Advances in Alternative Measures in Plant Protection

Advances in Alternative Measures in Plant Protection

Editors

Špela Mechora
Dragana Šunjka

MDPI • Basel • Beijing • Wuhan • Barcelona • Belgrade • Manchester • Tokyo • Cluj • Tianjin

Editors
Špela Mechora
Agency for Radwaste
Management
Slovenia

Dragana Šunjka
University of Novi Sad
Serbia

Editorial Office
MDPI
St. Alban-Anlage 66
4052 Basel, Switzerland

This is a reprint of articles from the Special Issue published online in the open access journal *Plants* (ISSN 2223-7747) (available at: https://www.mdpi.com/journal/plants/special_issues/Alternative_Measures).

For citation purposes, cite each article independently as indicated on the article page online and as indicated below:

LastName, A.A.; LastName, B.B.; LastName, C.C. Article Title. *Journal Name* **Year**, *Volume Number*, Page Range.

ISBN 978-3-0365-7396-0 (Hbk)
ISBN 978-3-0365-7397-7 (PDF)

Cover image courtesy of Dragan Mocevic

Contents

About the Editors

Špela Mechora

Špela Mechora gained her Ph.D. in the field of selenium metabolism in aquatic plants and selected agricultural plants. After that she was elected to Associate Professor for Ecology and lectured several courses and was the supervisor to several bachelor's and master's theses. Later she was employed in a business company where she was a research project manager for the field of ecology and health. She was involved in many research collaborations in national and international activities on the field of alternative measures in plant protection. She is currently employed at Agency for Radwaste management as a team member on a project of disposal facility in Slovenia, where she is responsible for radioactive safety. She became topical advisory panel member for Plants MDPI in year 2020. She serves as a peer reviewer for many journals with high Impact Factors and is a Guest Editor of Special Issue of Plants.

Dragana Šunjka

Dragana Šunjka is an Associate Professor of Phytopharmacy at the Faculty of Agriculture, University of Novi Sad, Serbia. Dragana obtained her Ph.D. in the field of phytopharmacy from the University of Novi Sad, and this was followed by postdoc research at EPFL (Lausanne, Switzerland) in the Central Environmental Laboratory. Her main research interest is oriented toward food safety and environment protection; pesticide behavior in agricultural products and the environment; pesticide quality; instrumental analysis. Recently, a significant part of her research activity focuses on the development and formulation of biopesticides and bioactive compounds. She is an author of more than 300 scientific papers published in international and national journals and conference proceedings, as well as books and book chapters. Dragana is involved in many research collaborations and has participated as a researcher in 20 projects in national and international scientific activities. She serves as a peer reviewer for many journals with high Impact Factors. She is the supervisor of three doctoral dissertations and several bachelor's and master's theses.

Preface to "Advances in Alternative Measures in Plant Protection"

Contemporary agricultural production demands increased applications of agrochemicals (pesticides and fertilizers), in order to obtain healthy crops with optimal nutritional value and high yields, as major challenges in food production. However, intensive and/or inappropriate use of agrochemicals leads to their accumulation in the agricultural products and environment, posing a risk to human health. They can also cause side effects on non-target organisms, contribute to the development of pests' resistance, etc. Due to increasing concern, the use of environmental-friendly methods for the control of weeds, diseases, and pests is being considered.

The book Advances in Alternative Measures in Plant Protection contains twenty-two scientific articles, sixteen original research papers and six original reviews. With these research and reviews, authors contribute to the development of new and biological ways of controlling pests in agricultural production. Scientists from different backgrounds have provided state-of-art research on alternative measures of plant protection, including biological control agents (predators, parasitoids, competitors, plant-plant interactions) plant and microorganisms-based biopesticides, and microelements. Some of the main advantages of the presented measures are the absence of toxicity to humans and vertebrates, beneficial organisms and the environment, without risk of pest resistance.

With this book, we tried to open new possibilities for the improvement of plant protection and present insight into recent research in the field. Hope that, together with authors and reviewers, we have contributed to the understanding of the importance of alternative measures in plant protection and reducing the use of synthetic pesticides. The data provided in this book will inspire researchers and enable future research. The book gives new strategies and technologies that can be utilized as plant protection measures, perfectly fitting the concept of sustainable development.

We are grateful to all the authors for their contributions and the reviewers, for their valuable recommendations leading to the improvement of all manuscripts. We would like to acknowledge the help and support provided by Managing Editor and MDPI Plants team.

Špela Mechora and Dragana Šunjka
Editors

Editorial

Advances in Alternative Measures in Plant Protection

Dragana Šunjka [1,*] and Špela Mechora [2,*]

[1] Faculty of Agriculture, University of Novi Sad, Trg Dositeja Obradovića 8, 21000 Novi Sad, Serbia
[2] Agency for Radwaste Management, Litostrojska 58A, 1000 Ljubljana, Slovenia
* Correspondence: dragana.sunjka@polj.edu.rs (D.Š.); spela.mechora@gmail.com (Š.M.)

Food production, along with the constant demand for higher yields, is an imperative of contemporary agricultural production. In order to accomplish this goal, it is necessary to control pests, weeds, and the causative agents of plant diseases, which implies the application of plant protection products. However, the intensive and/or unsuitable application of synthetic pesticides has led to a series of consequences for agricultural production itself and for the environment, non-target organisms, and for human health as well. The presence of resistant populations, side effects on pollinators and other beneficial organisms, and pesticide residues in food, water, and soil are just some of the consequences. The fact that in recent years the number of active substances approved for use in agricultural production is rapidly decreasing, all of the above requires continuous and comprehensive studies of the response of the living organism (beneficial and harmful effects) to various new agrochemicals and to products of biotechnology, followed by the mandatory risk assessment of the effect on humans and animals as well as the environment.

In order to minimize the use of chemical pesticides, the Special Issue "Advances in Alternative Measures in Plant Protection" intends to offer an insight into the alternative measures of plant protection in contemporary agriculture. Special Issue contains twenty-two scientific articles: sixteen original research articles and six original reviews. With these research and reviews, authors contribute to the development of new, biological ways of controlling pests in agricultural production.

As an alternative to synthetic pesticides for the control of plant diseases, especially for the management of plant pathogenic fungi, bioactive metabolites derived from algae/plant extracts, essential oils, and other materials can be used. The extract of seaweed *Gracilariopsis persica* has drawn attention for its biologically active compounds which show antifungal activities against *Botrytis cinerea*, *Aspergillus niger*, *Penicillium expansum*, and *Pyricularia oryzae* on the mycelial growth. The antifungal activity is most likely associated with its phenolic compounds, such as rosmarinic, oleic, and palmitic acid and quercetin, whose antifungal activity has already been reported [1]. The antifungal effect of phenolic compounds is potentially reflected in the reduction in the wood mass loss and restriction of wood biodegradation caused by *Ganoderma boninense* in inoculated oil palm woodblocks [2]. The hydromethanolic extract of *Rubia tinctorum* root also shows antifungal activity, alone or in combination with chitosan oligomers or with stevioside, against three *Botryosphaeriaceae* taxa by the inhibition of their mycelial growth. Considering the complexity of the extracts' chemical composition, the activity most likely should be ascribed to the combination of a group of them [3]. In addition to the antifungal effects, phenolic compounds play a notable role in the resistance of plants to diseases. The phenolic compound content usually increases after powdery mildew disease occurs, which leads to the conclusion that hybrids/cultivars with high phenolic contents should be recommended for the development of new superior cultivars as a source of resistance to fungal grape diseases [4].

An important part of sustainable pest management is natural enemies. Root fungal endophytes isolated from solanaceous plant species expressed activity in the control of a late blight of potato incited by *Phytophthora infestans* (Mont.) de Bary [5]. When it comes to the natural regulation of pests, it is particularly important to provide suitable conditions

Citation: Šunjka, D.; Mechora, Š. Advances in Alternative Measures in Plant Protection. *Plants* **2023**, *12*, 805. https://doi.org/10.3390/plants12040805

Received: 3 January 2023
Accepted: 21 January 2023
Published: 10 February 2023

for beneficial organisms. Field margin plant species are considered to be food and shelter for specific pests; therefore, it is crucial to study the biology of the host plants and how they interact with pests [6].

For the management of plant pathogens, as an environmentally friendly method of plant disease control, microorganisms can be used usually because of their capability of counteracting the growth. Species belonging to the bacterial genus *Pseudomonas* have a high potential in the control of plant pathogens. They can be used as an alcoholic extract derived from bacterial isolate, which manifests exceptional activity against some of the most common pre- and post-harvest fungal diseases [7].

Indoor agricultural production is particularly challenging due to conditions affordable for pests. As an alternative to chemical products, the control of causal agents of the major tomato diseases in greenhouse production was undertaken by *Bacillus* spp. or *B. subtilis* and the foliar application of *Reynoutria sachalinensis*, *Melaleuca alternifolia*, harpin αβ proteins, and bee honey. The use of these biorational products in the control of these diseases in greenhouse production has the potential to be incorporated into an integrated program for the management of the examined diseases in tomatoes [8]. Moreover, in the same conditions, the control of *Sclerotium rolfsii*, a destructive disease for many plants, including tomatoes, could be successfully undertaken using some antiseptic and disinfectant agents [9].

Plant extracts can indicate an herbicide effect as well. *Thymus vulgaris* hydrolate showed an inhibitory effect on the weed species, including *Amaranthus retroflexus* (L.), *Chenopodium album* (L.), *Portulaca oleracea* (L.), *Echinochloa crus-galli* (L.) P. Beauv., *Sorghum halepense* (L.) Pers., and *Solanum nigrum* L., through the inhibition of the seed germination and seedling growth. *T. vulgaris* hydrolate had the least negative phytotoxic effect on the germination of soybean, sunflower, and maize [10].

Understanding the plant–plant interaction is also one of the Advances in Alternative Measures in Plant Protection. The phytotoxic substances released by *Ambrosia trifida* L. have an allelopathic influence on the oxidative stress parameters and phenolic compounds in maize, soybean, and sunflower. The sunflower was the most sensitive crop to *A. trifida* allelochemicals, maize showed a mild sensitivity, while the soybean did not demonstrate sensitivity [11]. The phytotoxic effect could perhaps be reduced using some biostimulants. L-α-amino acid-based biostimulants show a protective effect against the stress caused by the application of the imazamox herbicide in sunflower plants, followed by the protection of the photosynthetic activity and reduction in the oxidative stress in the plant [12]. Furthermore, even some strains of entomopathogenic fungi can act as biostimulants for some crops by their ability to infect insect pests and to promote plant growth. For example, the isolate of *Metarhizium robertsii* (2693) causes the death of 73% *Tenebrio molitor* larvae, but also significantly increases the maize root length [13].

Aside from plants, microorganisms, insects, and nematodes, micronutrients have to be considered as a source of bioactive compounds. Recently, some micronutrients are recognized as an alternative to synthetic fungicides. Due to the influence of Selenium on the inhibition of the development of a *Fusarium proliferatum*, the addition of low concentrations of Se could improve conventional fungicides and decrease their side effects [14]. One of the most significant economic and ecologic forest tree species in Europe, the sessile oak, is highly threatened by the powdery mildew caused by *Erysiphe alphitoides* (Griffon and Maubl.). Evaluating the influence of the irrigation levels (overhead sprinklers) on the damage caused by powdery mildew to *Quercus petraea* in a nursery setting, it was showed that controlling the irrigation rate can become an effective component of integrated protection strategies against this pathogen [15]. The negative impact on the citrus production caused by pests *Tetranychus urticae* Koch 1836 could be successfully controlled by eco-friendly approaches, such as black soap and detergents, without having a negative impact on predators [16].

Furthermore, the authors review the latest knowledge regarding the importance of plant-parasitic nematodes in agricultural production [17] and genetically based plant resistance underlying plant–nematode interactions [18] as a safe alternative to chemical nematocides and describe the potential of using *Photorhabdus* spp. as biocontrol agents against a broad range of insect pests [19]. As an effective source of biofungicide, the authors provide recent knowledge of actinomycetes and their potential as biocontrol agents of phytopathogenic fungi [20].

Finally, the utility of biocontrol agents composed of microorganisms, plant-based compounds, as well RNAi-based technology has to be emphasized [21], not only in an organic agriculture, where they represent the exclusive measure of plant protection, but also in integrated and conventional agriculture. However, the main task in the field of plant protection, and the most challenging at the same time, is still finding alternative sources of compounds with pesticide activity and the improvement of their formulation [22].

As Guest Editors, we are grateful to all the authors for choosing this Special Issue to publish their research. We hope that together we have contributed to the understanding of the importance of alternative measures in plant protection and opened up new possibilities for the improvement of this field.

Author Contributions: The authors contributed equally to this article. All authors have read and agreed to the published version of the manuscript.

Acknowledgments: We would like to thank all authors for the excellent papers in Special Issue "Advances in Alternative Measures in Plant Protection". We also would like to express our gratitude to all of the reviewers, for their valuable recommendations leading to the improvement of all manuscripts. Finally, we would like to acknowledge the help and support provided by Managing Editor and Plants.

Conflicts of Interest: The authors declare no conflict of interest.

References

1. Pourakbar, L.; Moghaddam, S.S.; Enshasy, H.A.E.; Sayyed, R.Z. Antifungal Activity of the Extract of a Macroalgae, Gracilariopsis persica, against Four Plant Pathogenic Fungi. *Plants* **2021**, *10*, 1781. [CrossRef] [PubMed]
2. Surendran, A.; Siddiqui, Y.; Ahmad, K.; Fernanda, R. Deciphering the Physicochemical and Microscopical Changes in *Ganoderma boninense*-Infected Oil Palm Woodblocks under the Influence of Phenolic Compounds. *Plants* **2021**, *10*, 1797. [CrossRef] [PubMed]
3. Langa-Lomba, N.; Sánchez-Hernández, E.; Buzón-Durán, L.; González-García, V.; Casanova-Gascón, J.; Martín-Gil, J.; Martín-Ramos, P. Activity of Anthracenediones and Flavoring Phenols in Hydromethanolic Extracts of *Rubia tinctorum* against Grapevine Phytopathogenic Fungi. *Plants* **2021**, *10*, 1527. [CrossRef] [PubMed]
4. Atak, A.; Göksel, Z.; Yılmaz, Y. Changes in Major Phenolic Compounds of Seeds, Skins, and Pulps from Various *Vitis* spp. and the Effect of Powdery and Downy Mildew Diseases on Their Levels in Grape Leaves. *Plants* **2021**, *10*, 2554. [CrossRef]
5. El-Hasan, A.; Ngatia, G.; Link, T.I.; Voegele, R.T. Isolation, Identification, and Biocontrol Potential of Root Fungal Endophytes Associated with Solanaceous Plants against Potato Late Blight (*Phytophthora infestans*). *Plants* **2022**, *11*, 1605. [CrossRef]
6. Ndakidemi, B.J.; Mbega, E.R.; Ndakidemi, P.A.; Belmain, S.R.; Arnold, S.E.J.; Woolley, V.C.; Stevenson, P.C. Field Margin Plants Support Natural Enemies in Sub-Saharan Africa Smallholder Common Bean Farming Systems. *Plants* **2022**, *11*, 898. [CrossRef]
7. Librizzi, V.; Malacrinò, A.; Li Destri Nicosia, M.G.; Barger, N.; Luzzatto-Knaan, T.; Pangallo, S.; Agosteo, G.E.; Schena, L. Extracts from Environmental Strains of *Pseudomonas* spp. Effectively Control Fungal Plant Diseases. *Plants* **2022**, *11*, 436. [CrossRef]
8. Esquivel-Cervantes, L.F.; Tlapal-Bolaños, B.; Tovar-Pedraza, J.M.; Pérez Hernández, O.; Leyva Mir, S.G.; Camacho-Tapia, M. Efficacy of Biorational Products for Managing Diseases of Tomato in Greenhouse Production. *Plants* **2022**, *11*, 1638. [CrossRef]
9. Hussien, R.A.A.; Gnedy, M.M.A.; Sayed, A.A.S.; Bondok, A.; Alkhalifah, D.H.M.; Elkelish, A.; Tawfik, M.M. Evaluation of the Fungicidal Effect of Some Commercial Disinfectant and Sterilizer Agents Formulated as Soluble Liquid against *Sclerotium rolfsii* Infected Tomato Plant. *Plants* **2022**, *11*, 3542. [CrossRef]
10. Konstantinović, B.; Popov, M.; Samardžić, N.; Aćimović, M.; Šućur Elez, J.; Stojanović, T.; Crnković, M.; Rajković, M. The Effect of *Thymus vulgaris* L. Hydrolate Solutions on the Seed Germination, Seedling Length, and Oxidative Stress of Some Cultivated and Weed Species. *Plants* **2022**, *11*, 1782. [CrossRef]
11. Šućur, J.; Konstantinović, B.; Crnković, M.; Bursić, V.; Samardžić, N.; Malenčić, Đ.; Prvulović, D.; Popov, M.; Vuković, G. Chemical Composition of *Ambrosia trifida* L. and Its Allelopathic Influence on Crops. *Plants* **2021**, *10*, 2222. [CrossRef] [PubMed]
12. Navarro-León, E.; Borda, E.; Marín, C.; Sierras, N.; Blasco, B.; Ruiz, J.M. Application of an Enzymatic Hydrolysed L-α-Amino Acid Based Biostimulant to Improve Sunflower Tolerance to Imazamox. *Plants* **2022**, *11*, 2761. [CrossRef] [PubMed]
13. Praprotnik, E.; Lončar, J.; Razinger, J. Testing Virulence of Different Species of Insect Associated Fungi against Yellow Mealworm (*Coleoptera: Tenebrionidae*) and Their Potential Growth Stimulation to Maize. *Plants* **2021**, *10*, 2498. [CrossRef] [PubMed]
14. Troni, E.; Beccari, G.; D'Amato, R.; Tini, F.; Baldo, D.; Senatore, M.T.; Beone, G.M.; Fontanella, M.C.; Prodi, A.; Businelli, D.; et al. In Vitro Evaluation of the Inhibitory Activity of Different Selenium Chemical Forms on the Growth of a *Fusarium proliferatum* Strain Isolated from Rice Seedlings. *Plants* **2021**, *10*, 1725. [CrossRef] [PubMed]

15. Kasprzyk, W.; Baranowska, M.; Korzeniewicz, R.; Behnke-Borowczyk, J.; Kowalkowski, W. Effect of Irrigation Dose on Powdery Mildew Incidence and Root Biomass of Sessile Oaks (*Quercus petraea* (Matt.) Liebl.). *Plants* **2022**, *11*, 1248. [CrossRef] [PubMed]
16. Assouguem, A.; Kara, M.; Mechchate, H.; Al-Mekhlafi, F.A.; Nasr, F.; Farah, A.; Lazraq, A. Evaluation of the Impact of Different Management Methods on *Tetranychus urticae* (*Acari: Tetranychidae*) and Their Predators in Citrus Orchards. *Plants* **2022**, *11*, 623. [CrossRef]
17. Abd-Elgawad, M.M.M. Optimizing Safe Approaches to Manage Plant-Parasitic Nematodes. *Plants* **2021**, *10*, 1911. [CrossRef]
18. Abd-Elgawad, M.M.M. Understanding Molecular Plant–Nematode Interactions to Develop Alternative Approaches for Nematode Control. *Plants* **2022**, *11*, 2141. [CrossRef]
19. Abd-Elgawad, M.M.M. *Photorhabdus* spp.: An Overview of the Beneficial Aspects of Mutualistic Bacteria of Insecticidal Nematodes. *Plants* **2021**, *10*, 1660. [CrossRef]
20. Torres-Rodriguez, J.A.; Reyes-Pérez, J.J.; Quiñones-Aguilar, E.E.; Hernandez-Montiel, L.G. Actinomycete Potential as Biocontrol Agent of Phytopathogenic Fungi: Mechanisms, Source, and Applications. *Plants* **2022**, *11*, 3201. [CrossRef]
21. Kumar, J.; Ramlal, A.; Mallick, D.; Mishra, V. An Overview of Some Biopesticides and Their Importance in Plant Protection for Commercial Acceptance. *Plants* **2021**, *10*, 1185. [CrossRef] [PubMed]
22. Šunjka, D.; Mechora, Š. An Alternative Source of Biopesticides and Improvement in Their Formulation—Recent Advances. *Plants* **2022**, *11*, 3172. [CrossRef] [PubMed]

Review

An Overview of Some Biopesticides and Their Importance in Plant Protection for Commercial Acceptance

Jitendra Kumar [1,†], Ayyagari Ramlal [2,†], Dharmendra Mallick [3,‡] and Vachaspati Mishra [4,*]

1 Bangalore Bioinnovation Centre (BBC), Life Sciences Park, Electronics City Phase 1, Bengaluru, Karnataka 560100, India; director@bioinnovationcentre.com
2 Division of Genetics, Indian Agricultural Research Institute (IARI), Pusa Campus, New Delhi 110012, India; ramlal.ayyagari@gmail.com
3 Department of Botany, Deshbandhu College, University of Delhi, Delhi 110019, India; dkmallick@db.du.ac.in
4 Department of Botany, Dyal Singh College, University of Delhi, Delhi 110003, India
* Correspondence: mishravachaspati31@gmail.com
† Equally contributed to work.
‡ Deceased.

Abstract: Biopesticides are natural, biologically occurring compounds that are used to control various agricultural pests infesting plants in forests, gardens, farmlands, etc. There are different types of biopesticides that have been developed from various sources. This paper underscores the utility of biocontrol agents composed of microorganisms including bacteria, cyanobacteria, and microalgae, plant-based compounds, and recently applied RNAi-based technology. These techniques are described and suggestions are made for their application in modern agricultural practices for managing crop yield losses due to pest infestation. Biopesticides have several advantages over their chemical counterparts and are expected to occupy a large share of the market in the coming period.

Keywords: biopesticides; agriculture; food supply; microorganisms

Citation: Kumar, J.; Ramlal, A.; Mallick, D.; Mishra, V. An Overview of Some Biopesticides and Their Importance in Plant Protection for Commercial Acceptance. *Plants* **2021**, *10*, 1185. https://doi.org/10.3390/plants10061185

Academic Editors: Špela Mechora and Dragana Šunjka

Received: 23 May 2021
Accepted: 7 June 2021
Published: 10 June 2021

Publisher's Note: MDPI stays neutral with regard to jurisdictional claims in published maps and institutional affiliations.

1. Introduction

The global population is exploding at an exponential rate and is anticipated to reach approximately 9.7 billion by 2050, the largest share of which is in Africa and Asia [1]. This has imposed a large burden on agriculture and its allied sectors in terms of meeting food demands, which requires more inputs for crop production. Anthropogenic activities have affected people's surroundings and have also had negative impacts on the environment and ecosystems, including reductions in agricultural areas due to construction, the explosion of nutrient mining, degradation, and contamination of water resources (resulting in scarcity), aggregation of xenobiotics in the soils, and degeneration and deterioration of the quality, fertility, and efficiency of soil, with implications of soil erosion and climate change. In order to overcome these challenges and meet the requirements for food and supplies, the productivity and sustainability of agricultural practices should be improved and novel and improved strategies must be found. Enhanced agricultural productivity can be achieved in many ways, such as through increasing crop yield by providing manure and organic-based treatments, including biopesticides, or by limiting yield loss due to extreme environmental conditions (such as biotic and abiotic stresses) [2,3]. Abiotic stress can be largely controlled by the use of biostimulants and bioeffectors [4]. Biopesticides, which are pest management agents based on living microorganisms or natural products, offer a great promise in controlling yield loss without compromising the quality of the product.

The chemical pesticides used in crop protection, to reduce the damage caused by pathogens and pests in agricultural fields, pose many long-term threats and risks to living beings due to their harmful side effects. They are known to cause cancers [5] and foetal impairments [6] and they persist in the environment for many years (i.e., they are nonbiodegradable) [7]. Furthermore, based on their potential application and strong inhibitory activity against pests, these synthetic pesticides dominate the market and have a significant impact on the manufacture of products [8]. Based on a report by Business Communications Company (BCC), Inc., research on the global biopesticide and synthetic pesticide market showed that it was worth USD 61.2 billion in 2017 and is expected to rise to approximately USD 79.3 billion by 2022 [9,10]. Nutrient reduction and an increased disease incidence are quite common in crops grown on soils heavily subjected to chemical pesticides [11], and this is undesirable from the agricultural soil management for food and nutritional security standpoint. According to the Food and Agriculture Organization (FAO), United Nations (2017–2018), the top three leading pesticide-consuming countries are China, the USA, and Brazil [12]. In addition, pesticide consumption in India drastically increased from 50,410 tonnes (T) in 2016 to 58,160 T in 2018 [13]. The pesticides utilised for crops are as follows: fibre crops account for around 67%, fruits 50%, vegetables 46%, spices 43%, oilseeds 28%, and pulses 23% [14,15]. According to an annual report by the Ministry of Chemicals and Fertilizers, India (MoCF) (2019–2020), the production or manufacture of chemical pesticides increased from 186,000 metric tons (MT) in 2014–2015 to 217,000 MT in 2018–2019 [15]. The FAO also reported that from 2015 to 2018, the share of global pesticide consumption was 52.2% in Asia, 32.4% in the USA, 11.8% in Europe, 2% in Africa, and 1.6% in Oceania [12]. The per hectare consumption of pesticides by country is highest for China, followed by the UK, with the least in India [12]. Of the Indian states, Jammu and Kashmir had the highest chemical pesticide consumption, followed by Andhra Pradesh [13,14]. Based on the statistics of chemical pesticide consumption in India alone, it is imperative to seek alternative methods, especially to increase the use of biopesticides [15].

Biopesticides are naturally occurring compounds or agents that are obtained from animals, plants, and microorganisms such as bacteria, cyanobacteria, and microalgae and are used to control agricultural pests and pathogens. According to the US Environmental Protection Agency, biopesticides are 'derived from natural materials such as animals, plants, bacteria and certain minerals' [16]. Products such as genes or metabolites from these biocontrol agents can be used to prevent crop damage [16]. The use of biopesticides is, by far, more advantageous than the use of their counterparts, traditional chemical pesticides, as they are eco-friendly and host specific [17]. The use and application of agro-based chemicals in the agricultural sector to protect crop plants from invading and infecting pests can be greatly improved by employing biopesticides [17,18].

This paper provides up-to-date information on various important biopesticides, including types, advantages, and their utility in plant protection that would eventually lead to their commercial acceptance. Furthermore, various potential sources and technology involved in the production of biopesticides are briefly described.

2. Types of Biopesticides

There are many types of biopesticides, and they are classified according to their extraction sources and the type of molecule/compound used for their preparation [19]. The categories are listed below.

2.1. Microbial Pesticides

These are derived from microorganisms including bacteria, fungi, and viruses. The active molecules/compounds isolated from these organisms attack specific pest species or entomopathogenic nematodes. Those known as bioinsecticides, target insects that harm crops, while those that control weeds via microorganisms, such as fungi are referred to as bioherbicides. Over the last decade, extensive research activities on microbial biopesticides have led to the discovery and development of a good number of biopesticides and have paved the way for their marketability [19]. The successful use of *Bacillus thuringiensis* (Bt) and some other microbial species led to the discovery of many new microbial species and strains, and their valuable toxins and virulence factors that could be a boon for the biopesticide industry, and some of these have been translated into commercial products as well [19,20]. Major groups of bacterial entomopathogens include species of

Pseudomonas, *Yersinia*, *Chromobacterium*, etc., while fungi comprise species of *Beauveria*, *Metarhizium*, *Verticillium*, *Lecanicillium*, *Hirsutella*, *Paecilomyces*, etc. [18,21]. Other important microbial pesticide producers are baculoviruses that are species specific and their infectivity is associated with the crystalline occlusion bodies that are active against chewing insects (Lepidopteran caterpillars) [18]. The baculoviral occlusion body is basically a virion that is combined with the Bt toxin to produce recombinant baculovirus (ColorBtrus), producing occlusion bodies that incorporate the Bt insecticidal Cry1Ac toxin protein for enhancing the speed of action and pathogenicity with respect to its wild-type counterpart [18]. Entomopathogenic nematodes (EPNs) used as biocontrol agents belong primarily to species in the genera *Heterorhabditis* and *Steinernema*, associated with mutualistic symbiotic bacteria of the genera *Photorhabdus* and *Xenorhabdus* and are safe to mammals, environment, and nontarget organisms [18]. Their commercial development as biocontrol agents has been convenient because of their ease in mass production, using in vivo or in vitro techniques, and exemption from registration [18].

2.2. Biochemical Pesticides

Biochemical pesticides are naturally occurring products that are used to control pests through nontoxic mechanisms, whereas chemical pesticides use synthetic molecules that directly kill pests. Biochemical pesticides are further classified into different types depending upon whether they function in controlling infestations of insect pests by exploiting pheromones (semiochemicals), plant extracts/oils, or natural insect growth regulators.

2.2.1. Insect Pheromones

These are chemicals produced by insects which are mimicked for use in controlling insects in the integrated pest management programs. These chemicals are effective in disrupting insect mating to prevent the success of mating, thus reducing the number of insect progeny. The insects exploited in this process act as dispensers of pheromones that become confused due to the presence of pheromone flumes diffused in the surroundings. Insect pheromones are not true 'insecticides' since they do not kill insects but influence their olfactory system to affect behaviour [22]. A detailed account of the mode of action of pheromones is given by Ujváry [20]. In summary, the antennae of the perceiving insect adsorb pheromones, which then diffuse into the interior of the sensilla through microscopic pores in the cuticle. Once inside, these are transferred through the hydrophilic sensillum to the chemosensory membranes by pheromone-binding proteins (PBPs). Subsequently, the pheromone or pheromone–PBP complex interacts with a specific receptor protein, which transduces the chemical signal into an amplified electric signal by a second messenger system connected with neuronal machinery [23].

2.2.2. Plant-Based Extracts and Essential Oils

Over the last several years, plant-based extracts and essential oils have emerged as attractive alternatives to synthetic insecticides for insect pest management. These insecticides are naturally occurring insecticides as they are derived from plants and contain a range of bioactive chemicals [24]. Depending on physiological characteristics of insect species as well as the type of plant, plant extracts and essential oils (EOs) exhibit a wide range of action against insects: they can act as repellents, attractants, or antifeedants; they also may inhibit respiration, hamper the identification of host plants by insects, inhibit oviposition and decrease adult emergence by ovicidal and larvicidal effects [25–27]. Their composition varies greatly. Well-known examples in this regard are neem and lemongrass oil, which are very common in global herbal markets. A comprehensive study by Halder et al. [26] showed that a combination of neem oil with entomopathogenic microorganisms, including *Beauveria bassiana*, was very successful against vegetable sucking pests. However, it is very important to determine the dose of azadirachtin content in neem oil so as not to kill the nontarget organisms [28]. A similar strategy has to be established for the entomopathogenic fungi that need to be supported by complementary laboratory bioassays, station, and/or field experiments for effective management of the target pests without affecting nontarget insects [29]. As regards the marketability of essential oils, they in fact, represent a market estimated at USD 700.00 million and a total world production of 45,000 tons, and industries in the US are able to bring essential oil-based pesticides to market in a shortened time period, as compared to the time taken in conventional pesticide launch [30].

2.2.3. Insect Growth Regulators

Insect growth regulators (IGRs) inhibit certain fundamental processes required for the survival of insects, thereby killing them. Furthermore, these compounds are highly selective and less toxic to nontarget organisms [23]. Depending on the mode of action, IGRs had been recently grouped in chitin synthesis inhibitors (CSIs) and substances that interfere with the action of insect hormones (i.e., juvenile hormone analogues and ecdysteroids) [31]. IGRs can control many types of insects including fleas, cockroaches, and mosquitos even though they are not so fatal for adult insects [31]. Although low in toxicity to humans, they prevent reproduction, egg-hatch, and molting from one stage to the next in the young insects, while mixing them with other insecticides is able to kill even the adult insects [31].

2.3. GMO Products

These substances are produced through genetically modified organisms (GMOs). The genetic material is incorporated into the plant, which is then used as a source to produce pesticidal compounds, also referred to as plant-incorporated protectants (PIPs). Cry proteins are, by far the first-generation insecticidal PIPs that were introduced into the GM crops containing transgenes from the soil bacterium Bt. [30]. PIPs also demand the state of the research necessary for the ongoing environmental fate assessment of these molecules, primarily the RNAi-based PIPs [30,32] that would be discussed in a separate section.

3. Mode of Action of Biopesticides

Biopesticides act in a variety of ways on microorganisms depending on their type and nature. A few mechanisms through which biopesticides attack or kill pathogens are listed as follows [8].

3.1. Microbial Biopesticides

Fungicides and bactericides. These biopesticides generally inhibit or disrupt the process of translation and thus protein synthesis in numerous ways, including through binding of 50S ribosomes in prokaryotes, to prevent the transfer of peptides and inhibit chain elongation (such as blasticidin) [32,33]. Sometimes they interfere with the binding of aminoacyl tRNA to 30S and 70S ribosomal subunit complexes and inhibit translation (such as kasugamycin) [34]. In the case of streptomycin and mildiomycin, binding with the 30S ribosomal subunit causes abnormal synthesis of protein (nonfunctional) and blocks the activity of peptidyltransferase, respectively [35,36].

They can also disrupt plasma membrane permeability and cause leakage of substances (amino acids and electrolytes), thereby causing cell death (such as natamycin), and can inhibit chitin synthase activity (polyoxins) and inhibit trehalase, preventing the formation of glucose (validamycin) [31].

Insecticides upon reaching nerve endings, release gamma-aminobutyric acid (GABA), which causes GABA-gated Cl-ion channels to open, thus working by hyperpolarising the nerve membrane potential and blocking the electrical nerve conduction (avermectins and emamectin) [35,36]. Polynactins can cause leakage of potassium ions from mitochondria [36].

Herbicides inhibit phosphorylation in plants by blocking glutamine synthase, which causes an increase in ammonia (bilanafos) [36].

3.2. Biochemical Pesticides

These pesticides are derived from plants. Plants have evolved and developed many compounds, which can help to combat pathogenic microorganisms during the course of infection and attack. These compounds include steroids, alkaloids, phenylpropanoids, phenolics, terpenoids, and nitrogenated compounds. For instance, nicotine was the first insecticide obtained from tobacco leaves in the 17th-century that used to kill plum beetles [37,38]. Nicotine in tobacco is toxic to most herbivore insects and pesticides derived from them have been regarded as 'green pesticides' with high activity and low toxicity [38]. Duan et al. [38] have mentioned tobacco to be containing some useful ingredients, such as solanesol and nicotine, which exhibit potent inhibitory activity against *Staphylococcus aureus*, *Bacillus subtilis*, and *Micrococcus lysodeikticus*. Insecticides, such as azadirachtin and nicotine, function by either disrupting respiratory enzymes or inhibiting insect growth regulators, or by binding to sodium channels [39], while microbicides impair metabolic function and disrupt the integrity of plasma membrane and inhibit conidial formation [40].

3.3. GMO-Based Biopesticides

These are produced when genes are transferred into a plant, which allows it to produce compounds, such as Bt toxin, that can be used to combat pests. The delta endotoxins produced by the bacterium *B. thuringiensis* are broken down into smaller toxins in the insect gut by the action of proteases, which then bind to receptors in the midgut, causing cell expansion, rupture, and ion leakage leading to cell death [40].

4. Biopesticides from Algal and Cyanobacterial Sources

Microalgae can be used as an alternative technology to increase productivity in sustainable agricultural systems. A number of microalgae strains produce biologically active compounds that include antimicrobial compounds with the potential to act as biopesticides [40,41]. The biomass (extracts) can be applied as an alternative to chemical pesticides [40,42] since it can enhance plant growth and protect agricultural crops [42]. The filamentous cyanobacterium *Nostoc piscinale* and two single-celled green algae, *Chlamydopodium fusiforme* and *Chlorella vulgaris* are reported to have biopesticide activity against certain pathogens (Table 1). Some important microalgae have been exploited for their beneficial biopesticide activity in the cultivation of spices [43].

Table 1. A broad description of some common biopesticides, their types, sources, and target crops with the authors who published such reports.

Source	Type	Organism	Pest Type	Target Crop	Reference(s)
Bacteria	Insecticide	*Bacillus thuringiensis* var *kurstaki* *B. thuringiensis* var *tenebrionis*	caterpillars, fungi (*Botrytis*) Elm Leaf Beetle, Alfalfa weevil	vegetables, fruits, ornamentals, cereals Potato	Koul [44]; Bravo et al. [45] Saberi et al. [46]
	fungicide	*Bacillus subtilis*	*Botrytis* spp.	vegetables, fruits, and ornamentals	Koul [44]; Bravo et al. [45]
Fungi	insecticide	*Beauveria bassiana*	Whitefly	protected edible and ornamental plant production	McGuire and Northfield [47]
	fungicide	*Coniothyrium minitans* *Trichoderma harzianum*	*Sclerotinia* spp. *S sclerotiorum.*	outdoor edible and nonedible crops and protected crops Starwberry crops	Gams et al. [48] Dolatabadi et al. [49]
	herbicide	*Chondrostereum purpureum*	cut stumps of hardwood trees and shrubs	Forestry	Bailey [50]
	nematicide	*Paecilomyces lilacinus*	plant-parasitic nematodes in soil	vegetables, soft fruit, citrus, ornamentals, tobacco and turf	Moreno-Gavíra et al. [51]

Table 1. *Cont.*

Source	Type	Organism	Pest Type	Target Crop	Reference(s)
Virus	insecticide	*Cydia pomonella* granulovirus	codling moth	apples and pears	Kadoić Balaško et al. [52]
Oomycetes	herbicide	*Phytophthora palmivora*	*Morenia orderata*	citrus crops	Lala et al. [53]
Neem (*Azadirachta indica)*	insecticide	Azadirachtin	aphids, scale, thrips, whitefly, leafhoppers, weevils	vegetables, fruits, herbs, and ornamental crops	Chaudhary et al. [54]
Plant extracts	fungicide	*Reynoutria sachalinensis* (giant knotweed) extract	powdery mildew, downy mildew, *Botrytis*, late blight, citrus canker	protected ornamental and edible crops	Marrone [55]
	herbicide	Plant essential oils	Ragwort, many arthropods	Grassland	Isman [56]
	nematicide	*Quillaja saponaria*	plant parasitic nematodes	vineyards, orchards, field crops, ornamentals and turf	Guerra and Sepúlveda [57]
Talaromyces flavus; Clitoria ternatea (butterfly pea); *Trichoderma harzianum; Bacillus thuringiensis* var. *tenebrionis; Lactobacillus casei* fermentation products	biopesticides		*Glomerella cingulata* and *Colletotrichum acutatum; Helicoverpa* spp.; *Fusarium oxysporum Agelastica alni; Spodoptera litura, Helicoverpa armigera, Aphis gossypii; Xanthomonas fragariae; Spodoptera littoralis* and others	Strawberry, Cotton, Gladiolus hybrids, alder leaf, and hazelnut, other economically important plants and trees	Ishikawa [58]; Mensah et al. [59].; Kirk and Schafer [60] Eski et al. [61].; Dubois et al. [62]; Pavela et al. [63]; El-Abbassi et al. [64]
Semiochemical	attractant	Citronellol	tetranychid mites	apples, cucurbits, grapes, hops, nuts, pears, stone fruit, nursery, and ornamental crops	Mauchline et al. [65]; Mossa et al. [66]
	attractant	Multi-component sex pheromone, such as (E,E)-8,10-dodecadien-1-ol	codling moth	Fruits, such as apples and pears	El-Sayed et al. [67]
Arbuscular Mycorrhizal Fungi	Mutual inhabitant in the roots	Fungi	*Fusarium verticillioides;* pathogens affecting below ground plant organs	*Zea mays*	Olowe et al. [68]; Bharadwaj and Sharma [69]; Mukerji and Ciancio, [70]
Microalgae	Filamentous cyanobacterium; Single-celled green algae	*Nostoc piscinale; Chlamydopodium fusiforme; Chlorella vulgaris*	-	-	Ranglova et al. [41]
		Anabaena laxa and *Calothrix elenkinii*	Increase in fungicidal activity	Coriander, cumin, and fennel	Kumar et al. [43]
Nanobiopesticide	Silver nanobiopesticide	None	*Alternaria alternata, A. solani*	Alternaria leaf blight and leaf spot diseases in tomato, pepper, and potato	Narware et al. [71]
	Sargassum muticum derived NPs	None	*Ariadne merione*, a Lepidopteran pest	-	Narware et al. [71]; Rodrigues et al. [72]
	Caulerpa scalpelliformis and *Mesocyclops longisetus*-derived NPs	None	*Culex quinquefasciatus*	-	Narware et al. [71]

The use of chemical insecticides can result in numerous undesirable effects, including (i) killing of beneficial and nontargeted organisms and sometimes resurgence; (ii) rapid multiplication of secondary pests; (iii) development of pesticide resistance; (iv) contamination of the environment/ecosystem; (v) accumulation of pesticide residues in food materials; (vi) causing imbalanced ecological processes, such as pollination (pollinators affected by pesticides) and harm to living beings; (vii) carcinogenic and teratogenic effects in nature; and (viii) causing imbalances in hormone systems [8,73–75].

Several microorganisms have been explored for their potential in developing biopesticides. Microalgae have proved to be an excellent source owing to their advantages over traditional chemical pesticides. They produce a plethora of compounds with stimulating activities, including biomass and compounds, which can be used in the preparation of biopesticides, thereby enhancing crop protection [41]. Microalgae can be produced using wastewater, as they require nitrogen, phosphorus, and carbon and ammonium, which are abundant in wastewater, thus representing a nitrogen source. *Chlorella vulgaris* is generally used in the treatment of wastewater and is able to tolerate ammonium levels effectively. Ranglova et al. [41] assayed the efficacy of *C. vulgaris* against several phytopathogens, such as *Rhizoctonia solani*, *Fusarium oxysporum*, *Phytophthora capsica*, *Pythium ultimum*, *Clavibacter michiganensis*, *Xanthomonas campestris*, *Pseudomonas syringae*, and *Pectobacterium carotovorum*, while observing its antibacterial and antifungal activity, which were higher when cultivated in wastewater [41].

Gonçalves [3] argued that rice fields heavily sprayed with synthetic fertilisers to promote better productivity and yield left many detrimental effects on the environment and beneficial soil microflora, including decreased efficiency of fertiliser utilisation by the promotion of rice diseases, inhibition of microbiological nitrogen fixation, and increased nonpoint source pollution; importantly, they were also not cost effective. Furthermore, he added that in developing green rice, *Anabaena variabilis* could be a potent biofertiliser and biopesticide [3].

5. Biopesticide Activity from RNAi-Based Treatments

RNA interference technology is being used in the production of biopesticides due to the increased sensitivity towards pests and pathogens. Many transgenic crops (maize, soybean, and cotton) have been developed for resistance against particular pests [32]. Due to the limited consumption of genetically modified crops, RNA interference (RNAi) can be used as an alternative to overcome this problem. Studies carried out by Ratcliff et al. [76] and Ruiz et al. [77] demonstrated that transgenes had a significant impact on the functioning of plants upon viral infection through an RNAi mechanism. Similarly, Wang et al. [78] produced a barley crop completely resistant to barley yellow dwarf virus [76–78].

The mechanism of RNAi includes the expression of transgene dsRNA, which induces virus resistance and gene silencing in plants. Guide RNAs are formed as intermediaries; these are around 25 nt long and guide target RNAs for their degradation [79–81]. Dalmay et al. [81] reported that the process involves the use of RNA-dependent RNA polymerase RDR6 to generate double-stranded RNA (dsRNA) from target transcripts in plants, leading to the formation of small interfering RNA (siRNA) which, in turn, has silencing potential [81]. The RNase III domain-containing enzyme responsible for dsRNA cleavage, as observed in *Drosophila*, is called Dicer (also seen in plants and fungi) [82,83]. Following this, RNA-induced silencing complex (RISC)—a member of the conserved Argonaute family—is recruited, which mediates the cleavage of the target transcript [84,85], thus conferring resistance to the host [86].

RNAi technology has been used as a promising tool to overcome the ill effects of pests and pathogens. An RNAi method for oral application was developed by Baum et al. [85] using an artificial diet or transgenic maize against western corn rootworm (*Diabrotica virgifera*) to target V-ATPase subunits and alpha-tubulin [85]. Similarly, research conducted by Mao et al. showed the induction of growth defects in *Helicoverpa armigera*, the cotton bollworm, when given plant leaf material expressing a dsRNA specific to a cytochrome

P450 gene [87]. The first commercial, genetically modified variety showing the expression of dsRNA against an insect pest was developed in 2017 when Monsanto and Dow approved SmartStax PRO maize containing dsRNA against the western corn rootworm Snf7 gene [88]. Similarly, apple and potato expressing dsRNAs were approved for regulation of endogenous gene expression for quality enhancement [88,89]. Apart from insects and viruses, the mechanism of RNAi-mediated silencing has been used to control other plant pests and pathogens, including bacteria such as *Agrobacterium*, fungi such as powdery mildew, and root-knot nematodes [86,90]. The US environmental protection agency (EPA) approved the first PIP called SmartStax Pro in June 2017 that will help US farmers control corn rootworm, a devastating corn pest that has developed resistance to several other pesticides [91].

6. Bacteria-Based Biopesticide

Pesticides formulated using microorganisms and their products are highly effective, species specific, and eco-friendly, leading to acceptance of their use in pest management strategies worldwide [8–10,17]. Given their significance as stated [8–10,17], there is enough scope for further development in their marketing and profitability for the manufacturing industry.

The bacteria that are used as biopesticides can be divided into four categories [92], namely, crystalliferous spore formers (such as *Bacillus thuringiensis*), obligate pathogens (such as *B. popilliae*), potential pathogens (such as *Serratia marcescens*), and facultative pathogens (such as *Pseudomonas aeruginosa*). Of these, spore-forming bacteria are the most widely sought after for commercial use. The most commonly used bacteria, *B. thuringiensis* and *B. sphaericus*, are highly specific, safe, and effective organisms for insect control [92].

The Cry family of crystalline proteins are produced by *B. thuringiensis* in the parasporal crystals and encoded by the *cry* genes. The Cry proteins are globular molecules (65–145 kDa, depending on the strain) with three structural domains connected by single linkers. The Cry proteins belong to a single family that contains about 50 subgroups [92]. Further details of the Cry protein and its mechanism of action have been elaborately discussed by Koul [44]. Finally, pests are killed by lethal septicaemia and starvation. An example of a *Bacillus sphaericus*-based product has been known to contain a binary mosquito larvicidal toxin comprising BinB (51.4 kDa) and BinA (41.9 kDa), which is commonly used for mosquito control [44].

7. Biopesticides from Arbuscular Mycorrhizal Fungi (AMF)

Arbuscular mycorrhizal fungi (AMF) play a crucial role in enhancing the growth and yield of crops [93,94]. They enhance the resistance of crops against pathogens by raising their defences. The composition of AMF changes and its presence decreases depending upon the soil type and crop, as well as the application of fertilisers and tillage [95,96]. Plants have evolved many direct and indirect mechanisms to overcome herbivory; for example, they produce chemicals such as nicotine, gossypol, and many other such compounds, which can prevent herbivores from feeding on them. It has been observed that AMF colonisation on crop plants is extremely helpful in providing a good defensive ability in hosts by altering the gene expression patterns and directly or indirectly changing the nutritional status of crops [97,98]. Some examples of the utility of AMF application in plant biocontrol are mentioned in Table 1. More research needs to be carried out in this area.

8. Nanobiopesticides

The concept of 'nano' in biopesticides has revolutionised the field due to the size, structure, and nature of substances, which are formed in a size range of 1–100 nm. These small biologically active particles can prevent the growth of pathogens by either destroying or repelling them [97–99]. Nanoencapsulation, nanocontainers, and nanocages, because of their property of degradability, increase the stability and efficacy of pest control, and lower amounts are used when delivering nanobiopesticides [99]. The damages caused by the phytopathogens can also be overcome by the application of nanobiopesticides, primarily the metallic nanoparticles (NPs) of zinc, gold, silver, nickel, and titanium owing to their inherent antimicrobial properties. These have some added advantages over other biopesticides because of their increased solubilisation abilities and target-oriented delivery of the compound with enhanced efficiency. Bacterial, fungal, and plant extracts are used for the synthesis of NPs. It has been shown that silver nanobiopesticides (AgNPs) can be synthesised using marine organisms such as *Sargassum muticum*, *Mesocyclops longisetus*, and *Caulerpa scalpelliformis* [71]. The benefit of the use of microorganisms in the preparations of NPs is that microorganisms can withstand high concentrations of metals over plants and also their rate of production and management is much easier, as compared to the plants. Needless to stress here that microorganisms being very tiny, have better penetration ability than plants. Narware et al. [71] have mentioned a number of microorganism-derived NPs that are very useful in pest control (Table 1).

Bioherbicides have also been used in the formulations of nanobioherbicides. The efficacy of metabolites of *Photorhabdus luminescence*, an endosymbiotic bacterium of the *Heterorhabditis indica*, entomopathogenic and parasitic nematodes, are controlled [98]. Similarly, nanofungicides have also been prepared to control various pathogenic fungi which include *Bipolaris sorokiniana*, *Fusarium* sp., *Alternaria alternata*, and many others through AgNPs and *Magnaporthe grisea* and *B. sorokiniana* using metal nanoparticles. Apart from their ability of being readily soluble, the nanofungicides are very economical, eco-friendly, and safe [98].

9. Biopesticides from Aquatic Plants

Duckweed (*Lemna minor*), muskgrass (*Chara* spp.), water hyacinth (*Eichhornia crassipes*), hydrilla (*Hydrilla verticillate*), water lettuce (*Pistia stratiotes*), and filamentous algae (*Lyngbya wollei*) are some common aquatic plants. It is observed that some plants produce allelopathic compounds which have the potential to prevent the growth, germination, survival, and reproduction of surrounding organisms. Neem (*Azadirachta indica*) extract kills many insects, while *Eichhornia crassipes* has the ability to inhibit the growth of *Spodoptera litura*, a lepidopteran pest [100–103]. Similarly, *Chenopodium album* is inhibited by the presence of duckweed and water lettuce [100]. These examples illustrate that similar plants (or weeds) and their allelopathic chemicals have highly potent inhibitory properties against the pathogens and hence can be substituted for conventional chemical pesticides [103].

10. Merits of Biopesticides over Chemical Pesticides

Biopesticides have several merits over conventional chemical pesticides. They are environmentally friendly, target specific, and not deleterious to nontarget organisms and hence potent enough to replace synthetic pesticides for pest management [46]. Table 2 provides an overview of the disadvantages of using conventional chemical pesticides instead of biopesticides.

In recent years, the use of biopesticides is gaining momentum because they can be efficiently used in sustainable agricultural practices [2,3]. Biopesticides are highly effective in small amounts and decompose quickly without leaving problematic residues and hence can reduce the use of conventional pesticides as an integral component of IPM programs [102]. However, despite the merits of using biopesticides, their use has not been as widespread as expected, for the following reasons:

1. High cost of pesticide production due to the costs involved in screening, developing, and getting regulatory clearance for new biological agents;
2. Short shelf life due to the sensitivity of biopesticides to fluctuations in temperature and humidity;
3. Limited field efficacy due to climatic/regional variations in temperature, humidity, soil conditions, etc.;
4. Due to the high specificity of the biopesticides, i.e., they are only effective against target pathogens and pests, farmers are disinterested in them. They need to use

multiple biological agents to control different pathogens and pests in the field. These agents are confusing, costly, and cumbersome, and are also not available for every pest or pathogen.

Table 2. The various disadvantages of conventional chemical pesticides over biopesticides.

Conventional Chemical Pesticides	Biopesticides
Synthesised or produced from artificial/chemicals	Use naturally occurring compounds derived from living organisms for the production
They cause environmental pollution and are not eco-friendly	They do not cause environmental harm
Harmful to nontarget organisms	Do not cause harm to nontarget organisms
Cost ineffective	Cost efficient and cheaper, compared to chemical fertilisers
Microorganisms develop resistance gradually as the application increases	Pests do not develop resistance
High market value	Not preferred in the market
Contaminate water and soil	Cannot contaminate water sources
Lead to bioaccumulation	Do not lead to bioaccumulation

11. Commercial Exploitation of Biopesticides

Currently, a majority (about 90%) of microbial biopesticides on the market are derived from a single bacterium, *Bacillus thuringiensis* or Bt. Biopesticides make up a small share of the crop protection market, with a value of about USD 3 billion worldwide, accounting for just 5% of the total market [102]. In the United States market, more than 200 products are available, while the European Union market has only 60 analogues [43]. Biopesticide use at a global scale is increasing by almost 10% every year [43]. However, these pesticides are going to contribute noticeably to their global market consumption needs, which are to increase further in the future by substituting them for and thus reducing the over-reliance on chemical pesticides. Biopesticides are assessed in the EU by the same regulations used for the assessment of synthetic active substances, which require the addition of several new provisions in the current legislation, and the preparation of new guidelines facilitates the registration of prospective biopesticide products [78]. It is assumed that there are fewer active substances of biopesticides registered in the EU than in the USA, India, Brazil, or China [104]. It is expected that the use of biopesticides will be on par with synthetics by the early 2050s, but major uncertainties regarding the rates of uptake, especially in areas such as Africa and Southeast Asia, account for most of the flexibility in such projections [102].

12. Conclusions

The application of biofertilisers consisting of bacteria, cyanobacteria, or fungi can improve and restore the fertility of the soil and ensure sustainable agricultural production using green technology. Using microorganisms and microalgae as biopesticides can reduce the demand for energy and consumption of synthetic fertilisers and restore the efficiency of agroecosystems and wastelands. These organisms, when combined with the use of biotechnical innovations such as RNAi technology, can play a significant role in the production of secondary metabolites, biofertilisers, bioenergy, and bioprocessed products that would be also useful in pest control. RNAi-based biopesticides have gained enough momentum in recent years as a narrow-spectrum alternative to chemical-based control measures for specific and accurate targeting of pests and pathogens. In this regard, the use of bioinformatics-based dsRNA selection for effective RNAi design, coupled with adequate experimental testing, will likely eliminate the adverse impacts of RNAi-based biopesticides [86].

Considerable research on biological control agents, including biopesticides, is required for the development of the biopesticide market in the future. Scientists from diverse research institutes around the world are engaged in enormous research efforts in the field, but very few complete and systematic reports are available. Here, the utmost collaboration among enterprises and research institutes is needed, without which a scenario whereby biopesticides completely replace chemical pesticides seems impossible. In the current scenario, the agricultural sector needs to rely on both biopesticides and chemical pesticides. However, speeding up the practical application of laboratory results should facilitate large-scale industrial development. The inflow of biopesticides, however, has considerably reduced the use of synthetic chemicals because of stringent regulations [102]. Many substances have been researched to demonstrate their utility as biopesticides (Table 1), but extensive field research is required in order to assess their efficacy for precise pest problems under diverse cropping systems.

Farmers and society at large should benefit from the mixed and judicious use of both conventional chemical pesticides and biopesticides, while it is imperative to emphasise the research in the area of biopesticides for reaping greater benefits from it in the future.

Author Contributions: Writing—original draft preparation, A.R.; Conceptualisation, V.M.; writing—review and editing, V.M., J.K. and D.M.; funding acquisition, J.K. All authors have read and agreed to the published version of the manuscript.

Funding: The publication charges are paid from the funds of Bangalore Bioinnovation Centre, Karnataka Innovation and Technology Society (KITS), Department of Electronics, IT, BT and S&T, Government of Karnataka, India.

Institutional Review Board Statement: Not Applicable.

Informed Consent Statement: Not Applicable.

Data Availability Statement: Not Applicable.

Acknowledgments: The authors are thankful to the Indian Agricultural Research Institute, Pusa, New Delhi, for providing the library facility and the Bangalore Bioinnovation Centre, Karanataka Innovation and Technology Society, Department of Electronics, IT, BT and S&T, Government of Karnataka, India, Department of Biotechnology, Government of India.

Conflicts of Interest: The authors declare no conflict of interest.

References

1. United Nations Department of Economic and Social Affairs. *World Population Prospects: The 2015 Revision, Key Findings and Advance Tables*; Working Paper No ESA/P/WP; United Nations Department of Economic and Social Affairs: New York, NY, USA, 2015.
2. Pathak, J.; Maurya, P.K.; Singh, S.P.; Häder, D.P.; Sinha, R.P. Cyanobacterial farming for environment friendly sustainable agriculture practices: Innovations and perspectives. *Front. Environ. Sci.* **2018**, *6*, 7. [CrossRef]
3. Gonçalves, A.L. The Use of Microalgae and Cyanobacteria in the Improvement of Agricultural Practices: A Review on Their Biofertilising, Biostimulating and Biopesticide Roles. *Appl. Sci.* **2021**, *11*, 871. [CrossRef]
4. Van Oosten, M.J.; Pepe, O.; De Pascale, S.; Silletti, S.; Maggio, A. The role of biostimulants and bioeffectors as alleviators of abiotic stress in crop plants. *Chem. Biol. Technol. Agric.* **2017**, *4*, 5. [CrossRef]
5. Nicolopoulou-Stamati, P.; Maipas, S.; Kotampasi, C.; Stamatis, P.; Hens, L. Chemical Pesticides and Human Health: The Urgent Need for a New Concept in Agriculture. *Front. Public Health* **2016**, *4*, 148. [CrossRef]
6. Kalafati, L.; Barouni, R.; Karakousi, T.; Abdollahi, M.; Tsatsakis, A. Association of pesticide exposure with human congenital abnormalities. *Toxicol. Appl. Pharmacol.* **2018**, *346*, 58–75. [CrossRef]
7. Kalliora, C.; Mamoulakis, C.; Vasilopoulos, E.; Stamatiades, G.A.; Scholtz, M.T.; Voldner, E.; McMillan, A.C.; Van Heyst, B.J. A pesticide emission model (PEM). Part I: Model development. *Atmos. Environ.* **2002**, *36*, 5005–5013.
8. Liu, X.; Cao, A.; Yan, D.; Ouyang, C.; Wang, Q.; Li, Y. Overview of mechanisms and uses of biopesticides. *Int. J. Pest Manag.* **2021**, *67*, 65–72. [CrossRef]
9. Lehr, P. *Global Markets for Biopesticides*; Report Code CHM029E; BCC Research: Wellesley, MA, USA, 2014.
10. Chen, J. *Biopesticides: Global Markets to 2022*; Report Code CHM029G; BCC Research: Wellesley, MA, USA, 2018.
11. Tripathi, S.; Srivastava, P.; Devi, R.S.; Bhadouria, R. Influence of synthetic fertilizers and pesticides on soil health and soil microbiology. In *Agrochemicals Detection, Treatment and Remediation: Pesticides and Chemical Fertilisers*; Prasad, M.N.V., Ed.; Butterworth-Heinemann: Oxford, UK, 2020; pp. 25–54.
12. FAOSTAT. Pesticides. Food and Agriculture Organization, Rome. Available online: http://www.fao.org/faostat/en/#data/RP (accessed on 10 May 2021).

13. Consumption of Chemical Pesticides in Various States/UTs during 2014–2015 to 2019–2020. Department of Agriculture, Cooperation and Farmers Welfare, Ministry of Agriculture and Farmers Welfare, Government of India. Available online: http://ppqs.gov.in/statistical-database (accessed on 10 May 2021).
14. FICCI. *A Report on Indian Agrochemical Industry*; A Federation of Indian Chambers of Commerce and Industry Report; FICCI: New Delhi, India, 2016; p. 45. Available online: https://ficci.in/spdocument/23103/agrochemical-ficci.pdf (accessed on 3 May 2021).
15. Annual Report 2019–2020. Department of Chemicals and Petrochemicals, Ministry of Chemicals and Fertilisers, Government of India. Available online: http://chemicals.nic.in/sites/default/files/English%20Annual%20Report%20date%2024-2-2020.pdf (accessed on 10 May 2021).
16. EPA. Ingredients Used in Pesticide Products: Pesticides. What Are Biopesticides? Available online: https://www.epa.gov/ingredients-used-pesticide-products/what-are-biopesticides (accessed on 10 May 2021).
17. Essiedu, J.A.; Adepoju, F.O.; Ivantsova, M.N. Benefits and limitations in using biopesticides: A review. In Proceedings of the VII International Young Researchers' Conference—Physics, Technology, Innovations (PTI-2020), Ekaterinburg, Russia, 18–22 May 2020; Volume 2313, p. 080002. [CrossRef]
18. Chang, J.H.; Choi, J.Y.; Jin, B.R.; Roh, J.Y.; Olszewski, J.A.; Seo, S.J.; O'Reilly, D.R.; Je, Y.H. An improved baculovirus insecticide producing occlusion bodies that contain Bacillus thuringiensis insect toxin. *J. Invertebr. Pathol.* **2003**, *84*, 30–37. [CrossRef]
19. Ruiu, L. Microbial Biopesticides in Agroecosystems. *Agronomy* **2018**, *8*, 235. [CrossRef]
20. Ujváry, I. Chapter 3—Pest Control Agents from Natural Products. In *Handbook of Pesticide Toxicology*, 2nd ed.; Krieger, R.I., Krieger, W.C., Eds.; Academic Press: San Diego, CA, USA, 2001; pp. 109–179.
21. Sporleder, M.; Lacey, L.A. Biopesticides. In *Insect Pests of Potato*; Alyokhin, A., Vincent, C., Giordanengo, P., Eds.; Elsevier: Oxford, UK, 2021; pp. 463–497.
22. Gonzalez-Coloma, A.; Reina, M.; Diaz, C.E.; Fraga, B.M.; Santana-Meridas, O. Natural Product-Based Biopesticides for Insect Control. In *Reference Module in Chemistry, Molecular Sciences and Chemical Engineering*; Elsevier Inc.: Amsterdam, The Netherlands, 2013. [CrossRef]
23. Gurr, G.M.; Thwaite, W.G.; Nicol, H.I. Field evaluation of the effects of the insect growth regulator (tebufenozide) on entomophagous arthropods and pests of apples. *Austr. J. Entomol.* **1999**, *38*, 135–140. [CrossRef]
24. Magierowicz, K.; Górska-Drabik, E.; Golan, K. Effects of plant extracts and essential oils on the Inc.behavior of Acrobasis advenella (Zinck.) caterpillars and females. *J. Plant Dis. Prot.* **2020**, *127*, 63–71. [CrossRef]
25. Ali, M.A.; Doaa, S.M.; El-Sayed, H.S.; Asmaa, M.E. Antifeedant activity and some biochemical effects of garlic and lemon essential oils on *Spodoptera littoralis* (Boisduval) (Lepidoptera: Noctuidae). *J. Entomol. Zool.* **2017**, *5*, 1476–1482.
26. Halder, J.; Rai, A.B.; Kodandaram, M.H. Compatibility of Neem Oil and Different Entomopathogens for the Management of Major Vegetable Sucking Pests. *Natl. Acad. Sci. Lett.* **2013**, *36*, 19–25. [CrossRef]
27. Tripathi, A.K.; Upadhyay, S.; Bhuiyan, M.; Bhattacharya, P.R. A review on prospects of essential oils as biopesticide in insect-pest management. *J. Pharmacogn. Phytother.* **2009**, *1*, 52–63.
28. Mordue, A.J.; Morgan, E.D.; Nisbet, A.J. Azadirachtin, a Natural Product in Insect Control. In *Comprehensive Molecular Insect Science*; Gilbert, L.I., Ed.; Elsevier: Amsterdam, The Netherlands, 2005; pp. 117–135.
29. Dannon, H.F.; Dannon, A.E.; Douro-kpindou, O.K.; Zinsou, A.V.; Houndete, A.T.; Toffa-Mehinto, J.; Elegbede, I.A.T.M.; Olou, B.D.; Tamo, M. Toward the efficient use of *Beauveria bassiana* in integrated cotton insect pest management. *J. Cotton Res.* **2020**, *3*, 24. [CrossRef]
30. Parween, T.; Jan, S. Pesticides and environmental ecology. In *Ecophysiology of Pesticides*; Parween, T., Jan, S., Eds.; Academic Press: Cambridge, MA, USA, 2019; pp. 1–38.
31. Gwinn, K.D. Bioactive Natural Products in Plant Disease Control. In *Studies in Natural Products Chemistry*; Attaur, R., Ed.; Elsevier Inc.: Amsterdam, The Netherlands, 2018; Volume 56, pp. 229–246. [CrossRef]
32. Parker, K.M.; Barragán, B.V.; van Leeuwen, D.M.; Lever, M.A.; Mateescu, B.; Sander, M. Environmental Fate of RNA Interference Pesticides: Adsorption and Degradation of Double-Stranded RNA Molecules in Agricultural Soils. *Environ. Sci. Technol.* **2019**, *53*, 3027–3036. [CrossRef]
33. Svidritskiy, E.; Ling, C.; Ermolenko, D.N.; Korostelev, A.A. Blasticidin S inhibits translation by trapping deformed tRNA on the ribosome. *Proc. Natl. Acad. Sci. USA* **2013**, *110*, 12283–12288. [CrossRef] [PubMed]
34. Schuwirth, B.S.; Day, J.M.; Hau, C.W.; Janssen, G.R.; Dahlberg, A.E.; Cate, J.H.; Vila-Sanjurjo, A. Structural analysis of kasugamycin inhibition of translation. *Nat. Struct. Mol. Biol.* **2006**, *13*, 879–886. [CrossRef]
35. Arena, J.P.; Liu, K.K.; Paress, P.S.; Frazier, E.G.; Cully, D.F.; Mrozik, H.; Schaeffer, J.M. The mechanism of action of avermectins in Caenorhabditis elegans: Correlation between activation of glutamate-sensitive chloride current, membrane binding, and biological activity. *J. Parasitol.* **1995**, *2*, 286–294. [CrossRef]
36. Feduchi, E.; Cosín, M.; Carrasco, L. Mildiomycin: A nucleoside antibiotic that inhibits protein synthesis. *J. Antibiot.* **1985**, *38*, 415–419. [CrossRef]
37. Schorderet, W.S.; Kaminski, K.P.; Perret, J.-L.; Leroy, P.; Mazurov, A.; Peitsch, M.C.; Ivanov, N.V.; Hoeng, J. Antiparasitic properties of leaf extracts derived from selected Nicotiana species and Nicotiana tabacum varieties. *Food Chem. Toxicol.* **2019**, *132*, 110660. [CrossRef] [PubMed]
38. Duan, C.B.; Du, Y.; Hou, X.; Yan, N.; Dong, W.; Mao, X.; Zhang, Z. Chemical Basis of the Fungicidal Activity of Tobacco Extracts against *Valsa mali*. *Molecules* **2016**, *21*, 1743. [CrossRef] [PubMed]

39. Elgar, M.A.; Zhang, D.; Wang, Q.; Wittwer, B.; Thi Pham, H.; Johnson, T.L.; Freelance, C.B.; Coquilleau, M. Insect Antennal Morphology: The Evolution of Diverse Solutions to Odorant Perception. *Yale J. Biol. Med.* **2018**, *91*, 457–469. [PubMed]
40. Gomiero, T. Food quality assessment in organic vs. conventional agricultural produce: Findings and issues. *Appl. Soil Ecol.* **2018**, *123*, 714–728. [CrossRef]
41. Ranglová, K.; Lakatos, G.E.; Câmara Manoel, J.A.; Grivalský, T.; Suárez Estrella, F.; Acién Fernández, F.G.; Molnár, Z.; Ördög, V.; Masojídek, J. Growth, biostimulant and biopesticide activity of the MACC-1 Chlorella strain cultivated outdoors in inorganic medium and wastewater. *Algal Res.* **2021**, *53*, 102136. [CrossRef]
42. Costa, J.A.V.; Freitas, B.C.B.; Cruz, C.G.; Silveira, J.; Morais, M.G. Potential of microalgae as biopesticides to contribute to sustainable agriculture and environmental development. *J. Environ. Sci. Health Part B* **2019**, *54*, 366–375. [CrossRef] [PubMed]
43. Kumar, M.; Prasanna, R.; Bidyarani, N.; Babu, S.; Mishra, B.K.; Kumar, A.; Adak, A.; Jauhari, S.; Yadav, K.; Singh, R.; et al. Evaluating the plant growth promoting ability of thermotolerant bacteria and cyanobacteria and their interactions with seed spice crops. *Sci. Hortic.* **2013**, *164*, 94–101. [CrossRef]
44. Koul, O. Microbial biopesticides: Opportunities and challenges. *CAB Rev. Perspect. Agric. Vet. Sci. Nutr. Nat. Resour.* **2011**, *6*, 1–26.
45. Bravo, A.; Gill, S.S.; Soberón, M. Mode of action of *Bacillus thuringiensis* Cry and Cyt toxins and their potential for insect control. *Toxicon* **2007**, *49*, 423–435. [CrossRef]
46. Saberi, F.; Marzban, R.; Ardjmand, M.; Pajoum, S.F.; Tavakoli, O. Optimization of culture media to enhance the ability of local Bacillus thuringiensis var. tenebrionis. *J. Saudi Soc. Agric. Sci.* **2020**, *19*, 468–475. [CrossRef]
47. McGuire, A.V.; Northfield, T.D. Tropical Occurrence and Agricultural Importance of Beauveria bassiana and Metarhizium anisopliae. *Front. Sustain. Food Syst.* **2020**, *4*, 6. [CrossRef]
48. Gams, W.; Diederich, P.; Poldmaa, K. Fungicolous Fungi. In *Biodiversity of Fungi*; Mueller, G.M., Bills, G.F., Foster, M.S., Eds.; Academic Press: Cambridge, MA, USA, 2004; pp. 343–392. [CrossRef]
49. Dolatabadi, K.H.; Goltapeh, E.M.; Varma, A.; Rohani, N. In vitro evaluation of arbuscular mycorrhizal-like fungi and Trichoderma species against soil borne pathogens. *J. Agric. Technol.* **2011**, *7*, 73–84.
50. Bailey, K.L. The Bioherbicide Approach to Weed Control Using Plant Pathogens. In *Integrated Pest Management*; Abrol, D.P., Ed.; Academic Press: Cambridge, MA, USA, 2014; pp. 245–266. [CrossRef]
51. Moreno-Gavíra, A.; Huertas, V.; Diánez, F.; Sánchez-Montesinos, B.; Santos, M. Paecilomyces and Its Importance in the Biological Control of Agricultural Pests and Diseases. *Plants* **2020**, *9*, 1746. [CrossRef] [PubMed]
52. Kadoić, B.M.; Bažok, R.; Mikac, K.M.; Lemic, D.; Pajač, Ž.I. Pest Management Challenges and Control Practices in Codling Moth: A Review. *Insects* **2020**, *11*, 38. [CrossRef] [PubMed]
53. Lala, F.; Cahyaningrum, H.; Hadiarto, A.; Brahmantiyo, B. Implementation of Biopesticides to Control *Phytophthora palmivora* Butl. and *Conopomorpha cramerella* Snell. on Cacao in South Halmahera. In Proceedings of the 5th International Conference on Food, Agriculture and Natural Resources (FANRes, 2019), Ternate, Indonesia, 17–19 September 2019. [CrossRef]
54. Chaudhary, S.; Kanwar, R.K.; Sehgal, A.; Cahill, D.M.; Barrow, C.J.; Sehgal, R.; Kanwar, J.R. Progress on *Azadirachta indica* Based Biopesticides in Replacing Synthetic Toxic Pesticides. *Front. Plant Sci.* **2017**, *8*, 610. [CrossRef] [PubMed]
55. Marrone, P. An effective biofungicide with novel modes of action. *Pestic. Outlook* **2002**, *13*, 193–194. [CrossRef]
56. Isman, M.B. Bioinsecticides based on plant essential oils: A short overview. *Z Nat. C J. Biosci.* **2020**, *75*, 179–182. [CrossRef]
57. Guerra, F.; Sepúlveda, S. Saponin Production from Quillaja Genus Species. An Insight into Its Applications and Biology. *Sci. Agric.* **2021**, *78*, 305. [CrossRef]
58. Ishikawa, S. Integrated disease management of strawberry anthracnose and development of a new biopesticide. *J. Gen. Plant Pathol.* **2013**, *79*, 441–443. [CrossRef]
59. Mensah, R.; Moore, C.; Watts, N.; Deseo, M.A.; Glennie, P.; Pitt, A. Discovery and development of a new semiochemical biopesticide for cotton pest management: Assessment of extract effects on the cotton pest *Helicoverpa* spp. *Entom. Exp. Appl.* **2014**, *152*, 1–15. [CrossRef]
60. Kirk, W.W.; Schafer, R.S. Efficacy of new active ingredient formulations and new biopesticides for managing Fusarium root rot disease of gladiolus hybrids. *Acta Hortic.* **2015**, *1105*, 55–60. [CrossRef]
61. Eski, A.; Demir, D.; Sezen, K.; Demirbag, Z.A. New biopesticide from a local *Bacillus thuringiensis* var. tenebrionis (Xd3) against alder leaf beetle (Coleoptera: Chrysomelidae). *World J. Microbiol. Biotechnol.* **2017**, *33*, 95. [CrossRef] [PubMed]
62. Dubois, C.; Arsenault-Labrecque, G.; Pickford, J. Evaluation of a new biopesticide against angular leaf spot in a commercial operation system. *Acta Hortic.* **2017**, *1156*, 757–764. [CrossRef]
63. Pavela, R.; Waffo-Teguo, P.; Biais, B.; Richard, T.; Mérillon, J.M. *Vitis vinifera* canes, a source of stilbenoids against *Spodoptera littoralis* larvae. *J. Pest Sci.* **2017**, *90*, 961–970. [CrossRef]
64. El-Abbassi, A.; Saadaoui, N.; Kiai, H.; Raiti, J.; Hafidi, A. Potential applications of olive mill wastewater as biopesticide for crops protection. *Sci. Total Environ.* **2017**, *576*, 10–21. [CrossRef]
65. Mauchline, A.L.; Hervé, M.R.; Cook, S.M. Semiochemical-based alternatives to synthetic toxicant insecticides for pollen beetle management. *Arthropod-Plant Interact.* **2018**, *12*, 835–847. [CrossRef]
66. Mossa, A.T.H.; Abdelfattah, N.A.H.; Mohafrash, S.M.M. Nanoemulsion of camphor (*Eucalyptus globules*) essential oil, formulation, characterization and insecticidal activity against wheat weevil, *Sitophilus granaries*. *Asian J. Crop Sci.* **2017**, *9*, 50–62. [CrossRef]
67. El-Sayed, A.; Bengtsson, M.; Rauscher, S.; Lofqvist, J.; Witzgall, P. Multicomponent Sex Pheromone in Codling Moth (Lepidoptera: Tortricidae). *Environ. Entomol.* **1999**, *28*, 775–779. [CrossRef]

68. Olowe, O.M.; Olawuyi, O.J.; Sobowale, A.A.; Odebode, A.C. Role of arbuscular mycorrhizal fungi as biocontrol agents against Fusarium verticillioides causing ear rot of *Zea mays* L. (Maize). *Curr. Plant Biol.* **2018**, *15*, 30–37. [CrossRef]
69. Bharadwaj, A.; Sharma, S. Role of Arbuscular Mycorrhizae in pest management. *Int. J. Ecol. Environ. Sci.* **2006**, *32*, 193–205.
70. Mukerji, K.G.; Ciancio, A. Mycorrhizae in the integrated pest and disease management. In *General Concepts in Integrated Pest and Disease Management. Integrated Management of Plants Pests and Diseases*; Ciancio, A., Mukerji, K.G., Eds.; Springer: Dordrecht, Switzerland, 2007; Volume 1. [CrossRef]
71. Narware, J.; Yadav, R.N.; Keswani, C.; Singh, S.P.; Singh, H.B. Silver nanoparticle-based biopesticides for phytopathogens: Scope and potential in agriculture. In *Nano-Biopesticides Today and Future Perspectives*; Koul, O., Ed.; Academic Press: Cambridge, MA, USA, 2019; pp. 303–314.
72. Rodrigues, D.; Costa-Pinto, A.R.; Sousa, S.; Vasconcelos, M.W.; Pintado, M.M.; Pereira, L.; Rocha-Santos, T.A.P.; Costa, J.P.D.; Silva, A.M.S.; Duarte, A.C.; et al. *Sargassum muticum* and *Osmundea pinnatifida* Enzymatic Extracts: Chemical, Structural, and Cytotoxic Characterization. *Mar. Drugs* **2019**, *17*, 209. [CrossRef]
73. Sponsler, D.B.; Grozinger, C.M.; Hitaj, C.; Rundlöf, M.; Botías, C.; Code, A.; Lonsdorf, E.V.; Melathopoulos, A.P.; Smith, D.J.; Suryanarayanan, S.; et al. Pesticides and pollinators: A socioecological synthesis. *Sci. Total Environ.* **2019**, *662*, 1012–1027. [CrossRef]
74. Dar, S.A.; Wani, S.H.; Mir, S.H.; Showkat, A.; Dolkar, T.; Dawa, T. Biopesticides: Mode of Action, Efficacy and Scope in Pest Management. *J. Adv. Res. Biochem. Pharmacol.* **2021**, *4*, 1–8.
75. Malinga, L.N.; Laing, M.D. Efficacy of three biopesticides against cotton pests under field conditions in South Africa. *Crop. Prot.* **2021**, *145*, 105578. [CrossRef]
76. Ratcliff, F.; Harrison, B.D.; Baulcombe, D.C. A similarity between viral defense and gene silencing in plants. *Science* **1997**, *276*, 1558–1560. [CrossRef] [PubMed]
77. Ruiz, M.T.; Voinnet, O.; Baulcombe, D.C. Initiation and maintenance of virus-induced gene silencing. *Plant Cell* **1998**, *10*, 937–946. [CrossRef]
78. Wang, M.B.; Abbott, D.C.; Waterhouse, P.M. A single copy of a virus-derived transgene encoding hairpin RNA gives immunity to barley yellow dwarf virus. *Mol. Plant Pathol.* **2000**, *1*, 347–356. [CrossRef]
79. Hamilton, A.J.; Baulcombe, D.C. A species of small antisense RNA in posttranscriptional gene silencing in plants. *Science* **1999**, *286*, 950–952. [CrossRef] [PubMed]
80. Zamore, P.D.; Tuschl, T.; Sharp, P.A.; Bartel, D.P. RNAi: Doublestranded RNA directs the ATP-dependent cleavage of mRNA at 21 to 23 nucleotide intervals. *Cell* **2000**, *101*, 25–33. [CrossRef]
81. Dalmay, T.; Hamilton, A.; Rudd, S.; Angell, S.; Baulcombe, D.C. An RNA-Dependent RNA polymerase gene in Arabidopsis is required for posttranscriptional gene silencing mediated by a transgene but not by a virus. *Cell* **2000**, *101*, 543–553. [CrossRef]
82. Jacobsen, S.; Running, M.; Meyerowitz, E. Disruption of an RNA helicase/RNAse III gene in Arabidopsis causes unregulated cell division in floral meristems. *Development* **1999**, *126*, 5231–5243. [CrossRef]
83. Schauer, S.E.; Jacobsen, S.E.; Meinke, D.W.; Ray, A. DICER-LIKE1: Blind men and elephants in Arabidopsis development. *Trends Plant Sci.* **2002**, *7*, 487–491. [CrossRef]
84. Liu, J.; Carmell, M.A.; Rivas, F.V.; Marsden, C.G.; Thomson, J.M.; Song, J.-J.; Hammond, S.M.; Joshua-Tor, L.; Hannon, G.J. Argonaute2 is the catalytic engine of mammalian RNAi. *Science* **2004**, *305*, 1437–1441. [CrossRef] [PubMed]
85. Baumberger, N.; Baulcombe, D.C. Arabidopsis ARGONAUTE1 is an RNA Slicer that selectively recruits rnicroRNAs and short interfering RNAs. *Proc. Natl. Acad. Sci. USA* **2005**, *102*, 11928–11933. [CrossRef] [PubMed]
86. Fletcher, S.J.; Reeves, P.T.; Hoang, B.T.; Mitter, N. A perspective on RNAi-based biopesticides. *Front. Plant Sci.* **2020**, *11*, 51. [CrossRef]
87. Mao, Y.B.; Cai, W.J.; Wang, J.W.; Hong, G.J.; Tao, X.Y.; Wang, L.J.; Huang, Y.-P.; Chen, X.-Y. Silencing a cotton bollworm P450 monooxygenase gene by plant-mediated RNAi impairs larval tolerance of gossypol. *Nat. Biotechnol.* **2007**, *25*, 1307–1313. [CrossRef]
88. Waltz, E. USDA approves next-generation GM potato. *Nat. Biotechnol.* **2015**, *33*, 12–13. [CrossRef]
89. Baranski, R.; Klimek-Chodacka, M.; Lukasiewicz, A. Approved genetically modified (GM) horticultural plants: A 25-year perspective. *Folia Hortic.* **2019**, *31*, 3–49. [CrossRef]
90. Rosa, C.; Kuo, Y.W.; Wuriyanghan, H.; Falk, B.W. RNA interference mechanisms and applications in plant pathology. *Annu. Rev. Phytopathol.* **2018**, *56*, 581–610. [CrossRef]
91. Head, G.P.; Carroll, M.W.; Evans, S.P.; Rule, D.M.; Willse, A.R.; Clark, T.L.; Storer, N.P.; Flannagan, R.D.; Samuel, L.W.; Meinke, L.J. Evaluation of SmartStax and SmartStax PRO maize against western corn rootworm and northern corn rootworm: Efficacy and resistance management. *Pest Manag. Sci.* **2017**, *73*, 1883–1899. [CrossRef]
92. Roh, J.Y.; Choi, J.Y.; Li, M.S.; Jin, B.R.; Je, Y.H. Bacillus thuringiensis as a specific, safe, and effective tool for insect pest control. *J. Microbiol. Biotechnol.* **2007**, *17*, 547–549. [PubMed]
93. Mishra, V.; Ellouze, W.; Howard, R.J. Utility of Arbuscular Mycorrhizal Fungi for Improved Production and Disease Mitigation in Organic and Hydroponic Greenhouse Crops. *J. Hortic.* **2018**, *5*, 1–10. [CrossRef]
94. Ellouze, W.; Hamel, C.; Singh, A.K.; Mishra, V.; DePauw, R.M.; Knox, R.E. Abundance of the arbuscular mycorrhizal fungal taxa associated with the roots and rhizosphere soil of different durum wheat cultivars in the Canadian prairie. *Can. J. Microbiol.* **2018**, *64*, 527–536. [CrossRef] [PubMed]

95. Vannette, R.L.; Hunter, M.D. Mycorrhizal fungi as mediators of defence against insect pests in agricultural systems. *Agric. For. Entomol.* **2009**, *11*, 351–358. [CrossRef]
96. Olowe, O.M.; Odebode, A.C.; Olawuyi, O.J.; Akanmu, A.O. Correlation, Principal Component Analysis and Tolerance of Maize Genotypes to Drought and Diseases in Relation to Growth Traits. *Am. Eurasian J. Agric. Environ. Sci.* **2003**, *13*, 1554–1561.
97. Nuruzzaman, M.; Rahman, M.M.; Liu, Y.; Naidu, R. Nanoencapsulation, Nano-guard for Pesticides: A New Window for Safe Application. *J. Agric. Food Chem.* **2016**, *64*, 1447–1483. [CrossRef]
98. Kremer, R.J. Bioherbicides and nanotechnology: Current status and future trends. In *Nano-Biopesticides Today and Future Perspectives*; Koul, O., Ed.; Academic Press: Cambridge, MA, USA, 2019; pp. 353–366.
99. Abd-Elsalam, K.A.; Al-Dhabaan, F.A.; Alghuthaymi, M.; Njobeh, P.B.; Almoammar, H. Nanobiofungicides: Present concept and future perspectives in fungal control. In *Nano-Biopesticides Today and Future Perspectives*; Koul, O., Ed.; Academic Press: Cambridge, MA, USA, 2019; pp. 315–351.
100. Fu, Y.; Bhadha, J.H.; Rott, P.; Beuzelin, J.M.; Kanissery, R. Investigating the use of aquatic weeds as biopesticides towards promoting sustainable agriculture. *PLoS ONE* **2020**, *15*, e0237258. [CrossRef]
101. Sundh, I.; Goettel, M.S. Regulating biocontrol agents: A historical perspective and a critical examination comparing microbial and macrobial agents. *BioControl* **2012**, *58*, 575–593. [CrossRef]
102. Damalas, C.A.; Koutroubas, S.D. Current Status and Recent Developments in Biopesticide Use. *Agriculture* **2018**, *8*, 13. [CrossRef]
103. Kumar, S.; Singh, A. Biopesticides: Present status and the future prospects. *J. Fertil. Pestic.* **2015**, *6*, e129. [CrossRef]
104. Czaja, K.; Góralczyk, K.; Strucinski, P.; Hernik, A.; Korcz, W.; Minorczyk, M.; Łyczewska, M.; Ludwicki, J.K. Biopesticides—Towards increased consumer safety in the European Union. *Pest Manag. Sci.* **2015**, *71*, 3–6. [CrossRef] [PubMed]

Review

Photorhabdus spp.: An Overview of the Beneficial Aspects of Mutualistic Bacteria of Insecticidal Nematodes

Mahfouz M. M. Abd-Elgawad

National Research Centre, Plant Pathology Department, Agricultural and Biological Research Division, El-behooth St., Dokki, Giza 12622, Egypt; mahfouzian2000@yahoo.com

Abstract: The current approaches to sustainable agricultural development aspire to use safer means to control pests and pathogens. *Photorhabdus* bacteria that are insecticidal symbionts of entomopathogenic nematodes in the genus *Heterorhabditis* can provide such a service with a treasure trove of insecticidal compounds and an ability to cope with the insect immune system. This review highlights the need of *Photorhabdus*-derived insecticidal, fungicidal, pharmaceutical, parasiticidal, antimicrobial, and toxic materials to fit into current, or emerging, holistic strategies, mainly for managing plant pests and pathogens. The widespread use of these bacteria, however, has been slow, due to cost, natural presence within the uneven distribution of their nematode partners, and problems with trait stability during in vitro culture. Yet, progress has been made, showing an ability to overcome these obstacles via offering affordable mass production and mastered genome sequencing, while detecting more of their beneficial bacterial species/strains. Their high pathogenicity to a wide range of arthropods, efficiency against diseases, and versatility, suggest future promising industrial products. The many useful properties of these bacteria can facilitate their integration with other pest/disease management tactics for crop protection.

Keywords: biocontrol; *Heterorhabditis*; pest and pathogen management; *Photorhabdus*; marketing

Citation: Abd-Elgawad, M.M.M. *Photorhabdus* spp.: An Overview of the Beneficial Aspects of Mutualistic Bacteria of Insecticidal Nematodes. *Plants* **2021**, *10*, 1660. https://doi.org/10.3390/plants10081660

Academic Editors: Dragana Šunjka and Špela Mechora

Received: 4 July 2021
Accepted: 3 August 2021
Published: 12 August 2021

1. Introduction

In the agricultural sector, the current production practices are not sufficient to control insect pests and pathogens safely and with 100% efficacy. Growing dissatisfaction with the chemicals that are used for plant pest control has increased, due to their negative impacts on human health and non-target organisms, contamination, toxicity to the environment, and resistance development [1–3]. Thus, it has become imperative to decrease the use of these unsafe pesticides and replace them with benign ecologically sound products for crop pest control, in the context of sustainable agriculture.

Recently, the importance of using beneficial bacteria in integrated pest and pathogen management programs has been emphasized. These bacteria can reduce the chemical inputs used for plant protection, and stabilize ecological changes [4]. Conceivably, entomopathogenic bacterial species of the genus *Photorhabdus* (Enterobacteriales: Morganellaceae) may be a favorable alternative for expanding the biocontrol of many plant pests and pathogens, via their secretion of various arrays comprising effective bioactive metabolites [2,5–9]. This concept is based on a huge number of *Photorhabdus* genes that encode for producing relevant compounds, e.g., enzymes, toxins, antibiotics, and bacteriocins. They are found in the form of pathogenicity islands on the bacterial chromosomes representing different groups, e.g., the group of toxins is assorted to four prime groups. Additionally, discoveries of novel species of *Photorhabdus*, with presumed additional genes encoding for beneficial traits, are being discovered, and novel genes in the *Photorhabdus* species that are already described are ongoing, and can further be exploited [1,8–11]. The goal of this paper is to provide an overview on the pros and cons of *Photorhabdus* spp. as an all-in-one resource for biocontrol. The aim is not only to focus on the related background of the state-of-the-art knowledge, but also to highlight the modern biotechnology that can open further avenues of discovery that can be leveraged towards enhancing sustainable agriculture. So, this review addresses recent findings on their diversity, new taxa, and use in the existing or emerging programs to manage plant pests and pathogens.

2. Taxa, Diversity, and Lifestyle of *Photorhabdus* spp.

2.1. Their Taxa and Diversity

The number of *Photorhabdus* species has recently doubled, from four [12] to twenty [13]. An updated list of *Photorhabdus*, comprising the valid species/subspecies, is presented (Table 1). Yet, they are expected to increase as, in addition to their importance, much focus is directed to employ morphological, biochemical, physiological, and molecular approaches to characterize and identify them as mutualistic/symbiotic bacteria of *Heterorhabditis* spp. Some of *Photorhabdus* spp. are further divided into subspecies (Table 1). Using the entire genome-established phylogenesis of the *Photorhabdus* taxa (species, subspecies, strains, and isolates), a novel phylogenetic tree of the genus *Photorhabdus*, with new taxa, has been reconstructed [13]. For instance, considering the recent sequence comparative investigations of the related taxa, *P. aegyptia* was erected as a new species. Its type strain is a symbiont of *H. indica*, originally collected from Egypt.

As heterorhabditid nematodes are found in many, and wide, geographical areas, their symbionts, *Photorhabdus* spp., must have global distribution. An overall heterorhabditid nematode geographic distribution was reviewed [5]. While the most widespread species, *H. bacteriophora*, is found in zones having continental and Mediterranean climates, *H. indica* is familiar to the subtropics and tropics. *Heterorhabditis megidis* generally has a more northerly and more limited distribution than *H. bacteriophora*. The EPN species are found worldwide; the only continent where these nematodes have not been found is Antarctica. The longstanding recognition concerning the species-specific identification for the complex of *Photorhabdus-Heterorhabditis* as a dyad in their mutualism, is still generally valid. Thus, it can gain more knowledge regarding their distribution in space and diversity [14].

Basically, researchers have been trying to optimize EPN surveys and extraction methods [15,16] to discover novel species/strains that have adapted to local conditions, and to boost the effective control of pests and pathogens. Therefore, it can be generally assumed that the distribution of the EPN species is just an artefact of such sampling trials. However, increased attention is mainly paid to these bacteria when applied to kill insect pests and pathogens independently of their EPN partners.

Eventually, fueled by the gradual increase in the number of new *Photorhabdus* species/strains and novel technological approaches to identify them, the *Photorhabdus* taxa have become clearer and easier to study than before. Therefore, complete genome-constructed phylogenetic trees, along with accurate sequence comparative investigations, could correctly determine their relationship and diversity. This identification could also contribute to additional investigations of the relevant *Photorhabdus* spp.-bioactive compounds that can be applied in industrial and agricultural products [2,13,17].

Table 1. Updated list of species/subspecies in the genus *Photorhabdus* and their mutualistic heterorhabditid nematodes.

***Photorhabduss* Species**	***Heterorhabditis* Species**	**References**
P. akhurstii	*H. indica*	[18,19]
P. asymbiotica	undescribed species	[18,20]
P. australis subsp. *Thailandensis* subsp. *australis*	*H. gerrardi, H. indica*	[13,19,20]
P. bodei	*H. beicherriana*	[19]
P. caribbeanensis	*H. bacteriophora*	[19,21]

Table 1. *Cont.*

Photorhabduss Species	*Heterorhabditis* Species	References
P. cinerea	*H. downesi, H. megidis, H. bacteriophora*	[19,22]
P. hainanensis	undescribed species	[19,21]
P. heterorhabditis subsp. *Aluminescens* subsp. *heterorhabditis*	*H. zealandica*	[13,23]
P. kayaii	*H. bacteriophora*	[19,24]
P. khanii	*H. bacteriophora*	[19,21]
subsp. *guanajuatensis*	*H. atacamensis*	[25]
P. kleinii	*H. georgiana, H. bacteriophora*	[19,26]
P. laumondii subsp. *clarkei* subsp. *laumondii*	*H. bacteriophora*	[18,19]
P. luminescens	*H. bacteriophora, H. indica*	[27,28]
subsp. *sonorensis*	*H. sonorensis*	[29]
subsp. *mexicana*	*H. mexicana*	[25]
P. namnaonensis	*H. baujardi*	[19,30]
P. noenieputensis	*H. indica, Heterorhabditis* sp.	[19,31]
P. stackebrandtii	*H. bacteriophora, H. georgiana*	[19,32]
P. tasmanensis	*H. zealandica, H. marelatus*	[19,21]
P. temperata	*H. megidis, H. downesi, H. zealandica*	[18,19]
P. thracensis	*H. bacteriophora*	[19,21,24]
P. aegyptia	*H. indica*	[13]

2.2. The Lifestyle of Photorhabdus spp.

All the species of *Photorhabdus* are exclusively found as symbionts of the *Heterorhabditis* spp.-infective juvenile (IJ) stage [33]. As these bacteria have not been detected in free-living form in nature, they had previously raised doubts of their competence to live and infect insect pests and pathogens in the absence of their EPN partner. An exception is *P. asymbiotica*, which is known to be a human pathogen, although it also infects insects [34]. This species causes ulcerated skin lesions, both at the initial infection foci and, later, at disseminated distal sites [35]. Another featured difference to the other *Photorhabdus* spp. is the typical coloration of grey with pink spots, acquired by the *P. asymbiotica*-infected insect host, but that color for the other species of *Photorhabdus* is usually reddish [36].

During mutualism, these bacteria offer a few favors to the heterorhabditid nematodes, in terms of killing insects that are invaded by the nematodes, and supplying nutrition and defense to these entomopathogenic nematodes within the infected cadavers. The nematodes pay back such favors by protecting the mutualistic bacteria outside their insect host from harsh conditions, while enabling them to set off and multiply within the attacked insect hosts. Notwithstanding the EPNs, as originally soil-inhabiting creatures, their free-living stage, as IJ, harbors the bacteria in its intestinal lumen [33]. Due to their tripartite association [37], *Photorhabdus* spp. are naturally found either in their infected insect hosts or at the intestinal lumen (gut mucosa) of their nematode partner. As the IJs penetrate their insect hosts via natural openings (spiracles, anus, or mouth) or the cuticle, they move forward to the haemocoel, where they let out their mutualistic bacteria to multiply and produce secondary metabolites that kill their infected arthropods and turn their bodies into a nutrient soup. During this bioconversion, the nematode feed on the soup, in order to develop and reproduce while the bacteria produce antibiotics that keep the infected insect away from further microbial attacks.

Referring to *Photorhabdus* is frequently accompanied by its counterpart *Xenorhabdus* bacteria, because there is a great similarity between them in the general framework of living. *Xenorhabdus* and *Photorhabdus* are symbionts of EPN-IJs, of the genera *Steinernema* and *Heterorhabditis*, respectively. However, each of the two bacterial genera has its own attributes that distinguish its character. *Xenorhabdus* spp. live within a specialized receptacle at the front portion of the intestine, but *Photorhabdus* spp. are set at the gut mucosa of their respective nematodes [38,39]. The IJs of *Steinernema* spp. develop into amphimictic females and males. So, both sexes must infect the insect host and mate to reproduce in the insect hemocele, but the *Heterorhabditis*–IJs mature to hermaphroditic females and then to the two sexes. Thus, a single hermaphroditic *Heterorhabditis*–IJ can infect and multiply within the insect host. Also, the colonization behavior of the two bacterial genera differs within their nematode partners [39]. A significant trait that is used to distinguish between the two EPN genera, in case of their presence within their insect host, is the pigment of the *Photorhabdus* bacterium to color the host's body in a reddish shade. *Photorhabdus* is capable of fluorescing so often that the entire infected cadavers glow in unlighted places. *Photorhabdus* avoid the immune system of the insect host by adjusting the lipopolysaccharide to withstand the impact of the insect-derived antimicrobial produce of peptides, while *Xenorhabdus* disturbs the induction of such peptide expression [38]. Their 16S rRNA genes are more than 94% identical, though the genome of each genus may be disrupted by numerous deletions, inversions, insertions, and translocations [40]. *Photorhabdus* bacteria can go through major transcriptional reforming in the intestine of their EPN partner. They can induce general starvation mechanisms, turn into the pentose phosphate pathway to cope with oxidative stress and nutrition deficit, cellular acidification to slacken growth, and form biofilms to safely remain in the EPN intestine until transmitting to the insect hemolymph [32]. Such bacterial strategies can sustain them within the nematode gut and ensure felicitous transmission of the couple from one insect host to another. Interestingly, the *Xenorhabdus* and *Photorhabdus* species are able to grow in vitro as free-living organisms, without their partner EPNs, on artificial media under certain terms, i.e., adequate nutrient media with no competition.

Until the beginnings of the current century, it was believed that the relationship between EPNs and their mutualistic *Photorhabdus* bacteria was extremely specific, i.e., each EPN species carries only a corresponding bacterial species or subspecies. Contrary to the current findings [41–43], it was thought that the tight relatedness of the two taxonomic groups conduces co-speciation between each pair of *Photorhabdus* spp. and *Heterorhabditis* spp., i.e., the mutualism between *Heterorhabditis* and *Photorhabdus* had been previously thought to be strictly one-to-one, in terms of co-speciation [33,44]. However, an opposite opinion is exemplified in the cohabitation of mixed *Photorhabdus* species (*P. cinerea* and *P. temperata*) within a single nematode host; *Heterorhabditis downesi* [45]. The latter authors could further examine the aspects of competition between *P. cinerea* and *P. temperata*, associated with *H. downesi* at two levels. Apparently, *P. cinerea* can better protect the nematode against desiccation at the regional level, though *P. temperata* is superior to *P. cinerea* in protecting the scavengers that utilize vision in foraging. Moreover, *P. cinerea* surpasses *P. temperate* at the local level/within the infected insect host.

3. Pathogenicity of *Photorhabdus* spp.

3.1. *Range and Magnitude of Pathogenicity*

Basically, various *Heterorhabditis-Photorhabdus* partnerships, to infect and kill many insect pests, have been commercially applied as biocontrol agents, with continuous suggestions for developments, to prime them for effective alternative measures in plant protection [37,46–48]. In such cases of the natural EPN–bacterium complex, the bacterial host range is naturally confined to the ability of the IJs to find and penetrate the host, as a pre-requisite for the exponential growth of *Photorhabdus* spp. to attain high cell densities within the insect host. The bacterial cells can convert the insect tissues into a biomass that is necessary for the IJ development and reproduction. The pathogenicity relies on the bacterial growth. Therefore, the *Photorhabdus* growth rate is closely correlated with the time that is required for killing the insect. Admittedly, *Photorhabdus* spp. are highly virulent pathogens of a wide range of insect larvae [49].

Detecting the ability of *Photorhabdus* bacteria to survive in soil and in fresh water for 1 week, has probably fixed a time frame for their additional biocontrol applications, independent of their mutualistic EPNs [50]. Hence, various formulations (Figure 1), mainly based on just the bacteria and/or bacterial metabolites, have been recorded [2,4,12,51–57]. In this regard, increased pathogenicity islands of the *Photorhabdus* chromosome, with many genes encoding various insecticidal protein toxins, antibiotics, bacteriocins, and enzymes, were reviewed [5,58,59], but more have still been further identified, e.g., [55,56,60,61]. For instance, the insecticidal categories of protein toxins comprise toxin complexes (TCs), *Photorhabdus* insect-related (Pir) proteins, makes caterpillars floppy (Mcf) toxins, *Photorhabdus* virulence cassettes (Pvc), *Photorhabdus* insecticidal toxin (Pit), Photox, PaxAB, and Galtox [9]. While TCs are large multi-component toxins, others, such as Photox, are a smaller 46 kDa binary.

Clearly, the above-mentioned traits of the bacteria—such as the ability to circumvent the insect's immune system, reforming and adapting to the surroundings, especially under stressed conditions, and worldwide spread with a treasure trove of useful compounds—are compelling for their relevant usage. They may be a rationale to reflect their capacity in the broad and efficient biocontrol of plant pests and pathogens. Their impressive array of primary and secondary metabolites may be so virulent that a single bacterial cell is enough to kill the targeted host of some arthropods, within a relatively short period of time [49]. *Photorhabdus* bacteria are, independently, quite eligible to control a broad range of arthropod pests in numerous categories, especially in the orders Lepidoptera (butterflies and moths), Diptera (flies, including insects that transmit human and plant diseases), and Coleoptera (weevils and beetles) [2,7,62]. For instance, *P. luminescens* and its metabolites were found to be lethal to the greater wax moth, *Galleria mellonella*, when applied in sand media; the bacterium rapidly penetrated the *G. mellonella* hemocoele as it got contact with the larval body. Its toxic metabolites caused more larval death than the *P. luminescens* cells [63]. As the bacterial cells do have a free-living existence and can enter the insect haemocoele in the absence of the EPN vector, those authors stressed that the bacterium, or its toxic secretions, can be used for insect control, as an important component of various integrated pest management (IPM) programs.

Yet, *P. luminescens* received much research work, due to the type of species of its genus, with a worldwide distribution and high insecticidal activities. Such activities against the diamondback moth, *Plutella xylostella*, pupae surpassed that of *X. nematophila* [64]. The two bacterial species induced 60% and 40% mortality of *P. xylostella* pupae, with LC_{50} values of 5×10^4 and 5.5×10^5 cells/mL, respectively. The 48-h LC_{50} for toxin complex a (Tca) of *P. luminescens* against neonates of the Colorado potato beetle, *Leptinotarsa decemlineata*, was 2.7 ppm, and the second instar larval growth that was exposed to Tca for 72 h was almost fully inhibited at >0.5 ppm [65]. The strain K-1 of *P. luminescens akhurstii*, encapsulated in sodium alginate beads (2.5×10^7 cells/bead) and mixed with sterilized soil, killed 100% of the tobacco cutworm, *Spodoptera litura*, larvae in 48 h, though its partner (i.e., *H. indica*–IJs that possess the bacteria in their gut) scored only 40% mortality after 72 h. Koch's postulates for this strain, without the symbiont nematode, were evident, as the bacteria could be re-isolated from the dead insect. The LC_{50} dose of the strain K-1 was 1010 cells per *S. litura* sixth-instar larvae in 48 h [53]. A 100% mortality of both the fall armyworms, *Spodoptera frugiperda* and *G. mellonella*, was also attained after 48 h of treatment with the strain *P. luminescens akhurstii* SL0708 at 1×10^3–1×10^4 CFU/larva [66]. On the other hand, *P. temperata* showed oral toxicity to the olive moth, *Prays oleae*, with an LC_{50} of 58.1×10^6 cells/mL [67].

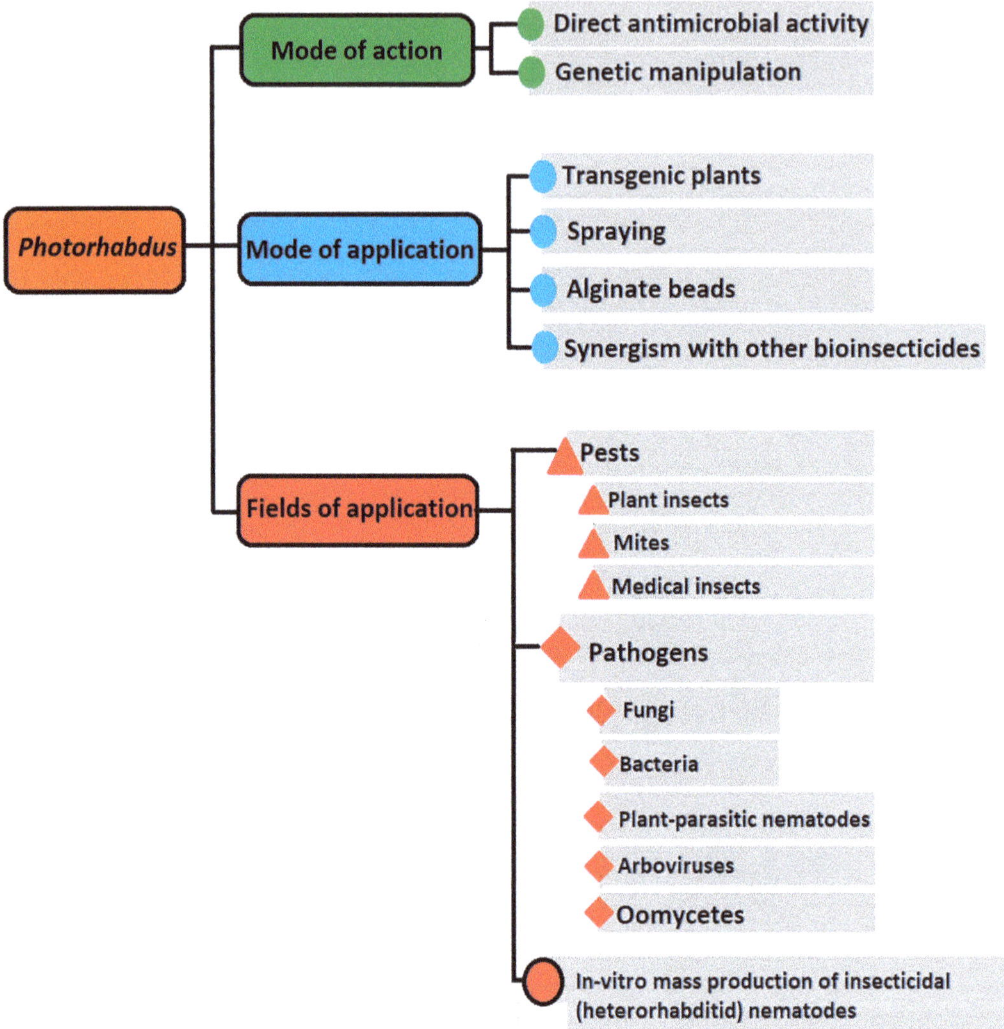

Figure 1. Simplified illustration for possible scopes of application, mode of application and mechanism of action of *Photorhabdus* bacteria to control various pests and pathogens.

Mohan et al. [51] tested the potential of *P. luminescens* to control the cabbage butterfly (*Pieris brassicae*), which is a polyphagous pest of crops in the family Brassicae. The bacterial cultures were grown overnight, in nutrient broth at 28 °C, up to a concentration of 10^8 colony-forming units (CFU)/mL. The bacterial culture was mixed with paraffin oil at 10 l/mL, Tween-20 at 0.5 l/mL, and sucrose at 0.5%, as a phagostimulant adjuvant using sterile water as the base before spraying. Initially, Mohan et al. [51] tested the survival and retention of bacterial pathogenicity in combination with various components used in the formulation. When the formulated *P. luminescens*, stored at 28 °C, was sprayed uniformly on the foliage of ornamental nasturtium, *Tropaeolum majus*, which was heavily infested with the 3–4 instar of *P. brassicae*, 100% mortality of the larvae was recorded within 24 h, compared to no mortality in the control plot. Re-isolation of the bacteria from the dead insects, and comparison with the original culture, proved Koch's postulates. Such results proved the direct toxicity of *P. luminescens* to the insects, under natural conditions, when used as a foliar spray. Also, Jallouli et al. [68] recorded promising insecticidal activity of *P. temperate* K122 against a stored grain pest, the Mediterranean flour moth, *Ephestia kuehniella*. At a high concentration of 12 × 10^8 cells/mL, 100% mortality of *E. kuehniella* larvae could be reached. Jallouli et al. [68] concluded that the insect mortality was due to toxaemia, as confirmed by the absence of variant small colonies, or *P. temperata* colonies, in *E. kuehniella* tissue. The histopathological effect of *P. temperata* toxins on the gut of infected *E. kuehniella* larvae indicated destruction of the gut epithelium, the appearance of large cavities, and cellular disintegration.

In addition to using *Photorhabdus*-concentrated metabolites or bacterial broth treatments, another avenue for pest or disease suppression is the development of bioactive compounds that are responsible for bacterial toxicity. Various active compounds in *Photorhabdus* spp. were reviewed [5]. For example, transcinnamic acid (TCA) was recently reported to be a major active compound in *P. luminescens'* suppressive activity against *Fusicladium effusum*, and thus further research to develop TCA, as a potential control agent, was suggested [69]. More research on the identification and activity of such bioactive compounds is warranted. Furthermore, regardless of the type of treatment (bioactive chemicals, metabolites and/or bacterial treatments), field testing and economic feasibility analysis will be needed, as various biotic and abiotic factors outside the laboratory may reduce the potency and longevity of the tested materials.

It appears that the arsenal of these bacteria still contains much that has yet to be discovered, against a broad range of various pathogens. The detection and cloning of other beneficial compounds from *Photorhabdus* bacteria are still ongoing [11]. Eventually, the bacteria in this genus possess metabolites with the main characteristics of common pesticides, i.e., their effect increases with an increase in the dose, and a negative correlation exists between the number of eggs laid/insects that are female, percentage of hatching, adult survival of the pest, and the bacterial dose [5]. Their toxins are so fatal that as low as 40 ng is sufficient to kill the tobacco hornworm *Manduca sexta* larvae [70].

3.2. Mode of Action of the Bacteria and Their Secreted Compounds

These bacteria are commonly known to kill insects via septicemia/toxemia, within the context of the natural *Photorhabdus–Heterorhabditis* complex [54]. However, stand-alone pathogenicity tests of the bacteria and/or their secreted compounds usually begin with injecting them directly into the insect haemocoel, by artificial means [71]. Bacterial protein toxins usually have oral and/or injectable toxicity to insects, with various modes of action [72]. The increasing ambition to exploit *Photorhabdus*-derived compounds in industry is due not only to their abundance, but also to their qualities that boost their functions. Complete genome-sequencing investigations, which started in 2003 using *P. luminescens* as a model [73,74], have been revealing the capacity of various *Photorhabdus* spp. to produce numerous secondary metabolites, such as peptides, polyketides, toxins, and hybrids. For instance, the parasiticidal material that is derived from *P. luminescens* metabolites is recognized as a small molecule that is stable at both pH changes in the range 2–12 and heating. This molecule can induce the trypanocidal activity via a mode of action that does not rely on nitric oxide [56]. Likewise, *P. luminescens* materials, such as isopropylstilbene [75] and the presumed GameXPeptides [76], demonstrated anti-*Leishmania donovani*, -*Trypanosoma cruzi*, and -*Plasmodium* leverage, respectively [55].

Each of the above-mentioned categories of toxins has a possible function as a biocontrol material, via a specific mechanism against arthropod pests, pathogens, and/or vector insects. While the Tcs damage epithelial cells at the insect intestine, for instance, Mcf enhances hemocytes apoptosis in the insect hemocoel [77]. On the other hand, Pvc can induce *G. mellonella* and *Manduca sexta* mortality, but Pir proteins of *P. luminescens laumondi* (TT01 strain) are responsible for insect death [2]. As Pvc can show a self-contained nanosyringe delivery mechanism, it can beat host cell membrane barriers and function independently from its bacterium, to disrupt the cytoskeleton of the insect host [78]. Some

Photorhabdus species can poison the intestinal epithelium of their insect hosts, via high-molecular-weight Mcf toxin. It can enable *Esherichia coli* both to persist within, and kill, the insect [79]. Thus, these toxins show bacterial strain-specific variations concerning toxicity to their insect hosts [9,80]. Different features regarding the detailed structure, mode of action, and putative function of the Tcs as 'polymorphic' toxins in the process of infection, have been discussed [81,82], but the modes of action of some recently detected metabolic compounds of *Photorhabdus* bacteria still need to be grasped, to enable their perfect use in the management of agricultural pathogens and insect pests [8,9]. *Photorhabdus* bacteria can also suppress important endoparasitic nematode species within plant roots, via their toxins and antibiotic compounds [5]. Furthermore, the Tc toxins could be cloned into *Arabidopsis* as a model plant, in order to offer protection from arthropod pests (Figure 1). Thus, these toxins are proposed as a potential substitution to the Bt toxin [49], for which insect resistance is developed [1].

The many examples of arthropod pest and pathogen mortalities that are caused by *Photorhabdus* spp. [3,5,48,53,83–85], do not negate the differences in the immune response between insect hosts. For instance, a cumulative percent mortality of 63% and 100% for *G. mellonella,* but 10% and 93% for *S. frugiperda,* was achieved after 72 h of injecting intra- or extra-cellular extracts of the strain *P. luminescens akhurstii* SL0708, respectively. These differences also indicate the impact of extracellular factors in pathogenicity [66]. Those authors detected proteases, esterases, ureases, hemolysins and siderophores as responsible for the high pathogenicity/extra-cellular activities. Moreover, the variation in immune response between host species may be due to both the evolutionary/environmental and biologic/genetic factors that are assigned to each host–pathogen system. The different system components, comprising induction, specificity, and memory of the immune system, can define the cognate resistance mechanism of the targeted insect species/population [86]. Abd El-Zaher et al. [87] reported that physical parameters (i.e., temperature, pH, and sodium chloride) variably affected the metabolite-induced mortality percentage for the *G. mellonella* larvae. Shapiro-Ilan et al. [88] reported that implementations of the bacteria and fermentation broth, to suppress pecan and peach pathogens, could be useful in reducing costs and/or avoiding regulatory issues compared to applying concentrated metabolites of these bacteria. However, they reported that the applications of bacterial broth could only inhibit the lesion growth that is induced by *Phytophthora cactorum* of pecan leaves, but its efficacy to the two other pathogens examined (*F. effusum* and *Armillaria tabescens*) was not apparent. Therefore, they perceived that more broth rates would stimulate a response. Alternatively, the toxicity of active compounds within the bacterial cell suspension, broth, or cell-free filtrates could be enhanced through medium optimization (e.g., [87,89]).

4. Pros and Cons of *Photorhabdus* spp.

4.1. The Positive Aspects

The above-mentioned common issues that are related to many chemical pesticides, as well as the costly and/or mixed performance of numerous existing biopesticides on one hand and the high levels of *Photorhabdus* virulence towards a wide variety of insects via just a few bacterial cells on the other, have been attracting commercial attention to use these bacteria and their bioactive compounds as new biopesticides [9,12,52,54]. Remarkably, the bacteria's recent inexpensive in vitro mass production [90] should increase the interest of many researchers in related fields (microbiology, nematology, molecular biology, pharmacology, etc.), for their beneficial applications. Collectively, *Photorhabdus* bacteria can be adopted for large-scale application, due to some of their characteristics discussed in the following section.

4.1.1. Cost-Effective Photorhabdus Mass Culture with Boosting Insecticidal Activity

In the recent past, many difficulties in the trait deterioration of *Photorhabdus* spp. were apparent during their in vitro mass culture. Furthermore, researchers are challenged by a cost and benefit tradeoff. Reduced costs with high bacterial yield, in terms of the scale-up of the mass production, are opposed by the requirement, to preserve beneficial *Photorhabdus* traits during the process of culturing [91]. Such bacterial attributes are significant for biocontrol programs, whether via *Photorhabdus* alone or the EPN–*Photorhabdus* complex. Fortunately, these bacterial traits could be characterized and maintained during their growth and metabolic phases of their inexpensive culturing [90]. Consequently, the full *Photorhabdus* capacity could be employed for the insect pathogenicity. Some significant advances in in vitro culture have been made. For example, Orozco-Hidalgo et al. [92] found that bacterial inoculation of the culture for 36 h could offer the highest *H. indica*–IJ yield. The authors stressed the merit of glucose as a substrate to increase/sustain the bacteria, so that higher nematode recovery could be achieved. Recently, an important discovery offers potentially the best conditions for enhancing the mass culture and the insecticidal efficacy of *P. temperata,* using the wastewater of the food industry as a basis for the medium [90]. The authors used both a special layout (Box–Behnken design) and response surface methodology (RSM) to optimize the bacterial mass production. This layout was based on three factors, carbon/nitrogen (C/N) ratio, sodium chloride concentration, and bacterial inoculum size. Both insecticidal efficacy and bacterial mass culture were tied to the media parameters. Furthermore, when wastewater from the food industry was optimized for utilization, as an inexpensive raw material for *P. temperate* culturing, its production costs were only USD 35 per kilogram medium [90], compared to the USD 679 per kilogram medium that used yeast extract and glucose as the main components [93]. About a 95% reduction in the total production cost was achieved, while the produced bacteria had high insecticidal activity [90]. Thus, this technique for optimizing the important combined factors via RSM, to yield a superior bioinsecticide with high bacterial mass production in a short time (48 h incubation), should be expanded to other species of *Photorhabdus*. This inexpensive technique should be considered for large-scale practice, as it also gives a share in getting low-cost formulations that are able to compete with conventional chemical compounds.

4.1.2. The Mounting Role of *Photorhabdus* Bacteria against Pests and Pathogens

The bacteria and their active metabolites show other merits that bode well for inclusion in promising approaches of modern farming. Consequently, broad biocontrol application is feasible against insects, fungi, oomycetes, mites, and bacteria infecting plants and, to a lesser degree, animals. Some examples of the expanding utility of *Photorhabdus* spp. include the following: (1) many *Photorhabdus* bacterial genes encode metabolites and toxins mostly with low molecular weight. These toxins proved to have insecticidal [94,95], antifungal, antibiotic [96], and antiparasitic [55,56] activities. They are effective, for instance, against the mushroom mite *Luciaphorus* sp. [97], *Venturia effusa* (a fungus causing pecan scab) [69,98], oomycetes that can severely limit the commercial productivity of pecans, peaches, and other fruit and nut trees [7,84], antibiotic-resistant bacteria [8], and even vector insects of human diseases, such as the mosquito species *Aedes aegypti* and *Ae. albopictus* [2]. Further, da Silva et al. [2] have reviewed the dengue virus as the most serious arbovirus, in reference to human morbidity and mortality. The viral serotypes can be transmitted mainly by females of *Ae. aegypti* mosquitoes, where the toxic effect of Cry4Ba that is derived from Bti against *Ae. Aegypti* could be enhanced by the *Xenorhabdus* and *Photorhabdus* bacteria; (2) ongoing field and laboratory assays can offer these bacteria and their nematode partners new positions for upgrading the control of additional pests via IPM programs [37,48,99,100]. Also, six *P. luminescens* isolates had antibacterial activities; each against one, two, or three of the tested bacterial species [101]. The metabolites or cell filtrates of some *Photorhabdus* isolates may have superior toxicity, relative to others tested as well [84]; (3) *Photorhabdus* species possess various toxins and compounds with various modes of action and sophisticated secretion systems that offer specificity of the cell surface receptor to dictate a specific interaction between the bacterial toxin and the insect midgut [82]; (4) *Photorhabdus luminescens* is so effective against the diamondback moth, *Plutella xylostella,* pupae that its pathogenic capability is superior to *X. nematophila,* though both bacterial species have a convergent lifestyle [64]; (5) *Photorhabdus* species could operate more effectively against pests via additive or synergistic incorporation with other advantageous inputs, such as another biocontrol agent [102,103]; (6) the toxic secretion of *P. luminescens* can be efficiently used to control various challenging pest populations, such as *Galleria mellonella* and subterranean termite *Macrotermis* spp. [85]. Shahina et al. [85] speculated that commercial use of *P. luminescens* cell suspensions as pesticides may overrule the problems of cost and reliability that are linked to the large-scale application of EPNs; (7) some *Photorhabdus* spp. showed miticidal activities against the economically important spider mite *Tetranychus urticae*, where the mortality rate increased as the bacterial dose or time elapsed increased [6]; (8) the supernatants of the *P. luminescens* culture blocked the nourishment of crickets, ants, and wasps [83,104]; (9) *Photorhabdus* toxin genes can generate transgenic plants for insect resistance [105,106]. There is potential to commercially produce an orally active agent from *Photorhabdus* bacteria for insect-resistant transgenic plant species. The prolonged monoculture of Bt transgenic plant cultivars has led to developing insects that are resistant against these cultivars, and the emergence of unforeseen pest problems [1]. Hence, *Photorhabdus* bacteria are being firmly suggested as an adequate alternative for Bt [2,9]. However, progress toward transgenic *Photorhabdus*-based plants is slow, as the current Bt transgenic strategies remain globally prevalent. Technical problems are usually linked to expressing a large, multi-subunit protein toxin into the needed transgenic plants, but small toxins of *Photorhabdus* and *Xenorhabdus* bacteria are among the best alternative sources of such insecticidal protein toxins [14,74]; and (10) these bacteria can favorably interact with plant roots as well [60]. The roots may attract EPNs that carry these bacteria to help control plant pests and pathogens [107,108]. Bacterial suspensions of *P. luminescens*, *Xenorhabdus* sp., and *X. szentirmaii* could significantly reduce the growth parameters of the root-knot nematode *Meloidogyne hapla*. They decreased its reproduction factor (RF) (55–62%), egg masses (48–68%), and number of galls (51–67%). Likewise, the cell-free supernatant of these bacteria reduced the number of egg masses (72–83%), galls (51–74%), and RF (62–72%) of the false root-knot nematode, *Nacobbus aberrans* [4,109]. Interestingly, on many economically important crops, the costs of controlling plant parasitic nematodes with these inexpensively produced bacteria [90] would generally be more economical than using common chemical nematicides, such as Cadusafos and Oxamyl. This is evidenced by the reported costs of using such chemicals on tomatoes [110], pepper [111], potatoes [112], and eggplant [113]. In addition, the application of *Photorhabdus* bacteria may also score a two-fold goal, i.e., control of the plant pathogens and insect pests simultaneously. Yet, achieving such a two-fold goal will necessitate optimal application tactics to maximize the field effectiveness of the *Photorhabdus* bacteria, e.g., a delivery system that is most conducive for the favorable and effective control of crop pests and pathogens via biocontrol agents. For example, various media/additions were found to potentiate *Photorhabdus* cell suspensions and cell-free filtrates against insect pests [87,89]. Ultimately, these attributes should be fully exploited by applying them in conventional and organic farming systems.

4.1.3. The Bacterial Metabolites as New Drugs for Diseases

Increased concern about controlling certain diseases has created much interest for safer and more effective drugs than the currently used ones. For instance, the only effective approach to control insect-borne diseases, such as Zika, chikungunya, Chagas, Leishmaniasis, and dengue, is to block the detrimental transmission of the disease-causal organisms. However, the control of these insect vectors has demonstrated issues, due to the expensive measures of their chemical and biological control. They frequently possess low specificity for the targeted vectors/organisms, and may also be toxic to non-target and beneficial organisms [2,56]. Additionally, evolved resistance was reported for common insecticides in global resistance-breaking populations of *Ae. albopictus* and *Ae. aegypti* [2]. Further, da Silva et al. [2], therefore, speculated that a broad variety of biologicals and chemicals should contribute to their control measure. They found entomopathogenic bacteria, e.g., *Photorhabdus* and *Xenorhabdus*, to rank high in this respect, as numerous studies have demonstrated their relevant efficiency. Within this frame, a peptide that is smaller than 3 kDa, secreted by *P. luminescens,* was aptly named as '*Photorhabdus*-derived leishmanicidal toxin'. This compound evidenced potent leishmanicidal action to suppress dimorphological forms (i.e., amastigote and promastigote forms) of *Leishmania amazonensis*. Leishmania-toxic peptide(s) could be new drugs concerning the remedy for leishmaniasis [55]. In another study, *Trypanosoma cruzi,* which gives rise to American trypanosomiasis (or Chagas disease of human beings) could be controlled via metabolites that are secreted by *P. luminescens* [56]. Recently, da Silva et al. [2] demonstrated the low specificity of the chemicals that are used against *Ae. aegypti* and *Ae. albopictus* that transmit diseases such as dengue, malaria, Zika, and chikungunya. On the contrary, they appreciated the value of toxin complexes and metabolites produced by *Photorhabdus* and *Xenorhabdus,* to effectively control such insect-borne diseases.

Some may mistakenly suspect that controlling these medical insects has no absolute, or even marginal, relationship with plant health. However, their related chemical insecticides, as the pyrethroids and organophosphate temephos, have become less effective, even against plant pests, due to their widespread use that includes these medical insects as well [2]. Thus, the non-use of such pesticides as organophosphates against medicinal insects will limit their over-application and avoid the pollution of the environment, plants, wildlife, and groundwater. Additionally, their non-excessive use may generally slow down the development of insect resistance.

4.1.4. *Photorhabdus*-Derived Natural Compounds as a Source for Industrial Products

The above-mentioned paradigms of *Photorhabdus*-derived insecticidal, fungicidal, pharmaceutical, parasiticidal, antimicrobial, and toxic materials may still be expanded to represent various types of expected industrial products. The richness of the relevant interesting compounds is reflected by an increasing number of materials that have been, and are still being, identified from these bacteria [5,8,11,55,56,61,69,94–96,98]. The targeted testing for the bioactivity of molecules reflecting the bacterial secondary metabolite genes and gene clusters are reported to be scarce, or mostly focused on medical applications [114]. Specifically, the molecule indole caused high levels of paralysis of plant-parasitic nematodes, such as *Meloidogyne incognita* and *Bursaphelenchus* spp., at a concentration of 100 to 300 mg/mL. Stock et al. [114] reviewed these bacterial metabolites in terms of their nematicidal, antimycotic, antibacterial, and insecticidal activities Thus, such metabolites may set up an inexhaustible mine of useful compounds, with multiple activities for new industrial products, mainly for crop protection.

4.2. Avoiding Negative Aspects

As *Photorhabdus* bacteria are originally mutualistic of insecticidal nematodes and highly efficient insect pathogens, via myriad toxins and small molecule effectors, cautious strategies should be set up to integrate these bacteria and/or their bioactive compounds into effective and safe management programs of crop insect pests. For instance, while some insect pests have developed resistance against Bt toxin, employing *Photorhabdus* toxins with alternate modes of action may resolve the issues of developed insect resistance. Practical use of *P. luminescens*, in conjunction with *B. thuringiensis kurstaki,* inhibited the growth of the Egyptian cotton leafworm *Spodoptera littoralis.* This synergism between the two bioinsecticides could use *B. thuringiensis* as a delivery means for *Photorhabdus* bacteria to infect the *S. littoralis* hemocoel and to reduce the risk of developing insect resistance [102]. Since Bt toxin occupies 90% of the bioinsecticide market [5], materializing this example can significantly raise *Photorhabdus* marketing. Other economically important insect pests that develop Bt resistance comprise *Ostrinia nubilalis* (European corn borer), *Heliothis virescens* (tobacco budworm), *Pectonophora gossypiella* (pink bollworm moth), *Culex quinquefasciatus* (mosquito), *Ae. aegypti* (yellow fever mosquito), *Trichloro plusiani* (tiger moth), *Leptinotarsa decemlineata* (Colorado potato beetle), *Spodoptera exigua* (beet armyworm), and *Chryosomela scripta* (cottonwood leaf beetle) [115,116]. Moreover, the toxin complex protein, TcaA toxin, showed toxicity against a wide range of agricultural pests, though the phylogenetic tree that was erected for TcaA indicated that this toxin did not have any ancestral relationship with BT toxins [116].

If so, the anticipated transgenic plants will be regulated by relevant legislation to be certain that such plants with genes of *Photorhabdus* bacteria can be produced without health or environmental risks [36]. No resistance to these bacteria has been reported in insect populations so far, but this line of thinking should be followed with any relevant toxins, as a means to avert or at least delay the development of insect/pathogen resistance. Therefore, many more investigations should be conducted before the exclusive use of these bacteria or their metabolites as a commercial tool to control crop insect pests and pathogens. Studies that are needed to enable its wise utilization should address both fundamentals, such as their molecular structure and mode of action, and applied aspects, such as their environmental stability, toxicity of different bacterial species/isolates against various insect pests and pathogens, and safety against non-target organisms. For example, Kumar et al. [117] found that two *P. luminescens* isolates failed to infect both the diamondback moth and the oriental leafworm moth (*Spodoptera litura*). On the contrary, Abdel-Razek [64] and Rajagopal et al. [53] found other isolates of the same bacterial species to be effective against the diamondback moth and the oriental leafworm moth, respectively. Likewise, technical issues, regarding further stand-alone formulations of the bacteria, their shelf life, and application, should be searched, particularly for their adequate integration into IPM programs. Examples of their usage have been reported [51,52,68], but their expansion and documentation should be attempted in earnest. Notably, a culture broth of *P. temperate temperata* mixed with *B. thuringiensis tenebrionis,* named "Col-Kill", proved efficacy in controlling the coleopteran *Phaedon brassicae* (Coleoptera: Chrysomelidae) [118], but it needs to be widely tested as a new, effective formulation. It could open another beneficial tactic of synergism and avoid the prolonged monoculture of Bt transgenic plants that have led to developing insect resistance.

Apart from the aforementioned concerns of *P. asymbiotica,* and occasional episodes of allergy processes that are experienced by people who have had close contact with *Photorhabdus* species and their symbiotic nematodes, these bacteria, similarly to EPNs, have a very low risk to human health [36]. Mohan and Sabir [119] and Fand et al. [120] reported other constraints, where *P. luminescens* harm *Trichogramma* parasitized insect eggs and do kill coccinellids (key mealybug predators), respectively. Therefore, the integration of the bacteria with other biocontrol agents should be utilized cautiously. Advances in biotechnological approaches, and consolidation of the *Photorhabdus* data with their regional and temporal efficacy and dynamics, may offer the holistic and sound knowledge that is needed to establish and evaluate the real benefits and risks of EPNs and/or their mutualistic bacteria in nature, in the long term. Given these basics, Abd-Elgawad [5] emphasized that management projects should be decided on a case-by-case basis for attaining the best *Photorhabdus* species or strain–pest (or pathogen) matching, without side effects to the fauna and flora of definite sites. For instance, the bacteria could be harmful to definite pollinators. *Photorhabdus* species may be able to digest the bee tissues effectively and offer a supply of nutrients to the mutualistic nematodes [121].

5. Conclusions

The potential of using *Photorhabdus* spp. as biocontrol agents in sustainable agriculture, against a broad range of insect pests and pathogens, is being boosted via various approaches. The discovery of novel species/strains worldwide will continue to broaden the pool of their bioactive compounds that are effective against economically important pests and pathogens. Developing inexpensive methods for their commercial production may expedite their use in the existing or emerging programs to manage plant pests and pathogens. Clearly, the toxins of these bacteria cause massive damage to the gut epithelium of insects, resulting in rupture of the gut integrity and crossing of the gut barrier. They could be used in foliar application or in the form of alginate beads; both as standalone insecticides for the most common management strategies. Furthermore, their long-acting strategies have been proved via incorporating their toxin genes into transgenic plants to control insect pests. *Photorhabdus*-derived insecticidal, fungicidal, parasiticidal, antimicrobial, and toxic materials should be leveraged to fit into holistic crop protection strategies. These may include their combination with other synergistic or additive compounds, to increase their efficacy while preventing, or reducing, the likelihood of developing pesticide-resistant pest/pathogen strains.

Funding: This research was funded by STDF, US-Egyptian project "Preparing and evaluating IPM tactics for increasing strawberry and citrus production" cycle 17 grant number 172 and The NRC in-house project No. 12050105 entitled "Pesticide alternatives against soil-borne pathogens and pests attacking economically important solanaceous crops", funded by The National Research Centre.

Institutional Review Board Statement: Not applicable.

Informed Consent Statement: Not applicable.

Data Availability Statement: Not applicable.

Acknowledgments: This article is supported in part by the US-Egypt Project cycle 17 (no. 172) entitled: Preparing and evaluating IPM tactics for increasing strawberry and citrus production. The study was also supported in part by the NRC In-house project No. 12050105 entitled "Pesticide alternatives against soil-borne pathogens and pests attacking economically important solanaceous crops". The author thanks David Shapiro-Ilan, Larry Duncan, and Zafar Handoo for their valuable comments on specific parts of the manuscript.

Conflicts of Interest: The author declares no conflict of interest. The funders had no role in the design of the study; in the collection, analyses, or interpretation of data; in the writing of the manuscript, or in the decision to publish the results.

References

1. Xiao, Y.; Wu, K. Recent progress on the interaction between insects and *Bacillus thuringiensis* crops. *Philos. Trans. R. Soc.* **2019**, *374*, 20180316. [CrossRef]
2. da Silva, W.J.; Pilz-Júnior, H.L.; Heermann, R.; da Silva, O.S. The great potential of entomopathogenic bacteria *Xenorhabdus* and *Photorhabdus* for mosquito control: A review. *Parasites Vectors* **2020**, *13*, 376. [CrossRef]
3. Coupland, J.; Abd-Elgawad, M.M.; Askary, T.H. Beneficial nematodes and the changing scope of crop protection. In *Biocontrol Agents: Entomopathogenic and Slug Parasitic Nematodes*; Abd-Elgawad, M.M.M., Askary, T.H., Coupland, J., Eds.; CAB International: Wallingford, UK, 2017; pp. 26–42.
4. Migunova, V.D.; Sasanelli, N. Bacteria as biocontrol tool against phytoparasitic nematodes. *Plants* **2021**, *10*, 389. [CrossRef] [PubMed]
5. Abd-Elgawad, M.M.M. Toxic secretions of *Photorhabdus* and their efficacy against crop insect pests. In *Biocontrol Agents: Entomopathogenic and Slug Parasitic Nematodes*; Abd-Elgawad, M.M.M., Askary, T.H., Coupland, J., Eds.; CAB International: Wallingford, UK, 2017; pp. 231–260.
6. Eroglu, C.; Cimenb, H.; Ulug, D.; Karagoz, M.; Hazir, S.; Cakmaka, I. Acaricidal effect of cell-free supernatants from *Xenorhabdus* and *Photorhabdus* bacteria against Tetranychus urticae (Acari: Tetranychidae). *J. Inver. Pathol.* **2019**, *106*, 61–66. [CrossRef]
7. Ffrench-Constant, R.; Waterfield, N.; Daborn, P. Insecticidal toxins from *Photorhabdus* and *Xenorhabdus*. In *Encyclopedia of Microbiology*, 4th ed.; Schmid, T.M., Ed.; Academic Press: New York, NY, USA, 2019; pp. 704–715. [CrossRef]
8. Muangpat, P.; Suwannaroj, M.; Yimthin, T.; Fukruksa, C.; Sitthisak, S.; Chantratita, N.; Vitta, A.; Thanwisai, A. Antibacterial activity of *Xenorhabdus* and *Photorhabdus* isolated from entomopathogenic nematodes against antibiotic-resistant bacteria. *PLoS ONE* **2020**, *15*, e0234129. [CrossRef]
9. Ahuja, A.; Kushwah, J.; Mathur, C.; Chauhan, K.; Dutta, T.K.; Somvanshi, V.S. Identification of Galtox, a new protein toxin from *Photorhabdus* bacterial symbionts of *Heterorhabditis*. *Toxicon* **2021**, *194*, 53–62. [CrossRef] [PubMed]
10. Park, H.; Sampathkumar, P.; Perez, C.E.; Lee, J.H.; Tran, J.; Bonanno, J.B.; Hallem, E.A.; Almo, S.C.; Crawford, J.M. Stilbene epoxidation and detoxification in a *Photorhabdus luminescens*-nematode symbiosis. *J. Biol. Chem.* **2017**, *292*, 6680–6694. [CrossRef]
11. Zhang, F.; Ren, J.; Zhan, J. Identification and characterization of an efficient phenylalanine ammonia-lyase from *Photorhabdus luminescens*. *Appl. Biochem. Biotech.* **2012**, *193*, 1099–1115. [CrossRef] [PubMed]
12. Kushwah, J.; Kumar, P.; Garg, V.; Somvanshi, V.S. Discovery of a highly virulent strain of *Photorhabdus luminescens* ssp. *akhurstii* from Meghalaya, India. *Indian J. Microbiol.* **2017**, *57*, 125–128. [CrossRef]
13. Machado, R.A.R.; Muller, A.; Ghazal, S.; Thanwisai, A.; Pagès, S.; Bode, H.B.; Hussein, M.A.; Khalil, K.M.; Tisa, L.S. *Photorhabdus heterorhabditis* subsp. *aluminescens* subsp. nov., *Photorhabdus heterorhabditis* subsp. *heterorhabditis* subsp. nov., *Photorhabdus australis*

subsp. *thailandensis* subsp. nov., *Photorhabdus australis* subsp. *australis* subsp. nov., and *Photorhabdus aegyptia* sp. nov. isolated from *Heterorhabditis* entomopathogenic nematodes. *Int. J. Syst. Evol. Microbiol.* **2021**, *71*. [CrossRef]
14. Ajnaga, E.; Kazimierczak, W. Evolution and taxonomy of nematode-associated entomopathogenic bacteria of the genera *Xenorhabdus* and *Photorhabdus*: An overview. *Symbiosis* **2020**, *80*, 1–13. [CrossRef]
15. Abd-Elgawad, M.M.M. Can rational sampling maximise isolation and fix distribution measure of entomopathogenic nematodes? *Nematology* **2020**, *22*, 907–916. [CrossRef]
16. Abd-Elgawad, M.M.M. Optimizing sampling and extraction methods for plant-parasitic and entomopathogenic nematodes. *Plants* **2021**, *10*, 629. [CrossRef]
17. Yooyangket, T.; Muangpat, P.; Polseela, R.; Tandhavanant, S.; Thanwisai, A.; Vitta, A. Identification of entomopathogenic nematodes and symbiotic bacteria from Nam Nao National Park in Thailand and larvicidal activity of symbiotic bacteria against *Aedes aegypti* and *Aedes albopictus*. *PLoS ONE* **2018**, *13*, e0195681. [CrossRef]
18. Fischer-Le Saux, M.; Viallard, V.; Brunel, B.; Normand, P.; Boemare, N.E. Polyphasic classification of the genus *Photorhabdus* and proposal of new taxa: *P. luminescens* subsp. *luminescens* subsp. nov., *P. luminescens* subsp. *akhurstii* subsp. nov., *P. luminescens* subsp. *laumondii* subsp. nov., *P. temperata* sp. nov., *P. temperate* subsp. *temperatas* ubsp. nov. and *P. asymbiotica* sp. nov. *Int. J. Syst. Bacteriol.* **1999**, *49*, 1645–1656.
19. Machado, R.A.R.; Wüthrich, D.; Kuhnert, P.; Arce, C.C.M.; Thönen, L.; Ruiz, C.; Zhang, X.; Robert, C.A.M.; Karimi, J.; Kamali, S.; et al. Whole-genome-based revisit of *Photorhabdus* phylogeny: Proposal for the elevation of most *Photorhabdus* subspecies to the species level and description of one novel species *Photorhabdus bodei* sp. nov., and one novel subspecies *Photorhabdus laumondii* subsp. *clarkei* subsp. nov. *Int. J. Syst. Evol. Microbiol.* **2018**, *68*, 2664–2681.
20. Akhurst, R.J.; Boemare, N.E.; Janssen, P.H.; Peel, M.M.; Alfredson, D.A.; Beard, C.E. Taxonomy of Australian clinical isolates of the genus *Photorhabdus* and proposal of *Photorhabdus asymbiotica* subsp. *asymbiotica* subsp. nov. and *P. asymbiotica* subsp. *australis* subsp. nov. *Int. J. Syst. Evol. Microbiol.* **2004**, *54*, 1301–1310. [CrossRef]
21. Tailliez, P.; Laroui, C.; Ginibre, N.; Paule, A.; Pages, S.; Boemare, N. Phylogeny of *Photorhabdus* and *Xenorhabdus* based on universally conserved protein-coding sequences and implications for the taxonomy of these two genera. Proposal of new taxa. *X. vietnamensis* sp. nov., *P. luminescens* subsp. *caribbeanensis* subsp. nov., *P. luminescens* subsp. *hainanensis* subsp. nov., *P. temperata* subsp. *khanii* subsp. nov., *P. temperata* subsp. *tasmaniensis* subsp. nov., and the reclassification of *P. luminescens* subsp. *thracensis* as *P. temperata* subsp. *thracensis* comb. nov. *Int. J. Syst. Evol. Microbiol.* **2010**, *60*, 1921–1937. [PubMed]
22. Tóth, T.; Lakatos, T. *Photorhabdus temperata* subsp. *cinerea* subsp. nov., isolated from *Heterorhabditis nematodes*. *Int. J. Syst. Evol. Microbiol.* **2008**, *58*, 2579–2581. [CrossRef]
23. Ferreira, T.; Reenen, C.A.; Endo, A.; Tailliez, P.; Pages, S.; Sproer, C.; Malan, A.P.; Dicks, L.M.T. *Photorhabdus heterorhabditis* sp. nov., a symbiont of the entomopathogenic nematode *Heterorhabditis zealandica*. *Int. J. Syst. Evol. Microbiol.* **2014**, *64*, 1540–1545. [CrossRef]
24. Hazir, S.; Stackebrandt, E.; Lang, E.; Schumann, P.; Ehlers, R.U.; Keskin, N. Two new subspecies of *Photorhabdus luminescens*, isolated from *Heterorhabditis bacteriophora* (Nematoda: Heterorhabditidae): *Photorhabdus luminescens* subsp. *Kayaii* subsp. nov. and *Photorhabdus luminescens* subsp. *thracensis* subsp. nov. *Syst. Appl. Microbiol.* **2004**, *27*, 36–42. [CrossRef]
25. Machado, R.A.R.; Bruno, P.; Arce, C.C.M.; Liechti, N.; Köhler, A.; Bernal, J.; Bruggmann, R.; Turlings, T.C.J. *Photorhabdus khanii* subsp. *guanajuatensis* subsp. nov., isolated from *Heterorhabditis atacamensis*, and *Photorhabdus luminescens* subsp. *Mexicana* subsp. nov., isolated from *Heterorhabditis mexicana* entomopathogenic nematodes. *Int. J. Syst. Evol. Microbiol.* **2019**, *69*, 652–661. [CrossRef]
26. An, R.; Grewal, P.S. *Photorhabdus luminescens* subsp. *kleinii* subsp. nov. (Enterobacteriales: Enterobacteriaceae). *Curr. Microbiol.* **2011**, *62*, 539–543. [CrossRef]
27. Thomas, G.M.; Poinar, G.O. *Xenorhabdus* gen. nov., a genus of entomopathogenic nematophilic bacteria of the family Enterobacteriaceae. *Int. J. Syst. Bacteriol.* **1979**, *29*, 352–360. [CrossRef]
28. Boemare, N.E.; Akhurst, R.J.; Mourant, R.G. DNA relatedness between *Xenorhabdus* spp. (Enterobacteriaceae), symbiotic bacteria of entomopathogenic nematodes, and a proposal to transfer *Xenorhabdus luminescens* to a new genus, *Photorhabdus* gen. nov. *Int. J. Syst. Bacteriol.* **1993**, *43*, 249–255. [CrossRef]
29. Orozco, R.A.; Hill, T.; Stock, S.P. Characterization and phylogenetic relationships of *Photorhabdus luminescens* subsp. *sonorensis* (γ-Proteobacteria: Enterobacteriaceae), the bacterial symbiont of the entomopathogenic nematode *Heterorhabditis sonorensis* (Nematoda: Heterorhabditidae). *Curr. Microbiol.* **2013**, *66*, 30–39. [CrossRef]
30. Glaeser, S.P.; Tobias, N.J.; Thanwisai, A.; Chantratita, N.; Bode, H.B.; Kampfer, P. *Photorhabdus luminescens* subsp. *namnaonensis* subsp. nov., isolated from *Heterorhabditis baujardi* nematodes. *Int. J. Syst. Evol. Microbiol.* **2017**, *67*, 1046–1051. [CrossRef]
31. Ferreira, T.; Reenen, C.A.; Pages, S.; Tailliez, P.; Malan, A.P.; Dicks, L.M.T. *Photorhabdus luminescens* subsp. *noenieputensis* subsp. nov., a symbiotic bacterium associated with a novel *Heterorhabditis* species related to *Heterorhabditis indica*. *Int. J. Syst. Evol. Microbiol.* **2013**, *63*, 1853–1858. [CrossRef]
32. An, R.; Grewal, P.S. Molecular mechanisms of persistence of mutualistic bacteria *Photorhabdus* in the entomopathogenic nematode host. *PLoS ONE* **2010**, *5*, e13154. [CrossRef]
33. Boemare, N. Biology, taxonomy and systematics of *Photorhabdus* and *Xenorhabdus*. In *Entomopathogenic Nematology*; Gaugler, R., Ed.; CAB International: Wallingford, UK, 2002; pp. 35–56.
34. Plichta, K.L.; Joyce, S.A.; Clarke, D.; Waterfield, N.; Stock, S.P. *Heterorhabditis gerrardi* n. sp. (Nematoda: Heterorhabditidae): The hidden host of *Photorhabdus asymbiotica* (Enterobacteriaceae:g-Proteobacteria). *J. Helminthol.* **2009**, *83*, 309–320. [CrossRef]

35. Gerrard, J.; Waterfield, N.; Vohra, R.; ffrench-Constant, R. Human infection with *Photorhabdus asymbiotica*: An emerging bacterial pathogen. *Microbes Infect./Inst. Pasteur.* **2004**, *6*, 229–237. [CrossRef]
36. Piedra-Buena, A.; López-Cepero, J.; Campos-Herrera, R. Entomopathogenic nematode production and application: Regulation, ecological impact and non-target effects. In *Nematodes Pathogenesis of Insects and Other Pests*; Campos-Herrera, R., Ed.; Springer International Publishing: Cham, Switzerland, 2015; pp. 255–282.
37. Askary, T.H.; Abd-Elgawad, M.M.M. Opportunities and challenges of entomopathogenic nematodes as bio control agents in their tripartite interactions. *Egypt. J. Biol. Pest Cont.* **2021**, *31*, 42. [CrossRef]
38. Goodrich-Blair, H.; Clarke, D.J. Mutualism and pathogenesis in *Xenorhabdus* and *Photorhabdus*: Two roads to the same destination. *Molecul. Microbiol.* **2007**, *64*, 260–268. [CrossRef] [PubMed]
39. Ciche, T.A.; Kim, K.S.; Kaufmann-Daszczuk, B.; Nguyen, K.C.Q.; Hall, D.H. Cell invasion and matricide during *Photorhabdus luminescens* transmission by *Heterorhabditis bacteriophora* nematodes. *Appl. Environ. Microbiol.* **2008**, *74*, 2275–2287. [CrossRef]
40. Chaston, J.M.; Suen, G.; Tucker, S.L.; Andersen, A.W.; Bhasin, A.; Bode, E. The entomopathogenic bacterial endosymbionts *Xenorhabdus* and *Photorhabdus*: Convergent lifestyles from divergent genomes. *PLoS ONE* **2011**, *6*, e27909. [CrossRef]
41. Maneesakorn, P.; An, R.S.; Daneshvar, H.; Taylor, K.; Bai, X.D.; Adams, B.J.; Grewal, P.S.; Chandrapatya, A. Phylogenetic and cophylogenetic relationships of entomopathogenic nematodes (Heterorhabditis: Rhabditida) and their symbiotic bacteria (Photorhabdus: Enterobacteriaceae). *Mol. Phylogenet. Evol.* **2011**, *59*, 271–280. [CrossRef]
42. Kazimierczak, W.; Skrzypek, H.; Sajnaga, E.; Skowronek, M.; Wasko, A.; Kreft, A. Strains of *Photorhabdus* spp. associated with polish *Heterorhabditis* isolates: Their molecular and phenotypic characterization and symbiont exchange. *Arch. Microbiol.* **2017**, *199*, 979–989. [CrossRef]
43. Maher, A.M.D.; Asaiyah, M.A.M.; Brophy, C.; Griffin, C.T. An entomopathogenic nematode extends its niche by associating with different symbionts. *Microb. Ecol.* **2017**, *73*, 211–223. [CrossRef]
44. Adams, B.J.; Fodor, A.; Koppenhofer, H.S.; Stackebrandt, E.; Patricia, S.; Klein, M.G. Biodiversity and systematics of nematode-bacterium entomopathogens. *Biol. Control* **2006**, *37*, 32–49. [CrossRef]
45. Maher, A.M.D.; Asaiyah, M.; Quinn, S.; Burke, R.; Wolff, H.; Bod, H.B.; Griffin, C.T. Competition and co-existence of two *Photorhabdus* symbionts with a nematode host. *Microb. Ecol.* **2021**, *81*, 223–239. [CrossRef]
46. Abd-Elgawad, M.M.M. Towards optimization of entomopathogenic nematodes for more service in the biological control of insect pests. Egypt. *J. Biol. Pest Cont.* **2019**, *29*, 77. [CrossRef]
47. Koppenhöfer, A.M.; Shapiro-Ilan, D.I.; Hiltpold, I. Entomopathogenic nematodes in sustainable food production. *Front. Sustain. Food Syst.* **2020**, *4*, 125. [CrossRef]
48. Shapiro-Ilan, D.; Hazir, S.; Glazer, I. Advances in use of entomopathogenic nematodes in integrated pest management. In *Integrated Management of Insect Pests: Current and Future Developments*; Kogan, M., Heinrichs, E.A., Eds.; Burleigh Dodds Science Publication: Cambridge, UK, 2020; pp. 1–30.
49. Clarke, D.J. *Photorhabdus*: A model for the analysis of pathogenicity and mutualism. *Cell Microbiol.* **2008**, *10*, 2159–2167. [CrossRef]
50. Morgan, J.A.W.; Kuntzelmann, V.; Tavernor, S.; Ousley, M.A.; Winstanley, C. Survival of *Xenorhabdus nematophilus* and *Photorhabdus luminescens* in water and soil. *J. Appl. Microbiol.* **1997**, *83*, 665–670. [CrossRef]
51. Mohan, S.; Raman, R.; Gaur, H.S. Foliar application of *Photorhabdus luminescens*, symbiotic bacteria from entomopathogenic nematode *Heterorhabditis indica*, to kill cabbage butterfly *Pieris brassicae*. *Curr. Sci.* **2003**, *84*, 1397.
52. Bhatnagar, R.K.; Rajagopal, R.; Rao, N.G.V. Biopesticidal Compositions. World Intellectual Property. Patent WO 2005/055724 A1. Available online: https://patentscope.wipo.int/search/en/detail.jsf?docId=WO2005055724 (accessed on 23 May 2021).
53. Rajagopal, R.; Mohan, S.; Bhatnagar, R.K. Direct infection of *Spodoptera litura* by *Photorhabdus luminescens* encapsulated in alginate beads. *J. Invert. Pathol.* **2006**, *93*, 50–53. [CrossRef] [PubMed]
54. Kushwah, J.; Somvanshi, V.S. *Photorhabdus*: A microbial factory of insect-killing toxins. In *Microbial Factories: Biodiversity, Biopolymers, Bioactive Molecules*; Kalia, V.C., Ed.; Springer: New Delhi, India, 2015; Volume 2, pp. 235–240. [CrossRef]
55. Antonello, A.M.; Sartori, T.; Folmer, C.A.P.; Brandelli, A.; Heermann, R.; Rodrigues, J.L.C.; Peres, A.; Romão, P.R.T.; Da Silva, O.S. Entomopathogenic bacteria *Photorhabdus luminescens* as drug source against *Leishmania amazonensis*. *Parasitology* **2018**, *145*, 1065–1074. [CrossRef] [PubMed]
56. Antonello, A.M.; Sartori, T.; Silva, M.B.; Prophiro, J.S.; Pinge-Filho, P.; Heermann, R. Anti-*Trypanosoma* activity of bioactive metabolites from *Photorhabdus luminescens* and *Xenorhabdus nematophila*. *Exp. Parasitol.* **2019**, *204*, 107724. [CrossRef]
57. An, R.; Grewal, P.S. *Photorhabdus temperata* subsp. *Stackebrandtii* subsp. nov. (Enterobacteriales: Enterobacteriaceae). *Curr. Microbiol.* **2011**, *61*, 291–297. [CrossRef]
58. Pidot, S.J.; Coyne, S.; Kloss, F.; Hertweck, C. Antibiotics from neglected bacterial sources. *Int. J. Med. Microbiol.* **2014**, *304*, 14–22. [CrossRef]
59. Ffrench-Constant, R.; Waterfield, N.; Daborn, P.; Joyce, S.; Bennett, H.; Au, C.; Dowling, A.; Boundy, S.; Reynolds, S.; Clarke, D. *Photorhabdus*: Towards a functional genomic analysis of a symbiont and pathogen. *FEMS Microbiol. Rev.* **2003**, *26*, 433–456. [CrossRef]
60. Regaiolo, A.; Dominelli, N.; Andresen, K.; Heermann, R. The biocontrol agent and insect pathogen *Photorhabdus luminescens* interacts with plant roots. *Appl. Environ. Microbiol.* **2020**, *86*, e00891-20. [CrossRef] [PubMed]
61. Eckstein, S.; Brehm, J.; Seidel, M.; Lechtenfeld, M.; Heermann, R. Two novel XRE-like transcriptional regulators control phenotypic heterogeneity in *Photorhabdus luminescens* cell populations. *BMC Microbiol.* **2021**, *21*, 63. [CrossRef] [PubMed]

62. Eliceche, D.P.; Rusconi, J.M.; Rosales, M.N.; Salas, A.; Achinelly, M.F. Infectivity by nematode-bacteria association on the potato weevil *Phyrdenus muriceus*. *Rev. Argent. Microbiol.* **2020**, *52*, 256–257. [CrossRef]
63. Mahar, A.N.; Munir, M.; Gowen, S.R.; Hague, N.G.M. Role of entomopathogenic bacteria, *Photorhabdus luminescens* and its toxic secretions against *Galleria mellonella* larvae. *J. Entomol.* **2005**, *2*, 69–76. [CrossRef]
64. Abdel-Razek, A.S. Pathogenic effects of *Xenorhabdus nematophilus* and *Photorhabdus luminescens* (Enterobacteriaceae) against pupae of the Diamondback Moth, *Plutella xylostella* (L.). *J. Pest Sci.* **2003**, *76*, 108–111. [CrossRef]
65. Blackburn, M.B.; Domek, J.M.; Gelman, D.B.; Hu, J.S. The broadly insecticidal *Photorhabdus luminescens* toxin complex a (Tca): Activity against the Colorado potato beetle, *Leptinotarsa decemlineata*, and sweet potato whitefly, *Bemisia tabaci*. *J. Insect Sci.* **2005**, *5*, 1–11. [CrossRef]
66. Salazar-Gutiérrez, J.D.; Castelblanco, A.; Rodríguez-Bocanegra, M.X.; Teran, W.; Sáenz-Aponte, A. *Photorhabdus luminescens* subsp. akhurstii SL0708 pathogenicity in *Spodoptera frugiperda* (Lepidoptera: Noctuidae) and *Galleria mellonella* (Lepidoptera: Pyralidae). *J. Asia-Pacific Entomol.* **2017**, *20*, 1112–1121. [CrossRef]
67. Tounsi, S.; Aoun, A.E.; Blight, M.; Rebai, A.; Jaoua, S. Evidence of oral toxicity of *Photorhabdus temperate* strain K122 against *Prays oleae* and its improvement by heterologous expression of *Bacillus thuringiensis* cry 1Aa and cry 1Ia genes. *J. Invert. Pathol.* **2006**, *91*, 131–135. [CrossRef] [PubMed]
68. Jallouli, W.; Abdelkefi-Mesrati, L.; Tounsi, S.; Jaoua, S.; Zouari, N. Potential of *Photorhabdus temperate* K122 bioinsecticide in protecting wheat flour against *Ephestia kuehniella*. *J. Stored Products Res.* **2013**, *53*, 61–66. [CrossRef]
69. Bock, C.H.; Shapiro-Ilan, D.I.; Wedge, D.E.; Cantrell, C.L. Identification of the antifungal compound, trans-cinnamic acid, produced by *Photorhabdus luminescens*, a potential bio-pesticide against pecan scab. *J. Pest Sci.* **2014**, *87*, 155–162. [CrossRef]
70. Bowen, D.J.; Ensign, J.C. Purification and characterization of a high molecular- weight insecticidal protein complex produced by the entomopathogenic bacterium *Photorhabdus luminescens*. *Appl. Environ. Microbiol.* **1998**, *64*, 3029–3035. [CrossRef]
71. Dunphy, G.B.; Webster, J.M. Virulence mechanisms of *Heterorhabditis heliothidis* and its bacterial associate, *Xenorhabdus luminescens*, in nonimmune larvae of the greater wax moth, *Galleria mellonella*. *Int. J. Parasitol.* **1988**, *18*, 729–737. [CrossRef]
72. English, L.; Slatin, S.L. Mode of action of delta-endotoxins from *Bacillus thuringiensis*: A comparison with other bacterial toxins. *Insect Biochem. Mol. Biol.* **1992**, *22*, 1–7. [CrossRef]
73. Williamson, V.M.; Kaya, H.K. Sequence of a symbiont. *Nat. Biotech.* **2003**, *21*, 1924–1925. [CrossRef]
74. Duchaud, E.; Rusniok, C.; Frangeul, L.; Buchrieser, C.; Givaudan, A.; Taourit, S. The genome sequence of the entomopathogenic bacterium *Photorhabdus luminescens*. *Nat. Biotechnol.* **2003**, *21*, 1307–1313. [CrossRef]
75. Kronenwerth, M.; Brachmann, A.O.; Kaiser, M.; Bode, H.B. Bioactive derivatives of isopropylstilbene from mutasynthesis and chemical synthesis. *ChemBioChem* **2014**, *15*, 2689–2691. [CrossRef] [PubMed]
76. Challinor, V.L.; Bode, H.B. Bioactive natural products from novel microbial sources. *Ann. N.Y. Acad. Sci.* **2015**, *1354*, 82–97. [CrossRef]
77. Jallouli, W.; Zouari, N.; Jaoua, S. Involvement of oxidative stress and growth at high cell density in the viable but nonculturable state of *Photorhabdus temperata* ssp. *temperata* strain K122. *Process. Biochem.* **2010**, *45*, 706–713. [CrossRef]
78. Vlisidou, I.; Hapeshi, A.; Healey, J.R.; Smart, K.; Yang, G.; Waterfield, N.R. The *Photorhabdus asymbiotica* virulence cassettes deliver protein effectors directly into target eukaryotic cells. *eLife* **2019**, *8*, e46259. [CrossRef]
79. Daborn, P.J.; Waterfield, N.; Silva, C.P.; Au, C.P.Y.; Sharma, S.; Ffrench-Constant, R.H. A single *Photorhabdus* gene, makes caterpillars floppy (mcf), allows *Escherichia coli* to persist within and kill insects. *Fixed Point Theory Appl.* **2002**, *99*, 10742–10747. [CrossRef]
80. Dutta, T.K.; Mathur, C.; Mandal, A.; Somvanshi, V.S. The differential strain virulence of the candidate toxins of *Photorhabdus akhurstii* can be correlated with their inter-strain gene sequence diversity. *Biotech* **2020**, *10*, 1–13. [CrossRef] [PubMed]
81. Zhang, D.; De Souza, R.F.; Anantharaman, V.; Iyer, L.M.; Aravind, L. Polymorphic toxin systems: Comprehensive characterization of trafficking modes, processing, mechanisms of action, immunity and ecology using comparative genomics. *Biol. Direct* **2012**, *7*, 18. [CrossRef]
82. Ffrench-Constant, R.H.; Dowling, A.J. Photorhabdus toxins. In *Advances in Insect Physiology*; Dhadialla, T.S., Gill, S.S., Eds.; Elsevier Academic Press: London, UK, 2014; Volume 47, pp. 343–388.
83. Zhou, X.; Kaya, H.K.; Heungens, K.; Goodrich-Blair, H. Response of ants to a deterrent factor(s) produced by the symbiotic bacteria of entomopathogenic nematodes. *Appl. Environ. Microbiol.* **2002**, *68*, 6202–6209. [CrossRef]
84. Shapiro-Ilan, D.I.; Reilly, C.C.; Hotchkiss, M.W. Suppressive effects of metabolites from *Photorhabdus* and *Xenorhabdus* spp. on phytopathogens of peach and pecan. *Arch. Phytopathol. Plant Protect.* **2009**, *42*, 715–728. [CrossRef]
85. Shahina, F.; Tabassum, K.A.; Salma, J.; Mahreen, G. Biopesticidal affect of *Photorhabdus luminescens* against *Galleria mellonella* larvae and subteranean termite (Termitidae: *Macrotermis*). *Pak. J. Nematol.* **2011**, *29*, 35–43.
86. Erler, S.; Popp, M.; Lattorff, M. Dynamics of immune system gene expression upon bacterial challenge and wounding in a social insect (*Bombus terrestris*). *PLoS ONE* **2011**, *6*, e18126. [CrossRef]
87. Abd El-Zaher, F.H.; Abd-Elgawad, M.M.M.; Abd El-Maksoud, H.K. Use of the entomopathogenic nematode symbiont *Photorhabdus luminescens* as a biocontrol agent. B- Factors affecting the cell-free filtrates from the bacterium. *J. Appl. Sci. Res.* **2012**, *8*, 4600–4614.
88. Shapiro-Ilan, D.I.; Bock, C.H.; Hotchkiss, M.W. Suppression of pecan and peach pathogens on different substrates using *Xenorhabdus bovienii* and *Photorhabdus luminescens*. *Biol. Cont.* **2014**, *77*, 1–6. [CrossRef]

89. Abd El-Zaher, F.H.; Abd El-Maksoud, H.K.; Abd-Elgawad, M.M.M. Use of *Photorhabdus* as a biopesticides. A-Cell suspension from *Photorhabdus* sp. against *Galleria mellonella* insect. *Catrina* **2008**, *3*, 125–129.
90. Keskes, S.; Jallouli, W.; Atitallah, I.B.; Driss, F.; Sahli, E.; Chamkha, M.; Touns, S. Development of a cost-effective medium for *Photorhabdus temperata* bioinsecticide production from wastewater and exploration of performance kinetic. *Sci. Rep.* **2021**, *11*, 779. [CrossRef]
91. Wang, Y.; Bilgrami, A.L.; Shapiro-Ilan, D.; Gaugler, R. Stability of entomopathogenic bacteria, *Xenorhabdus nematophila* and *Photorhabdus luminescens*, during in vitro culture. *J. Int. Microbiol. Biotechnol.* **2007**, *34*, 73–81. [CrossRef]
92. Orozco-Hidalgo, M.T.; Quevedo-Hidalgo, B.; Sáenz-Aponte, A. Growth kinetics and pathogenicity of *Photorhabdus luminescens* subsp. Akhurstii SL0708. *Egypt. J. Biol. Pest Cont.* **2019**, *29*, 71. [CrossRef]
93. Jallouli, W.; Hammami, W.; Zouari, N.; Jaoua, S. Medium optimization for biomass production and morphology variance overcome of *Photorhabdus temperata* ssp. *temperata* strain K122. *Process. Biochem.* **2008**, *43*, 1338–1344. [CrossRef]
94. Hinchliffe, S.J. Insecticidal toxins from the *Photorhabdus* and *Xenorhabdus* bacteria. *Open Toxinol. J.* **2010**, *3*, 101–118. [CrossRef]
95. Tobias, N.J.; Wolff, H.; Djahanschiri, B.; Grundmann, F.; Kronenwerth, M.; Shi, Y.M. Natural product diversity associated with the nematode symbionts *Photorhabdus* and *Xenorhabdus*. *Nat. Microbiol.* **2017**, *2*, 1676–1685. [CrossRef]
96. Bode, H.B. Entomopathogenic bacteria as a source of secondary metabolites. *Curr. Opin. Chem. Biol.* **2009**, *13*, 224–230. [CrossRef]
97. Bussaman, P.; Sermswan, R.W.; Grewal, P.S. Toxicity of the entomopathogenic bacteria *Photorhabdus* and *Xenorhabdus* to the mushroom mite (*Luciaphorus* sp.; Acari: Pygmephoridae). *Biocont. Sci. Tech.* **2006**, *16*, 245–256. [CrossRef]
98. Hazir, S.; Shapiro-Ilan, D.I.; Bock, C.H.; Hazir, C.; Leite, L.G.; Hotchkiss, M.W. Relative potency of culture supernatants of *Xenorhabdus* and *Photorhabdus* spp. on growth of some fungal phytopathogens. *Eur. J. Plant Pathol.* **2016**, *146*, 369–381. [CrossRef]
99. Shehata, I.E.; Hammam, M.M.A.; El-Borai, F.E.; Duncan, L.W.; Abd-Elgawad, M.M.M. Comparison of virulence, reproductive potential, and persistence among local *Heterorhabditis indica* populations for the control of *Temnorhynchus baal* (Reiche & Saulcy) (Coleoptera: Scarabaeidae) in Egypt. *Egy. J. Biol. Pest Cont.* **2019**, *29*, 32. [CrossRef]
100. Shehata, I.E.; Hammam, M.M.A.; El-Borai, F.E.; Duncan, L.W.; Abd-Elgawad, M.M.M. Traits of the entomopathogenic nematode, *Heterorhabditis bacteriophora* (Hb-EG strain), for potential biocontrol in strawberry fields. *Egy. J. Biol. Pest Cont.* **2020**, *30*, 40. [CrossRef]
101. Abd El-Zaher, F.H.; Abd-Elgawad, M.M.M. Assessment of the antagonistic potential for the bacterial symbionts of entomopathogenic nematodes. *J. Appl. Sci. Res.* **2012**, *8*, 4590–4599.
102. Jung, S.; Kim, Y. Synergistic effect of entomopathogenic bacteria (*Xenorhabdus* sp. and *Photorhabdus temperata* ssp. *temperata*) on the pathogenicity of *Bacillus thuringiensis* ssp. *Aizawai* against *Spodoptera exigua* (Lepidoptera: Noctuidae). *Environ. Entomol.* **2006**, *35*, 1584–1589. [CrossRef]
103. Benfarhat-Touzri, D.; Ben Amira, A.; Khedher, B.S.; Givaudan, A.; Jaoua, S.; Tounsi, S. Combinatorial effect of *Bacillus thuringiensis* kurstaki and *Photorhabdus luminescens* against *Spodoptera littoralis* (Lepidoptera: Noctuidae). *J. Basic Microbiol.* **2014**, *54*, 1160–1165. [CrossRef] [PubMed]
104. Gulcu, B.; Hazir, S.; Kaya, H.K. Scavenger deterrent factor (SDF) from symbiotic bacteria of entomopathogenic nematodes. *J. Invertebr. Pathol.* **2012**, *110*, 326–333. [CrossRef] [PubMed]
105. Liu, D.; Burton, S.; Glancy, T.; Li, Z.S.; Hampton, R.; Meade, T.; Merlo, D.J. Insect resistance conferred by 283-kDa *Photorhabdus luminescens* protein TcdA in *Arabidopsis thaliana*. *Nat. Biotechnol.* **2003**, *21*, 1222–1228. [CrossRef]
106. Petell, J.K.; Merlo, D.J.; Herman, R.A.; Roberts, J.L.; Guo, L.; Schafer, B.W.; Sukhapinda, K.; Merlo, A.O. Transgenic Plants Expressing *Photorhabdus* Toxin. US Patent No. 2004/6717035 B2. Available online: https://www.google.com/patents/US6717035 (accessed on 3 June 2021).
107. Rasmann, S.; Köllner, T.G.; Degenhardt, J.; Hiltpold, I.; Toepfer, S.; Kuhlmann, U.; Gershenzon, J.; Turlings, T.C.J. Recruitment of entomopathogenic nematodes by insect-damaged maize roots. *Nature* **2005**, *434*, 732–737. [CrossRef]
108. Wilschut, R.A.; Geisen, S. Nematodes as drivers of plant performance in natural systems. *Trends Plant Sci.* **2020**, *26*, 237–247. [CrossRef]
109. Caccia, M.; Marro, N.; Dueñas, J.R.; Doucet, M.E.; Lax, P. Effect of the entomopathogenic nematode-bacterial symbiont complex on *Meloidogyne hapla* and *Nacobbus aberrans* in short-term greenhouse trials. *Crop Prot.* **2018**, *114*, 162–166. [CrossRef]
110. Abd-Elgawad, M.M.M. Optimizing biological control agents for controlling nematodes of tomato in Egypt. *Egypt. J. Biol. Pest Cont.* **2020**, *30*, 58. [CrossRef]
111. Abd-Elgawad, M.M.M. Biological control agents in the integrated nematode management of pepper in Egypt. *Egypt. J. Biol. Pest Cont.* **2020**, *30*, 70. [CrossRef]
112. Abd-Elgawad, M.M.M. Biological control agents in the integrated nematode management of potato in Egypt. *Egypt. J. Biol. Pest Cont.* **2020**, *30*, 121. [CrossRef]
113. Abd-Elgawad, M.M.M. Biological control of nematodes infecting eggplant in Egypt. *Bull. Natl. Res. Cent.* **2021**, *45*, 6. [CrossRef]
114. Stock, S.P.; Kusakabe, A.; Orozco, R.A. Secondary metabolites produced by *Heterorhabditis* symbionts and their application in agriculture: What we know and what to do next. *J. Nematol.* **2017**, *49*, 373–383. [CrossRef]
115. Liu, Y.B.; Tabashnik, B.; Dennehy, T.; Patin, A.; Bartlet, A. Development time and resistance to Bt crops. *Nature* **1999**, *400*, 519. [CrossRef]
116. Maithri, S.K.; Ramesh, K.V.; Mutangana, D. Theoretical structure prediction of TcaA from *Photorhabdus luminescens* and aminopeptidase N receptor from *Helicoverpa armigera*. *Res. J. Recent Sci.* **2013**, *2*, 40–49.

117. Kumar, K.S.V.; Vendan, K.T.; Nagaraj, S.B. Isolation and characterization of entomopathogenic symbiotic bacterium, *Photorhabdus luminescens* of *Heterorhabditis indica* from soils of five agro climatic zones of Karnataka. *Biosci. Biotech. Res. Asia* **2014**, *11*, 129–139. [CrossRef]
118. Kim, E.; Jeoung, S.; Park, Y.; Kim, K.; Kim, Y. A novel formulation of *Bacillus thuringiensis* for the control of brassica leaf beetle, *Phaedon brassicae* (Coleoptera: Chrysomelidae). *J. Econ. Entomol.* **2015**, *108*, 2556–2565. [CrossRef] [PubMed]
119. Mohan, S.; Sabir, N. Biosafety concerns on the use of *Photorhabdus luminescens* as biopesticide: Experimental evidence of mortality in egg parasitoid *Trichogramma* spp. *Curr. Sci.* **2005**, *89*, 1268–1272.
120. Fand, B.B.; Gautam, R.D.; Kamra, A.; Suroshe, S.S.; Mohan, S. Bioefficacy of aqueous garlic extract and a symbiotic bacterium, *Photorhabdus luminescens* against *Phenacoccus solenopsis* Tinsley (Homoptera: Pseudococcidae). *Biopestic. Int.* **2012**, *8*, 38–48.
121. Nielsen-LeRoux, C.; Gaudriault, S.; Ramarao, N.; Lereclus, D.; Givaudan, A. How the insect pathogen bacteria *Bacillus thuringiensis* and *Xenorhabdus/Photorhabdus* occupy their hosts. *Curr. Opin. Microbiol.* **2012**, *15*, 220–231. [CrossRef] [PubMed]

MDPI

Review

Optimizing Safe Approaches to Manage Plant-Parasitic Nematodes

Mahfouz M. M. Abd-Elgawad

Plant Pathology Department, National Research Centre, El-Behooth St., Dokki, Giza 12622, Egypt; mahfouzian2000@yahoo.com

Abstract: Plant-parasitic nematodes (PPNs) infect and cause substantial yield losses of many foods, feed, and fiber crops. Increasing concern over chemical nematicides has increased interest in safe alternative methods to minimize these losses. This review focuses on the use and potential of current methods such as biologicals, botanicals, non-host crops, and related rotations, as well as modern techniques against PPNs in sustainable agroecosystems. To evaluate their potential for control, this review offers overviews of their interactions with other biotic and abiotic factors from the standpoint of PPN management. The positive or negative roles of specific production practices are assessed in the context of integrated pest management. Examples are given to reinforce PPN control and increase crop yields via dual-purpose, sequential, and co-application of agricultural inputs. The involved PPN control mechanisms were reviewed with suggestions to optimize their gains. Using the biologicals would preferably be backed by agricultural conservation practices to face issues related to their reliability, inconsistency, and slow activity against PPNs. These practices may comprise offering supplementary resources, such as adequate organic matter, enhancing their habitat quality via specific soil amendments, and reducing or avoiding negative influences of pesticides. Soil microbiome and planted genotypes should be manipulated in specific nematode-suppressive soils to conserve native biologicals that serve to control PPNs. Culture-dependent techniques may be expanded to use promising microbial groups of the suppressive soils to recycle in their host populations. Other modern techniques for PPN control are discussed to maximize their efficient use.

Keywords: nematode management; biological control; mechanisms; host plant resistance; synthetic nematicide; botanicals; optimizing strategies

Citation: Abd-Elgawad, M.M.M. Optimizing Safe Approaches to Manage Plant-Parasitic Nematodes. *Plants* **2021**, *10*, 1911. https://doi.org/10.3390/plants10091911

Academic Editors: Špela Mechora and Dragana Šunjka

Received: 24 July 2021
Accepted: 17 August 2021
Published: 14 September 2021

1. Introduction

Plant-parasitic nematodes (PPNs) can cause significant losses in the size and quality of a wide range of economically important crops. Previously, regulatory and sanitation measures entirely avoided their casual introduction, minimized their spread, and/or reduced their damage. However, the current widespread and severe damage of PPNs lead to the need for additional control measures. Synthetic nematicides have shown some nematode control with consequent yield increase—but many of them have been restricted or banned. This is due to their adverse effects on human health and the environment, as well as damage to the durability of many agricultural ecosystems. Increasing concern over such chemical nematicides has led to unprecedented and great efforts in various research areas to manage these pests safely and effectively.

Current nematode research has addressed genetics and molecular patterns associated with plant defense and damage in the event of nematode infection [1]. Research has also addressed microbial priming [2], which has achieved tremendous progress. Various techniques are being developed to fully grasp the interaction between PPNs and their host and non-host plants via the elicitor-receptor reciprocal action [3,4]. These substantial mechanisms are expected to provide us with the needed information to design durable nematode resistance in plants. Moreover, the processes engaged in plant defense and protection against PPN can be activated by beneficial microbes and synthetic elicitors that can be soundly and effectively exploited [4].

Various aspects of the current research focus on the fundamentals of the PPN–plant relationship. However, there have been related opportunities to exploit the available applications to safely control nematodes. Thus, benign alternative methods to chemical nematicides are expected to make up many of the the durable crop protection strategies. Abd-Elgawad [5] recently addressed the general strategies of using safe antagonists of PPNs. They are generally based on either augmentation (inundative and inoculative) or conservation biological control. This review extends and updates implementations of such strategies. It highlights the use and potential of various strategies and tactics that can contribute to PPN management. Such approaches may include biological control agents (BCAs), the use of botanicals (e.g., antagonistic plants), host plant resistance to nematodes with related crop rotations, and other advanced treatments. The desired outcome is not only to avoid plant damage and yield losses caused by the nematodes and contribute as best we can in sustainable agricultural ecosystems, but to summarize current progress made in the research and application of these techniques. It also presents key factors affecting their success and broader exploitation, as well as their merits and demerits, and discusses agricultural practices that optimize PPN control.

2. Biological Control Agents

2.1. Their General Categorization and Effects

Fungal and bacterial organisms are currently considered the most efficient and major biocontrol agents (BCAs) against PPNs [6,7]. Others, such as predaceous nematodes, mites, viruses, protozoans, oligochaetes, collembola, algae, and turbellarians, are also BCAs, but are less effective and less studied. Given the common occurrence of such numerous BCAs and their bioactive compounds, it is uncertain whether they can routinely limit PPN populations. To test their benefits, many hindrances must be overcome, including their mass culture, formulation, application techniques, and interactions, as they are applied to the seed, cultivated soil, or seedling medium for PPN control. The modes of action of these BCAs may be categorized into two major groups, i.e., either direct antagonism against PPNs or indirectly promoting operators of plant growth. However, BCAs and/or their bioactive metabolites are responsible for PPN management via various mechanisms. For instance, these fungi may be endoparasitic, toxin-producing, nematode-trapping, and/or parasites of eggs, juveniles, or adults. The other main taxon of BCAs was recently reviewed and shown to contain endophytic bacteria, rhizobacteria, obligate parasitic bacteria, symbiotic bacteria, opportunistic parasitic bacteria, and cry protein-forming bacteria [5]. These BCAs can also produce plant growth promotors for plant growth [8]. Directly, they can assist plants by easing resource possession and production of active compounds and hormones (e.g., gibberellins and cytokinin) necessary for plant growth. Indirectly, they can produce lytic enzymes and antibiotics to suppress pests and pathogens. These BCAs can also prime plants for PPN resistance. Molinari and Leonetti [2] have recently reported that BCAs can interact with roots to prime plants against infection by root-knot nematodes (RKNs), *Meloidogyne* spp., via upregulation of endogenous defense genes. They may comprise salicylic acid-dependent pathogenesis-related genes of the systemic acquired resistance, such as PR-1, PR-1b, PR-3, and PR-5. Moreover, related enzymes, e.g., endochitinase and glucanase, showed elevated activities in roots of pre-treated inoculated plants, which may open new avenues to novel PPN control.

2.2. Fungal and Bacterial Biocontrol

Fungal species related to genera, such as *Trichoderma*, *Purpureocillium*, *Catenaria*, *Actylellina*, *Dactylellina*, *Arthrobotrys*, *Aspergillus*, *Monacrosporium*, *Hirsutella*, and *Pochonia*, are outstanding BCAs against PPNs, especially for RKN control [6,9,10]. For example, endophytic fungi of the genera *Trichoderma*, *Fusarium*, *Alternaria*, *Purpureocillium*, and *Acremonium* can colonize plant roots and enhance plant defense via multiple factors [11]. They may repel

RKN second-stage juveniles (J_2) away from roots, retard or attenuate PPN development, or lower their fecundity. Species of *Trichoderma* and *Purpureocillium* can kill RKNs at different life stages in the root systems or soil. *Pochonia chlamydosporia* can induce systemic resistance against *M. incognita*. They can also target other important PPNs, such as the cyst nematodes, *Globodera* spp., especially when combined with additional BCA [7,12]. Silva S. et al. [13] screened 33 strains of *P. chlamydosporia* and *Purpureocillium lilacinum* and selected the most promising ones, e.g., *P. lilacinum* (CG1042, CG1101) and *P. chlamydosporia* (CG1006, CG1044) to be tested against *Meloidogyne enterolobii*, a relatively recently described but important RKN species, especially for tomato and banana. Both *P. lilacinum* and *P. chlamydosporia* caused 44 and 34% suppression in *M. enterolobii* eggs on tomato roots, respectively, whereas 34% suppression in *M. enterolobii* eggs was recorded on banana roots by *P. chlamydosporia*. However, such efficacies were noted when inoculation level of *M. enterolobii* eggs was as low as 500 eggs. Thus, applying both species, within the context of integrated pest management (IPM) programs, is suggested when *M. enterolobii* population levels are low. Another group of potential fungi is the arbuscular mycorrhizal fungi (AMF), which function as obligate plant root symbionts. The plant offers photosynthetic carbon for the symbionts, while the latter assist roots to uptake higher nutrients and boost both root growth and structure. Moreover, they usually compete for nutrition and space with PPNs and induce plant systemic resistance [14].

Likewise, numerous bacterial species of many genera, such as *Pseudomonas, Serratia, Bacillus, Pasteuria, Achromobacter, Variovorax, Rhizobium, Agrobacterium, Comamonas, Arthrobacter*, and *Burkholderia*, have shown nematicidal activities against PPNs [15–18]. Their modes of action against PPNs may vary even within a genus [7], but generally comprise antagonism, antibiotic production, and/or induced resistance. For instance, their nematicidal operations against RKNs are based on toxic particles of cry proteins in *Bacillus thuringiensis*, toxic antibiotics in *B. subtilis* and *B. cereus*, enzymatic activity in *B. firmus*, and repelling RKN J_2 by *B. cereus* after colonizing the roots. Strain efficacy may be more specific; the B. cereus strain BCM2 could reduce *M. incognita* invasion on tomato roots by 67.1% relative to the control. The *B. cereus* strain that colonizes tomato roots could affect the root exudates by raising the secretion of certain repellent *M. incognita* J_2 substances [19]. Lee and Kim [20] found that chitinase and protease, produced by *B. pumilus* strain L1, were responsible for *M. arenaria* antagonistic traits. They induced about 90% J_2 mortality and 88% inhibition of egg hatch.

The obligate parasites, *Pasteuria* spp., are extremely safe BCAs to manage PPNs. They can act on the nematodes under tough ecological conditions and with variable soil temperature, pH, and moisture. Their spores usually attach to the cuticle surface of the specific nematode species/race as they move about in the soil. Once adhered, they set up germ tubes that break into the nematode's interior body. The internal proliferation of these cells and sporulation suppresses nematode multiplication and causes nematode mortality. As they are species-specific, *Pasteuria* spp. do not hurt non-target organisms, e.g., as a RKN-specific parasite, *P. penetrans* can only infect the related J_2. The attached spores restrict the nematode movement and make them stick to the nearby nematodes. If the PPN can mature, the female may produce a few or no eggs in host plants. Abd-Elgawad [5] reviewed the attributes which allow *Pasteuria* spp. to integrate with other safe approaches, e.g., crop rotation, soil amendments, and nematode-resistant cultivars, to manage PPNs. Their endospores are resistant to mechanical shearing, drying, and heat. However, *Pasteuria* isolates should be screened to select the most adequate one(s) for biocontrol in specific agroecosystems because they are very specific and may only attack certain isolates of a given species.

2.3. Nematode-Suppressive Soils

Suppressive soils were reviewed as those in which harmful pathogens and parasites, herein PPNs, cannot set up or persist, found, but lead to no disease, or become established and initiate a mild disease that soon recedes [21]. The biological activity of such a specific soil is documented when its suppressiveness: (1) is removed by biocides; (2) can be conveyed to conducive soil with a modest volume of suppressive soil; (3) is specific to a nematode species; (4) can reduce multiplication in root-knot and cyst nematodes in the root zone; (5) can be detected by baiting methods; (6) is heat sensitive; (7) is density-dependent. To achieve these attributes, the BCAs in nematode-suppressive soils can act directly as nematode antagonists and/or they can indirectly prime plants and induce their defense responses against PPNs. Antibiosis and parasitism by BCAs were also suggested in a few soils with specific PPN suppressiveness. Topalović et al. [10] appraised fungi and bacteria that were characterized in PPN-suppressive soils via next-generation sequencing or extracted from dead or diseased PPNs. They noted that soil suppression may act against the relevant PPN species as the microbiome may vary from one soil to another. For instance, suppressive soil was more efficient in *M. hapla* than the *M. incognita* control. Additionally, soil properties and plant species/cultivar can also influence the magnitude of this suppression. Thus, to avoid the impact of soil physicochemical and nutritional features, Topalovic et al. [9] altered the approach of transferring soil suppressiveness to the conducive soil by using the microbial component of the suppressive soil only in a water suspension. However, the soilless suspension could not cause mortality of any PPN species, but it could in tomato-planted soil with the suspension [9].

The magnitudes of root colonization by BCAs and their possible metabolites and induced resistance are impacted by plant genotype. Nematode-susceptible plants will harbor more PPNs and need more BCAs to suppress them than poor host plants. Although two isolates of *P. chlamydosporia* prompted systemic resistance against RKNs, the induction was plant species-dependent. This reduced *M. incognita* female fecundity, infection, and reproduction of tomatoes, but not cucumbers [21]. Moreover, in a separate monoculture of different sugar beet cultivars in *Heterodera schachti*-infested soil, *H. schachtii*-tolerant cv. "Pauletta" enabled suppressiveness to be set up without the initial yield decrease noted in susceptible cv. "Beretta" [22]. Botelho et al. [23] speculated that the biological and physicochemical attributes of the coffee rhizosphere could dictate their impact on *Meloidogyne exigua* suppression under field conditions. Thus, such suppressive soils caused about 83% *M. exigua* J_2 mortality and attained the highest yields of coffee beans. Thus, further plant–nematode–microbe interactions in suppressive soils require additional study to be better understood to enable novel insights for the best exploitation of suppressiveness. Westphal [5] reported a few methods to examine the biology of PPN soil suppressiveness. They mostly rely on comparing PPN reproduction in sterilized vs. non-sterilized soils. A drawback in this approach is that the growth parameters of plants are usually better in sterilized soil, which impacts the PPN activities and other biologicals as well. Therefore, it may bias the results [24]. While culture-independent methods on the related microbiome have given a better understanding of the functional potential of many PPN suppressive BCAs, culture-dependent techniques enabled the use of some microbial groups in specific suppressive soils [10]. Both approaches should be timely and adequately used to adjust recycling of the relevant microbiome in their host populations and expanded long-term PPN suppression in other soils.

2.4. Evaluating Factors Affecting Their Success

2.4.1. Biological and Ecological Factors

Many factors can affect BCA–nematode interactions. Hence, the biology and ecology of these BCAs should be grasped from the standpoint of pest management so that they can be properly harnessed in biocontrol. Currently, facilitated-omics techniques should contribute to the better holistic perception of biocontrol mechanisms with related colonization processes at the rhizosphere and the relevant factors influencing them. In this vein, Cámara-Almirón et al. [25] reviewed the molecular basis used by many bacteria to antagonize plant pathogens and enhance plant growth and the structural units, necessary for their biofilm setting, e.g., the formation of the bacterial biofilm often influences and ensures a stable and effective biocontrol.

Basically, the biological strain, dose, time, and application method that likely achieve the best BCA(s)–nematode host matching should be optimized for the desired level of PPN management. The best BCA that properly fits the properties of the targeted habitats, cultivated crops, and PPNs to optimize the biocontrol gains should be selected. These factors can modify, to varying degrees, the effects of BCAs on the PPN control programs. Therefore, there is often a gap between BCA efficacy in the field and controlled experimental conditions. Careful manipulation of these factors can improve the effects of BCAs, especially in terms of their being more inconsistent, less effective, and/or slower-acting than pest control via chemicals. Admittedly, as soil food webs, including PPNs, have a cryptic nature, BCA ecology and their role in modifying nematode population levels and dynamics are largely unknown and deserve further study. Hence, significant variables can provide insights into how soil properties can be modulated to enhance biocontrol by conserving and favoring specific settings or BCAs [9,10,22,26–28]. For example, soil moisture and texture [29], salinity [30], mulching [31], and pH [32] were found to modulate nematode populations directly or indirectly by influencing their hosts or enemies [33].

Manufacturers of biocontrol products must consider these factors and their outputs [5]. Otherwise, a biocontrol product may contain several genera or groups of BCAs to increase its potency. Still, only one BCA in another product may be more effective than this combination (e.g., NemOut® (contains Bacillus subtilis + B. licheniformis + Trichoderma longibrachiatum) vs. Rizotec® (contains only *Pochonia chlamydosporia*)) [13].

Using BCAs is not an easy or routine task, but should be based on accurate, complementary data and a good conception of the possible involved factors. Therefore, sampling and primary tests are prerequisites to obtain the data on the related factors for effective IPM. Biological suppression may be assessed by comparing nematode reproduction in both untreated and treated soils with a proper biocide or heat to eliminate BCAs [24]. This test may take 1–3 months as targeted reproduction of PPNs is valued after at least one nematode generation. To shorten this period, survival of only free-living stages of the concerned PPNs may be assessed after several days in the untreated and treated soils. This alternative test offers rapid conclusions but, as such, is limited to measuring the effect of only BCAs on soil or migratory stages of PPNs. In the latter tests, other BCAs specialized in parasitizing PPN eggs and/or nematode-sedentary stages are mistakenly ignored. Furthermore, a PPN species not present in the field soil is preferably utilized in both tests to determine the level of biosuppression to avert the confusing impact of native nematodes, e.g., a host-specific parasite cannot be avoided. For instance, using the reniform nematode, *Rotylenchulus reniformis*, to evaluate its biosuppression in the sting nematode *Belonolaimus longicaudatus*-infested soil cannot detect a species-specific BCA, e.g., *Candidatus Pasteuria* usage, which is specific to parasitizing *B. longicaudatus* [34]. Mixing a small amount of the field soil into disinfested soil may resolve the issue. In this case, *B. longicaudatus* is conveyed with the field soil, and so other target PPNs can be added to the soil with few influences. Notwithstanding the utility of endemic *B. longicaudatus* to detect species-specific antagonists, the BCAs conveyed with the soil should have enough time to multiply to suppressive levels. The test does not negate that biocontrol of PPNs using an introduced BCA may not be as effective in various settings as that of indigenous BCA, due to ecological validity [35]. Eventually, relevant bioassays that validate PPN suppression in a specific agroecosystem should be carried out for the best BCA–PPN host matching.

2.4.2. Agricultural Practices

The positive or negative role of specific production practices should be assessed preferably in the context of IPM programs. Clearly, regulatory and phytosanitary measures should be exercised to avoid PPN contamination of plant materials, cultivated media, and used equipment [36,37]. Generally, fertilization and soil amendments can boost plant growth with consequent possible increases in population levels of both plant parasites and BCAs. However, significant exceptions should be considered [35]. Whenever possible, they must be manipulated to set up an adequate soil environment to boost and/or protect BCAs for conservation biological control. This is important for stable agroecosystems, e.g., alfalfa, turfgrass, and perennial crops, that can likely favor this conservation for long-term persistence of, especially indigenous, BCAs. In contrast, intensive production practices, soil infestations by polyspecific nematodes, introducing seedlings infected with new PPNs, and short crop sequences or sequential susceptible host plants may decrease the occurrence and persistence of these BCAs.

It is important not only to recognize promising BCAs, but also to combine them with conservation operations in the planted soil to enhance the efficacy, consistency, and duration of BCAs. These operations should manipulate the soil habitat to benefit the introduced BCA and any other useful organisms at the expense of PPNs. *Streptomyces*-treated soil showed a reduction in the total bacterial population that significantly changed the rhizosphere microbiome. However, the *Streptomyces* population in the treated rhizosphere was enhanced with less RKN damage to tomato roots [38]. Thus, the practices may comprise offering extra resources to BCAs, enhancing their habitat quality, and reducing or avoiding negative influences of pesticides on them. Resources, such as adequate organic matter, may be added to elevate the existing BCA efficacy and persistence. Specific soil amendments can boost the quality of particular soils [27]. On the contrary, pesticide usage, tillage, crop rotation, and fallow periods can directly disrupt BCA populations. Surprisingly, BCAs may be adversely affected if these practices can decrease their PPN hosts. Thus, these practices should also be adjusted for the effective use of the introduced BCAs in case of augmentation (inundative or inoculative) biological control. More selective pesticides and careful spatial (horizontal and vertical) and temporal usage of nematicides should be followed to evade or at least lessen their damage to BCAs [39].

The use of BCAs can also be optimized via sequential application or co-application with compatible agricultural inputs. The co-usage of *P. chlamydosporia* and chitosan enabled more root colonization by the fungi and less RKN damage than either alone [40]. Abd-Elgawad and Askary [7] reviewed effective strategies that combine BCAs with other cultural inputs to operate additively or synergistically in IPM. Optimistically, more recent operations are still being explored with other microbes. Admittedly, one of the present pressing issues is related to how to adequately address developing multi-dimensional biocontrol programs. One trend is to simultaneously use closely related or compatible species of BCAs that can offer optimum biocontrol efficacy, domination in the soil microbiome, and/or better root protection. Combining *B. subtilis* with *B. pumilus* gave more average increments in growth parameters of *M. incognita*-infected cowpea plants than either alone [41]. Thus, it is likely that different *Bacillus* species may present unlike anti-RKN mechanisms, which open new avenues to improve the biocontrol efficacy. Another trend is sequential usage. Dahlin et al. [42] found that fluopyram (a nematicide) reduced the *M. incognita* population on tomatoes at planting and that adding the *P. lilacinum* strain PL251 during the growing season could reinforce the reduction. The reductions in the number of *M. incognita*-J_2 were 56 and 68% when PL251 was applied alone and after fluopyram, respectively. Fewer *M. incognita* galls were found as *B. firmus* preceded synthetic nematicides in consecutive tomato cycles. The galling severity assessed by the galling index (scale of 0–10) was reduced from 7.7 in the control roots to 5.9, 4.3, and 4.0 for *B. firmus*, *B. firmus* + oxamyl, and *B. firmus* + fosthiazate, respectively [43]. These favorable results should be extended and documented as BCAs are compatible with other chemicals.

Dual-purpose BCAs should be utilized against more than one group of crop pests whenever possible. For instance, three populations of entomopathogenic nematodes, used as biocontrol agents against insect pests, could significantly ($p < 0.05$) reduce *M. incognita* populations on watermelon at 7 and 21 days after treatments [44]. Moreover, the entomopatogenic fungus *Lecanicillium muscarium* isolates (Lm1, Lm2, Lm3) were effective against *M. incognita* on tomatoes [45]. The isolate Lm1 reduced *M. incognita* egg and gall numbers by 90 and 80%, respectively. Thus, in these cases, it is assumed that BCAs could control both insect pests and PPNs. Sharma and Sharma [46] demonstrated suppression in various *M. incognita* stages and reproduction on tomato roots caused by individual or dual inoculations of AMF and plant growth-promoting rhizobacteria (PGPR). They found that colonization of AMF (*Rhizophagus irregularis*) and PGPR (*Pseudomonas jessenii* strain R62 and *Pseudomonas synxantha* strain R81) induced tomato resistance against RKN by upregulating the defense genes.

Nonetheless, it is suggested herein to examine the effects of BCAs on a case-by-case basis. This will allow us to monitor and optimize various techniques of PPN control according to the existing variables. It would enable us to unravel the complexity of the factors governing the activity and reproduction of PPNs and their associated BCAs in general or specific suppressive soils. In a soil microbiome, factors that positively affect specific BCAs that repel, suppress, or kill PPNs via competition, predation, or parasitism should be utilized for better plant growth. Likewise, adding amendments or related materials to the soil to release compounds that are toxic to the plant pathogens and/or inducing the defense systems of the host plants should be thoroughly examined and correctly adjusted [35,36]. Eventually, adopted recommendations to combat agricultural pests by decision makers and governmental bodies should guide and enlighten growers for optimizing practical use of bionematicides—especially in developing countries such as Egypt [47].

3. Botanicals as Bionematicides

3.1. Antagonistic Cultivated Plants

During their growth, antagonistic cultivated plants produce antihelminthic compounds that act as antagonists to the nematodes via various modes of action [48]. The nematostatic or nematicide compounds in the plant organs may be freed into the soil or operate within the plant to act as nematode traps or show unfavorable responses to PPNs. The broad conception of these plants may include different groups that can adversely affect various PPN populations, but poor and non-hosts will better be addressed separately [36] hereafter in more detail.

Many antagonistic plant species are found, but the most famous ones that are utilized against important PPNs include *Tagetes* spp., *Azadirahta indica*, *Brassica* spp., and *Crotalaria* spp. [48]. Various species of the genus *Tagetes* (marigold) may reduce PPN populations via different modes of action, e.g., by acting as a poor host or non-host, generate allelopathic compounds, trap the nematodes, or induce nematode antagonistic flora/fauna. Derivatives of bithienyl and alpha-terthienyl produced by marigold are toxic to the nematodes [49]. Species of marigold, such as *T. patula*, *T. erecta*, and *T. minuta*, are efficient, especially against two nematode genera, *Meloidogyne* and *Pratylenchus* [48]. Contrary to a tomato–tomato rotation, *T. erecta*, *T. patula*, and *T. signata* decreased RKN galling in subsequent susceptible tomato plants. Although the neem (*Azadirahta indica*) tree is generally considered to be antagonistic to many pests, the neem-based products are most commonly used in PPN control. They could demonstrate good nematicidal activities. Moreover, many species of *Brassica* (cruciferous plants), such as cabbage, broccoli, rape, canola, and mustard, can form glucosinolates (GSLs). Hydrolysis of secondary metabolite GSLs results in volatile and toxic isothiocynates (ITCs), which act as biofumigants against PPNs. Thus, the nematicidal properties of ITCs released into the soil may be attributed to aromatic GSLs (roots), indole GSLs (root and shoot), and aliphatic GSLs (seeds) when these repositories are rotated with PPN-susceptible plant species or grown as cover crops [50]. The biofumigation range for PPN control has been widened to include non-brassica antagonistic species. They can also form volatile pathogen-suppressing molecules. The hydrolysis of antecedent cyanogenic glycoside/dhurrin in the graminaceous plants (e.g., sudangrass (*Sorghum* × *drummondi*) and sorghum (*Sorghum bicolor*)) may produce cyanides that can kill PPNs [27]. Sunn hemp (*Crotalaria juncea*) as a cover crop and green manure is commonly utilized for its antagonistic impacts on RKNs in numerous crops. Moreover, *C. longirostrata* is is incorporated into the soil after growing as a cover crop to decrease RKN galling. Its effect concerning PPN control may be due to toxins produced during microbial degradation, not by toxic exudates from the plant [36]. Other species were tested for PPN control activity [51]. *Pratylenchus brachyurus* reproduction rates were lowered when maize (*Zea mays*) was intercropped or rotated with bristle oats and oilseed radish [52]. Growth of maize was enhanced (up to 34%) when intercropped with velvet bean (*Mucuna pruriens*) or jack bean (*Canavalia ensiformis*), while *Pratylenchus zeae* population levels were suppressed by 32%. Yields of maize intercropped with jack bean were raised (22–190%) under field conditions [51].

3.2. Plant-Related Materials and Compounds

Using the relevant compounds via extraction from the plants or incorporating plant parts into the soil is another and more common tactic for PPN control than using the entire plants. These materials are mainly extracted or formed from antagonistic plants. They may be grouped under various terms, such as natural compounds, organic acids, essential oils (EOs), and plant extracts and compounds. In contrast, not all of these groups are exclusively related to plants. For instance, acetic acid is produced as secondary plant metabolites [53] or as culture filtrates of the bacterium, *Lactobacillus brevis*, strain WiKim0069 [54]. This acid can damage the cuticle of RKN J_2, vacuolize the cytoplasm, and degrade the nuclei, causing death [54]. Numerous organic acids, such as amino, propionic, formic, and butyric acids, can exert toxic effects on PPN species [18]. They are formed via microbial decomposition of other compounds in the soil, mostly those related to plant materials/residues, but may also result from metabolites formed by soil organisms. Others, such as sesquiterpene heptalic acid produced by the fungus, *Trichoderma viride* [7], and hydroxamic acids from the grass, *Secale cereale* [55], have proved effective against important PPN species.

Neem's natural compounds, such as azadirachtin, kaempferol, thionemone, nimbidin, quercetin, nimbin, and salannin, also have nematicidal properties. Intercropped or treated plant roots, via soil application, can absorb these materials. Moreover, many natural nematicidal compounds have drawn the attention of the pesticide industry to develop their extraction and related processes. Hence, other effective treatments of neem may comprise root dipping in neem leaf extracts, soil amendment with its leaf extracts, mulching the soil with dried or fresh leaves, seed coating or drenching the soil with neem extract or oil, application of root exudates, or treating soil with seed or kernel powder [56]. Botanical extracts have conspicuous merits over synthetic nematicides. For example, they contain new compounds for PPN management [18,48]. Thus, nematodes are not yet able to develop resistance or inactivate them. In addition to originating from natural resources and having rapid biodegradation, they are always less concentrated and thus always less toxic than pure materials. These traits support their use as ecologically benign alternative nematicides.

Moreover, EOs have been tested for PPN control. The differences are clear in their related efficacies. Abd-Elgawad and Omer [57] tested the EOs of four medicinal plant species in the family Lamiaceae for PPN control. *Mentba spicata* caused the highest PPN mortality, followed by *Thymus vulgaris, Majorana hortensis*, and then *Mentha longifolia*. The corresponding major oils in these plants were carvone, P-cymene, terpinen-4-ol, and carvone, respectively. Generally, 0.1 oil solutions of each plant could inhibit more than 80% of *M. incognita* J_2 relative to 3.5% at the untreated check. Among the PPNs tested, *Rotylenchulus reniformis* was better controlled by the oil solutions than nematode species related to the genera *Criconemella* and *Hoplolaimus*. Recently, only 14 out of 29 EOs could have nematicidal efficacy in the range 8–100% at 1000 μg/mL, whereas the EO of Mexican tea (*Dysphania ambrosioides*) was the most efficient [58]. Mexican tea EO eliminated 99.5 and 100% of galls and eggs on susceptible tomato plants, respectively. They found that p-cymene (3.35%), E-ascaridole (8.45%), and (Z)-ascaridole (87.28%) formed 99.08% of the total composition in *D. ambrosioides* oil.

Kalaiselvi et al. [59] found that the elevation at which the plants are cultivated may affect their EOs in terms of quantity, chemical composition, nature, yield, and appearance. When *Artemisia nilagrica* was grown in low and high lands, its extracted EOs showed different lethal concentrations (LCs) against *M. incognita* on tomato. The LC_{50}/48 h was 10.23 and 5.75 μg/mL for EOs of low- and high-altitude plants, respectively. The EOs of low-

and high-level plants decreased *M. incognita* (J_2 and eggs) per 10 g root by approximately 68 and 87%, respectively. These oils also enhanced plant growth differently.

Many studies tested plant extracts against PPNs. Recent focus has been placed upon their compounds [9,10,60,61] as they are generally less toxic and safe alternatives to synthetic chemicals that can also efficiently control PPNs. A plant extract may have both acid and oil together, which can enhance nematode mortality as in vetiver grass *Vetiveria zizanioides* [62]. Extracts of *Paenoia rockii* and *Camellia oleifera* could inhibit egg hatching and J_2 activity of *M. incognita* [63]. Nematicidal activity of aqueous and methanolic extracts of *Ricinus communis*, *Taxus baccata*, *Raphanus raphanistrum*, *Sinapis arvensis*, and *Peganum harmala* on *M. incognita* J_2s was assessed at various exposure times and doses. Methanolic extracts showed more influence than the aqueous ones. Moreover, methanolic extracts of *S. arvensis*, *P. harmala*, and *T. baccata* had peak RKN mortality—86.6, 89.2, and 100%, respectively [60].

3.3. Safety, Reliability, and Economics of the Related Nematicidal Products

Some botanical-based nematicides are being commercially marketed, while others are still in the pipeline. For instance, numerous effective neem-based nematicidal formulations have been marketed, such as Neemrich, Neemix, Neemazal, Neemgold, and Neemax. In contrast, Nemastop is a commercial product with garlic (*Allium sativum*) extract (600 g ground garlic cloves/1 water). This has been marketed for PPN control but is not as effective as synthetic chemical nematicides or even other commercial biocontrol agents on eggplant [64]. However, this does not negate PPN suppression by allicin (diallyl thiosulfinate), the effective nematicidal compound of garlic. Allicin could control *M. incognita* and improve tomato yield [65]. Host range claims of bionematicidal products are often taken from the manufacturer's product labels. They have not necessarily been confirmed in neutral trials [66]. Under favorable conditions, PPN control usually increases with consequently elevated crop yield as a product concentration and/or exposure time is also enhanced.

Three basic elements are required for the bionematicides, in general, to be successful: (1) safety to the environment and human health, (2) reliable nematicidal effect, and (3) favorable economics. For instance, among synthetic chemical nematicides, ITCs are included as active ingredients. Notwithstanding the utility of natural ITCs as biofumigants against PPNs, they may share the same biochemical mechanism of action against the targeted PPNs. Thus, negative effects of ITCs as in mustard biofumigants have caused vulnerability and instability of soil food webs and suppression of beneficial organisms [27,67]. Ntalli and Caboni [50] speculated that non-target organisms are also adversely affected because both synthetic and natural components of ITCs interact in a non-specific and irreversible manner with amino acids and proteins. Hence, more studies harnessing their safe utilization as integrants in pest management programs should be conducted. On the contrary, azadirachtin compounds are relatively safe pesticides relevant to ecological issues and environmental risk. They have faint potential mobility in soil and degrade rapidly. Azadirachtin is non-mutagenic and pure azadirachtin is non-toxic to humans. Additionally, it possesses relative selectivity. Thus, it is safe for beneficial insects and can be utilized in IPM programs [48]. *Tagetes* spp. compounds in soil need to be further examined to establish their fate or degradation periods in field situations. On the one hand, researchers and stakeholders should consider that the notion of safety is relative and should be quantified as these materials are still chemicals, but not synthetic ones. On the other hand, several plant materials/extracts have been synthesized to ensure more safety for human application than the commonly known synthetic chemical nematicides.

Sikora et al. [51] suggested that antagonistic plants are very attractive tools for PPN control, but there are potentially new ones that could also be identified. Moreover, techniques should be sought for efficient and multi-purpose applications. Other merits of antagonistic plants are their effective operation in upgrading the soil characteristics. They are used as organic matter and green cover to raise soil quality [68]. Specific groups of antagonistic plants may possess additional merits. A striking example is to boost the activity of biocontrol agents against PPN in addition to their direct effect of reducing damage from pests. Contrary to the bacteria associated with soybean roots, the rhizobacteria isolated from the roots of antagonistic plant species *Ricinus communis*, *Mucuna deeringiana*, and *Canavalia ensiformis* could significantly decrease both *Meloidogyne incognita* and *Heterodera glycines* population densities on roots of soybean plants. Hence, Grubišić et al. [48] speculated that these plants may retain a selective action within each pest class as they possess multiple mechanisms with a wide spectrum. Additionally, these antagonists, related to legumes, can fix the atmospheric nitrogen, which boosts soil fertility.

More research is direly needed to determine the optimum conditions for these bionematicides in general. To optimize PPN control, their incorporation into the soil should target nematode-life stages and species that are most vulnerable. Variables, such as edaphic factors, tillage systems, proper planting date, favorable plant species, and suitable growth stage, should be examined to be best tailored for PPN control. Notwithstanding the nematicidal activity of brassicas cover crops to suppress PPN populations in soil, they may not provide consistent efficacy. Dutta et al. [27] stressed that temperatures may be too high for such plants to adequately show their nematicidal activities under greenhouse conditions.

A study of economic feasibility for applying any of the botanicals to PPN control should be conducted. If economic factors are not convenient, even a strategy involving good nematicidal properties to the targeted PPNs is doomed to recede. Components of economic success comprise the grower's sense to avoid crop losses caused by the nematode pests, the relative expenses of using this bionematicide compared to other options for PPN control, the value of the commodity (e.g., per acre), and its price in the relevant market. Thus, a grower should be enlightened about the indirect benefits of these safe nematicides, e.g., their use to avoid ecological pollution and health hazards, as well as to avoid nematode resistance of chemical nematicides. Such an agricultural extension would encourage growers to use them. Other policies may lower costs and improve the success of specific botanicals. Marigold seeds, for example, are costly relative to seeds of cover crops because marigold has a high value as an ornamental plant. Therefore, Grubišić et al. [48] suggested that the expenses of its seeds would be reasonably priced, or even lowered, if they were commercialized on a large scale as cover cropping for PPN management programs.

4. Exploiting Poor- or Non-Host Crops

As there are many nematode-susceptible plant species, we should make full use of poor- or non-host genotypes [69–71]. These are species/cultivars with genotypes that are immune, resistant, or tolerant to one or more of the PPN species. They are marketed based on their yield—not as cover crops utilized for pasturage or soil amendments/conservation [36]. Resistant or non-host plants can contribute to solving many PPN-related issues. Ensuring adequate crop sequences using non-host crops is the most effective method utilized for global RKN management. Such immune or resistant plant cultivars can be employed to enhance crop yields, suppress PPN population levels, and boost effective crop rotations [36]. Nematode-tolerant cultivars are usually called non-hosts—although PPN can reproduce on them. However, they can withstand nematode attack, and their crop yields are not significantly affected [71]. They can often enable less nematode reproduction than susceptible cultivars [22].

Older, related conceptions must be updated. Resistance-breaking pathotypes or populations of a nematode species may be able to reproduce on a cultivar that is known to be resistant to this species. These virulent populations can emerge in the field via repeated exposure to the resistance genes in the plants and may adversely affect resistance durability. However, such populations usually show faint competitiveness and less reproductive capacity on susceptible hosts than wild populations. In addition, virulent pathotypes reproduce only on plants with the gene on which their selection happened [72].

In a soil having a polyspecific nematode community, a non-host cultivar for one species may also be the susceptible host for another species. Fortunately, reasonable rotation crops for important PPN species with their host range size are listed [36]. It is justifiable to test any of the listed non-hosts in the targeted soil before recommending them. Moreover, the non-host cultivars would preferably be used in integrated PPN management strategies to avoid continuous resistant cultivar cultivation. As resistance-breaking PPN pathotypes may seriously result from this continuity, they can degrade sustainable crop production systems. Moreover, resistance found in some tomato cultivars failed, due to heat instability or sensitivity of the resistant *Mi* gene to high temperature. Incremental reductions in RKN resistance started as the soil temperature was raised above 25.6 °C. Thus, at 32.8 °C, resistance to RKN infection in these plants was fully broken [73]. Hence, such cultivars may be restricted to regions or seasons with cool soil temperatures.

Resistant cultivars of major crops, such as RKN-resistant tomato, are globally available, but emerging technical and economic issues with releasing others should be resolved [74]. Traditional techniques for host suitability designations of serious PPN species have been reviewed [74–76], but biochemical and DNA markers have merits to complement their phenotype screens [2,77]. Crucial factors affecting the phenotypic expression of the resistance should be adequately examined. Thus, a multi-disciplinary approach combining plant breeders, molecular biologists, and nematologists should investigate the level, nature, and inheritance of resistance traits. They should explore any DNA recombination during breeding cycles. Biochemical markers of genetic traits for PPN resistance should be sought in genome-assisted breeding strategies for their introgression into elite cultivars.

5. Other Methods of PPN Management

5.1. Additional Soil Amendments and Treatments

The broad concept of soil amendments is to use not only plant materials as a cover crop, compost, seed meal, and green manure, but also to mix them with other components. Contrary to the above-mentioned botanicals, they may include, for instance, various animal manures and/or nutrient salts to form different varieties of amendments. These additions are mostly organic matter and have been used in multi-purpose agricultural practices. They can suppress the population levels of many pathogens, pests, and weeds, enrich soil fertility, boost soil structure, increase communities of beneficial organisms, and/or induce systemic resistance of plant species [27]. Organic amendment herein refers to organic material brought from outside to the inside of the soil, e.g., industrial waste products or processing residues. This differs from the above-mentioned botanicals, which were added as fresh crop residue or grown in the rotation, e.g., break, cover, trap, antagonistic or green manure crops. Usually, merging large amounts of such organic material into the soil will reduce PPN densities. These may include many materials, such as oil cakes, sawdust, coffee husks, crustacean skeletons, chicken manure, paper waste, and crop residues, which showed various degrees of PPN control [36]. This action was mostly associated with corresponding increases in crop yields.

Moreover, an amendment that works well in soil with specific edaphic and biological factors may not work at all in another soil. Optimizing the PPN control efficacy relies on its compounds' compositions, quality, and quantity of its associated and interacting microbiome, and its ability to break down these compounds into elements that are suitable for plant growth and/or harmful to the nematodes. Fresh compost enriched with beneficial organisms and nutrients may show better efficacy against PPNs than aged compost. Abiotic factors, such as soil moisture and temperature, usually influence the microbiome and decomposition of these compounds. Soil amended with chicken manure and broccoli at $\geq$25 °C was superior to the same at 20 °C in reducing *M. incognita* galls on tomato roots [78]. Ntalli et al. [79] reviewed various soil amendments and their specific nematicidal activities. They categorized amendments as Brassicaceae and Asteraceae species (for cover-crop, biofumigation, rotation, and incorporation), biochars, composts, and vermicomposts (applied as recycling wastes), and other self-made products, such as canola or orange peel meals, dried leaves of *Canabis sativa*, and marigold or pennycress seed powder. Examples of safe strategies for applying various bionematicides or biocontrol methods against important nematode species are given (Table 1). Nonetheless, some examples may need continuous improvement to the above-mentioned aspects to improve their efficacy and reliability.

Table 1. Examples of various biocontrol agents and strategies against important nematode species.

Biological Control Agent	Nematode Species	Type of Study	Host Plant	Reference
Bacteria				
Bacillus firmus	*Meloidogyne incognita*	In vivo	tomato	[43]
Pasteuria penetranse	*Meloidogyne exigua*	In vivo	coffee	[23]
PGPR: Pseudomonas jessenii and *P. synxantha*	*M. incognita*	In vivo	tomato	[46]
Fungi				
(A) Filamentous: *Trichoderma* spp.	*Rotylenchulus reniformis, M. javanica, M. incognita, Heterodera cajani*	In vivo	tomato, brinjal, okra, soybean, sugarbeet, pigeonpea	[7]
AMF: *Rhizophagus irregularis*	*M. incognita*	In vivo	tomato	[46]
Endophyte: *Fusarium oxysporum*	*Radopholus similis*	In vivo	banana	[80]
(B) Mushrooms: *Lentinula edodes, Macrocybe titans, Pleurotus eryngii*	*M. javanica*	*In vitro*	tomato	[81]
(C) Yeasts: *Saccharomyces cerevisiae*	*M. incognita*	In vivo	eggplant	[82]
Co-application: *Pochonia chlamydosporia* and Chitosan	*M. javanica*	In vivo	tomato	[40]
Sequential application: Fluopyram and *Purpureocillium lilacinum*	*M. incognita*	In vivo	tomato	[42]
Dual-purpose: *Heterorhabditis bacteriophora* EGG	*M. incognita*	In vivo	watermelon	[44]
Algae: *Chlorella vulgaris*	*M. incognita*	In planta	potato	[83]
Nematode-suppressive soil	*M. hapla, Pratylenchus neglectus*	In vivo	tomato	[9]
Botanicals: *Tagetes* spp.	*M. incognita, M. javanica, M. acrita*	In vivo	tomato and *eggplant*	[48]
Soil amendments	*M. incognita, Heterodera glycines*	In vivo	tomato and soybean	[79]
RNA interference via stimulants of soil streptomycetes	*Heterodera avenae*	In planta	wheat	[84]

Using various composts as big sources of soil amendments should be further exploited. They could be manipulated via fermentation processes to make them enriched in the desired microbial species and PPN antagonistic compounds, such as phenolics and humic acids [85]. Composts can also enhance soil resident microbial antagonists, boost plant resistance or tolerance to various stresses, such as PPN infection, and change soil profiles to improper media for PPN reproduction. These gains should be optimized to improve PPN control by grasping the related edaphic factors as well. Eventually, their processes and components should be employed to obtain the desired PPN-suppressive soils.

Other treatments may be used when relevant factors and economic feasibility permit. For high cash crops, heating the soil can effectively manage RKN in protected cultivation [86]. Soil solarization could be effective against PPNs. Tarping the soil surface, especially in sunny regions, with transparent plastic sheets will raise soil temperatures enough to kill many pests and pathogens [87]. Solarization is more effective against PPNs in contained raised beds for cultivation in warm regions. Kokalis-Burelle et al. [88] found that the number of RKN galls on roots of sunflowers, snapdragon, and larkspur were less in steam-treated soil than in solarization alone. Steam treatment was as effective as methyl bromide in controlling *M. arenaria*. They concluded that soil steaming followed by solarization is so effective that it can be a safe alternative to chemical nematicides [88].

Biodisinfestation or biosolarization, that is, using soil amendments before solarization, could enhance the pest and pathogen suppression via rapid generation of harmful compounds, such as acetic and butyric acids, ultraviolet radiation, and lack of oxygen, due to microbial anaerobiosis [27]. However, lethal temperature and related duration may vary from one pathogen species to another [87]. Thus, sustained low PPN populations were sometimes not affected by fairly high temperatures ($\leq$45 °C). In these cases, lasting PPN populations at deeper depths away from the sun could recolonize and infect the plant roots.

Ozonated water, O_3wat, was reported to control *M. incognita* likely via modulated antioxidant systems without phytotoxicity. Tomato plants treated with O_3wat after or before *M. incognita* inoculation showed a root galling index (on a scale of 0–10) of 1.9 or 1.6, respectively, compared to 3.9 in the check [89]. As it degrades to water in a short time, O_3wat could suppress RKN populations early in the growing season without adverse effects.

5.2. Advanced Methods

Our targeted agroecosystems are facing real challenges that require advanced methods and innovative thinking for safe and effective PPN control. Some of the biggest challenges are the increased banning of numerous effective but synthetic chemical nematicides, vertical and horizontal agricultural expansion to raise and improve food production, the frequent appearance of resistance-breaking nematode pathotypes, global warming backing rapid PPN reproduction and spread and discovery of new PPN species [26] (to name but a few related to aggravated nematode damage). Hence, new PPN management tactics and strategies should detect more resilient BCAs and related materials that can best match these expected ecological windows of the pests and pathogens [90]. For instance, efficient methods for better understanding biological and ecological factors that affect BCAs should be employed [91]. This will enable bionematicides to effectively replace unsafe chemical nematicides for sustainable agriculture (Figure 1). Moreover, specific wavelengths via near-infrared spectroscopy could be used to detect soil nematode collections with different functions assigned to definite sets of soil organic matter [92]. Furthermore, developing bioactive compounds that have natural multifunctional derivatives, including nematicidal activity, are in progress. One such derivative is the chitin oligosaccharide dithicyclobutane (COSDTB) derivative. The 1, 3-dithicyclobutane-N-chitosan oligosaccharide could decrease *M. incognita* egg hatching by up to 90% at 2 mg/mL and cause 94% mortality of *M. incognita* J_2 at 4 mg/mL [93]. The role of silicon to support plant resistance against a variety of harmful bacterial and fungal invasions was recently reviewed [94]. As its salts can also suppress *Meloidogyne paranaensis* populations on coffee seedlings [95], silicates should be further tried in IPM programs in specific sites with various groups of pathogens.

Formulating industrial wastes as value-added products for PPN control will also optimize the gaining of the related industries. Waste such as orange bagasse, soybean hull, rice husk, poultry litter, and common bean hull were assessed for *M. javanica* control in the glasshouse [96]. Their mixtures, orange bagasse, soybean hulls, and powdered bean hulls, had a range of 55–100% RKN control. Other promising BCAs or their metabolites against PPNs are still under experimentation, e.g., *Rhodoblastus acidophilus* strain PSB-01 [97] and *Mortierella globalpina* [98]. On the other hand, nanoparticles [99] have proved to possess promising physical and chemical characteristics against nematodes. They have demonstrated effective PPN control with a few possible demerits.

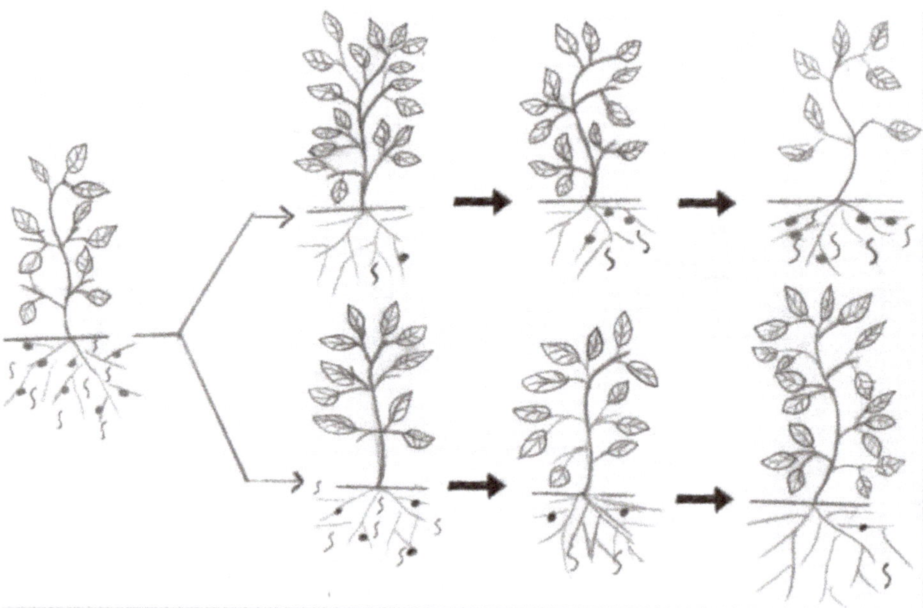

Figure 1. Effect of a chemical nematicide (upper trend) and a bionematicide (lower trend) on root-knot nematodes on susceptible plants. When both nematicides were applied, the chemical has a rapid and significant effect, reducing the nematode population. However, a few nematodes can escape its effect and reproduce to reach a damaging level, while the bionematicide can work continuously to keep the nematode below the economic threshold level [91].

Although optimizing PPN sampling and extraction methods to avoid their misuse and achieve cost-effective and efficient IPM programs are recently emphasized [100], such advances for PPN control should biotechnologically keep up in parallel to these improvements. The current research on genome sequencing technologies, small interfering RNA techniques (RNAi), and targeted genome editing should be harnessed to better grasp plant–nematode interaction mechanisms, and molecular enhancing of PPN-plant resistance should be used to boost these programs [1].

Likewise, other approaches may include expanding targeted biological seed treatment, remote sensing for specific nematicide applications, screening quarantine regulations, minimum tillage to potentiate PPN antagonists, biochemical marker-orientated selection for plant resistance, molecular monitoring and detection of PPNs, and indexing of PPN biodiversity via metagenomics [36]. Their expansion should not negate the continuous search for finding new BCA isolates or genetically engineered ones that are more persistent and compatible with beneficial rhizosphere organisms. There are application techniques that have not been tested on a large scale in earnest to develop them, e.g., spraying BCAs around the base of plants, practical use of a slow-release system, or dipping root plugs into BCA suspensions. Systematic experimentations and field trials testing the aforementioned techniques in various settings to show their worth with feasible, economical insights must be a way forward in crop protection/pest management.

Numerous biological products have demonstrated promising BCA efficacies against PPNs with low costs relative to other chemical nematicides. Table 2 lists examples of such costs for both types of products with their application rates as authorized by the Egyptian Ministry of Agriculture [47]. Thus, it is possible that using the BCA-listed products can offer control comparable to but more economical than these synthetic chemicals (Table 2). Hence, growers should be familiar with these products and their optimized usage, as well as the above-mentioned relevant agricultural practices via agricultural extensions. The end in view is that these bionematicides will ensure more safe applications, while promoting crop yields.

Table 2. Examples of bionematicidal product costs and used rates as compared to chemical nematicides in Egypt.

Active Ingredient	Product Name	Application Rate (Product Hectare $^{-1}$) +	Price per Hectare
Abamectin (soluble concentrate at 20 g/L) generated from the fermentation process of *Streptomyces avermitilis*	Tervigo 2% SC	5.95 L/Hectare	USD 319
10^9 CFU/mL of *Serratia* sp., *Pseudomonas* sp., *Azotobacter* sp., *Bacillus circulans* and *B. thuringiensis*	Micronema	71.4 L/Hectare (thrice)/year	USD 40
10^8 units/mL *Purpureocillium lilacinus*	Bio-Nematon	4.76 L/Hectare/year	USD 78
10^9 bacterium cells of *Serratia marcescens*/mL water	Nemaless	23.8 L/Hectare (thrice)/year	USD 95
Cadusafos (O-ethyl S,S-bis (1-methylpropyl) phosphorodithioate)	Rugby 10 G	57.14 Kg/Hectare	USD 1028
Oxamyl (methyl 2-(dimethylamino)-*N*-(methylcarbamoyloxy)-2 oxoethanimidothioate)	Vydate 24% SL	9.52 L/Hectare (twice)/year	USD 445

+ The rates are evenly applied to the soil (except oxamyl for foliar application). In some cases, these rates may be incorporated into potting mix, field soil, or applied in greenhouses for which other doses may be used according to the manufacturer's product labels on different crops [47,101–104].

6. Conclusions

The current literature on using safe approaches to manage PPNs is extensive given the considerable and negative effects of the synthetic chemical nematicides. These approaches may use various tactics and strategies, including different materials, such as BCAs, botanicals, poor- or non-host crops, and other advanced methods. However, such benign techniques mostly need to be further developed and/or optimized as many BCAs, for example, are more inconsistent, less effective, and/or slower-acting in nematode control than synthetic chemicals. Therefore, agricultural practices that favor the conservation biocontrol of PPNs should be recognized and earnestly applied. Moreover, bionematicides can be included in IPM programs in various ways that make them complementary or superior to these chemicals; they can exert synergistic or additive effects with other agricultural inputs. As numerous bionematicides are or are likely to become widely available soon, seeking their optimal performance is a continuous process. Hence, research priorities for harnessing such relevant and advanced methods should be identified to boost soil fertility within sustainable agricultural production systems. This will necessitate grasping the complex network of interactions among biotic and abiotic factors in intimate contact with these bionematicides to maximize their gains. Thus, the biology and ecology of these bionematicides can be seen as a research priority; they may even need to use previously developed, sophisticated methodologies. Meanwhile, stakeholders, such as nematologists and agronomists, can train, assist, and guide extension officers and farmers to optimize the quality of their produce. This can be achieved by minimizing the adverse effect of the pests in their crops via the improved and efficient application efficacy of these bionematicides.

Funding: This research was funded by STDF, US-Egyptian project cycle 17 grant number 172, and the NRC in-house project No. 12050105, entitled Pesticide Alternatives Against Soil-Borne Pathogens and Pests Attacking Economically Important Solanaceous Crops, funded by The National Research Centre.

Acknowledgments: This article is supported in part by the US-Egypt Project cycle 17 (no. 172) entitled "Preparing and evaluating IPM tactics for increasing strawberry and citrus production". The study was also supported in part by the NRC in-house project No. 12050105 entitled "Pesticide alternatives against soil-borne pathogens and pests attacking economically important solanaceous crops". The author thanks Zafar Handoo for his valuable comments on the manuscript.

Conflicts of Interest: The author declares no conflict of interest. The funders had no role in the design of the study; in the collection, analyses, or interpretation of data; in the writing of the manuscript, or in the decision to publish the results.

References

1. Ibrahim, H.M.M.; Ahmad, E.M.; Martínez-Medina, A.; Aly, M.A.M. Effective approaches to study the plant-root knot nematode interaction. *Plant Physiol. Biochem.* **2019**, *141*, 332–342. [CrossRef] [PubMed]
2. Molinari, S.; Leonetti, P. Bio-control agents activate plant immune response and prime susceptible tomato against root-knot nematodes. *PLoS ONE* **2019**, *14*, e0213230. [CrossRef]
3. Hallmann, J.; Subbotin, S.A. Methods for extraction, processing and detection of plant and soil nematodes. In *Plant Parasitic Nematodes in Subtropical and Tropical Agriculture*; Sikora, R.A., Coyne, D., Hallmann, J., Timper, P., Eds.; CABI: Wallingford, UK, 2018; pp. 87–119.
4. Abdul Malik, N.A.; Kumar, I.S.; Nadarajah, K. Elicitor and receptor molecules: Orchestrators of plant defense and immunity. *Int. J. Mol. Sci.* **2020**, *21*, 963. [CrossRef] [PubMed]
5. Abd-Elgawad, M.M.M. Plant-parasitic nematodes and their biocontrol agents: Current status and future vistas. In *Management of Phytonematodes: Recent Advances and Future Challenges*; Ansari, R.A., Rizvi, R., Mahmood, I., Eds.; Springer Nature: Singapore, 2020; pp. 171–204. [CrossRef]
6. Stirling, G.R. Biological control of plant-parasitic nematodes. In *Diseases of Nematodes*; Poinar, G.O., Jansson, H.-B., Eds.; CRC Press: Boca Raton, FL, USA, 2018; pp. 103–150. [CrossRef]
7. Abd-Elgawad, M.M.M.; Askary, T.H. Fungal and bacterial nematicides in integrated nematode management strategies. *Egypt. J. Biol. Pest Cont.* **2018**, *28*, 74. [CrossRef]
8. Glick, B.R. Plant growth-promoting bacteria: Mechanisms and applications. *Scientifica* **2012**, 1–15. [CrossRef] [PubMed]
9. Topalović, O.; Heuer, H.; Reineke, A.; Zinkernagel, J.; Hallmann, J. Antagonistic role of the microbiome from a *Meloidogyne hapla*-suppressive soil against species of plant-parasitic nematodes with different life strategies. *Nematology* **2019**, *22*, 75–86. [CrossRef]
10. Topalović, O.; Hussain, M.; Heuer, H. Plants and associated soil microbiota cooperatively suppress plant-parasitic nematodes. *Front. Microbiol.* **2020**, *11*, 313. [CrossRef]
11. Schouten, A. Mechanisms involved in nematode control by endophytic fungi. *Annu. Rev. Phytopathol.* **2016**, *54*, 121–142. [CrossRef]
12. Ghahremani, Z.; Escudero, N.; Saus, E.; Gabaldón, T.; Sorribas, F.J. *Pochonia chlamydosporia* induces plant-dependent systemic resistance to *Meloidogyne incognita*. *Front. Plant Sci.* **2019**, *10*, 945. [CrossRef]
13. Silva, S.D.; Carneiro, R.M.D.G.; Faria, M.; Souza, D.A.; Monnerat, R.G.; Lopes, R.B. Evaluation of *Pochonia chlamydosporia* and *Purpureocillium lilacinum* for suppression of *Meloidogyne enterolobii* on tomato and banana. *J. Nematol.* **2017**, *49*, 77–85. [CrossRef]
14. Schouteden, N.; Waele, D.D.; Panis, B.; Vos, C.M. Arbuscular mycorrhizal fungi for the biocontrol of plant-parasitic nematodes: A review of the mechanisms involved. *Front. Microbiol.* **2015**, *6*, 1280. [CrossRef] [PubMed]
15. Wolfgang, A.; Taffner, J.; Guimarães, R.A.; Coyne, D.; Berg, G. Novel strategies for soil-borne diseases: Exploiting the microbiome and volatile-based mechanisms toward controlling *Meloidogyne*-based disease complexes. *Front. Microbiol.* **2019**, *10*, 1296. [CrossRef] [PubMed]

16. Zhai, Y.L.; Shao, Z.Z.; Cai, M.M.; Zheng, L.Y.; Li, G.Y.; Yu, Z.N.; Zhang, J.B. Cyclo (l-Pro-l-Leu) of *Pseudomonas putida* MCCC 1A00316 isolated from Antarctic soil: Identification and characterization of activity against *Meloidogyne incognita*. *Molecules* **2019**, *24*, 768. [CrossRef] [PubMed]
17. Huang, D.; Yu, C.; Shao, Z.; Cai, M.; Li, G.; Zheng, L.; Yu, Z.; Zhang, J. Identification and characterization of nematicidal volatile organic compounds from deep-sea *Virgibacillus dokdonensis* MCCC 1A00493. *Molecules* **2020**, *25*, 744. [CrossRef]
18. Forghani, F.; Hajihassani, A. Recent advances in the development of environmentally benign treatments to control root-knot nematodes. *Front. Plant Sci.* **2020**, *11*, 1125. [CrossRef] [PubMed]
19. Li, X.; Hu, H.J.; Li, J.Y.; Wang, C.; Chen, S.L.; Yan, S.Z. Effects of the endophytic bacteria *Bacillus cereus* BCM2 on tomato root exudates and *Meloidogyne incognita* infection. *Pl. Dis.* **2019**, *103*, 1551–1558. [CrossRef]
20. Lee, Y.S.; Kim, K.Y. Antagonistic potential of *Bacillus pumilus* L1 against root-knot nematode, *Meloidogyne arenaria*. *J. Phytopathol.* **2016**, *164*, 29–39. [CrossRef]
21. Westphal, A. Detection and description of soils with specific nematode suppressiveness. *J. Nematol.* **2005**, *37*, 121–130. [PubMed]
22. Eberlein, C.; Heuer, H.; Westphal, A. Biological suppression of populations of *Heterodera schachtii* adapted to different host genotypes of sugar beet. *Front. Plant Sci.* **2020**, *11*, 812. [CrossRef]
23. Botelho, A.O.; Campos, V.P.; da Silva, J.C.P.; Freire, E.S.; de Pinho, R.S.C.; Barros, A.F.; Oliveira, D.F. Physicochemical and biological properties of the coffee (*Coffea arabica*) rhizosphere suppress the root-knot nematode *Meloidogyne exigua*. *Biocontrol Sci. Technol.* **2019**, *29*, 1181–1196. [CrossRef]
24. Stirling, G.R.; Rames, E.; Stirling, A.M.; Hamill, S. Factors associated with the suppressiveness of sugarcane soils to plant-parasitic nematodes. *J. Nematol.* **2011**, *43*, 135–148.
25. Cámara-Almirón, J.; Molina-Santiago, C.; Pérez-García, A.; de Vicente, A.; Cazorla, F.M.; Romero, D. Understanding bacterial physiology for improving full fitness. In *How Research Can Stimulate the Development of Commercial Biological Control Against Plant Diseases, Progress in Biological Control*; De Cal, A., Magan, N., Melgarejo, P., Eds.; Springer Nature: Cham, Switzerland, 2020; pp. 47–60. [CrossRef]
26. Abd-Elgawad, M.M.M. Comments on the use of biocontrol agents against plant-parasitic nematodes. *Int. J. PharmTech. Res.* **2016**, *9*, 352–359.
27. Dutta, T.K.; Khan, M.R.; Phani, V. Plant-parasitic nematode management via biofumigation using brassica and non-brassica plants: Current status and future prospects. *Curr. Plant Biol.* **2019**, *17*, 17–32. [CrossRef]
28. Poveda, J.; Abril-Urias, P.; Escobar, C. Biological control of plant-parasitic nematodes by filamentous fungi inducers of resistance: *Trichoderma*, mycorrhizal and endophytic fungi. *Front. Microbiol.* **2020**, *11*, 992. [CrossRef]
29. Duncan, L.W.; Stuart, R.J.; El-Borai, F.E.; Campos-Herrera, R.; Pathak, E.; Giurcanu, M.; Graham, J.H. Modifying orchard planting sites conserves entomopathogenic nematodes, reduces weevil herbivory and increases citrus tree growth, survival and fruit yield. *Biol. Cont.* **2013**, *64*, 26–36. [CrossRef]
30. Nielsen, A.L.; Spence, K.O.; Nakatani, J.; Lewis, E.E. Effect of soil salinity on entomopathogenic nematode survival and behaviour. *Nematology* **2011**, *3*, 859–867. [CrossRef]
31. Hussaini, S.S. Entomopathogenic nematodes: Ecology, diversity and geographical distribution. In *Biocontrol Agents: Entomopathogenic and Slug Parasitic Nematodes*; Abd-Elgawad, M.M.M., Askary, T.H., Coupland, J., Eds.; CAB International: Wallingford, UK, 2017; pp. 88–142.
32. Campos-Herrera, R.; Pathak, E.; El-Borai, F.E.; Schumann, A.; Abd-Elgawad, M.M.M.; Duncan, L.W. New citriculture system suppresses native and augmented entomopathogenic nematodes. *Biol. Cont.* **2013**, *66*, 183–194. [CrossRef]
33. Campos-Herrera, R.; Stuart, R.J.; Pathak, E.; EL-Borai, F.E.; Duncan, L.W. Temporal patterns of entomopathogenic nematodes in Florida citrus orchards: Evidence of natural regulation by microorganisms and nematode competitors. *Soil Biol. Biochem.* **2019**, *128*, 193–204. [CrossRef]
34. Giblin-Davis, R.M.; Williams, D.S.; Bekal, S.; Dickson, D.W.; Becker, J.O.; Preston, J.F. '*Candidatus* Pasteuria usgae' sp. nov., an obligate endoparasite of the phytoparasitic nematode, *Belonolaimus longicaudatus*. *Int. J. Syst. Evol. Microbiol.* **2003**, *53*, 197–200. [CrossRef]
35. Timper, P. Conserving and enhancing biological control of nematodes. *J. Nematol.* **2014**, *46*, 75–89. [PubMed]
36. Sikora, R.A.; Roberts, P.A. Management Practices: An overview of integrated nematode management technologies. In *Plant Parasitic Nematodes in Subtropical and Tropical Agriculture*; Sikora, R.A., Coyne, D., Hallmann, J., Timper, P., Eds.; CABI: Wallingford, UK, 2018; pp. 795–838.
37. Abd-Elgawad, M.M.M. Managing nematodes in Egyptian citrus orchards. *Bull. NRC* **2020**, *44*, 41. [CrossRef]
38. Ma, Y.; Li, Y.; Lai, H.; Guo, Q.; Xue, Q. Effects of two strains of *Streptomyces* on root-zone microbes and nematodes for biocontrol of root-knot nematode disease in tomato. *Appl. Soil. Ecol.* **2017**, *112*, 34–41. [CrossRef]
39. Gorny, A.M.; Hay, F.S.; Esker, P.; Pethybridge, S.J. Spatial and spatiotemporal analysis of *Meloidogyne hapla* and *Pratylenchus penetrans* populations in commercial potato fields in New York, USA. *Nematology* **2020**, *22*, 1–13. [CrossRef]
40. Escudero, N.; Lopez-Moya, F.; Ghahremani, Z.; Zavala-Gonzalez, E.A.; Alaguero-Cordovilla, A.; Ros-Ibañez, C.; Lacasa, A.; Sorribas, F.J.; Lopez-Llorca, L.V. Chitosan increases tomato root colonization by *Pochonia chlamydosporia* and their combination reduces root-knot nematode damage. *Front. Plant Sci.* **2017**, *8*, 1415. [CrossRef] [PubMed]
41. El-Nagdi, W.M.A.; Youssefi, M.M.A.; Abd-El-Khair, H.; Abd-Elgawad, M.M.M.; Dawood, M.G. Effectiveness of *Bacillus subtilis*, *B. pumilus*, *Pseudomonas fluorescens* on *Meloidogyne incognita* infecting cowpea. *Pak. J. Nematol.* **2019**, *37*, 35–43. [CrossRef]

42. Dahlin, P.; Eder, R.; Consoli, E.; Krauss, J.; Kiewnick, S. Integrated control of *Meloidogyne incognita* in tomatoes using fluopyram and *Purpureocillium lilacinum* strain 251. *Crop Prot.* **2019**, *124*, 104874. [CrossRef]
43. d'Errico, G.; Marra, R.; Crescenzi, A.; Davino, S.W.; Fanigliulo, A.; Woo, S.L.; Lorito, M. Integrated management strategies of *Meloidogyne incognita* and *Pseudopyrenochaeta lycopersici* on tomato using a *Bacillus firmus*-based product and two synthetic nematicides in two consecutive crop cycles in greenhouse. *Crop Prot.* **2019**, *122*, 159–164. [CrossRef]
44. Abd-Elgawad, M.M.M. Status of entomopathogenic nematodes in integrated pest management strategies in Egypt. In *Biocontrol Agents: Entomopathogenic and Slug Parasitic Nematodes*; Abd-Elgawad, M.M.M., Askary, T.H., Coupland, J., Eds.; CABI: Wallingford, UK, 2017; pp. 473–501.
45. Hussain, M.; Zouhar, M.; Rysanek, P. Suppression of *Meloidogyne incognita* by the entomopathogenic fungus *Lecanicillium muscarium*. *Plant Dis.* **2018**, *102*, 977–982. [CrossRef]
46. Sharma, I.P.; Sharma, A.K. Physiological and biochemical changes in tomato cultivar PT-3 with dual inoculation of mycorrhiza and PGPR against root-knot nematode. *Symbiosis* **2017**, *71*, 175–183. [CrossRef]
47. Anonymous. Adopted recommendations to combat agricultural pests. In *APC BooK 2018*; Commercial Al-Ahram Press: Qalioub, Egypt, 2018; (In Arabic) Agricultural Pesticide Committee, Ministry of Agriculture, Media Support Center Press. Available online: http://www.apc.gov.eg/Files/Releases/Recomm18/mobile/index.html#p=16 (accessed on 17 August 2021).
48. Grubišić, D.; Uroić, G.; Ivošević, A.; Grdiša, M. Nematode control by the use of antagonistic plants. *Agric. Consp. Sci.* **2018**, *83*, 269–275.
49. Hamaguchi, T.; Sato, K.; Vicente, C.S.L.; Hasegawa, K. Nematicidal actions of the marigold exudate α-terthienyl: Oxidative stress-inducing compound penetrates nematode hypodermis. *Biol. Open.* **2019**, *8*, bio038646. [CrossRef]
50. Ntalli, N.; Caboni, P. A review of isothiocyanates biofumigation activity on plant parasitic nematodes. *Phytochem. Rev.* **2017**, *16*, 827–834. [CrossRef]
51. Sikora, R.A.; Coyne, D.; Hallmann, J.; Timper, P. *Plant Parasitic Nematodes in Subtropical and Tropical Agriculture*, 3rd ed.; CABI: Cambridge, UK, 2018.
52. Chiamolera, F.M.; Dias-Arieira, C.R.; de Souto, E.R.; Biela, F.; da Cunha, T.P.L.; Santana, S.D.M.; Puerari, H.H. Susceptibility of winter crops to *Pratylenchus brachyurus* and effect on the nematode population in the maize crop. *Nematropica* **2012**, *42*, 267–275.
53. Ntalli, N.; Ratajczak, M.; Oplos, C.; Menkissoglu-Spiroudi, U.; Adamski, Z. Acetic acid, 2-undecanone, and (E)-2-decenal ultrastructural malformations on *Meloidogyne incognita*. *J. Nematol.* **2016**, *48*, 248–260. [CrossRef]
54. Seo, H.J.; Park, A.R.; Kim, S.; Yeon, J.; Yu, N.H.; Ha, S.; Chang, J.Y.; Palrk, H.W.; Kim, J.-C. Biological control of root-knot nematodes by organic acid-producing *Lactobacillus brevis* wikim0069 isolated from kimchi. *Plant Pathol. J.* **2019**, *35*, 662–673. [CrossRef] [PubMed]
55. Zasada, I.A.; Meyer, S.L.F.; Halbrendt, J.M.; Rice, C. Activity of hydroxamic acids from *Secale cereale* against the plant-parasitic nematodes *Meloidogyne incognita* and *Xiphinema americanum*. *Phytopathology* **2005**, *95*, 1116–1121. [CrossRef] [PubMed]
56. Ferraz, S.; Grassi de Freitas, L. Use of antagonistic plants and natural products. In *Nematology-Advances and Perspectives; Nematode Management and Utilisation*; Chen, Z.X., Chen, S.Y., Dickson, D.W., Eds.; CABI Publishing: Wallingford, UK, 2004; pp. 931–977. Volume II.
57. Abd-Elgawad, M.M.M.; Omer, E.A. Effect of essential oils of some medicinal plants on phytonematodes. *Anz. Schädlingskunde Pflanzenschutz. Umweltschutz/J. Pest Sci.* **1995**, *68*, 82–84. [CrossRef]
58. Barros, A.F.; Campos, V.P.; De Oliveira, D.F.; De Jesus Silva, F.; Jardim, I.N.; Costa, V.A.; Matrangolo, C.A.R.; Ribeiro, R.C.F.; Silval, G.H. Activities of essential oils from three Brazilian plants and benzaldehyde analogues against *Meloidogyne incognita*. *Nematology* **2019**, *21*, 1081–1089. [CrossRef]
59. Kalaiselvi, D.; Mohankumar, A.; Shanmugam, G.; Thiruppathi, G.; Nivitha, S.; Sundararaj, P. Altitude-related changes in the phytochemical profile of essential oils extracted from *Artemisia nilagirica* and their nematicidal activity against *Meloidogyne incognita*. *Ind. Crops Prod.* **2019**, *139*, 111472. [CrossRef]
60. Zaidat, S.A.E.; Mouhouche, F.; Babaali, D.; Abdessemed, N.; De Cara, M.; Hammache, M. Nematicidal activity of aqueous and organic extracts of local plants against *Meloidogyne incognita* (Kofoid and White) Chitwood in Algeria. *Egypt. J. Biol. Pest Cont.* **2020**, *30*, 46. [CrossRef]
61. Montiel-Rozas, M.; Díez-Rojo, M.; Ros, M.; Pascual, J.A. effect of plant extracts and metam sodium on the soilborne fungal pathogens, *Meloidogyne* spp., and soil microbial community. *Agronomy* **2020**, *10*, 513. [CrossRef]
62. Jindapunnapat, K.; Reetz, N.D.; MacDonald, M.H.; Bhagavathy, G.; Chinnasri, B.; Soonthornchareonnon, N.; Sasnarukkit, A.; Chauhan, K.R.; Chitwood, D.J.; Meyer, S.L. Activity of vetiver extracts and essential oil against *Meloidogyne incognita*. *J. Nematol.* **2018**, *50*, 147–162. [CrossRef] [PubMed]
63. Wen, Y.; Meyer, S.L.F.; MacDonald, M.H.; Zheng, L.; Jing, C.; Chitwood, D.J. Nematotoxicity of *Paeonia* spp. extracts and *Camellia oleifera* tea seed cake and extracts to *Heterodera glycines* and *Meloidogyne incognita*. *Plant Dis.* **2019**, *103*, 2191–2198. [CrossRef] [PubMed]
64. Abd-Elgawad, M.M.M.; Mohamed, M.M.M. Efficacy of selected biocontrol agents on *Meloidogyne incognita* on eggplant. *Nematol. Medit.* **2006**, *34*, 105–109.
65. Ji, X.; Li, J.; Meng, Z.; Dong, S.; Zhang, S.; Qiao, K. Inhibitory effect of allicin against *Meloidogyne incognita* and *Botrytis cinerea* in tomato. *Sci. Hortic.* **2019**, *253*, 203–208. [CrossRef]
66. Wilson, M.J.; Jackson, T.A. Progress in the commercialization of bionematicides. *BioControl* **2013**, *58*, 715–722. [CrossRef]

67. Ramirez, R.A.; Henderson, D.R.; Riga, E.; Lacey, L.A.; Synder, W.E. Harmful effects of mustard green manure on entomopathogenic nematodes, *Biol. Cont.* **2009**, *48*, 147–154.
68. Moreira, F.J.C.; Barbosa da Silva, M.C.; Araujo Rodrigues, A.; Neves Tavares, M.K. Alternative control of root-knot nematodes (*Meloidogyne javanica* and *M. enterolobii*) using antagonists. *Int. J. Agron. Agric. Res.* **2015**, *7*, 121–129.
69. Ndeve, N.D.; Matthews, W.C.; Santos, J.R.P.; Huynh, B.L.; Roberts, P.A. Broad-based root-knot nematode resistance identified in cowpea gene-pool two. *J. Nematol.* **2018**, 50. [CrossRef] [PubMed]
70. Zwart, R.S.; Thudi, M.; Channale, S.; Manchikatla, P.K.; Varshney, R.K.; Thompson, J.P. Resistance to plant-parasitic nematodes in chickpea: Current status and future perspectives. *Front. Plant Sci.* **2019**, *10*, 966. [CrossRef] [PubMed]
71. Abd-Elgawad, M.M.M. A new rating scale for screening plant genotypes against root-knot and reniform nematodes. Anz. Schadling. *Pflanzenschutz. Umweltschutz/J. Pest Sci.* **1991**, *64*, 37–39. [CrossRef]
72. Molinari, S. Resistance and virulence in plant-nematode interactions. In *Nematodes*; Boeri, F., Chung, J.A., Eds.; Nova Science Publisher Inc.: New York, NY, USA, 2012; pp. 59–82.
73. Noling, J.W. Nematode Management inTomatoes, Peppers, and Eggplant. 2019. Available online: https://edis.ifas.ufl.edu/publication/NG032 (accessed on 17 August 2021).
74. Starr, J.L.; Cook, R.; Bridge, J. *Plant Resistance to Parasitic Nematodes*; CABI: Wallingford, UK, 2002.
75. Starr, J.L.; Roberts, P.A. Resistance to plant-parasitic nematodes. In *Nematology, Advances and Perspective*; Chen, Z.X., Chen, S.Y., Dickson, D.W., Eds.; CABI: Wallingford, UK, 2004; Volume 2, pp. 879–907.
76. Abd-Elgawad, M.M.M. Towards sound use of statistics in nematology. *Bull. NRC.* **2021**, *45*, 13. [CrossRef]
77. Abd-Elgawad, M.M.M.; Molinari, S. Markers of plant resistance to nematodes: Classical and molecular strategies. *Nematol. Medit.* **2008**, *36*, 3–11.
78. Lopez-Perez, J.A.; Roubtsova, T.; Ploeg, A. Effect of three plant residues and chicken manure used as biofumigants at three temperatures on *Meloidogyne incognita* infestation of tomato in greenhouse experiments. *J. Nematol.* **2005**, *37*, 489–494.
79. Ntalli, N.; Adamski, Z.; Doula, M.; Monokrousos, N. Nematicidal amendments and soil remediation. *Plants* **2020**, *9*, 429. [CrossRef]
80. Vu, T.; Hauschild, R.; Sikora, R.A. *Fusarium oxysporum* endophytes induced systemic resistance against *Radopholus similis* on banana. *Nematology* **2006**, *8*, 847–852. [CrossRef]
81. Hahn, M.H.; May De Mio, L.L.; Kuhn, O.J.; Da Silva Silveira Duarte, H. Nematophagous mushrooms can be an alternative to control *Meloidogyne javanica*. *Biol. Cont.* **2019**, *138*, 104024. [CrossRef]
82. Youssef, M.M.A.; El-Nagdi, W.M.A. Population density of *Meloidogyne incognita* and eggplant growth vigour affected by sucrose-activated bread yeast (*Saccharomyces cerevisiae*). *Pak. J. Nematol.* **2018**, *36*, 117–122.
83. Ghareeb, R.Y.; Basyony, A.B.A.; Al-nazwani, M.S.; El-Saedy, M.A.M. Biocontrol of *Meloidogyne incognita* attacking potato plants using the extract of a new isolate of *Chlorella vulgaris* and molecular characterization. *Pl. Cell Biotech. Molecul. Biol.* **2021**, *13–14*, 47–62.
84. Blyuss, K.B.; Fatehi, F.; Tsygankova, V.A.; Biliavska, L.O.; Iutynska, G.O.; Yemets, A.I.; Blume, Y.B. RNAi-based biocontrol of wheat nematodes using natural poly-component biostimulants. *Front. Plant Sci.* **2019**, *10*, 483. [CrossRef] [PubMed]
85. Rouse-Miller, J.; Bartholomew, E.S.; St. Martin, C.C.G.; Vilpigue, P. Bioprospecting compost for long-term control of plant parasitic nematodes. In *Management of Phytonematodes: Recent Advances and Future Challenges*; Ansari, R., Rizvi, R., Mahmood, I., Eds.; Springer Nature: Singapore, 2020; pp. 35–50.
86. Hallmann, J.; Meressa, B.H. Nematode parasites of vegetables. In *Plant Parasitic Nematodes in Subtropical and Tropical Agriculture*; Sikora, R.A., Coyne, D., Hallmann, J., Timper, P., Eds.; CABI: Wallingford, UK, 2018; pp. 346–410.
87. Abd-Elgawad, M.M.M.; Elshahawy, I.E.; Abd-El-Kareem, F. Efficacy of soil solarization on black root rot disease and speculation on its leverage on nematodes and weeds of strawberry in Egypt. *Bull. NRC.* **2019**, *43*, 175. [CrossRef]
88. Kokalis-Burelle, N.; Rosskopf, E.N.; Butler, D.M.; Fennimore, S.A.; Holzinger, J. Evaluation of steam and soil solarization for *Meloidogyne arenaria* control in Florida floriculture crops. *J. Nematol.* **2016**, *48*, 183–192. [CrossRef]
89. Veronico, P.; Paciolla, C.; Sasanelli, N.; De Leonardis, S.; Melillo, M.T. Ozonated water reduces susceptibility in tomato plants to *Meloidogyne incognita* by the modulation of the antioxidant system. *Mol. Plant Pathol.* **2017**, *18*, 529–539. [CrossRef]
90. Magan, N. Importance of ecological windows for efficacy of biocontrol agents. In *How Research Can Stimulate the Development of Commercial Biological Control Against Plant Diseases, Progress in Biological Control*; De Cal, A., Magan, N., Melgarejo, P., Eds.; Springer Nature: Cham, Switzerland, 2020; pp. 1–14. [CrossRef]
91. Abd-Elgawad, M.M.M.; Askary, T.H. Factors affecting success of biological agents used in controlling plant-parasitic nematodes. *Egypt. J. Biol. Pest Cont.* **2020**, *30*, 17. [CrossRef]
92. Barthès, B.G.; Brunet, D.; Rabary, B.; Ba, O.; Villenave, C. Near infrared reflectance spectroscopy (NIRS) could be used for characterization of soil nematode community. *Soil Biol. Biochem.* **2011**, *43*, 1649–1659. [CrossRef]
93. Fan, Z.; Qin, Y.; Liu, S.; Xing, R.; Yu, H.; Chen, X.; Li, K.; Li, R.; Wang, X.; Li, P. The bioactivity of new chitin oligosaccharide dithiocarbamate derivatives evaluated against nematode disease (*Meloidogyne incognita*). *Carbohydr. Polym.* **2019**, *224*, 115155. [CrossRef]
94. Abd-El-Kareem, F.; Elshahawy, I.E.; Abd-Elgawad, M.M.M. Management of strawberry leaf blight disease caused by *Phomopsis obscurans* using silicate salts under field conditions. *Bull. NRC* **2019**, *43*, 1. [CrossRef]

95. Roldi, M.; Dias-Arieira, C.R.; Da Silva, S.A.; Dorigo, O.F.; Machado, A.C.Z. Control of *Meloidogyne paranaensis* in coffee plants mediated by silicon. *Nematology* **2017**, *19*, 245–250. [CrossRef]
96. Brito, O.D.C.; Ferreira, J.C.A.; Hernandes, I.; Silva, E.J.; Dias-Arieira, C.R. Management of *Meloidogyne javanica* on tomato using agro-industrial wastes. *Nematology* **2020**, 1–14. [CrossRef]
97. Cheng, F.; Wang, J.; Song, Z.; Cheng, J.; Zhang, D.; Liu, Y. Nematicidal effects of 5-Aminolevulinic acid on plant-parasitic nematodes. *J. Nematol.* **2017**, *49*, 295–303. [CrossRef]
98. DiLegge, M.J.; Manter, D.K.; Vivanco, J.M. A novel approach to determine generalist nematophagous microbes reveals *Mortierella globalpina* as a new biocontrol agent against *Meloidogyne* spp. nematodes. *Sci. Rep.* **2019**, *9*, 7521. [CrossRef]
99. Safeena, M.I.S.; Zakeel, M.C.M. Nanobiotechnology-driven management of phytonematodes. In *Management of phytonematodes: Recent Advances and Future Challenges*; Ansari, R.A., Rizvi, R., Mahmood, I., Eds.; Springer Nature: Singapore, 2020; pp. 1–33.
100. Abd-Elgawad, M.M.M. Optimizing sampling and extraction methods for plant-parasitic and entomopathogenic nematodes. *Plants* **2021**, *10*, 629. [CrossRef] [PubMed]
101. Abd-Elgawad, M.M.M. Optimizing biological control agents for controlling nematodes of tomato in Egypt. *Egypt. J. Biol. Pest Cont.* **2020**, *58*, 1–10. [CrossRef]
102. Abd-Elgawad, M.M.M. Biological control agents in the integrated nematode management of pepper in Egypt. *Egypt. J. Biol. Pest Cont.* **2020**, *30*, 70. [CrossRef]
103. Abd-Elgawad, M.M.M. Biological control agents in the integrated nematode management of potato in Egypt. *J. Biol. Pest Cont.* **2020**, *30*, 121. [CrossRef]
104. Abd-Elgawad, M.M.M. Biological control of nematodes infecting eggplant in Egypt. *Bull. NRC* **2021**, *45*, 6. [CrossRef]

MDPI

Review

Understanding Molecular Plant–Nematode Interactions to Develop Alternative Approaches for Nematode Control

Mahfouz M. M. Abd-Elgawad

National Research Centre, Plant Pathology Department, El-Behooth St., Dokki, Giza 12622, Egypt; mahfouzian2000@yahoo.com

Abstract: Developing control measures of plant-parasitic nematodes (PPNs) rank high as they cause big crop losses globally. The growing awareness of numerous unsafe chemical nematicides and the defects found in their alternatives are calling for rational molecular control of the nematodes. This control focuses on using genetically based plant resistance and exploiting molecular mechanisms underlying plant–nematode interactions. Rapid and significant advances in molecular techniques such as high-quality genome sequencing, interfering RNA (RNAi) and gene editing can offer a better grasp of these interactions. Efficient tools and resources emanating from such interactions are highlighted herein while issues in using them are summarized. Their revision clearly indicates the dire need to further upgrade knowledge about the mechanisms involved in host-specific susceptibility/resistance mediated by PPN effectors, resistance genes, or quantitative trait loci to boost their effective and sustainable use in economically important plant species. Therefore, it is suggested herein to employ the impacts of these techniques on a case-by-case basis. This will allow us to track and optimize PPN control according to the actual variables. It would enable us to precisely fix the factors governing the gene functions and expressions and combine them with other PPN control tactics into integrated management.

Keywords: plant–nematode interactions; nematode effectors and control; plant resistance

Citation: Abd-Elgawad, M.M.M. Understanding Molecular Plant–Nematode Interactions to Develop Alternative Approaches for Nematode Control. *Plants* **2022**, *11*, 2141. https://doi.org/10.3390/plants11162141

Academic Editors: Špela Mechora and Dragana Šunjka

Received: 11 July 2022
Accepted: 9 August 2022
Published: 17 August 2022

1. Introduction

With the ongoing nature of the socio-economic importance of agriculture, the global needs for sustainable and mounting food production to suffice the increased human population are evident. Thus, it is essential that issues associated with a full spectrum of crop production restrictions and losses are soundly solved. Plant–parasitic nematodes (PPNs) rank high among other crop pests and pathogens that constitute major constraints to agricultural production. Estimates of crop losses due to PPNs for the 20 life-sustaining crops averaged 12.6% of worldwide crop yield which equaled USD 215.77 billion of annual yield. An additional 20 crops with significant values for food and export have also a 14.45% annual yield loss which equaled USD 142.47 billion. The total 40 crops sustain an average of 13.5% losses which are estimated at USD 358.24 billion annually [1]. Clearly, these assessments will probably be elevated by adding other nematode-infected plant species worldwide to the list. Hence, adopting adequate and effective measures for optimizing PPN control tactics and strategies is a big challenge [2,3].

As PPNs are obligate parasites, they must feed on the roots or aerial parts of living plants to develop and reproduce either sexually or via parthenogenesis, i.e., reproduction without fertilization. While some PPN species have a restricted/narrow host range, others are polyphagous, but most PPNs are subterranean pests. Their second stage juvenile (J2) that hatches from the eggs is mostly the infective stage. It parasitizes plant roots of susceptible hosts, attracted to the roots via root exudates in certain host species. On the contrary, non-host/immune plants may have nematode-repellent materials in their roots. Hence, plant–nematode interactions have various aspects and may virtually occur even before the nematode touches the root of the plant (Figure 1). Eggs of *Globodera rostochiensis* need initial stimulation to hatch by chemical components in root exudates of their potato host [4]. Usually, many economically important and serious PPN species can attack and penetrate the plant roots and then migrate or develop feeding sites within them to secure their development and reproduction [5]. Therefore, a brief account of PPN categories is given in this review. Current PPN management measures to clarify their merits and demerits. Mounting concern about the demerits of PPN control methods has sparked broad interest in using resistant plant varieties/cultivars as safe, economic, and effective alternatives, especially to unsafe nematicides. This review addresses the mechanisms of natural resistance, especially against serious PPN species. It discusses current issues related to using both resistant plant genes and nematodes-effector proteins. These effectors have various functions, e.g., to detoxify enzymes in order to override the plant's antimicrobial compounds, master host immune signaling, and keep their sound feeding structure as well as the essential processes for PPN development [6,7].

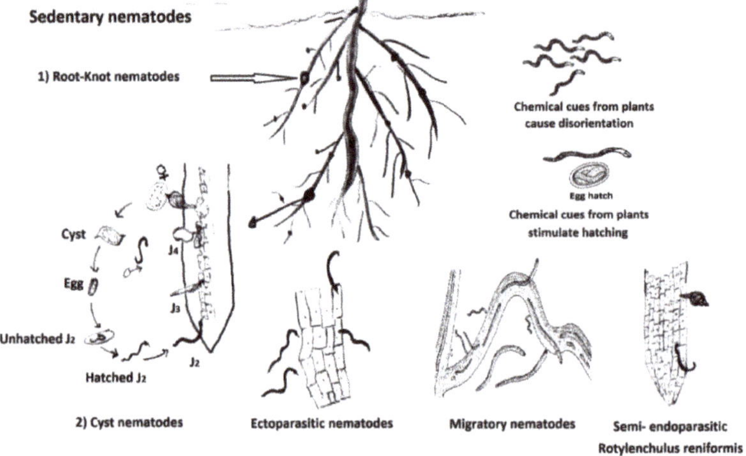

Figure 1. Categories of soil phytonematodes showing interactions with their host plants. Root metabolites can also function as PPN repellents, attractants, killers, or inhibitors.

As a consequence of these issues, there is a desperate need to exploit modern molecular technologies in developing such alternatives for PPN control strategies. Grasping molecular plant–nematode interactions enables us to adopt crucial factors and harness efficient tools and resources for use in these strategies. Although a few molecular mechanisms of nematode infection during both plant–nematode compatible and incompatible interactions have already been explored with favorable results, many gaps are still there that pose difficulties in employing the related strategies [8–13]. Therefore, examples are given herein to clarify future proposed directions of various approaches based on boosting resistance in plants and/or suppressing nematode effectors. They address the most recent developments regarding the molecular basis underlying plant–nematode interactions with various techniques utilized to enhance plant protection against serious PPNs [13–20]. Their expansion in an integrated approach is presented to attain effective and durable employment in nematode management.

2. General PPN Categories and Management Measures

Above-ground nematode parasites are less abundant and comprise stem and bulb nematodes, seed gall nematodes, and foliar nematodes. For the subterranean phytonematodes, the mode of their parasitism may group them into four wide categories (Figure 1): (i) Ectoparasites, such as species of the genus *Helicotylenchus* and other spiral nematodes.

They do not enter the roots but parasitize the root apex and/or peripheral cells via inserting long and robust feeding stylets. They usually move in soil searching for plant roots to parasitize. (ii) Migratory endoparasites, such as *Pratylenchus* spp. and *Ditylenchus dipsaci*. They can penetrate into and move within plant roots to feed and quit it to enter another one. After feeding, both ectoparasites and migratory endoparasites develop to J3, J4, and finally the adult without sexual dimorphism. (iii) Semi-endoparasites, such as *Tylenchulus semipenetrans* and *Rotylenchulus reniformis*, use only the nematode head to penetrate the root, but its posterior part remains in the soil. They are settled at one place on the root. The body of the female swells outside the roots. (iv) sedentary endoparasites, such as root-knot nematode (RKN) species (*Meloidogyne* spp.) and cyst nematode (CN) species (*Heterodera* and *Globodera* spp.). They enter and establish themselves within the roots. Both semi-endoparasites and sedentary endoparasites have sexual dimorphism. The swollen sedentary females sometimes protrude on the outside of the root. Migratory and sedentary endoparasites can damage plant tissues during their invasion, migration, and feeding on their susceptible hosts. Most of the studies on plant–nematode interactions have been centered on RKNs and CNs as their species are the most widespread and cause substantial crop losses in worldwide agricultural production. Launching and evolving of sedentary endoparasites-feeding sites for RKNs differ from that of CNs. The RKN J2 forms its feeding site on reaching the differentiating vascular tissue of plant roots via a few distinct giant/nurse cells, but cyst-forming J2 fixes it via setting syncytia close to the vascular bundle, where a few cells combine by resolving their cell walls. Having organized their feeding sites to transfer nutrients and solutes to the J2, the nematodes (RKN or CN) develop until reaching adult females via subsequent molts. These females lay eggs that hatch a new generation of J2s. Eventually, PPNs interact with their plant hosts in various courses ranging from transient ectoparasites to intimate involvement with their hosts, e.g., sedentary endoparasites.

Currently, PPNs are commonly managed via various production practices (chemical nematicides, bionematicides, resistant plants and crop rotation, soil amendments, fallowing, flooding, solarization, tillage, and use of certified transplants). Because most PPNs spend their lives within the soil or in plant roots, delivery of a chemical nematicide to the immediate surroundings of PPNs is generally difficult [21]. Yet, chemical nematicides are considered traditional means of effective PPN control (e.g., [22]). Unfortunately, the potential threat of these chemicals to wildlife, humans, and the environment, as well as the emergence of resistance-breaking nematode pathotypes/strains due to excessive use of these chemicals, has enforced the search for efficient and safe alternatives. Bionematicides are mostly safe alternatives, but they are frequently slower acting, less effective, and more inconsistent than these chemicals [23]. Using crop rotation is an effective and safe method for PPN control, were it not for the lack of PPN resistant/immune plant cultivars/varieties needed in the rotation. Soil amendments can enhance plant growth, but with the possible build-up in population densities of PPNs and BCAs, exceptions should be considered [24]. Related additions comprising botanical matrices and extracts, and purified secondary metabolites have received much research interest, but registration-processing and time-consuming issues have slowed their adoption [25]. Basic requirements for such materials are their safety, reliability, and favorable economics [26]. Fallowing and flooding may be used in PPN control but are not frequently economic for PPN control measures. Tillage is useful against many pests, weeds, and pathogens but can directly disrupt populations of PPN-antagonistic organisms and consequently increase nematode damage [24]. Certified transplants are excellent practices, but the plants should not grow into PPN-infested field soils, frequently an inevitable task. Eventually, the above-mentioned PPN chemical, cultural, and biological control techniques are not perfectly accepted and need deep revisions for safety and/or efficacy [19,27,28]. In addition, their demerits are frequently discouraging with regard to the generally low precision and accuracy in sampling the nematodes' subterranean and within plant life stages as well as a wide host range of PPNs and their diverse and clumped distribution [29].

Hada et al. [18] have recently emphasized that it is difficult to recommend a favorable PPN management tactic that is reliable, economical, safe, and harmless to the nontargets. Rather, farmers and stakeholders would turn to resistant varieties/cultivars and production practices for PPN control, but, for numerous crops, these methods and resources are mostly unavailable or unfavorable. Grasping molecular plant–nematode interactions may offer novel approaches and resources to fill these gaps and assist in nematode control. If so, the related multiplex mechanisms, especially for sedentary endoparasites, regarding their feeding sites within the plant roots as well as cellular and sub-cellular responses in the PPNs and their host plants should be fully understood and exploited.

3. The Mechanism of Natural Resistance

Contrary to compatible nematode–plant interactions in susceptible hosts, the single dominant resistance genes from plants interact specifically with corresponding avirulence (Avr) genes in the nematode, leading to an incompatible interaction. This incompatible interaction commences a cascade of plant responses against the nematode—defense strategies. Plants experience several modes of action for protection and immunity. A general innate/basal immune system can recognize nematode-associated molecular patterns (NAMPs) by pattern recognition receptors (PRRs) as the primary defense line 'layer' against plant parasites. The extracellular receptor proteins (receptor-like kinases and receptor-like proteins) may be initiation factors to elicit basal immunity, e.g., against RKNs [9]. A conserved ascaroside (Asc#18) is reported as a NAMP of *Heterodera glycines* whereas the *Arabidopsis* leucine-rich repeat (LRR) receptor-like kinase NILR1 is yet the only known cyst nematode PRR. The activation of NAMP-triggered immunity (NTI) leads to a series of immune responses such as the production of reactive oxygen species and secondary metabolites, cell death around the PPN-migratory tract, and/or reinforcement of cell walls [30]. Such plant responses may decelerate the early stages of PPN infection. They can share in effective defenses but only in non-host plants. In nematode-susceptible plants, PPNs can overcome NTI via secreting effector proteins known as effector-triggered suppression (ETS) to inhibit the basal immune responses. Cyst nematode effectors, such as Ha18764, GrVAP1, RHA1B, and GrCEP12, are synthesized in nematode-esophageal glands and secreted into the roots by the PPN stylet [13]. Suppressing the innate immune responses usually results in establishing feeding sites (e.g., giant cells for RKNs and syncytia for CNs) necessary for nematode development and multiplication on their susceptible hosts.

Contrary to the first line or innate immune defense system, another defense line 'layer' is found only in PPN-resistant plant genotypes. A widespread thought is that it is encoded by single dominant resistance genes (*R*-genes) or quantitative trait loci (QTL) to manifest a host-specific defense [7,13]. Yet, the thought should be boosted by the fact that the *R*-gene may include a small gene family with highly homologous copies clustered together. Relevant intracellular signaling pathways must also exist to enable the expression of the resistant response. Although a single gene in the cluster may determine resistance, multi-gene families are common for plant *R*-genes. For instance, *Mi-1* comprises a small gene family with seven highly homologous copies clustered together on the short arm of chromosome 6 on resistant tomato [31], but several other *Mi* genes have been found, different from *Mi-1* in genetic locations, functional characteristics, and specificity [32,33]. Although ten genes are recognized for resistance to *Meloidogyne* spp. In tomatoes, only seven genes (*Mi-2, Mi-3, Mi-4, Mi-5, Mi-6, Mi-9*, and *MI-HT*) can operate at high temperatures, e.g., above 32 °C [9]. Likewise, for the cyst nematode, genome sequencing combined with fine mapping could indicate that the *H1* locus harbors a cluster of intracellular nucleotide-binding (NB)-LRR proteins (NLR) candidate genes, revealing that the *H1* gene is also a classical single dominant *R*-gene [34]. The most common class of intracellular proteins related to these *R*-genes usually encodes NLR to activate the host-specific defense.

The *R*-genes operate via two modes of nematode interaction. The first mode has a direct pathway relying on a direct gene-for-gene interaction where the receptor protein of the resistant plant interacts with the nematode effectors mastered by avirulence (Avr) genes. Striking output of this incompatible reaction is a localized programmed cell death so that no nematode-feeding site is formed. For instance, Avr genes of RKN produce effectors that trigger the production and the expression of plant *Mi*-resistant genes in tomato plants resulting in a type of hypersensitive response (HR) after the nematode enters the plant root [35]. Likewise, single dominant *R*-gene *H1* from potato plant can award resistance against avirulent *Globodera rostochiensis* populations [36]. The second mode of the host-specific defense is named the guard hypothesis. Its mechanism starts as nematode effectors trigger the plant-virulence factors (protein) which stimulates *R*-gene [37]. The Avr genes of nematodes interact with tomato accessory protein, for example, leading to some modification of this plant protein, enabling the recognition by plant nucleotide-binding site (NBS)-LRR proteins that monitor for infection. Consequently, RKN development is indirectly prohibited via inhibiting the formation of feeding sites.

While nematode effectors are intra- and extracellularly recognized by immune receptors, these latter, encoded by *R*-genes, have the same structural type as PRRs. However, enforced by *R*-genes, these PRRs can activate higher (specific) defense responses, upon direct/indirect recognition of apoplastic effectors produced by designated nematode strain(s) than defense by basal immunity. Yet, there are various modifications of the resistance mechanisms. While HR-induced resistance can cause necrosis of the nematode-feeding sites within two days post-infection, another resistance mechanism is not based on HR but rather disintegration of these sites at almost two weeks post-infection. This latter, the delayed disintegration of the feeding site, is noted in a broad variety of incompatible plant–nematode interactions, e.g., *M. incognita*-pepper, *H. schachtii*-sugar beet, *H. glycines*-soybean, *H. avenae*-cereals, and *Globodera* spp.-potato [38].

Even the same crop, such as pepper, may carry two resistance genes, *Me-3* and *Me-1*, for a quick HR soon after nematode inoculation and for delayed degradation of giant cells, respectively. Fewer RKN juveniles develop and reproduce on *Me-1* than *Me-3*-resistant hosts. Interestingly, a large number of resistance genes to RKNs are recorded to be located on the P9 chromosome of pepper [34]. Therefore, Abd-Elgawad [39] noted the importance of resistant pepper varieties as they can suppress RKN populations to low levels in soil with high fruit yield under high initial RKN pressure. Yet, careful manipulation of RKN resistance in pepper should be based on the fact that the resistance response is the result of the specific *R*-gene-*Meloidogyne* species and the plant genotype together. In other words, there are diverse mechanisms of resistance and therefore the plant defenses rely on activating many known and unknown *R*-genes or QTLs, especially for the economically impactful RKNs and CNs [14,15,40].

4. Successes and Difficulties in Using *R*-Genes

Comprehensive references have addressed PPNs in temperate [41] and subtropical and tropical [5] agriculture materializing the successful use of naturally resistant plant species/cultivars. Although there are a good number of resistant genotypes, an urgent need is apparent for more ones to reduce PPN losses. Moreover, the majority of plant resistance genes used are effective against only the above-mentioned sedentary nematode category [30,33,42]. Hence, introgression of *R*-genes to confer nematode resistance to susceptible plants via classical genetic breeding can offer potent steps change in crop productivity [43–45]. Admittedly, plant genes responsible for PPN resistance are very useful in lowering PPN population levels, enhancing crop yields, and developing effective crop sequences.

In contrast to classical breeding for resistance, recognition and cloning of such genes found in a plant species can allow the transfer of resistance directly into other susceptible cultivar(s) with desirable traits of the same species, or even into cultivars of different species. Such genetic manipulations have the merits of avoiding linkage drag and scope to transfer resistance into genetic constitutions that prevent introgression by cross-breeding. Genes for nematode resistance could be cloned and transferred from some plant cultivars to others. The *Mi-1.2* from tomato against RKN (*Meloidogyne incognita*), $Hs1^{pro-1}$ from *Beta procumbens* against beet CN (*Heterodera schachtii*), *Gpa-2* from potato against potato CN (*Globodera pallida*) and *Hero A* from tomato against potato CNs (*G. pallida* and *G. rostochiensis*) and *Cre* loci from *Aegilops* spp. against cereal CN (*H. avenae*) in wheat are apparent examples [28,46]. The arsenal of nematode-resistant genes, especially for major PPNs, still has additional favorable ones, e.g., *Me* in pepper, *Rk* in cowpea, *Rhg1* in soybean, *Ma* in *Prunus* spp., and *Mex1* in coffee. Their benefits may be exemplified in the enhanced resistance to RKNs that was achieved via cloning and transferring the full genomic region of the *Mi-1* gene found in tomato into a distant plant species, lettuce, *Lactuca sativa* [47].

Conversely, the lack of novel resources to back certain resistant plant species in controlling a few species of key nematode pests is consistently increasing due to the slow decline that could be noticed in their *R*-gene effectiveness. A remarkable example is the current problem of using resistance derived from plant introduction accession 88788 in 95–98% of the soybean cyst nematode (*H. glycines*)-resistant soybean varieties cultivated in the USA. Although *H. glycines* is the most important pest of the soybean there, the related plant resistance encoded by a high copy number of the *rhg1-b* allele has already started to decrease. Therefore, Kahn et al. [48] added *Bacillus thuringiensis* delta-endotoxin (Cry14Ab) as a plant-incorporated protectant. Consequently, genetically engineered soybean plants expressing Cry14Ab showed a decrease in *H. glycines* cyst and egg counts relative to control plants, demonstrating excellent potential of Cry14Ab to control PPNs in soybean. Another type of issue is related to the gene construct itself, e.g., single or dual genes. Tomato plants genetically engineered using double structure (PjCHI-1 and CeCPI) genes with synthetic promoters could generate transgenic lines that displayed a better decrease in RKN infection and reproduction than transgenic tomatoes with a single gene [49].

Additional cases are related to elements and components mediating *R*-genes. It is well established that salicylic acid (SA) and jasmonic acid (JA) can play a critical role in the signaling/expression of both innate and *R*-gene-mediated defense responses against pests and pathogens [50]. Remarkably, SA is involved in PPN-plant resistance, especially against sedentary forms. Therefore, the suppression of plant defense by PPNs is usually accompanied by the downregulation of the genes involved in SA-mediated defense. However, the SA-dependent pathogenesis-related protein genes *PR-1* (*P6*) were elevated rapidly in plant roots of susceptible tomatoes to levels comparable to that in resistant tomatoes; plants infected by *Globodera rostochiensis* showed similar free SA levels in the incompatible and compatible interactions [51]. Notwithstanding the utility of SA to enhance plant resistance, free SA levels in roots of infected susceptible plants may be impacted differently according to the attacking PPN species/genus. Molinari [7] speculated that the early and abundant necrosis caused by *G. rostochiensis* may trigger the noticed early but transient rise of SA with stimulation of SA signaling in susceptible tomato. Clearly, this level of stimulation for SA signaling does not occur in *Meloidogyne*-plant compatible interaction as RKN move intercellularly, causing less tissue damage.

Ultimately, plants can still be immunized against nematode attacks via pre-treatments with auxins that mediate defense reactions, e.g., SA. The beneficial rhizosphere microorganisms, such as arbuscular mycorrhizal fungi and biocontrol agents, e.g., *Trichoderma* spp., can induce systemic acquired resistance-like responses against RKN [52–54]. This does not negate the fact that more investigations on recognition/signaling pathways interacting with components or genes required for *R* functions are direly needed.

5. Common Issues of Natural Plant Resistance

5.1. Resistance Breaking Nematode Pathotypes

The development of resistance-breaking pathotypes has been extensively studied and reported (e.g., [9,33,55,56]). Although the above-mentioned selection pressure is a common cause to generate these pathotypes or virulent populations, an intriguing study [57] partitioned virulent RKN populations into (a) populations extracted from a field with grown resistant tomatoes, (b) natural virulent populations isolated from fields without grown resistant tomatoes, and (c) virulent populations selected from laboratory-avirulent populations. They concluded that the genetic events resulting in the acquisition of virulence against the *Mi*-gene differ between selected and natural virulent populations. Moreover, selection pressure for virulence could accompany gaining additional function enabling these PPNs to circumvent the host response, e.g., by enhancing antioxidant enzyme activities [58]. These virulent populations are becoming of wide occurrence [9]. Although they are especially found in monoculture systems which may support the selection pressure events, the exact reasons for their occurrence are unclear. It may also be due to ecological factors, e.g., temperature and changes in PPN populations. Ultimately, such virulent PPN populations, which can develop on resistant crops, would turn nematode resistance in sustainable agriculture into elusive strategies.

5.2. Genetics of Virulence in Nematodes

Certain nematode reproduction usually undergoes obligate mitotic parthenogenesis (i.e., *M. javanica*, *M. incognita*, and *M. arenaria*) in the tropics. Others, such as *M. chitwoodi*, *M. hapla*, and *M. fallax*, generally reproduce by facultative meiotic parthenogenesis in temperate climates. Cyst nematodes are largely amphimictic. Their species with facultative reproduction usually have a narrower host range than the asexual species. However, sexual reproduction boosts adaptability and heterogeneity among and within PPN populations [33]. Accordingly, virulent populations may be more inducible in those species of sexual multiplication. These populations were detected from avirulent strains too in resistant tomato fields with a monocropping system [9]. On the other hand, caution should be exercised for these virulent nematode populations, as it is well known that natural nematode resistance may be encoded not only by single dominant genes but also in a polygenic manner [33]. In this vein, sound use of statistics in nematology could be a helping tool. Therefore, high-quality sequencing and assembly via joining long-read sequencing to utilize high-density genetic mapping can boost the detection and characterization of PPN-virulent genes. This novel scheme can support our grasp of the plant–PPN interaction.

5.3. The Temperature Factor

A remarkable example is the *Mi-1* gene of tomato used against RKNs. This gene cannot operate at temperatures above 28 °C for more than a few, maximum 48, hours after infection [35]. The RKN juveniles can establish their feeding site, relying on the temperature-dependent setback of resistance. Thereafter, resistance is not set any longer even at the permissive temperature. Therefore, HR-mediated resistance does not work to disrupt nematode growth and multiplication of the individuals that could form their feeding sites. Several factors were also reported to overcome *Mi-1*-mediated resistance. Populations of *M. javanica* and *M. incognita* that can infect and reproduce on tomato plants carrying *Mi-1* were documented [59]. Moreover, high population levels of *M. incognita* can seriously affect the resistance of the *Mi-1* gene [60]. On the contrary, *Mi 9* mediated resistance is operating at high temperatures and is localized to the short arm of chromosome 6 of tomato [32]. Temperature is a pivotal factor as it impacts tomato resistance and the metabolic and PPN multiplication rates.

5.4. Improper Research Methods and Tools

There are some molecular methods that should be dealt with carefully because they are based on materials that may be suitable for controlling a specific nematode genus but not others. Therefore, more studies with adequate tools and updated methods may be preferably directed towards nature, and structure of PPN-feeding tubes, the nematode-derived compounds, and consequent plant responses involved in such plant–nematode interactions for determining the molecular efficacy against the target nematode genus/species [9,10,12,16,27,61]. In this respect, as PPNs have a stylet orifice while feeding; it acts as a molecular sieve to uptake certain molecules and exclude others while feeding on tomato roots that express a nematicidal *Bacillus thuringiensis* crystal protein. The ultrastructure of these feeding tubes revealed that RKNs, but not CNs, can ingest larger transgenic proteins [62,63]. Thus, transgenic 54 kDa Cry6A and Cry5B proteins were ingested by and negatively affected *M. incognita* reproduction in tomato hairy roots [63,64]. On the contrary, resistance to cyst nematodes in roots expressing Cry5B protein from *Bacillus thuringiensis* is not conferred, i.e., the large 54 kDa Cry6A protein could not be ingested by *H. schachtii* due to the narrow orifice of the feeding tube; its size is limited to about 23 kDa [65]. This restriction severely limits the use of transgenic Cry proteins against some serious CNs.

Until not so long ago, there were many defects and flaws—now somewhat reduced—in the molecular tools and devices used. Remarkably, the quantitative polymerase chain reaction (qPCR) is superior to the frequently used PCR as the former enables not only the qualitative detection of target PPNs but also their quantification. It could be a faster and better alternative to the longstanding use of microscopy in PPN identification and counting during the study of nematode–host interactions, especially in developing countries. Although various qPCR diagnostic assays have been developed based on the internal transcribed spacer (ITS) of rDNA in many PPN species, related defects may arise. For instance, the high variability of ITS sequences in *Pratylenchus* spp. could enhance the risk of getting false-positive reactions (fragments from unidentified species) or false-negative reactions (variation existing between individuals). Moreover, imprecise quantification might also occur as some gene sequences are found in multiple copies in individual cells [66]. Furthermore, gene copy numbers can vary not only from one species to another but also amongst different PPN developmental stages [29]. Such confusing data may contribute to obtaining imprecise or unsound molecular nematode–host relationships. The main limit of qPCR is due to its failure to detect species that do not match the used primer/probes. Alternately, metagenomic methods can offer a reliable device, whether a PPN is found in databases, e.g., Genbank [67]. Based on the merits/demerits of each method, researchers should decide the approach that fulfills the intended goal(s).

Iqbal et al. [68] reviewed RNA interference (RNAi) of PPN genes as a now-common method. It involves engineering host plants to generate tall hairpin RNAs matching essential PPN genes. These genes are then processed into short interfering RNAs (siRNA) that trigger silencing as nematodes feed on cytoplasmic contents of the target plants [69]. They emphasized that the delivery of double-stranded RNA (dsRNA) to PPNs via host-induced gene silencing is more practical than spraying or any other method for a nematode-control strategy. Furthermore, they found that many of the tested genes reacted to RNAi knockdown differently [68]. Thus, they suggested that the original goal, types, R phenotypes of PPN strains, and current integration merits of RNAi should further be addressed; presumably, something more complex is occurring.

Common methods for transcriptome analysis of sedentary nematodes may rely on either isolating the nematodes from the plant tissue prior to RNA-sequencing (RNA-seq) or using dual RNA-seq where the plant roots and their invading nematodes are sequenced at the same time. The latter technique could have the merit of enabling PPN effector gene discovery and comparing the transcriptomic datasets between pre-parasitic and parasitic *Meloidogyne chitwoodi* juveniles on potato [62]. Thus, the dual RNA-seq could produce a substantial analysis of *M. chitwoodi* genes expressed during parasitism and encoded foreseen secreted proteins. This technique also considerably reduced the large list of genes in the *M. chitwoodi* secretome reported by the former method [70], isolating the nematodes from the roots led to recording genes not related to parasitism. While it is really difficult to functionally characterize ≥ 300 genes via a traditional method [70], dual RNA-seq could analyze the expression of fewer genes specifically at the early parasitic life stages of *M. chitwoodi* too [61].

6. General Approaches to Solve the Related Issues

Basically, genetic improvement of plants for nematode resistance to enhance their productivity via traditional breeding or genetic engineering is likely only if the desired alleles are present in the gene pools of the targeted plants. A notable example of RKN resistance in tomato is that all its current resistant varieties originated from just the *Mi* gene. Resistance resulted from hybridizing the wild tomato plant (as a single resistance gene source) with the commercial one [9]. Breeders and stakeholders have worked on enhancing the effectiveness of resistant strains. Some of the related genes could work at high temperatures, e.g., *Mi-HT*, *Mi-2*, *Mi-4*, *Mi-3*, *Mi-5*, *Mi-6*, and *Mi-9* are heat stable. Yet, further surveys of other diverse habitats may find new and indigenous PPN-resistance genes—*R*-genes that do not rely on *Mi*-genes.

Optimizing strategies for the efficient employment of durable resistant crops also requires a good knowledge of population genetics. As heat-stable resistance gene *Mi-9* is found in *Solannum arcanum*, resistance genes pyramiding in commercial varieties and genetic adjustments might enhance resistance durability. This could be done via manipulating plant metabolites that may comprise phenols, amino acids, and lipophilic molecules [71]. Furthermore, there is still much to grasp regarding resistance gene expression and function for various plant species and under different environments. Because there is great specificity of the virulent nematodes to the *R*-gene on which they were selected, the gene transfer or priming plants for immunization to counteract this virulence should be done using adequate molecular methods [7,33]. Moreover, durability could possibly be maintained via transferring multiple resistance genes to specific cultivar(s) within integrated nematode management systems. In such systems, using crop rotation and/or safe chemical nematicides can assist in reducing pressure on resistant cultivars/varieties to alleviate the emergence of virulent populations. BCAs can also offer a significant contribution to at least some of these systems. *Trichoderma asperellum* T34 reduced the number of eggs per plant of the virulent *M. incognita* population in both resistant and susceptible tomato cultivars. Fortunately, this fungal impact was additive with the *Mi-1.2* resistance gene of tomato [72]. Cloning and overexpressing the genes responsible for the biocontrol process from *Paecilomyces javanicus* may reinforce the plant immune response against RKN infection [16]. Likewise, engineered nanomaterials could show promising physical and chemical characteristics against nematodes [73].

Admittedly, examining the related biochemical, histological, and physiological aspects of plant–nematode interactions using sophisticated tools and devices may lead to novel and effective PPN management tools. A clear aim is to grasp molecular regulatory processes underlying PPN parasitism that could result in developing reliable PPN control strategies based on nematode genetic and plant-resistant backgrounds. In this respect, both the comprehensive secretome (different molecular proteins secreted via the nematode stylet that is repeatedly thrust into the cells of the plant roots) profiles and the whole-genome sequence of economically important PPN species have attained significant progress for important PPN species. For example, high-quality genome sequences of serious PPN species such as major RKN species [14,74–76] as well as less distributed ones, e.g., *Meloidogyne luci* [77], *M. enterolobii* [78], *M. exigua* [79], *M. chitwoodi* [80], and *M. graminicola* [81] are now available. Their availability should be harnessed not only to facilitate better comparative studies and phylogenomics on the related species but also help to recognize genomic variabilities and their main role in adaptability against different environmental factors and plant hosts, via examining the functional genomics. In this respect, a whole-genome shotgun study could reveal the long-read-based high-quality assembly of *M. arenaria* that may open new avenues to identify virulence-related genes [75]. These genes are frequently found in repeat-rich or highly variable regions in the genome. At hand, genome and transcriptome datasets are helpful in characterizing various PPN effector proteins and other genes involved in nematode parasitism. Additionally, more knowledge is still accumulating about these effector proteins to elucidate their significant roles during the penetration and migration within tissues of their plant hosts as well as parasitism comprising the adequate formation and maintenance of their feeding sites (e.g., nurse or giant cells for RKNs and syncytia for CNs), and deactivation of defense responses by their susceptible hosts [9,12,27].

Clearly, comparative secretome analyses among PPN species/strains/isolates are being investigated. They can determine which molecules are critical in inducing specific aspects of the disease and governing nematode virulence in the host plants. Thus, specific genes involved in the RNA interference pathways of the PPN species could be correctly targeted for nematode control [68]. Furthermore, combinatorial silencing of more than one functional gene at the same time could be more effective in PPN control [18]. Additionally, RNAi technology is being addressed to define specific PPN effectors to adapt them for effective nematode pest control. For instance, four isolates of the pinewood nematode, *Bursaphelenchus xylophilus*, with different levels of nematode virulence were recently compared to distinguish virulence determinants. These determinants, highly secreted by virulent *B. xylophilus* isolates, comprised Bx-CAT1 and Bx-CAT2 (as two C1A family cysteine peptidases), Bx-lip1 (lipase), and Bx-GH30 (glycoside hydrolase family 30). To quantitatively assess these four determinants at the transcript level at three stages, i.e., pre-inoculation, 3 days after inoculation (dai), and 7 dai into pine seedlings. Shinya et al. [20] used real-time reverse-transcription polymerase chain reaction analysis. They recorded significantly higher transcript levels of Bx-GH30, Bx-CAT2, and Bx-CAT1 in virulent isolates than in avirulent isolates at both pre-inoculation and 3 dai. While Bx-GH30 candidate virulent factor caused cell death in the plant, Bx-CAT2 was occupied in supplying nutrients for fungal feeding through soaking-mediated RNA interference. Shinya et al. [20] concluded that Bx-GH30 and Bx-CAT2 participate in the isolate virulence on host trees and may be engaged in pine wilt disease. Such nematode effectors can subsequently render themselves as potential candidate genes for nematode management. In this respect, RNAi may be utilized as a cellular procedure to degrade messenger RNA (mRNA), which plays the main role in protein synthesizing and consequently gene function. Thus, targeting 'candidate' effector genes of PPN species that cause successful infection of the host plant using the RNAi strategy could adequately suppress the genes responsible for this success [9,17–19,27,82]. The RNAi approach, for example, was utilized to knock down related effector genes of *Meloidogyne incognita* (e.g., *msp-16*, *msp-33*, *msp-20*, *msp-24*, and *msp-18*) that normally interact with plant transcription factors to express key cell wall-degrading enzymes (CWDE). The phenotypic plant data indicated that RNAi caused suppression of the targeted genes with a transcriptional shift in CWDE genes of the nematode [83].

7. Approaches to Strengthen Molecular PPN Control

Three main genetic classes for plant protection against PPNs have been used in a historical sequence, with overlapping between them. Traditional plant breeding for PPN resistance has long been used [84]. It has undoubtedly been progressing via genetic engineering too. This latter, the second class, aims at the general insertion of genetic material into a host genome. The latest class aims at genome editing in which DNA is inserted, deleted, modified, or replaced in the PPN genome. It aims at inserting genes to site-specific locations [85]. Ibrahim et al. [16] reported four main techniques of gene editing in order to boost the global breeding of cultivars resistant to RKN in a broad range of crops, namely recombinase-mediated site-specific gene integration, homologous recombination-dependent gene targeting, nuclease-mediated site-specific genome modifications, and oligonucleotide-directed mutagenesis. These techniques are expected to contribute to the rapid progress in grasping the plant–nematode interaction mechanisms and consequently ameliorate plant resistance against nematodes.

Rajput et al. [86] reviewed the related technologies, viz., the clustered regularly interspaced short palindromic repeat (CRISPR)/CRISPR-associated protein (CRISPR/Cas), as a strong device for accurately targeted modification of almost all crops' genomes to produce variation and expedite breeding plans. O'Halloran [87] provided a soft program, CRISPR-PN2, as a conclusive web-based stage that offers elastic use and control over the automated design of specific guide RNA sequences for CRISPR experiments in parasitic nematodes. The effective use of CRISPR/Cas9-directed genome editing in plant species has also been reviewed by Ibrahim et al. [16] in chickpea, the legume models *Medicago truncatula* and *Glycine max*. Its technology permits high-throughput gene editing at the genomic scale. The editing may assist in enhancing desired traits in plants with a restricted genetic pool and insufficient resistance sources.

7.1. Expanding the Use of Marker-Assisted Selection

Basically, marker-assisted selection (MAS) refers to utilizing a binding pattern of linked molecular (DNA) markers in order to indirectly select the desirable plant phenotype. Molecular markers are beneficial tools that can be used not only to set the introgression of genes related to economically desired traits but also to facilitate grasping molecular nematode–host interactions, as chromosome landmarks. Consequently, MAS are used for gene incorporation and stacking, as in tomato cultivars for multiple disease resistance traits. A striking example is *Mi-1* homologs that can grant resistance against a broad range of pests and pathogens, comprising the most common root-knot nematodes (*Meloidogyne javanica, M. incognita,* and *M. arenaria*), insects, i.e., potato aphids (*Macrosiphum euphorbiae*), and sweet potato whitefly (*Bemisia tabaci*), and oomycetes (*Phytophthora infestans*) in tomato plants [9,43]. Thus, various approaches relying on molecular markers for PPN resistance such as amplified fragment length polymorphisms (AFLPs), random amplified polymorphic DNA (RAPDs), restriction amplified length polymorphisms (RALPs), cleaved amplified polymorphic sequence (CAPS), reverse-transcription polymerase chain reaction (RT-PCR), single nucleotide polymorphisms (SNPs), sequenced characterized amplified regions (SCAR), sequence tagged site (STS), and simple sequence repeats (SSRs) are being developed [88,89]. They can be used to select a broad range of economically important plant species/cultivars for resistance against serious nematode pests (Table 1).

The readiness of marker application and affordability of marker genotyping, make MAS a good breeding option for many traits in most breeding programs. Moreover, marker development is advancing towards more reliable and efficient regeneration and genetic transformation systems with predictable and reproducible results. Very recently, the nematode resistance could be adequately addressed via using a genome-wide association study (GWAS) of single nucleotide polymorphism (SNP) markers with the PPN resistance. Thus, SNPs linked to resistance and the genes identified can establish a significant tool for introgression of resistance to *Heterodera glycines*, by marker-assisted selection in common bean (*Phaseolus vulgaris*) breeding programs [90]. Likewise, a locus on chromosome 13, comprising multiple TIR-NB-LRR genes and SNPs linked to *Meloidogyne javanica* resistance in soybean, was characterized by utilizing a combination of GWASs, resequencing, genetic mapping, and expression profiling [91]. Such technological progress would authorize a better understanding of gene function and expression with possible accredit of accurate genetic adjustment for PPN control and crop improvement [89].

Table 1. Examples of molecular markers for screening nematode resistance in main crops.

Crop	Nematode Species	Resistance Genes	Marker Type	References
Tomato	*Meloidogyne incognita*	*Mi 3*	RAPD and RFLP	[92]
Eggplant	*Meloidogyne javanica*	*Mi-1.2*	RT-PCR	[93]
Wheat	*Heterodera avenae*	*CreX* and *CreY*	SCAR	[94]
Pepper	*M. incognita, M. arenaria, M. javanica*	Me_3 and Me_4	RAPD and AFLP	[95]
Potato	*Globodera rostochinensis*	*H1*	RFLP	[96]
Soybean	*Heterodera glycines*	*Rhg1* and *Rhg4*	SNPs	[97]
Cucumber	*M. javanica*	*mj*	AFLP	[98]
Cotton	*M. incognita*	*qMi-C14*	SSR	[99]
Cotton	*Rotylenchulus reniformis*	Ren^{ari}	SSR	[100]
Peanut	*Meloidogyne arenaria*	*Rma*	CAPS, SSR, AFLP	[101]

7.2. Utilizing Proteinase Inhibitor Coding Genes

Proteinase inhibitors (PIs) can hinder the function of proteinases/proteases released by the nematodes. As PPNs invade plants, these PIs become active against all nematode proteinases; aspartic, cysteine, metalloproteinases, and serine. Ali et al. [27] reviewed various applications of PIs against PPNs. They emphasized that simultaneous use of different PIs could have an additive effect as it combines specificity with a broad range of resistance. Pyramiding genes of taro cystatin and fungal chitinase with a synthetic promoter could also increase resistance to RKNs in tomato [49]. These and similar approaches of combining more than one biocontrol measure/agent [27,102,103] can form featured bases for elevating transgenic plant resistance. Cystatins from various plant species rank high among other PIs in boosting nematode resistance in a variety of crops.

Abd-Elgawad [104] affirmed the importance of cystatins in increasing the nematode resistance within a plan that can upgrade eggplant production. Njom et al. [105] examined cysteine proteinases of the papain family (CPs) that attack nematodes and identified their specific molecular target(s). They concluded that multiple cuticle targets for these proteinases are found which probably make nematode resistance to these novel CPs slow to evolve. Thus, PIs have a future as a promising molecular control method against PPNs.

7.3. Use of RNA Interference

RNA interference is triggered by double-stranded RNA (dsRNA) inside the cell to degrade mRNA, the key to protein synthesis, and hence nematode-gene function. Therefore, the technique basically serves as a significant and robust device to analyze gene function in nematodes. Three classes of PPN-specific genes are being utilized as targets for RNAi techniques. These are genes enabling PPN parasitism, PPN developmental genes, and housekeeping genes [16]. As a genetically based approach, RNAi application has various aspects for effective and integrated control of PPNs. Its use to control plant infection with multiple plant pathogens proved to be promising [16,106]. This does not negate that the successful trials were solely based on single gene silencing for PPN control [82,83]. However, PPNs can masterly use several genes for accomplishing a specific function [18,19]. The nematode effectors found and expressed in subventral gland cells have many genes that can serve in nematode management as they are involved in the related nematode activities, e.g., penetration migration, and feeding within plant tissue. About 37 putative *M. incognita* esophageal gland secretory genes have been reported [83,107]. Nematode neuropeptides also serve in related processes such as host recognition, infection, and reproduction. This adds merit to the simultaneous silencing of genes as a promising tool in the control of PPNs. A dual gene construct of cysteine PI and a fungal chitinase with a synthetic promoter in transgenic tomato plants demonstrated considerably more reduction in RKN infection and reproduction than plants transformed with an individual gene [49]. Moreover, three *M. incognita* effectors, *Mi-msp1*, *Mi-msp16*, and *Mi-msp20* as fusion cassettes-1 and two FMRFamide-like peptides, *Mi-flp14*, *Mi-flp18*, and *Mi-msp20* as fusion cassettes-2 were successfully combined as targets of RNAi for nematode management. Their quantitative expression showed a significant decrease in mRNA abundance of target genes in *M. incognita* females in transgenic *Nicotiana tabacum* plants. The constructs, fusion 1 and fusion 2, granted up to an 85% decrease in *M. incognita* reproduction [18].

7.4. Nematicidal Proteins

Anti-nematode proteins such as some antibodies, lectins, and Bt Cry proteins can inhibit PPN development in plants. Yet, their mechanisms of inhibition vary and should be harnessed to optimize PPN control. Toxic lectins can block nematode-intestinal function [108]. The mechanism displayed by the lectins is crucial since several lectins bind with glycans. Overexpression of a *Galanthus nivalis* agglutinin (GNA)-related lectin driven by cauliflower mosaic virus promoter (CaMV35S) is exploited to offer anti-nematode efficacy in plants such as potato, oilseed rape (*Brassica napus*), and Arabidopsis concerning CNs, RKNs, and *Pratylenchus* spp. [27]. Some antibodies, known as plantibodies, are effective against PPNs in compatible plant–nematode interactions. They can oppose the active PPN-secreted proteins. They could, for example, react with secreted products of *G. pallida* and adversely affect the movement and invasion of this species to potato roots [109]. *Bacil-*

lus thuringiensis (Bt) toxins, known as Cry proteins, could directly reduce the *M. javanica* population on tomato roots by adding bacterial suspension or spore/crystal mixture [110] or indirectly induce resistance against *H. glycines* in transgenic soybean plants [48].

7.5. Chemodisruptive Peptides

Usually, PPNs use chemoreceptive neurons to approach their host plants or get away from their non-host. These neurons discern certain chemical stimuli for attacking the plants. Acetylcholinesterase (AChE) and/or nicotinic acetylcholine receptors are usually used for adequate operation of their nervous systems. A few peptides, at such low concentrations, can bind with these receptors and consequently disrupt the PPN ability of chemoreception by hindering their reaction to chemical signals [111]. A peptide secreted by transgenic potato plants could inhibit *G. pallida*-AchE resulting in the disorientation of attacking nematode. This led to a 52% reduction in the number of *G. pallida* females [112].

These peptides could also offer the prospect of an integrated nematode control strategy. A repellant peptide precisely directed at the sites of *G. pallida* invasion via a root tip-specific promoter from an Arabidopsis gene could be combined with the transgenic expression of a rice cystatin in potato to maintain a high degree of potato plant resistance against this CN [113]. Transgenic maize plants demonstrated good PPN control via combining digestive protease inhibitor cystatin with synthetic nematode repellent peptides [114,115]. Likewise, using chemo-disruptive peptides alone or integrated with cystatins into various plant species has been documented to show high levels of resistance against RKNs with a consequent increase in crop yields [27,116].

7.6. Employing Plant Resistance Mechanisms

Fundamentally, the above-mentioned two layers of plant-induced resistance and their related mechanisms against pests and pathogens should be fully employed. For instance, seeds or roots of some plants can release PPN-killing or repelling compounds in their exudates [3]. Proteomic methods detected 63 exuded proteins from soybean seeds, comprising a trypsin inhibitor, a β-1,3-glucanase, a lipoxygenase, a lectin, and a chitinase, all can contribute to plant defense. These exudates were able to suppress the hatching of *Meloidogyne incognita* eggs and to cause full mortality of the J2. Pretreatment of J2 with these exudates resulted in a 90% decrease in the gall number on plant roots [117]. While such findings should be exploited, caution should be exercised in other cases, e.g., a new chemo-attractant synthesized on *Arabidopsis* seeds could attract different RKN species to invade the freshly emerged seedling roots [118]. Phytohormones have significant roles in plant–nematode interactions since sedentary PPNs can alter auxin homeostasis via multiple strategies. Recent functional analyses indicated that PPNs have developed multiple approaches to manipulate indole-3-acetic acid (IAA) homeostasis to set an effective parasitic relation with their susceptible plants [119]. In contrast, the role of other hormones, such as salicylic acid, could also be exploited to boost plant resistance against PPNs [7,33].

As yet, more information needs to be generated for the related genes and defense mechanisms to encompass various aspects of host-specific resistance to optimize their efficacy and durability in field crops. Such information is expected to circumvent or overcome many of the above-mentioned issues causing a lack of developing sufficient nematode-resistant plant species/varieties. Otherwise, the two general schemes used to transfer a PPN-resistance gene, i.e., from one plant species to another or from a cultivar to another one within the same plant species, may face unexpected difficulties. For instance, the fact that backcrossing of an *Hs1*$^{pro-1}$ as a CN-resistant genotype and a susceptible sugarbeet plant did not lead to a resistance phenotype in the next generations is still raising an unsolved case [13]. Trials to transfer the *Mi-1* gene to *Arabidopsis* or tobacco were also ineffectual [33]. Conversely, favorable transfers of *R*-genes in heterologous species with monogenic resistance may result in resistance-breaking field pathotypes due to the imposed selection pressure. Other issues are related to the transfer of multiple disease resistance traits. These types are reflected in genetically transformed plants that share mixed characteristics of resistance and susceptibility. An outstanding example is the *Mi-1* gene transfer to eggplant from tomato. It could considerably lower RKN reproduction, but aphid resistance, displayed by the same transferred gene in tomato, was not attained [93]. This may raise the question of possible pleiotropic effects on the gene expressions. Another case of *R*-gene transformation raised the low level of resistance in the transformed plants as a result of the evident dosage impact of the *R*-gene copy number [33]. For such an impact, some authors assumed that expression of the resistance is more effective in homozygous than heterozygous genotypes of the tomato *Mi-1.2* gene, but others found the opposite for both the tomato *Mi-1.2* and the pepper *Me3* genes when the *R*-gene was introduced into homogeneous genetic backgrounds [120]. The authors assumed that transposable elements have a role in the creation and maintenance of *R*-genes-containing clusters in solanaceous crops as these elements are correlated with both large-scale genomic rearrangements and these genomic clusters. For instance, the sequencing of the P9 chromosome of pepper (carrying the *Me* gene cluster) showed how genome expansion due to these elements and duplication results in the advent of novel genes and functions or 'neofunctionalisation'. The frequent clustering of *R*-genes may ease the harmony of plant defenses against simultaneous pathogenic species and the development of new specificities to target an ever-changing array of pathogens [120,121].

Admittedly, the genetic background into which these genes are introduced is of supreme significance to the expression of PPN resistance and its durability, as shown with other pathosystems. Therefore, genetic constitutions of susceptible plants selected for the manipulations should possess an additional proper set of intracellular signaling pathways in order to employ the transferred *R*-gene(s) in proper resistance mechanisms. If so, molecular plant–nematode interactions can effectively serve this direction for other *R*-genes conferring PPN resistance for which the relevant signaling pathways are still insufficient.

7.7. Related Molecular Tools

Computational tools, bioinformatics technology, sound statistical methodology [121], and availability of increased molecular databases would ease grasping of the various types of nematode parasitism as well as the gene proteins, and recognizing pathways probably involved in plant–nematode interactions. These facilities, backed by reference genomes and novel genomic tools, comprising genotyping-by-sequencing (GBS), bulked-segregant analysis combined with whole-genome resequencing (BSA-seq), genome-wide association study (GWAS), and genomic selection (GS), will rapidly progress molecular PPN control to a high rank among other control measurements. This molecular PPN control can preferably be comprehensive in terms of addressing the management of polyspecific nematode populations [9]. Furthermore, the resistance to multiple PPN species should preferably be transferred into cultivars with resistance to other important pests. Clearly, researchers of relevant disciplines should better approach and apply the positive trends and standardization that serve this type for molecular control of PPNs. In this respect, novel plans to optimize nematode sampling [122] and focusing on recently recognized roles and tools to get better findings in the nematode realm are direly needed [16,123]. Such strategies can develop robust pest management programs able to efficiently replace unsafe nematicides while blocking the above-mentioned defects in the other control measures. Thus, sound integrated PPN management will combine molecular control and cultivation practices that lead to sustainable and high crop production techniques to keep the *R*-genes and sustain their durability.

Eventually, the regulatory picture for relevant transgenic plants is unclear and will stay so for the near future, as the scientific and social consequences of their release are debated. Meanwhile, there are recent references explaining the importance of using such modern methods to control PPNs on important crops such as tomatoes [124], and potatoes [125]. Although approaches to transgenic PPN control can be categorized as operating on nematode targets, nematode–plant interface, and plant response [126], the three classes might be interrelated to achieve the eventual control goal. Expression quantitative trait locus (eQTL) mapping-based approach to identify interacting sets of hosts and pathogen genes may further address how the PPN species can set gene expression differently on the account of their host's genotypes [127]. Moreover, on the nematode side, exploration of our current understanding of plant–pathogen molecular interactions and how they differ among different life strategies of various PPN genera and species should be further boosted. There are many putative control targets in the PPN-life cycles that can be exploited as illustrated by Perry and Moens [128]. Comparative genomics will upgrade our understanding of their parasitic strategies and lifestyles as well as the vulnerable life stages. Addressing incompatible nematode-host interactions is direly needed for crop species with limited availability of genetic and genomic resources, e.g., near-isogenic and mutant lines, completed genome and transcriptome sequences, and commercial full genome arrays [129]. Yet, practical use of this information for environmentally safe PPN management options is challenging.

Hence, favorable research avenues to overcome these difficulties should rationally forecast them to bring significant productivity to commercial agriculture. Ultimately, rational molecular control of the nematodes would be better integrated with other pest management measures to maximize crop production. Finally, the above-mentioned approaches should be integrated into PPN management in real time. For instance, there is no evidence that the novel strains of BCA have favorable traits without hard and tiring screening focusing on their virulence and versatility [130]. The alternate method is a directed search of mediums where BCAs will have had to develop the needed traits. This approach requires close academic–industry partnerships and a change in mindset away from the mold of using the traditional pesticide model to timely achieve tremendous strides.

In conclusion, PPNs are causing global crop losses while their classical control methods are not sufficient. Grasping the molecular basis of their interactions with plants can assist in developing new methods for PPN-molecular control for better nematode management. Omics technologies based on genes and proteins involved in the nematode activities and the plant resistance responses are key factors to evolve this control. Techniques such as next-generation sequencing of genomes and transcriptomes, RNAi, PIs, MAS, chemo-disruptive peptides, and genetic transformation systems with reproducible results must be available within PPN control strategies to enhance crop yields. Linking long-read sequencing to the use of high-density genetic mapping can also support the detection and characterization of PPN-virulent genes. Progress in computational biology, bioinformatics, and analyzing the omics large-scale data can efficiently boost these techniques to offer accurate recognition of components and pathways engaged in PPN parasitism and plant response. Updated genome-editing devices will serve classical plant breeding and precisely translate how gene actions are linked to phenotypic performances. Yet, these methods must be cautiously used to avoid unwanted effects such as PPN virulence and pleiotropic impacts on qualitative and quantitative crop yields. Definite issues such as those related to the transfer of multiple disease resistance traits, gene original construct, and specificity of the virulent nematodes to the *R*-gene may arise in particular cases. Therefore, employing the outcomes of these techniques on a case-by-case basis is proposed. This will enable us to monitor and improve PPN management according to the given variables. Eventually, molecular control methods of PPNs should be combined with other control tactics for integrated management as a way forward in crop protection/pest management.

Funding: This research was funded by The Science, Technology & Innovation Funding Authority (STDF), US-Egyptian project cycle 17 grant number 172 and The NRC help offered via In-house project headed by Mahfouz Abd-Elgawad in research plan no. 13 is acknowledged.

Institutional Review Board Statement: Not applicable.

Informed Consent Statement: Not applicable.

Data Availability Statement: Not applicable.

Acknowledgments: This article is supported in part by the US-Egypt Project cycle 17 (no. 172) entitled "Preparing and evaluating IPM tactics for increasing strawberry and citrus production."

The study was also supported in part by the NRC In-house project No. 12050105 entitled "Pesticide alternatives against soil-borne pathogens and pests attacking economically important solanaceous crops". The author thanks Mihail Kantor and Zafar Handoo for his valuable comments on the manuscript and Mostafa Hammam and Ibrahim Shehata for their help in the drawing.

Conflicts of Interest: The author declares no conflict of interest. The funders had no role in the design of the study; in the collection, analyses, or interpretation of data; in the writing of the manuscript, or in the decision to publish the results.

References

1. Abd-Elgawad, M.M.M.; Askary, T.H. Impact of phytonematodes on agriculture economy. In *Biocontrol Agents of Phytonematodes*; Askary, T.H., Martinelli, P.R.P., Eds.; CABI: Wallingford, UK, 2015; pp. 3–49.
2. Forghani, F.; Hajihassani, A. Recent advances in the development of environmentally benign treatments to control root-knot nematodes. *Front. Plant Sci.* **2020**, *11*, 1125. [CrossRef] [PubMed]
3. Abd-Elgawad, M.M.M. Optimizing safe approaches to manage plant-parasitic nematodes. *Plants* **2021**, *10*, 1911. [CrossRef] [PubMed]
4. Ochola, J.; Cortada, L.; Ng'ang'a, M.; Hassanali, A.; Coyne, D.; Torto, B. Mediation of potato–potato cyst nematode, *G. rostochiensis* interaction by specific root exudate compounds. *Front. Plant Sci.* **2020**, *11*, 649. [CrossRef] [PubMed]
5. Sikora, R.A.; Coyne, D.; Hallmann, J.; Timper, P. *Plant Parasitic Nematodes in Subtropical and Tropical Agriculture*, 3rd ed.; CABI: Cambridge, UK, 2018.
6. Quentin, M.; Abad, P.; Favery, B.; Rivas, S.; Reymond, P. Plant parasitic nematode effectors target host defense and nuclear functions to establish feeding cells. *Front. Plant Sci.* **2013**, *4*, 53. [CrossRef] [PubMed]
7. Molinari, S. Can the plant immune system be activated against animal parasites such as nematodes? *Nematology* **2020**, *22*, 481–492. [CrossRef]
8. Fuller, V.L.; Lilley, C.J.; Urwin, P.E. Nematode resistance. *New Phytol.* **2008**, *180*, 27–44. [CrossRef]
9. El-Sappah, A.H.; Islam, M.M.; El-Awady, H.H.; Yan, S.; Qi, S.; Liu, J.; Cheng, G.-T.; Liang, Y. Tomato natural resistance genes in controlling the root-knot nematode. *Genes* **2019**, *10*, 925. [CrossRef]
10. Bobardt, S.D.; Dillman, A.R.; Nair, M.G. The two faces of nematode infection: Virulence and immunomodulatory molecules from nematode parasites of mammals, insects and plants. *Front. Microbiol.* **2020**, *11*, 577846. [CrossRef]
11. Habash, S.S.; Brass, H.U.C.; Klein, A.S.; Klebl, D.P.; Weber, T.M.; Classen, T.; Pietruszka, J.; Grundler, F.M.W.; Schleker, A.S.S. Novel prodiginine derivatives demonstrate bioactivities on plants, nematodes, and fungi. *Front. Plant Sci.* **2020**, *11*, 579807. [CrossRef]
12. Cabral, D.; Forero Ballesteros, H.; de-Melo, B.P.; Lourenço-Tessutti, I.T.; Simões de Siqueira, K.M.; Obicci, L.; Grossi-de-Sa, M.F.; Hemerly, A.S.; de Almeida Engler, J. The armadillo BTB protein ABAP1 is a crucial player in DNA replication and transcription of nematode-induced galls. *Front. Plant Sci.* **2020**, *12*, 636663. [CrossRef]
13. Zheng, Q.; Putker, V.; Goverse, A. Molecular and cellular mechanisms involved in host-specific resistance to cyst nematodes in crops. *Front. Plant Sci.* **2021**, *12*, 641582. [CrossRef]
14. Blanc-Mathieu, R.; Perfus-Barbeoch, L.; Aury, J.-M.; Da Rocha, M.; Gouzy, J.; Sallet, E. Hybridization and polyploidy enable genomic plasticity without sex in the most devastating plant-parasitic nematodes. *PLoS Genet.* **2017**, *13*, e1006777. [CrossRef]
15. Jaubert-Possamai, S.; Noureddine, Y.; Favery, B. MicroRNAs, new players in the plant–nematode interaction. *Front. Plant Sci.* **2019**, *10*, 1180. [CrossRef]
16. Ibrahim, H.M.M.; Ahmad, E.M.; Martínez-Medina, A.; Aly, M.A.M. Effective approaches to study the plant-root knot nematode interaction. *Plant Physiol. Biochem.* **2019**, *141*, 332–342. [CrossRef]
17. Hada, A.; Kumari, C.; Phani, V.; Singh, D.; Chinnusamy, V.; Rao, U. Host-induced silencing of FMR Famide-like peptide genes, *flp-1* and *flp-12*, in rice impairs reproductive fitness of the root-knot nematode *Meloidogyne graminicola*. *Front. Plant Sci.* **2020**, *11*, 894. [CrossRef]
18. Hada, A.; Singh, D.; Papolu, P.K.; Banakar, P.; Raj, A.; Rao, U. Host-mediated RNAi for simultaneous silencing of different functional groups of genes in *Meloidogyne incognita* using fusion cassettes in *Nicotiana tabacum*. *Plant Cell Rep.* **2021**, *40*, 2287–2302. [CrossRef]
19. Hada, A.; Patil, B.L.; Bajpai, A.; Kesiraju, K.; Dinesh-Kumar, S.; Paraselli, B.; Sreevathsa, R.; Rao, U. Micro RNA-induced gene silencing strategy for the delivery of siRNAs targeting *Meloidogyne incognita* in a model plant *Nicotiana benthamiana*. *Pest Manag. Sci.* **2021**, *77*, 3396–3405. [CrossRef]
20. Shinya, R.; Kirino, H.; Morisaka, H.; Takeuchi-Kaneko, Y.; Futai, K.; Ueda, M. Comparative secretome and functional analyses reveal glycoside hydrolase family 30 and cysteine peptidase as virulence determinants in the pinewood nematode *Bursaphelenchus xylophilus*. *Front. Plant Sci.* **2021**, *12*, 640459. [CrossRef]
21. Chitwood, D.J. Nematicides. In *Encyclopedia of Agrochemicals*; Plimmer, J.R., Ed.; John Wiley & Sons: New York, NY, USA, 2003; Volume 3, pp. 1104–1115.
22. Abd-Elgawad, M.M.M.; Kour, F.F.H.; Montasser, S.A.; Hammam, M.M.A. Distribution and losses of *Tylenchulus semipenetrans* in citrus orchards on reclaimed land in Egypt. *Nematology* **2016**, *18*, 1141–1150. [CrossRef]

23. Abd-Elgawad, M.M.M.; Askary, T.H. Factors affecting success of biological agents used in controlling plant-parasitic nematodes. *Egypt. J. Biol. Pest. Cont.* **2020**, *30*, 17. [CrossRef]
24. Timper, P. Conserving and enhancing biological control of nematodes. *J. Nematol.* **2014**, *46*, 75–89.
25. Ntalli, N.; Adamski, Z.; Doula, M.; Monokrousos, N. Nematicidal amendments and soil remediation. *Plants* **2020**, *9*, 429. [CrossRef]
26. Dutta, T.K.; Khan, M.R.; Phani, V. Plant-parasitic nematode management via biofumigation using brassica and non-brassica plants: Current status and future prospects. *Curr. Plant Biol.* **2019**, *17*, 17–32. [CrossRef]
27. Ali, M.A.; Azeem, F.; Abbas, A.; Joyia, F.A.; Li, H.; Dababat, A.A. Transgenic strategies for enhancement of nematode resistance in plants. *Front. Plant Sci.* **2017**, *8*, 750. [CrossRef]
28. Sasanelli, N.; Konrat, A.; Migunova, V.; Toderas, I.; Iurcu-Straistaru, E.; Rusu, S.; Bivol, A.; Andoni, C.; Veronico, P. Review on control methods against plant parasitic nematodes applied in southern member states (C Zone) of the European Union. *Agriculture* **2021**, *11*, 602. [CrossRef]
29. Abd-Elgawad, M.M.M. Optimizing sampling and extraction methods for plant-parasitic and entomopathogenic nematodes. *Plants* **2021**, *10*, 629. [CrossRef] [PubMed]
30. Sato, K.; Kadota, Y.; Shirasu, K. Plant immune responses to parasitic nematodes. *Front. Plant Sci.* **2019**, *10*, 1165. [CrossRef] [PubMed]
31. Seah, S.; Telleen, A.C.; Williamson, V.M. Introgressed and endogenous *Mi-1* gene clusters in tomato differ by complex rearrangements in flanking sequences and show sequence exchange and diversifying selection among homologues. *Theoret. Appl. Gen.* **2007**, *114*, 1289–1302. [CrossRef]
32. Ammiraju, J.S.S.; Veremis, J.C.; Huang, X.; Roberts, P.A.; Kaloshian, I. The heat-stable, root-knot nematode resistance gene *Mi-9* from *Lycopersicum peruvianum* is localized on the short arm of chromosome 6. *Theoret. Appl. Gen.* **2003**, *106*, 478–484. [CrossRef]
33. Molinari, S. Resistance and virulence in plant-nematode interactions. In *Nematodes*; Boeri, F., Chung, J.A., Eds.; Nova Science Publisher, Inc.: New York, NY, USA, 2012; pp. 59–82.
34. Finkers-Tomczak, A.; Bakker, E.; de Boer, J.; van der Vossen, E.; Achenbach, U.; Golas, T. Comparative sequence analysis of the potato cyst nematode resistance locus H1 reveals a major lack of co-linearity between three haplotypes in potato (*Solanum tuberosum* ssp.). *Theor. Appl. Genet.* **2011**, *122*, 595–608. [CrossRef]
35. Dropkin, V.H.; Helgeson, J.P.; Upper, C.D. The hypersensitive reaction of tomatoes resistant to *Meloidogyne incognita*: Reversal by cytokinins. *J. Nematol.* **1969**, *1*, 55–60.
36. Janssen, R.; Bakker, J.; Gommers, F. Mendelian proof for a gene-for-gene relationship between virulence of *Globodera rostochiensis* and the *H1* resistance gene in *Solanum tuberosum* spp. andigena CPC 1673. *Rev. Nematol.* **1991**, *14*, 213–219.
37. Dangl, J.L.; Jones, J.D. Plant pathogens and integrated defence responses to infection. *Nature* **2001**, *411*, 826–833. [CrossRef]
38. Bakker, E.; Dees, R.; Bakker, J.; Goverse, A. Mechanisms involved in plant resistance to nematodes. In *Multi-Genic and Induced Systemic Resistance in Plants*; Tuzun, S., Bent, E., Eds.; Springer Science+Business Media, Inc.: Secaucus, NJ, USA, 2006; pp. 314–334.
39. Abd-Elgawad, M.M.M. Biological control agents in the integrated nematode management of pepper in Egypt. *Egypt. J. Biol. Pest Cont.* **2020**, *30*, 70. [CrossRef]
40. Williamson, V.M.; Roberts, P.A. Mechanisms and genetics of resistance. In *Root-Knot Nematodes;* Perry, R.N., Moens, M., Starr, J.L., Eds.; CAB International: Wallingford, UK, 2009; pp. 301–325.
41. Evans, K.; Trudgill, D.L.; Webster, J.M. *Plant Parasitic Nematodes in Temperate Agriculture*; CABI International: Wallingford, UK, 1993.
42. Sato, K.; Uehara, T.; Holbein, J.; Sasaki-Sekimoto, Y.; Gan, P.; Bino, T.; Yamaguchi, K.; Ichihashi, Y.; Maki, N.; Shigenobu, S.; et al. Transcriptomic analysis of resistant and susceptible responses in a new model root-knot nematode infection system using *Solanum torvum* and *Meloidogyne arenaria*. *Front. Plant Sci.* **2021**, *12*, 680151. [CrossRef]
43. Seifi, A.; Kaloshian, I.; Vossen, J.; Che, D.; Bhattarai, K.K.; Fan, J. Linked, if not the same, *Mi-1* homologues confer resistance to tomato powdery mildew and root-knot nematodes. *Mol. Plant-Microbe Interact.* **2011**, *24*, 441–450. [CrossRef]
44. Ndeve, N.D.; Matthews, W.C.; Santos, J.R.P.; Huynh, B.L.; Roberts, P.A. Broad-based root-knot nematode resistance identified in cowpea gene-pool two. *J. Nematol.* **2018**, *50*, 545–558. [CrossRef]
45. Bhuiyan, S.A.; Garlick, K. Evaluation of root-knot nematode resistance assays for sugarcane accession lines in Australia. *J. Nematol.* **2021**, *53*, 1–11. [CrossRef]
46. Khallouk, S.; Voisin, R.; Van Ghelder, C.; Engler, G.; Amiri, S.; Esmenjaud, D. Histological mechanisms of the resistance conferred by the *Ma* gene against *Meloidogyne incognita* in *Prunus* spp. *Phytopathology* **2011**, *101*, 945–951. [CrossRef]
47. Zhang, L.Y.; Zhang, Y.Y.; Chen, R.G.; Zhang, J.H.; Wang, T.T.; Li, H.X.; Ye, Z.B. Ectopic expression of the tomato *Mi-1* gene confers resistance to root knot nematodes in lettuce (*Lactuca sativa*). *Plant Mol. Biol. Rep.* **2010**, *28*, 204–211. [CrossRef]
48. Kahn, T.W.; Duck, N.B.; McCarville, M.T.A. *Bacillus thuringiensis* Cry protein controls soybean cyst nematode in transgenic soybean plants. *Nat. Commun.* **2021**, *12*, 3380. [CrossRef]
49. Chen, Y.L.; He, Y.; Hsiao, T.T.; Wang, C.J.; Tian, Z.; Yeh, K.W. Pyramiding taro cystatin and fungal chitinase genes driven by a synthetic promoter enhances resistance in tomato to root-knot nematode *Meloidogyne incognita*. *Plant Sci.* **2015**, *231*, 74–81. [CrossRef]

50. Glazebrook, J. Contrasting mechanisms of defense against biotrophic and nectrophic pathogens. *Annu. Rev. Phytopathol.* **2005**, *43*, 205–227. [CrossRef]
51. Uehara, T.; Sugiyama, S.; Matsuura, H.; Arie, T.; Masuta, C. Resistant and susceptible responses in tomato to cyst nematodes are differentially regulated by salicylic acid. *Plant Cell Physiol.* **2010**, *51*, 1524–1536. [CrossRef]
52. Zipfel, C.; Oldoryd, G.E.D. Plant signalling in symbiosis and immunity. *Nature* **2017**, *543*, 328–336. [CrossRef]
53. El-Nagdi, W.M.A.; Youssef, M.A.M.; Abd-El-Khair, H.; Abd-Elgawad, M.M.M. Effect of certain organic amendments and *Trichoderma* species on the root-knot nematode, *Meloidogyne incognita* infecting pea (*Pisum sativum* L.) plants. *Egypt. J. Biol. Pest Control* **2019**, *29*, 75. [CrossRef]
54. Gao, X.; Guo, H.; Zhang, Q. Arbuscular mycorrhizal fungi (AMF) enhanced the growth, yield, fiber quality and phosphorus regulation in upland cotton (*Gossypium hirsutum* L.). *Sci. Rep.* **2020**, *10*, 2084. [CrossRef]
55. Janssen, R.; Bakker, J.; Gommers, F. Selection of virulent and avirulent lines of *Globodera rostochiensis* for the *H1* resistance gene in *Solanum tuberosum* ssp. andigena CPC 1673. *Rev. Nématol.* **1990**, *13*, 265–268.
56. Castagnone-Sereno, P. Genetic variability of nematodes: A threat to the durability of plant resistance genes? *Euphytica* **2002**, *124*, 193–199. [CrossRef]
57. Xu, J.; Narabu, T.; Mizukubo, T.; Hibi, T. A molecular marker correlated with selected virulence against the tomato resistance gene *Mi* in *Meloidogyne incognita*, *M. javanica* and *M. arenaria*. *Phytopathology* **2001**, *91*, 377–382. [CrossRef]
58. Molinari, S. Antioxidant enzymes in (a)virulent populations of root-knot nematodes. *Nematology* **2009**, *11*, 689–697. [CrossRef]
59. Kaloshian, I.; Teixeira, M. Advances in plant−nematode interactions with emphasis on the notorious nematode genus *Meloidogyne*. *Phytopathology* **2019**, *109*, 1988–1996. [CrossRef]
60. Padilla-Hurtado, B.; Morillo-Coronado, Y.; Tarapues, S.; Burbano, S.; Soto-Suárez, M.; Urrea, R.; Ceballos-Aguirre, N. Evaluation of root-knot nematodes (*Meloidogyne* spp.) population density for disease resistance screening of tomato germplasm carrying the gene Mi-1. *Chil. J. Agric. Res.* **2022**, *82*, 157–166. [CrossRef]
61. Zhang, L.; Gleason, C. Transcriptome analyses of pre-parasitic and parasitic *Meloidogyne chitwoodi* race 1 to identify putative effector genes. *J. Nematol.* **2021**, *53*, e2021-84. [CrossRef] [PubMed]
62. Hussey, R.S.; Mims, C.W. Ultrastructure of esophageal glands and their secretory granules in the root-knot nematode *Meloidogyne incognita*. *Protoplasma* **1990**, *156*, 9–18. [CrossRef]
63. Li, X.Q.; Wei, J.Z.; Tan, A.; Aroian, R.V. Resistance to root-knot nematode in tomato roots expressing a nematicidal *Bacillus thuringiensis* crystal protein. *Plant Biotechnol. J.* **2007**, *5*, 455–464. [CrossRef] [PubMed]
64. Li, X.-Q.; Tan, A.; Voegtline, M.; Bekele, S.; Chen, C.-S.; Aroian, R.V. Expression of Cry5B protein from *Bacillus thuringiensis* in plant roots confers resistance to root-knot nematode. *Biol. Control* **2008**, *47*, 97–102. [CrossRef]
65. Urwin, P.E.; Mcpherson, M.J.; Atkinson, H.J. Enhanced transgenic plant resistance to nematodes by dual proteinase inhibitor constructs. *Planta* **1998**, *204*, 472–479. [CrossRef]
66. Orlando, V.; Grove, I.G.; Edwards, S.G.; Prior, T.; Roberts, D.; Neilson, R.; Back, M. Root-lesion nematodes of potato: Current status of diagnostics, pathogenicity and management. *Plant Pathol.* **2020**, *69*, 405–417. [CrossRef]
67. Dritsoulas, A.; El-Borai, F.E.; Shehata, I.E.; Hammam, M.M.; El-Ashry, R.M.; Mohamed, M.M.; Abd-Elgawad, M.M.; Duncan, L.W. Reclaimed desert habitats favor entomopathogenic nematode and microarthropod abundance compared to ancient farmlands in the Nile Basin. *J. Nematol.* **2021**, *53*, 1–13. [CrossRef]
68. Iqbal, S.; Fosu-Nyarko, J.; Jones, M.G.K. Attempt to silence genes of the RNAi pathways of the root-knot nematode, *Meloidogyne incognita* results in diverse responses including increase and no change in expression of some genes. *Front. Plant Sci.* **2020**, *11*, 328. [CrossRef]
69. Rosso, M.N.; Dubrana, M.P.; Cimbolini, N.; Jaubert, S.; Abad, P. Application of RNA interference to root-knot nematode genes encoding esophageal gland proteins. *Mol. Plant Microbe Interact.* **2005**, *18*, 615–620. [CrossRef]
70. Roze, E.; Hanse, B.; Mitreva, M.; Vanholme, B.; Bakker, J.; Smant, G. Mining the secretome of the root-knot nematode *Meloidogyne chitwoodi* for candidate parasitism genes. *Mol. Plant Pathol.* **2005**, *9*, 1–10. [CrossRef]
71. Wubie, M.; Temesgen, Z. Resistance mechanisms of tomato (*Solanum lycopersicum*) to root-knot nematodes (*Meloidogyne* species). *J. Plant Breed. Crop Sci.* **2019**, *11*, 33–40.
72. Pocurull, M.; Fullana, A.M.; Ferro, M.; Valero, P.; Escudero, N.; Saus, E.; Gabaldón, T.; Sorribas, F.J. Commercial formulates of *Trichoderma* induce systemic plant resistance to *Meloidogyne incognita* in tomato and the effect is additive to that of the mi-1.2 resistance gene. *Front. Microbiol.* **2020**, *10*, 3042. [CrossRef]
73. Zhao, Y.-L.; Wu, Q.-L.; Li, Y.-P.; Wang, D.-Y. Translocation, transfer, and in vivo safety evaluation of engineered nanomaterials in the non-mammalian alternative toxicity assay model of nematode *Caenorhabditis elegans*. *RSC Adv.* **2013**, *3*, 5741–5757. [CrossRef]
74. Szitenberg, A.; Salazar-Jaramillo, L.; Blok, V.C.; Laetsch, D.R.; Joseph, S.; Williamson, V.M.; Blaxter, M.L.; Lunt, D.H. Comparative genomics of apomictic root-knot nematodes: Hybridization, ploidy, and dynamic genome change. *Genome Biol. Evol.* **2017**, *9*, 2844–2861. [CrossRef]
75. Sato, K.; Kadota, Y.; Gan, P.; Bino, T.; Uehara, T.; Yamaguchi, K.; Ichihashi, Y.; Maki, N.; Iwahori, H.; Suzuki, T.; et al. High-quality genome sequence of the root-knot nematode *Meloidogyne arenaria* genotype A2-O. *Genome Announc.* **2018**, *6*, e00519-18. [CrossRef]
76. Mani, V.; Assefa, A.D.; Hahn, B.S. Transcriptome analysis and miRNA target profiling at various stages of root-knot nematode *Meloidogyne incognita* development for identification of potential regulatory networks. *Int. J. Mol. Sci.* **2001**, *22*, 7442. [CrossRef]

77. Susič, N.; Koutsovoulos, G.D.; Riccio, C.; Danchin, E.G.J. Genome sequence of the root-knot nematode *Meloidogyne luci*. *J. Nematol.* **2020**, *52*, 1–5. [CrossRef]
78. Koutsovoulos, G.D.; Poullet, M.; Elashry, A. Genome assembly and annotation of *Meloidogyne enterolobii*, an emerging parthenogenetic root-knot nematode. *Sci. Data* **2020**, *7*, 324. [CrossRef] [PubMed]
79. Phan, N.T.; Besnard, G.; Ouazahrou, R.; Sánchez, W.S.; Gil, L.; Manzi, S.; Bellafiore, S. Genome sequence of the coffee root-knot nematode *Meloidogyne exigua*. *J. Nematol.* **2001**, *53*, 1–6. [CrossRef] [PubMed]
80. Bali, S.; Hu, S.; Vining, K.; Brown, C.R.; Majtahedi, H.; Zhang, L.; Gleason, C.; Sathuvalli, V. Nematode genome announcement: Draft genome of *Meloidogyne chitwoodi*, an economically important pest of potato in the Pacific Northwest. *Mol. Plant-Microbe Interact.* **2021**, *34*, 981–986. [CrossRef] [PubMed]
81. Somvanshi, V.S.; Dash, M.; Bhat, C.G.; Budhwar, R.; Godwin, J.; Shukla, R.N.; Patrignani, A.; Schlapbach, R.; Rao, U. An improved draft genome assembly of *Meloidogyne graminicola* IARI strain using long-read sequencing. *Gene* **2021**, *793*, 145748. [CrossRef] [PubMed]
82. Papolu, P.K.; Gantasala, N.P.; Kamaraju, D.; Banakar, P.; Sreevathsa, R. Utility of host delivered RNAi of two FMRF amide like peptides, *flp-14* and *flp-18*, for the management of root knot nematode, *Meloidogyne incognita*. *PLoS ONE* **2013**, *8*, e80603. [CrossRef]
83. Shivakumara, T.N.; Papolu, P.K.; Dutta, T.K.; Kamaraju, D.; Rao, U. RNAi-induced silencing of an effector confers transcriptional oscillation in another group of effectors in the root-knot nematode, *Meloidogyne incognita*. *Nematology* **2016**, *18*, 857–870. [CrossRef]
84. Caromel, B.; Gebhardt, C. Breeding for nematode resistance: Use of genomic information. In *Genomics and Molecular Genetics of Plant-Nematode Interactions*; Jones, J., Gheysen, G., Fenoll, C., Eds.; Springer: Dordrecht, The Netherlands, 2011; pp. 465–492. [CrossRef]
85. Bak, R.O.; Gomez-Ospina, N.; Porteus, M.H. Gene editing on center stage. *Trends Genet.* **2018**, *34*, 600–611. [CrossRef]
86. Rajput, M.; Choudhary, K.; Kumar, M.; Vivekanand, V.; Chawade, A.; Ortiz, R.; Pareek, N. RNA interference and CRISPR/Cas gene editing for crop improvement: Paradigm shift towards sustainable agriculture. *Plants* **2001**, *10*, 1914. [CrossRef]
87. O'Halloran, D.M. CRISPR-PN2: A flexible and genome-aware platform for diverse CRISPR experiments in parasitic nematodes. *BioTechniques* **2001**, *71*, 495–498. [CrossRef]
88. Banu, J.G.; Meena, K.S.; Selvi, C.; Manickam, S.; Jainullabudeen, C.; Banu, G. Molecular marker-assisted selection for nematode resistance in crop plants. *J. Entomol. Zool. Stud.* **2017**, *5*, 1307–1311.
89. Simko, I.; Jia, M.; Venkatesh, J.; Kang, B.; Weng, Y.; Barcaccia, G.; Lanteri, S.; Bhattarai, G.; Foolad, M.R. Genomics and marker-assisted improvement of vegetable crops. *Crit. Rev. Plant Sci.* **2021**, *40*, 303–365. [CrossRef]
90. Shi, A.; Gepts, P.; Song, Q.; Xiong, H.; Michaels, T.E.; Chen, S. Genome-wide association study and genomic prediction for soybean cyst nematode resistance in USDA common bean (*Phaseolus vulgaris*) core collection. *Front. Plant Sci.* **2021**, *12*, 624156. [CrossRef]
91. Alekcevetch, J.C.; de Lima Passianotto, A.L.; Ferreira, E.G.C. Genome-wide association study for resistance to the *Meloidogyne javanica* causing root-knot nematode in soybean. *Theoret. Appl. Gen.* **2021**, *134*, 777–792. [CrossRef]
92. Williamson, V.M.; Ho, J.Y.; Wu, F.F.; Miller, N.; Kaloshian, I. A PCR based marker tightly linked to the nematode resistance gene, *Mi* in tomato. *Theoret. Appl. Gen.* **1997**, *87*, 757–763. [CrossRef]
93. Goggin, F.L.; Jia, L.L.; Shah, G.; Hebert, S.; Williamson, V.M.; Ullman, D.E. Heterologous expression of the *Mi-1.2* gene from tomato confers resistance against nematodes but not aphids in eggplant. *Mol. Plant-Microbe Interact.* **2006**, *19*, 383–388. [CrossRef]
94. Barloy, D.; Lemoine, J.; Abelard, P.; Tanguy, A.M.; Rivoal, R.; Jahier, J. Marker-assisted pyramiding of two cereal cyst nematode resistance genes from *Aegilops variabilis* in wheat. *Mol. Breed.* **2007**, *20*, 31–40. [CrossRef]
95. Djian-Caporalino, C.; Pijarowski, L.; Fazari, A.; Samson, M.; Gaveau, L.; O'Byrne, C.; Lefebvre, V.; Caranta, C.; Palloix, A.; Abad, P. High-resolution genetic mapping of pepper (*Capsicum annuum* L.) resistance loci Me_3 and Me_4 conferring heat-stable resistance root-knot nematodes (*Meloidogyne* spp.). *Theoret. Appl. Gen.* **2001**, *103*, 592–600. [CrossRef]
96. Gebhardt, C.; Mugniery, D.; Ritter, E.; Salamini, F.; Bonnel, E. Identification of RFLP markers closely linked to the H1 gene conferring resistance to *Globodera rostochiensis* in potato. *Theoret. Appl. Gen.* **1993**, *85*, 541–544. [CrossRef]
97. Kadam, S.; Tri, D.; Vuonga, Q.D.; Clinton, G.; Meinhardta Deshmukha, L.S.R.; Patila, G. Genomic-assisted phylogenetic analysis and marker development for next generation soybean cyst nematode resistance breeding. *Plant Sci.* **2016**, *242*, 342–350. [CrossRef]
98. Devran, Z.; Firat, A.F.; Tor, M.; Mutlu, N.; Elekçioglu, I.H. AFLP and SRAP markers linked to the *Mj* gene for root-knot nematode resistance in cucumber. *Sci. Agric.* **2011**, *68*, 115–119. [CrossRef]
99. Kumar, P.; He, Y.J.; Singh, R.; Davis, R.F.; Guo, H.; Paterson, A.H. Fine mapping and identification of candidate genes for a QTL affecting *Meloidogyne incognita* reproduction in Upland cotton. *BMC Genom.* **2016**, *17*, 567. [CrossRef]
100. Romano, G.B.; Sacks, E.J.; Stetina, S.R.; Robinson, A.F.; Fang, D.D.; Gutierrez, O.A.; Scheffler, J.A. Identification and genomic location of a reniform nematode (*Rotylenchulus reniformis*) resistance locus (Ren^{ari}) introgressed from *Gossypium aridum* into upland cotton (*G. hirsutum*). *Theoret. Appl. Gen.* **2009**, *120*, 139–150. [CrossRef]
101. Chu, Y.; Wu, C.L.; Holbrook, C.C.; Tillman, B.L.; Person, G.; Ozias-Akins, P. Marker-assisted selection to pyramid nematode resistance and the high oleic trait in peanut. *Plant Genome* **2011**, *4*, 110–117. [CrossRef]
102. Abd-Elgawad, M.M.M.; Askary, T.H. Fungal and bacterial nematicides in integrated nematode management strategies. *Egypt. J. Biol. Pest Control* **2018**, *28*, 74. [CrossRef]

103. Banakar, P.; Hada, A.; Papolu, P.K.; Rao, U. Simultaneous RNAi knockdown of three FMRF-amide-like peptide genes, Mi-flp1, Mi-flp12, and Mi-flp18 provides resistance to root-knot nematode, *Meloidogyne incognita*. *Front. Microbiol.* **2020**, *11*, 573916. [CrossRef] [PubMed]
104. Abd-Elgawad, M.M.M. Biological control of nematodes infecting eggplant in Egypt. *Bull. NRC* **2021**, *45*, 6. [CrossRef]
105. Njom, V.S.; Winks, T.; Diallo, O. The effects of plant cysteine proteinases on the nematode cuticle. *Parasites Vectors* **2021**, *14*, 302. [CrossRef] [PubMed]
106. Tripathi, L.; Atkinson, H.; Roderick, H.; Kubiriba, J.; Tripathi, J.N. Genetically engineered bananas resistant to *Xanthomonas* wilt disease and nematodes. *Food Energy Secur.* **2017**, *6*, 37–47. [CrossRef]
107. Huang, G.; Gao, B.; Maier, T.; Allen, R.; Davis, E.L.; Baum, T.J. A profile of putative parasitism genes expressed in the esophageal gland cells of the root-knot nematode *Meloidogyne incognita*. *Mol. Plant-Microbe Interact.* **2003**, *16*, 376–381. [CrossRef]
108. Vasconcelos, I.M.; Oliveira, J.T. Antinutritional properties of plant lectins. *Toxicon* **2004**, *44*, 385–403. [CrossRef]
109. Fioretti, L.; Porter, A.; Haydock, P.J.; Curtis, R. Monoclonal antibodies reactive with secreted-excreted products from the amphids and the cuticle surface of *Globodera pallida* affect nematode movement and delay invasion of potato roots. *Int. J. Parasitol.* **2002**, *32*, 1709–1718. [CrossRef]
110. Ravari, S.B.; Moghaddam, E.M. Efficacy of *Bacillus thuringiensis* Cry14 toxin against root knot nematode, *Meloidogyne javanica*. *Plant Prot. Sci.* **2015**, *51*, 46–51. [CrossRef]
111. Winter, M.D.; Mcpherson, M.J.; Atkinson, H.J. Neuronal uptake of pesticides disrupts chemosensory cells of nematodes. *Parasitology* **2002**, *125*, 561–565.
112. Liu, B.; Hibbard, J.K.; Urwin, P.E.; Atkinson, H.J. The production of synthetic chemodisruptive peptides in planta disrupts the establishment of cyst nematodes. *Plant Biotechnol. J.* **2005**, *3*, 487–496. [CrossRef]
113. Green, J.; Wang, D.; Lilley, C.J.; Urwin, P.E.; Atkinson, H.J. Transgenic potatoes for potato cyst nematode control can replace pesticide use without impact on soil quality. *PLoS ONE* **2012**, *7*, e30973. [CrossRef]
114. Roderick, H.; Tripathi, L.; Babirye, A.; Wang, D.; Tripathi, J.; Urwin, P.E. Generation of transgenic plantain (*Musa* spp.) with resistance to plant pathogenic nematodes. *Mol. Plant Pathol.* **2012**, *13*, 842–851. [CrossRef]
115. Tripathi, L.; Tripathi, J.N.; Roderick, H.; Atkinson, H.J. Engineering nematode resistant plantains for sub-Saharan Africa. *Acta Hortic.* **2013**, *974*, 99–107. [CrossRef]
116. Papolu, P.K.; Dutta, T.K.; Hada, A.; Singh, D.; Rao, U. The production of a synthetic chemodisruptive peptide in planta precludes *Meloidogyne incognita* multiplication in *Solanum melongena*. *Physiol. Mol. Plant Pathol.* **2020**, *112*, 101542. [CrossRef]
117. Rocha, R.O.; Morais, J.K.S.; Oliveira, J.T.A. Proteome of soybean seed exudates contains plant defense-related proteins active against the root-knot nematode *Meloidogyne incognita*. *J. Agric. Food Chem.* **2015**, *63*, 5335–5343. [CrossRef]
118. Tsai, A.Y.L.; Higaki, T.; Nguyen, C.N.; Perfus-Barbeoch, L.; Favery, B.; Sawa, S. Regulation of root-knot nematode behavior by seed-coat mucilage-derived attractants. *Mol. Plant.* **2019**, *12*, 99–112. [CrossRef]
119. Oosterbeek, M.; Lozano-Torres, J.L.; Bakker, J.; Goverse, A. Sedentary plant-parasitic nematodes alter auxin homeostasis via multiple strategies. *Front. Plant Sci.* **2021**, *12*, 668548. [CrossRef]
120. Barbary, A.; Djian-Caporalino, C.; Palloix, A.; Castagnone-Sereno, P. Host genetic resistance to root-knot nematodes, *Meloidogyne* spp., in Solanaceae: From genes to the field. *Pest Manag. Sci.* **2015**, *71*, 1591–1598. [CrossRef]
121. Hulbert, S.H.; Craig, A.W.; Shavannor, M.S.; Qing, S. Resistance gene complexes: Evolution and utilization. *Annu. Rev. Phytopathol.* **2001**, *39*, 285–312. [CrossRef]
122. Abd-Elgawad, M.M.M. Can rational sampling maximise isolation and fix distribution measure of entomopathogenic nematodes? *Nematology* **2020**, *22*, 907–916. [CrossRef]
123. Abd-Elgawad, M.M.M. Towards optimization of entomopathogenic nematodes for more service in the biological control of insect pests. *Egypt. J. Biol. Pest Cont.* **2019**, *29*, 77. [CrossRef]
124. Abd-Elgawad, M.M.M. Optimizing biological control agents for controlling nematodes of tomato in Egypt. *Egypt. J. Biol. Pest Cont.* **2020**, *30*, 58. [CrossRef]
125. Abd-Elgawad, M.M.M. Biological control agents in the integrated nematode management of potato in Egypt. *Egypt. J. Biol. Pest Cont.* **2020**, *30*, 121. [CrossRef]
126. McCarter, J.P. Molecular approaches toward resistance to plant-parasitic nematodes. In *Plant Cell Monographs: Cell Biology of Plant Nematode Parasitism*; Berg, R.H., Taylor, C.G., Eds.; Springer: Berlin, Germany, 2008; pp. 239–268. [CrossRef]
127. Guo, Y.; Fudali, S.; Gimeno, J.; Digennaro, P.; Chang, S.; Williamson, V.M.; Bird, D.M.; Nielsen, D. Networks underpinning symbiosis revealed through cross-species eQTL mapping. *Genetics* **2017**, *206*, 2175–2184. [CrossRef]
128. Perry, R.N.; Moens, M. Introduction to plant-parasitic nematodes: Modes of parasitism. In *Genomics and Molecular Genetics of Plant–Nematode Interactions*; Jones, J.T., Gheysen, L., Fenoll, C., Eds.; Springer: Berlin/Heidelberg, Germany, 2011; pp. 3–20.
129. Kaloshian, I.; Desmond, O.J.; Atamian, H.S. Disease resistance-genes and defense responses during incompatible interactions. In *Genomics and Molecular Genetics of Plant–Nematode Interactions*; Jones, J.T., Gheysen, L., Fenoll, C., Eds.; Springer: Berlin/Heidelberg, Germany, 2011; pp. 309–324. [CrossRef]
130. Sun, B.; Zhang, X.; Song, L.; Zheng, L.; Wei, X.; Gu, X.; Cui, Y.; Hu, B.; Yoshiga, T.; Abd-Elgawad, M.M.; et al. Evaluation of indigenous entomopathogenic nematodes in Southwest China as potential biocontrol agents against *Spodoptera litura* (Lepidoptera: Noctuidae). *J. Nematol.* **2021**, *53*, 1–17. [CrossRef]

MDPI

Review

An Alternative Source of Biopesticides and Improvement in Their Formulation—Recent Advances

Dragana Šunjka [1,*] and Špela Mechora [2]

[1] Faculty of Agriculture, University of Novi Sad, Trg Dositeja Obradovića 8, 21000 Novi Sad, Serbia
[2] Agency for Radwaste Management, Litostrojska 58A, 1000 Ljubljana, Slovenia
* Correspondence: dragana.sunjka@polj.edu.rs

Abstract: Plant protection in contemporary agriculture requires intensive pesticide application. Their use has enabled the increase in yields, simplifying cultivation systems and crop protection strategies, through successful control of harmful organisms. However, it has led to the accumulation of pesticides in agricultural products and the environment, contaminating the ecosystem and causing adverse health effects. Therefore, finding new possibilities for plant protection and effective control of pests without consequences for humans and the environment is imperative for agricultural production. The most important alternatives to the use of chemical plant protection products are biopesticides. However, in order to increase their application and availability, it is necessary to improve efficacy and stability through new active substances and improved formulations. This paper represents an overview of the recent knowledge in the field of biopesticides and discusses the possibilities of the use of some new active substances and the improvement of formulations.

Keywords: biopesticides; organic waste; formulation; improvement

Citation: Šunjka, D.; Mechora, Š. An Alternative Source of Biopesticides and Improvement in Their Formulation—Recent Advances. *Plants* **2022**, *11*, 3172. https://doi.org/10.3390/plants11223172

Academic Editor: Per Kudsk

Received: 31 October 2022
Accepted: 16 November 2022
Published: 20 November 2022

1. Introduction

It is estimated that by 2050, agricultural production will have to increase by 70% in order to feed the growing population [1]. However, food production is a great challenge due to climate change, lack of fresh water, reduction of arable land, and particularly the presence of pests and diseases [2–8]. In order to achieve these goals, appropriate plant protection is required. At the same time, growing concerns about food safety, the trend of organic agricultural production, the presence of resistant pest populations, and the disruption of biodiversity as a result of intensive use of chemical pesticides are some of the main challenges of modern science.

A particularly important issue in contemporary agriculture, from the aspect of plant protection, is the control of invasive species. Due to global travel and trade, pests appear earlier in the season and away from their native environments where they have not been introduced naturally [2]. In newly introduced habitats they can cause great damage to indigenous plant species and the environment. As a result of invasive expansion in the past, insects such as grape phylloxera (*Daktulosphaira vitifoliae* (Fitch)) [9] and the Colorado potato beetle (*Leptinotarsa decemlineata* (Say)) [10] were introduced from the United States into Europe, where they originally never appeared before. More recently, newly introduced species are *Drosophila suzukii*, native to East Asia [11], and the brown marmorated stink bug *Halyomorpha halys* (Stål) [12] and *Lycorma delicatula* (White), originating from northern China, which pose a serious threat to vines, apple, plum, and pear, but also to ornamental and forestry production [13]. Some of the examples of pathogens with a strong negative impact are the fungi *Phytophthora infestans* ([Mont] de Bary) [14] and *Puccinia graminis* f. sp. tritici (Pgt) [15]. Although a native species to Europe, the small bacteria, Flavescence Dorée phytoplasma [16], has in recent decades come to be considered an invasive and quarantined pest in European viticulture. The introduction of the American grapevine leafhopper (*Scaphoideus titanus* (Ball)), a vector whose life cycle is closely linked to the vine, has allowed rapid spread, and a negative impact on viticulture had been reported. The reduction of crop yield is also affected by weeds. Although weeds are a habitat for some beneficial organisms, a food source for pollinators, and contribute to reducing soil erosion, they are defined as "unwanted plants", which makes weeds necessary to eliminate from agricultural ecosystems. One of the typical examples is the spread of allochthonous species *Ambrosia artemisiifolia* L., *Cuscuta campestris* Yunk., *Erigeron canadensis* L., and others [17]. In addition to insects, plant pathogens, and weeds, extensive damage to crops can also be caused by nematodes.

Besides the direct impact on yield, the side effects of pests manifest after harvesting as well as a result of the presence of mycotoxins in food and feed, which can seriously endanger human and animal health [18]. The presence of pathogens, weed seeds, eggs, and larvae of insects, nematodes, etc., endanger agricultural products as well. Although these are just some examples, they confirm the significance and magnitude of the resulting damage.

All the above-mentioned require intensive crop protection. However, modern agriculture almost completely relies on the use of chemical compounds, i.e., pesticides. Their use has enabled the increase in yields, simplifying cultivation systems and crop protection strategies through the successful control of harmful organisms. If pesticides would be totally forbidden, in only one year diseases, pests, and weeds would reduce world food production by 17–20% [19]. Nevertheless, their long-term use has led to a situation in which farmers have almost completely stopped using traditional methods of plant protection and replaced them with pesticides.

However, the intensive and/or inappropriate use of chemicals causes ecosystem contamination and adverse health effects; pesticide residues in food endanger human health [20,21], along with their accumulation in the environment, affect non-target organisms in a negative way [22], and endanger the future of plant production by the development of pest resistance [23]. Moreover, the impossibility of applying synthetic pesticides during the ripening and harvesting period (especially in greenhouses) complicates plant protection even more. These are only some of the side effects of pesticide application. Despite this, the exporting of synthetic pesticides banned in the European Union (EU) to developing countries is still a big issue that needs to be addressed, especially given the fact that these chemicals pose serious and long-term risks to human health and the environment [24]. Therefore, there is a need to find new possibilities for plant protection and effective control of pests without consequences for humans and the environment, as imperative for agricultural production. The most important alternative to the use of chemical plant protection products are biopesticides, naturally originated compounds with pesticide activity.

Biopesticides are also an important part of sustainable agriculture, which fulfills the United Nations 2030 agenda [25]. However, in order to increase their application and availability, it is necessary to improve efficacy and stability through new active substances and advanced formulations.

This paper represents an overview of the recent knowledge in the field of biopesticides with the main objective being to highlight the possibilities of the use of some new active substances and formulations.

2. Biopesticides—Advantages and Disadvantages

The term biopesticides refers to the application of microbiological, biochemical, and macrobiological plant protection products for the control of pests, weeds, and disease-causing agents (Table 1). Microbiological pesticides contain selected genres of specific species or mixtures of different bacteria, fungi, viruses, or protozoa. These beneficial organisms produce toxins, vitamins, enzymes, and plant hormones that can act antagonistically on disease-causing, harmful insects, nematodes, and weeds. Additionally, beneficial microorganisms produce vitamins, enzymes, and plant hormones that can have an effect on plants' immune systems by increasing their resistance. They are widely used and account for about 30% of the total production and sale of biopesticides. The specific mode of action of microorganism-based biopesticides relies on the competence for space and food, direct antagonism in relation to the growth of the target organism, and immunization of the host plant [26]. Microorganisms in plant protection products must have strong power to compete with the autochthonous microbial population and a high degree of ability to survive and adapt to the newly created conditions in which they should achieve their best efficiency.

Table 1. Biopesticides.

Microbiological Pesticides	Biochemical Pesticides	Macrobiological Pesticides
Bacteria	Plants	Insects
Fungi	Animals	Mites
Virus	Minerals	Nematodes
Protozoa	Insects	

Biochemical pesticides are substances of natural origin that control harmful organisms through non-toxic mechanisms. They are produced by plants, animals, minerals, insects, etc. The most important biochemical pesticides are botanical pesticides, such as plant extracts and essential oils, i.e., plant derivatives. Plant extracts are chemicals or mixtures of chemicals obtained from higher plants. They usually contain various types of metabolites, including alkaloids, phenols, terpenoids, and secondary substances developed by plants in order to help them in the protection from the harmful insect. These products are characterized by a variety of compositions and modes of action, which have an influence on insects by repelling or exhibiting insecticidal effects. Plant extracts offer an unlimited resource of biodegradable, economical, and renewable alternative pest control measures. Essential oils are natural substances, a complex mixture of lipophilic, liquid, fragrant, and volatile components located in the secretory structures of aromatic plants, which are often responsible for the characteristic smell or taste. They exhibit physiological functions with hormonal action, keep coenzymes in a reduced form, and represent a source of energy. They also have an ecological function that is reflected in the reduction of respiration, creating a specific microclimate that protects plants from excessive transpiration, reflection, and refraction of light and participates in the plant–plant, plant–animal, and plant–insect interactions. These compounds are responsible for attracting insects, which is important for pollination. They can inhibit the germination of seeds of other and their own species, protect the plant from insect and animal attacks, and protect the plant from infection by microorganisms. Essential oils accumulate in all types of vegetative organs such as flowers, leaves, bark, tree, roots, rhizomes, fruits, and seeds. So far, thousands of compounds from the terpene class in essential oils have been identified [27].

Even macrobiological pesticides must not be left out. Macrobiological pesticides include zoophagous species such as insects, mites, and nematodes, which, depending on their way of life, are divided into predators, parasites, and parasitoids, also called natural enemies. Although natural enemies are already present in the environment, often for their successful activity in agroecosystems, it is necessary to introduce or apply them regularly when the need arises.

Can biopesticides suppress the use of chemical pesticides as the dominant strategy for the control of harmful pests in the future? Unrealistic expectations, problems with quality control, short shelf life, lack of awareness of their importance, and relatively high costs compared to the pesticide used in conventional agriculture are the main reasons for their insufficient application.

Despite this, the market of biopesticides has grown significantly in recent years due to increased awareness of their potential and focused attention on environmental and health risks associated with the use of synthetic pesticides. Biopesticides are only 4–5% of the global pesticide market, but it is estimated that this percentage could increase up to 20% in the near future [28], while according to some authors, by 2050 the importance of biopesticides in agricultural production will be equal to chemical pesticides [29]. It is difficult to achieve a complete replacement of chemical pesticides with biopesticides, considering the large number of harmful pests, invasive species, pesticide resistance, as well as climate change. However, it is reasonable to expect that biopesticides along with conventional pesticides will be included in the Integrated Plant Protection System (IPM). One example of Good Agricultural Practices is the integration of microbiological biopesticides into IPM with the aim of increasing the efficacy of natural agents, known as bio-intensive plant protection (BIPM) [30]. It has been proven that their efficacy increases in combination with other methods of integrated plant protection. To protect plants with the lowest risk, the use of biopesticides as an alternative to chemical pesticides is required. Plants, microorganisms, insects, and certain minerals are well-known sources of biopesticides, and according to the EPA, nowadays biopesticides include plant-incorporated protectants (PIPs) as well [31].

The most used biopesticides belong to the neem derivatives (azadirachtin) and *Bacillus thuringiensis* (Berliner), a Gram-positive bacteria with the capacity to produce a toxin that is completely harmless to humans, plants, and other animals but has great potential to be fatal to a wide range of insect pests, and the latest data indicate that 75% of biopesticides are based on *B. thuringiensis* [32]. Neem is a widespread botanical biopesticide [33]; however, due to its instability in the environment, its effectiveness is temporary.

It is estimated that due to difficulties in isolating new organisms/compounds, the types of biopesticides will change. The intensive use of biopesticides in plant protection is limited by a number of risks—limited spectrum, short supply of products, interaction with non-target organisms, the virulence of strains, etc. Registration of bioinsecticides requires demanding protocols, causing a low rate of research in the field of biopesticides.

In the European Union, the development of non-chemical alternatives to synthetic pesticides has been recognized and defined by the law with the aim of achieving their sustainable use and reduction of side effects on human health and the environment (2009/128/EC) [34]. However, only a few countries recognize biopesticides as a separate group of pesticides, which has led to inappropriate methods for assessing their efficacy and safety [35].

The main steps in biopesticides development are the selection of biocontrol agents and formulation technology [36]. This requires a continuous and comprehensive study of the living organisms' responses (beneficial and side effects), both on various agrochemicals and biotechnology products, the discovery of new bioactive compounds, improvements of formulations, and mandatory risk assessment, not only for humans and animals but also for the environment.

3. Sources of Biopesticides

Recently, renewable resources of biopesticides have been recognized as the potential to overcome resource limitations and environmental pollution. Within a circular economy, organic solid wastes are considered to be a useful resource for obtaining value-added products [37–39].

3.1. Agricultural and Forest Waste as a Source of Biopesticide Compounds

The use of agricultural and forest waste as a potential bio-resource for the production of pesticide compounds has been enabled by thermochemical processing (Table 2). From the aspect of sustainable use of waste and biomass, pyrolysis technology is being actively studied. This technology is not novel, since the use of pyrolysis products dates way back to the Middle Paleolithic times [40]. Depending on the process, pyrolysis can be slow or fast, resulting in liquid (bio-oil), solid (bio-char), and gaseous (syngas) products. However, there is insufficient scientific evidence for supporting claims regarding the efficacy and toxicological assessment of the used products.

Bio-oil obtained by lignin pyrolysis at 450 and 550 °C shows insecticidal, fungicidal, and bactericidal activity [41]. Birchwood, pyrolyzed at 380 °C, in the form of birch tar oil mixed with Vaseline® has a strong repellent effect against mollusks [42], which makes it an effective, cheap, and simple method for controlling mollusks. Furthermore, the pesticide activity of liquids obtained from slow pyrolysis (pyroligneous acids) and hydrothermal carbonization (HTC) products has been stated [40,41,43,44]. However, there are numerous challenges in the development of new biocontrol technologies and barriers to their commercialization. The characterization of pyrolysis liquid showed that polycyclic aromatic hydrocarbons (PAHs) mainly contribute to pesticide activity [41,43]. They are major environmental pollutants with toxic, mutagenic, and carcinogenic effects on various organisms; thus, their release into the environment is highly restricted. Nevertheless, by reducing the temperature for the separation of liquids during pyrolysis below 300 °C, the appearance of PAHs and tar is reduced, which can be one of the possible directions for development and potential application [45,46]. Studying the pesticidal activity of various liquids obtained by slow pyrolysis of pine bark, pine forest waste, wheat straw, willow, and hydrothermally carbonized (HTC) willow, at temperatures of 280 °C and 260 °C, showed high efficiency of slow pyrolysis liquid from willow on *Arianta arbustorum* (L.) (repellent action), *Brassica rapa* (L.) (herbicidal action), and *Rhopalosiphum padi* (L.) (insecticidal action) [47]. A slightly lower efficacy was found in the liquid from wheat, bark, and forest residues. The pesticide activity of PAH-free liquids from slow pyrolysis originates from simple organic volatile compounds and numerous phenolic compounds. The main reasons for the higher pesticide activity of willow-derived pyrolysis liquid are high levels of acetic and other carboxylic acids, as well as the presence of dozens of different phenolic compounds. Thus, slow pyrolysis liquids represent a potential source of biopesticides, but there is a necessity for the optimization of technology production.

As already stated, an important source of bioactive compounds are plants' secondary metabolites. The biological activities of phenolic and polyphenolic compounds in olive mill wastewater have already been confirmed for the control of several insect pests, mostly due to the presence of polyphenol oleuropein [48–52]. These compounds have potent bioactive properties as insecticides and growth regulators and can be successfully used for the control of the Mediterranean fruit fly *Ceratitis capitata* (Wiedemann) [53], a highly invasive species.

A valuable agricultural by-product is grape pomace. In wine production, at least 20% of the fruit weight is discarded as this substance [54]. Grape pomace is used for the production of spirits and liquors, as fertilizer or animal feed, and even as composting material [55]. However, the presence of bioactive compounds, such as stilbenes resveratrol, in pomace extracts, indicates that grape pomace is a valuable source of biopesticides [56]. Due to its ability to inhibit the growth of bacteria, fungi, and viruses, resveratrol has been intensively studied [57,58]. Recently, resveratrol was proven to be effective in the control of *Botrytis cinerea* (Pers.), suppressing mycelial growth and conidia germination [59,60].

Table 2. Organic waste as a source of biopesticide compounds.

Source of Biopesticide	Authors
Bio-oil (fast pyrolysis of biomass)	[41]
Birch tar oil (fast pyrolysis of biomass)	[42]
Pyroligneous acids (slow pyrolysis of pine bark, pine forest waste, wheat straw)	[40,41,43,44]
Olive mill wastewater	[48–53]
Grape pomace	[54,56–60]

3.2. Solid Waste as a Medium for Microbial-Based Pesticides

Microbial-based biopesticides require a rich nutrient commercial medium for developing microbial species. In the process of production, this is a cost-consuming step. Production of microbial biopesticides using nonhazardous solid biodegradable waste, instead of commercial media, can contribute to solving the problem of increased waste production [61]. As an alternative to conventional media, kitchen waste could be used as a substrate for bio-pesticide production. Results showed that kitchen waste in a combination with wheat bran, soybean cake power, grain hulls, and mixed ions is suitable for the growth of *B. thuringiensis* [62,63], the most used biocontrol agent worldwide.

Biodegradable wastes from agriculture, industry, and households are being studied in order to obtain biopesticides through solid-state fermentation. At the laboratory level, it was confirmed that biowaste (source-selected organic fraction of municipal solid waste) is a substrate suitable for *B. thuringiensis* growth under non-sterile conditions [39,64]. Some authors provided results of using chicken feathers as a medium for the production of entomopathogenic bacteria (*Bacillus thuringiensis* serovar *israelensis* (Bti) and *Bacillus sphaericus* (Meyer and Neide) (Bs)) [65].

Furthermore, fungal-based biocontrol agents, such as the genus *Beauveria* and *Trichoderma*, are also highly promising alternatives to traditional chemical pesticides. Optimization of the fermentation process would allow the use of agroindustrial waste, rice husk, as a source of fungal production [66].

As a potential source of biopesticides and biofertilizers, some authors state the use of microalgae due to the production of biologically active compounds with antimicrobial activity [67,68]. An added value represents the possibility of using wastewater as a growing medium given that they require nitrogen, phosphorus, carbon, and ammonium [69].

The release of huge amounts of bioorganic waste from the food industry, poultries, and fisheries [65] is an opportunity for biopesticide production, either as a medium or as a source of bioactive compounds.

4. Advances in Formulations

Biopesticides are considered a powerful tool for developing a more rational strategy for the use of pesticides, which should lead to the improvement of the balance between efficacy, production costs, and application [70]. The history of biopesticides, classification, and mode of action are well known and extensively studied [71–74]. However, in order to ensure its successful application, the efficacy and achievement of higher formulation stability still have to be improved.

The formulation of biopesticides needs to ensure the stability of the organism, i.e., the compounds included in the preparation composition during production, distribution, storage, handling, and application of the preparation. Moreover, it is necessary to protect the biological agent/compound from the external environment influence, as well as enable an increase of the organism's activity during its reproduction, contact, or interaction with the target organisms. All of the above are achieved by adding proper non-pesticide compounds [31].

The formulation of such products poses a challenge, considering that such formulations must meet a number of objectives such as satisfactory efficiency, environmental acceptability, constancy after application, and uniform distribution throughout the treated object.

The fact that inert ingredients enhance biopesticide activity has opened new opportunities for further development in this field. A suitable formulation can improve the stability of the plant protection product and increase or expand the activity with a reduction of the influence of external factors [75].

For example, essential oils, regardless of origin, can be phytotoxic if applied to plants in high concentrations. Compared to conventional insecticides, insecticides based on essential oils are less effective and therefore must be applied at relatively high concentrations, often in the range of 0.5–1.5%. However, the procedure for bioinsecticide registration must include an assessment of the phytotoxic effects on the crops with special care. Little or no residues are expected in the agricultural product after the application of essential oil-based pesticides, and the potential impact on the organoleptic quality of the treated products has not been studied in detail to date [28].

A relatively new area that finds its use in agriculture and the food industry is nanotechnology [76]. This technology has enabled the development of controlled-release formulations, such as nanoemulsions, nanosuspensions, nanocapsule suspensions, etc. [77].

Besides the improvement of the bioactivity of synthetic pesticides, the use of nanotechnologies in plant protection enables the limitations of biopesticides to be overcome [78]. Due to their unique properties, nanomaterials are considered suitable carriers for stabilizing fertilizers and pesticides, as well as for facilitating the controlled transfer of nutrients and increasing plant protection. Thus, stability, persistence in the environment, and toxicity to target organisms would be improved, while the side effects and phytotoxicity would be reduced [79]. At the same time, nanotechnology-based pesticides, especially biosynthesized and bioinspired materials, make a huge contribution to sustainable development [80] due to their controlled release of the active ingredients [81], ensuring their efficacy in long-term use with the possibility of resolving the issue of accumulation of pesticide residues [82].

Despite the promising and well-known pesticidal activity of botanical pesticides, only a few essential oil-based biopesticides are available [28,82,83]. The main disadvantage of the commercialization of these biopesticides as plant protection products is their high volatility, which limits application with a relatively short activity in the field, requiring repeated applications [84]. The possible ways to overcome these limitations can be achieved with nanotechnologies; the essential oil needs to be enclosed within nanoparticles or nanoemulsions, which allow their stability and dispersibility in water [85]. Nanoencapsulation is based on encapsulating EOs in materials in the order of nanometers.

Along with botanical pesticides, microbial pesticides could express better pesticide activity for some pests in the field; however, they show activity against only one type of pest. This is one of the biggest disadvantages of microbial pesticides [86]. Environmental factors such as desiccation, heat, light, and UV reduce the activity of microbial pesticides, causing continuous crop destruction [87]. Among the newer technologies for the production of biopesticides based on microorganisms is bioencapsulation technology. Encapsulation of microorganisms in microcapsules has significant survival benefits, while also ensuring the controlled release of these bacteria throughout the growing season [75,88–90]. Encapsulation involves the active ingredient being enclosed within the polymer. The size of the capsule, which provides controlled release of the active ingredient after plant protection product application, varies from 2–50 µm, or 1–2 µm [91,92]. For encapsulation of microorganisms, various materials are used, including natural and synthetic polymers such as agar and agarose, starch, corn syrup, polyacrylamide, and polyurethane from artificial materials [93,94]. The capsule does not erode during the release process. The pores close again when the capsule is exposed to osmotic stress/dehydration. Capsules can be stored at room temperature, and storage time can be significantly extended by adding nutrients to the capsule.

However, sustainable nanotechnology requires a science-based environmental risk assessment [95]. The production of nanoparticles needs hazardous materials and advanced, modern equipment and exhibit side effects on the environment. Based on several toxicological studies, concerns have emerged about the safety of nanomaterials and the side effects on the environment [96]. Micro/nanoemulsions are stable systems with the possibility of spreading on the plant surface, which increases bioactivity, but at the same time, this increases phytotoxicity [97]. Therefore, over the last decade, research has shifted towards environmentally friendly and sustainable, more economical, "green" synthesis with the aim of supporting the growing use of nanoparticles in various industries. Green synthesis, as part of bioinspired protocols, provides reliable and sustainable methods for nanoparticle bio-synthesis using a wide range of microorganisms rather than current synthetic processes [98].

A good example is a formulation of nanoemulsions based on commercial essential oils (EO) using polyoxyethylene sorbitan monooleate and water [82]. By adding distilled water and agarose to the EO-based nano-formulations, followed by the addition of sodium polyacrylate, EO-based nanogels can be formulated. The results proved optimal physical characteristics of essential oil-based nanoemulsions and insecticidal activity against the test insect *Tribolium confusum* (Jacquelin du Val).

The application of biosynthesis as a tool for the design and development of many nanomaterials has become a sustainable and eco-friendly method [80]. Myco-nanotechnology, based on the development of nanoparticles employing fungi [94], and many other biological systems, such as bacteria, yeast, actinomycetes, and plant parts, have shown promising suitability for nanomaterial biosynthesis and represent a greener alternative to chemically synthesized nanoparticles [99].

One of the main limiting factors of biopesticide application is UV sensitivity. In the formulation of biopesticides based on baculovirus as an active ingredient, this limitation was overcome using titanium dioxide as a UV absorbent. Sensitive viral DNA was protected by the ENTOSTAT wax capsule, which dissolved in the alkaline intestines of insects to release the virus. Moreover, this prolongs the efficacy and stability of biopesticides without side effects on crops [75].

Improvements in the formulation of biopesticides based on *Bacillus thuringiensis* using starch-producing industry wastewater could be successfully carried out with the addition of a soybean medium [100].

5. Perspectives

In order to make the agricultural system safer and more sustainable, the European Commission announced plans to reduce the use and risk of chemical and more hazardous pesticides by 50% by 2030 [101]. In order to reduce the use of pesticides, in addition to numerous preventive measures, it is necessary to introduce biopesticides in plant protection. Currently, 60 biopesticide active substances are available on the EU market, while the number in the United States market is over 200 [102]. Such a small number of registered biopesticide plant protection products is a consequence of limited resources of bioactive agents, demanding registration procedures, etc.

In order to fulfill the goal, it is necessary to intensify the development of biopesticides and overcome their main shortcomings by choosing proper active agents and improving formulations. This primarily refers to increasing the specificity and longevity of the plant protection product, reducing the effective dose, and improving the speed of activity. A shorter shelf-life not only reduces the efficacy of biopesticides but also reduces their competitiveness with chemicals.

Although biopesticides include a wide range of living organisms, products of their metabolism, and compounds of plant origin, they have completely different characteristics and activities in specific ecosystems. This requires extensive scientific research in order to create the conditions for the transition to the phase of commercialization and wider use of such preparations [103]. At the same time, the main limitation of biopesticides commercialization is the strict regulations concerning their placement on the market. Long-term demanding procedures with insufficiently defined or maladapted principles are the main obstacles to the more significant development of the biological products industry for plant protection. In addition, not recognizing the difference between living organisms and bioactive compounds represents one of the biggest problems when registering biopesticides [104].

In view of the above stated reasons, the imperative is finding an optimal solution that would simplify the procedures and enable the development of new biopesticides, followed by the release of plant protection products on the market with a competitive price acceptable to agricultural producers [102], and thus an increase in their application [77].

Until the categories of "basic substances" and "low-risk substances" were introduced, biopesticides did not have a regulatory category [105]. Today, for basic substances not primarily intended for plant protection products, but which may provide crop protection possibilities, an application authorization is granted for the whole EU for an indefinite period. Specific to low-risk substances is that their application can be partially approved based on data from the literature and scientifically justified opinions. In this case, it is assessed whether microbiological and semiochemical products (e.g., pheromones) comply with low-risk criteria, in an assessment that only lasts for 120 days, while the approval may be valid for 15 years instead of 10 years. For many biopesticides, the components of the formulation are inert or not toxicologically significant, and the risk assessment can be based only on the active substance and on scientific evidence.

Limited sources of biopesticides are one of the most significant challenges. Finding new sources of bioactive compounds with an emphasis on renewable sources is imperative for modern plant protection. However, most of the research conducted in the field of biocontrol has provided only empirical results at the laboratory level. Many of these research studies will never be continued in real environmental conditions, primarily because the commercialization of biopesticide products requires a continuous and easily accessible source of bioactive components and a suitable formulation.

Furthermore, the formulation of biopesticides represents another crucial challenge. Production of biopesticides is more expensive, and their use is often more technically complex than the chemical ones. Thus, it is the main barrier to biopesticide placement on the market. However, research on their production, formulation, and application could greatly help in the commercialization of biopesticides. With the development and improvement of biopesticide formulations, the balance between efficiency, production costs, and application will be improved [70], which will lead to intensifying their application in the future.

6. Conclusions

The finding of alternative sources of compounds with pesticide activity is the most challenging issue in the field of plant protection. To overcome resource limitations and environmental pollution, renewable resources play an important role. Aside from plants as a source of bioactive compounds, the use of different types of waste as a medium for microbial growth is a significant source of biopesticides. This also could contribute to solving the problem of increased waste production. Although significant progress has been made in the development of the formulations and methods of application, further research is necessary regarding the use of biopesticides in plant protection. This is especially aimed at the need for a legal framework that would regulate nanomaterials placed on the food market. Future research will aim at the improvement of techniques and multidisciplinary research that will provide good, safe, effective, and inexpensive plant protection products.

Author Contributions: Writing—original draft preparation, D.Š.; Conceptualization, D.Š. and Š.M.; writing—review and editing, D.Š. and Š.M. All authors have read and agreed to the published version of the manuscript.

Funding: This research was funded by the Ministry of Education, Science, and Technological Development of the Republic of Serbia, Grant No. 451 03 68/2022-14/200117.

Data Availability Statement: Not applicable.

Conflicts of Interest: The authors declare no conflict of interest.

References

1. Diouf, J. FAO's Director-General on How to Feed the World in 2050. *Popul. Dev. Rev.* **2009**, *35*, 837–839.
2. IPPC Secretariat. *Scientific Review of the Impact of Climate Change on Plant Pests—A Global Challenge to Prevent and Mitigate Plant Pest Risks in Agriculture, Forestry and Ecosystems*; FAO on behalf of the IPPC Secretariat: Rome, Italy, 2021.
3. Mancosu, N.; Snyder, R.L.; Kyriakakis, G.; Spano, D. Water Scarcity and Future Challenges for Food Production. *Water* **2015**, *7*, 975–992. [CrossRef]
4. Jagadeesan, P. Factors affecting food security and contribution of modern technologies in food sustainability. *J. Sci. Food Agric.* **2011**, *91*, 2707–2714. [CrossRef]
5. Savary, S.; Willocquet, L.; Pethybridge, S.J.; Esker, P.; McRoberts, N.; Nelson, A. The global burden of pathogens and pests on major food crops. *Nat. Ecol. Evol.* **2019**, *3*, 430–439. [CrossRef]
6. Arora, N.K. Impact of climate change on agriculture production and its sustainable solutions. *Environ. Sustain.* **2019**, *2*, 95–96. [CrossRef]

7. Skendžić, S.; Zovko, M.; Živković, I.P.; Lešić, V.; Lemić, D. The impact of climate change on agricultural insect pests. *Insects* **2021**, *12*, 440. [CrossRef]
8. Adler, C.; Athanassiou, C.; Carvalho, M.O.; Emekci, M.; Gvozdenac, S.; Hamel, D.; Riudavets, J.; Stejskal, V.; Trdan, S.; Trematerra, P. Changes in the distribution and pest risk of stored product insects in Europe due to global warming: Need for pan-European pest monitoring and improved food-safety. *J. Stored Prod. Res.* **2022**, *97*, 101977. [CrossRef]
9. Eitle, M.W.; Carolan, J.C.; Griesser, M.; Forneck, A. The salivary gland proteome of root-galling grape phylloxera (*Daktulosphaira vitifoliae* Fitch) feeding on *Vitis* spp. *PLoS ONE* **2019**, *14*, e0225881. [CrossRef]
10. Wang, C.; Xu, H.; Pan, X. Management of Colorado potato beetle in invasive frontier areas. *J. Integr. Agric.* **2020**, *19*, 360–366. [CrossRef]
11. EPPO. Global Database. Last Updated: 22 July 2020. Available online: https://gd.eppo.int/taxon/DROSSU/distribution (accessed on 14 May 2022).
12. Ludwick, D.; Morrison, W.R.; Acebes-Doria, A.L.; Agnello, A.M.; Bergh, J.C.; Buffington, M.L.; Hamilton, G.C.; Harper, J.K.; Hoelmer, K.A.; Krawczyk, G.; et al. Invasion of the Brown Marmorated Stink Bug (Hemiptera: Pentatomidae) into the United States: Developing a national response to an invasive species crisis through collaborative research and outreach efforts. *J. Integr. Pest Manag.* **2020**, *11*, 4. [CrossRef]
13. Nakashita, A.; Wang, Y.; Lu, S.; Shimada, K.; Tsuchida, T. Ecology and genetic structure of the invasive spotted lanternfly *Lycorma delicatula* in Japan where its distribution is slowly expanding. *Sci. Rep.* **2022**, *12*, 1543. [CrossRef] [PubMed]
14. Grunwald, N.; Brown, C.; Ip, H.S.; Chang, J.H. Genetic processes facilitating pathogen emergence. In *Tactical Sciences for Biosecurity in Animal and Plant Systems*; IGI Global: Hershey, PA, USA, 2022; pp. 32–53.
15. Figueroa, M.; Upadhyaya, N.M.; Sperschneider, J.; Park, R.F.; Szabo, L.J.; Steffenson, B. Changing the game: Using integrative genomics to probe virulence mechanisms of the stem rust pathogen *Puccinia graminis f.* sp. tritici. *Front. Plant. Sci.* **2016**, *7*, 205. [CrossRef] [PubMed]
16. Bertaccini, A.; Duduk, B.; Paltrinieri, S.; Contaldo, N. Phytoplasmas and phytoplasma diseases: A severe threat to agriculture. *Am. J. Plant Sci.* **2014**, *5*, 1763–1788. [CrossRef]
17. Vrbničanin, S.; Božić, D. Biological invasions: The example of weed species. *Acta Herbol.* **2014**, *23*, 97–112. [CrossRef]
18. Magan, N.; Medina, A.; Aldred, D. Possible climate change effects on mycotoxin contamination of food crops pre- and post-harvest. *Plant. Pathol.* **2011**, *60*, 150–163. [CrossRef]
19. Popp, J.; Pető, K.; Nagy, J. Pesticide productivity and food security. A review. *Agron. Sustain. Dev.* **2013**, *33*, 243–255. [CrossRef]
20. Verger, P.; Agamy, N.; Anshasi, M.; Al-Yousfi, A.B. Occurrence of pesticide residues in fruits and vegetables for the Eastern Mediterranean Region and potential impact on public health. *Food Control* **2021**, *119*, 107457.
21. Calderon, R.; García-Hernández, J.; Palma, P.; Leyva-Morales, J.B.; Zambrano-Soria, M.; Bastidas-Bastidas, P.J.; Godoy, M. Assessment of pesticide residues in vegetables commonly consumed in Chile and Mexico: Potential impacts for public health. *J. Food Compos. Anal.* **2022**, *108*, 104420. [CrossRef]
22. Iyaniwura, T.T. Non-target and environmental hazards of pesticides. *Rev. Environ. Health* **1991**, *9*, 161–176. [CrossRef]
23. Barzman, M.; Bàrberi, P.; Birch, A.N.E. Eight principles of integrated pest management. *Agron. Sustain. Dev.* **2015**, *35*, 1199–1215. [CrossRef]
24. Sarkar, S.; Dias Bernardes Gil, J.; Keeley, J.; Möhring, N.; Jansen, K. *The Use of Pesticides in Developing Countries and Their Impact on Health and the Right to Food*; Directorate-General for External Policies Policy Department: Strasbourg, France, 2021.
25. Fenibo, E.O.; Ijoma, G.N.; Matambo, T. Biopesticides in sustainable agriculture: A Critical Sustainable development driver governed by green chemistry principles. Front. Sustain. *Food. Syst.* **2021**, *5*, 619058. [CrossRef]
26. Grahovac, M. Biological Control of Colletotrichum spp. Parasites of Stored Apple fruits. Ph.D. Dissertation, Faculty of Agriculture, University of Novi Sad, Novi Sad, Serbia, 2014.
27. Barka, E.A.; Lahlali, R.; Mauch-Mani, B. Editorial: Elicitors, secret agents at the service of the plant kingdom. *Front. Plant Sci.* **2022**, *13*, 1060483. [CrossRef]
28. Isman, M.B. Botanical insecticides in the twenty-first century-fulfilling their promise? *Annu. Rev. Entomol.* **2020**, *65*, 233–249. [CrossRef] [PubMed]
29. Olson, S. An analysis of the biopesticide market now and where it is going. *Outlooks. Pest. Manag.* **2015**, *26*, 203–206. [CrossRef]
30. Reddy, P. *Biointensive Integrated Pest Management in Horticultural Ecosystems*; Springer: Berlin/Heidelberg, Germany, 2014.
31. Šunjka, D.; Vuković, S. Biopesticides in plant protection chapter. In *Agricultural Studies on Different Subjects*; ÇIĞ, A., Ed.; IKSAD Publishing House: Ankara, Turkey, 2021; pp. 381–412.
32. Samada, L.H.; Tambunan, U.S.F. Biopesticides as Promising Alternatives to Chemical Pesticides: A Review of Their Current and Future Status. *J. Biol. Sci.* **2020**, *20*, 66–76. [CrossRef]
33. Chaudhary, S.; Kanwar, R.K.; Sehgal, A.; Cahill, D.M.; Barrow, C.J.; Sehgal, R.; Kanwar, J.R. Progress on Azadirachta indica Based Biopesticides in Replacing Synthetic Toxic Pesticides. *Front. Plant. Sci.* **2017**, *8*, 610. [CrossRef]
34. Official Journal of the European Union, L 309/71, DIRECTIVE 2009/128/EC OF THE EUROPEAN PARLIAMENT AND OF THE COUNCIL of 21 October 2009, 24.11.2009, European Parliament, Council of the European Union, Bruxelles/Brussel, Belgium. (accessed on 14 May 2022).
35. Sundh, I.; Goettel, M.S. Regulating biocontrol agents: A historical perspective and a critical examination comparing microbial and microbial agents. *BioControl* **2013**, *58*, 575–593. [CrossRef]

36. Mishra, J.; Dutta, V.; Arora, N.K. Biopesticides in India: Technology and sustainability linkages. *3 Biotech* **2020**, *10*, 210. [CrossRef]
37. Galanakis, C.M. Recovery of high added-value components from food wastes: Conventional, emerging technologies and commercialized applications. *Trends Food Sci. Technol.* **2012**, *26*, 68–87. [CrossRef]
38. Galanakis, C.M.; Markouli, E.; Gekas, V. Recovery and fractionation of different phenolic classes from winery sludge using ultrafiltration. *Sep. Purif. Technol.* **2013**, *107*, 245–251. [CrossRef]
39. Ballardo, C.; Barrena, R.; Artola, A.; Sánchez, A. A novel strategy for producing compost with enhanced biopesticide properties through solid-state fermentation of biowaste and inoculation with Bacillus thuringiensis. *Waste Manag.* **2017**, *70*, 53–58. [CrossRef] [PubMed]
40. Tiilikkala, K.; Fagernäs, L.; Tiilikkala, J. History and use of wood pyrolysis liquids as biocide and plant protection product. *Open Agric. J.* **2010**, *4*, 111–118. [CrossRef]
41. Hossain, M.M.; Scott, M.I.; McGarvey, B.D.; Conn, K.; Ferrante, L.; Berruti, F.; Briens, C. Insecticidal and anti-microbial activity of bio-oil derived from fast pyrolysis of lignin, cellulose, and hemicellulose. *J. Pest. Sci.* **2015**, *88*, 171–179. [CrossRef]
42. Lindqvist, I.; Lindqvist, B.; Tiilikkala, K. Birch tar oil is an effective mollusc repellent: Field and laboratory experiments using *Arianta arbustorum* (Gastropoda: Helicidae) and *Arion lusitanicus* (Gastropoda: Arionidae). *Agric. Food Sci.* **2010**, *19*, 1–12. [CrossRef]
43. Hagner, M.; Kuoppala, E.; Fagernäs, L.; Tiilikkala, K.; Setälä, H. Using the copse snail *Arianta arbustorum* (Linnaeus) to detect repellent compounds and the quality of wood vinegar. *Int. J. Environ. Res.* **2015**, *9*, 53–60.
44. Yatagai, E.; Nishimoto, M.; Hori, K.; Ohira, T.; Shibata, A. Termiticidal activity of wood vinegar, its components and their homologues. *J. Wood. Sci.* **2002**, *48*, 338–342. [CrossRef]
45. Fagernäs, L.; Kuoppala, E.; Simell, P. Polycyclic aromatic hydrocarbons in birch wood slow pyrolysis products. *Energy Fuels* **2012**, *26*, 6960–6970. [CrossRef]
46. Fagernäs, L.; Kuoppala, E.; Arpiainen, V. Composition, utilization and economic assessment of torrefaction condensates. *Energy Fuels* **2015**, *29*, 3134–3142. [CrossRef]
47. Hagner, M.; Tiilikkala, K.; Lindqvist, I.; Klaus, N.; Hanne, W.; Anssi, K.; Kimmo, R. Performance of liquids from slow pyrolysis and hydrothermal carbonization in plant protection. *Waste Biomass Valor.* **2020**, *11*, 1005–1016. [CrossRef]
48. Lo Scalzo, R.; Scarpati, M.L.; Verzegnassi, B.; Vita, G. Olea europaea chemical repellent to Dacus oleae females. *J. Chem. Ecol.* **1994**, *2*, 1813–1923. [CrossRef]
49. Barbouche, N.; Dhouib, S.; Ammar, M.; Ben Hamouda, M.H. Action de trois substrats alimentaires: Bersim, faux poivrier et olivier Sur le développement de la cuticule chez le criquet pèlerin Schistocerca gregaria. *Ann. L'institut Natl. Rech. Agron. Tunis* **1996**, *69*, 131–146.
50. Hemming, J.D.C.; Lindroth, R.L. Effects of phenolic glycosides and protein on gypsy moth (Lepidoptera: Lymantriidae) and Forest tent *caterpillar* (Lepidoptera: Lasiocampidae) performance and detoxication activities. *Environ. Entomol.* **2000**, *29*, 1108–1115. [CrossRef]
51. Ouguas, Y.; Hilal, A.; Elhadram, I. Effet biocide des extraits phénoliques oléicoles Sur les adultes du psylle de l'olivier Euphyllura olivina costa (Homoptera: Psyllidae) Sur deux variétés d'oliviers. *Menara Arbequine Maroc Rev. Ezzaitouna* **2010**, *11*, 1–15.
52. Golawska, S.; Kapusta, I.; Lukasik, I.; Wojcicka, A. Effect of phenolics on the pea aphid, *Acyrthosiphon pisum* (Harris) population on *Pisum sativum* L. (Fabaceae). *Pestycydy/Pesticides* **2008**, *3–4*, 71–77.
53. Di Ilio, V.; Cristofaro, M. Polyphenolic extracts from the olive mill wastewater as a source of biopesticides and their effects on the life cycle of the Mediterranean fruit fly *Ceratitis capitata* (Diptera, Tephriditae). *Int. J. Trop. Insect. Sci.* **2021**, *41*, 359–366. [CrossRef]
54. Beres, C.; Costa, G.N.S.; Cabezudo, I.; da Silva-James, N.K.; Teles, A.S.C.; Cruz, A.P.G.; Mellinger-Silva, C.; Tonon, R.V. Towards integral utilization of grape pomace from winemaking process: A review. *Waste Manag.* **2017**, *68*, 581–594. [CrossRef]
55. Muhlack, R.A.; Potumarthi, R.; Jeffery, D.W. Sustainable wineries through waste valorisation: A review of grape marc utilisation for value-added products. *Waste Manag.* **2018**, *72*, 99–118. [CrossRef]
56. Ferri, M.; Vannini, M.; Ehrnell, M.; Eliasson, L.; Xanthakis, E.; Monari, S.; Sisti, L.; Marchese, P.; Celli, A.; Tassoni, A. From winery waste to bioactive compounds and new polymeric biocomposites: A contribution to the circular economy concept. *J. Adv. Res.* **2020**, *24*, 1–11. [CrossRef]
57. Campagna, M.; Rivas, C. Antiviral activity of resveratrol. *Biochem. Soc. Trans.* **2010**, *38*, 50–53. [CrossRef]
58. Abba, Y.; Hassim, H.; Hamzah, H.; Noordin, M.M. Antiviral activity of resveratrol against human and animal viruses. *Adv. Virol.* **2015**, *2015*, 184241. [CrossRef]
59. Adrian, M.; Jeandet, P. Effects of resveratrol on the ultrastructure of Botrytis cinerea conidia and biological significance in plant/pathogen interactions. *Fitoterapia* **2012**, *83*, 1345–1350. [CrossRef] [PubMed]
60. Xu, D.; Yu, G.; Xi, P.; Kong, X.; Wang, Q.; Gao, L.; Jiang, Z. Synergistic effects of resveratrol and pyrimethanil against *Botrytis cinerea* on grape. *Molecules* **2018**, *23*, 1455. [CrossRef] [PubMed]
61. Balasubramanian, S.; Tyagi, R.D. 3–Biopesticide Production From Solid Wastes. In *Tyagi, Ashok Pandey, Current Developments in Biotechnology and Bioengineering*; Jonathan, W.-C., Wong, R.D., Eds.; Elsevier: Amsterdam, The Netherlands, 2017; pp. 43–58.
62. Zhang, W.; Qiu, L.; Gong, A.; Cao, Y.; Wang, B. Solid-state fermentation of kitchen waste for production of *Bacillus thuringiensis*-based bio-pesticide. *BioResources* **2013**, *8*, 1124–1135. [CrossRef]
63. Zhang, W.; Zou, H.; Jiang, L.; Yao, J.; Liang, J.; Wang, Q. Semi-solid state fermentation of food waste for production of Bacillus thuringiensis biopesticide. *Biotechnol. Bioproc. E* **2015**, *20*, 1123–1132. [CrossRef]

64. Molina-Peñate, E.; Sánchez, A.; Artola, A. Enzymatic hydrolysis of the organic fraction of municipal solid waste: Optimization and valorization of the solid fraction for Bacillus thuringiensis biopesticide production through solid-state fermentation. *Waste Manag.* **2022**, *137*, 304–311. [CrossRef]
65. Poopathi, S.; Murugan, K.; Selvakumari, J.; Mani, C.; Bala, P. Production of Microbial Bio-Pesticides from Waste Disposal of Chicken Feathers. *Ferment. Technol.* **2016**, *5*, e123.
66. Sala, A.; Artola, A.; Sánchez, A.; Barrena, R. Rice husk as a source for fungal biopesticide production by solid-state fermentation using B. bassiana and T. harzianum. *Bioresour. Technol.* **2020**, *296*, 122322. [CrossRef]
67. Gomiero, T. Food quality assessment in organic vs. conventional agricultural produce: Findings and issues. *Appl. Soil Ecol.* **2018**, *123*, 714–728. [CrossRef]
68. Ranglová, K.; Lakatos, G.E.; Câmara Manoel, J.A.; Grivalský, T.; Suárez Estrella, F.; Acién Fernández, F.G.; Molnár, Z.; Ördög, V.; Masojídek, J. Growth, biostimulant and biopesticide activity of the MACC-1 Chlorella strain cultivated outdoors in inorganic medium and wastewater. *Algal Res.* **2021**, *53*, 102136. [CrossRef]
69. Kumar, P.; Pandhi, S.; Mahato, D.K.; Kamle, M.; Mishra, A. Bacillus-based nano-bioformulations for phytopathogens and insect–pest management. *Egypt. J. Biol. Pest Control* **2021**, *31*, 128. [CrossRef]
70. Khater, H.F. Prospects of Botanical Biopesticides in Insect Pest Management. *Pharmacologia* **2012**, *12*, 641–656.
71. Lengai, G.; Muthomi, J. Biopesticides and their role in sustainable agricultural production. *J. Biosci. Med.* **2018**, *6*, 7–41. [CrossRef]
72. Gupta, S.; Dikshit, A.K. Biopesticides: An ecofriendly approach for pest control. *J. Biopestic.* **2010**, *3*, 186.
73. Copping, L.G.; Menn, J.J. Biopesticides: A review of their action, applications and efficacy. *Pest Manag. Sci. Former. Pestic. Sci.* **2000**, *56*, 651–676. [CrossRef]
74. Pavela, R.; Benelli, G. Essential oils as ecofriendly biopesticides? Challenges and constraints. *Trends Plant Sci.* **2016**, *21*, 1000–1007. [CrossRef] [PubMed]
75. Wilson, K.; Grzywacz, D.; Curcic, I.; Scoates, F.; Harper, K.; Rice, A.; Paul, N.; Dillon, A. A novel formulation technology for baculoviruses protects biopesticide from degradation by ultraviolet radiation. *Sci. Rep.* **2020**, *10*, 13301. [CrossRef] [PubMed]
76. Lowry, G.V.; Avellan, A.; Gilbertson, L.M. Opportunities and challenges for nanotechnology in the agri-tech revolution. *Nat. Nanotechnol.* **2019**, *14*, 517–522. [CrossRef]
77. Glare, T.; Caradus, J.; Gelernter, W.; Jakson, T.; Keyhani, N.; Kohl, J.; Stewart, A. Have biopesticides come to age. *Trend Biotechnol.* **2012**, *30*, 250–258. [CrossRef]
78. de Oliveira, J.L. 1-Nano-biopesticides: Present concepts and future perspectives in integrated pest management. In *Advances in Nano-Fertilizers and Nano-Pesticides in Agriculture*; Woodhead Publishing Series in Food Science, Technology and Nutrition: Sawston, UK, 2021; pp. 1–27.
79. Ali, S.S.; Al-Tohamy, R.; Koutra, E.; Moawad, M.S.; Kornaros, M.; Mustafa, A.M.; Mahmoud, Y.A.; Badr, A.; Osman, M.E.H.; Elsamahy, T.; et al. Nanobiotechnological advancements in agriculture and food industry: Applications, nanotoxicity, and future perspectives. *Sci. Total Environ.* **2021**, *792*, 148359. [CrossRef]
80. He, X.; Deng, H.; Hwang, H.M. The current application of nanotechnology in food and agriculture. *J. Food. Drug. Anal.* **2019**, *27*, 1–21. [CrossRef]
81. Manchikanti, P. Bioavailability and environmental safety of nanobiopesticides. In *Nano-Biopesticides Today and Future Perspectives*; Elsevier: Amsterdam, The Netherlands, 2019; pp. 207–222.
82. Palermo, D.; Giunti, G.; Laudani, F.; Palmeri, V.; Campolo, O. Essential oil-based nano-biopesticides: Formulation and bioactivity against the confused flour beetle Tribolium confusum. *Sustainability* **2021**, *13*, 9746. [CrossRef]
83. Jadhav, K.; Nagarkar, J. Development of karanj oil and castor oil nanoemulsions and its encapsulation into beads for controlled release larvicidal activity. *J. Adv. Sci. Res.* **2022**, *13*, 93–100. [CrossRef]
84. Pascual-Villalobos, M.J.; Guirao, P.; Díaz-Baños, F.G.; Cantó-Tejero, M.; Villora, G. Oil in water nanoemulsion 9 formulations of botanical active substances. In *Nano-Biopesticides Today and Future Perspectives*; Koul, O., Ed.; Academic Press: London, UK, 2019; pp. 223–247.
85. Werdin González, J.O.; Gutiérrez, M.M.; Ferrero, A.A.; Fernández Band, B. Essential oils nanoformulations for stored-product pest control-Characterization and biological properties. *Chemosphere* **2014**, *100*, 130–138. [CrossRef]
86. Hazafa, A.; Murad, M.; Masood, M.U.; Bilal, S.; Khan, M.N.; Farooq, Q.; Iqbal, M.O.; Shakir, M.; Naeem, M. *Nano-Biopesticides as an Emerging Technology for Pest Management*; Insecticides, Ed.; IntechOpen: London, UK, 2021.
87. Essiedu, J.A.; Adepoju, F.O.; Ivantsova, M.N. (Eds.) Benefits and limitations in using biopesticides: A review. In *AIP Conference Proceedings*; AIP Publishing LLC: College Park, MD, USA, 2020.
88. Cassidy, M.B.; Leung, K.T.; Lee, H.; Trevors, J.T. Survival of lac-lux marked Pseudomonas aeruginosa UG2Lr cells encapsulated in k-carrageenan and alginate. *J. Microbiol Methods* **1995**, *23*, 281–290. [CrossRef]
89. Bashan, Y.; Hernandez, J.P.; Leyva, L.A.; Bacilio, M. Alginate microbeads as inoculant carriers for plant growth-promoting bacteria. *Biol Fertil. Soils* **2002**, *35*, 359–368. [CrossRef]
90. Vemmer, M.; Patel, A.V. Review of encapsulation methods suitable for microbial biological control agents. *Biol. Control* **2013**, *67*, 380–389. [CrossRef]
91. Bashir, O.; Claverie, J.P.; Lemoyne, P.; Vincent, C. Controlled-release of Bacillus thurigiensis formulations encapsulated in light-resistant colloidosomal microcapsules for the management of lepidopteran pests of Brassica crops. *Peer. J.* **2016**, *4*, e2524. [CrossRef]

92. do Nascimento Junior, D.R.; Tabernero, A.; Cabral Albuquerque, E.C.d.M.; Vieira de Melo, S.A.B. biopesticide encapsulation using supercritical CO2: A comprehensive review and potential applications. *Molecules* **2021**, *26*, 4003. [CrossRef]
93. Vassilev, N.; Vassileva, M.; Martos, V.; Garcia del Moral, L.F.; Kowalska, J.; Tylkowski, B.; Malusá, E. Formulation of microbial inoculants by encapsulation in natural polysaccharides: Focus on beneficial properties of carrier additives and derivatives. *Front. Plant Sci.* **2020**, *11*, 270. [CrossRef]
94. Saberi Riseh, R.; Skorik, Y.A.; Thakur, V.K.; Moradi Pour, M.; Tamanadar, E.; Noghabi, S.S. Encapsulation of plant biocontrol bacteria with alginate as a main polymer material. *Int. J. Mol. Sci.* **2021**, *22*, 11165. [CrossRef]
95. Svendsen, C.; Walker, L.A.; Matzke, M.; Lahive, E.; Harrison, S.; Crossley, A.; Park, B.; Lofts, S.; Lynch, I.; Vázquez-Campos, S.; et al. Key principles and operational practices for improved nanotechnology environmental exposure assessment. *Nat. Nanotechnol.* **2020**, *15*, 731–742. [CrossRef] [PubMed]
96. Sekhon, B.S. Nanotechnology in agri-food production: An overview. *Nanotechnol. Sci. Appl.* **2014**, *20*, 31–53. [CrossRef] [PubMed]
97. López, M.D.; Cantó-Tejero, M.; Pascual-Villalobos, M.J. New Insights Into Biopesticides: Solid and liquid formulations of essential oils and derivatives. *Front. Agron.* **2021**, *3*, 763530. [CrossRef]
98. Bahrulolum, H.; Nooraei, S.; Javanshir, N.; Tarrahimofrad, H.; Mirbagheri, V.S.; Easton, A.J.; Ahmadian, G. Green synthesis of metal nanoparticles using microorganisms and their application in the agrifood sector. *J. Nanobiotechnol.* **2021**, *19*, 86. [CrossRef] [PubMed]
99. Patil, A.G.; Kounaina, K.; Aishwarya, S.; Harshitha, N.; Satapathy, P.; Hudeda, S.P.; Reddy, K.R.; Alrafas, H.; Yadav, A.N.; Raghu, A.V.; et al. Myco-Nanotechnology for Sustainable Agriculture: Challenges and Opportunities. In *Recent Trends in Mycological Research*; Biology, F., Yadav, A.N., Eds.; Springer: Cham, Switzerland, 2021.
100. Ndao, A.; Kumar, L.R.; Tyagi, R.D.; Valéro, J. Biopesticide and formulation processes based on starch industrial wastewater fortified with soybean medium. *J. Environ. Sci. Health* **2020**, *55*, 115–126. [CrossRef]
101. Available online: https://food.ec.europa.eu/plants/pesticides/sustainable-use-pesticides/farm-fork-targets-progress_en (accessed on 11 October 2022).
102. United Nations Department of Economic and Social Affairs. *World Population Prospects: The 2015 Revision, Key Findings and Advance Tables*; Working Paper No ESA/P/WP; United Nations Department of Economic and Social Affairs: New York, NY, USA, 2015.
103. Damalas, C.A.; Koutroubas, S.D. Current Status and Recent Developments in Biopesticide Use. *Agriculture* **2018**, *8*, 13. [CrossRef]
104. Chandler, D.; Bailey, A.S.; Tatchell, G.M.; Davidson, G.; Greaves, J.; Grant, W.P. The development, regulation and use of biopesticides for integrated pest management. *Philos. Trans. R. Soc. Lond. B Biol. Sci.* **2011**, *366*, 1987–1998. [CrossRef]
105. COMMISSION REGULATION (EU) 2017/1432, Official Journal of the European Union L 205/59, 8.8.2017. European Parliament, Council of the European Union, Bruxelles/Brussel, Belgium. Available online: https://eur-lex.europa.eu/legal-content/EN/ALL/?uri=CELEX%3A32017R1432 (accessed on 21 September 2022).

MDPI

Review

Actinomycete Potential as Biocontrol Agent of Phytopathogenic Fungi: Mechanisms, Source, and Applications

Juan A. Torres-Rodriguez [1], Juan J. Reyes-Pérez [2,*], Evangelina E. Quiñones-Aguilar [3] and Luis G. Hernandez-Montiel [1,*]

[1] Nanotechnology and Microbial Biocontrol Group, Centro de Investigaciones Biológicas del Noroeste, Av. Politécnico Nacional 195, Col. Playa Palo de Santa Rita Sur, La Paz 23090, Mexico
[2] Facultad de Ciencias Pecuarias, Universidad Técnica Estatal de Quevedo, Av. Quito km 1.5 vía a Santo Domingo, Quevedo 120501, Ecuador
[3] Centro de Investigaciones y Asistencia en Tecnología y Diseño del Estado de Jalisco, Camino Arenero, El Bajío del Arenal, Guadalajara 45019, Mexico
* Correspondence: jreyes@uteq.edu.ec (J.J.R.-P.); lhernandez@cibnor.mx (L.G.H.-M.)

Abstract: Synthetic fungicides have been the main control of phytopathogenic fungi. However, they cause harm to humans, animals, and the environment, as well as generating resistance in phytopathogenic fungi. In the last few decades, the use of microorganisms as biocontrol agents of phytopathogenic fungi has been an alternative to synthetic fungicide application. Actinomycetes isolated from terrestrial, marine, wetland, saline, and endophyte environments have been used for phytopathogenic fungus biocontrol. At present, there is a need for searching new secondary compounds and metabolites of different isolation sources of actinomycetes; however, little information is available on those isolated from other environments as biocontrol agents in agriculture. Therefore, the objective of this review is to compare the antifungal activity and the main mechanisms of action in actinomycetes isolated from different environments and to describe recent achievements of their application in agriculture. Although actinomycetes have potential as biocontrol agents of phytopathogenic fungi, few studies of actinomycetes are available of those from marine, saline, and wetland environments, which have equal or greater potential as biocontrol agents than isolates of actinomycetes from terrestrial environments.

Keywords: antifungal activity; marine; saline; wetland; post-harvest

Citation: Torres-Rodriguez, J.A.; Reyes-Pérez, J.J.; Quiñones-Aguilar, E.E.; Hernandez-Montiel, L.G. Actinomycete Potential as Biocontrol Agent of Phytopathogenic Fungi: Mechanisms, Source, and Applications. *Plants* **2022**, *11*, 3201. https://doi.org/10.3390/plants11233201

Academic Editors: Špela Mechora and Dragana Šunjka

Received: 6 October 2022
Accepted: 17 November 2022
Published: 23 November 2022

1. Introduction

One of the key problems in agriculture is the damage caused by phytopathogenic fungi [1]. Conventional methods to control phytopathogenic fungi have been carried out by using synthetic fungicides; however, their application causes resistance in microorganisms and harm to human, animal, and environmental health [2]. With the objective of achieving food production efficiency from an ecological and economic points of view, the search for an alternative to decrease the use of synthetic fungicides in agriculture is a global priority [3].

In recent years, the use of actinomycetes as a biocontrol agent on phytopathogenic fungi has been an alternative to the application of synthetic fungicides [4]. Actinomycetes are Gram-positive bacteria found in different habitats, humidity, pH, and temperature [5]. Actinomycetes have been isolated from different environments, such as terrestrial, marine, hypersaline, wetlands, and plant endophytes, among others [6,7].

The main antagonistic mechanisms of actinomycetes to control phytopathogenic fungi are competence for space and nutrients [8], antibiotics [9], siderophores [10], lytic enzymes [11], volatile organic compounds (VOCs) [12], and host resistance induction [13]. Additionally, actinomycetes promote plant growth and development through the synthesis of phytohormones, atmospheric nitrogen fixation, and mineral solubilization, among others [14]. Several studies of actinomycetes have been reported; however, few studies have focused on actinomycetes, isolated from different environments, used as biocontrol agents due to their effect in agriculture and antagonistic mechanisms to phytopathogenic fungi.

2. General Characteristics of Actinomycetes

Actinomycetes form vegetative or aerial mycelia and are capable of reproducing by binary fission [15]. In vitro culture and the natural environment have a typical smell of humid soil because of the production of two geosmin and 2-methylisoborneol volatile organic compounds [16]. Spore production is a result of nutrient depletion, allowing actinomycetes to remain latent until they find favorable conditions for growth [14]. Additionally, filamentous and sporulating natures allow them to compete more efficiently against other organisms found in the rhizosphere [17]. Their cell wall is a rigid structure formed by complex compounds, such as peptidoglycan, teichoic and teichuronic acids, and polysaccharides. Actinomycetes also have a high guanine and cytosine content in DNA [18].

3. Actinomycetes as Biocontrol Agents

The importance of the use of actinomycetes as biocontrol agents is explained by inherent positive characteristics: (1) they are not harmful to human and animal health; (2) they are not toxic to plants; (3) they improve plant yield; and (4) they decrease the use of synthetic fungicides [19,20]. Among the different genera, *Streptomyces* has been investigated extensively because it is easy to isolate [21]. Actinomycetes have a slower growth than bacteria. Thus, growth improvement techniques should be applied to obtain desirable actinomycetes in culture media. These techniques are based on selective isolation media and the pretreatment of samples, such as: soil with calcium carbonate, both by drying and heating, wet, and chemical pretreatments, among others [22]. One of the ways to stimulate actinomycete populations in soil is by adding biostimulants and organic fertilizers, such as compost and vermicompost. *S. sampsonii* and *S. flavovariabilis* isolates from soil amended with vermicompost showed the highest antagonistic activity towards *Rhizoctonia solani, Alternaria tenuissima, Aspergillus niger*, and *Penicillium expansum* [23]. In addition, soil amended with *Brassica napus* and *Brassica rapa* leaf residues promoted the increase in actinomycete populations in the soil. The increase in the actinomycete population showed a strong correlation with the suppression of the *R. solani* wilt disease [24]. Different actinomycetes have been studied as biocontrol agents on phytopathogenic fungi and as mechanisms of action (Table 1).

Table 1. Antagonistic mechanisms of actinomycetes for the control of phytopathogenic fungi.

Actinomycete	Phytopathogen	Host	In Vivo Inhibition	Antagonistic Mechanisms	Reference
Streptomyces sp.	*Colletotrichum fragariae*	Strawberry	100%	Secondary metabolites	[19]
S. sampsonii	*Sclerotinia sclerotiorum*	Green bean	100%	Secondary metabolites	[11]
Streptomyces sp.	*Ralstonia solanacearum*	Tomato	97%	Induction of host resistance	[13]
S. sichuanensis	*Fusarium oxysporum*	Banana	51%	Siderophores	[25]
Amycolatopsis sp.	*F. graminearum*	Maize	79%	Lytic enzyme	[26]
Arthrobacter humicola	*A. alternata*	Tomato	31%	Secondary metabolites	[27]
Nocardiopsis dassonvillei	*Bipolaris sorokiniana*	Wheat	72%	Siderophores and lytic enzyme	[28]
S. rameus	*R. bataticola*	Bean	70%	Siderophores and lytic enzyme	[10]
S. globisporous	*R. solani*	Tomato	50%	Induction of host resistance	[29]

4. Main Actinomycete Antagonistic Mechanisms to Phytopathogenic Fungi

Biocontrol agents use a combination of several antagonistic mechanisms of action to control phytopathogenic fungi [30]. The main antagonistic mechanisms of actinomycetes are their competence for space and nutrients, antibiotics, siderophores, lytic enzymes, and induction of host resistance, among others [1,13,19] (Figure 1). Understanding the mechanisms of action of biocontrol agents is essential in order to improve their viability and increase their potential [31].

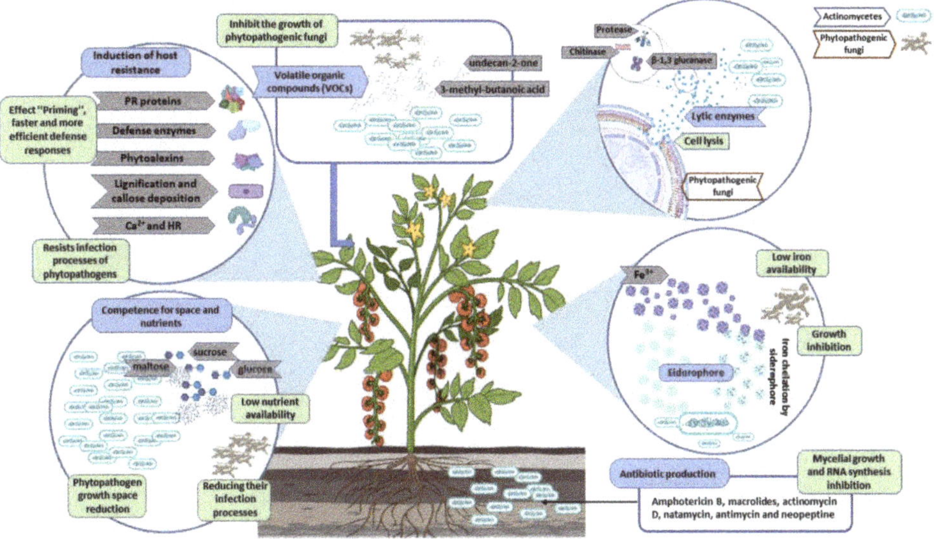

Figure 1. Main actinomycete antagonistic mechanisms to phytopathogenic fungi.

4.1. Competence for Space and Nutrients

Competition is an indirect mechanism of actinomycetes for the growth inhibition of phytopathogenic fungi [8]. Competence between two or more microorganisms begins for the same carbon source (carbohydrates such as sucrose, glucose, maltose, and fructose) or space for their growth [32,33]. The ecological plasticity and fast growth of antagonistic microorganisms allow them to assimilate the available nutrients in the host at a greater amount than phytopathogenic fungi; thus, the spore germination stage and infection processes to the host are reduced [34]. Competence is also an effective biocontrol mechanism when the antagonist is found in sufficient volumes and assimilates nutrients faster and in greater quantity than phytopathogenic fungi [30].

4.2. Antibiotic Production

Actinomycetes produce secondary metabolites with antifungal properties [1]. Approximately 80% of antibiotics, such as streptomycin, spectinomycin, neomycin, tetracycline, erythromycin, and nystatin, are produced by actinomycetes [35].

Furthermore, many metabolites have been discovered with antimicrobial properties similar to phytopathogenic fungi, such as amphotericin B, macrolides, actinomycin D, natamycin, antimycin, and neopeptine [36–38]. Macrolides are a group of antibiotics produced by actinomycetes that inhibit fungus protein synthesis [39]. Amphotericin B joins selectively to ergosterol in the fungal cell membrane, producing changes in permeability and inducing cell lysis [40]. Moreover, actinomycin D production by *Streptomyces* sp. strains limit microbial growth and RNA synthesis [9]. Antimycin inhibits the mitochondrial electron transport chain between cytochromes *b* and *c* [41]. Natamycin blocks fungal growth when it joins to the ergosterol of the fungus cell membrane [42]. Neopeptine is an inhibitor of the microbial cell wall biosynthesis at the enzymatic level [37]. Another important process that involves antibiotic production is symbiosis between actinomycetes and plants, as the antibiotic protects the plant from phytopathogenic fungi, and the plant exudates allow actinomycete development [43].

4.3. Siderophore Production

Siderophores are molecules that perform sequestration on low-molecular weight irons (500–1000 Da) and link with Fe^{3+} ions to be transported to the cell and secreted in response to low Fe^{3+} availability [44]. Siderophores are classified as: phenolate, catecholate, hydroxamate, and carboxylate; some have a group mix (mixed types) [45]. Siderophore production has been demonstrated by *Streptomyces* strains that produce hydroxamate-type siderophores known as deferoxamine [20]. Moreover, heterobactins are catecholate-hydroximate mixed-type siderophores that have been found in *Rhodococcus erythropolis* [46], and albisporachelin is a hydroxamate-type siderophore produced by *Amycolatopsis albispora* [47]. A sufficient amount of siderophore production by biocontrol agents limits Fe^{3+} availability for phytopathogenic fungi. Thus, growth and virulence are limited because microorganisms without iron in their environment cannot perform vital processes, such as synthesis and repair of nucleic acids, respiration, photosynthetic transport, and nitrate reduction or free radical detoxification [48].

4.4. Lytic Enzyme Production

Actinomycetes produce lytic enzymes, such as chitinase, β-1,3-glucanase, and protease that degrade the fungal cell wall [26] and cause loss of membrane integrity, set intracellular material free, and cell death [1]. The fungus cell wall is responsible for a cell's physical integrity, formed by chitin, β-1,3-glucan, and protein [49]. The β-1,3 glucanase hydrolyze β-D-glycosidic bonds of β-1,3 glucan, and chitinases hydrolyze chitin β-1,4 N-acetyl-β-D-glycosamide bonds, breaking fungal cell walls [31,50]. Proteases hydrolyze proteins, specifically mannoproteins, make up the phytopathogenic fungi cell wall [51].

4.5. Volatile Organic Compounds (VOCs)

Volatile organic compounds (VOCs) are low molecular weight compounds that evaporate easily at a normal temperature and pressure, which gives them the ability to diffuse through the atmosphere and soil [52]. Most VOCs are lipid-soluble and thus have low water solubility. These organic compounds travel great distances in structurally heterogeneous environments, as well as in solid, liquid, or gaseous compounds [31]. VOCs produced by actinomycetes inhibit the growth of phytopathogenic fungi, promote plant growth, possess nematocidal activity, and induce systemic resistance in plants [11,12]. They inhibit the mycelia, causing swelling, conidia collapse, and structural alterations in the fungal cell wall [53]. The *Streptomyces* species produces 2-ethyl-5-methylpyrazine and dimethyl disulfide that inhibit mycelial growth and spore germination [12]. VOCs such as S-methyl ethanethioate, 1,2-dimethyldisulfane, 2-methyl propanoic acid, acetic acid, 3-methyl-butanoic acid, undecan-2-one, nonan-2-one, and 2-isopropyl-5-methylcyclohexan-1-ol have been reported from the actinomycetes *Nocardiopsis* sp., which inhibit mycelial growth of fungi [54].

4.6. Induction of Host Resistance

Induced resistance in plants is activated by antagonist actinomycetes that cause a defense response in the host through several chemical or biochemical reactions [13]. Systemic acquired resistance (SAR) and induced systemic resistance (ISR) are two forms of induced resistance, characterized based on signaling pathways [30]. SAR stimulates a rapid response in the phytopathogens and actinomycetes, stimulating a special ISR state called "priming", for faster and stronger defense responses [55,56]. Actinomycetes are capable of inducing defense responses in plants through the overproduction of: (1) enzymes related to defense, which strengthen the cell wall structure, avoiding the entrance of phytopathogenic fungi, their colonization toward the plant, and catalyzing phenolic compound oxidation to quinones that are toxic for fungi [29]; (2) proteins (PR) related to pathogenesis, such as chitinase hydrolytic enzymes, and β-1,3-glucanase that break the phytopathogenic fungi cell wall structure [57]; (3) phytoalexins, which are toxic for phytopathogenic fungi, inhibit germ tube elongation and growth, decrease mycelial growth and limit glucose absorption [30,58]; (4) lignification promotion that contributes to plant cell wall hardening [59]; and (5) callus formation induction that isolates stress (biotic and abiotic) in the tissue, locally, by depositing a physical barrier [56,60].

5. Actinomycete Isolation from Different Environments

Actinomycetes have been isolated from different environments, such as terrestrial, marine, hypersaline, wetland, as well as plant endophytes, among others. Marine environments cover more than 67% of terrestrial surface, and only 1% of the microorganisms have been studied [61]. Marine actinomycetes living in extreme environmental conditions are ideal for the synthesis of new secondary metabolites because of their adaptation to reproduce, grow, and feed [5]. Endophyte actinomycetes are microorganisms that inhabit plant tissues during the totality or part of the life cycle and do not cause negative effects in the host [17]. Additionally, molecules produce functions as growth promoter metabolites, antimicrobials to phytopathogenic fungi, improve gene expression of plant defense that codify enzymes, such as phenylalanine ammonia-lyase (PAL), and improve nutrient absorption [62]. Wetlands are biologically important ecosystems that provide habitat, food, and spawning areas for a number of plants and animals [7]. Hypersaline environments are extreme habitats with high concentrations of salt, alkalinity, and low oxygen. Actinomycetes have been isolated from different hypersaline environments, such as salt lakes, salt flats, salt mines, and brine wells; however, these environments remain unexplored [6]. Compounds and secondary metabolites of terrestrial microorganisms have been studied extensively, hence the importance of searching for new isolation sources for actinomycetes [33].

6. Antifungal In Vitro Activity of Actinomycetes Isolated in Different Environments

The *Streptomyces*, *Micromonospora*, and *Nocardiopsis* species are within the main actinomycetes that have been studied for their antifungal in vitro activity [63,64] and isolated from different environments, such as terrestrial, marine, saline, and wetland (Figure 2).

Actinomycetes of terrestrial origin, such as *Streptomyces* sp., have demonstrated to reduce the mycelial growth of *R. bataticola* by 65.3% [10]. Similar results were obtained for *Streptomyces* sp. isolated from a terrestrial environment, reducing the mycelial growth of *Botrytis cinerea* by 77% [59]. The antifungal activity of *Streptomyces* sp. VOCs of terrestrial origin inhibited the mycelial growth of *F. solani* by 69% [53]. In addition, in another investigation, *Streptomyces* sp. VOCs reduced the mycelial growth of *C. acutatum* by 77% [64]. Endophytic actinomycetes from marine and wetland environments have also inhibited the growth of phytopathogenic fungi under in vitro conditions. A study of *S. polychromogenes* endophytes from date palm roots inhibited the mycelial growth of *F. solani;* the in vitro antifungal activity was associated with the production of lytic enzymes that degrade the cell wall [1]. A *Streptomyces* sp. Extract of marine origin containing oligomycin A inhibited the growth of *Pyricularia oryzae* hyphae by 83%, which damaged the fungal membrane, inhibited conidial germination and appressoria formation [65]. *Streptomyces* spp. From marine environments have also inhibited the growth of *Penicillum digitatum*, *A. niger* and *F. solani* by 92, 73, and 72%, respectively [66]. Actinomycetes from marine environments, such as *Streptomyces* sp. And *N. lucentensis*, inhibited the mycelial growth of *F. solani* by 72 and 68%, respectively, and *Streptomyces* sp. showed no significant differences with the synthetic fungicide [67].

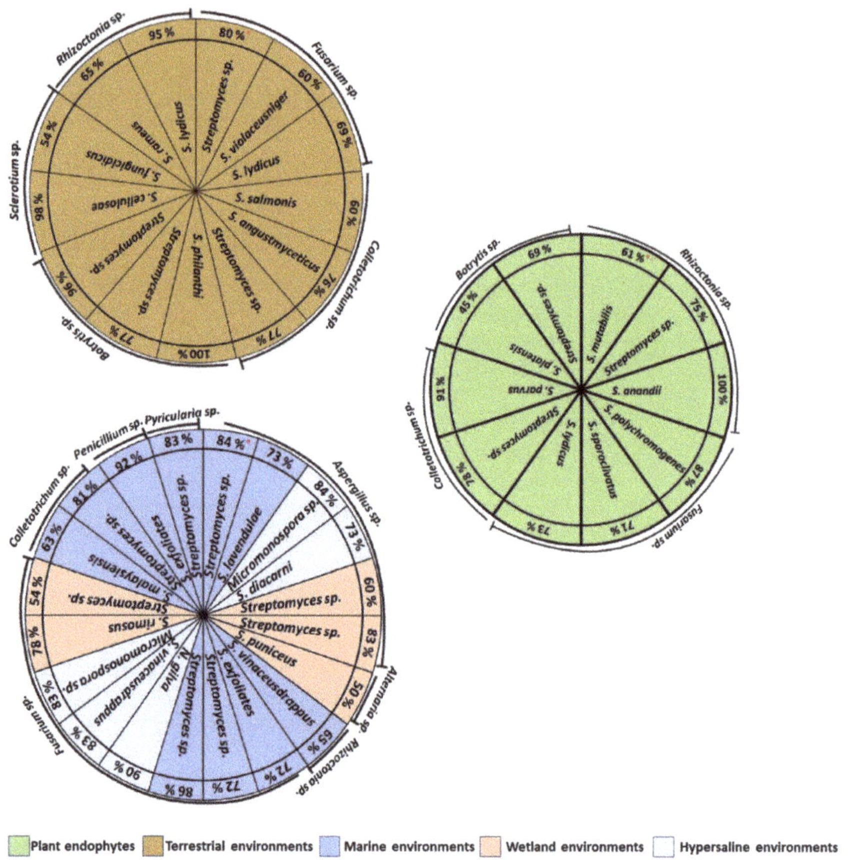

Figure 2. Antifungal activity (% growth inhibition *) in vitro of actinomycetes isolated from different environments [1,3,4,6,10,19–21,53,59,60,64–85].

7. Antifungal Activity of Actinomycetes In Vivo Isolated from Different Environments

The diseases transmitted by soil phytopathogenic fungi are difficult to control with synthetic fungicides [21]. Plant diseases cause a yield loss of 50%, particularly in developing countries [86]. The antifungal activity of actinomycetes isolated from different environmental conditions has been demonstrated in vitro conditions. However, research in actinomycetes as biocontrol agents in vivo conditions has been limited to the study of terrestrial actinomycete isolates (Figure 3) due to difficulties in sampling and culturing microorganisms of marine, saline, and wetland environments, among others [5]. Nevertheless, interest still exists in finding more efficient strains that differ considerably with respect to their biocontrol efficiency [19].

Streptomyces species from terrestrial environments have been shown to significantly reduce the incidence of *B. cinerea* disease on chickpea plants by 47%, compared to the control, and induce resistance in the host plant through antioxidant enzymes and phenolic compounds [59]. The antifungal activity of the *S. sichuanensis* strain from terrestrial environments towards *F. oxysporum* was associated with siderophore production and whose extracts induced apoptosis of phytopathogen cells. In the greenhouse experiment, the *S. sichuanensis* strain significantly inhibited *F. oxysporum* infection in roots and bulbs of banana seedlings and reduced the disease index by 51% [25].

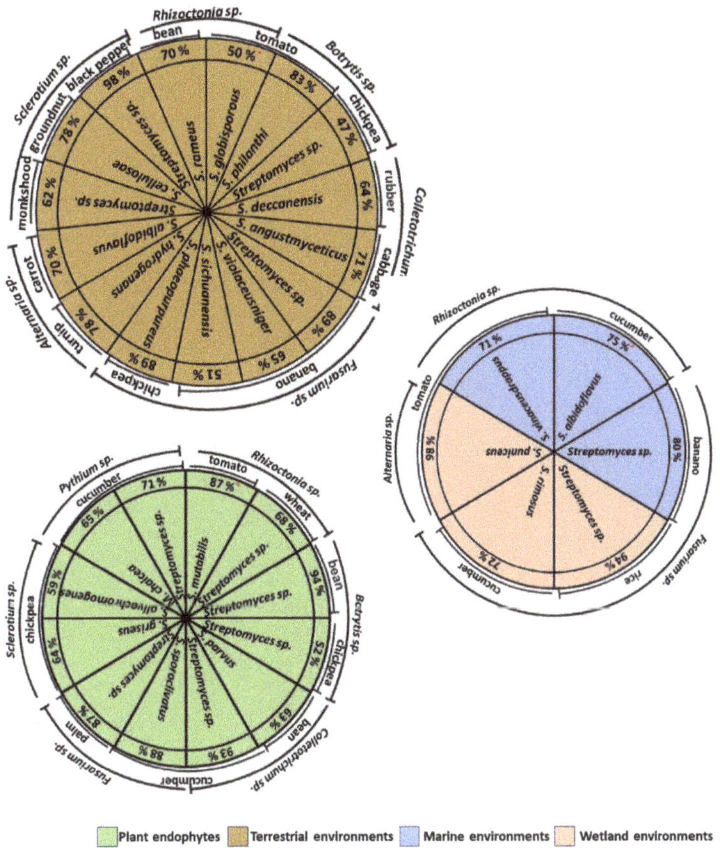

Figure 3. Antifungal activity (% growth inhibition *) of actinomycetes in vivo isolated from different environments [1,4,10,21,25,29,59,60,63,69–71,73,74,77,78,80,83,85,87–97].

Moreover, the actinomycete endophytes of date palms decreased the sudden decline syndrome (SDS) disease, caused by *F. solani*, by 86% under greenhouse conditions; these effects are related to the production of antifungal metabolites of the *S. coeruleoprunus* strain [1]. Studies of *Streptomyces* sp. extracts from marine environments have shown that the disease index of *F. oxyspoum* significantly decreased by 80%. This effect could have been associated with secondary metabolites causing the loss of osmotic balance, cell membrane rupture and leakage of cellular components of *F. oxyspoum* [85]. Similarly, the application of *S. vinaceusdrappus* from the marine environment on tomato plants showed a disease reduction (71%) of root rot caused by *R. solani* compared to the untreated control [73]. In detached tomato leaves, co-inoculation of *A. solani* with *S. puniceus* extract from wetland environments reduced the disease by 98%, relative to the control, due to the presence of antifungal metabolites, such as *Alteramide A* [77]. These investigations confirm the potential of actinomycetes isolated from different environments, not only terrestrial, in plant disease management.

8. Antifungal Activity of Actinomycetes Isolated in Different Environments in Postharvest Fruit

The main losses in post-harvest fruit are caused by phytopathogenic fungi, which represent more than 50% of agricultural production [98]. In post-harvest fruit management, antagonists are subjected to changes in pH, temperature, and humidity because in these conditions the efficiency of biocontrol agents can be affected [99]. Actinomycetes from marine, saline, hypersaline, and wetland environments are subjected to extreme environmental conditions that allow them to adapt to the changes in temperature, pH, and humidity that occur post-harvest [19,100]. However, in most of the studies of actinomycetes and biocontrol of phytopathogenic fungi in post-harvest fruit, the isolates provided are from terrestrial environments (Figure 4). More studies should be performed with these microorganisms isolated from different environments.

In the post-harvest trial on strawberries inoculated with *B. cinerea*, VOCs from *Streptomyces* sp. isolated from terrestrial environment inhibited the development of gray mold symptoms on fruit by more than 87% compared to untreated control strawberries. In addition, *B. cinerea* conidia showed symptoms of swelling and crumbling and the fungal mycelium showed structural alterations [53]. Moreover, incubation of apples infected with *C. acutatum* in semi-closed boxes with *Streptomyces* sp. strains showed that the VOCs produced by *Streptomyces sp.* reduced the rotting areas of the apples by 66% in relation to the control treatment [64].

Marine actinomycetes, such as *S. chumphonensis*, reduced citrus green mold disease caused by *P. digitatum* by 93%. The authors suggest that this effect may be related to the production of antimicrobial substances [101]. Furthermore, *Streptomyces* sp. species from marine environments and their metabolites showed high efficacy in the control of *C. fragariae* in strawberry fruit, reducing the severity of anthracnose disease by 76%, in addition, fruit hardness and color were maintained [80].

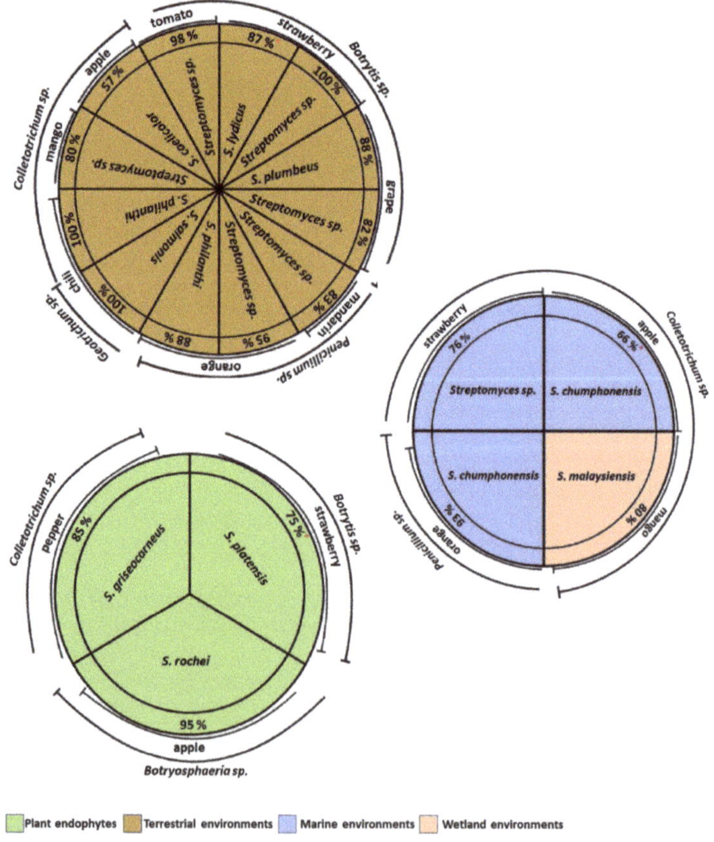

Figure 4. Antifungal activity (% growth inhibition *) of actinomycetes in post-harvest isolated from different environments [3,19,20,53,64,68,84,102–109].

9. Commercial Products Based on Actinomycetes

The main problem in obtaining commercial products based on microorganisms is that their biocontrol capacity is different in in vitro trials and field experiments. In addition, developing a commercial microorganism product is a complex, time-consuming, costly and interactive process. The success of a biocontrol agent is its formulation, which must include a specific concentration of the new microorganisms and a set of other inert ingredients to produce a commercial product for its use in field conditions, and must show repeated positive results, reasonable prices and easy handling [17]. The efficiency of these biocontrol agents is affected by environmental factors, such as temperature, humidity, precipitation, among other abiotic aspects which synthetic fungicides have overcome [33].

The factors outlined above all make the transfer of an effective biocontrol agent under controlled laboratory conditions to a commercially available product for application under field conditions difficult. Although the use of microorganisms as a biocontrol agent is a current option to reduce synthetic fungicides, the ratio of actinomycetes registered as biocontrol agents for commercial availability is still low [110]. From the commercial products based on actinomycetes, Mycostop is the only product registered in Canada, the European Union, and the United States of America (Table 2). Overall, an open field for the industry is envisaged for actinomycete-based products in agriculture.

Table 2. Commercial products based in actinomycetes.

Commercial Product	Actinomycete	Registered Countries	Phytopathogen Species/Target Disease	Main Effects	Reference
Mycostop	*S. griseovirids*	Canada, UE countries, and USA	*Alternaria*, *R. solani*, *Fusarium*, *Botrytis*, *Phytophthora*, and *Pythium*	Space and nutrient competence and produces polyenic antibiotics	[111]
Actinovate	*S. lydicus*	Canada and USA	*Pythium*, *Fusarium*, *Phytophthora*, *Rhizoctonia*, and *Verticillium*, powdery and downy mildew, and *Botrytis*, *Alternaria*, *Geotrichum*, and *Sclerotinia*	Induces resistance in plants and produces extracellular chitinases	[112]
Mycocide KIBC	*S. colombiensis*	South Korea	Powdery mildews, grey mold, and brown patch	Produces enzymes and antibiotics	[113]
Safegrow KIBC	*S. kasugaensis*	South Korea	Sheath blight and large patch	Produces enzymes and antibiotics	[113]
Kasugamycin, Kasumin	*S. kasugaensis*	Ukraine	Leaf spot, scab, and root rot	Inhibit protein biosynthesis	[114]
Agrimycin, Paushak, Cuprimicin 17, Astrepto 17	*S. griseus*	India, USA, New Zealand, China, Ukraine and Canada	Bacterial rots, *Xanthomona*, and *Pseudomonas*	Inhibit protein biosynthesis	[112]
Polyoxorim (Endorse, Polyoxin Z and Stopit)	*S. cacaoi* var. *asoensis*	UE countries	*Sphaerotheca*, powdery mildews, *Botrytis*, *Sclerotium*, *Corynespora*, *Cochliobolus*, *Alternaria*, sheath blight, and *Helminthosporium*	Inhibit cell wall biosynthesis and causes abnormal spore germ tube swelling and hypha points	[115]
Validacin, Valimun, Dantotsupadan-valida, Mycin Hustler, Valida	*S. hygroscopicus*	-	*Rhizoctonia*	Inhibit trehalase in *Rhizoctonia*	[116]

10. Conclusions

Actinomycetes are an option to control phytopathogenic fungi in agriculture and their application reduces the use of synthetic fungicides. Marine, saline, and wetland environments are important sources for actinomycete isolation and in the discovery of new compounds and secondary metabolites. Biocontrol studies have focused on isolates of actinomycetes from terrestrial environments. Nevertheless, actinomycetes from marine, saline, and wetland environments have equal or greater antifungal activity than those from terrestrial environments.

Author Contributions: Conceptualization, J.A.T.-R., L.G.H.-M. and J.J.R.-P.; methodology, J.A.T.-R., L.G.H.-M. and J.J.R.-P.; validation, E.E.Q.-A.; writing, J.A.T.-R., L.G.H.-M. and J.J.R.-P.; review and editing, L.G.H.-M., J.J.R.-P. and E.E.Q.-A.; visualization, L.G.H.-M. and J.A.T.-R.; supervision, L.G.H.-M. and J.A.T.-R.; project administration, L.G.H.-M. All authors have read and agreed to the published version of the manuscript.

Funding: This research received no external funding.

Institutional Review Board Statement: Not applicable.

Informed Consent Statement: Not applicable.

Data Availability Statement: Not applicable.

Acknowledgments: The first author is grateful to Consejo Nacional de Ciencia y Tecnologia (CONACYT), Mexico for Grant No. 2021910046 to carry out the Doctoral thesis project and to the State Technical University of Quevedo, for the support granted through the Competitive Fund for Scientific and Technological Research (FOCICYT) 8th Call, through the project PFOC8-10-2021 "Stimulation of biological and agricultural productivity by the application of silicon in horticultural crops.

Conflicts of Interest: The authors declare no conflict of interest.

References

1. Alblooshi, A.A.; Purayil, G.P.; Saeed, E.E.; Ramadan, G.A.; Tariq, S.; Altaee, A.S.; El-Tarabily, K.A.; AbuQamar, S.F. Biocontrol potential of endophytic actinobacteria against *Fusarium solani*, the causal agent of sudden decline syndrome on date palm in the UAE. *J. Fungi* **2022**, *8*, 8. [CrossRef] [PubMed]
2. Torres-Rodriguez, J.A.; Reyes-Pérez, J.J.; Castellanos, T.; Angulo, C.; Quiñones-Aguilar, E.E.; Hernandez-Montiel, L.G. A biopolymer with antimicrobial properties and plant resistance inducer against phytopathogens: Chitosan. *Not. Bot. Horti Agrobot. Cluj-Napoca* **2021**, *49*, 12231. [CrossRef]
3. Xiao, L.; Niu, H.-J.; Qu, T.-L.; Zhang, X.-F.; Du, F.-Y. *Streptomyces* sp. FX13 inhibits fungicide-resistant *Botrytis cinerea in vitro* and *in vivo* by producing oligomycin A. *Pestic. Biochem. Physiol.* **2021**, *175*, 104834. [CrossRef] [PubMed]
4. Boukaew, S.; Yossan, S.; Cheirsilp, B.; Prasertsan, P. Impact of environmental factors on *Streptomyces* spp. metabolites against *Botrytis cinerea*. *J. Basic Microbiol.* **2022**, *62*, 611–622. [CrossRef]
5. Jagannathan, S.V.; Manemann, E.M.; Rowe, S.E.; Callender, M.C.; Soto, W. Marine actinomycetes, new sources of biotechnological products. *Mar. Drugs* **2021**, *19*, 365. [CrossRef]
6. Tatar, D. Isolation, phylogenetic analysis and antimicrobial activity of halophilic actinomycetes from different saline environments located near Çorum province. *Biologia* **2021**, *76*, 773–780. [CrossRef]
7. Maua, J.O.; Mbuvi, M.T.E.; Matiku, P.; Munguti, S.; Mateche, E.; Owili, M. The difficult choice-to conserve the living filters or utilizing the full potential of wetlands: Insights from the Yala swamp, Kenya. *Environ. Chall.* **2022**, *6*, 100427. [CrossRef]
8. Fadhilah, Q.G.; Santoso, I.; Maryanto, A.E.; Abdullah, S.; Yasman, Y. Evaluation of the antifungal activity of marine actinomycetes isolates against the phytopathogenic fungi *Colletotrichum siamense* KA: A preliminary study for new antifungal compound discovery. *Pharmacia* **2021**, *68*, 837–843. [CrossRef]
9. Xia, X.; Liu, J.; Huang, L.; Zhang, X.; Deng, Y.; Li, F.; Liu, Z.; Huang, R. Molecular details of actinomycin D-Treated MRSA revealed via high-dimensional data. *Mar. Drugs* **2022**, *20*, 114. [CrossRef]
10. Meena, L.I.; Rajeswari, E.; Ahiladevi, P.; Kamalakannan, A.; Kalaiselvi, T. Antifungal potential of *Streptomyces rameus* GgS 48 against mungbean root rot [*Rhizoctonia bataticola* (Taub.) Butler]. *J. Biosci.* **2022**, *47*, 10. [CrossRef] [PubMed]
11. Gebily, D.A.S.; Ghanem, G.A.M.; Ragab, M.M.; Ali, A.M.; Soliman, N.E.K.; Abd El-Moity, T.H. Characterization and potential antifungal activities of three *Streptomyces* spp. as biocontrol agents against *Sclerotinia sclerotiorum* (Lib.) de Bary infecting green bean. *Egypt. J. Biol. Pest Control* **2021**, *31*, 33. [CrossRef]
12. Gong, Y.; Liu, J.-Q.; Xu, M.-J.; Zhang, C.-M.; Gao, J.; Li, C.-G.; Xing, K.; Qin, S. Antifungal volatile organic compounds from *Streptomyces setonii* WY228 control black spot disease of sweet potato. *Appl. Environ. Microbiol.* **2022**, *88*, e02317-21. [CrossRef] [PubMed]
13. Kaari, M.; Joseph, J.; Manikkam, R.; Sreenivasan, A.; Venugopal, G.; Alexander, B.; Krishnan, S. Anti-Biofilm activity and biocontrol potential of streptomyces cultures against *Ralstonia solanacearum* on tomato plants. *Indian J. Microbiol.* **2022**, *62*, 32–39. [CrossRef] [PubMed]

14. Mitra, D.; Mondal, R.; Khoshru, B.; Senapati, A.; Radha, T.K.; Mahakur, B.; Uniyal, N.; Myo, E.M.; Boutaj, H.; Sierra, B.E.G.; et al. Actinobacteria-enhanced plant growth, nutrient acquisition, and crop protection: Advances in soil, plant, and microbial multifactorial interactions. *Pedosphere* **2022**, *32*, 149–170. [CrossRef]
15. van Bergeijk, D.A.; Terlouw, B.R.; Medema, M.H.; van Wezel, G.P. Ecology and genomics of actinobacteria: New concepts for natural product discovery. *Nat. Rev. Microbiol.* **2020**, *18*, 546–558. [CrossRef]
16. Cheng, Z.; McCann, S.; Faraone, N.; Clarke, J.-A.; Hudson, E.A.; Cloonan, K.; Hillier, N.K.; Tahlan, K. Production of plant-associated volatiles by select model and industrially important *Streptomyces* spp. *Microorganisms* **2020**, *8*, 1767. [CrossRef] [PubMed]
17. Vurukonda, S.S.K.P.; Giovanardi, D.; Stefani, E. Plant growth promoting and biocontrol activity of *Streptomyces* spp. as endophytes. *Int. J. Mol. Sci.* **2018**, *19*, 952. [CrossRef]
18. Gomes, E.d.B.; Dias, L.R.L.; Miranda, R.d.C.M. Actinomycetes bioactive compounds: Biological control of fungi and phytopathogenic insect. *Afr. J. Biotechnol.* **2018**, *17*, 552–559. [CrossRef]
19. Li, X.; Jing, T.; Zhou, D.; Zhang, M.; Qi, D.; Zang, X.; Zhao, Y.; Li, K.; Tang, W.; Chen, Y.; et al. Biocontrol efficacy and possible mechanism of *Streptomyces* sp. H4 against postharvest anthracnose caused by *Colletotrichum fragariae* on strawberry fruit. *Postharvest Biol. Technol.* **2021**, *175*, 111401. [CrossRef]
20. Zhou, D.; Jing, T.; Chen, Y.; Yun, T.; Qi, D.; Zang, X.; Zhang, M.; Wei, Y.; Li, K.; Zhao, Y.; et al. Biocontrol potential of a newly isolated Streptomyces sp. HSL-9B from mangrove forest on postharvest anthracnose of mango fruit caused by *Colletotrichum gloeosporioides*. *Food Control* **2022**, *135*, 108836. [CrossRef]
21. Zou, N.; Zhou, D.; Chen, Y.; Lin, P.; Chen, Y.; Wang, W.; Xie, J.; Wang, M. A novel antifungal actinomycete *Streptomyces* sp. strain H3-2 effectively controls banana *Fusarium* wilt. *Front. Microbiol.* **2021**, *12*, 706647. [CrossRef]
22. Tiwari, K.; Gupta, R.K. Diversity and isolation of rare actinomycetes: An overview. *Crit. Rev. Microbiol.* **2013**, *39*, 256–294. [CrossRef] [PubMed]
23. Javoreková, S.; Kovácsová, S.; Medo, J.; Maková, J.; Petrová, J.; Hleba, L.; Košťálová, D.; Cinkocki, R. Soil amended with organic fertilizers as a source of actinomycetes with high potential as biocontrol agents. *J. Microbiol. Biotechnol. Food Sci.* **2021**, *8*, 1352–1359. [CrossRef]
24. Ascencion, L.C.; Liang, W.J.; Yen, T.B. Control of *Rhizoctonia solani* damping-off disease after soil amendment with dry tissues of Brassica results from increase in Actinomycetes population. *Biol. Control* **2015**, *82*, 21–30. [CrossRef]
25. Qi, D.; Zou, L.; Zhou, D.; Zhang, M.; Wei, Y.; Li, K.; Zhao, Y.; Zhang, L.; Xie, J. Biocontrol potential and antifungal mechanism of a novel *Streptomyces sichuanensis* against *Fusarium oxysporum* f. sp. cubense tropical race 4 *in vitro* and *in vivo*. *Appl. Microbiol. Biotechnol.* **2022**, *106*, 1633–1649. [CrossRef]
26. Cabrera, R.; García-López, H.; Aguirre-von-Wobeser, E.; Orozco-Avitia, J.A.; Gutiérrez-Saldaña, A.H. *Amycolatopsis* BX17: An actinobacterial strain isolated from soil of a traditional milpa agroecosystem with potential biocontrol against *Fusarium graminearum*. *Biol. Control* **2020**, *147*, 104285. [CrossRef]
27. Ramlawi, S.; Abusharkh, S.; Carroll, A.; McMullin, D.R.; Avis, T.J. Biological and chemical characterization of antimicrobial activity in *Arthrobacter* spp. isolated from disease-suppressive compost. *J. Basic Microbiol.* **2021**, *61*, 745–756. [CrossRef]
28. Allali, K.; Goudjal, Y.; Zamoum, M.; Bouznada, K.; Sabaou, N.; Zitouni, A. *Nocardiopsis dassonvillei* strain MB22 from the Algerian Sahara promotes wheat seedlings growth and potentially controls the common root rot pathogen *Bipolaris sorokiniana*. *J. Plant Pathol.* **2019**, *101*, 1115–1125. [CrossRef]
29. Ebrahimi-Zarandi, M.; Bonjar, G.H.S.; Riseh, R.S.; El-Shetehy, M.; Saadoun, I.; Barka, E.A. Exploring two *Streptomyces species* to control *Rhizoctonia solani* in tomato. *Agronomy* **2021**, *11*, 1384. [CrossRef]
30. Lahlali, R.; Ezrari, S.; Radouane, N.; Kenfaoui, J.; Esmaeel, Q.; El Hamss, H.; Belabess, Z.; Ait Barka, E. Biological control of plant pathogens: A global perspective. *Microorganisms* **2022**, *10*, 596. [CrossRef]
31. Carmona-Hernandez, S.; Reyes-Pérez, J.J.; Chiquito-Contreras, R.G.; Rincon-Enriquez, G.; Cerdan-Cabrera, C.R.; Hernandez-Montiel, L.G. Biocontrol of postharvest fruit fungal diseases by bacterial antagonists: A review. *Agronomy* **2019**, *9*, 121. [CrossRef]
32. Dutta, S.; Lee, Y.H. High-throughput identification of genes influencing the competitive ability to obtain nutrients and performance of biocontrol in *Pseudomonas putida* JBC17. *Sci. Rep.* **2022**, *8*, 14336. [CrossRef] [PubMed]
33. Farda, B.; Djebaili, R.; Vaccarelli, I.; Del Gallo, M.; Pellegrini, M. Actinomycetes from caves: An overview of their diversity, biotechnological properties, and insights for their use in soil environments. *Microorganisms* **2022**, *10*, 453. [CrossRef] [PubMed]
34. Köhl, J.; Kolnaar, R.; Ravensberg, W.J. Mode of action of microbial biological control agents against plant diseases: Relevance beyond efficacy. *Front. Plant Sci.* **2019**, *10*, 845. [CrossRef]
35. Palla, M.S.; Guntuku, G.S.; Muthyala, M.K.K.; Pingali, S.; Sahu, P.K. Isolation and molecular characterization of antifungal metabolite producing actinomycete from mangrove soil. *Beni-Suef Univ. J. Basic Appl. Sci.* **2018**, *7*, 250–256. [CrossRef]
36. Wang, H.; Tian, R.; Tian, Q.; Yan, X.; Huang, L.; Ji, Z. Investigation on the antifungal ingredients of *Saccharothrix yanglingensis* Hhs. 015, an antagonistic endophytic actinomycete isolated from cucumber plant. *Molecules* **2019**, *24*, 3686. [CrossRef]
37. Zhang, D.; Lu, Y.; Chen, H.; Wu, C.; Zhang, H.; Chen, L.; Chen, X. Antifungal peptides produced by actinomycetes and their biological activities against plant diseases. *J. Antibiot.* **2020**, *73*, 265–282. [CrossRef]
38. Polyak, Y.M.; Sukharevich, V.I. Detection and regulation of antagonistic properties of the soil actinomycete *Streptomyces* sp. 89. *Biol. Bull.* **2021**, *48*, 626–634. [CrossRef]
39. Vázquez-Laslop, N.; Mankin, A.S. How macrolide antibiotics work. *Trends Biochem. Sci.* **2018**, *43*, 668–684. [CrossRef]

40. Hartmann, D.O.; Shimizu, K.; Rothkegel, M.; Petkovic, M.; Ferraz, R.; Petrovski, Ž.; Branco, L.C.; Lopes, J.N.C.; Pereira, C.S. Tailoring amphotericin B as an ionic liquid: An upfront strategy to potentiate the biological activity of antifungal drugs. *RSC Adv.* **2021**, *11*, 14441–14452. [CrossRef]
41. Young, D.H.; Wang, N.X.; Meyer, S.T.; Avila-Adame, C. Characterization of the mechanism of action of the fungicide fenpicoxamid and its metabolite UK-2A. *Pest Manag. Sci.* **2018**, *74*, 489–498. [CrossRef] [PubMed]
42. Meena, M.; Prajapati, P.; Ravichandran, C.; Sehrawat, R. Natamycin: A natural preservative for food applications-a review. *Food Sci. Biotechnol.* **2021**, *30*, 1481–1496. [CrossRef]
43. Sharma, S.; Fulke, A.B.; Chaubey, A. Bioprospection of marine actinomycetes: Recent advances, challenges and future perspectives. *Acta Oceanol. Sin.* **2019**, *38*, 1–17. [CrossRef]
44. Ghosh, S.K.; Bera, T.; Chakrabarty, A.M. Microbial siderophore–A boon to agricultural sciences. *Biol. Control* **2020**, *144*, 104214. [CrossRef]
45. Roskova, Z.; Skarohlid, R.; McGachy, L. Siderophores: An alternative bioremediation strategy? *Sci. Total Environ.* **2022**, *819*, 153144. [CrossRef] [PubMed]
46. Retamal-Morales, G.; Senges, C.H.R.; Stapf, M.; Olguín, A.; Modak, B.; Bandow, J.E.; Tischler, D.; Schlömann, M.; Levicán, G. Isolation and characterization of arsenic-binding siderophores from *Rhodococcus erythropolis* S43: Role of heterobactin B and other heterobactin variants. *Appl. Microbiol. Biotechnol.* **2021**, *105*, 1731–1744. [CrossRef]
47. Wu, Q.; Deering, R.W.; Zhang, G.; Wang, B.; Li, X.; Sun, J.; Chen, J.; Zhang, H.; Rowley, D.C.; Wang, H. *Albisporachelin*, a new hydroxamate type siderophore from the deep ocean sediment-derived actinomycete *Amycolatopsis albispora* WP1T. *Mar. Drugs* **2018**, *16*, 199. [CrossRef]
48. Hernández-Montiel, L.G.; Rivas-García, T.; Romero-Bastidas, M.; Chiquito-Contreras, C.J.; Ruiz-Espinoza, F.H.; Chiquito-Contreras, R.G. Antagonistic potential of bacteria and marine yeasts for the control of phytopathogenic fungi. *Rev. Mex. Cienc. Agríc.* **2018**, *9*, 4311–4321. [CrossRef]
49. Mottola, A.; Ramírez-Zavala, B.; Hünniger, K.; Kurzai, O.; Morschhäuser, J. The zinc cluster transcription factor Czf1 regulates cell wall architecture and integrity in *Candida albicans*. *Mol. Microbiol.* **2021**, *116*, 483–497. [CrossRef]
50. Dimkić, I.; Janakiev, T.; Petrović, M.; Degrassi, G.; Fira, D. Plant-associated *Bacillus* and *Pseudomonas* antimicrobial activities in plant disease suppression via biological control mechanisms-A review. *Physiol. Mol. Plant Pathol.* **2022**, *117*, 101754. [CrossRef]
51. Jadhav, H.P.; Shaikh, S.S.; Sayyed, R.Z. Role of hydrolytic enzymes of rhizoflora in biocontrol of fungal phytopathogens. An overview. In *Rhizotrophs: Plant Growth Promotion to Bioremediation*; Mehnaz, S., Ed.; Microorganisms for Sustainability; Springer: Singapore, 2017; pp. 183–203. ISBN 978-981-10-4862-3.
52. del Rosario Cappellari, L.; Chiappero, J.; Banchio, E. Invisible signals from the underground: A practical method to investigate the effect of microbial volatile organic compounds emitted by rhizobacteria on plant growth. *Biochem. Mol. Biol. Educ.* **2019**, *47*, 388–393. [CrossRef] [PubMed]
53. Ayed, A.; Kalai-Grami, L.; Ben Slimene, I.; Chaouachi, M.; Mankai, H.; Karkouch, I.; Djebali, N.; Elkahoui, S.; Tabbene, O.; Limam, F. Antifungal activity of volatile organic compounds from *Streptomyces* sp. strain S97 against *Botrytis cinerea*. *Biocontrol Sci. Technol.* **2021**, *31*, 1330–1348. [CrossRef]
54. Widada, J.; Damayanti, E.; Alhakim, M.R.; Yuwono, T.; Mustofa, M. Two strains of airborne *Nocardiopsis alba* producing different volatile organic compounds (VOCs) as biofungicide for *Ganoderma boninense*. *FEMS Microbiol. Lett.* **2021**, *368*, fnab138. [CrossRef]
55. Alipour Kafi, S.; Karimi, E.; Akhlaghi Motlagh, M.; Amini, Z.; Mohammadi, A.; Sadeghi, A. Isolation and identification of *Amycolatopsis* sp. strain 1119 with potential to improve cucumber fruit yield and induce plant defense responses in commercial greenhouse. *Plant Soil* **2021**, *468*, 125–145. [CrossRef]
56. Yu, Y.; Gui, Y.; Li, Z.; Jiang, C.; Guo, J.; Niu, D. Induced systemic resistance for improving plant immunity by beneficial microbes. *Plants* **2022**, *11*, 386. [CrossRef] [PubMed]
57. Vilasinee, S.; Toanuna, C.; McGovern, R.J.; Nalumpang, S. Expression of pathogenesis-related (PR) genes in tomato against *Fusarium* wilt by challenge inoculation with *Streptomyces* NSP3. *Int. J. Agric. Technol.* **2019**, *15*, 157–170.
58. Bizuneh, G.K. The chemical diversity and biological activities of phytoalexins. *Tradit. Med.* **2021**, *21*, 31–43. [CrossRef]
59. Vijayabharathi, R.; Gopalakrishnan, S.; Sathya, A.; Vasanth Kumar, M.; Srinivas, V.; Mamta, S. *Streptomyces* sp. as plant growth-promoters and host-plant resistance inducers against *Botrytis cinerea* in chickpea. *Biocontrol Sci. Technol.* **2018**, *28*, 1140–1163. [CrossRef]
60. Li, Y.; He, F.; Guo, Q.; Feng, Z.; Zhang, M.; Ji, C.; Xue, Q.; Lai, H. Compositional and functional comparison on the rhizosphere microbial community between healthy and *Sclerotium rolfsii*-infected monkshood (*Aconitum carmichaelii*) revealed the biocontrol potential of healthy monkshood rhizosphere microorganisms. *Biol. Control* **2022**, *165*, 104790. [CrossRef]
61. Ameen, F.; AlNadhari, S.; Al-Homaidan, A.A. Marine microorganisms as an untapped source of bioactive compounds. *Saudi J. Biol. Sci.* **2021**, *28*, 224–231. [CrossRef]
62. Vijayabharathi, R.; Gopalakrishnan, S.; Sathya, A.; Srinivas, V.; Sharma, M. Deciphering the tri-dimensional effect of endophytic *Streptomyces* sp. on chickpea for plant growth promotion, helper effect with *Mesorhizobium ciceri* and host-plant resistance induction against *Botrytis cinerea*. *Microb. Pathog.* **2018**, *122*, 98–107. [CrossRef] [PubMed]
63. Jing, T.; Zhou, D.; Zhang, M.; Yun, T.; Qi, D.; Wei, Y.; Chen, Y.; Zang, X.; Wang, W.; Xie, J. Newly isolated *Streptomyces* sp. JBS5-6 as a potential biocontrol agent to control banana Fusarium wilt: Genome sequencing and secondary metabolite cluster profiles. *Front. Microbiol.* **2020**, *11*, 3036. [CrossRef]

64. Jepsen, T.; Jensen, B.; Jørgensen, N.O.G. Volatiles produced by *Streptomyces* spp. delay rot in apples caused by *Colletotrichum acutatum*. *Curr. Res. Microb. Sci.* **2022**, *3*, 100121. [CrossRef]
65. Buatong, J.; Rukachaisirikul, V.; Sangkanu, S.; Surup, F.; Phongpaichit, S. Antifungal metabolites from marine-derived *Streptomyces* sp. AMA49 against *Pyricularia oryzae*. *J. Pure Appl. Microbiol.* **2019**, *13*, 653–665. [CrossRef]
66. Uba, B.O.; Okoye, E.B.; Anyaeji, O.J.; Ogbonnaya, O.C. Antagonistic potentials of actinomycetes isolated from coastal area of Niger Delta against *Citrus sinensis* (Sweet Orange) and *Lycopersicum esculentum* (Tomato) fungal pathogens. *Res. Rev. A J. Biotech.* **2019**, *9*, 4–15.
67. Torres-Rodriguez, J.; Reyes-Pérez, J.; Castellanos, T.; Angulo, C.; Quiñones-Aguilar, E.; Hernandez-Montiel, L. Identication and morphological characterization of marine actinomycetes as biocontrol agents of *Fusarium solani* in tomato. *Rev. Fac. Agron. Univ. Zulia* **2022**, *39*, e223915. [CrossRef]
68. Wan, M.; Li, G.; Zhang, J.; Jiang, D.; Huang, H.C. Effect of volatile substances of *Streptomyces platensis* F-1 on control of plant fungal diseases. *Biol. Control* **2008**, *46*, 552–559. [CrossRef]
69. Shimizu, M.; Yazawa, S.; Ushijima, Y. A promising strain of endophytic *Streptomyces* sp. for biological control of cucumber anthracnose. *J. Gen. Plant Pathol.* **2009**, *75*, 27–36. [CrossRef]
70. Gholami, M.; Khakvar, R.; AliasgarZad, N. Application of endophytic bacteria for controlling anthracnose disease (*Colletotrichum lindemuthianum*) on bean plants. *Arch. Phytopathol. Plant Prot.* **2013**, *46*, 1831–1838. [CrossRef]
71. Goudjal, Y.; Toumatia, O.; Yekkour, A.; Sabaou, N.; Mathieu, F.; Zitouni, A. Biocontrol of *Rhizoctonia solani* damping-off and promotion of tomato plant growth by endophytic actinomycetes isolated from native plants of Algerian Sahara. *Microbiol. Res.* **2014**, *169*, 59–65. [CrossRef]
72. Malviya, N.; Yandigeri, M.S.; Yadav, A.K.; Solanki, M.K.; Arora, D.K. Isolation and characterization of novel alkali-halophilic actinomycetes from the Chilika brackish water lake, India. *Ann. Microbiol.* **2014**, *64*, 1829–1838. [CrossRef]
73. Yandigeri, M.S.; Malviya, N.; Solanki, M.K.; Shrivastava, P.; Sivakumar, G. Chitinolytic *Streptomyces vinaceusdrappus* S5MW2 isolated from Chilika lake, India enhances plant growth and biocontrol efficacy through chitin supplementation against *Rhizoctonia solani*. *World J. Microbiol. Biotechnol.* **2015**, *31*, 1217–1225. [CrossRef] [PubMed]
74. Lu, D.; Ma, Z.; Xu, X.; Yu, X. Isolation and identification of biocontrol agent *Streptomyces rimosus* M527 against *Fusarium oxysporum* f. sp. *cucumerinum*. *J. Basic Microbiol.* **2016**, *56*, 929–933. [CrossRef]
75. Mingma, R.; Duangmal, K. Characterization, antifungal activity and plant growth promoting potential of endophytic actinomycetes isolated from rice (*Oryza sativa* L.). *Chiang Mai J. Sci.* **2018**, *45*, 2652–2665.
76. Adlin Jenifer, J.S.C.; Michaelbabu, M.; Eswaramoorthy Thirumalaikumar, C.L.; Jeraldin Nisha, S.R.; Uma, G.; Citarasu, T. Antimicrobial potential of haloalkaliphilic *Nocardiopsis* sp. AJ1 isolated from solar salterns in India. *J. Basic Microbiol.* **2019**, *59*, 288–301. [CrossRef]
77. Hao, L.; Zheng, X.; Wang, Y.; Li, S.; Shang, C.; Xu, Y. Inhibition of tomato early blight disease by culture extracts of a *Streptomyces puniceus* isolate from mangrove soil. *Phytopathology* **2019**, *109*, 1149–1156. [CrossRef]
78. Wonglom, P.; Suwannarach, N.; Lumyong, S.; Ito, S.; Matsui, K.; Sunpapao, A. *Streptomyces angustmyceticus* NR8-2 as a potential microorganism for the biological control of leaf spots of *Brassica rapa* subsp. *pekinensis* caused by *Colletotrichum* sp. and *Curvularia lunata*. *Biol. Control* **2019**, *138*, 104046. [CrossRef]
79. Benhadj, M.; Metrouh, R.; Menasria, T.; Gacemi-Kirane, D.; Slim, F.Z.; Ranque, S. Broad-spectrum antimicrobial activity of wetland-derived *Streptomyces* sp. ActiF450. *Excli J.* **2020**, *19*, 360–371. [CrossRef]
80. Cao, P.; Li, C.; Wang, H.; Yu, Z.; Xu, X.; Wang, X.; Zhao, J.; Xiang, W. Community structures and antifungal activity of root-associated endophytic actinobacteria in healthy and diseased cucumber plants and *Streptomyces* sp. HAAG3-15 as a promising biocontrol agent. *Microorganisms* **2020**, *8*, 236. [CrossRef] [PubMed]
81. Ghadhban, W.I.; Abd Burghal, A. Identification of antibiotic producing actinomycetes isolated from sediment in Basra, Iraq. *Int. J. Pharm. Sci.* **2020**, *11*, b92–b98. [CrossRef]
82. Peng, C.; An, D.; Ding, W.-X.; Zhu, Y.-X.; Ye, L.; Li, J. Fungichromin production by *Streptomyces* sp. WP-1, an endophyte from *Pinus dabeshanensis*, and its antifungal activity against *Fusarium oxysporum*. *Appl. Microbiol. Biotechnol.* **2020**, *104*, 10437–10449. [CrossRef]
83. Abo-Zaid, G.; Abdelkhalek, A.; Matar, S.; Darwish, M.; Abdel-Gayed, M. Application of bio-friendly formulations of chitinase-producing *Streptomyces* cellulosae Actino 48 for controlling peanut soil-borne diseases caused by *Sclerotium rolfsii*. *J. Fungi* **2021**, *7*, 167. [CrossRef]
84. Boukaew, S.; Cheirsilp, B.; Prasertsan, P.; Yossan, S. Antifungal effect of volatile organic compounds produced by *Streptomyces salmonis* PSRDC-09 against anthracnose pathogen *Colletotrichum gloeosporioides* PSU-03 in postharvest chili fruit. *J. Appl. Microbiol.* **2021**, *131*, 1452–1463. [CrossRef]
85. Li, X.; Li, K.; Zhou, D.; Zhang, M.; Qi, D.; Jing, T.; Zang, X.; Qi, C.; Wang, W.; Xie, J. Biological control of banana wilt disease caused by *Fusarium oxyspoum* f. sp. *Cubense* using *Streptomyces* sp. H4. *Biol. Control* **2021**, *155*, 104524. [CrossRef]
86. Ziedan, E.S.H. A review of the efficacy of biofumigation agents in the control of soil-borne plant diseases. *J. Plant Prot. Res.* **2022**, *62*, 1–11. [CrossRef]
87. El-Tarabily, K.A.; Nassar, A.H.; Hardy, G.S.J.; Sivasithamparam, K. Plant growth promotion and biological control of *Pythium aphanidermatum*, a pathogen of cucumber, by endophytic actinomycetes. *J. Appl. Microbiol.* **2009**, *107*, 1765–1766. [CrossRef]

88. Costa, F.G.; Zucchi, T.D.; Melo, I.S.D. Biological control of phytopathogenic fungi by endophytic actinomycetes isolated from maize (*Zea mays* L.). *Braz. Arch. Biol. Technol.* **2013**, *56*, 948–955. [CrossRef]
89. Singh, S.P.; Gupta, R.; Gaur, R.; Srivastava, A.K. Antagonistic actinomycetes mediated resistance in *Solanum lycopersicon* Mill. against *Rhizoctonia solani* Kühn. *Proc. Natl. Acad. Sci. India Sect. B Biol. Sci.* **2017**, *87*, 789–798. [CrossRef]
90. Anusha, B.G.; Gopalakrishnan, S.; Naik, M.K.; Sharma, M. Evaluation of *Streptomyces* spp. and *Bacillus* spp. for biocontrol of *Fusarium* wilt in chickpea (*Cicer arietinum* L.). *Arch. Phytopathol. Plant Prot.* **2019**, *52*, 417–442. [CrossRef]
91. Barnett, S.J.; Ballard, R.A.; Franco, C.M. Field assessment of microbial inoculants to control *Rhizoctonia* root rot on wheat. *Biol. Control* **2019**, *132*, 152–160. [CrossRef]
92. Pathom-aree, W.; Kreawsa, S.; Kamjam, M.; Tokuyama, S.; Yoosathaporn, S.; Lumyong, S. Potential of selected mangrove *Streptomyces* as plant growth promoter and rice bakanae disease control agent. *Chiang Mai J. Sci.* **2019**, *46*, 261–276.
93. Zhang, J.; Gao, Z.; Li, N.; Zhang, X. The antifungal activity of biocontrol actinomycetes G-1 against *Alternaria radicina*. *J. Henan Agric. Sci.* **2019**, *48*, 99–104.
94. El-Shatoury, S.A.; Ameen, F.; Moussa, H.; Wahid, O.A.; Dewedar, A.; AlNadhari, S. Biocontrol of chocolate spot disease (*Botrytis cinerea*) in faba bean using endophytic actinomycetes *Streptomyces*: A field study to compare application techniques. *PeerJ* **2020**, *8*, e8582. [CrossRef]
95. Gu, L.; Zhang, K.; Zhang, N.; Li, X.; Liu, Z. Control of the rubber anthracnose fungus *Colletotrichum gloeosporioides* using culture filtrate extract from *Streptomyces deccanensis* QY-3. *Antonie Van Leeuwenhoek* **2020**, *113*, 1573–1585. [CrossRef]
96. Saeed, I.; Khan, S.H.; Rasheed, A.; Jahangir, M.M.; Jabbar, A.; Shaheen, H.M.F.; Din, W.U.; Mazhar, K. Assessment of antagonistic potential of bacteria as biocontrol agent against *Alternaria* leaf spot of turnip. *Pak. J. Phytopathol.* **2021**, *33*, 401–409. [CrossRef]
97. Yao, X.; Zhang, Z.; Huang, J.; Wei, S.; Sun, X.; Chen, Y.; Liu, H.; Li, S. Candicidin isomer production is essential for biocontrol of cucumber *Rhizoctonia* rot by *Streptomyces albidoflavus* W68. *Appl. Environ. Microbiol.* **2021**, *87*, e03078-20. [CrossRef]
98. Shewa, A.G.; Gobena, D.A.; Ali, M.K. Review on postharvest quality and handling of apple. *J. Agric. Sci. Food Technol.* **2022**, *8*, 028–032. [CrossRef]
99. Li, Y.; Xia, M.; He, P.; Yang, Q.; Wu, Y.; He, P.; Ahmed, A.; Li, X.; Wang, Y.; Munir, S.; et al. Developing *Penicillium digitatum* management strategies on post-harvest citrus fruits with metabolic components and colonization of *Bacillus subtilis* L1-21. *J. Fungi* **2022**, *8*, 80. [CrossRef]
100. Sarkar, G.; Suthindhiran, K. Diversity and biotechnological potential of marine actinomycetes from India. *Indian J. Microbiol.* **2022**, *Vol.*, 1–19. [CrossRef]
101. Hu, X.; Cheng, B.; Du, D.; Huang, Z.; Pu, Z.; Chen, G.; Peng, A.; Lu, L. Isolation and identification of a marine actinomycete strain and its control efficacy against citrus green and blue moulds. *Biotechnol. Biotechnol. Equip.* **2019**, *33*, 719–729. [CrossRef]
102. Najmeh, S.; Hosein, S.B.G.; Sareh, S.; Bonjar, L.S. Biological control of citrus green mould, *Penicillium digitatum*, by antifungal activities of Streptomyces isolates from agricultural soils. *Afr. J. Microbiol. Res.* **2014**, *8*, 1501–1509. [CrossRef]
103. Yong, D.; Wang, C.; Li, G.; Li, B. Control efficiency of endophytic actinomycetes A-1 against apple fruit ring rot and its influence on the activity of defense-related enzymes. *Acta Phytophylacica Sin.* **2014**, *41*, 335–341.
104. Lyu, A.; Liu, H.; Che, H.; Yang, L.; Zhang, J.; Wu, M.; Chen, W.; Li, G. Reveromycins A and B from *Streptomyces* sp. 3-10: Antifungal activity against plant pathogenic fungi *in vitro* and in a strawberry food model system. *Front. Microbiol.* **2017**, *8*, 3–10. [CrossRef]
105. Boukaew, S.; Petlamul, W.; Bunkrongcheap, R.; Chookaew, T.; Kabbua, T.; Thippated, A.; Prasertsan, P. Fumigant activity of volatile compounds of *Streptomyces philanthi* RM-1-138 and pure chemicals (acetophenone and phenylethyl alcohol) against anthracnose pathogen in postharvest chili fruit. *Crop Prot.* **2018**, *103*, 1–8. [CrossRef]
106. Liotti, R.G.; da Silva Figueiredo, M.I.; Soares, M.A. *Streptomyces griseocarneus* R132 controls phytopathogens and promotes growth of pepper (*Capsicum annuum*). *Biol. Control* **2019**, *138*, 104065. [CrossRef]
107. Boukaew, S.; Petlamul, W.; Prasertsan, P. Comparison of the biocontrol efficacy of culture filtrate from *Streptomyces philanthi* RL-1-178 and acetic acid against *Penicillium digitatum*, *in vitro* and *in vivo*. *Eur. J. Plant Pathol.* **2020**, *158*, 939–949. [CrossRef]
108. Kim, J.D.; Kang, J.E.; Kim, B.S. Postharvest disease control efficacy of the polyene macrolide lucensomycin produced by *Streptomyces plumbeus* strain CA5 against gray mold on grapes. *Postharvest Biol. Technol.* **2020**, *162*, 111115. [CrossRef]
109. Nguyen, H.T.T.; Park, A.R.; Hwang, I.M.; Kim, J.C. Identification and delineation of action mechanism of antifungal agents: Reveromycin E and its new derivative isolated from *Streptomyces* sp. JCK-6141. *Postharvest Biol. Technol.* **2021**, *182*, 111700. [CrossRef]
110. Chaurasia, A.; Meena, B.R.; Tripathi, A.N.; Pandey, K.K.; Rai, A.B.; Singh, B. Actinomycetes: An unexplored microorganisms for plant growth promotion and biocontrol in vegetable crops. *World J. Microbiol. Biotechnol.* **2018**, *34*, 132. [CrossRef]
111. Lahdenperä, M.L.; Simon, E.; Uoti, J. Mycostop-a novel biofungicide based on *Streptomyces* bacteria. In *Developments in Agricultural and Managed Forest Ecology*; Beemster, A.B.R., Bollen, G.J., Gerlagh, M., Ruissen, M.A., Schippers, B., Tempel, A., Eds.; Elsevier: Amsterdam, The Netherlands, 1991; Volume 23, pp. 258–263.
112. Tian, X.; Zheng, Y. Evaluation of biological control agents for *Fusarium* wilt in Hiemalis begonia. *Can. J. Plant Pathol.* **2013**, *35*, 363–370. [CrossRef]
113. Kabaluk, J.T.; Svircev, A.M.; Goettel, M.S.; Woo, S.G. *The Use and Regulation of Microbial Pesticides in Representative Jurisdiction Worldwide*; IOBC Global: Hong Kong, China, 2010; p. 99.

114. Okuyama, A.; Machiyama, N.; Kinoshita, T.; Tanaka, N. Inhibition by kasugamycin of initiation complex formation on 30S ribosomes. *Biochem. Biophys. Res. Commun.* **1971**, *43*, 196–199. [CrossRef] [PubMed]
115. Copping, L.G.; Duke, S.O. Natural products that have been used commercially as crop protection agents. *Pest Manag. Sci.* **2007**, *63*, 524–554. [CrossRef] [PubMed]
116. Copping, L.G.; Menn, J.J. Biopesticides: A review of their action, applications and efficacy. *Pest Manag. Sci.* **2000**, *56*, 651–676. [CrossRef]

 *plants*

Article

Activity of Anthracenediones and Flavoring Phenols in Hydromethanolic Extracts of *Rubia tinctorum* against Grapevine Phytopathogenic Fungi

Natalia Langa-Lomba [1,2], Eva Sánchez-Hernández [3], Laura Buzón-Durán [3], Vicente González-García [2], José Casanova-Gascón [1], Jesús Martín-Gil [3] and Pablo Martín-Ramos [1,*]

1 Instituto Universitario de Investigación en Ciencias Ambientales de Aragón (IUCA), EPS, Universidad de Zaragoza, Carretera de Cuarte, s/n, 22071 Huesca, Spain; natalialangalomba@gmail.com (N.L.-L.); jcasan@unizar.es (J.C.-G.)

2 Agrifood Research and Technology Centre of Aragón, Plant Protection Unit, Instituto Agroalimentario de Aragón—IA2 (CITA-Universidad de Zaragoza), Avda. Montañana 930, 50059 Zaragoza, Spain; vgonzalezg@aragon.es

3 Department of Agricultural and Forestry Engineering, ETSIIAA, Universidad de Valladolid, 34004 Palencia, Spain; eva.sanchez.hernandez@uva.es (E.S.-H.); laura.buzon@uva.es (L.B.-D.); mgil@iaf.uva.es (J.M.-G.)

* Correspondence: pmr@unizar.es

Citation: Langa-Lomba, N.; Sánchez-Hernández, E.; Buzón-Durán, L.; González-García, V.; Casanova-Gascón, J.; Martín-Gil, J.; Martín-Ramos, P. Activity of Anthracenediones and Flavoring Phenols in Hydromethanolic Extracts of *Rubia tinctorum* against Grapevine Phytopathogenic Fungi. *Plants* **2021**, *10*, 1527. https://doi.org/10.3390/plants10081527

Academic Editors: Dragana Šunjka and Špela Mechora

Received: 8 July 2021
Accepted: 23 July 2021
Published: 26 July 2021

Publisher's Note: MDPI stays neutral with regard to jurisdictional claims in published maps and institutional affiliations.

Abstract: In this work, the chemical composition of *Rubia tinctorum* root hydromethanolic extract was analyzed by GC–MS, and over 50 constituents were identified. The main phytochemicals were alizarin-related anthraquinones and flavoring phenol compounds. The antifungal activity of this extract, alone and in combination with chitosan oligomers (COS) or with stevioside, was evaluated against the pathogenic taxa *Diplodia seriata*, *Dothiorella viticola* and *Neofusicoccum parvum*, responsible for the so-called Botryosphaeria dieback of grapevine. In vitro mycelial growth inhibition tests showed remarkable activity for the pure extract, with EC_{50} and EC_{90} values as low as 66 and 88 $\mu g \cdot mL^{-1}$, respectively. Nonetheless, enhanced activity was attained upon the formation of conjugate complexes with COS or with stevioside, with synergy factors of up to 5.4 and 3.3, respectively, resulting in EC_{50} and EC_{90} values as low as 22 and 56 $\mu g \cdot mL^{-1}$, respectively. The conjugate with the best performance (COS-*R. tinctorum* extract) was then assayed ex situ on autoclaved grapevine wood against *D. seriata*, confirming its antifungal behavior on this plant material. Finally, the same conjugate was evaluated in greenhouse assays on grafted grapevine plants artificially inoculated with the three aforementioned fungal species, resulting in a significant reduction in the infection rate in all cases. This natural antifungal compound represents a promising alternative for developing sustainable control methods against grapevine trunk diseases.

Keywords: antifungal; *Botryosphaeriaceae*; chitosan; GTDs; madder; stevioside; *Vitis vinifera*

1. Introduction

The joint presence of compounds of quinone and phenol categories in plant extracts and, specifically, the differential content of anthracenediones and 4-*tert*-butyl-2-phenylphenol, which might be responsible for the chromatic aberration of teak (difference between heartwood and sapwood), has been the object of attention in the bibliography [1].

Anthracenediones are a class of molecules based on the 9,10-anthracenedione parent (Figure 1a), which–among others–include purpurin (Figure 1a) and those synthesized by the American Cyanamid Laboratories in the late 1970s [2]. Although mitoxantrone, which has a dihydroxyanthraquinone central chromophore with two symmetrical aminoalkyl side chains (Figure 1a), is considered the biologically most active anthracenedione [3], other anthracenediones have also been reported to have antimicrobial activities: for instance, anthraquinone aglycones have been found to have a remarkable in vitro activity against clinical strains of dermatophytes [4]; anthraquinone derivatives exhibit antifungal activity against *Candida albicans* (C.P. Robin) Berkhout, *Cryptococcus neoformans* (San Felice) Vuill., *Trichophyton mentagrophytes* (C.P. Robin) R. Blanch., *Aspergillus fumigatus* Fresen. and *Sporothrix schenckii* Hektoen and C.F. Perkins [5,6]; purpurin possesses remarkable antifungal activity against *Candida* spp. [7]; and alizarin or 1,2-dihydroxyanthraquinone (Figure 1b) show antifungal behavior against *Aspergillus niger* Tieghem and *A. ochraceus* K. Wilhelm [8].

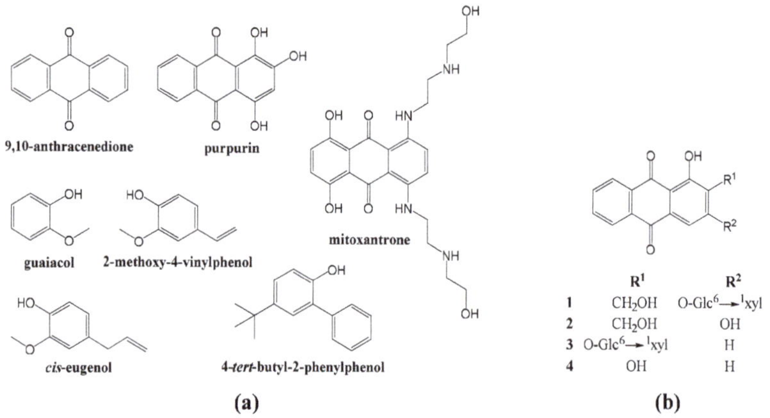

	R^1	R^2
1	CH_2OH	O-Glc^6→^{1}xyl
2	CH_2OH	OH
3	O-Glc^6→^{1}xyl	H
4	OH	H

Figure 1. (**a**) Structures of different anthracenediones and phenols; (**b**) structures of alizarin-3-*O*-*β*-primeveroside, 3; lucidin-3-*O*-*β*-primeveroside, 1; and their aglycons (alizarin, 4; lucidin, 2). Glc, D-glucose; xyl, D-xylose.

Flavoring phenols is a category that includes small free phenolic compounds (Figure 1a), such as 2-methoxy-phenol (or guaiacol), 2-methoxy-4-vinylphenol (or 4-vinyl-guaiacol), *cis*-2-methoxy -4-(1-propenyl)-phenol (or *cis*-eugenol) and 4-*tert*-butyl-2-phenyl-phenol, which participate in the aroma of wine. Guaiacol and eugenol are characterized by spice, clove, and smoke notes (guaiacol provides a roasted aroma and eugenol confers a clove aroma); and 4-vinyl-guaiacol has an odor reminiscent of carnation (*Dianthus* flowers). 4-((1E)-3-hydroxy-1-propenyl)-2-methoxyphenol (or coniferyl alcohol) is a precursor of grape and wine volatiles [9]. All of them are present in oak, but 4-*tert*-butyl-2-phenyl-phenol has been referred as a constituent of *Rubia cordifolia* L. essential oil [10,11]. Regarding their antifungal activities, a strong antifungal activity against *Botrytis cinerea* Pers. has been referred for eugenol [12], and guaiacol has been found to be effective against sap-staining fungi (*Ophiostoma* spp.) [13].

In this paper, the possibility of a joint presence of both anthracenediones and flavoring phenols in *Rubia tinctorum* L. (*Rubiaceae*) has been explored, given that the presence of 9,10-anthraquinones and other biologically active compounds has been reported for other members of the genus *Rubia*, mainly for *R. cordifolia*, as summarized in the review paper by Singh, et al. [14].

R. tinctorum is widely distributed in southern and southeastern Europe, in the Mediterranean area, and in central Asia. Its reddish roots contain hydroxyanthraquinones, such as alizarin (used for the dyeing of textiles [15] and in the treatment of kidney and bladder stones), purpurin (1,2,4-trihydroxyanthraquinone), and lucidin (Figure 1b.4) [16,17]; and flavoring phenols such as 4-vinyl-guaiacol [18].

The interest in the joint presence of anthracenediones and phenols (as 2-methoxyphenols and 4-*tert*-butyl-2-phenyl-phenol) lies in the possibility of synergies that enhance their microbiological activity. In particular, this work focuses on their potential application for the control of grapevine trunk diseases (GTDs), currently considered one of the most relevant challenges in Viticulture, as these pathologies cause significant economic losses in grape growing areas all over the world. Under this generic concept, a series of mycoses are grouped, which affect the wood of grapevine throughout its entire life cycle [19,20]. Among them, those that affect young plants coming from the nursery and in the first years after planting are especially important from the economic point of view, being responsible for numerous losses derived from the removal and replacement of plants in hundreds of thousands of hectares around the world [21]. Some of these include the so-called "Black Foot" disease, caused by different species belonging to soil-borne genera like *Ilyonectria*, *Campylocarpon*, *Cylindrocladiella*, *Dactylonectria*, etc.; the etiological agents responsible for Petri disease (mainly species of the genus *Phaeoacremonium*, and *Phaeomoniella chlamydospora* (W. Gams, Crous, M.J. Wingf. and Mugnai) Crous and W. Gams) that for many authors would be part of the first stages of the complex esca syndrome; or some species of the ascomycete family *Botryosphaeriaceae*, especially certain aggressive taxa in the early years of the plant such as *Neofusicoccum parvum* (included in the present study). In addition to these pathologies, other complex syndromes have been described, such as the aforementioned esca (attributable to certain species of lignicolous basidiomycetes), Eutypiosis (caused in Europe by *Eutypa lata* (Pers.) Tul. and C. Tul.), or the so-called Botryosphaeria decay of grapevine plants (also known as "Black Dead Arm" disease) caused by various genera and species of this family such as the aforementioned *N. parvum*, *Diplodia* spp., *Dothiorella* spp., *Lasiodiplodia* spp. or *Botryosphaeria* spp.

Given that the prohibition of active ingredients such as sodium arsenite and benzimidazoles, which were used to control GTDs, has worsened the impact of these diseases, they have become the subject of intense research efforts. Unfortunately, due to the breadth and complexity of the problem, no single effective control measure against these mycoses has been developed to date. Current strategies and future prospects for the management of GTDs are thoroughly discussed in the review papers by Fontaine, et al. [22], Bertsch, et al. [20], Mondello, et al. [23] and Gramaje, et al. [24], but the use of active ingredients of natural origin, instead of conventional chemicals, poses an especially interesting approach, aligned with the criteria of European legislation currently in force (Article 14 in European Directive 2009/128/EC).

Taking into consideration that many phytochemicals have solubility and bioavailability problems, in this work the bioactivity of the hydromethanolic extracts of *R. tinctorum* against GTDs has also been assayed after the formation of conjugate complexes, either with chitosan oligomers (COS) or with stevioside [a terpene glycoside obtained from *Stevia rebaudiana* (Bertoni) Bertoni extract], which also have antifungal properties and which may lead to a synergistic fungicide behaviour [25,26].

2. Material and Methods

2.1. Plant Material and Chemicals

The specimens of *Rubia tinctorum* under study were collected on the banks of the Carrión river as it passes through the town of Palencia (Spain). The roots were shade-dried and pulverized to fine powder in a mechanical grinder. Samples from different specimens (n = 25) were thoroughly mixed to obtain composite samples.

Chitosan (CAS 9012-76-4; high MW: 310,000–375,000 Da) was supplied by Hangzhou Simit Chem. & Tech. Co. (Hangzhou, China). Neutrase^{TM} 0.8 L enzyme was supplied by Novozymes A/S (Bagsværd, Denmark). Stevioside (CAS 57817-89-7, 99%) was purchased from Wako Chemicals GmbH (Neuss, Germany). Quantities of 4-*tert*-butyl-2-phenylphenol (CAS 98-27-1, 97%), 1,2-dihydroxyanthraquinone (CAS 72-48-0, 97%), sodium alginate (CAS 9005-38-3), calcium carbonate (CAS 471-34-1, $\geq$99.0%) and methanol (CAS 67-56-1, UHPLC, suitable for mass spectrometry) were acquired from Sigma-Aldrich Química (Madrid, Spain). Agar (CAS 9002-18-0) and PDA (potato dextrose agar) were supplied by Becton Dickinson (Bergen County, NJ, USA).

2.2. Preparation and Physicochemical Characterization of the of *R. tinctorum* Extracts

Rubia tinctorum samples were mixed (1:20, *w*/*v*) with a methanol/water solution (1:1 *v*/*v*) and heated in a water bath at 50 °C for 30 min, followed by sonication for 5 min in pulse mode with a 1 min stop for each 2.5 min, using a 1000 W probe-type ultrasonicator operated at 20 kHz (model UIP1000hdT, Hielscher Ultrasonics, Teltow, Germany). The solution was then centrifuged at 9000 rpm for 15 min and the supernatant was filtered through Whatman No. 1 paper. Aliquots were lyophilized for the vibrational spectroscopy analysis.

The infrared vibrational spectra of both dried and ground roots and the lyophilized extract were registered using a Thermo Scientific (Waltham, MA, USA) Nicolet iS50 Fourier-transform infrared spectrometer, equipped with an in-built diamond attenuated total reflection (ATR) system. The spectra were collected with a 1 cm^{-1} spectral resolution over the 400–4000 cm^{-1} range, taking the interferograms that resulted from co-adding 64 scans. The spectra were then corrected using the advanced ATR correction algorithm [27] available in OMNIC^{TM} software suite.

The hydroalcoholic plant extract was studied by gas chromatography–mass spectrometry (GC–MS) at the Research Support Services (STI) at Universidad de Alicante (Alicante, Spain), using a gas chromatograph model 7890A coupled to a quadrupole mass spectrometer model 5975C (both from Agilent Technologies). The chromatographic conditions were: 3 injections/vial, injection volume = 1 µL; injector temperature = 280 °C, in splitless mode; initial oven temperature = 60 °C, 2 min, followed by ramp-up of 10 °C/min to a final temperature of 300 °C, 15 min. The chromatographic column used for the separation of the compounds was an Agilent Technologies HP-5MS UI of 30 m length, 0.250 mm diameter and 0.25 µm film. The mass spectrometer conditions were: temperature of the electron impact source of the mass spectrometer = 230 °C and of the quadrupole = 150 °C; ionization energy = 70 eV. Test mixture 2 for apolar capillary columns according to Grob (Supelco 86501) and PFTBA tuning standards were used for equipment calibration. NIST11 library and the monograph by Adams [28] were used for compound identification.

2.3. Preparation of Chitosan Oligomers and Bioactive Formulations

Chitosan oligomers (COS) were prepared according to the procedure reported by Santos-Moriano, et al. [29], with the modifications indicated in [30], obtaining oligomers with a molecular weight <2000 Da.

The COS-*R. tinctorum* and stevioside–*R. tinctorum* conjugate complexes were obtained by mixing the respective solutions in a 1:1 (*v*/*v*) ratio. The mixtures were then sonicated for 15 min in five 3-minute periods (so that the temperature did not exceed 60 °C) using a probe-type ultrasonicator.

For the assays carried out on autoclaved wood, the conjugate complex was dispersed in an agar matrix (15 g/L in Milli-Q water), using a procedure analogous to the one described below for the in vitro tests.

For the in vivo assays, the bioactive product was dispersed in a calcium alginate matrix. Hydrogel beads were prepared as follows: the control product was added to a 3% sodium alginate solution in a 2:8 ratio (20 mL compound/80 mL sodium alginate). Subsequently, this solution was dispensed drop by drop onto a 3% calcium carbonate solution to polymerize (30 min curing), obtaining beads with diameters in the 4–6 mm range.

2.4. Fungal Isolates

The three fungal isolates used (Table 1) were supplied as lyophilized vials (later reconstituted and refreshed as PDA subcultures) by the Agricultural Technological Institute of Castilla and Leon (ITACYL, Valladolid, Spain) [31].

Table 1. Fungal isolates used in the study.

Code	Isolate	Binomial Nomenclature	Geographical Origin	Host/Date
ITACYL_F098	Y-084-01-01a	*Diplodia seriata* De Not.	Spain (DO Toro)	Grapevine ('Tempranillo') 2004
ITACYL_F118	Y-103-08-01	*Dothiorella viticola* A.J.L.Phillips and J.Luque	Spain (Extremadura)	Grapevine 2004
ITACYL_F111	Y-091-03-01c	*Neofusicoccum parvum* (Pennycook and Samuels) Crous, Slippers and A.J.L.Phillips	Spain (Navarra, nursery)	Grapevine ('Verdejo') 2006

2.5. Antifungal Activity Assessment

2.5.1. In vitro Tests of Mycelial Growth Inhibition

The antifungal activity of the different treatments was determined using the agar dilution method according to EUCAST standard antifungal susceptibility testing procedures [32], by incorporating aliquots of stock solutions onto the PDA medium to obtain concentrations in the 15.62–1500 μg·mL^{-1} range. Mycelial plugs (∅ = 5 mm) from the margin of 1-week-old PDA cultures of *D. seriata*, *D. viticola* or *N. parvum* were transferred to plates incorporating the above-mentioned concentrations for each treatment (3 plates per treatment/concentration, with 2 replicates each). Plates were then incubated at 25 °C in the dark for a week. PDA medium without any amendment was used as control. Mycelial growth inhibition was estimated according to the formula: $((d_c - d_t)/d_c) \times 100$, where d_c and d_t represent the average diameters of the fungal colony of the control and of the treated fungal colony, respectively. Effective concentrations (EC_{50} and EC_{90}) were estimated using PROBIT analysis in IBM SPSS Statistics v.25 (IBM; Armonk, NY, USA) software. The level of interaction (i.e., synergy factors) was determined according to Wadley's method [33].

2.5.2. Assays on Autoclaved Grape Wood

The formulation (COS-*R. tinctorum* conjugate) that showed the best performance in the in vitro assays was then tested on autoclaved grapevine wood to assess its behaviour on plant material against the least sensitive fungus in the previous plate tests. One-year-old dormant canes (*Vitis vinifera* L. cv. 'Tempranillo') were cut into 16 cm (length) and 0.8–1 cm (diameter) segments and autoclaved twice at 121 °C (20 min) to eliminate any microbial contamination. Inoculation was performed by first making two approximately 3 mm deep slits with a scalpel (without reaching the medullary tissue) per shoot, 8–10 cm apart and located in the internodes. A 3 mm diameter plug of PDA agar coming from the margin of a 10-day colony of the pathogen (*D. seriata*) was placed in each slit, flanked by 2 plugs (∅ = 3 mm) of bacteriological agar that contained the tested conjugate complex. After this, the wounds were covered with autoclaved cotton moistened with sterile bi-distilled water and sealed with Parafilm^{TM} tape. Inoculated shoots were placed in transparent culture boxes on a bed of sterile filter paper, periodically moistened (with sterile double distilled water), and incubated for 21 days in a climatic chamber at 26 °C, with 70% RH and a 12/12 h photoperiod. A total of 5 boxes with 3 replicates/box each were arranged, together with a positive control inoculated only with *D. seriata* (1 box with 3 replicates) and a negative control without pathogen, inoculated only with the conjugate (also 1 box with 3 replicates).

After the incubation period, segments were recovered from the boxes, and each of them was divided into two halves of approximately 8 cm, before longitudinal cuts were made in each half. Finally, the length of the vascular necroses produced was measured longitudinally on upper and lower directions from the inoculation point for both halves, and compared with those of controls.

2.5.3. Greenhouse Bioassays on Grafted Plants

Bioassays with COS (chosen as a reference) and COS-*R. tinctorum* conjugate complexes were performed in living plants in order to scale the protective capabilities of these compounds against the three selected *Botryosphaeriaceae* species in young grapevine plants. As summarized in Table S1, plant material consisted of 30 plants of 'Tempranillo' (CL. 32 clone) (2 year old) cultivar and 30 plants of 'Garnacha' (VCR3 clone) (1 year old) cultivar, grafted on 775P and 110R rootstocks, respectively. Each plant was cultured on a 3.5 L plastic pot containing a mixed substrate of moss peat and sterilized natural soil (75:25), incorporating slow release fertilizer when needed along the culture cycle. Plants were maintained in the greenhouse with drip irrigation and anti-weed ground cover from June to December 2020 (6 months). One week after placing them in pots, young, grafted plants were artificially inoculated with the pathogens and the COS-*R. tinctorum* treatment. Five repetitions (plants) were arranged for each pathogen*cultivar combination, together with 4 positive controls/(pathogen*cultivar) plus 3 negative controls (incorporating only the bioactive product) for each cultivar. Inoculations of both pathogens and the control product were carried out directly on the trunk of the living plants at two sites per stem (separated >5 cm) below the grafting point and not reaching the root crown. For the different fungi, agar plugs from the margin of 5 day old fresh PDA cultures of each species were used as fungal inoculum. In the aforementioned two inoculation points of each grapevine plant, slits of approx. 15 mm in diameter and 5 mm deep were made with a scalpel. Subsequently, 5 mm diameter agar plugs were placed directly into contact with vascular tissue in the stem; simultaneously, calcium alginate hydrogel beads containing the bioactive product were placed at both sides of the agar plug; and the whole set was covered with cotton soaked in sterile bi-distilled water and sealed with Parafilm^{TM} tape. During the culture period, application of copper (cuprous oxide 75%, Cobre Nordox^{TM} 75 WG) to control downy mildew outbreaks was performed in mid-July, accompanied with a first sprouting (followed by periodic sprouting). Plants were visually examined weekly for the presence of foliar symptoms. After six months in the greenhouse, the plants were removed, two sections of the inoculated stems between the grafting point and the root crown were prepared and sectioned longitudinally. The length of the vascular necroses was scored longitudinally on upper and lower directions from the inoculation point for both halves of the longitudinal cut, and the average measures of these were statistically analysed and compared depending on the type of pathogen. All the data were compared with positive and negative controls. Finally, grapevine plants removed and measured at the end of the assay were also processed to re-isolate the different pathogenic taxa previously inoculated. Thus, 5 mm long wood chips exhibiting vascular necroses (1–2 cm around the wounds) were washed, their surface sterilized, then placed in PDA plates amended with streptomycin sulphate (to avoid bacterial contamination) and incubated in a culture chamber at 26 °C in the dark for 2–3 days.

2.6. *Statistical Analyses*

The results of the in vitro inhibition of mycelial growth were statistically analyzed using one-way analysis of variance (ANOVA), followed by post hoc comparison of means through Tukey's test at $p < 0.05$ (provided that the homogeneity and homoscedasticity requirements were satisfied, according to the Shapiro–Wilk and Levene tests). In the case of autoclaved grapevine wood and greenhouse assay results, since the normality and homoscedasticity requirements were not met, the Kruskal–Wallis non-parametric test was used instead, with the Conover–Iman test for post hoc multiple pairwise comparisons. R statistical software was used for all the statistical analyses [34].

3. Results

3.1. *Vibrational Characterization*

The assignment of the main absorption bands in the infrared spectra of the *R. tinctorum* root powder and root extracts is shown in Table 2. The most prominent band, attributed to the benzene ring in aromatic compounds, occurs at ca. 1500 cm^{-1}. The bands at 1592 cm^{-1} and 1676 cm^{-1} can be assigned to the in-phase C=O and symmetrical C=C vibrations from anthraquinone.

Table 2. Main bands in the infrared spectra of root and lyophilized *R. tinctorum* extract and of two of its main constituents.

R. tinctorum		Anthraquinone	4-tert-butyl-2-phenylphenol	Assignment
Root Powder	Extract			
3334	3335			Bonded O–H stretching (cellulose)
2920	2920	2964 2925		sp^3 C–H =C–H groups of aromatic rings
		2856		aliphatic C–H asymmetrical stretching
		2724		β–OH, typical of α-hydroxy anthraquinone
1727	1733			C=O from esters
		1704		ester C=O
		1676		C=O in anthraquinones
1639	1620	1633		C=O in anthraquinones
1602	1605	1592	1585	phenyl ring (aromatic skeletal vibration) >C=C< in anthraquinones
1552			1545	carboxylate stretches/C=C aromatic
	1511		1480	methylene C–H bend
		1461	1470	methyl C–H asymmetrical
	1435		1430	=C–H in plane bending
1414	1416		1420	vinyl C–H in plane bending
1370	1406 1370	1377 1366	1385	C–C asymmetrical stretching phenolic hydroxyl groups
			1355	C–O stretching/methylene C–H bending
		1333 1329	1325	C–H in-plane deformation methylene C–H bending
1316	1316	1306		vinylidene C–H in plane bending
1255	1255	1287	1270	C–O stretching/C=C symmetric stretching
		1207	1215	C–O stretching/C–H in plane bending
		1171	1180	–C–O–C– stretching
	1142	1153	1135	
	1100	1099		
		1087	1080	
1020			1025	C–C stretching
	951	969		C–H out-of-plane bending

3.2. Gas Chromatography–Mass Spectrometry Analysis of the Extract

In *R. tinctorum* root hydromethanolic extracts, the main analyzed components (Table 3) were: the anthraquinone family (19.4%) consisting of 2-methyl-9,10-anthracenedione (or *β*-methylanthraquinone) (15.5%), 1,2-dihydroxyanthraquinone (or alizarin), 1,8-dihydroxy-3-methylanthraquinone, 1-hydroxy-9,10-anthracenedione (or *α*-hydroxyanthraquinone) and 1-hydroxy-4-methylanthraquinone; cyclopentenones (2.3%), such as 2-hydroxy-2-cyclopenten-

1-one and 4-cyclopentene-1,3-dione; and the phenol category (7.5%), constituted by *cis*-2-methoxy-4-(1-propenyl)-phenol (or *cis*-eugenol), 2-methoxy-phenol (or guaiacol), 2-methoxy-4-vinylphenol (or 4-vinyl-guaiacol), 4-*tert*-butyl-2-phenylphenol, and coniferyl alcohol. Other phytochemicals of interest were 4-methoxy-4',5'-methylenedioxybiphenyl-2-carboxylic acid (8.6%), 1,4-diacetyl-3-acetoxymethyl-2,5-methylene-l-rhamnitol (8.3%) and guanosine (5.8%).

Table 3. Phytochemicals identified in *R. tinctorum* root hydromethanolic extract by GC–MS.

Peak	R_t (min)	Area (%)	Assignments
1	4.6369	1.99	4-pentenoic acid, ethyl ester
2	4.7440	0.26	l-gala-l-ido-octose
3	4.8414	0.52	4-cyclopentene-1,3-dione
4	5.0021	1.09	oxime-, methoxy-phenyl-
5	5.1968	0.46	1-(2-furanyl)-ethanone
6	5.2942	1.36	2,5-diethenyltetrahydro-2-methyl-furan
7	5.3770	1.82	2-hydroxy-2-cyclopenten-1-one
8	6.0781	0.70	2,4-dihydroxy-2,5-dimethyl-3(2H)-furan-3-one
9	6.3312	2.91	2-hydroxy-γ-butyrolactone
10	6.6331	2.47	glycerin
11	6.7694	1.93	1,2-cyclopentanedione, 3-methyl-
12	6.9690	1.19	2-acetamido-2-deoxy-α-D-glucopyranose
13	7.1345	0.76	butyronitrile, 4-ethoxy-3-hydroxy-
14	7.2952	1.55	2,5-dimethyl-4-hydroxy-3(2H)-furanone
15	7.4267	0.77	trimethyl(tetrahydrofuran-2-ylperoxy)silane
16	7.6798	1.01	2-methoxy-phenol (or guaiacol)
17	7.7869	2.20	L-alanine, methyl ester
18	8.2008	2.65	dimethyl dl-malate
19	8.3955	0.64	ethanamine, N-ethyl-N-nitroso-
20	8.5270	2.47	4H-pyran-4-one, 2,3-dihydro-3,5-dihydroxy-6-methyl-
21	9.1453	0.87	4H-pyran-4-one, 3,5-dihydroxy-2-methyl-
22	9.2865	0.97	catechol
23	9.4471	0.94	1,4:3,6-dianhydro-α-d-glucopyranose
24	9.6857	0.35	5-hydroxymethylfurfural
25	10.3965	0.31	2-acetoxy-5-hydroxyacetophenone
26	10.5718	0.50	p-cymen-7-ol
27	10.8882	2.80	2-methoxy-4-vinylphenol (or 4-vinylguaiacol)
28	11.2193	0.98	DL-arabinose
29	11.7451	1.09	DL-proline, 5-oxo-, methyl ester
30	12.0470	0.74	vanillin
31	12.6702	1.96	2-methoxy-4-(1-propenyl)-phenol (Z)- (or *cis*-isoeugenol)
32	13.1473	1.05	1-[4-(methylthio)phenyl]-ethanone
33	13.4492	0.86	butylated hydroxytoluene
34	13.6877	0.57	benzeneacetic acid, 4-hydroxy-3-methoxy-, methyl ester
35	13.9458	0.82	1,4-diacetyl-3-acetoxymethyl-2,5-methylene-l-rhamnitol
36	14.3693	2.44	α-methyl-l-sorboside
37	15.0461	5.78	guanosine
38	15.5865	8.31	1,4-diacetyl-3-acetoxymethyl-2,5-methylene-l-rhamnitol
39	16.0734	1.65	4-((1E)-3-hydroxy-1-propenyl)-2-methoxyphenol (or coniferyl alcohol)
40	17.4561	0.43	5-amino-1-(4-amino-furazan-3-yl)-1H-[1–3]triazole-4-carbonitrile
41	17.9088	1.18	hexadecanoic acid, methyl ester
42	18.2594	1.17	n-hexadecanoic acid
43	19.1552	0.76	5-(1,1-dimethylethyl)[1,1′-biphenyl]-2-ol (or 4-tert-butyl-2-phenylphenol)
45	19.4278	0.69	cyclopentadecane
46	19.6421	0.15	1-hydroxy-9,10-anthracenedione (or α-hydroxyanthraquinone)
47	19.8709	15.54	9,10-anthracenedione, 2-methyl- (or β-methylanthraquinone)
48	20.7278	1.43	1-hydroxy-4-methylanthraquinone
49	21.1659	1.75	1,2-dihydroxyanthraquinone (or alizarin)

Table 3. *Cont.*

Peak	R_t (min)	Area (%)	Assignments
50	21.8038	1.40	azacyclotridecan-2-one, 1-(3-aminopropyl)-
51	23.0550	0.81	glycerol 1-palmitate
52	23.3812	0.73	bis(2-ethylhexyl) phthalate
53	23.8973	0.24	1,8-dihydroxy-3-methyl-9,10-anthracenedione (or 1,8-dihydroxy-3-methyl anthraquinone)
54	24.2673	8.57	4-methoxy-4′,5′-methylenedioxybiphenyl-2-carboxylic acid
55	24.4474	0.42	9-octadecenoic acid (Z)-, 2-hydroxy-1-(hydroxymethyl)ethyl ester
56	25.4260	0.35	squalene
57	29.2333	0.62	octasiloxane, 1,1,3,3,5,5,7,7,9,9,11,11,13,13,15,15-hexadecamethyl-
58	29.9051	0.40	γ-sitosterol

It is worth noting that the flavoring phenols found in the hydroalcoholic extracts from *R. tinctorum* (guaiacol, 4-vinyl-guaiacol and *cis*-eugenol) were the same present in oak, which are used to confer aroma to wine.

3.3. *Antifungal Activity*

3.3.1. In vitro Tests of Mycelial Growth Inhibition

The results of the mycelial growth inhibition tests for the hydromethanolic *R. tinctorum* root extract, alone or forming a conjugate complex with COS or stevioside, are presented in Figure 2 and Figures S2–S4. The antifungal activity of the extract was found to be much higher than those of COS and stevioside alone, reaching full inhibition at concentrations in the 93.8–250 $\mu g \cdot mL^{-1}$ range, depending on the pathogen (vs. 1500 $\mu g \cdot mL^{-1}$ for COS and stevioside). Upon conjugation with stevioside, some improvement in the germicide effect could be observed: for instance, the inhibition of *D. seriata* was higher at the 78.1 $\mu g \cdot mL^{-1}$ concentration (76.3% vs. 45.9%), and the full inhibition of *D. viticola* and *N. parvum* was attained at a lower concentration (93.8 vs. 125 $\mu g \cdot mL^{-1}$, and 125 vs. 250 $\mu g \cdot mL^{-1}$, respectively). Nonetheless, the best results were obtained for the COS–*R. tinctorum* extracts, for which full inhibition was recorded at the lowest concentrations (in the 70.3–78.1 $\mu g \cdot mL^{-1}$ range).

In order to provide a tentative explanation for the strong antifungal activity observed in the extracts, three of the presumably bioactive constituents were also assayed (an anthracenedione, a phenol and a purine nucleoside) separately. The results, presented in Figures S5–S8, showed that 4-*tert*-butyl-2-phenylphenol was the most active (full inhibition of the three fungi was attained at concentrations in the 78.1–93.8 $\mu g \cdot mL^{-1}$ range), but 1,2,4-trihydroxyanthraquinone and guanosine were also effective (full inhibition was reached at concentrations in the 187.5–500 and 250–375 $\mu g \cdot mL^{-1}$ ranges, respectively). Such values are comparable to those found for the whole *R. tinctorum* extract, suggesting that the activity cannot be ascribed to a single constituent, but rather to the combination of several of them.

To quantify the synergistic behavior observed for the conjugate complexes, effective concentrations were estimated (Table 4) and synergy factors (SF) were then calculated according to Wadley's method (Table 5). As expected, the synergism between COS and *R. tinctorum* extract was noticeably higher than the one observed between stevioside and *R. tinctorum* extract, with SF values in the 2.23–5.35 and 1.36–3.29 range, respectively.

3.3.2. Assays on Autoclaved Grapevine Wood

The results from the ex situ experiment conducted on autoclaved grapevine canes for the most promising treatment (COS-*R. tinctorum* conjugate complex) and the least sensitive fungus (*D. seriata*), presented in Table 6, showed that the application of the bioactive product led to statistically significant differences in terms of vascular necroses vs. the positive control. Nonetheless, it did not lead to full inhibition, given that there were statistically significant differences in the length of vascular lesions compared with the negative control (shoots inoculated only with the bioactive compound). This could be tentatively attributed to the chosen dispersion medium (agar), which was replaced with calcium alginate in subsequent in vivo experiments.

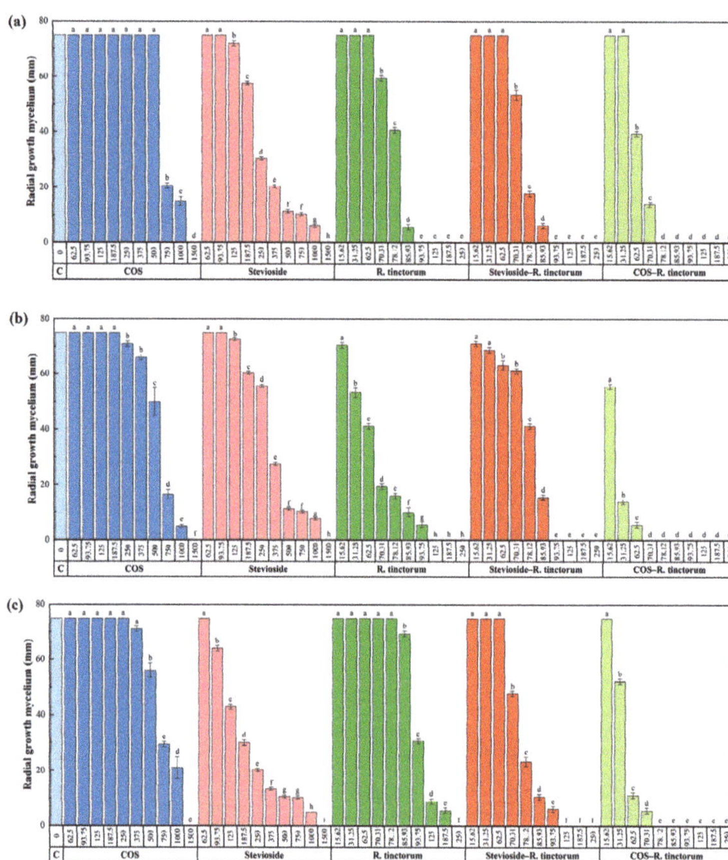

Figure 2. Colony growth measures of (**a**) *D. seriata*, (**b**) *D. viticola* and (**c**) *N. parvum* strains when cultured in PDA plates containing the various control products (viz. chitosan oligomers (COS), stevioside, *R. tinctorum* hydromethanolic extract, stevioside–*R. tinctorum* and COS-*R. tinctorum* conjugate complexes) at concentrations in the 62.5–1500 and 15.62–250 μg·mL^{-1} range ordered according to the least and the most active products, respectively. The same letters above concentrations indicate that they are not significantly different at $p < 0.05$. Error bars represent standard deviations.

Table 4. EC_{50} and EC_{90} effective concentrations. Values are expressed in μg·mL^{-1}, and are followed by the standard errors of fit.

Pathogen	EC	COS	Stevioside	*R. tinctorum*	COS—*R. tinctorum*	Stevioside—*R. tinctorum*	4-*tert* …	1,2,4-trihydro …	Guanosine
D. seriata	EC_{50}	744.4 ± 43.9	288.1 ± 15.3	78.0 ± 0.8	63.1 ± 0.3	73.6 ± 0.3	53.0 ± 2.1	45.4 ± 3.4	130.4 ± 12.8
	EC_{90}	1179.9 ± 58.2	840.5 ± 62.3	87.8 ± 1.9	73.4 ± 0.9	82.4 ±0.7	73.2 ± 2.3	171.4 ± 18.7	249.9 ± 28.5
D. viticola	EC_{50}	554.3 ± 27.4	306.9 ± 26.6	66.2 ± 2.9	22.1 ± 1.4	80.0 ± 0.7	25.7 ± 3.6	37.2 *	182.7 ± 7.7
	EC_{90}	1138.7 ± 75.0	917.0 ± 74.3	90.2 ± 8.7	55.5 ± 4.6	90.7 ± 1.5	71.2 ± 9.0	74.9 *	308.1 ± 23.7
N. parvum	EC_{50}	680.2 ± 43.1	194.8 ± 13.4	92.3 ± 0.5	38.2 ± 1.4	75.1 ± 0.8	62.2 ± 0.7	72.0 ± 14.8	95.1 ± 22.6
	EC_{90}	1326.6 ± 83.2	723.8 ± 56.7	184.0 ± 1.1	66.3 ± 4.2	89.2 ± 1.9	70.6 ± 2.2	338.4 ± 37.9	317.8 ± 33.9

* Could not be reliably calculated (lack of points).

Table 5. Synergy factors, estimated according to Wadley's method.

Pathogen	EC	Synergy Factor	
		COS-*R. tinctorum*	Stevioside—*R. tinctorum*
D. seriata	EC_{50}	2.24	1.67
	EC_{90}	2.23	1.93
D. viticola	EC_{50}	5.35	1.36
	EC_{90}	3.01	1.81
N. parvum	EC_{50}	4.26	1.67
	EC_{90}	4.87	3.29

Table 6. Kruskal–Wallis test and multiple pairwise comparisons using the Conover–Iman procedure for the lengths of the vascular necroses scored for *D. seriata* in the ex situ autoclaved grapevine canes assay.

Sample	Frequency	Sum of Ranks	Mean of Ranks	Groups		
COS-*R. tinctorum* negative control	24	300.000	12.500	A		
COS-*R. tinctorum-D. seriata*	120	10,193.000	84.942		B	
Positive control	24	3703.000	154.292			C

Treatments/controls labelled with the different letters are significantly different at $p < 0.05$.

3.3.3. Greenhouse Bioassays on Grafted Plants

When the best treatment (COS-*R. tinctorum* conjugate complex) was further assayed in vivo, significant differences were found against the positive controls in all cases (Table 7), confirming its antifungal behavior on the plant material. Nonetheless, complete inhibition was not reached against any of the three pathogens for the assayed dose (100 $\mu g \cdot mL^{-1}$) comparing with non-infected controls, suggesting that a higher concentration than the EC_{90} values found in the in vitro tests (and/or a different dispersion medium) should be assayed when the treatment is used in future field trials.

Table 7. Kruskal-Wallis test and multiple pairwise comparisons using the Conover–Iman procedure for the lengths of the vascular necroses for the three phytopathogen in greenhouse in vivo assays.

Pathogen	Sample	Frequency	Sum of Ranks	Mean of Ranks	Groups		
D. seriata	COS-*R. tinctorum* negative control	32	725.500	22.672	A		
	COS-*R. tinctorum-D. seriata*	72	6124.000	85.056		B	
	Positive control	56	6030.500	107.688			C
D. viticola	COS-*R. tinctorum* negative control	32	1295.000	40.469	A		
	COS-*R. tinctorum-D. viticola*	72	4885.000	67.847		B	
	Positive control	64	8016.000	125.250			C
N. parvum	COS-*R. tinctorum* negative control	32	572.000	17.875	A		
	COS-*R. tinctorum-N. parvum*	48	3695.000	76.979		B	
	Positive control	64	6173.000	96.453			C

Treatments/controls labelled with the different letters are significantly different at $p < 0.05$.

4. Discussion

4.1. On the Constituents of *R. tinctorum* Extracts

The composition here reported was different from that found by Derksen and Van Beek [35] (using LC–DAD and HPLC–MS(/MS) with ESI or APCI), where lucidin primeveroside and ruberythric acid were the major anthraquinone components in an ethanolic-water extract, and from the one reported by Jalill [18] for a methanolic extract, which was rich in 9,12-octadecadienoic acid (29.75%), 9-octadecenoic acid hexadecyl ester (26.1%) and 2-ethyl-2-(hydroxymethyl)-1,3-propanediol, (10.1%), but poor in anthracenediones (4.0%) and 2-methoxy-4-vinylphenol (0.5%).

Significant differences in composition were also observed in comparison with the *Rubia cordifolia* essential oil characterized by GC–MS, in which mollugin (rubimaillin or methyl 6-hydroxy-2,2-dimethylbenzo[h]chromene-5-carboxylate) was found to be the major component, followed by 3-methyl-2-cyclopenten-1-one, eugenol, anethole and 4-*tert*-butyl-2-phenylphenol [10,11].

Although the geographical location, time of year and age of the plant are known to influence the composition [15], the observed differences should be mainly ascribed to differences in both the extractive chemicals and in the extraction process (nature of the alcoholic solvent, alcohol:water ratio and mechanical enhancers such as sonication [36,37]), and to the characterization technique, provided that previous studies on *R. tinctorum* extracts [36,38–41] were conducted by HPLC and LC–HRMS (instead of GC–MS) and generally focused only on anthraquinones, anthraquinone glycosides and aglycones.

4.2. *On the Combined Effect of Anthraquinones and Phenols*

It is known that increasing the activity of a parent molecule can be pursued either by testing multiple substituent changes on the base core (the impact of the number, nature, and location of substituents on the anthraquinone moiety on its inhibitory potency against pathogenic fungi has been studied in [42]), or by testing the effect of coexistence with other molecules with which synergistic behavior may occur. In general, anthraquinone per se is a relatively inert compound, but in the presence of glucose, anthrahydroquinone units (formed by reduction of anthraquinone) reduce the quinone–methide units (issued by dehydration of phenolic β-*O*-4 lignin) mainly by electron transfer leading to guaiacol [43]. Thus, the presence of 4-vinyl-guaiacol, *cis*-eugenol, coniferyl alcohol or 4-*tert*-butyl-2-phenylphenol phytochemicals in the *R. tinctorum* hydromethanolic extract should be referred to the same origin. As regards a subsequent interaction of these phenols with anthraquinones, it cannot be excluded: Maurino et al. [44] have demonstrated that quinonoid compounds excited by sunlight react with phenols, transforming them into tetrasubstituted dihydroxybiphenyls and phenoxyphenols. Nevertheless, in the absence of induced sunlight, no reaction between anthraquinones and methoxy- and phenyl-phenols has been described in the literature (to the best of the authors' knowledge), so at this point it is not possible to establish whether the activity of *R. tinctorum* extracts may be referred to an additive effect of both families of components or to a synergistic one.

4.3. *Comparison with Efficacies Reported in the Literature*

An overview of the antimicrobial activities reported for *R. tinctorum* in the literature is presented in Table S2. Concerning its antifungal behavior, full inhibition of *Aspergillus flavus* Link and *Fusarium oxysporum* Schltdl. at a concentration of 100 $\mu g \cdot mL^{-1}$ has been reported by Kalyoncu, et al. [45], and inhibition percentages in the 18–43% range were reported against *Trichoderma viride* Pers., *Doratomyces stemonitis* (Pers.) Nees, *Aspergillus niger*, *Penicillium verrucosum* Dierckx, *Alternaria alternate* (Fr.) Keissl., *Aureobasidium pullulans* (de Bary) G. Arnaud and *Mucor mucedo* L. by Manojlovic et al. [46], although the assayed concentration was not reported. Activity against other fungi (e.g., *Penicillium expansum* Link, *Geotrichum candidum* Link, *Fusarium solani* (Mart.) Sacc., *Postia placenta* (Fr.) M.J. Larsen and Lombard, *Trametes versicolor* (L.) Lloyd) has also been reported, albeit not in a quantitative manner [47,48].

The contribution of anthraquinones to antifungal activity is well-established, given that anthracenediones from other plants have proven to be effective against a wide variety of phytopathogenic fungi. For instance, anthraquinones isolated from *Cassia tora* L., *Coccoloba mollis* Casar., *Rheum palmatum* L., *Morinda lucida* Benth. or *Aegle marmelos* (L.) Corrêa, to name a few, showed antifungal behavior against phytopathogenic fungi such as *Botrytis cinerea*, *Blumeria graminis* (DC.) Speer, *Phytophthora infestans* (Mont.) de Bary, *Puccinia recondita* Roberge ex Desm., *Pyricularia grisea* Sacc., *Rhizoctonia solani* J.G. Kühn, *Botryospheria ribis* Grossenbacher and Duggar, *B. rhodina* (Berk. and M.A. Curtis) Arx, *Lasiodiplodia theobromae* (Pat.) Griffon and Maubl., *Fusarium* sp., *Fusarium graminearum*

Schwabe, *Mycosphaerella melonis* (Pass.) W.F. Chiu and J.C. Walker, *Fusarium oxysporum* f. sp. *vasinfectum* (G.F. Atk.) W.C. Snyder and H.N. Hansen, *Phyllosticta zeae* Stout, *Sclerotinia sclerotiorum* (Lib.) de Bary, *Cladosporium cucumerinum* Ellis and Arthur and *Aspergillus* spp. [49–53]. The underlying mechanism of action has been studied, for example, for purpurin against *Candida* spp., finding that it elevates intracellular ROS levels, depolarizes the mitochondrial membrane potential, downregulates of the expression of hypha-specific genes and the central morphogenetic regulator Ras1p and degrades DNA [54,55].

On the other hand, the antifungal activities of 2-methoxy- and 2-*tert*-butyl-substituted phenols against phytopathogens have been less studied, although a strong antifungal activity of 2-methoxy-4-(1-propenyl)-phenol against *Botrytis cinerea* was reported by Wang et al. [12]; against *B. rhodina, Rhizoctonia* sp. and *Alternaria* sp. by de Oliveira Pereira et al. [56]; and against *A. alternata* (Fr.) Keissl., *Sarocladium oryzae* (Sawada) W. Gams and D. Hawksw., *F. graminearum, F. equiseti* (Corda) Sacc. and *F. verticillioides* (Sacc.) Nirenberg by Pilar Santamarina et al. [57]. Likewise, 2-methoxy-phenol was effective against sap-staining fungi (*Ophiostoma* spp.), according to Velmurugan et al. [13]. Regarding their mechanism of action, it has been proposed that, for instance, eugenol acts on cell membrane by a mechanism that seems to involve the inhibition of ergosterol biosynthesis, and the lower ergosterol content interferes with the integrity and functionality of the cell membrane [56]. It has also been suggested that, taking into consideration that it induced the generation of H_2O_2 and increased free Ca^{2+} in the cytoplasm, its activity may also be referred to membrane binding and permeability alteration, leading to the destabilization and disruption of the plasma membrane [12].

4.4. On the Synergistic Behaviour of R. tinctorum Extracts with COS and Stevioside

To date, it has been verified that chitosan acts as an elicitor on *R. tinctorum*, stimulating anthraquinone synthesis [58]; chitosan/poly (lactic acid) nanoparticles have been evaluated as a novel carrier for the delivery of anthraquinone [59]; and chitosan-based hydrogels have been studied for the adsorption of anthraquinone dyes [60]. Nonetheless, after a thorough bibliographical survey, no previous examples of the use of chitosan or stevioside for the formation of conjugate complexes with anthraquinones could be found.

On the other hand, examples of synergistic behaviour have been reported, for instance, for chitosan combined with *Cinnamomum zeylanicum* Blume essential oils, rich in eugenol [61]. These authors hypothesized that eugenol alters the surface and structure of the fungal cell wall, and COS acts as a potentiator by reducing cell wall synthesis and facilitating death in an energy-dependent manner. In this regard, the accepted and potential mechanisms of action behind the antimicrobial properties of chitosan have been thoroughly discussed in the review paper by Ma et al. [62]. Those of stevioside have been discussed in [63], and are related to the uncoupling of mitochondrial oxidative phosphorylation and the permeabilization of the cell membrane.

Nonetheless, taking into consideration that the antifungal activity of both COS and stevioside alone was substantially lower than that of the *R. tinctorum* extract, and given that the use of most free anthraquinones in pharmaceutical industries is limited by their poor water solubility and low bioavailability [64], the observed strong synergistic behavior with COS and stevioside should probably be referred to a solubility and bioavailability enhancement through the formation of inclusion compounds or conjugate complexes (discussed, in the case of chitosan, in the recent review paper by Detsi et al. [65] and, for steviol glycosides, in the works by Nguyen et al. [66,67]). Examples of antifungal activity enhancement via the formation of conjugate complexes against GTDs have been previously reported in [25,26,68], albeit with worse EC_{50} and EC_{90} values than those reported in this work.

5. Conclusions

The GC–MS analysis of *R. tinctorum* hydroalcoholic extracts revealed that, apart from members of the anthraquinone family (19.4%), flavoring phenols similar to those found in oak (used to confer aroma to wine) and guanosine were also present. *R. tinctorum* extract, alone and forming conjugate complexes with COS and stevioside, along with three of its constituents, were assayed in vitro against three *Botryosphaeriaceae* taxa. *R. tinctorum* extract led to a strong mycelial growth inhibitory effect in all cases, with EC_{90} values as 88 $\mu g \cdot mL^{-1}$. Although 4-*tert*-butyl-2-phenylphenol was its most active constituent, 1,2,4-trihydroxyanthraquinone and guanosine were also effective, suggesting the activity cannot be ascribed to a single constituent, but rather to the combination of several of them. As regards the strong synergistic behavior observed upon conjugation with COS, which resulted in EC_{90} values in the 56–73 $\mu g \cdot mL^{-1}$ range, it may be ascribed to solubility and bioavailability enhancement, rather than to the antifungal activity of chitosan (which is much weaker than that of *R. tinctorum*). The treatment for which the best results were attained in plate tests (COS-*R. tinctorum* conjugate complex) was then tested ex situ on autoclaved grapevine twigs and in young, grafted plants in greenhouse assays. A significant reduction in the infection rate was found in all cases. Hence, this natural antifungal compound may deserve further examination in larger field trials, as it may be hold promise for the sustainable control of GTDs.

Supplementary Materials: The following are available online at https://www.mdpi.com/article/10.3390/plants10081527/s1, Table S1. Repetitions for each of the plant/treatment combinations in the greenhouse bioassay. Each grafted plant was inoculated at two sites below grafting point; Table S2. Examples of application of *R. tinctorum* extracts against microorganisms reported in the literature; Figure S1. GC–MS spectrum of *R. tinctorum* root hydromethanolic extract; Figures S2–S4. Mycelial growth inhibition of *D. seriata*/*D. viticola*/*N. parvum* upon treatment with: chitosan oligomers, stevioside, *R. tinctorum* hydromethanolic extract, stevioside–*R. tinctorum* conjugate complex, and COS-*R. tinctorum* conjugate complex at different concentrations; Figure S5. Colony growth measures of *D. seriata*, *D. viticola* and *N. parvum* strains when cultured in PDA plates containing the main phytochemicals found in *R. tinctorum* hydromethanolic extracts at concentrations in the 62.5–1500 and 15.62–250 $\mu g \cdot mL^{-1}$ range for the least and the most active products, respectively; Figures S6–S8. Mycelial growth inhibition of *D. seriata*/*D. viticola*/*N. parvum* upon treatment with the main phytochemicals found in *R. tinctorum* hydromethanolic extracts: purpurin, guanosine, and 4-*tert*-butyl-2-phenylphenol, at different concentrations.

Author Contributions: Conceptualization, J.M.-G., P.M.-R. and V.G.-G.; methodology, J.M.-G., J.C.-G. and V.G.-G.; validation, J.C.-G., V.G.-G. and P.M.-R.; formal analysis, J.C.-G., V.G.-G. and P.M.-R.; investigation, N.L.-L., E.S.-H., L.B.-D., V.G.-G., J.C.-G., J.M.-G. and P.M.-R.; resources, J.M.-G. and P.M.-R.; data curation, J.C.-G.; writing—original draft preparation, N.L.-L., E.S.-H., L.B.-D., V.G.-G., J.C.-G., J.M.-G. and P.M.-R.; writing—review and editing, V.G.-G. and P.M.-R.; visualization, N.L-L. and E.S.-H.; supervision, V.G.-G. and P.M.-R.; project administration, V.G.-G., J.M.-G. and P.M.-R.; funding acquisition, J.M.-G. and P.M.-R. All authors have read and agreed to the published version of the manuscript.

Funding: This research was funded by Junta de Castilla y León under project VA258P18, with FEDER co-funding; by Cátedra Agrobank under the "IV Convocatoria de Ayudas de la Cátedra AgroBank para la transferencia del conocimiento al sector agroalimentario" program and by Fundación Ibercaja-Universidad de Zaragoza under the "Convocatoria Fundación Ibercaja-Universidad de Zaragoza de proyectos de investigación, desarrollo e innovación para jóvenes investigadores" program.

Data Availability Statement: The data presented in this study are available on request from the corresponding author. The data are not publicly available due to their relevance to an ongoing Ph.D. thesis.

Acknowledgments: V.G.-G thanks C. Julián (Plant Protection Unit, CITA) for her technical assistance. The authors gratefully acknowledge the support of Pilar Blasco and Pablo Candela at the Servicios Técnicos de Investigación, Universidad de Alicante, for conducting the GC–MS analyses.

Conflicts of Interest: The authors declare no conflict of interest. The funders had no role in the design of the study; in the collection, analyses, or interpretation of data; in the writing of the manuscript, or in the decision to publish the results.

References

1. Qiu, H.; Liu, R.; Long, L. Analysis of chemical composition of extractives by acetone and the chromatic aberration of teak (Tectona Grandis L.F.) from China. *Molecules* **2019**, *24*, 1989. [CrossRef] [PubMed]
2. Murdock, K.C.; Child, R.G.; Fabio, P.F.; Angier, R.D.; Wallace, R.E.; Durr, F.E.; Citarella, R.V. Antitumor agents. 1. 1,4-Bis[(aminoalkyl)amino]-9,10-anthracenediones. *J. Med. Chem.* **1979**, *22*, 1024–1030. [CrossRef] [PubMed]
3. Coufal, N.; Farnaes, L. Anthracyclines and anthracenediones. In *Cancer Management in Man: Chemotherapy, Biological Therapy, Hyperthermia and Supporting Measures*; Minev, B., Ed.; Springer: Dordrecht, The Netherlands, 2011; pp. 87–102. [CrossRef]
4. Wuthi-udomlert, M.; Kupittayanant, P.; Gritsanapan, W. In vitro evaluation of antifungal activity of anthraquinone derivatives of Senna alata. *J. Health Res.* **2010**, *24*, 117–122.
5. Agarwal, S.; Singh, S.S.; Verma, S.; Kumar, S. Antifungal activity of anthraquinone derivatives from Rheum emodi. *J. Ethnopharmacol.* **2000**, *72*, 43–46. [CrossRef]
6. Singh, D.; Verma, N.; Raghuwanshi, S.; Shukla, P.; Kulshreshtha, D. Antifungal anthraquinones from Saprosma fragrans. *Bioorg. Med. Chem. Lett.* **2006**, *16*, 4512–4514. [CrossRef]
7. Singh, J.; Hussain, Y.; Luqman, S.; Meena, A. Purpurin: A natural anthraquinone with multifaceted pharmacological activities. *Phytother. Res.* **2020**, *35*, 2418–2428. [CrossRef]
8. Essaidi, I.; Snoussi, A.; Koubaier, H.B.H.; Casabianca, H.; Bouzouita, N. Effect of acid hydrolysis on alizarin content, antioxidant and antimicrobial activities of Rubia tinctorum extracts. *Pigment. Resin Technol.* **2017**, *46*, 379–384. [CrossRef]
9. Ilc, T.; Werck-Reichhart, D.; Navrot, N. Meta-analysis of the core aroma components of grape and wine aroma. *Front. Plant Sci.* **2016**, *7*, 1472. [CrossRef]
10. Miyazawa, M.; Kawata, J. Identification of the key aroma compounds in dried roots of Rubia cordifolia. *J. Oleo Sci.* **2006**, *55*, 37–39. [CrossRef]
11. Li, W.-Q.; Quan, M.-P.; Li, Q. Chemical composition and antibacterial activity of the essential oil from Qiancao (Rubia cordifolia Linn.) roots against selected foodborne pathogens. *Asian J. Agric. Food Sci.* **2019**, *7*. [CrossRef]
12. Wang, C.; Zhang, J.; Chen, H.; Fan, Y.; Shi, Z. Antifungal activity of eugenol against Botrytis cinerea. *Trop. Plant Pathol.* **2010**, *35*, 137–143. [CrossRef]
13. Velmurugan, N.; Han, S.S.; Lee, Y.S. Antifungal activity of neutralized wood vinegar with water extracts of Pinus densiflora and Quercus serrata saw dusts. *Int. J. Environ. Res.* **2009**, *3*, 167–176. [CrossRef]
14. Singh, R.; Chauhan, S.M.S.G. 9,10-Anthraquinones and other biologically active compounds from the GenusRubia. *Chem. Biodivers.* **2004**, *1*, 1241–1264. [CrossRef]
15. Angelini, L.G.; Pistelli, L.; Belloni, P.; Bertoli, A.; Panconesi, S. Rubia tinctorum a source of natural dyes: Agronomic evaluation, quantitative analysis of alizarin and industrial assays. *Ind. Crop. Prod.* **1997**, *6*, 303–311. [CrossRef]
16. Boldizsár, I.; Szücs, Z.; Füzfai, Z.; Molnár-Perl, I. Identification and quantification of the constituents of madder root by gas chromatography and high-performance liquid chromatography. *J. Chromatogr. A* **2006**, *1133*, 259–274. [CrossRef]
17. Nakanishi, F.; Nagasawa, Y.; Kabaya, Y.; Sekimoto, H.; Shimomura, K. Characterization of lucidin formation in *Rubia tinctorum* L. *Plant Physiol. Biochem.* **2005**, *43*, 921–928. [CrossRef]
18. Jalill, R.D.A. GC-MS analysis of extract of Rubia tinctorum having anticancer properties. *Int. J. Pharmacogn. Phytochem. Res.* **2017**, *9*, 286–292.
19. Úrbez-Torres, J.R.; Hrycan, J.; Hart, M.; Bowen, P.; Forge, T. Grapevine trunk disease fungi: Their roles as latent pathogens and stress factors that favour disease development and symptom expression. *Phytopathol. Mediterr.* **2020**, *59*, 395–424. [CrossRef]
20. Bertsch, C.; Ramírez-Suero, M.; Magninrobert, M.; Larignon, P.; Chong, J.; Mansour, E.A.; Spagnolo, A.; Clément, C.; Fontaine, F. Grapevine trunk diseases: Complex and still poorly understood. *Plant Pathol.* **2012**, *62*, 243–265. [CrossRef]
21. Gramaje, D.; Armengol, J. Fungal Trunk pathogens in the grapevine propagation process: Potential inoculum sources, detection, identification, and management strategies. *Plant Dis.* **2011**, *95*, 1040–1055. [CrossRef]
22. Fontaine, F.; Gramaje, D.; Armengol, J.; Smart, R.; Nagy, Z.A.; Borgo, M.; Rego, C.; Corio-Costet, M.-F. *Grapevine Trunk Diseases: A Review*; OIV Publications: Paris, France, 2016; p. 25.
23. Mondello, V.; Songy, A.; Battiston, E.; Pinto, C.; Coppin, C.; Trotel-Aziz, P.; Clément, C.; Mugnai, L.; Fontaine, F. Grapevine trunk diseases: A review of fifteen years of trials for their control with chemicals and biocontrol agents. *Plant Dis.* **2018**, *102*, 1189–1217. [CrossRef] [PubMed]
24. Gramaje, D.; Úrbez-Torres, J.R.; Sosnowski, M.R. Managing grapevine trunk diseases with respect to etiology and epidemiology: Current strategies and future prospects. *Plant Dis.* **2018**, *102*, 12–39. [CrossRef] [PubMed]
25. Langa-Lomba, N.; Buzón-Durán, L.; Sánchez-Hernández, E.; Martín-Ramos, P.; Casanova-Gascón, J.; Martín-Gil, J.; González-García, V. Antifungal activity against *Botryosphaeriaceae* fungi of the hydro-methanolic extract of *Silybum marianum* capitula conjugated with stevioside. *Plants* **2021**, *10*, 1363. [CrossRef]
26. Langa-Lomba, N.; Buzón-Durán, L.; Martín-Ramos, P.; Casanova-Gascón, J.; Martín-Gil, J.; Sánchez-Hernández, E.; González-García, V. Assessment of conjugate complexes of chitosan and *Urtica dioica* or *Equisetum arvense* extracts for the control of grapevine trunk pathogens. *Agronomy* **2021**, *11*, 976. [CrossRef]
27. Nunn, S.; Nishikida, K. *Advanced ATR Correction Algorithm—Application Note 50581*; ThermoScientific: Madison, WI, USA, 2008; p. 4.
28. Adams, R.P. *Identification of Essential Oil Components by Gas Chromatography/Mass Spectroscopy*, 4th ed.; Allured Publishing Corp.: Carol Stream, IL, USA, 2007; p. 804.

29. Santos-Moriano, P.; Fernandez-Arrojo, L.; Mengíbar, M.; Belmonte-Reche, E.; Peñalver, P.; Acosta, F.N.; Ballesteros, A.O.; Morales, J.C.; Kidibule, P.; Fernandez-Lobato, M.; et al. Enzymatic production of fully deacetylated chitooligosaccharides and their neuroprotective and anti-inflammatory properties. *Biocatal. Biotransform.* **2017**, *36*, 57–67. [CrossRef]
30. Buzón-Durán, L.; Martín-Gil, J.; Pérez-Lebeña, E.; Ruano-Rosa, D.; Revuelta, J.L.; Casanova-Gascón, J.; Ramos-Sánchez, M.C.; Martín-Ramos, P. Antifungal agents based on chitosan oligomers, ε-polylysine and streptomyces spp. secondary metabolites against three botryosphaeriaceae species. *Antibiotics* **2019**, *8*, 99. [CrossRef]
31. Martin, M.T.; Cobos, R. Identification of fungi associated with grapevine decline in Castilla y León (Spain). *Phytdopathol. Mediterr.* **2007**, *46*, 18–25. [CrossRef]
32. Arendrup, M.C.; Cuenca-Estrella, M.; Lass-Flörl, C.; Hope, W. EUCAST technical note on the EUCAST definitive document EDef 7.2: Method for the determination of broth dilution minimum inhibitory concentrations of antifungal agents for yeasts EDef 7.2 (EUCAST-AFST)*. *Clin. Microbiol. Infect.* **2012**, *18*, E246–E247. [CrossRef]
33. Levy, Y.; Benderly, M.; Cohen, Y.; Gisi, U.; Bassand, D. The joint action of fungicides in mixtures: Comparison of two methods for synergy calculation. *EPPO Bull.* **1986**, *16*, 651–657. [CrossRef]
34. R Core Team. *R: A Language and Environment for Statistical Computing*; R Foundation for Statistical Computing: Vienna, Austria, 2020.
35. Derksen, G.C.H.; Van Beek, T.A. Rubia tinctorum L. In *Studies in Natural Products Chemistry*; Ur-Rahman, A., Ed.; Elsevier: Amsterdam, The Netherlands, 2002; Volume 26, pp. 629–684.
36. DE Santis, D.; Moresi, M. Production of alizarin extracts from Rubia tinctorum and assessment of their dyeing properties. *Ind. Crop. Prod.* **2007**, *26*, 151–162. [CrossRef]
37. Dulo, B.; Phan, K.; Githaiga, J.; Raes, K.; De Meester, S. Natural quinone dyes: A review on structure, extraction techniques, analysis and application potential. *Waste Biomass Valoriz.* **2021**, 1–36. [CrossRef]
38. Derksen, G.; van Beek, T.A.; de Groot, Æ.; Capelle, A. High-performance liquid chromatographic method for the analysis of anthraquinone glycosides and aglycones in madder root (*Rubia tinctorum* L.). *J. Chromatogr. A* **1998**, *816*, 277–281. [CrossRef]
39. Derksen, G.C.H.; Lelyveld, G.P.; Van Beek, T.A.; Capelle, A. Two validated HPLC methods for the quanti?cation of alizarin and other anthraquinones inRubia tinctorum cultivars. *Phytochem. Anal.* **2004**, *15*, 397–406. [CrossRef]
40. Krizsán, K.; Szókán, G.; Tóth, Z.A.; Hollósy, F.; Laszlo, M.; Khlafulla, A. HPLC Analysis of Anthraquinone Derivatives in Madder Root (*Rubia Tinctorum*) and Its Cell Cultures. *J. Liq. Chromatogr. Relat. Technol.* **1996**, *19*, 2295–2314. [CrossRef]
41. Eltamany, E.E.; Nafie, M.S.; Khodeer, D.M.; El-Tanahy, A.; Abdel-Kader, M.S.; Badr, J.M.; Abdelhameed, R.F.A. Rubia tinctorum root extracts: Chemical profile and management of type II diabetes mellitus. *RSC Adv.* **2020**, *10*, 24159–24168. [CrossRef]
42. Friedman, M.; Xu, A.; Lee, R.; Nguyen, D.N.; Phan, T.A.; Hamada, S.M.; Panchel, R.; Tam, C.C.; Kim, J.H.; Cheng, L.W.; et al. The inhibitory activity of anthraquinones against pathogenic protozoa, bacteria, and fungi and the relationship to structure. *Molecules* **2020**, *25*, 3101. [CrossRef]
43. Megiatto, J.J.D.; Cazeils, E.; Ham-Pichavant, F.; Grelier, S.; Gardrat, C.; Castellan, A. Styrene-spaced copolymers including anthraquinone and β-O-4 lignin model units: Synthesis, characterization and reactivity under alkaline pulping conditions. *Biomacromolecules* **2012**, *13*, 1652–1662. [CrossRef]
44. Maurino, V.; Borghesi, D.; Vione, D.; Minero, C. Transformation of phenolic compounds upon UVA irradiation of anthraquinone-2-sulfonate. *Photochem. Photobiol. Sci.* **2007**, *7*, 321–327. [CrossRef]
45. Kalyoncu, F.; Çetin, B.; Saglam, H. Antimicrobial activity of common madder (*Rubia tinctorum* L.). *Phytother. Res.* **2006**, *20*, 490–492. [CrossRef]
46. Manojlovic, N.; Solujic, S.; Sukdolak, S.; Milosev, M. Antifungal activity of Rubia tinctorum, Rhamnus frangula and Caloplaca cerina. *Fitoterapia* **2005**, *76*, 244–246. [CrossRef]
47. Mehrabian, S.; Majd, A.; Majd, I. Antimicrobial effects of three plants (rubia tinctorum, carthamus tinctorius and juglans regia) on some airborne microorganisms. *Aerobiologia* **2000**, *16*, 455–458. [CrossRef]
48. Ozen, E.; Yeniocak, M.; Goktas, O.; Alma, M.H.; Yilmaz, F. Antimicrobial and antifungal properties of madder root (Rubia tinctorum) colorant used as an environmentally-friendly wood preservative. *Bioresources* **2014**, *9*, 1998–2009. [CrossRef]
49. Kim, Y.-M.; Lee, C.-H.; Kim, A.H.-G.; Lee, H.-S. Anthraquinones isolated from Cassiatora (Leguminosae) seed show an antifungal property against phytopathogenic fungi. *J. Agric. Food Chem.* **2004**, *52*, 6096–6100. [CrossRef]
50. De Barros, I.B.; Daniel, J.F.D.S.; Pinto, J.P.; Rezende, M.I.; Filho, R.B.; Ferreira, D.T. Phytochemical and antifungal activity of anthraquinones and root and leaf extracts of Coccoloba mollis on phytopathogens. *Braz. Arch. Biol. Technol.* **2011**, *54*, 535–541. [CrossRef]
51. Shang, X.-F.; Zhao, Z.-M.; Li, J.-C.; Yang, G.-Z.; Liu, Y.-Q.; Dai, L.-X.; Zhang, Z.-J.; Yang, Z.-G.; Miao, X.-L.; Yang, C.-J.; et al. Insecticidal and antifungal activities of Rheum palmatum L. anthraquinones and structurally related compounds. *Ind. Crop. Prod.* **2019**, *137*, 508–520. [CrossRef]
52. Rath, G.; Ndonzao, M.; Hostettmann, K. Antifungal anthraquinones from Morinda lucida. *Int. J. Pharmacogn.* **1995**, *33*, 107–114. [CrossRef]
53. Mishra, B.B.; Kishore, N.; Tiwari, V.K.; Singh, D.D.; Tripathi, V. A novel antifungal anthraquinone from seeds of Aegle marmelos Correa (family Rutaceae). *Fitoterapia* **2010**, *81*, 104–107. [CrossRef]
54. Tsang, P.W.-K.; Bandara, H.; Fong, W.-P. Purpurin suppresses candida albicans biofilm formation and hyphal development. *PLoS ONE* **2012**, *7*, e50866. [CrossRef]

55. Tsang, P.W.-K.; Wong, A.P.-K.; Yang, H.-P.; Li, N.-F. Purpurin triggers caspase-independent apoptosis in candida dubliniensis biofilms. *PLoS ONE* **2013**, *8*, e86032. [CrossRef]
56. Pereira, F.D.O.; Mendes, J.M.; Lima, E.D.O. Investigation on mechanism of antifungal activity of eugenol againstTrichophyton rubrum. *Med. Mycol.* **2013**, *51*, 507–513. [CrossRef]
57. Santamarina, M.P.; Roselló, J.; Giménez, S.; Blázquez, M.A. Commercial Laurus nobilis L. and Syzygium aromaticum L. Merr. & Perry essential oils against post-harvest phytopathogenic fungi on rice. *LWT* **2016**, *65*, 325–332. [CrossRef]
58. Vasconsuelo, A.; Giulietti, A.M.; Boland, R. Signal transduction events mediating chitosan stimulation of anthraquinone synthesis in Rubia tinctorum. *Plant Sci.* **2004**, *166*, 405–413. [CrossRef]
59. Jeevitha, D.; Amarnath, K. Chitosan/PLA nanoparticles as a novel carrier for the delivery of anthraquinone: Synthesis, characterization and in vitro cytotoxicity evaluation. *Coll. Surf. B Biointerf.* **2013**, *101*, 126–134. [CrossRef] [PubMed]
60. Oladipo, A.A.; Gazi, M.; Yilmaz, E. Single and binary adsorption of azo and anthraquinone dyes by chitosan-based hydrogel: Selectivity factor and Box-Behnken process design. *Chem. Eng. Res. Des.* **2015**, *104*, 264–279. [CrossRef]
61. Mohammadi, A.; Hashemi, M.; Hosseini, S. The control of Botrytis fruit rot in strawberry using combined treatments of Chitosan with Zataria multiflora or Cinnamomum zeylanicum essential oil. *J. Food Sci. Technol.* **2015**, *52*, 7441–7448. [CrossRef]
62. Ma, Z.; Garrido-Maestu, A.; Jeong, K.C. Application, mode of action, and in vivo activity of chitosan and its micro- and nanoparticles as antimicrobial agents: A review. *Carbohydr. Polym.* **2017**, *176*, 257–265. [CrossRef]
63. Buzón-Durán, L.; Martín-Gil, J.; Ramos-Sánchez, M.D.C.; Pérez-Lebeña, E.; Marcos-Robles, J.L.; Fombellida-Villafruela, Á.; Martín-Ramos, P. Antifungal activity against *Fusarium culmorum* of stevioside, *Silybum marianum* seed extracts, and their conjugate complexes. *Antibiotics* **2020**, *9*, 440. [CrossRef]
64. Xu, H.; Lu, Y.; Zhang, T.; Liu, K.; Liu, L.; He, Z.; Xu, B.; Wu, X. Characterization of binding interactions of anthraquinones and bovine β-lactoglobulin. *Food Chem.* **2019**, *281*, 28–35. [CrossRef]
65. Detsi, A.; Kavetsou, E.; Kostopoulou, I.; Pitterou, I.; Pontillo, A.R.N.; Tzani, A.; Christodoulou, P.; Siliachli, A.; Zoumpoulakis, P. Nanosystems for the encapsulation of natural products: The case of chitosan biopolymer as a matrix. *Pharmaceutics* **2020**, *12*, 669. [CrossRef]
66. Nguyen, T.T.H.; Si, J.; Kang, C.; Chung, B.; Chung, D.; Kim, D. Facile preparation of water soluble curcuminoids extracted from turmeric (*Curcuma longa* L.) powder by using steviol glucosides. *Food Chem.* **2017**, *214*, 366–373. [CrossRef]
67. Nguyen, T.T.H.; Yu, S.-H.; Kim, J.; An, E.; Hwang, K.; Park, J.-S.; Kim, D. Enhancement of quercetin water solubility with steviol glucosides and the studies of biological properties. *Funct. Foods Health Dis.* **2015**, *5*, 437. [CrossRef]
68. Buzón-Durán, L.; Langa-Lomba, N.; González-García, V.; Casanova-Gascón, J.; Martín-Gil, J.; Pérez-Lebeña, E.; Martín-Ramos, P. On the applicability of chitosan oligomers-amino acid conjugate complexes as eco-friendly fungicides against grapevine trunk pathogens. *Agronomy* **2021**, *11*, 324. [CrossRef]

MDPI

Article

In Vitro Evaluation of the Inhibitory Activity of Different Selenium Chemical Forms on the Growth of a *Fusarium proliferatum* Strain Isolated from Rice Seedlings

Elisabetta Troni [1], Giovanni Beccari [1,*], Roberto D'Amato [1,*], Francesco Tini [1], David Baldo [2], Maria Teresa Senatore [2], Gian Maria Beone [3], Maria Chiara Fontanella [3], Antonio Prodi [2], Daniela Businelli [1] and Lorenzo Covarelli [1]

[1] Department of Agricultural, Food and Environmental Sciences, University of Perugia, 06121 Perugia, Italy; elisabetta8891@gmail.com (E.T.); francesco.tini@collaboratori.unipg.it (F.T.); daniela.businelli@unipg.it (D.B.); lorenzo.covarelli@unipg.it (L.C.)

[2] Department of Agricultural and Food Sciences, Alma Mater Studiorum University of Bologna, 40127 Bologna, Italy; david.baldo@unibo.it (D.B.); mariateresa.senatore@unibo.it (M.T.S.); antonio.prodi@unibo.it (A.P.)

[3] Department for Sustainable Food Process, Catholic University of the Sacred Heart of Piacenza, 29122 Piacenza, Italy; gian.beone@unicatt.it (G.M.B.); mariachiara.fontanella@unicatt.it (M.C.F.)

* Correspondence: giovanni.beccari@unipg.it (G.B.); roberto.damato@unipg.it (R.D.)

Citation: Troni, E.; Beccari, G.; D'Amato, R.; Tini, F.; Baldo, D.; Senatore, M.T.; Beone, G.M.; Fontanella, M.C.; Prodi, A.; Businelli, D.; et al. In Vitro Evaluation of the Inhibitory Activity of Different Selenium Chemical Forms on the Growth of a *Fusarium proliferatum* Strain Isolated from Rice Seedlings. *Plants* **2021**, *10*, 1725. https://doi.org/10.3390/plants10081725

Academic Editors: Dragana Šunjka and Špela Mechora

Received: 2 August 2021
Accepted: 17 August 2021
Published: 20 August 2021

Publisher's Note: MDPI stays neutral with regard to jurisdictional claims in published maps and institutional affiliations.

Abstract: In this study, the in vitro effects of different Se concentrations (5, 10, 15, 20, and 100 mg kg^{-1}) from different Se forms (sodium selenite, sodium selenate, selenomethionine, and selenocystine) on the development of a *Fusarium proliferatum* strain isolated from rice were investigated. A concentration-dependent effect was detected. Se reduced fungal growth starting from 10 mg kg^{-1} and increasing the concentration (15, 20, and 100 mg kg^{-1}) enhanced the inhibitory effect. Se bioactivity was also chemical form dependent. Selenocystine was found to be the most effective at the lowest concentration (5 mg kg^{-1}). Complete growth inhibition was observed at 20 mg kg^{-1} of Se from selenite, selenomethionine, and selenocystine. Se speciation analysis revealed that fungus was able to change the Se speciation when the lowest Se concentration was applied. Scanning Electron Microscopy showed an alteration of the fungal morphology induced by Se. Considering that the inorganic forms have a higher solubility in water and are cheaper than organic forms, 20 mg kg^{-1} of Se from selenite can be suggested as the best combination suitable to inhibit *F. proliferatum* strain. The addition of low concentrations of Se from selenite to conventional fungicides may be a promising alternative approach for the control of *Fusarium* species.

Keywords: fungi; *Fusarium*; selenium; micronutrient; inhibition; bioactivity

1. Introduction

Selenium (Se) is an essential micronutrient for humans and animals, and is involved in numerous biological processes, such as cellular response to oxidative stress, cellular differentiation, redox signaling, and protein folding [1–3]. More than 25 Se-containing proteins have been identified in mammals, having a role in the regulation of redox processes. Among Se proteins, Se is a crucial component of glutathione peroxidase, whose main biological role is to protect against oxidative damage by reducing free hydrogen peroxide to water and lipid hydroperoxides to their corresponding alcohols [4].

In addition to humans and animals, Se is also beneficial to plants when applied at low concentrations [5–8]. For example, Se contributes to the control of water status [9], prevents oxidative stress, delays senescence, and promotes growth [10]. Due to this experimental evidence, numerous studies have investigated Se-biofortification strategies for providing plant protection against abiotic stresses and, at the same time, when possible, beneficial food for human health [11–13]. As demonstrated for many other nutrients, Se also shows a U-shaped relationship between concentration in living organisms and the risk of deficiency toxicity that occurs both below and above the physiological range, which is very narrow [3,14]. Furthermore, Se effects on living organisms are not only concentration dependent but are also related to its chemical form and its bioavailability [15].

Depending on certain experimental conditions (i.e., application mode, concentration, form, and application timing), Se is beneficial to plants and, at the same time, detrimental to plant pathogens [16–18]. For this reason, the activity and role of Se within a plant–pathogen interaction are worthy of more in-depth elucidation. To date, several studies concerning the use of Se salt treatment for the control of a range of plant pathogens have been undertaken, such as, *Aspergillus funiculosus, Alternaria tenuis, Fusarium* spp. and *Fusarium graminearum* in artificial media [19–21]; *Fusarium* spp. and *Alternaria brassicicola* in Indian mustard [22]; *Fusarium oxysporum* f. sp. *lycopersici* in tomato [23]; *Penicillium expansum* in artificial media [18]; *Botrytis cinerea* in tomato [24]; and *F. graminearum* in wheat [21]. Other studies showed the Se protective effect in plants against the activity of mycotoxins, such as zearalenone [25,26] and aflatoxin B_1 [27]. The in vitro and in vivo ability of Se to reduce the deoxynivalenol (DON) production by *F. graminearum* was also explored [21]. Additionally, in soil, Se was found to enhance the microbiome diversities and the relative abundance of Plant Growth Promoting Bacteria (PGPB), while reducing the number of pathogenic fungi [28].

These results indicate that Se might serve as a potential alternative to synthetic fungicides for the control of certain plant diseases caused by several fungal pathogens [18].

In a previous study [29] on rice (*Oryza sativa* L.) seedlings (variety Selenium) cultivated in a hydroponic system in the presence of half-strength Hoagland solution [30], browning of stem bases was detected in more than 50% of total plants. Interestingly, the rice seedlings grown with the same method but in the presence of a Se salt (sodium selenite at a concentration of 20 mg L^{-1} of Se) showed a noticeably lower presence of these symptoms (observed only in approximately 5% of total plants). This observation, as successively described in the present paper, led us to identify the fungal microorganism associated with these symptoms as belonging to the species *Fusarium proliferatum*, a member of the *Fusarium fujikuroi* species complex (FFSC), a group of 40 closely related *Fusarium* species defined by morphological traits, sexual compatibility, and DNA-based phylogenetic analysis [31–33]. *F. proliferatum* is a globally widespread causal agent of diseases of various economically important plants including staple crops such as cereals. *F. proliferatum* is mostly found to colonize maize [34], but has also been isolated from rice [35,36], wheat [33], sorghum, millet [37,38], asparagus [39], garlic [40], and date palm [41]. In rice, *F. proliferatum* is a well-known pathogen associated with Bakanea disease. This seed-borne disease is also caused, in addition to *F. proliferatum*, by other species belonging to FFSC, such as *Fusarium fujikuroi* and *Fusarium verticillioides*. The typical symptoms of Bakanae disease are seedling blight, root and crown rot, pale green to yellowing of foliage, chlorotic leaves, and abnormal elongation [36,42,43].

The presence of *F. proliferatum* in plants is also a potential risk to animal and human health because of its ability to biosynthesize several mycotoxins, such as fumonisins [44]. In particular, maize and, to a lesser extent, rice, are the matrices in which natural contaminations of this mycotoxin are more common [45].

Due to the global importance of both *F. proliferatum* and the host (rice) from which it has been isolated, and because, to the best of our knowledge, no studies have been conducted on the effect of Se against *F. proliferatum*, in the present study we investigated the in vitro effect of various Se concentrations (5, 10, 15, 20, and 100 mg kg^{-1} of Se) from four different Se chemical forms (sodium selenite, sodium selenate, selenomethionine (Se-Met), and selenocystine (Se-Cys)) on the development of a *F. proliferatum* strain isolated from rice seedlings. Additionally, a Se speciation analysis was performed to obtain information on the occurrence of Se metabolites and their distribution, as a consequence of Se bioconversion operated by the fungus. Finally, Scanning Electron Microscopy (SEM) analysis was also carried out to investigate the possible modifications induced by Se on the hyphal morphology of *F. proliferatum*.

As a result, the best combination of Se form and Se concentration suitable for the in vitro inhibition of the development of *F. proliferatum* is described. In addition, the obtained results were found to be useful for the hypothesis of new and alternative approaches to manage *F. proliferatum* infections of rice and other cereal crops, perhaps coupled with a Se-biofortification strategy [46] of grains destined for human and animal food products.

2. Results

2.1. Identification of the *F. proliferatum* Strain PG-CH1 Isolated from Rice Seedlings

The strain PG-CH1, obtained from rice seedlings, was identified as *F. proliferatum* showing a similarity score of >99% with the reference sequences of the same species deposited on NCBI and *Fusarium MLST* databases. The identification was also confirmed by phylogenetic analysis (Figure 1) based on a single locus dataset of *translation elongation factor 1α* (*tef1α*) partial sequences (631 bp). The strain PG-CH1 clustered together with the reference strain *F. proliferatum* G18SXS9-2 (Accession Number MK952837) and *Gibberella intermedia* S1S (sexual stage of *F. proliferatum*; Accession Number JN092349) showed high bootstrap support (95%).

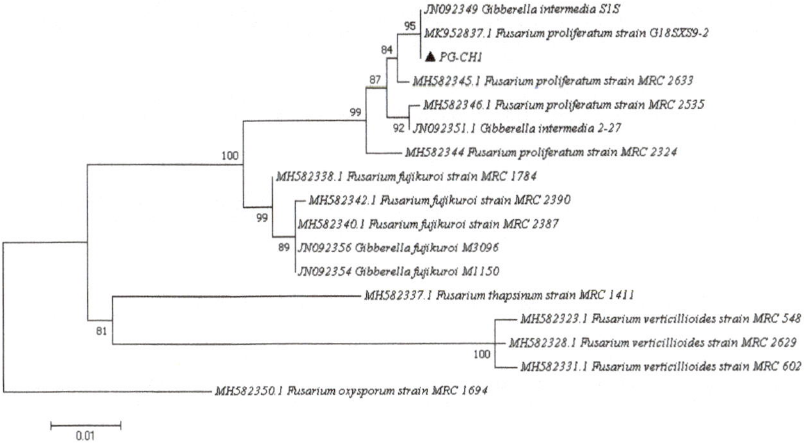

Figure 1. Phylogenetic relationship of PG-CH1 strain (triangle) and members of the *Fusarium fujikuroi* species complex shown in a maximum likelihood dendrogram based on the Kimura 2-parameter model. The tree with the highest log likelihood (−1567.09) is shown. Bootstrap values are indicated above the branch nodes. A discrete Gamma distribution was used to model evolutionary rate differences among sites.

2.2. The In Vitro Inhibitory Activity of Different Se Forms on *F. proliferatum* Strain PG-CH1 Growth

One representative image for each treatment (untreated or treated with Se from different forms and concentrations) is shown in Figure 2, in which the in vitro inhibitory activity of Se on *F. proliferatum* strain PG-CH1 colony growth on Potato Dextrose Agar (PDA) is reported. In combination with the growth reduction, as can be inferred from Figure 2, an alteration induced by Se on the morphology of *F. proliferatum* strain PG-CH1 colonies was also observed. In particular, a lower cottony texture and a lower mycelium density, in addition to a more intense red pigmentation in the central area of the colony, were visible in the presence of Se in comparison to the untreated control. These morphological modifications were macroscopically more appreciable in the presence of lower Se concentrations (5–10 mg kg^{-1}) that allowed a certain colony development. Interestingly, the presence of Se from selenate induced a considerable mycelium ramification (Figure 2b) that was not detected in the presence of Se from other chemical forms. To improve clarity, please note that the ramification was considered for radial growth measurement. Data relative to radial growth reduction in *F. proliferatum* strain PG-CH1 colonies concerning the untreated control (0 mg kg^{-1} of Se), measured after 10 days of incubation at 22 ± 2 °C in the dark in the presence of various Se concentrations (5, 10, 15, 20, and 100 mg kg^{-1}) from different Se forms (selenite, selenate, Se-Met, and Se-Cys), are summarized in Figure 3.

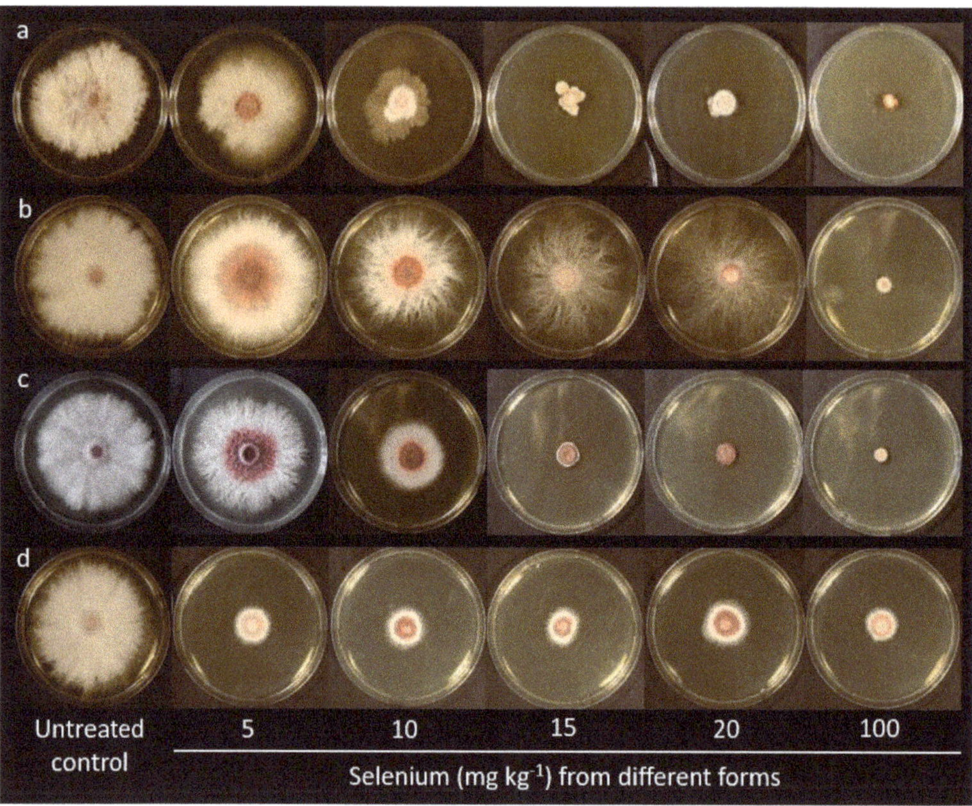

Figure 2. Effect of increasing selenium concentrations from different selenium forms on the colony development of *Fusarium proliferatum* strain PG–CH1 after 10 days of incubation at 22 ± 2 °C in the dark in comparison to the untreated control. Selenite (**a**), selenate (**b**), selenomethionine (**c**), selenocystine (**d**).

Significant differences in radial growth reduction in *F. proliferatum* strain PG-CH1 colonies relative to the untreated control (0 mg kg^{-1} of Se) were detected both within a Se form for different concentrations and within a Se concentration for different Se forms. Focusing the attention on Se from selenite (Figure 3), the inhibitory activity of 5 mg kg^{-1} concentration was significantly lower ($p < 0.05$) than those caused by all other concentrations, whereas Se concentrations of 15, 20, and 100 mg kg^{-1} showed a significantly higher inhibitory activity ($p < 0.05$) than the concentration of 10 mg kg^{-1}. The same significant ($p < 0.05$) gradient described for selenite was observed for Se-Met (Figure 3). In contrast, regarding Se from selenate (Figure 3), the inhibitory activity caused by Se concentrations of 5, 10, 15, and 20 mg kg^{-1} was not significantly different ($p > 0.05$). Only the highest Se concentration (100 mg kg^{-1}) showed a significant effect ($p < 0.05$) on the reduction in *F. proliferatum* strain PG-CH1 growth in comparison to the inhibition caused by all other concentrations.

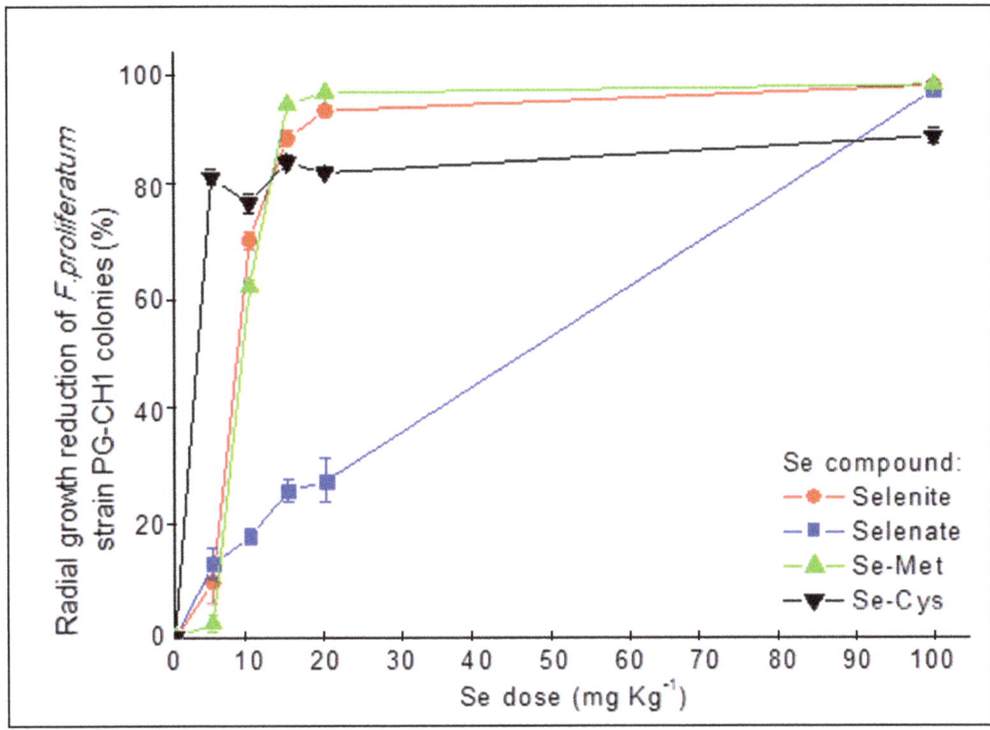

Se dose (mg kg^{-1})	Radial growth reduction of *F. proliferatum* strain PG-CH1 colonies (%)			
	Selenite	Selenate	Se-Met	Se-Cys
5	9.98 ± 7.44 Aa	13.18 ± 5.57 Aa	2.48 ± 2.91 Aa	82.21 ± 2.78 Ab
10	70.94 ± 3.05 Bac	18.00 ± 2.83 Ab	62.76 ± 1.95 Bc	77.54 ± 3.36 Aa
15	89.07 ± 2.59 Ca	26.14 ± 4.04 Ab	95.17 ± 0.60 Ca	84.89 ± 2.22 Aa
20	94.09 ± 0.30 Ca	28.06 ± 8.01 Ab	97.27 ± 0.14 Ca	83.02 ± 0.68 Aa
100	98.45 ± 0.04 Ca	97.55 ± 0.28 Bb	98.64 ± 0.04 Ca	89.36 ± 2.71 Aa

Figure 3. Effect of five selenium (Se) concentrations (5, 10, 15, 20, and 100 mg kg^{-1}) from sodium selenite (selenite; red circles), sodium selenate (selenate; blue squares), selenomethionine (Se–Met; green up–triangles), and selenocystine (Se-Cys; black down-triangles) on *Fusarium proliferatum* strain PG-CH1 colony development. The inhibitory activity was measured after 10 days of incubation at 22 ± 2 °C in the dark and expressed as the mean (± standard error) of colony radial growth reduction (%) relative to the untreated control (0 mg kg^{-1} of Se), calculated according to the equation: (radial growth$_{control}$ − radial growth$_{treatment}$): radial growth$_{control}$ × 100. One-way ANOVA was used to determine statistically significant differences in radial growth of the tested strain. In the table, within a Se form (A–C) or a Se concentration (a–b), means with the same letter are not significantly different at $p < 0.05$ based on the Tukey Honestly Significant Difference multiple comparison test.

The growth reduction performed by Se from Se-Cys (Figure 3) was not significantly different ($p < 0.05$) across the five Se concentrations tested; that is, the inhibitory activity detected at the lowest concentration of 5 mg kg^{-1} of Se was not different from that observed following the treatment with higher Se concentrations, including that of 100 mg kg^{-1}. The concentration of 5 mg kg^{-1} Se from Se-Cys showed a significantly higher ($p < 0.05$) inhibitory activity with respect to Se from all other chemical forms. Increasing the concentration to 10 mg kg^{-1} of Se from Se-Cys and selenite showed a similar ($p > 0.05$) effect on the growth reduction in *F. proliferatum* strain PG-CH1, followed by Se from Se-Met. Se inhibition activity from this last compound was, in turn, significantly lower ($p < 0.05$) than Se from Se-Cys but not significantly different ($p > 0.05$) from that observed for Se from selenite. Conversely, at this concentration, Se from selenate showed the significantly lowest ($p < 0.05$) inhibitory activity compared with that of Se from the other three forms. At the concentrations of 15 and 20 mg kg^{-1}, Se from selenite, Se-Cys, and Se-Met showed a similar ($p > 0.05$) effect on fungal growth reduction, whereas Se from selenite showed, in both cases, a significantly lower ($p < 0.05$) inhibitory activity than those detected for Se from the other three chemical forms. Finally, at the highest tested concentration (100 mg kg^{-1}), the inhibitory activity of Se from Se-Cys was significantly lower ($p < 0.05$) than those detected for Se from selenite, selenate, and Se-Met.

Summarizing, all Se treatments from different Se forms showed a certain effect on *F. proliferatum* strain PG-CH1. In particular, the highest inhibitory activity at the lowest concentration of Se (5 mg kg^{-1}) tested in this experiment was caused by Se from Se-Cys, whereas increasing the concentrations (10, 15, and 20 mg kg^{-1}) of Se from selenite, Se-Cys, and Se-Met showed a higher reduction in fungal colony growth than Se from selenate. Se from selenite, selenate, and Se-Met inhibited *F. proliferatum* strain PG-CH1 growth more than Se-Cys at 100 mg kg^{-1}. The control treatment (PDA amended with Na^+ 100 mg kg^{-1} from Sodium chloride (NaCl)) showed a negligible effect on the growth of *F. proliferatum* strain PG-CH1 colonies, because only a radial growth reduction up to 1.8% was observed (Supplementary Figure S1).

2.3. Se Speciation

Se speciation analysis was performed on Se-amended PDA in the presence of *F. proliferatum* strain PG-CH1, and on Se-amended PDA in the absence of *F. proliferatum* strain PG-CH1. In the presence of fungus, Se-speciation data showed a transformation of the applied Se form (selenite, selenate, Se-Cys, and Se-Met) in other Se chemical forms (Figure 4). Conversely, no transformation of the applied Se form was observed in the absence of *F. proliferatum* strain PG-CH1 (data not reported).

In general, in the presence of fungus, the observed Se transformation in other Se chemical forms was reduced by increasing Se concentration to such an extent that no conversion was detected at the highest Se concentration (100 mg kg^{-1}). This observation suggests the capacity of the fungus to metabolize Se at low concentrations, whereas accumulation of the applied Se form occurred at the highest Se concentration. For that reason, in the following, we discuss the Se speciation results in the presence of *F. proliferatum* strain PG-CH1, focusing on the lowest Se concentrations applied of 5 and 10 mg kg^{-1}. When inorganic Se from selenite was applied, the conversions in Se-Cys (up to 36%) and Se-Met (up to 16%) were mainly observed (Figure 4a); when inorganic Se from selenate was applied, the conversions in Se-Met (up to 58%), Se-Cys (up to 16%), and selenite (up to 8%) occurred (Figure 4b). When organic Se from Se-Met was applied, Se speciation analysis revealed mainly the conversion in Se-Cys (up to 72%) and a trace of selenite (up to 4%) (Figure 4c); when Se-Cys was applied, it was converted to selenite (up to 24%), selenate (up to 11%), and Se-Met (up to 8%) (Figure 4d).

To ascribe with certainty the Se transformation in other Se forms to *F. proliferatum* strain PG-CH1, Se speciation was also performed on PDA amended with Se in the absence of the fungus. Based on the above results (Figure 4), the lowest Se concentration (as the concentration at which the greater Se conversion into other Se forms occurred) was selected and Se speciation analysis was performed on PDA amended with 5 mg kg^{-1} of Se from selenite, selenate, Se-Cys, and Se-Met in the absence of *F. proliferatum* strain PG-CH1. The obtained data did not show the transformation of the applied Se form in other Se chemical forms, thus confirming that the observed Se transformations in other chemical forms were undertaken by *F. proliferatum* strain PG-CH1.

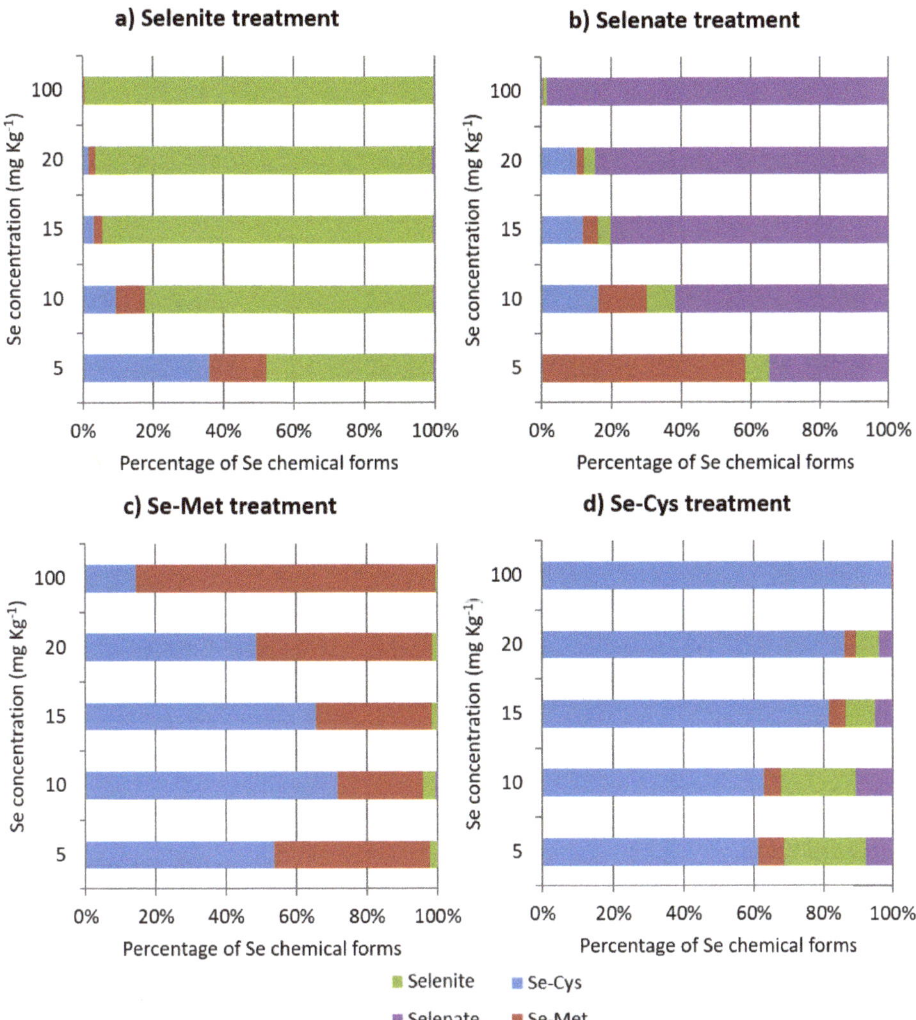

Figure 4. Selenium (Se) speciation data showing different Se chemical forms and their distribution after 10 days of incubation in the growth medium (potato dextrose agar) in the presence of *Fusarium proliferatum* strain PG-CH1 and following the application of increasing Se concentrations (5, 10, 15, 20, and 100 mg kg^{-1}) as sodium selenite (selenite; **a**), sodium selenate (selenite; **b**), selenomethionine (Se-Met; **c**), and selenocystine (Se-Cys; **d**).

2.4. Selenium (20 mg kg^{-1}) from Sodium Selenite Effect Observed by SEM

Based on the results obtained for the in vitro inhibitory activity of Se from different Se forms on *F. proliferatum* strain PG-CH1 growth, for SEM analysis we focused our attention on the effect of 20 mg kg^{-1} of Se from selenite (Figure 5).

In general, a lower mycelium density with shorten hyphae was observed in the treated samples (Figure 5d) in comparison to the untreated ones (Figure 5a). In detail, a strong hyphal collapse was also noticed in the treated samples (Figure 5e,f) compared to the untreated samples (Figure 5b,c). This observation showed additional evidence that Se from selenite at the concentration of 20 mg kg^{-1} was able to determine the inhibition of *F. proliferatum* strain PG-CH1 growth, which was also manifested by the alteration of fungal hyphae morphology and density.

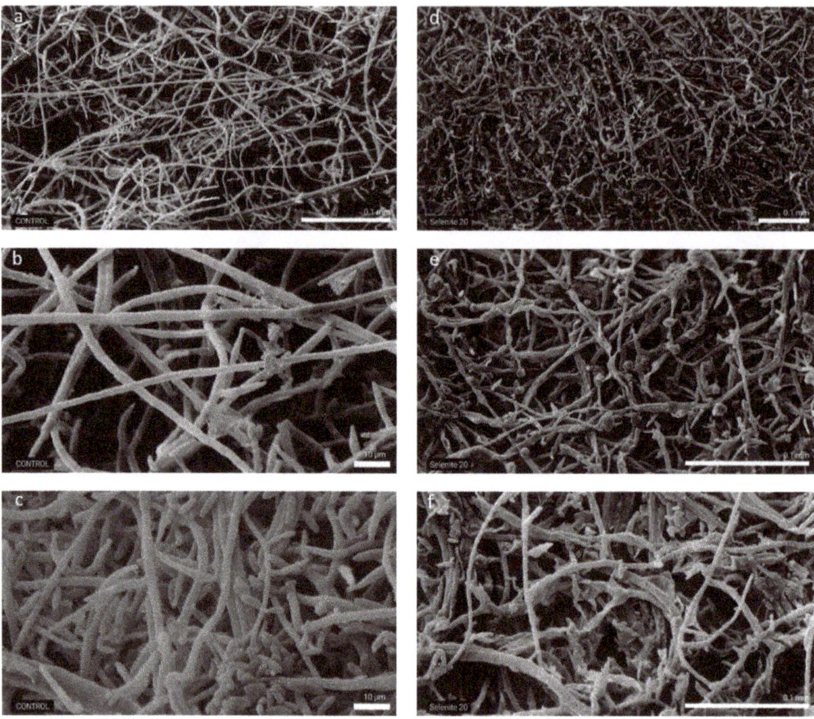

Figure 5. Scanning Electron Microscopy images showing *Fusarium proliferatum* strain PG-CH1 hyphae grown for 10 days on potato dextrose agar (PDA) not amended with selenium (Se) (labelled as control) (**a–c**) and on PDA amended with 20 mg kg^{-1} of Se as sodium selenite (selenite; labelled as selenite 20) (**d–f**). The effect of 20 mg kg^{-1} of Se as selenite caused appreciable modifications in the morphology and density of *F. proliferatum* strain PG-CH1 hyphae (**d–f**) in comparison to the untreated control (**a–c**).

3. Discussion

This study aimed to assess the inhibitory activity of Se toward *F. proliferatum* strain PG-CH1 isolated from rice seedlings. Several studies [3,14,15] showed that Se effects on living organisms depend on Se concentration, in addition to the Se chemical form and bioavailability. Se is one of the few non-metals exhibiting variable oxidation states—selenate (+6), selenite (+4), elemental Se (0), and selenide (−2). Moreover, the distribution of the valence states in a given environment strongly depends on several parameters, such as biological interactions, pH, redox conditions, the solubility of its salts, the chemical complex of soluble and solid ligands, reaction kinetics, and temperature [47,48]. Consequently, Se naturally occurs in different chemical forms. Among these, in living organisms, Se is found both in inorganic forms (mainly as selenite and selenate) and organic forms (mainly as Se-Cys and Se-Met) [49]. In general, it has been reported that organic Se forms have less toxicity compared to inorganic forms in living organisms [3].

Based on the above, in this study, four different Se chemical forms (inorganic Se as selenite and selenate; organic Se as Se-Met and Se-Cys) in the concentration range 5–100 mg kg^{-1} of Se were tested. This concentration range was selected after a preliminary test in which *F. proliferatum* strain PG-CH1 was grown on PDA amended with Se concentration (from selenite) in the range 1–200 mg kg^{-1}. A negligible effect on the reduction in the fungal growth was observed up to 5 mg kg^{-1} of Se, whereas the fungal growth was almost completely inhibited at 100 mg kg^{-1} of Se.

In general, in our experimental conditions, a significant in vitro concentration-dependent Se bioactivity from different Se forms towards *F. proliferatum* strain PG-CH1 was observed. For example, with the exception of Se from selenate, Se reduced (>60%) the *F. proliferatum* strain PG-CH1 growth relative to the untreated control starting from the concentration of 10 mg kg^{-1}. Additionally, by increasing Se concentration to 15 and 20 mg kg^{-1}, the growth reduction caused by Se from selenite, Se-Met, and Se-Cys exceeded 80%. Finally, at the concentration of 100 mg kg^{-1} of Se, *F. proliferatum* strain PG-CH1growth was almost completely inhibited (growth reduction of $\geq$89%), regardless of the applied Se form. An in vitro concentration-dependent activity of Se was previously reported to inhibit *A. funiculosis*, *A. tenuis*, and *Fusarium* spp. (Se from selenite) [19], *P. expansum* (Se from selenite) [18], *B. cinerea* (Se from selenite) [50], *A. brassicicola*, *Fusarium* spp. (Se from selenate) [22], and *F. graminearum* (Se from selenite, selenate, Se-Met, and Se-Cys) [21] growth. In addition to reporting that the effect of Se (from selenite) against a strain of *Fusarium* sp. was concentration dependent, Yin et al. 2017 [51], showed that Se (from selenite) in the range 0.1–1.0 mg kg^{-1} promoted the growth of another strain belonging to the *F. tricinctum* species.

Moreover, the results obtained in this study showed that Se bioactivity was not only concentration dependent but, as mentioned above, also related to the Se chemical form deployed and its bioavailability. For example, Se from selenate exhibited the lowest bioactivity in up to 20 mg kg^{-1} of Se (radial growth reduction in *F. proliferatum* strain PG-CH1 colonies of up to 28%) compared to the other Se forms (percentage of radial growth reduction in *F. proliferatum* colonies ranged from 83 to 97% at 20 mg kg^{-1} of Se from selenite, Se-Cys, and Se-Met). Conversely, Se from Se-Cys was found to be the most effective at the lowest Se concentration (radial growth reduction in *F. proliferatum* colonies at 5 mg Kg^{-1} of 79.3%, whereas Se from other forms reduced the growth by up to 13.1% at 5 mg kg^{-1} of Se). However, it should be recalled that, at the concentrations of 10, 15, and 20 mg kg^{-1}, Se from selenate altered the morphology of *F. proliferatum* strain PG-CH1 colonies by inducing hyphae ramification and decreasing mycelium density. In addition, in vitro (20–80 mg L^{-1}) and in planta (wheat, 20 mg L^{-1}) effects of Se from selenite, selenate, Se-Met, and Se-Cys on fungal growth (in vitro), symptoms (in planta), and DON accumulation (in vitro and in planta) was compound dependent [21].

For further investigation of the process of growth inhibition resulting from Se supplementation to the growth medium, we performed Se speciation analysis to obtain information on the occurrence of the process of Se conversion in other chemical forms operated by *F. proliferatum* strain PG-CH1, or on the process of Se accumulation. No transformation of the applied Se forms occurred in the absence of *F. proliferatum*, whereas transformation of the applied Se forms was observed in the presence of the fungal microorganism, indicating that *F. proliferatum* strain PG-CH1 can metabolize Se.

In particular, the fungal capacity to transform the applied Se form was reduced by increasing Se concentration (Figure 4). Thus, we can hypothesize that the complete growth inhibition in *F. proliferatum* strain PG-CH1 colonies observed at 100 mg kg^{-1} of Se may be related to the accumulation (and consequent toxicity) of the applied Se forms. Concerning the lowest Se concentration applied, the observed inhibition of *F. proliferatum* strain PG-CH1 growth is more difficult to explain. Based on the collected speciation data and considering that the biochemistry of Se resembles that of sulfur (S) [52], we hypothesize several mechanisms as follows:

(i) The occurrence of redox processes mediated by Se (e.g., selenate reduction to selenite, see Figure 4b), which may have affected the regular physiological processes of the fungal microorganism.

(ii) The formation of organic Se compounds caused by the fungus (e.g., inorganic Se is converted to Se-Met and Se-Cys, see Figure 4a,b; Se-Met is converted to Se-Cys, see Figure 4c; Se-Cys is converted to Se-Met in small amounts, see Figure 4d), which may lead to direct incorporation of Se-Cys and Se-Met into proteins, rather than the analog sulfur-containing amino acids S-Cys and S-Met, a process previously reported in the literature [53,54]. In this case, the growth inhibitory effect may be ascribed to a different conformation of the Se proteins concerning analog S proteins to such an extent that a modification in protein activity should not be excluded.

(iii) The direct incorporation of the applied Se-Cys and Se-Met into proteins, a process which may explain the higher bioavailability of organic Se forms compared to inorganic Se forms.

(iv) The relationship of Se with oxidative stress [5]. For example, it is known that in certain filamentous fungi, mycelium metamorphosis, in structures that ensure fungal propagation, is induced by increased oxidative stress. Conversely, decreased oxidative stress causes a permanence of undifferentiated mycelia, and inhibition of metamorphosis and fungal propagation [55]. The extension of this theory predicts that any antioxidant (such as Se at certain doses) can stop fungal propagation by inhibiting its metamorphosis, thus acting as a natural fungicide.

To support hypotheses (ii) and (iii) for explaining the Se toxic effect, it has been reported that an excess of Se-containing proteins can have adverse effects on cellular metabolism [56]. Finally, we also noted a strong smell in the culture medium amended with Se, which may be due to the production of the volatile compound dimethyl selenide, $(CH_3)_2Se$, as a consequence of the detoxification mechanism of the fungus. As reported in the literature [57], the methylation of inorganic Se by microorganisms produces volatile inorganic compounds that are less toxic than inorganic forms.

Se speciation analysis is a powerful tool for detecting the final products of the Se bioconversion operated by fungi; nevertheless, Se speciation data alone are not sufficient to provide an in-depth explanation of the metabolic pathway of Se in fungal microorganisms. Several authors attempted to explain the bioavailability and toxicity of the Se forms for fungal microorganisms at the molecular and/or physiological levels. For example, Reference [58] reports that selenite and selenate are assimilated through oxyanion transporters, and, once inside the cells, they are transformed into selenide through a reductive pathway that may involve the enzyme glutathione peroxidase. In the presence of oxygen, selenide can promote the formation of reactive oxygen species (ROS), which may damage DNA, proteins, and other cellular macromolecules. Selenide is also the intermediate for the formation of selenocystine, from which selenomethionine can then be formed in organisms with a functional transsulfuration pathway. Several metabolomic studies demonstrated that selenocysteine can also be formed from selenomethionine [59,60]. However, the reason for the differences in bioavailability and toxicity possessed by different Se forms remains unknown and further studies are needed to better understand the critical metabolic processes that determine Se tolerance or toxicity. Based on the obtained results and considering that the inorganic forms of Se (selenite and selenate) have a higher solubility in water and are cheaper than organic forms (Se-Cys and Se-Met), we concluded that 20 mg kg^{-1} of Se from selenite can be suggested as the best combination of Se form and concentration suitable to in vitro inhibit the development of the considered *F. proliferatum* strain PG-CH1.

Even if the use of a single strain of *F. proliferatum* does not allow a general conclusion to be drawn at the species level, the results obtained may provide a basis for further investigations concerning Se effects against this polyphagous and global fungal pathogen, and for other *Fusarium* or fungal species. Further developments of this research can be undertaken by screening a higher number of *F. proliferatum* strains to assess the inhibitory activity of Se in a wider population context. This because not all isolates of a species may be sensitive to Se activity, and the phenomenon of Se tolerance by fungi has been previously described [61].

In addition, the activity of Se on the biosynthesis of fumonisins mycotoxins by *F. proliferatum* may also be interesting to explore, in addition to the impact of Se on the growth of other *Fusarium* species and their mycotoxins. Finally, these results may be also useful for evaluating the ability of Se to control *F. proliferatum* infections of rice in the field, in addition to this or other *Fusarium* species in different cereal hosts. Moreover, the addition of low concentrations of Se from selenite to conventional fungicides may be a promising alternative approach for the control of *Fusarium* species in different cereal hosts [62,63].

Many fungicides are categorized as polluting, and it can be reasonably assumed that the addition of Se may help in reducing the concentration of the chemical active ingredient by maintaining the same fungicidal efficacy and decreasing the potentially hazardous effect on the environment and human health [17,19] simultaneously realizing a Se-biofortification of cereal grains [64]. Within the U-shaped range [14], Se is an essential micronutrient with beneficial effects in animals and humans [65].

4. Materials and Methods

4.1. Chemicals and Reagents

Sodium selenite (Na_2SeO_3; selenite), sodium selenate (Na_2SeO_4; selenate), Se-L-Cystine ($C_6H_{12}N_2O_4Se_2$; Se-Cys), Se-DL-methionine ($C_5H_{11}NO_2Se$; Se-Met), Hoagland stock solution, NaCl, streptomycin sulfate, protease, ethanol 95%, agarose, trizma base-glacial acid acetic-ethylene diamine tetra acetic acid disodium salt dihydrate (TAE), EF1 and EF2 primers, glutaraldehyde, and phosphate buffer were all purchased from Sigma-Aldrich (Saint Louis, MO, USA). PDA was purchased from Biolife Italiana (Milan, Italy). Sodium hypochlorite (NaClO) 7% was purchased from Carlo Erba Reagents (Milan, Italy). RedSafe was purchased from iNtRON Biotechnology (Burlington, MA, USA). Gene Ruler1 kb was purchased from Thermo Fisher Scientific (Walthman, MA, USA). Dnase free sterile water was purchased from 5prime (Hilden, Germany). HyperLadder 100–1000 bp was purchased by Bioline (Cincinnati, OH, USA).

4.2. Obtainment of *Fusarium proliferatum* Strain PG-CH1 from Rice Seedlings

The *F. proliferatum* strain PG-CH1 used in this study was isolated from rice seedlings (variety Selenium). To assess the fungal microorganisms affecting rice seedlings, portions of symptomatic material were surface disinfected for 1 min with water-ethanol 95%-NaClO (7%, solution) (82:10:8% vol.) and rinsed with sterile water for 1 min. After the disinfection process, small pieces (0.5 cm) of seedling tissue were placed onto PDA supplemented with streptomycin sulfate (0.16 g L^{-1}) into 5 Petri dishes (90 mm diameter) containing 7 pieces each (7 pieces from one single seedling, 1 seedling per plate, 5 seedlings in total). The dishes were incubated at 22 °C in the dark, and after 5 days a combination of visual and stereomicroscope (SZX9, Olympus, Tokyo, Japan) observations were carried out on each piece to assess fungal development. After visual observations, fungal colonies having similar morphology on PDA showed a higher (95%) isolation incidence.

According to colony color and shape on PDA from visual examination, in addition to the morphology of reproductive structures by microscope analysis (Axiophot, Zeiss, Oberkochen, Germany), they were considered to potentially belong to the genus *Fusarium* and a representative isolate (named PG-CH1) of all of those obtained was transferred onto new plates containing PDA and grown at 22 °C in the dark. After the obtainment of monosporic culture, the isolate was placed onto new PDA plates at 22 °C in the dark for two weeks. Successively, the mycelium was scraped from the PDA surface and placed into 2 mL sterile plastic tubes (Eppendorf, Hamburg, Germany) and stored at −80 °C. Following freeze drying with a lyophilizer (Heto Powder Dry LL3000; Thermo Fisher Scientific, Walthman, MA, USA), the mycelium was finely ground with a grinding machine (MM200, Retsch, Dusseldorf, Germany) for 6 min with a frequency of 25 Hz.

DNA extraction was carried out using the method previously described in [66]. Extracted genomic DNA was visualized on a 1% agarose, TAE gel in TAE buffer (1X) containing 500 µL L^{-1} of RedSafe. DNA fragments were separated in 10 cm long agarose gels, with an electrophoresis apparatus (Eppendorf) applying a tension of 110 V for ~30 min. Electrophoretic runs were visualized using an ultraviolet transilluminator (Euroclone, Milan, Italy). DNA concentration was estimated by comparison with Gene Ruler 1 kb included in each gel as a control. DNA was diluted with Dnase free sterile water for molecular biology use to obtain a concentration of ~30 ng µL^{-1} and stored at −20 °C until use. The DNA extracted from the *Fusarium* isolate was subjected to *tef1α* gene amplification, purification, and sequencing. PCR protocol is described in [67] adopting EF1

(ATGGGTAAGGA(A/G)GACAAGAC) and EF2 (GGA(G/A)GTACCAGT(G/C)ATCATGTT) primers [68]. PCR assays were performed on a T-100 thermal cycler (Bio-Rad, Hercules, CA, USA). The PCR fragment was visualized on TAE 1X agarose gel (2%) containing 500 μL L^{-1} of RedSafe. The DNA fragment was separated by an electrophoresis apparatus applying a tension of 110 V for ~40 min. Electrophoretic runs were observed with an ultraviolet transilluminator. The size of the amplified fragment was obtained by comparison with HyperLadder 100–1000 bp. The PCR fragment was purified and sequenced by an external sequencing service (Genewiz Genomics Europe, Takeley, UK). The sequence obtained was verified by Chromatogram Explorer Lite v 4.0.0 (HeracleBiosoft srl, Mioveni, Romania, 2011) and compared to those deposited on the BLAST database (National Center for Biotechnology Information (NCBI) Basic Local Search Tool (BLAST)), available online at http://blast.ncbi.nlm.nih.gov (accessed on 1 November 2018) [69], and the *Fusarium* MLST database (available online at http://fusarium.mycobank.org/, accessed on 1 November 2018).

To further confirm the species identity, the strain PG-CH1 was subjected to phylogenetic analysis based on *tef1α* sequences. The sequence of the PG-CH1 strain was aligned with 15 reference *tef1α* sequences of most common pathogens associated with seedling rice diseases such as *F. fujikuroi*, *F. verticillioides*, and *F. proliferatum* (Table 1) [36,70]. The sequences were aligned using the Muscle Algorithm implemented in the MEGA 7 software package [71]. Phylogenetic analysis was conducted in MEGA, via maximum likelihood following the best fit model of molecular evolution as determined by Bayesian information criterion (BIC) scores. A discrete Gamma distribution was used to model evolutionary rate differences among sites. All positions with less than 90% of coverage were eliminated. Statistical support of branches was evaluated using bootstrap analysis of 1000 replicates. The out-group isolate *F. oxysporum* MRC 1694 (accession number MH582350) was used for rooting the tree [72].

Table 1. *Fusarium* spp. strains used in the phylogenetic analysis and related information.

Strain	Species	Species Complex	Host	Geographic Origin/Substrate	GenBank Accession Number	References
MRC 548	*F. verticillioides*	*Fujikuroi*	Maize	South Africa	MH582323.1	[72]
MRC 602	*F. verticillioides*	*Fujikuroi*	Maize	South Africa	MH582331.1	[72]
MRC 1411	*F. tapsinum*	*Fujikuroi*	Maize	North Carolina-USA	MH582337.1	[72]
MRC 1784	*F. fujikuroi*	*Fujikuroi*	Rawcotton	Georgia, USA	MH582338.1	[72]
MRC 2324	*F. proliferaum*	*Fujikuroi*	Cotton boll	Alabama, USA	MH582344.1	[72]
MRC 2387	*F. fujikuroi*	*Fujikuroi*	Rice	Japan	MH582340.1	[70]
MRC 2390	*F. fujikuroi*	*Fujikuroi*	Unknown	Unknown	MH582342.1	[72]
MRC 2535	*F. proliferatum*	*Fujikuroi*	River sediment	Japan	MH582346.1	[72]
MRC 2629	*F. verticillioides*	*Fujikuroi*	Maize	Iowa, USA	MH582328.1	[72]
MRC 2633	*F. proliferatum*	*Fujikuroi*	Wheat	India	MH582345.1	[72]
S1S	*F. proliferatum*	*Fujikuroi*	Rice (seed)	Italy	JN092349	[70]
2–27	*F. proliferatum*	*Fujikuroi*	Rice (seed)	Italy	JN092351	[70]
M3096	*F. fujikuroi*	*Fujikuroi*	Rice	Georgia, USA	JN092356	[70]
M1150	*F. fujikuroi*	*Fujikuroi*	Rice	Taiwan	JN092354	[70]
G18SXS9-2	*F. proliferatum*	*Fujikuroi*	Unknown	Unknown	MK952837.1	Unknown
MRC 1694	*F. oxysporum*	*Oxysporum*	Unknown	Human	MH582350.1	[72]

4.3. *In Vitro Evaluation of the Inhibitory Activity of Se Forms on* F. proliferatum *Strain PG-CH1*

PDA was supplemented with four different Se forms (selenite, selenate, Se-Cys, and Se-Met) at five different Se concentrations (5, 10, 15, 20, and 100 mg kg^{-1} of Se). A weighed amount of each Se form was dissolved in sterile distilled water to obtain a concentrated solution of 7 g L^{-1} of Se. A predetermined volume (64.3, 128.6, 192.8, 257.1, and 1280.0 μL) of the Se-concentrated obtained solutions was added to 20 g of PDA so that

Se concentrations in the growth medium were 5, 10, 15, 20, and 100 mg kg^{-1}. Successively, mycelium plugs (5 mm diam.) were taken from the edge of one week old *F. proliferatum* strain PG-CH1 colonies, developed on PDA at 22 °C, in the dark. The plugs were placed onto Se amended PDA, at the center of the plate, with the mycelium side facing upwards. Untreated controls (0 mg kg^{-1} of Se), and control treatments with NaCl (as counter ions of selenite and selenate), were also included in the experiment to ascribe the observed growth reduction of *F. proliferatum* strain PG-CH1 to Se with certainty. For this purpose, NaCl was added to the growth medium at the concentration of 254 mg kg^{-1} to fix the amount of Na^+ at 100 mg kg^{-1}, which is the same as the maximum Se concentration used in the experiment. Furthermore, the selected concentration of 254 mg kg^{-1} of NaCl (100 mg kg^{-1} of Na^+) allowed the concentration of Na^+ combined with inorganic Se (selenite and selenate) to be exceeded. After 10 days of incubation at 22 ± 2 °C in the dark, radial growth of fungal colonies was measured by ImageJ software (https://imagej.nih.gov/ij/, accessed on 20 January 2019), and the inhibitory activity was expressed as the percentage of radial growth reduction concerning the untreated control, according to Equation (1):

$$\begin{gathered}\textit{Radial growth reduction of F. proliferatum strain PG}-\textit{CH1colonies}\ (\%) = \\ (\text{radial growth}_{\text{control}} - \text{radial growth}_{\text{treatment}}) : \text{radial growth}_{\text{control}} \times 100\end{gathered} \quad (1)$$

Four replicates per treatment were performed. This experiment was repeated twice and the results of the second experiment, representative of the entire study, are shown. The preliminary first experiment showed a colony growth pattern, which was confirmed by the results obtained in the second experiment. The following analyses were realized exclusively using the materials of the second experiment.

4.4. Se Speciation Analysis

The Se speciation analysis was performed on 10 day old *F. proliferatum* colonies grown on amended PDA. About 16 g of sample (the sample comprises both mycelium and growth medium) was added with 10 mL of distilled water to 50 mL centrifuge tubes, accurately stirred to disperse the PDA gel, and sonicated for 2 min with an ultrasound probe. Then, protease was added up to 2.0 mg mL^{-1} and the obtained sample was stirred in a water bath at 37 °C for 15 h. Because PDA contained 2% by weight of agar, and considering that agar forms a gel in the concentration range of 0.5–2% by weight, the obtained samples were 10-fold diluted with distilled water, then cooled at room temperature and centrifuged at 5000 rpm for 10 min. Then, the supernatant was collected and filtered through 0.22 μm Millex GV filters (Millipore Corporation). The Se standards selenite, selenate, Se-Met, and Se-Cys were prepared in ultrapure (>18 MΩ) water (Supplementary Figure S2). Speciation of Se was performed using a Liquid Chromatography-Inductively Coupled Plasma Mass Spectrometry (LC-ICP-MS/MS) system consisting of an Agilent 1260 Infinity II LC system and an Agilent 8900 ICP-tandem mass spectrometer (Agilent Technologies Japan, Ltd., Tokyo, Japan). Details of the mobile phases, column, and MS/MS conditions are available in Supplementary Table S1.

4.5. Observation by Scanning Electron Microscopy

For this analysis, based on the results obtained from in vitro evaluation of the inhibitory activity of different Se forms on *F. proliferatum* strain PG-CH1 growth, we focused our attention on Se activity from selenite. In detail, a mycelium plug (5 mm diameter) of *F. proliferatum* developed in the presence of 0, 5, 10, 15, 20, and 100 mg kg^{-1} of Se from selenite was sampled from 10 day old colonies to perform SEM analysis. A mycelium plug was collected from two replicates of the previously described experiment for a total of 12 samples (5 different Se concentrations from selenite plus the untreated control). Samples were prepared for SEM observations following the protocol described in [73] with slight modifications. Samples were fixed in 5% glutaraldehyde for 24 h; washed three times with 0.1 M phosphate buffer, pH 7.2; rinsed three times in distilled water; and dehydrated in ethanol series (25%, 50%, 75%, 90%, and 100%) for 7 min each. Samples were then transferred to a critical point dryer (Emitech K850, Quorum Technologies Ltd., Laughton, UK) to complete the drying process with carbon dioxide as a transition fluid. Specimens were then mounted on aluminum stubs with double-sided carbon tape, coated with gold in a sputter (Emitech K500, Quorum Technologies Ltd., Tokyo, Japan), and observed with a SEM 515 (Philips, Amsterdam, The Netherlands) at 12 kV. Observations were conducted for all 12 samples.

4.6. Statistical Analysis

Data of radial growth reduction of *F. proliferatum* strain PG-CH1 colonies after 10 days of incubation at 22 ± 2 °C in the dark in the presence of different Se concentrations from different Se forms were subject to one-way ANOVA. Data were analyzed considering the "radial growth reduction of *F. proliferatum* strain PG-CH1 colonies" (variable) for each "Se concentration" or "Se form" (factors). The results were expressed as the mean of four biological replicates (±standard error). To check for pairwise contrasts, Tukey Honestly Significant Difference multiple comparison tests were performed ($p < 0.05$) using the Microsoft Excel (Microsoft Corporation, Redmond, WA, USA) Macro "DSAASTAT ver. 1.0192" (macro developed by University of Perugia, Italy) [74].

Supplementary Materials: The following are available online at https://www.mdpi.com/article/10.3390/plants10081725/s1, Figure S1. Comparison between *Fusarium proliferatum* strain PG-CH1 development on untreated potato dextrose agar (PDA) (untreated control; a,d), on PDA amended with 100 mg kg^{-1} of Selenium from sodium selenite (b), sodium selenate (e), and on PDA amended with 100 mg kg^{-1} of Na^+ from NaCl (c,f) (control treatment). Figure S2. Liquid chromatography-inductively coupled plasma mass spectrometry (LC-ICP-MS/MS) chromatogram of 1 μg L^{-1} of selenium standards. Peak identities are selenocystine (a), selenomethionine (b), sodium selenite (c), and sodium selenate (d). Table S1: Liquid chromatography-inductively coupled plasma mass spectrometry (LC–ICP-MS/MS) conditions.

Author Contributions: Conceptualization, G.B., R.D., D.B. (Daniela Businelli) and L.C.; methodology, E.T., G.B., R.D., G.M.B. and M.C.F.; formal analysis, E.T., G.B. and R.D.; investigation, E.T., G.B., R.D., F.T., D.B. (David Baldo), M.T.S., G.M.B., M.C.F. and A.P.; resources, D.B. (Daniela Businelli) and L.C.; data curation, E.T., G.B., R.D. and F.T.; writing—original draft preparation, E.T., G.B. and R.D.; writing—review and editing, F.T., D.B. (Daniela Businelli) and L.C.; visualization, E.T., G.B., R.D., D.B. (David Baldo), M.T.S. and A.P.; supervision, D.B. (Daniela Businelli) and L.C. All authors have read and agreed to the published version of the manuscript.

Funding: This research received no external funding.

Acknowledgments: Graphical abstract was created with Biorender.com, accessed on 1 July 2021.

Conflicts of Interest: The authors declare no conflict of interest.

References

1. Steinbrenner, H.; Speckmann, B.; Klotz, L.O. Selenoproteins: Antioxidant selenoenzymes and beyond. *Arch. Biochem. Biophys.* **2016**, *595*, 113–119. [CrossRef]
2. Roman, M.; Jitaru, P.; Barbante, C. Selenium biochemistry and its role for human health. *Metallomics* **2014**, *6*, 25–54. [CrossRef]
3. D'Amato, R.; Regni, L.; Falcinelli, B.; Mattioli, S.; Benincasa, P.; Dal Bosco, A.; Pacheco, P.; Proietti, P.; Troni, E.; Santi, C.; et al. Current knowledge on selenium biofortification to improve the nutraceutical profile of food: A comprehensive review. *J. Agric. Food Chem.* **2020**, *68*, 4075–4097. [CrossRef]
4. Ursini, F.; Maiorino, M. Glutathione peroxidases. In *Encyclopedia of Biological Chemistry*, 2nd ed.; Elsevier Inc.: Amsterdam, The Netherlands, 2013; pp. 399–404.
5. Gupta, M.; Gupta, S. An overview of selenium uptake, metabolism, and toxicity in plants. *Front Plant. Sci.* **2017**, *7*, 2074. [CrossRef]
6. Vats, S. *Biotic and Abiotic Stress Tolerance in Plants*; Vats, S., Ed.; Springer: Singapore; Springer Nature Singapore Pte Ltd.: Singapore, 2018.
7. Hasanuzzaman, M.; Bhuyan, M.H.M.B.; Raza, A.; Hawrylak-Nowak, B.; Matraszek-Gawron, R.; Mahmud, J.A.; Nahar, K.; Fujita, M. Selenium in plants: Boon or bane? *Environ. Exp. Bot.* **2020**, *178*, 104170. [CrossRef]

8. Hasanuzzaman, M.; Borhannuddin Bhuyan, M.H.M.; Raza, A.; Hawrylak-Nowak, B.; Matraszek-Gawron, R.; Nahar, K.; Fujita, M. Selenium toxicity in plants and environment: Biogeochemistry and remediation possibilities. *Plants* **2020**, *9*, 1711. [CrossRef] [PubMed]
9. Kuznetsov, V.; Kuznetsov, V. Selenium regulates the water status of plants exposed to drought. *Dokl. Biol. Sci.* **2003**, *390*, 266–268. [CrossRef] [PubMed]
10. Gilmara, P.d.S.; Leonardo, C.C.; Victor, V.C.; Sylvia, L.O.S.; Edilaine, I.F.T. Selenium and agricultural crops. *Afr. J. Agric. Res.* **2017**, *12*, 2545–2554. [CrossRef]
11. Bocchini, M.; D'Amato, R.; Ciancaleoni, S.; Fontanella, M.C.; Palmerini, C.A.; Beone, G.M.; Onofri, A.; Negri, V.; Marconi, G.; Albertini, E.; et al. Soil Selenium (Se) biofortification changes the physiological, biochemical and epigenetic responses to water stress in *Zea mays* L. by inducing a higher drought tolerance. *Front. Plant Sci.* **2018**, *9*, 389. [CrossRef] [PubMed]
12. Proietti, P.; Nasini, L.; Del Buono, D.; D'Amato, R.; Tedeschini, E.; Businelli, D. Selenium protects olive (*Olea europaea* L.) from drought stress. *Sci. Hortic.* **2013**, *164*, 165–171. [CrossRef]
13. Feng, R.; Wei, C.; Tu, S. The roles of selenium in protecting plants against abiotic stresses. *Environ. Exp. Bot.* **2013**, *87*, 58–68. [CrossRef]
14. Rayman, M.P. Selenium intake, status, and health: A complex relationship. *Hormones* **2019**, *19*, 9–14. [CrossRef] [PubMed]
15. Thiry, C.; Ruttens, A.; De Temmerman, L.; Schneider, Y.J.; Pussemier, L. Current knowledge in species-related bioavailability of selenium in food. *Food Chem.* **2012**, *130*, 767–784. [CrossRef]
16. Bhatia, P.; Aureli, F.; D'Amato, M.; Prakash, R.; Cameotra, S.S.; Nagaraja, T.P.; Cubadda, F. Selenium bioaccessibility and speciation in biofortified *Pleurotus* mushrooms grown on selenium-rich agricultural residues. *Food Chem.* **2013**, *140*, 225–230. [CrossRef]
17. Hasanuzzam, M.; Hossain, M.A.; Fujita, M. Selenium in higher plants: Physiological role, antioxidant metabolism and abiotic stress tolerance. *J. Plant Sci.* **2010**, *5*, 354–375. [CrossRef]
18. Wu, Z.-L.; Yin, X.-B.; Lin, Z.-Q.; Bañuelos, G.S.; Yuan, L.-X.; Liu, Y.; Li, M. Inhibitory effect of selenium against *Penicillium expansum* and its possible mechanisms of action. *Curr. Microbiol.* **2014**, *69*, 192–201. [CrossRef] [PubMed]
19. Razak, A.A.; El-Tantawy, H.; El-Sheikh, H.H.; Gharieb, M.M. Influence of selenium on the efficiency of fungicide action against certain fungi. *Biol. Trace Elem. Res.* **1991**, *28*, 47–56. [CrossRef] [PubMed]
20. Ramadan, S.E.; Razak, A.A.; Yousseff, Y.A.; Sedky, N.M. Selenium metabolism in a strain of *Fusarium*. *Biol. Trace Elem. Res.* **1988**, *18*, 161–170. [CrossRef] [PubMed]
21. Mao, X.; Hua, C.; Yang, L.; Zhang, Y.; Sun, Z.; Li, L.; Li, T. The effects of selenium on wheat fusarium head blight and DON accumulation were selenium compound-dependent. *Toxins* **2020**, *12*, 573. [CrossRef] [PubMed]
22. Hanson, B.; Garifullina, G.F.; Lindblom, S.D.; Wangeline, A.; Ackley, A.; Kramer, K.; Norton, A.P.; Lawrence, C.B.; Pilon-Smits, E.A.H. Selenium accumulation protects *Brassica juncea* from invertebrate herbivory and fungal infection. *New Phytol.* **2003**, *159*, 461–469. [CrossRef] [PubMed]
23. Companioni, B.; Medrano, J.; Torres, J.A.; Flores, A.; Rodríguez, E.; Benavides, A. Protective action of sodium selenite against *Fusarium* wilt in tomato: Total protein contents, levels of phenolic compounds and changes in antioxidant potential. *Acta Hortic.* **2012**, *947*, 321–327. [CrossRef]
24. You, J.; Zhang, J.; Wu, M.; Yang, L.; Chen, W.; Li, G. Multiple criteria-based screening of *Trichoderma* isolates for biological control of *Botrytis cinerea* on tomato. *Biol. Control.* **2016**, *101*, 31–38. [CrossRef]
25. Filek, M.; Łabanowska, M.; Kurdziel, M.; Sieprawska, A. Electron paramagnetic resonance (EPR) spectroscopy in studies of the protective effects of 24-epibrasinoide and selenium against zearalenone-stimulation of the oxidative stress in germinating grains of wheat. *Toxins* **2017**, *9*, 178. [CrossRef] [PubMed]
26. Kornaś, A.; Filek, M.; Sieprawska, A.; Bednarska-Kozakiewicz, E.; Gawrońska, K.; Miszalski, Z. Foliar application of selenium for protection against the first stages of mycotoxin infection of crop plant leaves. *J. Sci. Food Agric.* **2019**, *99*, 482–485. [CrossRef] [PubMed]
27. Agar, G.; Alpsoy, L.; Bozari, S.; Erturk, F.A.; Yildirim, N. Determination of protective role of selenium against aflatoxin B1-induced DNA damage. *Toxicol. Ind. Health* **2013**, *29*, 396–403. [CrossRef] [PubMed]
28. Liu, X.; Zhao, Z.; Duan, B.; Hu, C.; Zhao, X.; Guo, Z. Effect of applied sulphur on the uptake by wheat of selenium applied as selenite. *Plant Soil* **2015**, *386*, 35–45. [CrossRef]
29. D'Amato, R.; Fontanella, M.C.; Falcinelli, B.; Beone, G.M.; Bravi, E.; Marconi, O.; Benincasa, P.; Businelli, D. Selenium biofortification in rice (*Oryza sativa* L.) sprouting: Effects on Se yield and nutritional traits with focus on phenolic acid profile. *J. Agric. Food Chem.* **2018**, *66*, 4082–4090. [CrossRef]
30. Kane, C.D.; Jasoni, R.L.; Peffley, E.P.; Thompson, L.D.; Green, C.J.; Pare, P.; Tissue, D. Nutrient solution and solution pH influences on onion growth and mineral content. *J. Plant Nutr.* **2006**, *29*, 375–390. [CrossRef]
31. O'Donnell, K.; Nirenberg, H.I.; Aoki, T.; Cigelnik, E. A Multigene phylogeny of the *Gibberella fujikuroi* species complex: Detection of additional phylogenetically distinct species. *Mycoscience* **2000**, *41*, 61–78. [CrossRef]
32. O'Donnell, K.; Rooney, A.P.; Proctor, R.H.; Brown, D.W.; McCormick, S.P.; Ward, T.J.; Frandsen, R.J.N.; Lysøe, E.; Rehner, S.A.; Aoki, T.; et al. Phylogenetic analyses of RPB1 and RPB2 support a middle Cretaceous origin for a clade comprising all agriculturally and medically important fusaria. *Fungal Genet. Biol.* **2013**, *52*, 20–31. [CrossRef]

33. Palacios, S.A.; Susca, A.; Haidukowski, M.; Stea, G.; Cendoya, E.; Ramírez, M.L.; Chulze, S.N.; Farnochi, M.C.; Moretti, A.; Torres, A.M. Genetic variability and fumonisin production by *Fusarium proliferatum* isolated from durum wheat grains in Argentina. *Int. J. Food Microbiol.* **2015**, *201*, 35–41. [CrossRef]
34. Chulze, S.N.; Ramirez, M.L.; Farnochi, M.C.; Pascale, M.; Visconti, A.; March, G. *Fusarium* and fumonisin occurrence in Argentinian corn at different ear maturity stages. *J. Agric. Food Chem.* **1996**, *44*, 2797–2801. [CrossRef]
35. Amatulli, M.T.; Spadaro, D.; Gullino, M.L.; Garibaldi, A. Molecular identification of *Fusarium* spp. associated with bakanae disease of rice in Italy and assessment of their pathogenicity. *Plant Pathol.* **2010**, *59*, 839–844. [CrossRef]
36. Wulff, E.G.; Sørensen, J.L.; Lübeck, M.; Nielsen, K.F.; Thrane, U.; Torp, J. *Fusarium* spp. associated with rice Bakanae: Ecology, genetic diversity, pathogenicity and toxigenicity. *Environ. Microbiol.* **2010**, *12*, 649–657. [CrossRef] [PubMed]
37. Leslie, J.F.; Zeller, K.A.; Lamprecht, S.C.; Rheeder, J.P.; Marasas, W.F.O. Toxicity, pathogenicity, and genetic differentiation of five species of *Fusarium* from sorghum and millet. *Phytopathology* **2005**, *95*, 275–283. [CrossRef]
38. Vismer, H.F.; Shephard, G.S.; van der Westhuizen, L.; Mngqawa, P.; Bushula-Njah, V.; Leslie, J.F. Mycotoxins produced by *Fusarium proliferatum* and *F. pseudonygamai* on maize, sorghum and pearl millet grains in vitro. *Int. J. Food Microbiol.* **2019**, *296*, 31–36. [CrossRef]
39. Logrieco, A.; Doko, B.; Moretti, A.; Frisullo, S.; Visconti, A. Occurrence of fumonisin B1 and B2 in *Fusarium proliferatum* infected asparagus plants. *J. Agric. Food Chem.* **1998**, *46*, 5201–5204. [CrossRef]
40. Gálvez, L.; Urbaniak, M.; Waśkiewicz, A.; Stępień, Ł.; Palmero, D. *Fusarium proliferatum*–Causal agent of garlic bulb rot in Spain: Genetic variability and mycotoxin production. *Food Microbiol.* **2017**, *67*, 41–48. [CrossRef]
41. Abdalla, M.Y.; Al-Rokibah, A.; Moretti, A.; Mulè, G. Pathogenicity of toxigenic *Fusarium proliferatum* from date palm in Saudi Arabia. *Plant Dis.* **2000**, *84*, 321–324. [CrossRef] [PubMed]
42. Desjardins, A.E.; Manandhar, H.K.; Plattner, R.D.; Manandhar, G.G.; Poling, S.M.; Maragos, C.M. *Fusarium* species from nepalese rice and production of mycotoxins and gibberellic acid by selected species. *Appl. Environ. Microbiol.* **2000**, *66*, 1020–1025. [CrossRef] [PubMed]
43. Mohd Zainudin, N.A.I.; Razak, A.A.; Salleh, B. Bakanae disease of rice in malaysia and indonesia: Etiology of the causal agent based on morphological, physiological and pathogenicity characteristics. *J. Plant Prot. Res.* **2008**, *48*, 475–485. [CrossRef]
44. Ji, F.; He, D.; Olaniran, A.O.; Mokoena, M.P.; Xu, J.; Shi, J. Occurrence, toxicity, production and detection of *Fusarium* mycotoxin: A review. *Food Prod. Process. Nutr.* **2019**, *1*, 1–14. [CrossRef]
45. Braun, M.S.; Wink, M. Exposure, occurrence, and chemistry of fumonisins and their cryptic derivatives. *Compr. Rev. Food Sci. Food Saf.* **2018**, *17*, 769–791. [CrossRef]
46. Schiavon, M.; Nardi, S.; dalla Vecchia, F.; Ertani, A. Selenium biofortification in the 21st century: Status and challenges for healthy human nutrition. *Plant Soil* **2020**, *453*, 245–270. [CrossRef] [PubMed]
47. Reddy, C.C.; Massaro, E.J. Biochemistry of selenium: A brief overview. *Toxicol. Sci.* **1983**, *3*, 431–436. [CrossRef]
48. McNeal, J.M.; Balistrieri, L.S. Geochemistry and Occurrence of Selenium: An Overview. In *Selenium in Agriculture and the Environment*; Jacob, L.W., Ed.; Soil Science Society of America: Madison, WI, USA, 1989; pp. 1–13.
49. Boyd, R. Selenium stories. *Nat. Chem.* **2011**, *3*, 570. [CrossRef]
50. Wu, Z.; Yin, X.; Bañuelos, G.S.; Lin, Z.Q.; Zhu, Z.; Liu, Y.; Yuan, L.; Li, M. Effect of selenium on control of postharvest gray mold of tomato fruit and the possible mechanisms involved. *Front. Microbiol.* **2016**, *6*, 1–11. [CrossRef] [PubMed]
51. Yin, H.; Zhang, Y.; Zhang, F.; Hu, J.T.; Zhao, Y.M.; Cheng, B.L. Effects of selenium on *Fusarium* growth and associated fermentation products and the relationship with chondrocyte viability. *Biomed. Environ. Sci.* **2017**, *30*, 134–138. [PubMed]
52. Jacob, C.; Giles, G.I.; Giles, G.I.; Sies, H. Sulfur and selenium: The role of oxidation state in protein structure and function. *Angew. Chem. Int. Ed.* **2003**, *42*, 4742–4758. [CrossRef] [PubMed]
53. Läuchli, A. Selenium in plants: Uptake, functions, and environmental toxicity. *Bot. Acta* **1993**, *106*, 455–468. [CrossRef]
54. Patching, S.G.; Gardiner, P.H.E. Recent developments in selenium metabolism and chemical speciation: A review. *J. Trace Elem. Med. Biol.* **1999**, *13*, 193–214. [CrossRef]
55. Georgiou, C.D.; Patsoukis, N.; Papapostolou, I.; Zervoudakis, G. Sclerotial metamorphosis in filamentous fungi is induced by oxidative stress. *Integr. Comp. Biol.* **2006**, *46*, 691–712. [CrossRef]
56. Brown, T.A.; Shrift, A. Selenium: Toxicity and tolerance in higher plants. *Biol. Rev.* **1982**, *57*, 59–84. [CrossRef]
57. da Silva, M.d.C.S.; da Luz, J.M.R.; Paiva, A.P.S.; Mendes, D.R.; Carvalho, A.A.C.; Naozuka, J.; Kasuya, M.C.M. Growth and Tolerance of *Pleurotus ostreatus* at Different Selenium Forms. *J. Agric. Sci.* **2019**, *11*, 151. [CrossRef]
58. Herrero, E.; Wellinger, R.E. Yeast as a model system to study metabolic impact of selenium compounds. *Microb. Cell* **2015**, *2*, 139–149. [CrossRef] [PubMed]
59. Rao, Y.; McCooeye, M.; Windust, A.; Bramanti, E.; D'Ulivo, A.; Mester, Z. Mapping of selenium metabolic pathway in yeast by Liquid Chromatography−Orbitrap Mass Spectrometry. *Anal. Chem.* **2010**, *82*, 8121–8130. [CrossRef]
60. Arnaudguilhem, C.; Bierla, K.; Ouerdane, L.; Preud'homme, H.; Yiannikouris, A.; Lobinski, R. Selenium metabolomics in yeast using complementary reversed-phase/hydrophilic ion interaction (HILIC) liquid chromatography-electrospray hybrid quadrupole trap/Orbitrap mass spectrometry. *Anal. Chim. Acta* **2012**, *757*, 26–38. [CrossRef] [PubMed]
61. Rosenfield, C.E.; Kenyon, J.A.; James, B.R.; Santelli, C.M. Selenium (IV, VI) reduction and tolerance by fungi in an oxic environment. *Geobiology* **2016**, *15*, 441–452. [CrossRef] [PubMed]

62. Ma, Y.; Liu, R.; Gong, X.; Li, Z.; Huang, Q.; Wang, H.; Song, G. Synthesis and herbicidal activity of *N,N*-diethyl-3-(arylselenonyl)-1 H-1,2,4-triazole-1 -carboxamide. *J. Agric. Food Chem.* **2006**, *54*, 7724–7728. [CrossRef] [PubMed]
63. Singh, P.K. Synthesis and fungicidal activity of novel 3-(substituted/unsubstituted phenylselenonyl)-1-ribosyl/deoxyribosyl-1 H -1,2,4-triazole. *J. Agric. Food Chem.* **2012**, *60*, 5813–5818. [CrossRef]
64. Zhu, Y.G.; Pilon-Smits, E.A.H.; Zhao, F.J.; Williams, P.N.; Meharg, A.A. Selenium in higher plants: Understanding mechanisms for biofortification and phytoremediation. *Trends Plant Sci.* **2009**, *14*, 436–442. [CrossRef] [PubMed]
65. Hartikainen, H. Biogeochemistry of selenium and its impact on food chain quality and human health. *J. Trace Elem. Med. Biol.* **2005**, *18*, 309–318. [CrossRef] [PubMed]
66. Beccari, G.; Senatore, M.T.; Tini, F.; Sulyok, M.; Covarelli, L. Fungal community, *Fusarium* head blight complex and secondary metabolites associated with malting barley grains harvested in Umbria, central Italy. *Int. J. Food Microbiol.* **2018**, *273*, 33–42. [CrossRef]
67. Beccari, G.; Prodi, A.; Senatore, M.T.; Balmas, V.; Tini, F.; Onofri, A.; Pedini, L.; Sulyok, M.; Brocca, L.; Covarelli, L. Cultivation area affects the presence of fungal communities and secondary metabolites in Italian durum wheat grains. *Toxins* **2020**, *12*, 97. [CrossRef] [PubMed]
68. O'Donnell, K.; Kistler, H.C.; Cigelnik, E.; Ploetz, R.C. Multiple evolutionary origins of the fungus causing panama disease of banana: Concordant evidence from nuclear and mitochondrial gene genealogies. *Proc. Natl. Acad. Sci. USA* **1998**, *95*, 2044–2049. [CrossRef]
69. National Center for Biotechnology Information (NCBI) Basic Local Search (BLAST) Tool. Available online: http://blast.ncbi.nlm.nih.gov (accessed on 1 November 2018).
70. Amatulli, M.T.; Spadaro, D.; Gullino, M.L.; Garibaldi, A. Conventional and real-time PCR for the identification of *Fusarium fujikuroi* and *Fusarium proliferatum* from diseased rice tissues and seeds. *Eur. J. Plant Pathol.* **2012**, *134*, 401–408. [CrossRef]
71. Kumar, S.; Stecher, G.; Tamura, K. MEGA7: Molecular Evolutionary Genetics Analysis Version 7.0 for Bigger Datasets. *Mol. Biol. Evol.* **2016**, *33*, 1870–1874. [CrossRef]
72. O'Donnell, K.; McCormick, S.P.; Busman, M.; Proctor, R.H.; Ward, T.J.; Doehring, G.; Geiser, D.M.; Alberts, J.F.; Rheeder, J.P.; Marasas; et al. 1984 "Toxigenic *Fusarium* species: Identity and mycotoxicology" revisited. *Mycologia* **2018**, *110*, 1058–1080. [CrossRef]
73. Alves, E.; Lucas, G.C.; Pozza, E.A.; de Carvalho Alves, M. Scanning electron microscopy for fungal sample examination. In *Laboratory Protocols in Fungal Biology*; Springer: New York, NY, USA, 2013; pp. 133–150.
74. Onofri, A.; Pannacci, E. Spreadsheet tools for biometry classes in crop science programmes. *Commun. Biometry Crop. Sci.* **2014**, *9*, 3–13.

MDPI

Article

Antifungal Activity of the Extract of a Macroalgae, *Gracilariopsis persica*, against Four Plant Pathogenic Fungi

Latifeh Pourakbar [1,*], Sina Siavash Moghaddam [2], Hesham Ali El Enshasy [3,4,5] and R. Z. Sayyed [6]

1 Department of Biology, Faculty of Science, Urmia University, Urmia 5756151818, Iran
2 Department of Plant Production and Genetics, Faculty of Agriculture, Urmia University, Urmia 5756151818, Iran; sinamoghaddam2003@gmail.com
3 Institute of Bioproduct Development (IBD), Universiti Teknologi Malaysia (UTM), Skudai, Johor Bahru 81310, Johor, Malaysia; henshasy@ibd.utm.my
4 School of Chemical and Energy Engineering, Faculty of Engineering, Universiti Teknologi Malaysia (UTM), Skudai, Johor Bahru 81310, Johor, Malaysia
5 City of Scientific Research and Technology Applications (SRTA), New Burg Al Arab, Alexandria 21934, Egypt
6 Department of Microbiology, PSGVP Mandal's, Arts, Science & Commerce College, Shahada 425409, India; sayyedrz@gmail.com
* Correspondence: l.pourakbar@urmia.ac.ir

Abstract: Nowadays, the extract of seaweeds has drawn attention as a rich source of bioactive metabolites. Seaweeds are known for their biologically active compounds whose antibacterial and antifungal activities have been documented. This research aimed to study the profile of phenolic compounds using the HPLC method and determine biologically active compounds using the GC-MS method and the antifungal activity of *Gracilariopsis persica* against plant pathogenic fungi. *G. persica* was collected from its natural habitat in Suru of Bandar Abbas, Iran, dried, and extracted by methanol. The quantitative results on phenolic compounds using the HPLC method showed that the most abundant compounds in *G. persica* were rosmarinic acid (20.9 ± 0.41 mg/kg DW) and quercetin (11.21 ± 0.20 mg/kg DW), and the least abundant was cinnamic acid (1.4 ± 0.10 mg/kg DW). The GC-MS chromatography revealed 50 peaks in the methanolic extract of *G. persica*, implying 50 compounds. The most abundant components included cholest-5-en-3-ol (3 beta) (27.64%), palmitic acid (17.11%), heptadecane (7.71%), and palmitic acid methyl ester (6.66%). The antifungal activity of different concentrations of the extract was determined in vitro. The results as to the effect of the alga extract at the rates of 200, 400, 600, 800, and 1000 µL on the mycelial growth of four important plant pathogenic fungi, including *Botrytis cinerea, Aspergillus niger, Penicillium expansum,* and *Pyricularia oryzae*, revealed that the mycelial growth of all four fungi was lower at higher concentrations of the alga extract. However, the extract concentration of 1000 µL completely inhibited their mycelial growth. The antifungal activity of this alga may be related to the phenolic compounds, e.g., rosmarinic acid and quercetin, as well as compounds such as palmitic acid, oleic acid, and other components identified using the GC-MS method whose antifungal effects have already been confirmed.

Keywords: biologically active compounds; fungistatic effects; quercetin; red alga; rosmarinic acid

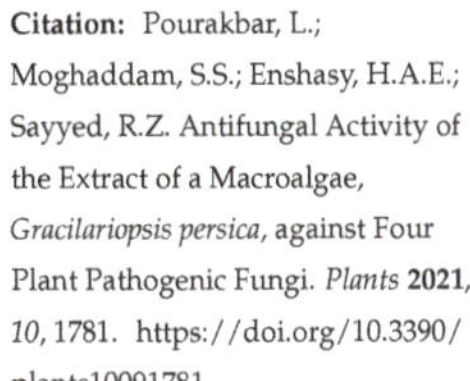

Citation: Pourakbar, L.; Moghaddam, S.S.; Enshasy, H.A.E.; Sayyed, R.Z. Antifungal Activity of the Extract of a Macroalgae, *Gracilariopsis persica*, against Four Plant Pathogenic Fungi. *Plants* **2021**, *10*, 1781. https://doi.org/10.3390/plants10091781

Academic Editors: Špela Mechora and Dragana Šunjka

Received: 27 July 2021
Accepted: 24 August 2021
Published: 26 August 2021

1. Introduction

Algae are autotrophic organisms that contain chlorophyll but lack flowering organs or real roots, stems, and leaves and can convert solar energy into chemical energy by photosynthesis [1]. Macroalgae, or seaweeds, have long been used as a source of food, forage, fertilizer, and medication. Seaweeds provide many raw materials, including agar, algin, and carrageenan used in many industrial sectors. Moreover, they are consumed as food in many Asian countries [2] because they contain carotenoids, diet fibers, proteins, necessary fatty acids, vitamins, and minerals required for human nutrition [3]. Depending on their phylum, growth stage, and environmental conditions, algae can contain different amounts of bioactive compounds, including secondary metabolites, that could exhibit antiviral, antibacterial, and antifungal activities [4–7].

Plant diseases, especially those caused by plant pathogenic fungi, are key factors in the production and quality of crops [8–10] and in the postharvest life of fresh fruit and vegetables [11]. Among these pathogens, *Pyricularia oryzae* (*P. oryzae*) and *Botrytis cinerea* (*B. cinerea*) are two globally distributed fungi with a wide range of hosts, ranked as the first and second most important plant pathogens in the world based on their economic and scientific importance [12]. Known as the cause of rice blast disease, *P. oryzae* is the most damaging rice disease in the world. It is reported that it annually destroys 10–30% of the rice crop, sometimes reaching 100% in the case of an epidemic [12,13]. *B. cinerea* is responsible for various diseases, e.g., blossom blight, leaf spot, and fruit and bulb rot, both on farms and post-harvest crops. Since the organs of the infectious agent grow on the surface of the infected tissues, the developed infection is called gray mold [14]. *Aspergillus niger* is one of the most dominant species from the genus *Aspergillus*, which is responsible for black mold disease in fruits and vegetables [15,16]. It is also known as the common contaminant of foods, especially in sun-dried foods, grains, and nuts, and is the primary agent of postharvest rot of fruits and vegetables. In addition, it is regarded as an opportunistic pathogen of humans [17]. *Penicillium expansum* is known as the cause of blue mold or soft rot in many vegetables and fruits in the world. Its name originated from the blue-color conidium masses that it produces in the infected parts [18,19]. This species is the most important postharvest pathogen of apples and pears as it not only damages them extensively during storage but is also vital, as it produces the carcinogen toxin patulin in the infected fruits used to make fruit juice [20].

There are different ways to control diseases, e.g., the use of resistant cultivars, agronomic practices such as the use of healthy seeds, the balanced use of fertilizers, and biological and chemical control methods [21]. As a result of the frequent use of fungicides, some fungal isolates have become resistant to different fungicide groups [22]; therefore, these fungicides can be used limitedly in disease control [23,24]. It is, therefore, necessary to find new chemicals or use effective natural substances to control the disease.

Moisture on plants is a major factor involved in the fungal infections of these four fungal diseases. The current strategies to control fungal diseases include preventing moisture on leaf area for a prolonged period, the development of resistance in the host plants, and the application of fungicides. Nowadays, various artificial fungicides are applied for the protection of plants from these diseases [21]. Interest is growing in research on the industrial application of medicinal plants due to increasing attention to the use of natural antioxidants and fungicides as an alternative to synthetic ones, which are unsafe and toxic. Recently, many studies have reported the antifungal and antimicrobial activities of algae extracts, essential oils, and other materials. The number of compounds derived from various families of macroalgae, including green algae, brown algae, and red algae, is estimated at 40,000, which plays a key role in plant protection and improvement and is indeed a new approach for pest management [25,26].

Gracilaria is a genus of red algae with a global spread found in polar, moderate, and hot regions, but most of its species have been reported in tropical waters [27]. Research on the coastal areas in Southern Iran has detected a species of these algae in the Persian Gulf called *Gracilariopsis persica* (Rhodophyta) [28].

Most synthetic fungicides can potentially be harmful to the environment and can leave toxic residues in soil and/or crops [29]. Therefore, it is of crucial significance to find new biodegradable natural and eco-friendly bioactive compounds that possess potential biorational activity [30]. Accordingly, a goal we pursued in this research was to assess and examine natural active ingredients in the extract of *G. persica* using HPLC and GC-MS methods with a focus on its biorational effect (relatively non-toxic with few environmental side effects) against four pathogenic fungi.

2. Materials and Methods

2.1. Sampling from *G. persica*

Algae seedlings were collected from their natural habitat in Suru of Bandar Abbas, Iran at high water on November 16, 2018. In the habitat of *G. persica* on the transect, three transects were selected parallel to the coast. Samples were randomly taken by throwing 50 × 50 cm quadrats in five replications and cutting the seedlings in each quadrat with a spatula from their joining to the sand bed. The samples were placed within moist Kenaf fibers (damped by seawater) in multiple layers. They were then transferred to a laboratory with coolers [31]. The thalli collected from the natural habitat were placed in an aquarium for 24 h. The aquarium for the seawater was filled with 40-ppt saline water and kept at 25–27 °C with aeration. The thalli of the algae were rinsed with the water of the related aquarium to remove mud and unwanted epiphytes.

2.2. Extraction to Assess Antifungal Activity

The collected algae were dried for 1 week in the ambient shade drying conditions at an average temperature of 27 °C and after grinding, their powder was used for extraction. Therefore, 50 g of the ground alga's dry matter was poured into a 500-milliliter Erlenmeyer flask where 200 mL of methanol was added. It was then shaken at 25 °C for 10 days, and its extract was filtered through a grade 2 Whatman® filter paper. The extract was dried with an evaporator and kept in a sterilized sealed glass container in darkness at 4 °C [32]. Then, to check the antifungal effects, the dried extract was dissolved in water.

2.3. Preparation of Methanolic Extract for HPLC and GC-MS Analysis

Two g of the powdered alga sample were added to 25 mL of methanol containing 1% acetic acid, placed in a magnetic shaker for 3 h, and filtered by grade 1 Whatman® filter paper.

2.4. Analysis of Methanol Extract by Gas Chromatograph-Mass Spectrometry (GC-MS)

The methanolic extract of the alga was analyzed using Agilent 7890A Gas Chromatography/Mass Spectrometry (GC/MS) (Agilent 7890A, Agilent Technologies Inc., Santa Clara, CA, USA) and 5975 A mass spectrophotometer using an HP-5 MS capillary column (polydimethylsiloxane with a length of 30 m, an internal diameter of 0.25 mm, and a thickness of 0.25 μm). The initial temperature of the oven was set to increase from 80 to 180 °C at a rate of 8 °C/min. Helium was used as the carrier gas whose speed was 1 mm/min along the column length. The injection valve was set in the split mode at a ratio of 1:500 and an injection temperature of 250 °C. The mass spectrum was 40–500 mass/load and ionization energy of 70 eV. The whole run time was 55 min. The mass libraries Wily 2007 and NIST were employed to identify the compounds. Data were processed in the Windows-based Chemstation software. The relative percentage of the extract constituents was expressed in percent based on the peak level.

2.5. Measurement of Phenol Compounds by the HPLC Method

The phenolic compounds were isolated, detected, and quantified using High-Performance Liquid Chromatography HPLC (Agilent 1100, Agilent Technologies Inc., Santa Clara, CA, USA) equipped with a 20-microliter injection loop, a four-solvent gradient pump, a degasser, a column oven (set at 25 °C), and a diode array detector set at 250, 272, and 310 nm.

The isolation was carried out by an Octadecyl Saline Column (ODC) (with a length of 25 cm, an internal diameter of 4.6 mm, and a particle size of 5 μm, ZOR BAX Eclipse XDB, Germany). Data were processed using the Chemstation software. The mobile phase consisted of acetic acid (1%) (A) and acetonitrile (B) at a flow rate of 1 mL/min. The phenolic compounds were eluted under the following conditions: 1 mL/min of flow rate, the temperature of 25 °C, isocratic conditions from 0 to 10 min with 10% B, gradient conditions from 10 to 25% B in 5 min, from 25 to 65% B in 10 min, from 65 to 100% B in 15 min, followed by washing and reconditioning the column. The injection volume was

10 μL and phenolic compounds were detected at wavelengths of 250, 272, and 310 nm. The phenolic compounds were identified by comparing their relative retention times and compared with standards [33].

2.6. Preparation of Potato Dextrose Agar (PDA)

Fifty g of peeled Irish potatoes were rinsed after slicing and boiled for 1 h with distilled water in a conical flask. A muslin cloth was applied to filter the boiled potatoes and the volume was increased to 250 mL with water along with dextrose (5 g) and agar (3.5 g). The suspension was dissolved with heat and shaking, then left to cool down at room temperature for two min, then sterilized in an autoclave at 121 °C for one h.

2.7. In Vitro Antifungal Activity of the Extract

The antifungal effect of the extract of *G. persica* was checked using the following method. The fungi *B. cinerea*, *P. oryzae*, *A. niger*, and *P. expansum* were purchased from the Fungi and Bacteria Collection Center of Iran and cultured in a laboratory. To evaluate the antifungal activity of the alga extract, 200, 400, 600, 800, and 1000 μL of the extract were poured on sterile PDA plates. Distilled water was chosen as the control. Then, a ring with a diameter of 5 mm was taken from the margin of the actively growing colonies of the fungi and transferred to the center of the culture medium. All inoculated plates were kept in an incubator at 25 °C and were checked daily until the surface of the culture medium was entirely covered with the fungi in the control treatments. Then, the radial growth of the fungi was measured. The extract inhibitory effect on the mycelial growth of the fungi was calculated for each concentration by using the following formula [34]:

$$\text{MGI}\ (\%) = \frac{dc - dt}{dc} \times 100 \quad (1)$$

where MGI is the mycelial growth inhibition, dc is the mean diameter (mm) of the fungal mycelial growth in the control treatment, and dt is the mean diameter (mm) of the fungal mycelial growth extract-containing treatment.

2.8. Fungistatic or Fungicide Effect of the Extract

At the end of the trials, another experiment was conducted on the treatments in which no growth was observed to determine whether the extracts had killed the fungi (fungicidal effect) or had inhibited their growth temporarily (fungistatic effect). For this experiment, a PDA culture medium that contained no additive was prepared and poured into Petri dishes. The fungi-containing rings from the treatments that had no growth were transferred to the new culture medium. The Petri dishes were placed in an incubator at 25 °C and were re-checked after 7 days. The treatments in which the mycelial growth was observed after 7 days showed the fungistatic activity of the extract, while the lack of mycelial growth in the fungi-containing ring in the culture medium exhibited its fungicide activity.

2.9. Data Analysis

The data were subjected to the analysis of variance (ANOVA) and the comparison of means in the SPSS (Ver. 16) software package. The mean values were compared with Duncan's multiple range test. In all graphs, the results were expressed in average values of three replications ± standard deviation (SD). Additionally, the graphs were drawn in MS-Excel (2016) software package.

3. Results

3.1. Results of GC-MS

The biologically active components in the methanolic extract of *G. persica* were detected using GC-MS with a running time of 55 min. The GC-MS chromatogram showed 50 peaks in methanolic extract, implying 50 compounds (Figure 1). The spectra of these compounds were compared with Wiley 7.0 and National Institute of Standards and Technology libraries.

The detected compounds are listed in Table 1. The main fractions included cholest-5-en-3-ol (3. beta), palmitic acid, heptadecane, palmitic acid methyl ester, bicyclo [3.1.1] heptane, 2,6,6-trimethyl or pinane, isobutyl phthalate, *N*-cyano-*N′*, *N′*, *N″*, *N″*-tetramethyl-1,3,5-triazinetriamine, 5-thiazoleethanol, 4-methyl, and phytol.

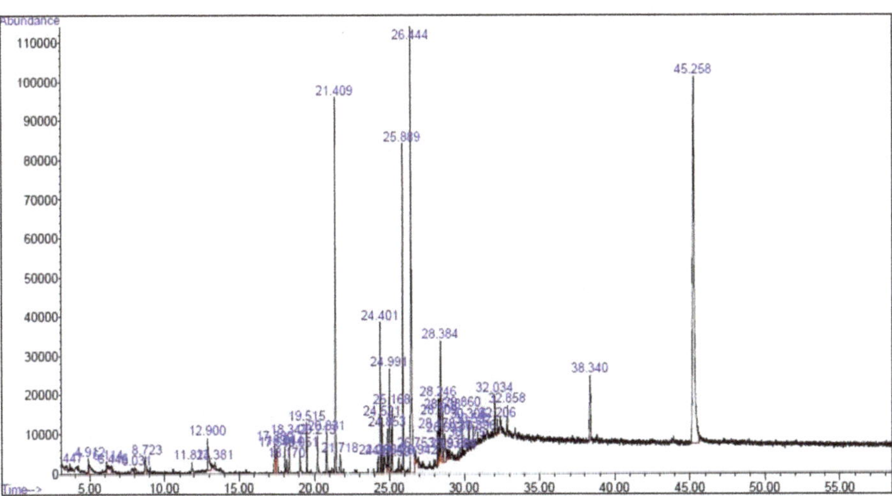

Figure 1. Gas chromatography and mass spectroscopy chromatogram of the methanolic extract of the *Gracilariopsis persica* alga.

Table 1. The phytocomponents were detected in the methanolic extract of the *Gracilariopsis persica* using gas chromatography and mass spectroscopy.

Retension Time	Name of the Compound	Area %	Bioactivity	Reference
3.447	Boron, trihydro (N-methylmethanamine-, (T-4)-	0.36		
4.912	2 (3H)-Furanone, dihydro-4-hydroxy-	0.45	Antibacterial	[35]
6.114	1-Hepten-3-ol	0.21	-	
6.646	D-Ribonic acid, 2,3-O-(ethoxymethylene)-	0.37	-	
8.031	2-Octyldodecan-1-ol	0.19	-	
8.723	Docosane	0.88	Antifangal	[36]
11.824	2-Decyne	0.46	-	
12.900	5-Thiazoleethanol, 4-methyl-	2.12	Antifungal, anti-inflammatory, anti-allergic	[37]
13.381	Carvacrol	0.25	Anti-inflammatory, antioxidant, antitumor, antibacterial	[38]
17.380	2-(1,4,4-Trimethyl-cyclohex-2-enyl)-ethanol	1.21		
17.541	Decamethylpentasiloxane	0.83	Antibacterial, antifungal	[39]
18.044	2-cyclohexene-1-one	0.62	Antibacterial	[40]
18.170	10-Methyl-9-oxabicyc lo[6.4.0]dodecan-1(8)-ene	0.43	-	
18.342	2-(4H)-Benzofuranone, 5,6,7,7a-tetrahydro-4,4,7a-trimethyl-	0.76	Antimicrobial preservative, antifungal, antibacterial	[41]
19.051	2-Propenoic acid, 3-(1-aziridinyl)-,methyl ester	0.85		
19515	Diethyl Phthalate	1.06	Antioxidant	[42]
20.213	Methanone, diphenyl-	0.98	Antibacterial, antifungal	[43]
20.831	Cyclododecasiloxane, tetracosamethyl-	1.37	Antimicobial, antirheumatic antispasmodic	[44]
21.409	Heptadecane	7.71	Anticancer, anti-inflammatory	[38]
21.718	n-Dodecanal	0.66	Antibacterial	[41]
24.230	Cyclododecasiloxane, Tetracosamethyl-	0.39	Antimicobial, antirheumatic antispasmodic	[44]
24.662	Caffeine	0.97	Antibacterial, antifungal	[45]
24.401	Bicyclo[3.1.1] heptane, 2,6,6-trimethyl Pinane	3.17	Antifungal	[46]
24.521	11-Dodecen-2-one	1.13		

Table 1. *Cont.*

Retension Time	Name of the Compound	Area %	Bioactivity	Reference
24.853	E-10-Methyl-11-tetradecen-1-ol acetate	0.76		
24.991	Isobutyl phthalate	2.25	Antibacterial, antifungal	[42]
25.168	Neophytadine	1.66	Antimicrobial, anti-inflammatory	[47]
25.591	2-Propenoic acid, 2-methyl-	0.43	Antibacterial, antifungal	[48]
25.889	Palmitic acid methyl ester	6.66	Antibacterial, antifungal	[41]
26.444	Palmitic acid	17.11	Antibacterial, antifungal	[49]
26.753	Oleic acid	0.24	Antibacterial, antifungal	[49]
26.942	1-Decanol, 2-hexyl-	0.46	Antimicrobial	[50]
28.178	Linolelaidic acid, methyl ester	0.66	Antibacterial, antifungal	[41]
28.246	Oleic acid methyl ester	1.18	Antibacterial, antifungal	[51]
28.309	Oleic acid methyl ester	0.94	Antibacterial, antifungal	[41]
28.384	Phytol	2.34	Antibacterial, antifungal	[52]
28.538	Stearic acid methyl ester	1.01	Antibacterial, antifungal	[41]
28.693	9-Hexadecenoic acid	1.73	Antibacterial, inflammator	[52]
28.933	9-Hexadecenoic acid	0.32	Antibacterial, inflammator	[52]
29.608	E-11-Tetradecenoic acid	0.25	Antibacterial, antifungal	[52]
29.860	Hexadecanedioic acid	0.93	Antimicrobial, anti-inflammatory	[52]
30.175	9-Borabicyclo[3.3.1]nonane, dimethylamino)propyl]	0.27	-	
0.306	(-)-18-noramborx	0.45	-	
30.592	Trisiloxane, 1,1,1,3,5,5,5-heptamethyl	0.53	Antifungal	[53]
30.896	Hexamethylcyclotrisiloxane	0.28	Antimicrobial	[54]
32.034	1,3-Bis(trimethylsilyl)benzene	0.79	-	
32.206	p-Bis(trimethylsilyl)benzene	0.47	-	
32.858	Phthalic acid,bis(2-ethylhexyl) ester	0.75	Antibacterial, antifungal	[35]
38.340	N-Methyl-1-adamantane acetamide	2.52	Antimicrobial	[55]
45.258	Cholest-5-en-3-ol (3.beta.)-	27.64	Antibacterial	[56,57]

3.2. Quantitative Analysis of Phenolic Compounds Determined Using HPLC

Figure 2 displays the standard HPLC chromatogram of all the recorded compounds at 250, 272, and 310 nm. According to Figure 2, all the studied compounds responded to the three spectra and were isolated successfully. The compounds of the alga extract were also separated using the same method at the same wavelengths, and they were identified using the standard graph (Figure 3a–c).

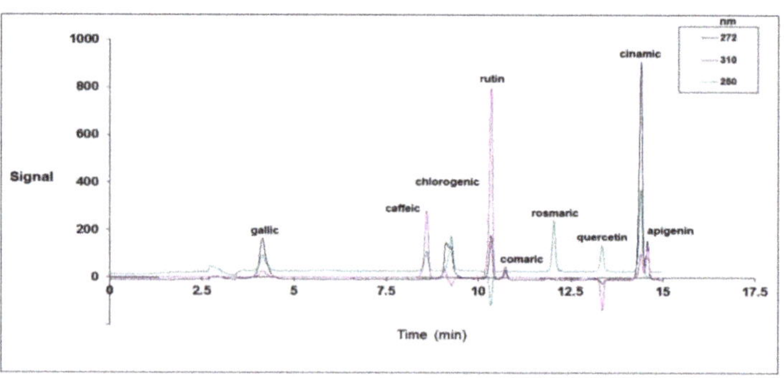

Figure 2. The standard graph of the phenolic compounds in the HPLC chromatogram.

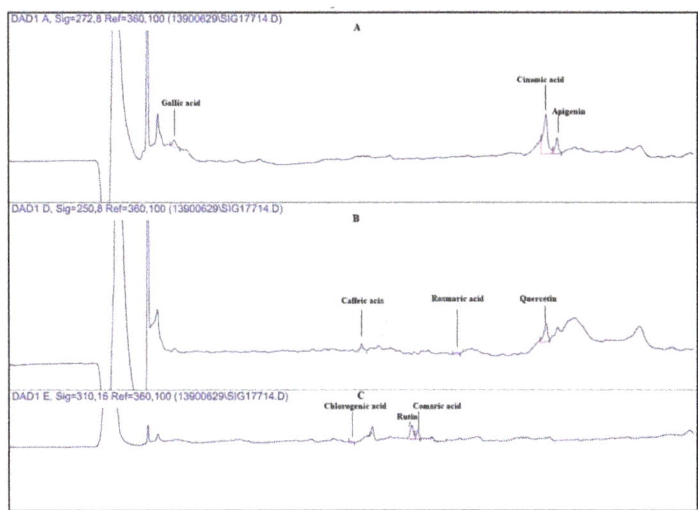

Figure 3. The diagram of identifying the phenolic compounds using the HPLC method at three wavelengths of (**A**) 272 nm, (**B**) 250 nm, and (**C**) 310 nm.

The quantitative results for the phenolic compounds using the HPLC method showed that the most abundant phenols in *G. Persica* were rosmarinic acid (20.9 ± 0.41 mg/kg DW) and quercetin (11.21 ± 0.20 mg/kg DW), and the least abundant was cinnamic acid (1.4 ± 0.10 mg/kg DW) (Figures 3 and 4).

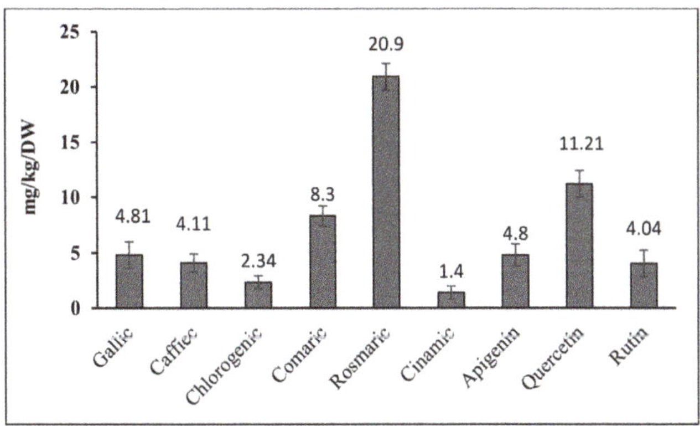

Figure 4. The quantity of the phenolic compounds identified using the HPLC method in *G. persica* (columns are mean SD ± 3 replications and vertical bars show standard deviation).

3.3. Antifungal Activity

Based on the results, as the concentration of the *G. persica* extract was increased, the growth of *B. cinerea*, *P. oryzae*, *A. niger*, and *P. expansum* was inhibited to a significantly higher extent (Figure 5). The lowest MGI in all four fungi was 35.18, 16.88, 24.50, and 37.21% recorded at the extract volume of 200 µL, respectively. The mycelial growth of *B. cinerea* and *P. expansum* was completely inhibited at an extract volume of 800 µL, and that of *P. oryzae* and *A. niger* was fully inhibited at the extract rate of 1000 µL (Figure 5).

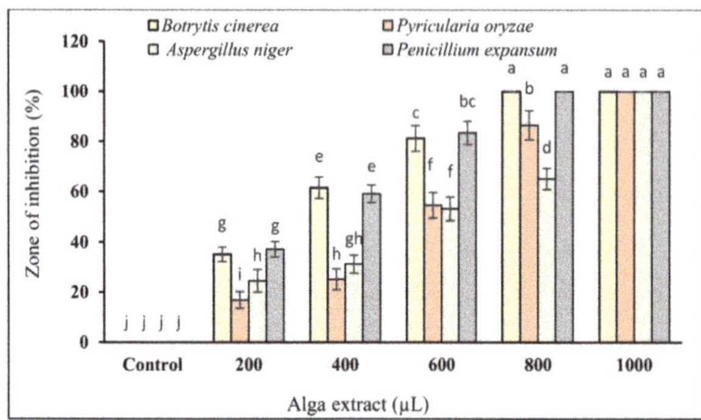

Figure 5. The effect of the *G. persica* extracts at volumes, 200, 400, 600, 800, and 1000 μL on the mycelial growth inhibition of *B. cinerea*, *P. oryzae*, *A. niger*, and *P. expansum* (columns are mean SD ± 3 replications, vertical bars show standard deviation, and different letters imply significant differences at the $p < 0.05$ level based on Duncan's test.).

The results as to the effect of the alga extract on the mycelial growth of the studied fungi revealed a clear dose–response of the extract. The 100% inhibition was achieved at 800 and 1000 μL for *B. cinerea* and *P. expansum* (Figure S1) and 1000 μL for *P. oryzae* and *A. niger* (Figure S2).

Fungistatic

The fungistatic effect of various concentrations of the extract was observed as inhibition of mycelial growth. However, re-culturing the mycelial growth following the re-inoculation into the fresh medium resulted in the re-growth of the fungal mycelium in the new medium, implying that the alga extract could only inhibit the growth of the fungi (fungistatic).

In the study of fungicidal or fungistatic effects of extracts at concentrations that completely inhibited mycelium growth, the re-cultures resulting from the transfer of the fungi-containing ring to the PDA culture medium showed that the fungi reappeared and grew after 5–7 days on the fresh culture medium. The results revealed that *P. oryzae* and *A. niger* started to grow after 5 days and *P. expanstum* and *B. cinerea* started to grow after 7 days. During the study period (7 days), the growth of mycelium in the re-cultures of the control fungi was completely evident from the second day, and it was much higher (up to 90%) at the end of 7 days than the re-cultures performed at concentrations in which mycelium growth was completely inhibited. These results indicate that the algae extract only inhibited the growth of fungi and fungicidal properties were observed in none of the extract concentrations.

4. Discussion

The results obtained from the GC-MS analysis of this study specify that *G. persica* showed a large number of bioactive compounds with antioxidant, antibacterial, and antifungal properties. There are various reports about the compounds derived from macroalgae as a potential source of biochemical and medicinal properties including antibacterial [57], antifungal [58], antiviral [59], antioxidant [60], and anti-inflammation activities [61].

The results of the present study revealed that the extract of *G. persica* had a high potential to inhibit the mycelial growth of plant pathogenic fungi. Previous research findings have indicated that alga extracts have agents for the biological control of the growth of hyphae and germination, an increase in intracellular holes (vacuolization), and the disruption of the functioning of fungal cells [62]. On the other hand, research on different species of *Gracilaria* has revealed this genus' antibacterial and antifungal activities. In this regard, Singh and Raadha [63] studied the *G. corticata* extract. They found that this species, at a rate of 1000 μL, could inhibit the growth of human pathogenic bacteria and fungi, including *Salmonella typhimurium*, *Escherichia coli*, *Staphylococcus aureus*, and *Candida albicans*, and was a natural source of antibiotics. Dayuti [64] also reported the antibacterial activity of *G. verrucosa* against *S. typhimurium* and *E. coli*. A study on the antifungal and antibacterial properties of the *G. confervoides* extract showed that 100 μL of the *G. confervoid* extract could prevent the aerial mycelial growth of the cucumber pathogens *Rhizoctonia solani* and *Macrophomina phaseolinae* [58]. This is consistent with our findings as to the fungistatic activity of the *G. persica* extract. In an experiment, Kolanjinathan and Stella [65] found that *G. corticata* could inhibit the growth of *Aspergillus flavus*, *A. fumigatus*, and *A. niger*, as well as the human pathogen *albicans*. The minimum inhibitory rate of this extract was estimated at 2–16 mg/mL. This is in agreement with our findings as to the antifungal property of *G. persica* in inhibiting the growth of *A. niger*.

The results of HPLC analysis in our study showed that *G. persica* is rich in polyphenolic compounds. It is now well established that the antifungal activity of alga extracts may be related to the presence of phytochemicals, e.g., tannins and phenols. Phenolic compounds are likely to influence the growth and metabolism of fungi [66]. Sea resources are the most enormous remaining reservoirs of natural molecules, which are assessed for therapeutic activities and provide valuable ideas for developing new medications against cancer, microbial infections, and inflammations [67]. Although terrestrial biodiversity constitutes the basis of the pharmaceutical industry, oceans have rich biodiversity and can produce commercially invaluable modern compounds. A comparison of our findings with those of other studies confirms that *G. persica* is a rich source of biological components.

Some phytocompounds, which combine alkaloids, flavonoids, and saturated and unsaturated fatty acids, have antimicrobial, anti-inflammation, anti-cancer, anti-coagulant, and anti-arrhythmic activities. According to the results of GC-MS, most of the compounds detected in the methanolic extract of *G. persica* (36 out of the 50 detected components) are compounds whose antimicrobial and antifungal activities have already been documented. These interpretations corroborate the findings of current research (Table 1).

The results of HPLC revealed that *G. persica* had invaluable phenolic compounds, e.g., rosmarinic acid and quercetin, implying its antioxidant effect. It has been reported that the difference in plant extracts' biological activity depends on their constituents; therefore, a single compound may be responsible for an extract's effects alone or in synergy with other compounds [68].

Polyphenols are a remarkable group of plant metabolites that have an efficient antimicrobial performance. A number of studies have postulated about the interaction of the synergy of polyphenols with antibiotics against microbial resistance, e.g., epigallocatechin gallate of green tea [69], tellimagrandin I, and rugosin B of *Rosa floribunda* 'Dubline Bay' (aka 'MACdub') [70], and the synergy of rosmarinic acid and antibiotics against methicillin-resistant *Staphylococcus aureus* [71]. Likewise, the antimicrobial effect of quercetin as the second dominant phenolic compound in *G. persica* against *Escherichia coli*, *Staphylococcus aureus*, and *Pseudomonas fluorescens* [72], and the antifungal effects of quercetin and rutin against *Cryptococcus* spp. [73] have been documented. It has been reported that these compounds can change the structure of the cell membrane, thereby destroying the plasma membrane integrity in fungal cells and increasing its permeability, which results in an increase in the K^+ outflow from the cytoplasm of fungal cells. These effects may cause the polarity of the membrane to be lost by altering ion transport, or they may reduce energy production (ATP) by altering the membrane structure through impairing glucose uptake or inhibiting the enzymes involved in oxidative stress or phosphorylation precursor. The increase in cytoplasm membrane permeability ultimately leads to cell death due to the dispersion and loss of the cell pH gradient, a decline in the ATP level, and the failure of the proton driving force. Indeed, nutrient uptake, nucleic acid synthesis, and ATP*ase* activity sections are most damaged in the tissues of the fungi [74,75].

The experimental evidence of current research regarding the examination of polyphenolic compounds in *G. persica* showed that quercetin was the highest among the flavonoid compounds. The possible activity of flavonoids may also be involved in mitochondrial damage and ROS production by inducing the transcription factors related to apoptosis and increasing the level of proptose proteins [76]. Hwang et al. [77] revealed that flavonoids disturbed the performance of mitochondria in the *C. albicans* strain by increasing the ROS level. The elevated intracellular ROS level and the disturbed mitochondrial performance play a significant role in apoptosis induction [78,79]. Quercetin has two aromatic rings in its structure and can penetrate the phospholipid membrane [80], where it damages DNA by inducing oxidative damage and, finally, cause cell death by apoptosis, which is an irreversible process [81].

5. Conclusions

Herbal medicines play a significant role in human health and are an inspiring source of new medicinal compounds. It can be concluded from the results of the present study that *G. persica* has a high potential to be used in the pharmaceutical industry to improve health owing to its different compounds with antimicrobial activity.

The inhibitory effect of the alga on the growth of plant pathogenic fungi *in vitro* is apparent from our experimental evidence. Thus, this alga might be used to produce an environmentally friendly, reliable, and economic antifungal agent to control *B. cinerea*, *P. oryzae*, *A. niger*, and *P. expansum*. This can be a high potential alternative to highly toxic chemical fungicides in plant disease management. The fungicidal efficacy still has to be shown in vivo. Above all, our results suggest that the algae in the Persian Gulf and Oman Sea can be a rich source of different macroalgae species with unique antimicrobial activities, which may have various applications in agriculture and the control of plant disease in the future.

Supplementary Materials: The following are available online at https://www.mdpi.com/article/10.3390/plants10091781/s1, Figure S1: The effect of different concentrations of the *G. persica* extract (control, 400, and 1000 µL) on the mycelial growth of *B. cinerea* and *P. expansum.* Figure S2: The effect of different concentrations of the *G. persica* extract (control, 400, and 1000 µL) on the mycelial growth of *P. oryzae* and *A. niger.*

Author Contributions: Writing—original draft, methodology, and formal analysis, L.P.; conceptualization, supervision, and project administration, S.S.M.; writing—editing and proofreading and formal analysis, R.Z.S. and H.A.E.E. All authors have read and agreed to the published version of the manuscript.

Funding: This work was funded by RMC, Universiti Teknologi Malaysia (UTM), Malaysia through grant No. R.J130000.7609.4C336 and R.J130000.7609.4C359.

Institutional Review Board Statement: Not applicable.

Informed Consent Statement: Not applicable.

Data Availability Statement: Not applicable.

Acknowledgments: H.A.E. would like to thank RMC, Universiti Teknologi Malaysia (UTM), Malaysia for financial support through grant No. R.J130000.7609.4C336 and R.J130000.7609.4C359.

Conflicts of Interest: The authors declare no conflict of interest.

References

1. Lee, R.E. *Phycology;* Cambridge University Press: Cambridge, UK, 2008.
2. Mishra, V.K.; Temelli, F.; Ooraikul, B.; Shacklock, P.F.; Craigie, J.S. Lipids of the red alga, Palmaria palmata. *Bot. Mar.* **1993**, *36*, 169–174. [CrossRef]
3. Mohamed, F.; Namitha, K.K.; Chidambara Murthy, K.N.; Mahadeva Swamy, M.; Sarada, R.; Khanam, S.; Subbarao, P.V.; Ravishankar, G.A. Chemical composition, iron bioavailability and antioxidant activity of *Kappsphycus alvarezi*. *J. Agric. Chem.* **2005**, *53*, 792–797.
4. Plaza, M.; Santoyo, S.; Jaime, L.; Reina, G.G.B.; Herrero, M.; Senorans, F.J.; Lbanez, E. Screening for bioactive compounds from algae. *J. Pharmac. Biomed. Anal.* **2010**, *51*, 450–455. [CrossRef] [PubMed]

5. Al-Saif, S.S.A.; Abd-Alraouf, N.; El-Wazanani, H.A.; Aref, I.A. Antibacterial substances from marine algae isolated from Jeddah coast of Red sea, Saudi Arabia. *Sau. J. Biolog. Sci.* **2014**, *14*, 57–64. [CrossRef] [PubMed]
6. Righini, H.; Roberti, R.; Baraldi, E. Use of algae in strawberry management. *J. Appl. Phyco.* **2018**, *30*, 3551–3564. [CrossRef]
7. Soares, A.R.; Robaina, M.C.S.; Mendes, G.S.; Silva, T.S.L.; Gestinari, L.M.S.; Pamplona, O.S.; Yoneshigue-Valentin, Y.; Kaiser, C.R.; Romanos, M.T.V. Antiviral activity of extracts from Brazilian seaweeds against herpes simplex virus. *Rev. Bras. Farmacogn. Braz. J. Pharmacogn.* **2012**, *22*, 714–723. [CrossRef]
8. Sagar, A.; Sayyed, R.Z.; Ramteke, P.W.; Sharma, S.; Marraiki, N.; Elgorban, A.E.; Syed, A. ACC deaminase and antioxidant enzymes producing halophilic *Enterobacter* sp. PR14 promotes the growth of rice and millets under salinity stress. *Physiol. Mol. Biol. Plants* **2020**, *26*, 1847–1854. [CrossRef] [PubMed]
9. Basu, A.; Prasad, P.; Das, S.N.; Kalam, S.; Sayyed, R.Z.; Reddy, M.S.; Enshasy, H.E. Plant Growth Promoting Rhizobacteria (PGPR) as Green Bioinoculants: Recent Developments, Constraints, and Prospects. *Sustainability* **2021**, *13*, 1140. [CrossRef]
10. Hamid, B.; Zaman, M.; Farooq, S.; Fatima, S.; Sayyed, R.Z.; Baba, Z.A.; Sheikh, T.A.; Reddy, M.S.; El Enshasy, H.; Gafur, A.; et al. Bacterial Plant Biostimulants: A Sustainable Way towards Improving Growth, Productivity, and Health of Crops. *Sustainability* **2021**, *21*, 2856. [CrossRef]
11. Saharan, V.; Sharma, G.; Yadav, M.; Choudhary, M.K.; Sharma, S.S.; Pal, A.; Biswas, P. Synthesis and in vitro antifungal efficacy of Cu–chitosan nanoparticles against pathogenic fungi of tomato. *Int. J. Biol. Macromol.* **2015**, *75*, 346–353. [CrossRef]
12. Dean, R.; Van Kan, J.A.; Pretorius, Z.A.; Hammond-Kosack, K.E.; Di Pietro, A.; Spanu, P.D.; Rudd, J.J.; Dickman, M.; Kahmann, R.; Ellis, J.; et al. The Top 10 fungal pathogens in molecular plant pathology. *Mol. Plant. Pathol.* **2012**, *13*, 414–430. [CrossRef] [PubMed]
13. Wang, X.; Lee, S.; Wang, J.; Ma, J.; Bianco, T.; Jia, Y. Current advances on genetic resistance to rice blast disease. In *Rice-Germplasm, Genetics, and Improvement*; Yan, W., Bao, J., Eds.; InTech: London, UK, 2014; pp. 195–217.
14. Bautista-Baños, S. *Postharvest Decay: Control Strategies*; Elsevier: Amsterdam, The Netherlands; Academic Press: San Diego, CA, USA, 2014; p. 383.
15. Pitt, J.I.; Hocking, A.D. *Fungi and Food Spoilage*, 3rd ed.; Springer Science + Business Media, LLC: New York, NY, USA, 2009; p. 519.
16. Palencia, E.R.; Hinton, D.M.; Bacon, C.W. The black *Aspergillus* species of maize and peanuts and their potential for myctoxin production. *Toxins* **2010**, *2*, 399–416. [CrossRef]
17. Ziani, K.; Fernandez-Pan, I.; Royo, M.; Maté, J.I. Antifungal activity of films and solutions based on chitosan against typical seed fungi. *Food Hydrocolloids* **2009**, *23*, 2309–2314. [CrossRef]
18. Kwon, J.H.; Jeong, S.G.; Hong, S.B.; Chae, Y.S.; Park, C.S. Occurrence of blue mold on sweet persimmon (*Diospyros kaki*) caused by *Penicillium expansum*. *Res. Plant Dis.* **2006**, *12*, 290–293. [CrossRef]
19. Gao, L.L.; Zhang, Q.; Sun, X.Y.; Jiang, L.; Zhang, R.; Sun, G.Y.; Zha, Y.L.; Biggs, A.R. Etiology of moldy core, core browning, and core rot of Fuji apple in China. *Plant Dis.* **2013**, *97*, 510–516. [CrossRef]
20. Tannous, J.; Atoui, A.; El Khoury, A.; Francis, Z.; Oswald, I.P.; Puel, O.; Lteif, R. A study on the physicochemical parameters for *Penicillium expansum* growth and patulin production: Effect of temperature, pH, and water activity. *Food Sci. Nutr.* **2016**, *4*, 611–622. [CrossRef]
21. Kalam, S.; Basu, A.; Ahmad, I.; Sayyed, R.Z.; Enshasy, D.; Dailin, J.; Suriani, N.L. Recent understanding of soil Acidobacteria and their ecological significance: A critical review. *Front. Microbiol.* **2020**, *11*, 580024. [CrossRef]
22. Arora, H.; Sharma, A.; Sharma, S.; Haron, F.F.; Gafur, A.; Sayyed, R.Z.; Datta, R. Pythium damping-off and root rot of *Capsicum annuum* L.: Impacts, Diagnosis, and Management. *Microorganisms* **2021**, *9*, 823. [CrossRef] [PubMed]
23. Miah, G.; Rafii, M.Y.; Ismail, M.R.; Puteh, A.B.; Rahim, H.A.; Asfaliza, R.; Latif, M.A. Blast resistance in rice: A review of conventional breeding to molecular approaches. *Mol. Biol. Rep.* **2013**, *40*, 2369–2388. [CrossRef] [PubMed]
24. Chauhan, B.S.; Jabran, K.; Mahajan, G. *Rice Production Worldwide*; Springer International Publishing AG: Berlin/Heidelberg, Germany, 2017; p. 563.
25. Raven, P.H.; Evert, R.F.; Eichom, S.E. *Biology of Plants*, 5th ed.; Worth Publishers: New York, NY, USA, 1992.
26. Saidani, K.; Bedjou, F.; Benabdesslam, F.; Touati, N. Antifungal activity of methanolic extracts of four Algerian marine algae species. *Afr. J. Biotechnol.* **2012**, *11*, 9496–9500. [CrossRef]
27. Sahoo, D.; Yarish, C. Mariculture of seaweeds. In *Phycological Methods: Algal Culturing Techniques*; Andersen, R., Ed.; Academic Press: New York, NY, USA, 2005; pp. 219–237.
28. Sohrabipour, J.; Rabii, R. A list of marine algae of sea shores of the Persian Gulf and Oman Sea in the Hormozgan province. *Iran. J. Bot.* **1999**, *8*, 131–162.
29. Reshma, P.; Naik, M.K.; Aiyaz, M.; Niranjana, S.R.; Chennappa, G.; Shaikh, S.S.; Sayyed, R.Z. Induced systemic resistance by 2,4diacetylphloroglucinol positive fluorescent Pseudomonas strains against rice sheath blight. *Indian J. Exp. Biol.* **2018**, *56*, 207–212.
30. Zope, V.P.; Jadhav, H.P.; Sayyed, R.Z. Neem cake carrier prolongs shelf life of biocontrol fungus *Trichoderma viridae*. *Indian J. Exp. Biol.* **2019**, *57*, 372–375.
31. Liu, S. Study on the commercial cultivation of *Gracilaria* in south. *Chi. J. Oce. Lim.* **1987**, *5*, 281–283.
32. Kumar, C.S.; Dronamraju, V.; Sarada, L.; Ramasamy, R. Seaweed extracts control the leaf spot disease of the medicinal plant *Gymnema sylvestre*. *Ind. J. Sci. Technol.* **2008**, *1*, 1–5. [CrossRef]

33. Seal, T. Quantitative HPLC analysis of phenolic acids, flavonoids, and ascorbic acid in four different solvent extracts of two wild edible leaves, *Sonchus arvensis* and *Oenanthe linearis* of North-Eastern region in India. *J. Appl. Pharmacol. Sci.* **2016**, *6*, 157–166. [CrossRef]
34. Shahi, S.K.; Patra, M.; Shukla, A.C.; Dikshit, A. Use of essential oil as botanical-pesticide against postharvest spoilage in *Malus pumilo* fruits. *Biocontrol* **2003**, *48*, 223–232. [CrossRef]
35. Husain, A.; Mumtaz, A.; Siddiqui, N. Synthesis, reactions and biological activity of 3-arylidene-5-(4-methylphenyl)-2(3H)-furanones. *J. Serb. Chem. Soc.* **2009**, *74*, 103–115. [CrossRef]
36. Mansur Ali, S.; Ahmed Khan, N.; Sagathevan, K.; Anwar, A.; Siddiqui, R. Biologically active metabolite(s) from haemolymph of red-headed centipede Scolopendra subspinipes possess broad spectrum antibacterial activity. *AMB Express* **2019**, *9*, 1–17.
37. Paerl, R.W.; Bertrand, E.M.; Rowland, E.; Schatt, P.; Mehiri, M.; Niehaus, T.D.; Hanson, A.D.; Riemann, L.; Bouget, F.Y. Carboxythiazole is a key microbial nutrient currency and critical component of thiamin biosynthesis. *Sci. Rep.* **2018**, *8*, 8876. [CrossRef] [PubMed]
38. Magi, G.; Marini, E.; Facinelli, B. Antimicrobial activity of essential oils and carvacrol, and synergy of carvacrol and erythromycin, against clinical, erythromycin-resistant Group A *Streptococci*. *Front. Microbiol.* **2015**, *6*, 1–7. [CrossRef]
39. Mahmud, P.I.A.M.; Wan Ahmad, W.Y.; Ibrahim, N.; Abu Bakar, M. Antibacterial activity and major constituents of *polyalthia cinnamomea* basic fraction. *Sains Malays.* **2018**, *47*, 2063–2071. [CrossRef]
40. Vyas, D.H.; Tala, S.D.; Akbari, J.D.; Dhaduk, M.F.; Joshi, H.S. Synthesis, antimicrobial, and antitubercular activity of some cyclohexenone and indazole derivatives. *Ind. J. Chem.* **2009**, *48*, 1405–1410.
41. Mujeeb, F.; Bajpai, P.; Pathak, N. Phytochemical evaluation, antimicrobial activity, and determination of bioactive components from leaves of *Aegle marmelos*. *Biol. Med. Res. Int.* **2014**, *2014*, 1–11.
42. Al-Bari, M.A.A.; Sayeed, M.A.; Rahman, M.S.; Mossadik, M.A. Characterization and antimicrobial activities of a phthalic acid derivative produced by *Streptomyces bangladeshiensis*—A novel species collected in Bangladesh. *Res. J. Medic. Med. Sci.* **2006**, *1*, 77–81.
43. Uma, M.; Jothinayaki, S.; Kumaravel, S.; Kalaiselvi, P. Determination of bioactive components of *Plectranthus amboinicus* Lour by GC–MS analysis. *N.Y. Sci. J.* **2011**, *4*, 66.
44. Singh, H.B.; Handique, A.K. Antifungal activity of the essential oil of *Hyptis suaveolens* and its efficacy in biocontrol measures in combination with *Trichoderma harzianum*. *J. Essent. Oil Res.* **2007**, *9*, 683–687. [CrossRef]
45. Kim, Y.W.; Chun, H.J.; Kim, I.; Liu, H.; Shick Ahn, W. Antimicrobial and antifungal effects of green tea extracts against microorganisms causing vaginitis. *Food Sci. Biotechnol.* **2013**, *22*, 713–719. [CrossRef]
46. Nikitina, L.E.; Startseva, V.A.; Dorofeeva, L.; Artemova, N.P.; Kuznetsov, I.V.; Lisovskaya, S.A.; Glushko, N.P. Antifungal activity of bicylic monoterpenoids and terpeneslfides. *Chem. Nat. Comp.* **2010**, *46*, 28–32. [CrossRef]
47. Kumara Swamy, M.; Arumugam, G.; Kaur, R.; Ghasemzadeh, A.; Yusoff, M.M.; Sinniah, U.R. GC-MS-based metabolite profiling, antioxidant and antimicrobial properties of different solvent extracts of Malaysian *Plectranthus amboinicus* leaves. *Evidence Based Compl. Altern. Medi.* **2017**, *2017*, 1–10. [CrossRef]
48. Qi, S.H.; Xu, Y.; Gao, J.; Qian, P.Y.; Zhang, S. Antibacterial and antilarval compounds from the marine bacterium *Pseudomonas rhizosphaerae*. *Ecol. Environ. Microbiol.* **2009**, *59*, 229–233.
49. Chandrasekaran, M.; Senthilkumar, A.; Venkatesalu, V. Antibacterial and antifungal efficacy of fatty acid methyl esters from the leaves of *Sesuvium portulacastrum* L. *Eur. Rev. Medicol. Pharmacol. Sci.* **2011**, *15*, 775–780.
50. Togashi, N.; Shiraishi, A.; Nishizaka, M.; Matsuoka, K.; Endo, K.; Hamashima, H.; Inoue, Y. Antibacterial activity of long-chain fatty alcohols against *Staphylococcus aureus*. *Molecules* **2007**, *12*, 139–148. [CrossRef] [PubMed]
51. Pejin, B.; Savic, A.; Sokovic, M.; Glamoclija, J.; Ciric, A.; Nikolic, M. Further in vitro evaluation of the antiradical and antimicrobial activity of phytol. *Nat. Prod. Res.* **2014**, *28*, 372–376. [CrossRef] [PubMed]
52. Swamy, M.K.; Sinniah, U.R. A comprehensive review on the phytochemical constituents and pharmacological activities of Pogostemon cablin Benth.: An aromatic medicinal plant of industrial importance. *Molecules* **2015**, *20*, 8521–8547. [CrossRef]
53. Shobier, A.H.; Abdel, S.A.; Barakat, G.M. GC/MS spectroscopic approach and antifungal potential of bioactive extracts produced by marine macroalgae. *Egy. J. Aqu. Res.* **2016**, *42*, 289–299. [CrossRef]
54. Keskin, D.; Ceyhan, N.; Ugur, A.; Durgan Dbeys, A. Antimicrobial activity and chemical constituents of West Anatolia olive (*Olea europaea* L.) leaves. *J. Food. Env.* **2012**, *10*, 99–102.
55. Rateb, H.S.; Ahmed, H.E.; Ahmed, S.; Ihmaid, S.; Afifi, H.S. Discovery of novel phthalimide analogs: Synthesis, antimicrobial and antitubercular screening with molecular docking studies. *EXCLI J.* **2016**, *15*, 781–796.
56. Saeidniaa, S.; Permehb, P.; Goharia, A.R.; Mashinchian-Moradib, A. *Gracilariopsis persica* from Persian Gulf contains bioactive sterols. *IJPR* **2012**, *11*, 845–849.
57. Nirupama, R. Fungal disease of white muscardiane in silkworm, *Bombyx mori* L. *Mun. Ent. Zool.* **2014**, *9*, 870–875.
58. Soliman, A.S.; Ahmed, A.Y.; Abdel-Ghafour, A.; El-Sheekh, M.; Sobhy, H. Antifungal bio-efficacy of the red algae *Gracilaria* confervoides extracts against three pathogenic fungi of the cucumber plant. *Middle East. J. Appl.* **2018**, *8*, 727–735.
59. Kim, S.K.; Karadeniz, F. Anti-HIV activity of extracts and compounds from marine algae. *Adv. Food. Nutr. Res.* **2011**, *64*, 255–265. [PubMed]
60. Zarei Jeliani, Z.; Mashjoor, S.; Soleimani, S.; Pirian, K.; Sedaghat, F. Antioxidant activity and cytotoxicity of organic extracts from three species of green macroalgae of Ulvaceae from Persian Gulf. *Modares J. Biotech.* **2018**, *9*, 59–67.

61. Abdelhamid, A.; Jouini, M.; Bel Haj Amor, H.; Mzoughi, Z.; Dridi, M.; Ben Said, R.; Bouraoui, A. Phytochemical analysis and evaluation of the antioxidant, anti-inflammatory, and anti-nociceptive potential of phlorotannin-rich fractions from three Mediterranean brown seaweeds. *Mar. Biotechnol.* **2018**, *20*, 60–74. [CrossRef]
62. Komari, S.S.; Kumar, V.D.; Priyanka, B. Antifungal efficacy of seaweed extracts against fungal pathogen of silkworm, *Bombyx mori* L. *Int. J. Agric. Res.* **2017**, *12*, 123–129. [CrossRef]
63. Singh, M.J.; Raadha, C.K. Studies on the antimicrobial potency of the marine algae-*Gracilaria corticata* var cylindrical and *Hydroclathrus clathratus*. *Adv. Life Sci. Health* **2015**, *2*, 50–55.
64. Dayuti, S. Antibacterial activity of red algae (*Gracilaria verrucosa*) extract against Escherichia coli and *Salmonella typhimurium*. *Earth Environ. Sci.* **2018**, *137*, 1–5.
65. Kolanjinathan, K.; Stella, D. Pharmacological effect of *Gracilaria corticata* solvent extracts against human pathogenic bacteria and fungi. *Int. J. Pharmacol. Biol. Arch.* **2011**, *2*, 1722–1728.
66. Perry, N.B.; Blunt, J.W.; Munro, M.H. Cytotoxic and antifungal 1, 4-naphthoquinone and related compounds from a New Zealand brown algae. *Landsburgia quercifolia*. *J. Nat. Prod.* **1991**, *54*, 978–985. [CrossRef]
67. Elena, M.; Francisco, Y.; Erickson, K.L. Mailiohydrin, a cytotoxic chamigrene dibromohydrin from a Phillippine *Laurencia* species. *J. Nat. Prod.* **2003**, *64*, 790–791.
68. Smit, A.J. Medicinal and pharmaceutical uses of seaweed natural products: A review. *J. Appl. Phycol.* **2004**, *16*, 245–262. [CrossRef]
69. Zhao, W.H.; Hu, Z.Q.; Hara, Y.; Shimamura, T. Inhibition of penicillinase by epigallocatechin gallate resulting in restoration of antibacterial activity of penicillin against penicillinase-producing *Staphylococcus aureus*. *Antim. Agents Chemo.* **2002**, *46*, 2266–2268. [CrossRef]
70. Shiota, S.; Shimizu, M.; Mizusima, T.; Ito, H.; Hatano, T.; Yoshida, T.; Tsuchiya, T. Restoration of the effectiveness of beta-lactams on methicillin-resistant *Staphylococcus aureus* by tellimagrandin I from rose red. *FEMS Microbiol. Lett.* **2000**, *185*, 135–138.
71. Ekambaram, P.S.; Perumal, S.S.; Balakrishnan, A.; Marappan, N.; Gajendran, S.S.; Viswanathan, V. Antibacterial synergy between rosmarinic acid and antibiotics against methicillin-resistant *Staphylococcus aureus*. *J. Int. Ethnopharm.* **2016**, *5*, 358–363. [CrossRef] [PubMed]
72. Arima, H.; Ashida, H.; Danno, G. Rutinen hanced antibacterial activities of flavonoids against *Bacillus cereus* and *Salmonella enteritidis*. *Biosci. Biotechnol. Biochem.* **2002**, *66*, 1009–1014. [CrossRef] [PubMed]
73. Oliveira, V.M.; Carraro, E.; Auler, M.E.; Khalil, N.M. Quercetin and rutin as potential agents antifungal against *Cryptococcus* spp. *Braz. J. Biol.* **2016**, *76*, 1029–1034. [CrossRef]
74. Skocibusic, M.; Bezic, N.; Dunkic, V. Food Chemistry. Phytochemical composition and antimicrobial activities of the essential oil from *Satureja subspicata* Vis. *Grow. Croatia* **2006**, *96*, 20–28.
75. Borghi, E.; Morace, G.; Borgo, F.; Rajendran, R.; Sherry, L.; Nile, C.; Ramage, G. New strategic insights into managing fungal biofilms. *Front. Microbiol.* **2015**, *6*, 1077. [CrossRef]
76. Lee, J.; Lee, D.G. Novel antifungal mechanism of resveratrol: Apoptosis inducer in *Candida albicans*. *Curr. Microbiol.* **2015**, *70*, 383–389. [CrossRef]
77. Hwang, I.; Lee, J.; Jin, H.G.; Woo, E.R.; Lee, D.G. Amentoflavone stimulates mitochondrial dysfunction and induces apoptotic cell death in *Candida albicans*. *Mycopathologia* **2012**, *173*, 207–218. [CrossRef]
78. Simon, H.U.; Haj-Yehia, A.; Levi-Schaffer, F. Role of reactive oxygen species (ROS) in apoptosis induction. *Apoptosis* **2000**, *5*, 415–418. [CrossRef] [PubMed]
79. Cho, J.; Lee, D.G. The antimicrobial peptide arenicin-1 promotes generation of reactive oxygen species and induction of apoptosis. *Biochim. Biophys. Acta* **2011**, *1810*, 1246–1250. [CrossRef] [PubMed]
80. Daglia, M. Polyphenols as antimicrobial agents. *Curr. Opin. Biotechnol.* **2012**, *23*, 174–181. [CrossRef] [PubMed]
81. Salvador, V.A.G. Evaluation of Apoptosis and Necrosis in *Saccharomyces cerevisiae* during Wine Fermentations. Ph.D. Thesis, Universidade Técnica de Lisboa, Lisbon, Portugal, 2009.

 MDPI

Article

Deciphering the Physicochemical and Microscopical Changes in *Ganoderma boninense*-Infected Oil Palm Woodblocks under the Influence of Phenolic Compounds

Arthy Surendran [1,*], Yasmeen Siddiqui [2,*], Khairulmazmi Ahmad [2,*] and Rozi Fernanda [2]

1 School of Life Sciences, University of Warwick Wellesbourne, Warwick CV35 9EF, UK
2 Sustainable Agronomy and Crop Protection, Institute of Plantation Studies, Universiti Putra Malaysia, Serdang 43400, Malaysia; rozifernanda86@gmail.com
* Correspondence: Arthy.Surendran@warwick.ac.uk (A.S.); yasmeen@upm.edu.my (Y.S.); khairulmazmi@upm.edu.my (K.A.); Tel.: +60-3-9769-4135 (K.A.)

Abstract: The threat of *Ganoderma boninense*, the causal agent of basal stem rot disease, in the oil palm industry warrants finding an effective control for it. The weakest link in the disease management strategy is the unattended stumps/debris in the plantations. Hence, this study aimed to determine whether the selected phenolic compounds could control *G. boninense* in inoculated oil palm woodblocks and restrict wood biodegradation. Results indicated a significant reduction in the wood mass loss when treated with all the phenolic compounds. Surprisingly, syringic and vanillic acids behaved ambivalently; at a lower concentration, the wood mass loss was increased, but it decreased as the concentrations were increased. In all four phenolic compounds, the inhibition of mass loss was dependent on the concentration of the compounds. After 120 days, the mass loss was only 31%, with 63% relative degradation of lignin and cellulose, and 74% of hemicellulose and wood anatomy, including silica bodies, were intact in those woodblocks treated with 1 mM benzoic acid. This study emphasizes the physicochemical and anatomical changes occurring in the oil palm wood during *G. boninense* colonization, and suggests that treating oil palm stumps with benzoic acid could be a solution to reducing the *G. boninense* inoculum pressure during replantation in a sustainable manner.

Keywords: basal stem rot (BSR); phenolic compounds; lignin; cellulose; hemicellulose; silica body; crystalline cellulose; biodegradation

Citation: Surendran, A.; Siddiqui, Y.; Ahmad, K.; Fernanda, R. Deciphering the Physicochemical and Microscopical Changes in *Ganoderma boninense*-Infected Oil Palm Woodblocks under the Influence of Phenolic Compounds. *Plants* **2021**, *10*, 1797. https://doi.org/10.3390/plants10091797

Academic Editors: Špela Mechora and Dragana Šunjka

Received: 2 July 2021
Accepted: 19 July 2021
Published: 28 August 2021

1. Introduction

The rapid expansion of plantations of oil palms (*Elaeis guineensis* Jacq.) has led to its emergence as a commodity of strategic global importance. It is one of the most economically important oil crops in Southeast Asia. Palm oil is very useful, and is used extensively in food and as a precursor for biodiesel. The palm kernel cake resulting after oil extraction is also used as animal feed and bio-fertilizer. In the decades since 1980, the production of oil palm has increased from 5 Mt to more than 45 Mt, with an annual average growth rate of 7.8% [1], and is still increasing.

Trees in the tropical rain forest are hosts to a wide range of stem and root pathogens, typically belonging to the basidiomycete genera. One such predominant pathogen for the oil palm is *Ganoderma boninense*, which causes basal stem rot (BSR). Losses in the fruit bunch yield of between 0.04 and 4.34 t ha^{-1} from 10 to 22 years of planting, due to the BSR in oil palms, have been reported, and it is predicted that more than 60 million mature oil palms could be infected in Malaysia [2–4]. It was estimated if a tree dies at the age of 10 years, it can cause a loss of 15 years of production, which is 600 kg of crude palm oil at a monetary loss of approximately USD 675 [5].

The management of fungal stem and root rot is universally difficult [1], and BSR disease is no exception, if not more difficult than most. This fungus is a saprophyte that can survive for a long time in the debris, stumps and leftover roots of unattended logs in the plantation [1]. The initial spread of BSR in oil palm could be via basidiospore dispersal, root to root contact, and also via root contact with the unattended infected debris [6,7]. The unattended debris acts as a source of the inoculum of *G. boninense* during the replantation. The reports suggest that the disease incidence (DI) has been increasing in successive replantation. It has been reported that in the first replantation, oil palm plants of 10 years old have been affected with less than 2% of DI; however, at the third replantation, DI (70%) has increased alarmingly [8]. In his study, Khairudin (1993) has concluded that 93% of seedlings that are growing around the diseased stumps have also been infected with BSR disease [9].

The newly cut stumps and the unattended infected trunks are the most vulnerable links in the disease management chain. This is because the fresh-cut stumps and the trunks are still alive, and serve as a nutrient reservoir for the pathogens to flourish. The treatment of the cut stumps with chemicals is not effective because of the larger size and the anatomy of oil palm roots. Oil palm roots form around an orthotropic taproot, and the horizontal lateral roots occupy about 16 m^3 of soil mass [10], which makes it very tedious and expensive to excavate as a way to control BSR disease.

To control BSR disease in oil palms, the *G. boninense* surviving in cut stumps must be eradicated. To protect the surface of the stump from attack by *G. boninense*, there is an urgent need to find some antimicrobial compounds that can travel to the site of infection (roots) efficiently and act cost-effectively. Through a series of experiments and screening, four naturally occurring phenolic compounds, namely, benzoic, salicylic, syringic and vanillic acids, were selected based on their antifungal efficiency, along with their inhibitory potential toward the ligninolytic and cellulolytic enzymes [11,12]. These naturally occurring phenolic compounds are involved in the lignin synthesis pathway in oil palm and hence could be potential agents to eliminate *G. boninense*. To the best of our knowledge, little or no study has been conducted to evaluate the degradation pattern of *G. boninense* in the presence or absence of selected phenolic compounds, which could be important to our understanding for the development of an alternative BSR management technique.

2. Materials and Methods

2.1. Chemicals

Potato dextrose agar (PDA), salicylic acid, syringic acid and vanillic acid were purchased from Friedemann Schmidt (Germany), while benzoic acid was obtained from R&M Chemicals and Reagents (Malaysia). Four phenolic compounds with a concentration of 1, 5, 10 and 15 mM were used.

2.2. Microorganism, Culture Conditions and Treatments

The isolate *Ganoderma boninense* (PER 71) was obtained from the Malaysian Palm Oil Board (MPOB). The culture was maintained by sub-culturing the fungus at regular intervals on PDA at 28 °C. Four phenolic compounds, namely, benzoic acid, salicylic acid, syringic acid and vanillic acid, were tested to see whether they would inhibit the biodegradation ability of *G. boninense* on oil palm woodblocks.

2.3. Biodegradation of Woodblocks

The rate of woodblock decay was investigated using the method described by Schirp and Wolcott (2005), with slight modifications [13]. The healthy oil palm trunks were collected from the MPOB, and were cut into 20 × 20 × 40 mm blocks. The blocks were oven-dried at 60 °C for 48 h to obtain a constant dry weight. The dried woodblocks were weighed and labeled individually, to obtain the initial dry mass, followed by double sterilization for 30 min at 121 °C, as described by Bucher et al. (2004) [14]. Tissue-culture jars (350 mL capacity, 15 cm in length and 75 mm in diameter) containing 20 mL of PDA media were autoclaved at 121 °C for 20 min, followed by cooling and solidification of the media. The jars were then inoculated with *G. boninense* culture (two mycelial plugs of 3 mm Ø) and incubated at 28 ± 2 °C until full coverage of the media by fungal mycelium was reached. The phenolic stock solutions were prepared according to Sudrendran et al. (2017) [11]. The sterilized woodblocks were dipped for 30 min in the respective phenolic compounds of each concentration (1, 5, 10 and 15 mM) and air-dried in the laminar hood until no dripping was observed [14]. After complete colonization of PDA by *G. boninense* in the jars, the sterilized healthy and phenolic compound-treated oil palm woodblocks were placed individually on the top of the *G. boninense* culture. The jars containing woodblocks without any phenolic compound treatment served as the controls. All the jars were then incubated at 28 ± 2 °C for 10, 30, 45 and 120 days. For each sampling period, five woodblocks were harvested. The surface of each block was carefully wiped to remove any mycelium, followed by weighing. The experiments were repeated twice, and the mass loss was calculated using Equation (1):

$$Percentage\ mass\ loss\ (\%) = \frac{Initial\ weight\ after\ treatment - Final\ weight}{Initial\ drymass} \times 100 \quad (1)$$

2.4. Anatomical Characterization during Biodegradation of Oil Palm Woodblocks

Scanning Electron Microscopy (SEM) Analysis

At the end of each incubation time, the *G. boninense*-colonized wood samples were collected. The mycelia from the samples were wiped off. The radial and vertical samples, with a dimension of 0.5 × 0.5 cm, were sputter-coated with gold–palladium alloy. The SEM (S-3400N Hitachi, UPM, Serdang Malaysia) was performed to determine the morphological changes of the wood and *G. boninense* during the degradation period.

2.5. Chemical Characterization during Biodegradation of Oil Palm Woodblocks

2.5.1. Fourier Transform Infrared (FT-IR) Spectroscopy Analysis

At the end of each incubation period, the *G. boninense*-colonized woodblocks were sampled out of the jars, and the mycelium was carefully wiped off from the woodblocks. Later, the woodblocks were impregnated with water to remove the remaining mycelium. The wood samples were powdered in the mortar and pestle using liquid nitrogen. The powder was sieved, and those fractions with an average size less than 0.5 mm were retained to develop KBr pellets. The FTIR spectra for the wood samples were measured by direct transmittance using the KBr pellet technique. The amount of sample in the pellets was constant (5 mg/500 mg KBr). The spectra were recorded in the range of 4000–400 cm^{-1} using a Spectrum 100 FTIR Spectrometer (Perkin Elmer, Inc., Waltham, MA, USA). Five recordings were performed for each sample, and evaluation was made from the average of the five using the software [15]. Healthy wood without any treatment was also analyzed to compare with the degraded wood. From the FT-IR spectra, the lower order index (LOI, A_{1430}/A_{898}) was used to estimate the crystal structure of cellulose material [16], and the total crystallinity index (TCI, A_{1372}/A_{2900}) was used to estimate the infrared crystallinity index [16]. The ratio between the syringyl and guaiacyl S/G (S/G, A_{1327}/A_{1271}) units in the lignin was also estimated [17].

2.5.2. Thermogravimetry (TGA) Analysis

At each sampling time, the wood samples were removed from the jars and cleaned of mycelium, as mentioned above. The wood samples were then powdered, using a mortar and pestle and liquid nitrogen. The powder was sieved; the fraction with an average size of less than 0.5 mm was retained. TGA analysis was performed using a Perkin Elmer thermogravimetric analyzer (Pyris 1 TGA, Pyris software 7.0—Perkin Elmer) at a constant nitrogen flow of 20 mL min^{-1} and constant heating of 15 °C min^{-1}. The heating scans were performed with 5 mg of wood samples in the temperature range of 25–890 °C [15]. Healthy wood without any treatment was analyzed for comparison with the degraded wood. The amount of lignin, cellulose and hemicellulose was quantified according to Yang et al. [18]. Oil palm lignin degradation takes place at 700–800 °C, hemicellulose and cellulose at 0–300 °C and 330–340 °C, respectively. The relative percentage of lignin, cellulose and hemicellulose was calculated according to Equation (2):

$$Relative\ lignin\ degraded\ (\%) = \frac{Initial\ lignin\ content - Final\ lignin\ content}{Initial\ lignin\ content} \times 100 \quad (2)$$

For the relative percentage of cellulose and hemicellulose when degraded, the lignin in Equation (2) is replaced with cellulose and hemicellulose, respectively.

2.6. Statistical Analyses

The data were analyzed using SAS statistical software (PC-SAS software V8.2, SAS Institute, Cary, NC, USA). A p-value of ≤ 0.01 was considered significant. All the experiments were repeated twice, with six replicates for each treatment. The means were compared using Tukey's range test.

3. Results

3.1. Mass Loss

Initially, the colonization of *G. boninense* in the control woodblocks was very slow and steady for the first ten days. Eventually, *G. boninense* rapidly colonized the control woodblocks after 30 days, with a weight loss of 37%. With the progression of the incubation period, a significantly higher ($p \leq 0.05$) weight loss was observed in the control woodblocks, compared to the woodblocks treated with 1 mM benzoic acid. The mass loss at various time intervals, i.e., 10, 30, 45 and 120 days, are shown in Figure 1a–d, respectively. As the colonization of *G. boninense* increased with time, a decrease in the weight of the oil palm woodblocks was observed. The colonization rates in those treated woodblocks with a higher concentration of phenolic compounds were prominently less when compared to the control woodblocks, except for the syringic and vanillic acid treatments. On the 120th day, the mass loss due to the colonization of *G. boninense* on the woodblocks that were treated with 1 mM of syringic (74.9%) and vanillic (83%) acids was more marked when compared to the control (71.8%) woodblocks. These two phenolic compounds enhanced the growth of *G. boninense* at their lower concentration, that is, 1 mM. However, the same compounds proved to be poisonous when the concentrations increased further. As the concentration of the syringic acid was increased to 5 mM in the woodblocks, a significant decrease in mass loss due to the colonization of *G. boninense* was observed. The weight losses in the woodblocks treated with 5 mM syringic acid (25.5%) were almost half when compared to the 1 mM syringic acid treatment (Figure 1d), whereas, on the 120th day, the 10 mM vanillic acid-treated woodblocks had a significant reduction in weight loss of about 19.6%. At the end of the study, among the four phenolic compounds, benzoic acid was the best inhibitor against *G. boninense* colonization at 1 mM concentration, followed by salicylic acid, with a 40.3% weight loss in the treated woodblocks. The woodblocks treated with 5 mM and above of benzoic acid, 10 mM and above of salicylic and syringic acids, and 15 mM of vanillic acids retained their physical properties, with no growth of *G. boninense*, as in the healthy oil palm woodblock, until the end of the study. At the end of the decay period, the degraded wood turned from brown-black to whitish tan with some black spots. The woodblocks were soft, fragile and easily shredded into longitudinal fragments.

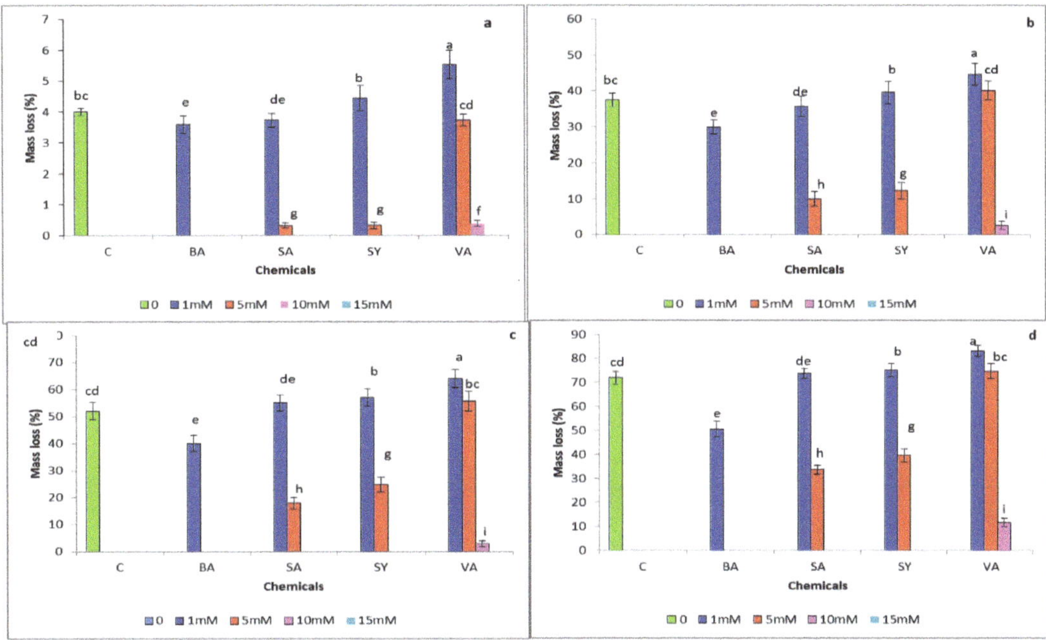

Figure 1. Degradation of oil palm wood by *G. boninense* with and without phenolic treatments at various time intervals. (**a**) 10th-day sampling of degraded oil palm wood; (**b**) 30th-day sampling of degraded oil palm wood; (**c**) 45th-day sampling of degraded oil palm wood; (**d**) 120th-day sampling of degraded oil palm wood. Vertical bars indicate standard deviation (S.D.) $n = 6$. C—control, BA—benzoic acid, SA—salicylic acid, SY—syringic acid and VA—vanillic acid. No data due to absolute inhibition for BA (5, 10 and 15 mM), SA (10 and 15 mM), SY (10 and 15 mM) and VA (15 mM).

3.2. Anatomical Characterization during Biodegradation of Oil Palm Wood

Scanning Electron Microscopy (SEM) Analysis

The SEM micrographs of the healthy oil palm woodblocks revealed many silica bodies embedded on the surface of the fiber strands. The silica bodies were attached to circular craters and were arranged in a pattern of flowers that typifies oil palm wood (Figure 2a). Silica bodies were spread uniformly over the strand's surface. In general, the SEM observations indicated that, initially, *G. boninense* colonized the substrate surfaces via apical hyphal extension (Figure 2b). The initial colonization on the 10th day was up to the cell lumen, which provoked erosion. As the degradation progressed, the eroded zones coalesced, and the fungal hyphae filled in the voids at 30 days. At the end of the degradation (120 days), silica bodies on the surface of the oil palm woodblocks were completely dissociated, and extensive deterioration of the wood cell walls was observed in the control (Figure 2c).

The SEM of the woodblocks treated with 1 mM vanillic and syringic acid displayed a similar pattern, with denser hyphae than the control (Figure 2d). Although flower-shaped silica bodies were surrounded by the *G. boninense*, the silica bodies looked intact (Figure 2d). After incubation for 30 days, the hyphae of the *G. boninense* were abundant and were spread across the pits of the vessels, as well as the fibers, in a bridge-like arrangement. As the degradation time was prolonged, very few boreholes were observed in some of the parenchymal tissue, and *G. boninense* intervention was evident (Figure 2e). At the end (120 days) of this study, the woodblocks treated with 1 mM vanillic and syringic acids displayed more disintegrated parenchymal tissues, compared to the control (Figure 2f). No flower-patterned silica bodies were observed in the woodblocks treated with 1 mM of syringic and vanillic acids. Similar observations were also observed in the woodblocks treated with 5 and 10 mM vanillic acid, and also in those treated with 5 mM syringic acid, but the intensity of the colonization was less marked. No *G. boninense* growth was observed in the woodblocks treated with 10 and 15 mM syringic acid and the 15 mM vanillic acid-treated woodblocks. The woodblocks treated with 1 and 5 mM salicylic acid were colonized with *G. boninense*, but with low visible intensity. However, the woodblocks with 5 mM concentration were significantly better, with less colonization. The colony of hyphae in 5 mM salicylic acid-treated woodblocks displayed a cleft circle at the end of the hyphae and, after a closer look, it was confirmed that the hyphae were almost emptied (Figure 2g,h). No growth of *G. boninense* was observed on the woodblocks treated with a 10 mM concentration of salicylic acid. *G. boninense* colonized the least heavily in the woodblocks treated with 1 mM benzoic acid until the end of the study. *G. boninense* hyphae were thin on the 10th day, when compared to the *G. boninense* grown on the control woodblocks. At the end of the degradation process (120th day), dense colonization was observed, with very thin hyphae of *G. boninense* on those woodblocks treated with 1 mM benzoic acid (Figure 2i); the flower-shaped silica components embedded in the circular craters were intact on the surface of the woodblocks.

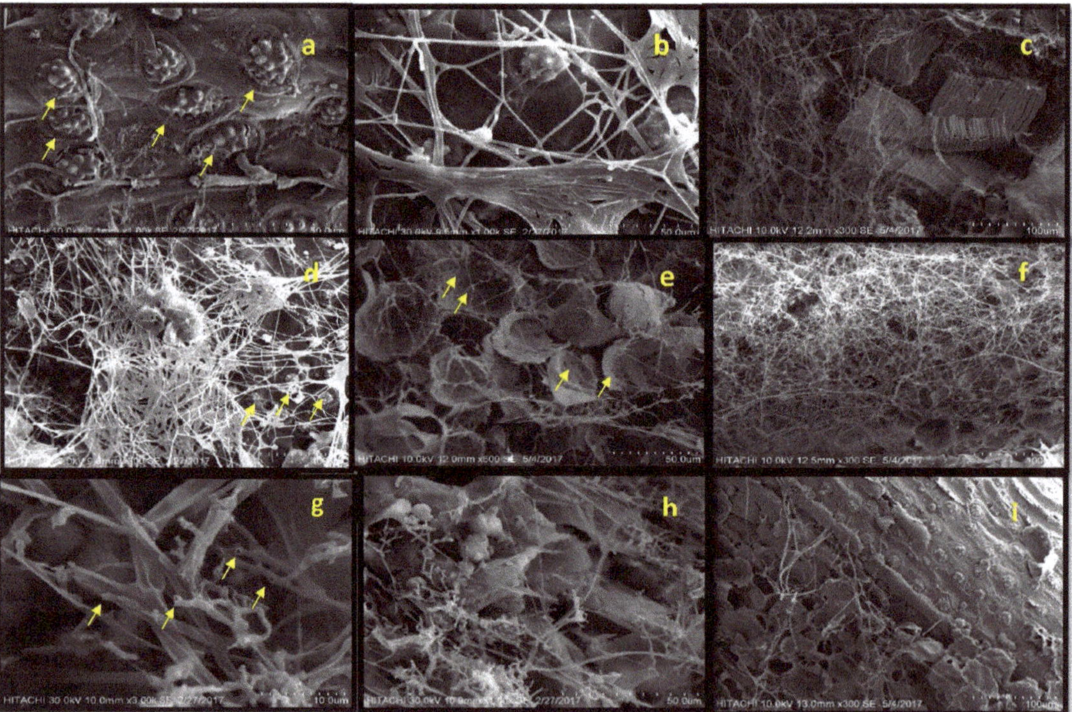

Figure 2. Scanning electron microscope (SEM) analysis of healthy and infected oil palm wood sample subjected to various treatments. (**a**) Intact silica bodies in healthy oil palm wood; (**b**) the initial stage of apical colonization of *G. boninense* in oil palm wood; (c) the control wood samples with silica bodies on the surface of oil palm woodblocks were completely dissociated, and show extensive deterioration in the wood cell walls; (**d**) dense hyphae on oil palm wood treated with 1 mM vanillic acid; (**e**) woodblocks treated with 1 mM syringic acid at 30th day; (**f**) colonization of *G. boninense* on woodblocks treated with 1 mM vanillic acid at the120th day; (**g**) a closer look at the hyphae with clefts grown on woodblocks treated with 5 mM salicylic acid; (**h**) *G. boninense* with deformities on woodblocks treated with 1 mM salicylic acid; (**i**) intact silica bodies on the oil palm woodblocks treated with 1 mM benzoic acid at the 120th day.

3.3. Chemical Characterization during Biodegradation of Oil Palm Woodblocks

3.3.1. Fourier Transform Infrared (FT-IR) Spectroscopy Analysis

FT-IR spectroscopy was performed to compare the changes in the functional groups in the wood components in both healthy and treated wood samples that have been degraded by *G. boninense*. This study's focus was on the band modifications related to the carbohydrates (cellulose, hemicellulose and starch) and the lignin in the oil palm wood. The FT-IR spectra in the fingerprint region between 1800 to 500 cm^{-1} were analyzed. Tables S1–S3 represent the LOI, TCI and S/G of the wood components. The TCI represents the crystallinity degree of the cellulose, and LOI represents the overall degree of order in the cellulose. This analysis has given a clear demonstration of the structural degradation of cellulose. Not many changes were observed in the first ten days of the degradation process between the control and the phenolic compound-treated woodblocks. As the samples were degraded by *G. boninense*, a 3340 cm^{-1} band increased in its intensity and broadening. This was observed in both the control and the treated samples. As the band broadened, it shifted to the lower frequency. However, the intensity of this band in the wood samples that were treated with 10 mM vanillic, 5 mM salicylic, and 1 mM of benzoic acids was less intense when compared to the control samples. The TCI of the control samples increased until the 30-day point and started decreasing as the degradation period continued, but the LOI remained constant for the control samples for the first 30 days and decreased as the degradation continued (Tables S1 and S2). However, woodblocks treated with 1 mM benzoic acid maintained almost the same LOI until the end of the study, but a constant decrement was observed in the TCI values. The LOI values of the woodblocks treated with salicylic acid were constant for the first 30 days, and decreased with the increase in the degradation period. On the other hand, the TCI values of the woodblocks treated with 1 mM salicylic acid decreased for the first 30 days, and started increasing until the 45th day, again decreasing from then onward. The LOI and TCI values of the wood samples treated with vanillic acid continued reducing throughout the degradation period, except for the TCI values on the 45th day.

As shown in Table 1, *G. boninense* invaded the lignin, cellulose and xylene bonds simultaneously. Only stretching in various bonds was observed, and no bending was visible in the FT-IR analysis. The stretching of the C-O bonds of the syringyl and guaiacyl units of lignin, and also the stretching of the C=O bonds of aryl ketone and C=C in the aromatic group of lignin, occurred on the 30th day. Besides lignin, C-H and C-O stretching in cellulose and xylene were detected. A band at 1700 cm^{-1} was observed in the wood samples treated with 1 mM benzoic, 5 mM salicylic and 15 mM vanillic acids. This band is associated with the C=O stretching of conjugated or aromatic ketones (1700 cm^{-1}) [15] at days 30 and 45. As degradation progressed further, this peak disappeared (day 120). The intensity of the band at 1670 cm^{-1} corresponds to the C=O stretching of lignin, which is more intense in the samples treated with 1 mM syringic and vanillic acids. The LOI and TCI values also dropped. This indicates the extensive degradation of cellulose and hemicellulose components, which increased the residual lignin concentration. These findings were in agreement with the TGA studies (results below). Similar bands appeared for the remaining treatments, with lesser intensity. In the case of 1 mM benzoic acid, as the degradation progressed further, the band at 1670 cm^{-1} disappeared. A constant decrement was observed in the S/G ratio in the control wood samples. This indicates that *G. boninense* degraded more G units in the lignin compared to the S unit. In only the wood samples treated with syringic and vanillic acids, the S/G ratio has slightly increased by the 30th and 45th days, and dropped at the 120th day of biodegradation.

Table 1. Characteristic bands in the FT-IR spectra of the studied oil palm wood samples in the 400–4000 cm^{-1} region.

Wave Number (cm^{-1})	Assignment	Source	Observation	References
		Lignin Bands		
1700	C=O stretching of conjugated or aromatic ketones	Lignin	Appeared in wood samples treated with 1 mM benzoic, 5 mM salicylic and 15 mM vanillic acids. Later disappeared on the 120th day.	[19]
1612	Unconjugated carboxyl stretch of both lignin and cellulose	Lignin	A high-intensity band appeared in woodblocks treated with 1 mM syringic and vanillic acids. A medium-intensity band was observed in 1 mM benzoic acid initially, and was not observed in the later stages of degradation.	[20,21]
1620	C=C stretching in the aromatic groups of lignin	Lignin	This band was observed in all the treated as well as the control wood samples.	[21,22]
1670	C=O stretching in the conjugated p-substituted aryl ketone	Lignin	This band was observed in all the treated as well as the control wood samples.	[19,23]
1270	C-O stretching in xylene and hemicellulose and guaiacyl structure in lignin	Lignin	The intensity of the band decreased when the lignin and the adjacent hemicellulose degraded appeared in the control wood samples, along with wood samples treated with 1 mM phenolic compounds. It appeared in all the other wood samples as the degradation proceeded.	[24]
1034	Deformation vibration of C-H bond in aromatic rings	Lignin	This band was observed in all the treated as well as the control wood samples.	[25]
1247	C-O stretching in lignin (Guaiacyl units) and hemicellulose	Lignin	This band was observed in all the treated as well as the control wood samples. However, the highest intensity of this band was observed in wood samples treated with 1 mM vanillic acid.	[19,23]
		Carbohydrate Bands		
1336	OH in-plane bending cellulose	Cellulose	A weak band appeared only at the end of the degradation period.	[26,27]
1320	C-H variation in cellulose and C-O stretching in syringyl unit of lignin	Cellulose and lignin	The intensity of the band decreased when the lignin and the adjacent hemicellulose degraded.	[28]
128	C-H stretching	Starch	A weak band appeared only at the end of the degradation period.	[14]
		Hydroxy Bands		
3427	Intra-molecular OH stretching in cellulose	Cellulose	Increased intensity of this band indicates that more hydroxyl groups are available resulting from the hydrolysis. The increment in these two bands was observed throughout the degradation process.	[29]
3340	Bonded OH stretching	Carbohydrate and lignin		

3.3.2. Thermogravimetry (TGA) Analysis

The chemical composition of healthy oil palm woodblocks that were treated with phenolic compounds and infected with *G. boninense* are shown in Tables 2 and S4. The three distinct stages of weight loss were observed in both the treated and the control oil palm woodblocks. In TGA, the first is the dehydration stage, followed by active pyrolysis and passive pyrolysis; these were in the ranges of 0–100 °C, 100–300 °C and 350–800 °C. A negligible mass loss was observed both in the control and in the treated samples at the dehydration stage. A considerable mass loss had occurred in between the temperature ranges of 110–500 °C. The degradation of the cellulose and hemicellulose had occurred in this range, which was the start of the active pyrolysis step. The estimation of hemicellulose, cellulose and lignin were recorded at temperature ranges of 220–300 °C, 330–340 °C and 700–800 °C, respectively [18]. As the degradation progressed further, an increase in the thermal degradation rate was observed in the woodblocks colonized with *G. boninense*. Table S4 represents the lignin, cellulose and hemicellulose contents of the healthy oil palm woodblocks before the treatment.

The relative percentages of lignin, hemicellulose and cellulose degradation in control woodblocks colonized with *G. boninense* were elucidated according to TGA; on the 10th day, *G. boninense* had degraded the hemicellulose (68.8%), followed by lignin (49.6%), and then cellulose (37.1%). As the degradation progressed further, the rate of degradation of the hemicellulose had decreased gradually and, simultaneously, the rate of degradation of the lignin had increased in parallel on the 120th day. Hence, at the end of the degradation period (120th day), *G. boninense* had degraded the lignin and hemicellulose almost equally, but at the same time, the degradation in the cellulose was the least. This result indicated that *G. boninense* had utilized all the three components of the oil palm wood during colonization. During the initial stages of colonization, *G. boninense* preferred more hemicellulose components to degrade, but at a later stage, there was a change in preference toward lignin rather than hemicellulose.

Table 2 indicates the difference between the treated woodblocks with that of the control woodblocks, and the relative degradation of lignin, hemicellulose and cellulose components. An increase in the concentration of phenolic compounds in the treatment woodblocks had decreased the degradation rate, as informed by the lesser colonization of *G. boninense* and the lesser utilization of carbon sources.

Table 2. Percentage of the chemical composition of oil palm wood (with and without treatment) degraded by *G. boninense* at various time intervals.

Phenolic Compounds (mM)	Biodegradation (Days)											
	10th Day			30th Day			45th Day			120th Day		
	Relative (%) Degradation											
	L	C	H	L	C	H	L	C	H	L	C	H
Control	49.6	37.1	68.9	55.6	46.5	69.9	64.7	46.5	70.6	72.2	60.9	71.1
BA 1	2.3	14.4	18.7	54.9	44.5	35.5	56.4	47.5	49.2	63.2	63.9	74.2
SA 1	47.4	15.1	33.4	54.9	40.1	48.5	57.1	49.5	66.6	72.2	61.9	70.6
SA 5	32.3	9.0	22.4	45.7	27.1	39.8	49.3	29.8	41.1	68.0	35.8	45.8
SY 1	54.9	34.8	49.2	62.4	47.5	50.5	66.9	53.8	52.5	71.4	66.2	69.2
SY 5	10.5	5.4	23.7	25.6	13.4	27.8	32.3	25.4	33.1	49.6	29.1	49.5
VA 1	62.4	40.5	64.9	69.9	48.2	67.6	72.2	52.2	71.2	85.0	69.6	76.9
VA 5	54.9	57.5	51.2	63.9	60.9	60.9	66.2	61.5	63.9	69.9	65.2	77.9
VA10	1.5	2.5	22.7	24.8	17.1	31.4	33.1	19.7	32.8	62.4	20.4	42.5

BA—benzoic acid, SA—salicylic acid, SY—syringic acid, VA—vanillic acid, L—lignin, C—cellulose, and H—hemicellulose.

The maximum destruction of lignin was recorded in the woodblocks that were treated with 1 mM vanillic acid, and this tendency continued until the end of the study. On the 10th day of colonization of the *G. boninense* on oil palm woodblock treated with 1 mM vanillic acid, degradation ratios of 62.4, 64.9 and 40.5% of lignin, hemicellulose and cellulose, respectively, were recorded. This was followed by 1 mM syringic acid-treated woodblocks, and on the 10th day, 54.9% of the lignin, 49.2% of the hemicellulose, and 34.8% of the cellulose components were found to have degraded. On the 45th day, 72.2% of the lignin was found to have degraded in the 1 mM vanillic acid-treated woodblock, as against 64.7% of lignin degradation in the control woodblocks. The degradation pattern in oil palm wood treated with vanillic acid appeared similar with hemicellulose and lignin to that of the control woodblocks, initially, but the destruction of lignin was found to be discernibly higher in the 1 mM vanillic acid-treated woodblocks (Table 2). However, in the case of woodblocks treated with syringic acid, *G. boninense* degraded more lignin and cellulose components than it did the hemicellulose components.

Overall, the woodblocks treated with 1 mM benzoic acid recorded a minimum degradation rate when compared to other phenolic compounds at the minimal concentration. The degradation in those woodblocks treated with 1 mM benzoic acid was accelerated after the 10th day. The lignin degradation had increased from 2.3% on day 10 to 54.95 on day 30; similarly, hemicellulose destruction increased from 18.7% to 55.5%, and in the case of cellulose, from 14.4% to 44.5%, respectively. At the end of this study, the relative percentages of lignin, hemicellulose and cellulose that had degraded in the woodblocks treated with 1 mM benzoic acid were 63.2%, 63.9% and 74.2%, respectively. Slow and steady rates in the degradation of wood components were observed in those woodblocks treated with 1 and 5 mM salicylic acid, but the ratio of intensity varied between them temporally. On day 10, the ratios of the destruction of lignin, hemicellulose and cellulose were at 47.4–32.3%, 33.4–22.4% and 15–9%, respectively. On the final day, *G. boninense* degraded a higher percentage of lignin, cellulose and hemicellulose, at 72.2%, 60.9% and 71.1%, respectively.

4. Discussion

Abiding by the "zero burnings" policy, instead of controlled burning, the felled oil palm trunks and the stumps (with roots) are left to degrade in field conditions. This serves as a host for the saprophytic, pathogenic *G. boninense* to survive for a long time until they come in contact with healthy oil palm roots. Increases in disease incidence and disease severity were observed across the consecutive replantation. This is directly linked to the increase in size of the inoculum of *G. boninense* [30]. Nearly 60% of the estates in Southeast Asia were reported for the presence of this disease [1]. Although the stumps and felled oil palm trunks are considered to be the weakest link in the BSR disease management strategy, it is one that has been overlooked for some time.

Recently, researchers have initiated research on the identification of biocontrols of *G. boninense* in the felled trunks [31]. Despite the theoretical fact that biocontrol agents are superior to other control agents, their efficiency in the field is always questionable. This issue apart, the introduction of new biocontrols into the environment can also cause an unwanted change to the biodiversity of the ecosystem [25]. Hence, to overcome the above-mentioned issues, naturally occurring phenolic compounds might be a potential candidate to control the *G. boninense* that resides saprophytically in the oil palm trunks.

Hence, an in vitro study was conducted to evaluate the effects of phenolic compounds on controlling the growth of *G. boninense* in oil palm woodblocks. The wood can be studied in the following three ways, microscopic, spectroscopic, and thermodynamics, providing useful information on the surface morphology and chemistry of wood samples during their degradation by *G. boninense* when in the presence of phenolic compounds. In the current study, the analyses were made using SEM, TGA and FTIR.

G. boninense is a white-rot fungus (WRF) that is known for its remarkable ability to produce both oxidative and hydrolytic enzymes for the degradation of lignin and other cellular components [32]. In our study, the mass loss at the end of the degradation by *G. boninense* was 70% on the 120th day. The observed mass loss in this study is 10–20% higher than the *Populus deltoides* wood degraded by *Pycnoporus sanguineus* and *Ganoderma. lucidium* at 120 days [24]. Both *P. sanguineus* and *G. lucidium* are considered to be extensive degraders of poplar wood. From the point of view of mass loss, we can state that *G. boninense* can be potentially destructive under favorable conditions. Although this in vitro study cannot be considered as absolute evidence of the behavior of *G. boninense*, still, it is useful to predict the trajectory of degradation of oil palm wood.

During the degradation period, black spots appeared on the woodblocks, which could be associated with the deposition of manganese peroxidase by the fungus that colonizes the woodblocks [20]. *G. boninense* is known to produce three ligninolytic enzymes, namely, laccase, lignin peroxidase and manganese peroxidase [12]. At the end of the degradation process, the woodblocks were soft, spongy and pale in color, which is the characteristic feature of wood degraded by WRF [33].

In this study, a higher concentration of phenolic compounds was observed to reduce the growth of *G. boninense* on oil palm woodblocks. The rate of mass loss due to the colonization of *G. boninense* is inversely pronominally connected to the concentration of the phenolic compounds. However, the two para-methylated phenolic compounds (syringic and vanillic acid) behaved ambivalently, and at the lower concentration of phenolic compounds, increased the colonization of *G. boninense,* thus leading to increased mass loss. A higher mass loss of about 80% and 71% was also observed in the woodblocks treated with phenolic compounds, at 1 mM of the vanillic and syringic acids, respectively. This behavior is not surprising because these two phenolic compounds are not only metabolized by *G. boninense* but also by a wide range of white-rot fungi (WRF), such as *Phanerochaete chrysoporium, Prognathodes dichrous* and *Pleurotus ostreatus,* at lower concentrations [28,34]. These para-methylated aromatic compounds are less inhibitory to the growth of the WRF, as well as being easy to degrade. The conversion of phenol into para-methylated forms is considered to be one of the stratagems of the WRF while colonizing lignocellulose materials [35].

G. boninense produces the laccase enzyme [12], which acts as a detoxifying enzyme, as well as being responsible for the degradation of aromatic structures in the lignin. Hence, the phenolic compounds at lower concentrations in this study would have been oxidized by the laccase enzymes produced by *G. boninense*. The enhanced laccase enzyme produced by *Coriolus versicolor* has also detoxified various phenolic compounds, such as ferulic acid, xylidine, vanillic acid, cinnamic acid and guaiacol at 1 mM, but, as the concentration increased further, the phenolic compounds inhibited the growth of *C. versicolor* [36].

The presence of phenolic compounds at minimal concentration, along with the oil palm wood chips in the broth, not only increased the production of oxidative enzymes but also hydrolytic enzymes [11,12,37]. From this, we can infer that the phenolic compounds at lower concentrations induced the production of the oxidative and hydrolytic enzymes; as a result, more substrate (lignin, cellulose and hemicellulose) has been utilized by the enzymes [11,37]. This, in turn, led to a substantial mass loss in the oil palm woodblocks.

When the wood samples were viewed under SEM, the images indicated that *G. boninense* is a WRF that adapts to a simultaneous degradation pattern; that is, the degradation of the lignin and cellulose components of the oil palm woodblocks. The wood cell walls are attacked from the lumen, and the degradation is associated with the hyphae and not with the diffusion mechanism [33]. The erosion troughs observed beneath the hyphae indicate that *G. boninense* adopts a simultaneous degradation of lignin and cellulose components. This result was similar to *G. lucidium,* which degrades both lignin and cellulose [24]. Some of the species of WRF are known for their ability to degrade the wood components simultaneously. The rate of lignin and cellulose compound removal depends on various factors, such as the type of wood substrate, as well as the fungal species [21]. The surface recalcitrance of healthy oil palm wood was evident in the micrographs. The silica bodies serve as a major physical barrier for the enzymatic hydrolyzing process, due to the difficulty in penetrating the wood surface to access the cellulose and hemicellulose for sugar production. Although benzoic acid-treated wood at lower concentrations could not inhibit the *G. boninense* growth, it prevented the disintegration of silica bodies and could have aided in strengthening the structural barrier. The presence of phenolic compounds did not relatively alter the degradation pattern; however, it modified the extent of colonization by *G. boninense* on the surface of the woodblocks.

In the process of degradation, the WRF considerably affects the lignin structure. The demethoxylation of lignin takes place at the initial processes of degradation, followed by the formation of oxidized products [22]. This makes the lignin even more complex to detect [38]. Hence, FT-IR and TGA analyses were performed to further analyze the chemical changes in the wood structure during the biodegradation, as well as to quantify the remaining wood components. In FT-IR spectroscopy, the modification occurs in the cellulose as well as the lignin components of the oil palm woodblocks during the degradation process. A great reduction was observed at 1410 and 1415 cm^{-1} (cellulose), 1375 cm^{-1} (cellulose and hemicellulose) and also at 1028 cm^{-1} (starch) [29]. As the degradation proceeded further, these peaks disappeared. The peaks corresponding to the lignin (166, 1634, 1628 cm^{-1}) had increased initially and later disappeared, as was evident in the current findings [19,20]. Again, these suggest that *G. boninense* is a simultaneous degrader of lignin, hemicellulose and cellulose components, almost equally. When comparing the wood components, the highly crystalline cellulose components are the least degraded. This could be due to the presence of the strong covalent bonds between them, as well their interaction with the lignin [23].

The FT-IR spectra showed that the crystallinity indexes (TCI, LOI) increased initially and then decreased. The initial increment and later decrement in the crystallinity index were observed in various wood-rotting fungi [39]. This indicated the higher accessibility of the cellulase enzyme initially; thus, the substrate (cellulose) has degraded, leading to a decrease in the crystallinity index [40]. As the degradation of oil palm wood continued, the ordered crystalline structure of the cellulose was disturbed, and more amorphous domains were also introduced. The S/G ratio constantly decreased throughout the study of the control, indicating that *G. boninense* preferred S-type lignin more than the G-type. This could be due to the high prevalence of ether-type linkages, compared to G-type lignin [26]. However, the S/G ratio increased in all the woodblocks treated with syringic and vanillic acid from 30 to 45 days, and a great decrement was observed on the 75th day. This indicates that the presence of phenolic compounds in the oil palm wood can influence the preference of the type of lignin that *G. boninense* utilizes.

According to our previous study, the predominant oxidative enzyme of *G. boninense* is laccase [12]. Peroxidases (lignin and manganese) are produced predominantly by *G. boninense,* but no significant difference was observed in the production of hydrolytic enzymes in the presence of syringic and vanillic acids [12,37]. The results revealed that the presence of phenolic compounds only alters the preference of lignin subunits (S and G) and not the degradation pattern or the degradation ratio of lignin-hemicelluloses and cellulose. However, an increase in the concentration of phenolic compounds reduced the quantity of the degraded components.

The phenolic compounds are the key compounds in signal transduction in the plant defense system, and are also involved in the elimination of pathogens. Hence, an increase in the content of phenolic compounds in the plant is linked to induced resistance in plants [27,41]. The phenolic compounds used in this study are associated with lignin and lignin biosynthesis pathways. Syringic and vanillic acids are involved in the lignin structure, whereas benzoic acid and salicylic acids are involved in the synthesis of lignin. The cells of cut stumps and the felled trunks are intact and alive for a considerable period. Hence, the phenolic treatment of felled oil palm trunks and stumps can control *G. boninense* and can develop resistance against it.

A three-year study conducted by the Malaysian Palm Oil Board (MPOB) on live BSR-infected oil palm stumps treated with 1.2 kg Dazoment revealed a 10% survival rate of *Ganoderma* inoculum [42]. This practice can greatly reduce the inoculum survival rate. Nevertheless, the cost of the amount of Dazoment required to eradicate the inoculum should not be overlooked. The cost of the above-mentioned phenolic will comparatively be less, and be used in reduced quantities to treat the palm; hence, it could be a cost-effective method. Besides this, the salient feature of phenolic compounds is that they are resistant to autoxidation [43]. Hence, they could be directly applied to the stump via soil drenching. However, the effect of these phenolic compounds on the environment should be evaluated before their application in the field.

5. Conclusions

The results of this study verify and conclude that naturally occurring phenolic compounds can be an effective controller of BSR disease, and, hence, they are considered to be potential compounds to treat the infected stumps and debris. Benzoic acid is the most effective compound to control *G. boninense* growth in the wood samples at the minimal concentration (5 mM), followed by salicylic acid. As a recommendation, it is essential to determine the threshold concentration of phenolic compounds before their application, because a lower concentration of phenolic compounds has enhanced the colonization of *G. boninense*, which defeats the purpose of treatment. This study could serve as a stepping-stone towards the development of an effective stump treatment, with the ultimate aim of minimizing BSR infection in oil palm estates in a sustainable manner.

Supplementary Materials: The following are available online at https://www.mdpi.com/article/10.3390/plants10091797/s1, Table S1: Lower Order Index (LOI) of oil palm wood (with and without treatment) degraded by *G. boninense* at various time intervals; Table S2: Total Crystallinity Index (TCI) of oil palm wood (with and without treatment) degraded by *G. boninense* at various time intervals; Table S3: S/G ratio of oil palm wood (with and without treatment) degraded by *G. boninense* at various time intervals; Table S4: Percentage of the chemical composition of healthy oil palm wood.

Author Contributions: Conceptualization, Y.S.; methodology, A.S. and Y.S.; validation, Y.S. and K.A.; formal analysis, A.S.; investigation, A.S.; resources, Y.S. and K.A.; data curation, A.S.; writing—original draft preparation, A.S.; writing—review and editing, Y.S., K.A. and R.F.; visualization, Y.S.; supervision, Y.S.; project administration, Y.S.; funding acquisition, Y.S. All authors have read and agreed to the published version of the manuscript.

Funding: This research was funded by Ministry of Education, Malaysia under the Fundamental Research Grant Scheme (FRGS/1/2019/STG05/UPM/02/22).

Institutional Review Board Statement: Not applicable.

Informed Consent Statement: Not applicable.

Data Availability Statement: Not applicable.

Acknowledgments: The authors would like to thank the Ministry of Education, Malaysia for providing financial support under the Fundamental Research Grant Scheme (FRGS/1/2019/STG05/UPM/02/22) and Universiti Putra Malaysia.

Conflicts of Interest: The authors declare no conflict of interest.

References

1. Mohammed, C.; Rimbawanto, A.; Page, D. Management of basidiomycete root-and stem-rot diseases in oil palm, rubber and tropical hardwood plantation crops. *For. Pathol.* **2014**, *44*, 428–446. [CrossRef]
2. Olaniyi, O.N.; Szulczyk, K.R. Estimating the economic damage and treatment cost of basal stem rot striking the Malaysian oil palms. *For. Policy Econ.* **2020**, *116*, 102163. [CrossRef]
3. Midot, F.; Lau, S.Y.L.; Wong, W.C.; Tung, H.J.; Yap, M.L.; Lo, M.L.; Jee, M.S.; Dom, S.P.; Melling, L. Genetic Diversity and Demographic History of Ganoderma boninense in Oil Palm Plantations of Sarawak, Malaysia Inferred from ITS Regions. *Microorganisms* **2019**, *7*, 464. [CrossRef]
4. Roslan, A.; Idris, A. Economic impact of Ganoderma incidence on Malaysian oil palm plantation—A case study in Johor. *Oil Palm Ind. Econ. J.* **2012**, *12*, 24–30.
5. Chung, G. Management of Ganoderma diseases in oil palm plantations. *Planter* **2011**, *87*, 325–339.
6. Flood, J.; Hasan, Y.; Turner, P.; O'Grady, E. The spread of Ganoderma from infective sources in the field and its implications for management of the disease in oil palm. In *Ganoderma Diseases of Perennial Crops*; CABI: Wallingford, UK, 2000; pp. 101–112.
7. Pilotti, C.; Gorea, E.; Bonneau, L. Basidiospores as sources of inoculum in the spread of Ganoderma boninense in oil palm plantations in Papua New Guinea. *Plant Pathol.* **2018**, *67*, 1841–1849. [CrossRef]
8. Susanto, A.; Sudharto, P.; Purba, R. Enhancing biological control of basal stem rot disease (*Ganoderma boninense*) in oil palm plantations. *Mycopathologia* **2005**, *159*, 153–157. [CrossRef]
9. Khairudin, H. *Basal Stem Rot of Oil Palm Caused by Ganoderma boninense: An Update, Proceedings of the International Palm Oil Congress 'Update and Vision' (Agriculture), Bangi, Selangor, Malaysia, 20–25 September 1993*; Paper No. 46; Palm Oil Research Institute of Malaysia: Bangi, Selangor, Malaysia, 1993.
10. Jourdan, C.; Rey, H. Architecture and development of the oil-palm (Elaeis guineensis Jacq.) root system. *Plant Soil* **1997**, *189*, 33–48. [CrossRef]
11. Surendran, A.; Siddiqui, Y.; Saud, H.M.; Manickam, N. The antagonistic effect of phenolic compounds on ligninolytic and cellulolytic enzymes of Ganoderma Boninense, causing basal stem rot in Oil Palm. *Int. J. Agric. Biol.* **2017**, *19*, 1437–1446.

12. Surendran, A.; Siddiqui, Y.; Saud, H.M.; Ali, N.S.; Manickam, S. Inhibition and kinetic studies of lignin degrading enzymes of Ganoderma boninense by naturally occurring phenolic compounds. *J. Appl. Microbiol.* **2018**, *125*, 876–887. [CrossRef]
13. Schirp, A.; Wolcott, M.P. Influence of fungal decay and moisture absorption on mechanical properties of extruded wood-plastic composites. *Wood Fiber Sci.* **2005**, *37*, 643–652.
14. Bucher, V.; Hyde, K.; Pointing, S.; Reddy, C. Production of wood decay enzymes, mass loss and lignin solubilization in wood by marine ascomycetes and their anamorphs. *Fungal Divers.* **2004**, *15*, 1–4.
15. Popescu, C.-M.; Popescu, M.-C.; Vasile, C. Structural changes in biodegraded lime wood. *Carbohydr. Polym.* **2010**, *79*, 362–372. [CrossRef]
16. Nelson, M.L.; O'Connor, R.T. Relation of certain infrared bands to cellulose crystallinity and crystal lattice type. Part II. A new infrared ratio for estimation of crystallinity in celluloses I and II. *J. Appl. Polym. Sci.* **1964**, *8*, 1325–1341. [CrossRef]
17. Fan, M.; Dai, D.; Huang, B. Fourier transform infrared spectroscopy for natural fibres. In *Fourier Transform-Materials Analysis*; Intechopen: London, UK, 2012.
18. Yang, H.; Yan, R.; Chin, T.; Liang, D.T.; Chen, H.; Zheng, C. Thermogravimetric analysis– Fourier transform infrared analysis of palm oil waste pyrolysis. *Energy Fuels* **2004**, *18*, 1814–1821. [CrossRef]
19. Pandey, K.K.; Pitman, A. FTIR studies of the changes in wood chemistry following decay by brown-rot and white-rot fungi. *Int. Biodeterior. Biodegrad.* **2003**, *52*, 151–160. [CrossRef]
20. Blanchette, R.A. A review of microbial deterioration found in archaeological wood from different environments. *Int. Biodeterior. Biodegrad.* **2000**, *46*, 189–204. [CrossRef]
21. Kirk, T.K.; Moore, W.E. Removing lignin from wood with white-rot fungi and digestibility of resulting wood. *Wood Fiber Sci.* **2007**, *4*, 72–79.
22. Alfredsen, G.; Bader, T.K.; Dibdiakova, J.; Filbakk, T.; Bollmus, S.; Hofstetter, K. Thermogravimetric analysis for wood decay characterisation. *Eur. J. Wood Wood Prod.* **2012**, *70*, 527–530. [CrossRef]
23. Su, Y.; Yu, X.; Sun, Y.; Wang, G.; Chen, H.; Chen, G. Evaluation of screened lignin-degrading fungi for the biological pretreatment of corn stover. *Sci. Rep.* **2018**, *8*, 5385. [CrossRef]
24. Luna, M.L.; Murace, M.A.; Keil, G.D.; Otaño, M.E. Patterns of decay caused by Pycnoporus sanguineus and Ganoderma lucidum (Aphyllophorales) in poplar wood. *IAWA J.* **2004**, *25*, 425–433. [CrossRef]
25. Kües, U.; Mai, C.; Militz, H.; Kües, U. 14. Biological Wood Protection against Decay, Microbial Staining, Fungal Moulding and Insect Pests. In *Wood Production, Wood Technology, and Biotechnological Impacts*; 2007; pp. 273–294. Available online: https://www.researchgate.net/publication/262179411_Biological_Wood_Protection_against_Decay_Microbial_Staining_Fungal_Moulding_and_Insect_Pests.
26. Del Rıo, J.; Speranza, M.; Gutiérrez, A.; Martınez, M.; Martınez, A. Lignin attack during eucalypt wood decay by selected basidiomycetes: A Py-GC/MS study. *J. Anal. Appl. Pyrolysis* **2002**, *64*, 421–431. [CrossRef]
27. Chappell, J.; Hahlbrock, K. Transcription of plant defence genes in response to UV light or fungal elicitor. *Nature* **1984**, *311*, 76–78. [CrossRef]
28. Bari, E.; Nazarnezhad, N.; Kazemi, S.M.; Ghanbary, M.A.T.; Mohebby, B.; Schmidt, O.; Clausen, C.A. Comparison between degradation capabilities of the white rot fungi Pleurotus ostreatus and Trametes versicolor in beech wood. *Int. Biodeterior. Biodegrad.* **2015**, *104*, 231–237. [CrossRef]
29. Darwish, S.S.; El Hadidi, N.; Mansour, M. The effect of fungal decay on *Ficus sycomoru* wood. *Int. J. Conserv. Sci.* **2013**, *4*, 271–282.
30. Susanto, A.; Suprianto, E.; Purba, A.; Prasetyo, A. Oil palm breeding for tolerance to G. boninense. In *Disease Management Strategies in Plantations, Yagyakarta*; Indonesian Oil Palm Research Institute: Medan, Indonesia, 2009; pp. 4–8.
31. Naidu, Y.; Siddiqui, Y.; Rafii, M.Y.; Saud, H.M.; Idris, A.S. Inoculation of oil palm seedlings in Malaysia with white-rot hymenomycetes: Assessment of pathogenicity and vegetative growth. *Crop Prot.* **2018**, *110*, 146–154. [CrossRef]
32. Cragg, S.M.; Beckham, G.T.; Bruce, N.C.; Bugg, T.D.; Distel, D.L.; Dupree, P.; Etxabe, A.G.; Goodell, B.S.; Jellison, J.; McGeehan, J.E. Lignocellulose degradation mechanisms across the Tree of Life. *Curr. Opin. Chem. Biol.* **2015**, *29*, 108–119. [CrossRef]
33. Guillén, F.; Martínez, M.J.; Gutiérrez, A.; Del Rio, J. Biodegradation of lignocellu-losics: Microbial, chemical, and enzymatic aspects of the fungal attack of lignin. *Int. Microbiol.* **2005**, *8*, 195–204.
34. Buswell, J.; Ander, P.; Pettersson, B.; Eriksson, K.-E. Oxidative decarboxylation of vanillic acid by Sporotrichum pulverulentum. *FEBS Lett.* **1979**, *103*, 98–101. [CrossRef]
35. Gupta, J.K.; Hamp, S.G.; Buswell, J.A.; Eriksson, K.-E. Metabolism of trans-ferulic acid by the white-rot fungus Sporotrichum pulverulentum. *Arch. Microbiol.* **1981**, *128*, 349–354. [CrossRef]
36. Mishra, V.; Jana, A.K.; Jana, M.M.; Gupta, A. Fungal pretreatment of sweet sorghum bagasse with supplements: Improvement in lignin degradation, selectivity and enzymatic saccharification. *3 Biotech* **2017**, *7*, 110. [CrossRef] [PubMed]
37. Surendran, A.; Siddiqui, Y.; Ali, N.S.; Manickam, S. Inhibition and kinetic studies of cellulose-and hemicellulose-degrading enzymes of Ganoderma boninense by naturally occurring phenolic compounds. *J. Appl. Microbiol.* **2018**, *124*, 1544–1555. [CrossRef] [PubMed]
38. Carrier, M.; Loppinet-Serani, A.; Denux, D.; Lasnier, J.-M.; Ham-Pichavant, F.; Cansell, F.; Aymonier, C. Thermogravimetric analysis as a new method to determine the lignocellulosic composition of biomass. *Biomass Bioenergy* **2011**, *35*, 298–307. [CrossRef]
39. Hastrup, A.C.S.; Howell, C.; Larsen, F.H.; Sathitsuksanoh, N.; Goodell, B.; Jellison, J. Differences in crystalline cellulose modification due to degradation by brown and white-rot fungi. *Fungal Biol.* **2012**, *116*, 1052–1063. [CrossRef] [PubMed]

40. Hall, M.; Bansal, P.; Lee, J.H.; Realff, M.J.; Bommarius, A.S. Cellulose crystallinity–a key predictor of the enzymatic hydrolysis rate. *FEBS J.* **2010**, *277*, 1571–1582. [CrossRef]
41. Dixon, R.A.; Paiva, N.L. Stress-induced phenylpropanoid metabolism. *Plant Cell* **1995**, *7*, 1085. [CrossRef]
42. Idris, A.; Maizatul, S. Stump treatment with dazomet for controlling Ganoderma disease in oil palm. *MPOB TS Inf. Ser.* **2012**, *107*, 615–616.
43. Srivastava, P.; Andersen, P.C.; Marois, J.J.; Wright, D.L.; Srivastava, M.; Harmon, P.F. Effect of phenolic compounds on growth and ligninolytic enzyme production in Botryosphaeria isolates. *Crop Prot.* **2013**, *43*, 146–156. [CrossRef]

MDPI

Article

Chemical Composition of *Ambrosia trifida* L. and Its Allelopathic Influence on Crops

Jovana Šućur [1], Bojan Konstantinović [2], Marina Crnković [1], Vojislava Bursić [2], Nataša Samardžić [2,*], Đorđe Malenčić [1], Dejan Prvulović [1], Milena Popov [2] and Gorica Vuković [3]

1 Department of Field and Vegetable Crops, Faculty of Agriculture, University of Novi Sad, Trg Dositeja Obradovića 8, 21000 Novi Sad, Serbia; jovana.sucur@polj.uns.ac.rs (J.Š.); m.crnkovic.95@gmail.com (M.C.); djordje.malencic@polj.uns.ac.rs (Đ.M.); dejanp@polj.uns.ac.rs (D.P.)

2 Department of Plant and Environmental Protection, Faculty of Agriculture, University of Novi Sad, Trg Dositeja Obradovića 8, 21000 Novi Sad, Serbia; bojank@polj.uns.ac.rs (B.K.); bursicv@polj.uns.ac.rs (V.B.); milena.popov@polj.uns.ac.rs (M.P.)

3 Department of Pesticides and Herbology, Faculty of Agriculture, University of Belgrade, Nemanjina 6, 11080 Belgrade, Serbia; goricavukovic@yahoo.com

* Correspondence: natasa.samardzic@polj.uns.ac.rs; Tel.: +381-214853359

Abstract: Phytotoxic substances released by invasive plants have been reported to have anti-pathogen, anti-herbivore, and allelopathic activity. The aim of this study was to determine the allelopathic influence of the *Ambrosia trifida* L. on oxidative stress parameters (the lipid peroxidation process; reduced glutathione (GSH) content; and activity of antioxidant enzymes catalase (CAT), superoxide dismutase (SOD), and peroxidase (PX)) and phenolic compounds (total phenolic and tannin content) in maize (*Zea mays* L.), soybean (*Glycine max* L.), and sunflower (*Helianthus annuus* L.) crops to explore the effect of released allelochemicals through *A. trifida* root on crops. An analysis by HPLC confirmed the presence of protocatechuic acid, p-hydroxybenzoic acid, vanillic acid, and syringic acid as major components in the *A. trifida*. Based on the obtained results for oxidative stress parameters, it can be concluded that the sunflower was the most sensitive species to *A. trifida* allelochemicals among the tested crops. The other two crops tested showed a different sensitivity to *A. trifida*. The soybean did not show sensitivity, while the maize showed sensitivity only 10 days after the sowing.

Keywords: allelopathy; oxidative stress; maize; soybean; sunflower

Citation: Šućur, J.; Konstantinović, B.; Crnković, M.; Bursić, V.; Samardžić, N.; Malenčić, Đ.; Prvulović, D.; Popov, M.; Vuković, G. Chemical Composition of *Ambrosia trifida* L. and Its Allelopathic Influence on Crops. *Plants* **2021**, *10*, 2222. https://doi.org/10.3390/plants10102222

Academic Editors: Špela Mechora and Dragana Šunjka

Received: 15 September 2021
Accepted: 12 October 2021
Published: 19 October 2021

1. Introduction

Ambrosia trifida L., commonly named giant ragweed, is a flowering plant belonging to the *Asteraceae* family. The giant ragweed is American (especially North American) in origin and is spreading as a pioneer species worldwide, invading cultivated fields, particularly wheat, corn, and soybean fields. The earliest germination and emergence, as well as the largest seeds and seedlings, give *A. trifida* a decisive advantage over the other plants [1]. Thus, *A. trifida* is listed, along with *Ageratum conyzoides* L. and *Lantana camara* L., as the most economically destructive weed in the world [2]. The same authors concluded that the competition is an important interference mechanism that is responsible for *A. trifida* infestation, but the allelopathy of *A. trifida* would also be an important mechanism.

Allelopathy is defined as "The biochemical interactions among all types of plants, including microorganisms" [3]. One organism produces chemicals that affect another organism and those chemicals are called allelochemicals [4]. Plant phenolic compounds are considered as a major source of allelochemicals [5]. They can be involved in the production of reactive oxygen species (ROS) and can suppress antioxidant enzyme activity-inducing oxidative stress in plants [6,7]. High production of ROS that exceeds the capacity of antioxidant defence enzymes results in oxidative stress and plant cell death [8].

Thus, the phenolic compounds act as antioxidant, antimutagenic and leaf movement-regulating agents, in turn protecting the organism that produces them from the oxidative stress created by the metabolism and their physical environment [9]. Phenolic compounds present in plants are very important constituents that defend plants from infection and injury [10,11]. It has been found that p-hydroxy benzoic acid increases resistance of wheat (*Triticum aestivum* L.) against a pathogen infection, increasing abiotic stress tolerance and impermeability of the cell wall [12]. Furthermore, it was found that syringic acid and p-hydroxy benzoic acid possess antimicrobial activity [13].

Phenolic compounds synthesized by plants are also involved in plant allelopathy inhibiting the growth of the other plant competitors [14,15]. Some weeds inhibit crop growth and development by interfering with them through released allelochemicals. The *Ambrosia* genus biosynthesizes and releases several types of secondary metabolites (phenolics, flavonoids, sesquiterpenes, ambrosin, isabelin, psilostachyin, coronopilin, thiarubrines, and thiophenes) with a broad spectrum of biological activities including allelopathic reactions [2,16]. Specific chemical compounds, i.e., carotane sesquiterpenes, thiarubrines, and thiophenes, were identified from *A. trifida* [2].

Another important non-enzymatic antioxidant molecule is reduced glutathione (GSH), a thiol tripeptide containing γ-glutamate, glycine and, the most important amino acid, cysteine. It has several roles in animal and plant cells and one of them is to buffer the ROS molecules by donating a hydrogen (H) molecule [17]. The donator is a thiol (SH) group from cysteine, which oxidizes to the glutathione disulfide (GSSG) after the donation. This reaction prevents the oxidative stress and the cell death. The ratio of the reduced glutathione to oxidized glutathione (GSH/GSSG) within cells is a measure of the cellular oxidative stress [18].

Interactions between plants and the environment are an important consideration for understanding allelopathy. One of the studies demonstrated that the temperature stress enhances allelochemical inhibition, which indicates that interactions between plants and the environment are important for understanding allelopathy [19].

Phytotoxic substances released by invasive plants have been reported to have anti-pathogen, anti-herbivore, and allelopathic activity [20]. Bearing in mind that *A. trifida* has the ability to spread, becoming an invasive weed, the aim of this study was to determine the allelopathic influence of *A. trifida* on maize (*Zea mays* L.), soybean (*Glycine max* L.), and sunflower (*Helianthus annuus* L.) plants to explore the effect of released allelochemicals through *A. trifida* root on crops. The effect of released allelochemicals of *A. trifida* on oxidative stress parameters (the lipid peroxidation process, reduced glutathione (GSH) content, as well as the activity of antioxidant enzymes (catalase (CAT)), superoxide dismutase (SOD), and peroxidase (PX)) and phenolic compounds (total phenolic and tannin content) in crop plants were examined. The chemical composition of the *A. trifida* was achieved by the HPLC.

Ambrosia trifida is locally present in the Central Bačka (Vojvodina) and is expected to spread in the future [21]. The results of the study that dealt with the antioxidant potential of the ragweeds showed that the range of the total phenolic compounds was between 30.0 and 111.1 mg gallic acid equivalents (GAE) per gram of the dry weight of leaves. The contents of all of the measured phenolic compounds in *A. trifida* were pronounced when compared with *A. artemisiifolia* and *Iva xanthifolia*. When it comes to the antioxidant activity, the radical scavenging abilities of the hydroxyl and 1,1-diphenyl-2-picrylhydrazyl (DPPH) were notably higher in the case of *A. trifida* leaves, compared with the other two investigated species [22].

2. Results

2.1. Phenolic Compounds

The main constituent of the phenolic components (Table 1.) was protocatechuic acid (8.90 ± 0.20 µg/g), followed by p-hydroxybenzoic, vanillic, and syringic acid (4.50 ± 0.01, 3.55 ± 0.01, and 1.75 ± 0.09 µg/g, respectively). The p-Coumaric acid and ferulic acid were represented at the concentration levels of 0.96 ± 0.01 and 0.70 ± 0.01 µg/g, respectively.

Table 1. The identified and quantified phenolic compounds in the *Ambrosia trifida*.

Phenolic Compounds	mean ± SD µg/g *
gallic acid	nd
ferulic acid	0.70 ± 0.01
2-hydroxycinnamic acid	nd
trans-cinnamic acid	nd
caffeic acid	nd
p-coumaric acid	0.96 ± 0.00
chlorogenic acid	nd
quercetin	nd
(+)-catechin	nd
protocatechuic acid	8.90 ± 0.28
p-hydroxybenzoic acid	4.50 ± 0.00
vanillic acid	3.55 ± 0.00
epicatechin	nd
syringic acid	1.75 ± 0.09
kaempferol	nd

* Data represent the means of the three replicates ± standard deviation; nd = not defined.

2.2. The Effects of Ambrosia trifida on Sunflower, Soybean and Maize Total Phenolics and Tannins

The highest amount of the total phenolics was obtained in sunflower leaves at the weed:crop ratio of 3:1, at 10 days after the sowing (30.50 ± 5.18 mg GAE/g DW) (Table 2). There were no significant differences in total tannins in sunflower leaves. Soybean leaves at the weed:crop ratio was 3:1, at 14 days after the sowing had the highest total phenolic content (14.17 ± 0.44 mg GAE/g DW), while the content of tannins was lower in soybean leaves at the weed:crop ratio of 3:1, at 10 days after the sowing (0.41 ± 0.30 mg GAE/g DW), as well as at the weed:crop ratio of 3:1, at 14 days after the sowing (2.69 ± 1.10 mg GAE/g DW). In maize leaves, the highest amount of the total phenolic compounds was obtained at the weed:crop ratio of 3:1, at 10 days after the sowing (22.72 ± 1.49 mg GAE/g DW). The content of the tannins in maize leaves was lower in the treatments compared with the control group 14 days after the sowing.

Table 2. Total phenolics (TP) and total tannins (TT) in leaves of sunflower, soybean and maize plants of the three-term ratio 1:3, 1:1, 3:1 (weed-crop).

Time	Ratio	Sunflower		Soybean		Maize	
		TP	TT	TP	TT	TP	TT
7 days	C	11.67 ± 0.90 ^{a}	4.36 ± 0.46 ^{ab}	9.93 ± 1.18 ^{a}	3.09 ± 1.12 ^{b}	14.82 ± 2.13 ^{ac}	5.65 ± 1.99 ^{bc}
	1:3	7.95 ± 0.92 ^{a}	3.35 ± 0.34 ^{ab}	11.23 ± 0.89 ^{a}	7.03 ± 0.67 ^{d}	14.34 ± 1.74 ^{ac}	8.00 ± 1.67 ^{c}
	1:1	11.74 ± 0.89 ^{a}	4.76 ± 0.07 ^{ab}	10.12 ± 0.64 ^{a}	5.90 ± 0.60 ^{cd}	15.52 ± 1.20 ^{ac}	6.77 ± 0.82 ^{c}
	3:1	7.18 ± 1.14 ^{a}	3.23 ± 0.17 ^{ab}	10.59 ± 1.09 ^{a}	5.19 ± 1.11 ^{bcd}	13.75 ± 0.96 ^{b}	5.86 ± 1.00 ^{bc}
10 days	C	11.20 ± 1.02 ^{a}	3.70 ± 0.22 ^{ab}	10.23 ± 0.78 ^{a}	4.30 ± 0.53 ^{bc}	15.86 ± 0.48 ^{ac}	7.98 ± 0.22 ^{c}
	1:3	6.83 ± 0.78 ^{a}	2.89 ± 0.15 ^{a}	11.08 ± 0.47 ^{a}	4.93 ± 0.77 ^{bcd}	17.28 ± 0.57 ^{c}	8.40 ± 0.42 ^{c}
	1:1	9.09 ± 1.34 ^{a}	3.88 ± 0.26 ^{ab}	10.90 ± 0.45 ^{a}	4.33 ± 1.01 ^{bc}	14.76 ± 1.09 ^{ac}	5.52 ± 0.92 ^{bc}
	3:1	30.50 ± 5.18 ^{b}	5.01 ± 0.17 ^{b}	11.05 ± 0.94 ^{a}	0.41 ± 0.30 ^{a}	22.72 ± 1.49 ^{a}	5.70 ± 0.61 ^{bc}
14 days	C	12.81 ± 0.55 ^{a}	5.04 ± 0.38 ^{b}	10.32 ± 0.51 ^{a}	6.10 ± 0.31 ^{c}	16.55 ± 0.12 ^{c}	12.85 ± 0.22 ^{a}
	1:3	14.26 ± 0.51 ^{a}	4.42 ± 0.44 ^{ab}	10.60 ± 0.35 ^{a}	4.22 ± 0.18 ^{bc}	15.05 ± 0.18 ^{ac}	6.91 ± 0.16 ^{c}
	1:1	13.67 ± 1.19 ^{a}	5.00 ± 0.51 ^{b}	10.39 ± 0.53 ^{a}	3.66 ± 0.47 ^{bc}	14.64 ± 0.73 ^{ac}	5.70 ± 0.76 ^{bc}
	3:1	16.14 ± 1.35 ^{a}	5.10 ± 0.14 ^{b}	14.17 ± 0.44 ^{b}	2.69 ± 1.10 ^{b}	12.72 ± 0.80 ^{b}	3.12 ± 0.17 ^{b}

TP (mgGAE/gDW); TT (mgGAE/gDW). The data are mean values ± standard error. $^{a–d}$ Values without the same superscripts within each column differ significantly ($p < 0.05$).

2.3. The Effects of Ambrosia trifida on Sunflower, Soybean and Maize Antioxidant Enzyme Activity, GSH Content and MDA Content

The results of the *A. trifida* effects on the antioxidant enzyme activity, as well as the GSH and MDA content of sunflower, soybean, and maize, are shown in Tables 3–5, respectively.

Table 3. The activity of the antioxidant enzymes, reduced glutathione (GSH) content, and malondialdehyde (MDA) content in leaves of sunflower plants of the three-term ratio 1:3, 1:1, 3:1 (weed-crop).

Time	Ratio	CAT	SOD	PX	GSH	MDA
7 days	C	0.29 ± 0.16 [a]	1241.71 ± 1150.06 [d]	0.77 ± 0.65 [c]	0.20 ± 0.20 [a]	35.04 ± 13.26 [a]
	1:3	0.06 ± 0.07 [b]	1428.85 ± 1337.20 [abc]	0.47 ± 0.35 [a]	0.16 ± 0.15 [b]	32.16 ± 10.38 [a]
	1:1	0.17 ± 0.04 [c]	1467.84 ± 1376.18 [a]	0.55 ± 0.43 [ab]	0.16 ± 0.16 [b]	25.21 ± 3.43 [a]
	3:1	0.09 ± 0.04 [c]	888.89 ± 797.24 [e]	0.62 ± 0.50 [abc]	0.17 ± 0.16 [b]	35.26 ± 13.48 [a]
10 days	C	0.10 ± 0.03 [a]	1396.19 ± 1304.54 [abc]	0.63 ± 0.51 [abc]	0.180.17 [a]	42.73 ± 20.96 [ab]
	1:3	0.12 ± 0.01 [a]	1443.81 ± 1352.16 [ab]	0.61 ± 0.49 [abc]	0.20 ± 0.19 [b]	34.83 ± 13.05 [a]
	1:1	0.07 ± 0.06 [a]	1304.76 ± 1213.11 [c]	0.62 ± 0.50 [abc]	0.17 ± 0.16 [c]	55.66 ± 33.88 [ab]
	3:1	0.15 ± 0.02 [a]	1329.52 ± 1237.87 [b]	0.70 ± 0.58 [b]	0.17 ± 0.16 [a]	45.62 ± 23.84 [ab]
14 days	C	0.11 ± 0.02 [a]	1505.24 ± 1413.59 [a]	0.81 ± 0.69 [d]	0.21 ± 0.20 [a]	40.49 ± 18.71 [ab]
	1:3	0.09 ± 0.04 [a]	1456.86 ± 1365.21 [ab]	0.57 ± 0.49 [ab]	0.17 ± 0.16 [b]	39.53 ± 17.75 [ab]
	1:1	0.17 ± 0.04 [a]	1472.55 ± 1380.90 [a]	0.89 ± 0.77 [e]	0.26 ± 0.25 [c]	71.69 ± 49.91 [b]
	3:1	0.12 ± 0.01 [a]	1488.23 ± 1396.58 [a]	0.91 ± 0.79 [e]	0.20 ± 0.19 [a]	90.38 ± 68.61 [c]

Activity of antioxidant enzymes (U/g FW); CAT (catalase, U/g FW); SOD (superoxide dismutase, U/g FW); PX (peroxidase, U/g FW); GSH content (reduced glutathione, µmol GSH/g FW); MDA content (nmol MDA/g FW). The data are mean values ± standard error. [a–e] Values without the same superscripts within each column differ significantly ($p < 0.05$).

Table 4. The activity of the antioxidant enzymes, reduced glutathione (GSH) and malondialdehyde (MDA) content in leaves of soybean plants of the three-term ratio 1:3, 1:1, and 3:1 (weed-crop).

Time	Ratio	CAT	SOD	PX	GSH	MDA
7 days	C	0.13 ± 0.03 [a]	1062.38 ± 956.17 [a]	1.50 ± 1.20 [a]	0.25 ± 0.24 [a]	91.13 ± 84.96 [a]
	1:3	0.06 ± 0.04 [a]	877.19 ± 770.99 [c]	0.83 ± 0.53 [b]	0.20 ± 0.19 [b]	80.34 ± 74.17 [b]
	1:1	0.13 ± 0.03 [a]	1027.29 ± 921.09 [ac]	1.05 ± 0.76 [ab]	0.20 ± 0.19 [b]	55.66 ± 49.49 [f]
	3:1	0.08 ± 0.02 [a]	1109.16 ± 1002.96 [a]	1.27 ± 0.97 [ab]	0.31 ± 0.30 [c]	74.68 ± 68.51 [be]
10 days	C	0.07 ± 0.03 [a]	1308.57 ± 1202.37 [b]	1.20 ± 0.91 [ab]	0.24 ± 0.23 [a]	109.19 ± 103.01 [cd]
	1:3	0.09 ± 0.01 [a]	1369.52 ± 1263.32 [b]	1.05 ± 0.76 [ab]	0.20 ± 0.19 [b]	101.39 ± 95.22 [c]
	1:1	0.09 ± 0.01 [a]	1022.86 ± 916.65 [ac]	1.41 ± 1.12 [a]	0.26 ± 0.25 [c]	67.63 ± 61.45 [e]
	3:1	0.20 ± 0.10 [a]	1099.05 ± 992.84 [a]	2.34 ± 2.05 [cd]	0.31 ± 0.30 [d]	85.36 ± 79.19 [ab]
14 days	C	0.08 ± 0.02 [a]	1392.16 ± 1285.95 [b]	2.39 ± 2.09 [cd]	0.31 ± 0.30 [a]	114.10 ± 107.93 [c]
	1:3	0.11 ± 0.01 [a]	1447.06 ± 1340.86 [b]	2.01 ± 1.72 [c]	0.32 ± 0.31 [a]	100.32 ± 94.15 [d]
	1:1	0.15 ± 0.05 [a]	1433.33 ± 1327.13 [b]	2.51 ± 2.21 [d]	0.29 ± 0.28 [b]	103.74 ± 97.57 [d]
	3:1	0.11 ± 0.01 [a]	623.53 ± 517.33 [d]	2.36 ± 2.06 [cd]	0.24 ± 0.23 [c]	100.21 ± 94.04 [d]

Activity of antioxidant enzymes (U/g FW); CAT (catalase, U/g FW); SOD (superoxide dismutase, U/g FW); PX (peroxidase, U/g FW); GSH content (reduced glutathione, µmol GSH/g FW); MDA content (nmol MDA/g FW). The data are mean values ± standard error. [a–e] Values without the same superscripts within each column differ significantly ($p < 0.05$).

According to Duncan's multiple range tests, a significant decrease in the activity of the antioxidant enzymes CAT and PX was detected in sunflower leaves 7 days after the sowing at all examined ratios (PX from 20 to 39%, CAT from 43 to 79%). The activity of the SOD was lower only in the weed:crop ratio of 3:1 (28%). Lower production of GSH was detected in the sunflower plants leaves 7 days after the sowing in all examined ratios. After 14 days, there were no significant differences in the GSH production in the weed:crop ratio of 3:1, statistically lower GSH was measured in the weed:crop ratio of 1:3 (19%), while, in the weed:crop ratio of 1:1, significantly higher GSH content was measured (24%). The MDA content, the main end product of the lipid peroxidation process, is used as a biomarker for the oxidative stress. A statistically significant increase in MDA accumulation was recorded in sunflower leaves 14 days after the sowing whereby the weed:crop ratios were 3:1 and 1:1 (123% and 77%, respectively).

Table 5. The activity of the antioxidant enzymes, reduced glutathione (GSH) and malondialdehyde (MDA) content in leaves of maize plants of the three-term ratio 1:3, 1:1, 3:1 (weed-crop).

Time	Ratio	CAT	SOD	PX	GSH	MDA
7 days	C	0.33 ± 0.16^{a}	1335.28 ± 1258.50^{e}	1.12 ± 0.96^{c}	0.20 ± 0.19^{a}	49.25 ± 33.69^{a}
	1:3	0.23 ± 0.07^{a}	1085.77 ± 1008.99^{b}	1.00 ± 0.83^{d}	0.22 ± 0.20^{a}	43.80 ± 28.24^{a}
	1:1	0.72 ± 0.55^{b}	1042.88 ± 966.10^{b}	1.32 ± 1.16^{abc}	0.20 ± 0.19^{a}	39.53 ± 23.96^{a}
	3:1	0.20 ± 0.04^{a}	807.02 ± 730.23^{d}	1.04 ± 0.88^{d}	0.21 ± 0.20^{a}	51.17 ± 35.61^{a}
10 days	C	0.22 ± 0.06^{a}	754.29 ± 677.50^{d}	1.43 ± 1.27^{ab}	0.23 ± 0.22^{a}	50.43 ± 34.86^{a}
	1:3	0.22 ± 0.06^{a}	923.81 ± 847.02^{a}	1.21 ± 1.05^{a}	0.25 ± 0.24^{b}	44.23 ± 28.67^{a}
	1:1	0.34 ± 0.18^{a}	910.48 ± 833.69^{ac}	1.53 ± 1.37^{b}	0.19 ± 0.17^{c}	48.50 ± 32.94^{a}
	3:1	0.21 ± 0.04^{a}	935.24 ± 858.45^{abc}	1.42 ± 1.26^{ab}	0.33 ± 0.32^{d}	54.06 ± 38.49^{a}
14 days	C	0.14 ± 0.02^{a}	851.15 ± 774.37^{c}	1.33 ± 1.17^{abc}	0.25 ± 0.24^{a}	58.01 ± 42.45^{a}
	1:3	0.12 ± 0.04^{a}	1012.58 ± 935.79^{ab}	1.35 ± 1.19^{abc}	0.29 ± 0.27^{b}	58.65 ± 43.09^{a}
	1:1	0.31 ± 0.15^{a}	1018.87 ± 942.08^{ab}	1.61 ± 1.45^{e}	0.28 ± 0.27^{b}	58.01 ± 42.45^{a}
	3:1	0.23 ± 0.07^{a}	1008.39 ± 931.60^{ab}	1.78 ± 1.62^{f}	0.30 ± 0.29^{b}	52.88 ± 37.32^{a}

Activity of antioxidant enzymes (U/g FW); CAT (catalase, U/g FW); SOD (superoxide dismutase, U/g FW); PX (peroxidase, U/g FW); GSH content (reduced glutathione, µmol GSH/g FW); MDA content (nmol MDA/g FW). The data are mean values ± standard error. [a–e] Values without the same superscripts within each column differ significantly ($p < 0.05$).

A significant decrease in the activity of the SOD was detected in soybean leaves at the weed:crop ratio of 3:1, at 10 and 14 days after the sowing (16% and 55%, respectively). No significant difference was detected in the activity of CAT, while a significant increase in PX activity was detected at 10 days after the sowing at the weed:crop ratio of 3:1 when the highest increase in the activity of the PX was measured (95%). Furthermore, significant production of the GSH was detected in the soybean plants leaves at 7 and 10 days after the sowing at the weed:crop ratio of 3:1 (25% and 28%, respectively), while, in the same ratio at the end of the experimental time, the production of GSH was lower compared to the control (22%). In the soybean leaves, the amount of MDA was lower in the treatments compared with the control during the experimental period (e.g., from 9 to 12%, 14 days after the sowing).

The activity of the SOD showed an increase from 18 to 24% in maize leaves at 10 and 14 days after the sowing at all examined ratios. The highest activity of the CAT was measured at the weed:crop ratio of 1:1, at 7 days after the sowing (119%). The highest activity of PX was detected in maize leaves at the weed:crop ratios of 1:1 and 3:1, at 14 days after the sowing (21% and 34%, respectively). Significant induction of the GSH production was detected in the maize plants leaves 14 days after the sowing in all examined weed:crop ratios (1:3, 1:1, and 3:1) and accumulation of the GSH compared with the control was 16%, 12%, and 20%, respectively. A significant increase in the LP intensity was recorded in maize leaves 10 days after the sowing at the weed:crop ratio of 3:1 (8%). On the other hand, 14 days after the sowing, there was no significant increase in the LP intensity in the treatments compared with the control.

3. Discussion

The allelopathic effect of *A. trifida* was investigated in a number of studies. It was reported that *A. trifida* aqueous extracts were significantly inhibitory to the sorghum seedling growth [23]. The effects of the *A. trifida* volatile oil on other plants were examined and it was found that *A. trifida* volatile oil significantly inhibited the germination and growth of maize and wheat but these volatiles also significantly stimulated the germination and growth of barnyard grass weed (*Echinochloa crus-galli*) [24]. Primarily, the volatile allelochemicals of *A. trifida* acting against other plants and stimulating the germination of barnyard grass weed were terpenoids alcanfor, borneol, and borneol acetate.

The main constituent of the phenolic components in *Ambrosia trifida* was protocatechuic acid, followed by p-hydroxy benzoic, vanillic, and syringic acid. Protocatechuic and p-hydroxy benzoic acid formation result from reactions of hydroxylation and methylation that can occur in the aromatic ring of benzoic acid [14].

The effect of *A. trifida* root exudates on wheat (*T. aestivum)* growth was examined and it was found that soil phytotoxicity did not result primarily from *A. trifida* root exudates, but from the residues in the soil [16]. It was also found that 1α-angeloyloxycarotol and 1α-(2-methylbutyroyloxy)-carotol released by *A. trifida* act as allelochemicals and inhibit the growth of wheat. Another study reported that metabolites with carotene skeletons have strong biological activity [25].

The allelochemical stress is a phenomenon when allelochemical compounds suppress the plant growth. The accepted mode of action of many allelochemical compounds is the production of reactive oxygen species (ROS) and induction of oxidative stress. The excessive production of ROS is accompanied by the activation of enzymatic defenses. The activity of the antioxidant enzymes is frequently used as an indicator of the oxidative stress in plants, while an increase in the lipid peroxidation is a widely used stress indicator of the plant membranes [26]. Taking into account that the activity of the antioxidant enzymes in plants can be changed under oxidative stress [26], the changes in the activity of the antioxidant enzymes in sunflower, soybean, and maize leaves could occur as a response to the oxidative stress induced by the allelochemicals released by *A. trifida*. In addition, if allelopathy-provoked stress is strong enough, the activity of the antioxidant enzymes could not prevent the oxidative stress when lipid peroxidation is increased.

The study results showed that the presence of *A. trifida* strongly affected lipid peroxidation in sunflower leaves, particularly at 14 days after the sowing at the weed:crop ratio was 3:1 when the MDA accumulation was the highest, which preceded a significant decrease in the activity of the antioxidant enzymes, i.e., CAT, SOD, and PX (7 days after the sowing), and significantly higher total phenolic content (10 days after sowing). The level of the GSH molecule production differed between the three tested plant species. In sunflower and maize leaves, increased production of GSH was detected at 14 days after the sowing, while, in case of soybean leaves, the production decreased. The differences in the GSH production among plant species were previously reported [27], where the lower production of GSH compared with the control in *Lactuca sativa* cv. Phillipus and a higher production compared with the control in *Brassica oleracea* cv. Bronco were observed. The obtained results showed that the GSH production is in correlation with the MDA production. Specifically, in sunflower and maize leaves, high levels of the GSH and MDA production were detected at 14 days after the sowing, while, in soybean leaves, the levels were low. The reason lies in the fact that the GSH has the role to reduce the occurred lipid peroxides, after which it oxidizes to the GSSG [28].

It was reported that the stress-sensitive plants under the stress conditions accumulate flavonoids, i.e., one group of the plant phenolics which are effective scavengers of ROS and flavonoid productions under stress conditions. This represents negative correlations with an increase in the antioxidant enzyme activity [29]. A significantly higher accumulation of the MDA in sunflower leaves points to the fact that allelopathy-provoked stress by *A. trifida* plants was strong enough and the scavenging effects of the antioxidant enzymes, with decreased activity, could not prevent oxidative burst and induction of the lipid peroxidation.

The maize plants were mildly affected; an increase in LP intensity was recorded at 10 days after the sowing, but at 14 days after the sowing, there was no significant increase in the LP intensity. On the other hand, soybean plants were not affected by the presence of *A. trifida* plants. The amount of MDA was lower in the treatments compared with the control at 14 days after the sowing. The obtained results for the MDA content are in accordance with the results for the activity of antioxidant enzymes when an increase in the activity of antioxidant enzymes (CAT, SOD and PX) was detected. The increased activity of the antioxidant enzymes in soybean and maize plants suggests that the defensive system of the plants prevailed. It was reported that the presence of the neighboring weeds caused an accumulation of the hydrogen peroxide (H_2O_2) and the reduction in anthocyanin content in the first leaf of the maize seedlings [30].

Total phenols were produced in higher amounts in all of the examined plants, but the total tannins, which are antioxidants, had lower production compared with the control. This could be correlated with the increased level of the lipid peroxidation in sunflower and maize, since there are bigger amounts of phenolic compounds without the antioxidant role.

In this research, the sunflower was the most sensitive species to the *A. trifida* allelochemicals among the investigated crops. The other two tested crops showed a different sensitivity to *A. trifida*. The soybean did not show any sensitivity, while maize showed sensitivity only at 10 days after the sowing. These findings are in agreement with the results of the study which reported that different crops showed a different sensitivity to the common ragweed (*Ambrosia artemisiifolia* L.) [31]. Among the tested crops (alfalfa, barley, maize, lettuce, tomato, and wheat), the tomato was the most sensitive indicator species to the *A. artemisiifolia* allelopathic residues.

4. Materials and Methods

4.1. The Validation of the Method

The LC-MD/MS validation parameters were set in accordance with the AOAC guidelines [32]. The validation parameters are shown in Table 6. The validation parameters included retention time (Rt) (expressed in min), precursor product ion, the correlation coefficient (R^2), repeatability (expressed in %RSD), as well as the limit of quantification (LOQ) (expressed in µg/kg).

Table 6. The LC-MS/MS validation parameters.

Phenolic Compounds	Rt (min)	Precursor Product Ion	R^2	Repeatability (RSD, %)	LOQ (µg/kg)
gallic acid	4.12	169→125	0.9967	9.5	0.1
ferulic acid	12.62	193→134 193→177.5	0.9988	12.1	0.1
2-hydroxycinnamic acid	7.25	163→117 163→119	0.9954	5.9	0.1
trans-cinnamic acid	13.27	147→147	0.9816	8.1	0.1
caffeic acid	10.86	179→135	0.9995	11.7	0.1
p-coumaric acid	12.20	163→93 163→119	0.9990	8.6	0.1
chlorogenic acid	10.05	353→191	0.9990	9.2	0.1
quercetin	15.13	301→151 301→179	0.9969	7.4	0.1
(+)-catechin	9.79	289→205 289→245	0.9980	4.3	0.1
protocatechuic acid	7.53	153→109	0.9995	4.3	0.1
p-hydroxybenzoic acid	9.62	137→93	0.9980	9.2	0.1
vanillic acid	10.68	167→108	0.9996	11.7	0.1
epicatechin	11.14	189→245	0.9986	5.9	0.1
syringic acid	11.17	197→182	0.9998	4.3	0.1
kaempferol	15.65	285→169 285→285	0.9968	4.2	0.1

4.2. Plant Material and Plant Extraction

The giant ragweed was collected in Kosančić village in Serbia (19°28′30.07″ E, 45°30′30.20″ N) in the six-leaf stage in May 2017. The plants were dried at 30 °C for two weeks. Dried leaves were powdered in the mill and stored at +4 °C.

The powdery material (1 g) was soaked in 10 mL of solvent (methanol and HPLC—grade water) and sonicated for 60 min at 55 °C. The supernatant was removed after the centrifugation (4000 rpm for 5 min). The aliquot was evaporated to dryness and reconstituted in 0.5 mL of the mobile phase. After that, the extract was ready for the LC-MS/MS analyses. The method was performed with the modifications compared with the previous studies [33,34].

4.3. *LC-MS/MS Analysis*

The LC was performed with an Agilent 1200 HPLC system equipped with a G1379B degasser, a G1312B binary pump, a G1367D autosampler, and a G1316B column oven. The chromatographic separation was achieved by the Zorbax Eclipse XDB C18 column (150 × 4.6 mm, 1.8 μm) maintained at 30 °C. The mobile phases were 0.1% formic acid in methanol (solvent A) and 0.1% formic acid in Milli-Q water (solvent B). The gradient was 0 min (80% B), 10 min (50% B), 20 min (5% B), 24 min (0% B), 25 min (80% B), with the flow rate of 0.6 mL/min. For the MS analysis, an Agilent 6410 Triple-Quad LC/MS system was applied. Agilent MassHunter data acquisition, qualitative analysis, and quantitative analysis software were used for method development and data acquisition [34]. All of the used solvents were of a chromatography grade and were obtained from J. T. Baker (Deventer, The Netherlands). The gallic acid, ferulic acid, 2-hydroxycinnamic acid, *trans*-cinnamic acid, caffeic acid, p-coumaric acid, chlorogenic acid, quercetin, (+)-catechin, protocatechuic acid, p-hydroxybenzoic acid, vanillic acid, epicatechin, syringic acid, and kaempferol were used as analytical standards. The stock solutions of individual phenol compound were prepared at the concentration of 1.0 mg/mL in methanol. The stock mixture standard (working solution containing all 15 phenol compounds) was obtained by mixing and diluting the stock standards with a mobile phase in the final concentration of 100 μg/mL. The composite mixtures of all phenolic compounds at the appropriate concentrations were used to spike samples in the data validation settings. The acetic acid was of a *p.a.* grade (Carl Roth).

4.4. *Crop Plant Growth*

A competition trial was conducted in vitro at the Laboratory of Biochemistry, Faculty of Agriculture, Novi Sad, Serbia. The crop (maize (*Zea mays* L.), sunflower (*Helianthus annuus* L.), soybean (*Glycine max* (L.) Merr.), and the weed (*Ambrosia trifida* L.) seeds were surface-sterilized with 3% H_2O_2 (v/v) and washed with deionised water. The seeds of crops and *A. trifida* were sown in pots 40 × 30 cm, at a three-term ratio of 1:3, 1:1, and 3:1 (weed:crop), for each crop separately, while the control pots contained crops only. The sowing substrate consisted of garden humus and sterilized sand at a ratio 2:1. After the sowing, the pots were covered with aluminum foil and put in a growing chamber under the controlled conditions (28 °C, 60% relative humidity, a photoperiod of 18 h, and a light intensity of 10.000 lx). The irrigation was carried out with the sterilized water every day. The evaluation was conducted at 7, 10, and 14 days after the emergence.

4.5. *Determination of Total Phenolics and Tannins*

The air-dried crop leaves (from each growth condition), collected at 7, 10, and 14 days after the emergence, were extracted with 10 mL of 70% ethanol. After 24 h, the extracts were filtered through Whatman No. 4 filter paper and stored at 4 °C until analyzed.

The total phenolics (TP) and tannins (TT) were determined according to the Folin–Ciocalteu method [35]. The leaf extracts were mixed with deionized water, 20% sodium carbonate, and Folin–Ciocalteu reagent diluted with distilled water in proportion 1:2. The absorbance of the reaction mixture was measured after incubation at the room temperature for 30 min at 720 nm using an UV/VIS spectrophotometer (Thermo Scientific Evolution 220, Waltham, MA, USA). The standard curve for TP and TT contents was plotted using gallic acid (GA) solution (0.1–2.0 mg/mL). The data were expressed as mg of gallic acid equivalent per gram of dry weight (mg GAE/g DW).

4.6. Determination of the Oxidative Stress Parameters

For the determination of the oxidative stress parameters, i.e., activity of antioxidant enzyme activity, content of reduced glutathione (GSH), and intensity of lipid peroxidation, 2 g of fresh crop plant material (leaves from each growth condition: control and three ratios 1:3, 1:1, 3:1 (weed:crop)), collected at 7, 10, and 14 days after the emergence, were crushed and homogenized in 10 mL of phosphate buffer (0.1 M, pH 7.0) and prepared in-house. After centrifugation, clear supernatants were used for further biochemical analyses. Biochemical assays were carried out spectrophotometrically using an UV/VIS spectrophotometer (Thermo Scientific Evolution 220 (USA)).

The catalase (CAT) (EC 1.11.1.6) activity was determined according to the following method [36]. The decomposition of the H_2O_2 was followed by a decrease in absorbance at 240 nm. The enzyme extract was added to the assay mixture containing 50 mM of potassium phosphate buffer (pH 7.0) prepared in-house and 10 mM of H_2O_2. The activity of the enzyme was expressed as U per gram of fresh weight (U/g FW).

The assay of the superoxide dismutase (SOD) (EC 1.15.1.1) activity [37] is based on the ability of the enzyme extracts to inhibit the photochemical reduction in the nitro blue tetrazolium (NBT) chloride. The reaction medium was prepared by mixing 50 mM of phosphate buffer (pH 7.8), 75 μM of NBT, 13 mM of L-methionine, 0.1 mM of EDTA, 2 μM of riboflavin, and enzyme extract. It was kept under a fluorescent lamp for 30 min, and the absorbance was read at 560 nm. One unit of the SOD activity was defined as the amount of enzymes required to inhibit reduction in NBT by 50%. The activity of the enzyme was expressed as U per gram of fresh weight (U/g FW).

The peroxidase (PX) (EC 1.11.1.7) activity was measured using pyrogallol as a substrate. This method [38] is based on the purpurogallin content measurement, i.e., a product of the pyrogallol oxidation. The enzyme extract was added to the assay mixture containing 180 mM of pyrogallol and 2 mM of H_2O_2. The absorbance was recorded at 430 nm. The activity of the enzyme was expressed as U per gram of fresh weight (U/g FW).

The reduced glutathione (GSH) was determined [39] and expressed as μmol GSH per gram of fresh weight (μmol GSH/g FW).

The intensity of the lipid peroxidation (LP) was determined using the thiobarbituric acid (TBA) test at 532 nm [34]. The enzyme extract was incubated with 20% trichloroacetic acid (TCA) containing 0.5% thiobarbituric acid for 40 min at 95 °C. The reaction was stopped by cooling on ice for 10 min, after which the product was centrifuged at 10,000 rpm for 15 min. The total amount of TBA-reactive substances was given as nmol of malondialdehyde (MDA) equivalents per gram of fresh weight (nmol MDA/g FW).

4.7. Statistical Analysis

All measurements were performed in triplicates. The values of the biochemical parameters were expressed as a mean ± standard error of mean and tested by ANOVA, followed by a comparison of the means by Duncan's multiple range test ($p < 0.05$). The data were analyzed using STATISTICA for Windows, version 11.0. Comparable percentage was calculated with the formula:

$$\Delta\ (\%) = (100 \times \text{sample}/\text{control}) - 100$$

5. Conclusions

According to the obtained results for the oxidative stress parameters, it can be concluded that the tested crops showed different sensitivity to *A. trifida*. The highest amount of MDA was detected in sunflower leaves, particularly 14 days after the sowing at the weed:crop ratio of 3:1, thus the sunflower was the most sensitive crop tested. Maize showed mild sensitivity, while the soybean did not show sensitivity in these conditions.

Author Contributions: Conceptualization, J.Š.; methodology, Đ.M. and D.P.; software, N.S.; validation, N.S. and V.B.; formal analysis, V.B. and G.V.; investigation, J.Š. and M.C.; resources, M.P.; data curation, J.Š.; writing—original draft preparation, J.Š.; writing—review and editing, B.K.; visualization, M.C.; supervision, B.K.; project administration, J.Š.; funding acquisition, B.K. All authors have read and agreed to the published version of the manuscript.

Funding: This research was supported by the the Ministry of Education, Science and Technological Development of the Republic of Serbia, grant number: 451-03-9/2021-14/200117.

Institutional Review Board Statement: Not applicable.

Informed Consent Statement: Not applicable.

Data Availability Statement: The data is contained within this manuscript. All data, tables and figures in this manuscript are original.

Acknowledgments: The authors would like to acknowledge Tijana Stojanović, Teaching Assistant, Faculty of Agriculture, University of Novi Sad, for the support and the preparation of the manuscript.

Conflicts of Interest: The authors declare no conflict of interest. The funders had no role in the design of the study; in the collection, analyses, or interpretation of data; in the writing of the manuscript, nor in the decision to publish the results.

References

1. Abul-Fatih, H.A.; Bazzaz, F.A. The biology of *Ambrosia trifida* L. I. Influence of species removal on the organization of the plant community. *New Phytol.* **1979**, *83*, 813–816. [CrossRef]
2. Kong, C.H. Ecological pest management and control by using allelopathic weeds (*Ageratum conyzoides*, *Ambrosia trifida*, and *Lantana camara*) and their allelochemicals in China. *Weed Biol. Manag.* **2010**, *10*, 73–80. [CrossRef]
3. Einhellig, F.A. Allelopathy: Current status and future goals. In *Allelopathy: Organisms, Processes, and applications*; Inderjit, Dakshini, K.M.M., Einhellig, F.A., Eds.; ACS Symposium Series; American Chemical Society: Washington, DC, USA, 1994; pp. 1–24.
4. Kamal, J. Allelopathy: A brief review. *Jordan J. Appl. Sci.* **2020**, *9*, 1–12.
5. Šežiene, V.; Baležentienė, L.; Maruška, A. Identification and allelochemical activity of phenolic compounds in extracts from the dominant plant species established in clear-cuts of Scots pine stands. *iFor.-Biogeosci. For.* **2017**, *10*, 310. [CrossRef]
6. Inzé, D.; Montagy, M.V. Oxidative stress in plants. *Curr. Opin. Biotechnol.* **1995**, *6*, 153–158. [CrossRef]
7. Gniazdowska, A.; Bogatek, R. Allelopathic interactions between plants. Multi site action of allelochemicals. *Acta Physiol. Plant.* **2005**, *27*, 395–407. [CrossRef]
8. Chaki, M.; Begara-Morales, J.; Barroso, J. Oxidative stress in plants. *Antioxidants* **2020**, *9*, 481–484. [CrossRef]
9. Khadem, S.; Marles, R.J. Monocyclic phenolic acids; hydroxy- and polyhydroxybenzoic acids: Occurrence and recent bioactivity studies. *Molecules* **2010**, *15*, 7985–8005. [CrossRef] [PubMed]
10. Ćetković, G.S.; Čanadanović-Brunet, J.M.; Djilas, S.M.; Tumbas, V.T.; Markov, S.L.; Cvetković, D.D. Antioxidant potential, lipid peroxidation inhibition and antimicrobial activities of *Satureja montana* L. subsp. *kitaibelii* extracts. *Int. J. Mol. Sci.* **2007**, *8*, 1013–1027. [CrossRef]
11. Smirnof, N. *Antioxidants and Reactive Oxygen Species in Plants*; John Wiley & Sons: Oxford, UK, 2005; pp. 1–320.
12. Horvath, E.; Pal, M.; Szalai, G.; Paldi, E.; Janda, T. Exogenous 4-hydroxybenzoic acid and salicylic acid modulate the effect of short term drought and freezing stress on wheat plants. *Biologia Plant.* **2007**, *51*, 480–487. [CrossRef]
13. Chong, K.P.; Rossall, S.; Atong, M. In Vitro antimicrobial activity and fungitoxicity of syringic acid, caffeic acid and 4-hydroxybenzoic acid against *Ganoderma boninense*. *J. Agric. Sci.* **2009**, *1*, 15–20. [CrossRef]
14. Heleno, S.A.; Martins, A.; Queiroz, M.J.R.P.; Ferreira, I.C.F.R. Bioactivity of phenolic acids: Metabolites *versus* parent compounds: A review. *Food Chem.* **2015**, *173*, 501–513. [CrossRef] [PubMed]
15. Li, Z.H.; Wang, Q.; Ruan, X.; Pan, C.; Jiang, D.A. Phenolics and plant allelopathy. *Molecules* **2010**, *15*, 8933–8952. [CrossRef] [PubMed]
16. Kong, C.H.; Wang, P.; Xu, X. Allelopathic interference of *Ambrosia trifida* with wheat (*Triticum aestivum*). *Agric. Ecosyst. Environ.* **2007**, *119*, 416–420. [CrossRef]
17. Diaz Vivancos, P.; Wolff, T.; Markovic, J.; Pallardó, F.V.; Foyer, C.H. A nuclear glutathione cycle within the cell cycle. *Biochem. J.* **2010**, *431*, 169–178. [CrossRef]
18. Bounous, G.; Molson, J.H. The antioxidant system. *Anticancer Res.* **2003**, *23*, 1411–1416.
19. Einhellig, F.A.; Eckrich, P.C. Interactions of temperature and ferulic acid stress on grain sorghum and soybeans. *J. Chem. Ecol.* **1984**, *10*, 161–170. [CrossRef]
20. Kato-Noguchi, H.; Kurniadie, D. Allelopathy of *Lantana camaraas* an invasive plant. *Plants* **2021**, *10*, 1028–1038. [CrossRef]
21. Savić, B.G.; Stanković, D.M.; Živković, S.M.; Ognjanović, M.R.; Tasić, G.S.; Mihajlović, I.J.; Brdarić, T.P. Electrochemical oxidation of a complex mixture of phenolic compounds in the base media using PbO_2-GNRs anodes. *Appl. Surf. Sci.* **2020**, *529*, 147120. [CrossRef]

22. Kiprovski, B.; Miladinović, J.; Koren, A.; Malenčić, Đ.; Mikulič-Petkovšek, M. Black and yellow soybean: Contribution of seed quality to oxidative stress response during plant development. *Genetika* **2019**, *51*, 495–510. [CrossRef]
23. Rasmussen, J.A.; Einhellig, F.A. Synergistic inhibitory effects of *p*-coumaric and ferulic acids on germination and growth of grain sorghum. *J. Chem. Ecol.* **1977**, *3*, 197–205. [CrossRef]
24. Wang, P.; Liang, W.; Kong, C.; Jiang, Y. Allelopathic potential of volatile allelochemicals of *Ambrosia trifida* L. on other plants. *Allelopath. J.* **2005**, *15*, 131–136.
25. Macías, F.A.; Varela, R.M.; Simonet, S.M.; Cutler, H.G.; Cutler, S.J.; Eden, M.A.; Hill, R.A. Bioactive carotanes from *Trichoderma virens*. *J. Nat. Prod.* **2000**, *63*, 1197–1200. [CrossRef]
26. Awasthi, R.; Bhandari, K.; Nayyar, H. Temperature stress and redox homeostasis in agricultural crops. *Front. Environ. Sci.* **2015**, *3*, 1–24. [CrossRef]
27. Barrameda-Medina, Y.; Montesinos-Pereira, D.; Romero, L.; Blasco, B.; Ruiz, J.M. Role of GSH homeostasis under Zn toxicity in plants with different Zn tolerance. *Plant Sci.* **2014**, *227*, 110–121. [CrossRef] [PubMed]
28. Lu, S.C. Glutathione synthesis. *Biochim. Biophys. Acta (BBA)–Gen. Subj.* **2013**, *1830*, 3143–3153. [CrossRef] [PubMed]
29. Agati, G.; Azzarello, E.; Pollastri, S.; Tattini, M. Flavonoids as antioxidants in plants: Location and functional significance. *Plant Sci.* **2012**, *196*, 67–76. [CrossRef] [PubMed]
30. Afifi, M.; Swanton, C. Early physiological mechanisms of weed competition. *Weed Sci.* **2012**, *60*, 542–551. [CrossRef]
31. Vidotto, F.; Tesio, F.; Ferrero, A. Allelopathic effects of *Ambrosia artemisiifolia* L. in the invasive process. *Crop Prot.* **2013**, *54*, 161–167. [CrossRef]
32. AOAC Guidelines for Single Laboratory Validation of Chemical Methods for Dietary Supplements and Botanicals. Harmonized guidelines for single laboratory validation of methods of analysis (IUPAC Technical Report). *Pure Appl. Chem.* **2002**, *74*, 835–855. [CrossRef]
33. Generalić, I.; Skorza, D.; Šurjak, J.; Možina, S.; Ljubenkov, I.; Katalinić, A.; Šimat, V.; Katalinić, V. Seasonal variations of phenolic compounds and biological properties in sage (*Salvia officinalis* L.). *Chem. Biodivers.* **2012**, *9*, 441–457. [CrossRef]
34. Malenčić, Đ.; Kiprovski, B.; Bursić, V.; Vuković, G.; Ćupina, B.; Mikic, A. Dietary phenolics and antioxidant capacity of selected legumes seeds from the Central Balkans. *Acta Aliment.* **2018**, *47*, 340–349. [CrossRef]
35. Hagerman, A.; Harvey-Mueller, I.; Makkar, H.P.S. *Quantification of Tannins in Tree Foliage—A Laboratory Manual, FAO/IAEA*; Joint FAO/IAEA Division of Nuclear Techniques in Food and Agriculture: Vienna, Austria, 2000; pp. 1–31.
36. Sathya, E.; Bjorn, M. Spectrophotometric Assays for Antioxidant Enzymes in Plants. In *Plant Stress Tolerance: Methods and Protocols*, 2nd ed.; Sathya, E., Bjorn, M., Eds.; Humana Press: New York, NY, USA, 2010; Volume 639, pp. 273–280.
37. Mandal, S.; Mitra, A.; Mallick, N. Biochemical characterization of oxidative burst during interaction between *Solanum lycopersicum* and *Fusarium oxysporum* f. sp. *lycopersici*. *Physiol. Mol. Plant Pathol.* **2008**, *72*, 56–61. [CrossRef]
38. Morkunas, I.; Gmerek, J. The possible involvement of peroxidase in defense of yellow lupine embryo axes against *Fusarium oxysporum*. *J. Plant Physiol.* **2007**, *164*, 185–194. [CrossRef] [PubMed]
39. Sedlak, J.; Lindsay, H. Estimation of total protein bound and non protein sulphydryl groups in tissue with Ellman's reagent. *Anal. Biochem.* **1968**, *25*, 192–205. [CrossRef]

MDPI

Article

Testing Virulence of Different Species of Insect Associated Fungi against Yellow Mealworm (Coleoptera: Tenebrionidae) and Their Potential Growth Stimulation to Maize

Eva Praprotnik *, Jernej Lončar and Jaka Razinger

Plant Protection Department, Agricultural Institute of Slovenia, 1000 Ljubljana, Slovenia; jernej.loncar@kis.si (J.L.); jaka.razinger@kis.si (J.R.)
* Correspondence: eva.praprotnik@kis.si

Abstract: This paper investigates 71 isolates of two genera of entomopathogens, *Metarhizium* and *Beauveria*, and a biostimulative genus *Trichoderma*, for their ability to infect yellow mealworms (*Tenebrio molitor*) and to stimulate maize (*Zea mays*) growth. Fungal origin, host, and isolation methods were taken into account in virulence analysis as well. Isolates *Metarhizium brunneum* (1154) and *Beauveria bassiana* (2121) showed the highest mortality (100%) against *T. molitor*. High virulence seems to be associated with fungi isolated from wild adult mycosed insects, meadow habitats, and Lepidopteran hosts, but due to uneven sample distribution, we cannot draw firm conclusions. *Trichoderma atroviride* (2882) and *Trichoderma gamsii* (2883) increased shoot length, three *Metarhizium robertsii* isolates (2691, 2693, and 2688) increased root length and two *M. robertsii* isolates (2146 and 2794) increased plant dry weight. Considering both criteria, the isolate *M. robertsii* (2693) was the best as it caused the death of 73% *T. molitor* larvae and also significantly increased maize root length by 24.4%. The results warrant further studies with this isolate in a tri-trophic system.

Keywords: entomopathogenic fungi; *Tenebrio molitor*; virulence; pathogenicity; growth stimulation; plant–microbe–pest interactions; rhizosphere competence

Citation: Praprotnik, E.; Lončar, J.; Razinger, J. Testing Virulence of Different Species of Insect Associated Fungi against Yellow Mealworm (Coleoptera: Tenebrionidae) and Their Potential Growth Stimulation to Maize. *Plants* **2021**, *10*, 2498. https://doi.org/10.3390/plants10112498

Academic Editors: Špela Mechora and Dragana Šunjka

Received: 14 October 2021
Accepted: 15 November 2021
Published: 18 November 2021

1. Introduction

Entomopathogenic fungi are primarily known for their ability to parasitize insects and kill or severely harm them [1–3]. Fungi of the hypocrealean family Cordycipitaceae include important entomopathogens, of which certain species of *Metarhizium*, *Beauveria* and *Isaria* are most studied. The use of these typically facultative parasitic fungi as biopesticides is prevalent due to the wide range of target hosts and their ability to complete their life cycles also independently from insect hosts [4]. Entomopathogens can also colonize the rhizosphere and plant tissues as endophytes and act as plant growth promoters [5,6]. The occurrence of entomopathogenic endophytes is reported in more than 50 host plants, including cereals, legumes, oil and fiber crops, herbs, deciduous and coniferous trees, and others (reviewed in [7]). Their association with plants allows them to interact closely with insect herbivores in a tri-trophic system [8,9], ultimately impacting economic aspects, particularly in agriculture [10]. However, to design successful pest management strategies, it is necessary to fully understand the ecological role of implemented microbes.

Coating seeds with plant beneficial entomopathogens is a viable method for delivering microbes to germinating crops. It can be a cost-reducing alternative to soil inoculation, which requires large amounts of microbial inoculum, which could be an economic disadvantage if applied on a larger scale [11]. Seed coating can improve plant defenses by adding a certain concentration of beneficial organisms to the soil in the immediate vicinity of the germinated seed, which promotes seedling development and acts against plant pathogens or insect pests. For example, reducing wireworm pressure during the first three weeks of maize growth can significantly minimize crop loss [12,13]; therefore, a specific treatment that would enhance seed germination or speed up early-stage growth would be highly beneficial as it would give the plant an advantage to oppose soil pests. Secondly, the level of defense can be improved by direct insect interfering activities of entomopathogens. If present as endophytes and rhizosphere colonizers, they could directly protect plants at later physiological stages. Coating bean *Phaseolus vulgaris* L. seeds with *Beauveria bassiana* (Balsamo) Vuillemin and *Metarhizium robertsii* J.F.Bisch., Rehner and Humber significantly reduced the population of the spider mite *Tetranychus urticae* Koch while improving plant growth within five weeks after inoculation [14]. Similar effects were observed when maize and tobacco seeds were coated with *M. robertsii* [1,15] and white jute seeds with *B. bassiana* [16].

However, biotic and abiotic conditions, as well as genotypic and phenotypic plasticity of host plants and fungi, can significantly affect the insect associated fungi–host interactions [17]. Evolutionary theory supports the important role of grass endophytes in defense against herbivores in a mutualistic manner. However, this relationship is not fixed and may, under certain conditions, turn into a neutral or even antagonistic interaction [18]. As plants influence the chemical and nutritional properties of their rhizospheres, the entomopathogenic fungi living there are under strong selection pressure to utilize the specific rhizodeposits and might be subject to habitat selection rather than the presence/absence of an insect host [19,20]. Therefore, these factors should be considered, especially at tri (multi)-trophic levels when considering entomopathogens for commercial use.

Our question was whether the origin of the fungus, host, or the isolation method could affect the fungal virulence of the fungus and whether biostimulative properties are common among highly virulent fungal isolates. Therefore, we looked at 66 strains belonging to the entomopathogenic genera *Metarhizium* and *Beauveria*, as well as five from genus *Trichoderma*, primarily known as a biostimulative fungus [21,22], however, also a proven insects' facultative pathogen [23–26] in order to evaluate their ability to infect yellow mealworms (*Tenebrio molitor* Linnaeus, 1758) and stimulate maize (*Zea mays* L.) growth. Mealworms are known for their susceptibility to entomopathogenic fungal infection and are a suitable test organism for assessing fungal virulence. Although they are mostly known as storage pests, their natural environment is dark and moist earth's floor, mostly under rocks or in leaf-litter [27]. Maize served as a model plant because most isolates were isolated from maize fields or from the rhizosphere of wild Poaceae species growing in dry Karst meadows.

2. Results

2.1. Virulence Bioassay

Altogether 71 fungal isolates were analyzed for their virulence against *T. molitor* (Table 1). All isolates, with the exception of *Trichoderma atroviride* P.Karst. (number of strains tested: n = 2), *Trichoderma harzianum* Rifai (n = 1) and *Trichoderma gamsii* Samuels and Druzhin. (n = 1), showed pathogenicity against *T. molitor*. After 14 days *Metarhizium brunneum* Petch (n = 4) caused mortality ranging from 21.43 to 100.00%, while *M. robertsii* (n = 53) caused mortality ranging from 5.27 to 84.62%, and *Metarhizium guizhouense* Q.T. Chen and H.L. Guo (n = 3) caused mortality ranging from 3.70 to 32.14%. The isolates of *B. bassiana* (n = 5) caused mortality ranging from 53.33 to 100.00%. Isolates *B. bassiana* (2121) and *M. brunneum* (1154) had the highest Abbott's corrected mortality after 7- and 14-days post inoculation. Three isolates of *B. bassiana*, two of *M. brunneum*, and two of *M. robertsii* caused mortality of at least 75% after 14 days.

Table 1. Origin and virulence of selected fungal isolates against larvae of *Tenebrio molitor*. ACM–Abbott's corrected mortality 7 and 14 days after inoculation; LT50–Median lethal time of *T. molitor* in days; Green fill indicates most virulent isolates with ACM 14 days after inoculation >75%, yellow fill indicates moderately virulent isolates with ACM 14 days after inoculation between 50% and 75%; * Asterisk indicates significance for survival curve analysis; Italic font indicates unreliable ACM results due to high control mortality (sterile 0.1% Tween 80).

Isolate	Taxon	Habitat or Origin	Isolation Type/ Host Organism	Host Developmental Stage	Host Origin	ACM 7.00 (%)	ACM 14.00 (%)	LT_{50} (d)
1154	*MB*	soil	*Galleria mellonella*	larvae	reared	61.54	100.00	6.00 *
1868	*MB*	meadow	*Agriotes* sp.	adult	wild	26.92	86.96	8.00 *
2121	*BB*	cauliflower field	Curculionidae	adult	wild	65.38	100.00	6.00 *
2631	*MR*	maize field	*Tenebrio molitor*	larvae	reared	10.34	52.00	12.5 *
2632	*MR*	maize field	*Tenebrio molitor*	larvae	reared	10.34	32.00	
2245	*MR*	maize field	*Tenebrio molitor*	larvae	reared	6.90	28.00	
2246	*MR*	maize field	*Tenebrio molitor*	larvae	reared	0.00	20.00	
2215	*MR*	maize field	*Tenebrio molitor*	larvae	reared	11.11	37.50	14.00
2216	*MR*	maize field	*Tenebrio molitor*	larvae	reared	7.41	54.17	10.00 *
2299	*BB*	meadow	*Galleria mellonella*	larvae	reared	18.52	54.17	10.00 *
2300	*BB*	meadow	*Galleria mellonella*	larvae	reared	33.33	79.17	10.00 *
2635	*MR*	maize field	*Tenebrio molitor*	larvae	reared	21.43	65.38	11.50 *
2637	*MR*	maize field	*Tenebrio molitor*	larvae	reared	17.86	84.62	11.00 *
2641	*MR*	maize field	*Tenebrio molitor*	larvae	reared	21.43	65.38	11.00 *
2697	*ND*	maize field	*Tenebrio molitor*	larvae	reared	*−18.18*	*−23.08*	
2298	*BB*	meadow	*Galleria mellonella*	larvae	reared	*42.11*	*88.24*	6.00 *
2243	*MR*	maize field	*Tenebrio molitor*	larvae	reared	0.00	37.50	14.00
2151	*MR*	maize field	*Tenebrio molitor*	larvae	reared	*−18.18*	*20.00*	13.00
2703	*MB*	soil	ND	ND	ND	0.00	21.43	*
2009	*MR*	soil	selective medium	-	-	10.34	32.14	14.00 *
2010	*MG*	soil	selective medium	-	-	3.45	32.14	*
2011	*MR*	soil	*Galleria mellonella*	larvae	reared	16.67	31.03	
2686	*MR*	maize field	*Tenebrio molitor*	larvae	reared	7.41	43.48	13.00 *
2687	*MR*	maize field	*Tenebrio molitor*	larvae	reared	11.11	56.52	11.00 *
2690	*MB*	soil	ND	ND	ND	7.41	47.83	12.00 *
2692	*MR*	maize field	*Diabrotica v. virgifera*	adult	wild	7.41	39.13	14.00 *
2152	*MR*	maize field	*Tenebrio molitor*	larvae	reared	−3.45	26.92	*
2146	*MR*	maize field	*Tenebrio molitor*	larvae	reared	20.00	54.17	11.00 *
2147	*MR*	maize field	*Tenebrio molitor*	larvae	reared	23.33	58.33	11.00 *
2251	*MR*	maize field	*Tenebrio molitor*	larvae	reared	16.67	75.00	11.00 *
2789	*MR*	maize field	*Diabrotica v. virgifera*	larvae	wild	16.67	41.67	13.00 *
2793	*MR*	maize field	selective medium	-	-	20.00	70.83	11.00 *
2794	*MR*	maize field	selective medium	-	-	10.00	20.83	
2795	*MR*	maize field	selective medium	-	-	36.67	66.67	11.00 *
2645	*MR*	maize field	*Tenebrio molitor*	larvae	reared	3.33	26.92	
2691	*MR*	blueberry field	*Tenebrio molitor*	larvae	reared	6.67	53.85	11.00 *
2693	*MR*	blueberry field	*Tenebrio molitor*	larvae	reared	6.67	73.08	11.00 *
2790	*MR*	maize field	*Diabrotica v. virgifera*	larvae	wild	0	19.23	14.00
2634	*MR*	maize field	*Tenebrio molitor*	larvae	reared	10.34	8.33	
2214	*MR*	maize field	*Tenebrio molitor*	larvae	reared	10.34	33.33	14.00
2702	*MR*	maize field	*Tenebrio molitor*	larvae	reared	0	8.33	
2791	*MR*	maize field	selective medium	-	-	37.93	62.50	12.00 *
2250	*MG*	maize field	*Tenebrio molitor*	larvae	reared	3.33	3.70	
2685	*MR*	maize field	*Tenebrio molitor*	larvae	reared	3.33	14.81	
2640	*MR*	maize field	*Tenebrio molitor*	larvae	reared	17.24	29.63	
2694	*MR*	strawberry field	*Tenebrio molitor*	larvae	reared	−3.45	25.93	
2695	*MR*	strawberry field	*Tenebrio molitor*	larvae	reared	6.90	22.22	
2788	*MR*	maize field	*Diabrotica v. virgifera*	larvae	wild	10.34	29.63	
2792	*MR*	maize field	selective medium	-	-	3.45	29.63	
2796	*MR*	maize field	selective medium	-	-	3.45	22.22	
2688	*MR*	maize field	*Tenebrio molitor*	larvae	reared	3.33	5.27	*
2154	*MR*	maize field	*Tenebrio molitor*	larvae	reared	3.57	*−23.81*	
2153	*MR*	maize field	*Tenebrio molitor*	larvae	reared	3.33	23.33	*
2217	*MR*	maize field	*Tenebrio molitor*	larvae	reared	3.33	10.00	
2698	*MR*	basil leaf	unknown larva	larvae	wild	3.33	13.33	

Table 1. *Cont.*

Isolate	Taxon	Habitat or Origin	Isolation Type/ Host Organism	Host Developmental Stage	Host Origin	ACM 7.00 (%)	ACM 14.00 (%)	LT_{50} (d)
2699	*MR*	blueberry field	*Tenebrio molitor*	larvae	reared	3.33	10.00	
2700	*MR*	maize field	*Tenebrio molitor*	larvae	reared	10	20.00	
2701	*MR*	maize field	*Tenebrio molitor*	larvae	reared	3.33	13.33	
2239	*MR*	maize field	*Tenebrio molitor*	larvae	reared	0	20.00	
2704	*BB*	meadow	unknown larva	larvae	wild	0	53.33	14.00 *
2247	*MG*	maize field	*Tenebrio molitor*	larvae	reared	3.33	13.33	
2752	*TA*	decaying corn ear	natural substratum	-	-	0	0.00	
2815	*TB*	maize field	selective medium	-	-	0	3.33	
2878	*TH*	maize field	selective medium	-	-	0	0.00	
2882	*TA*	maize field	selective medium	-	-	0	0.00	
2883	*TG*	maize field	selective medium	-	-	0	0.00	
2150	*MR*	maize field	*Galleria mellonella*	larvae	reared	16.67	46.67	*
2240	*MR*	maize field	*Tenebrio molitor*	larvae	reared	23.33	50.00	14.00 *
2148	*MR*	maize field	*Galleria mellonella*	larvae	reared	23.33	53.33	14.00 *
2636	*MR*	maize field	*Tenebrio molitor*	larvae	reared	16.67	63.33	12.50 *
2642	*MR*	maize field	*Tenebrio molitor*	larvae	reared	3.33	23.33	*
Actara 25 WG	-	-	-	-	-	11.86	47.37	13.50 *
Mycotal	*LM*	-	-	-	-	−1.69	−3.51	
Force	-	-	-	-	-	1.69	1.75	
Met52 EC	*MB*	-	-	-	-	2.54	11.18	

Note: Actara 25 WG–insecticide based on the active ingredient Thiamethoxam (25% *w*/*w*); Mycotal–biological insecticide based on the active ingredient *L. muscarium* strain Ve6; Force 1.5G–insecticide based on the active ingredient Tefluthrin (0.15% *w*/*w*); Met52 EC–biological insecticide based on the active ingredient *M. brunneum* strain F52. *MB: Metarhizium brunneum; MR: Metarhizium robertsii; MG: Metarhizium guizhouense; BB: Beauveria bassiana; LM: Lecanicillium muscarium; TA: Trichoderma atroviride; TB: Trichoderma brevicompactum; TG: Trichoderma gamsii; TH: Trichoderma harzianum; ND*: No data.

2.2. Influence of Fungal Origin and Isolation Method on Mortality Rate

For exploratory data analysis, we illustrated different parameters in correlation with ACM of 67 fungal isolates (Figure 1). The results indicate positive correlation between ACM and the genus *Beauveria*, adult Lepidoptera insect host, and meadows. Conversely, a negative correlation is shown between ACM and the genus *Trichoderma*.

We detected no significant difference in ACM on 14th day between isolates isolated from a wild host versus a reared host ($F_{1,50} = 0.897$, $p = 0.348$), from a live organism versus a selective medium ($\chi^2(1) = 1.8712$, $p = 0.1713$) and from bulk soil versus rhizosphere soil ($F_{1,56} = 0.144$, $p = 0.706$). On the other hand, we detected a significant difference in ACM on 14th day between isolates of different genera ($\chi^2(2) = 18.423$, $p = 0.0001$), isolates isolated from a meadow versus a field ($F_{1,57} = 7.182$, $p = 0.0096$), and marginally significant differences in ACM of isolates from an adult insect host versus larvae ($\chi^2(1) = 4.0098$, $p = 0.0452$) and from a Lepidoptera insect host versus a Coleoptera ($\chi^2(1) = 4.1391$, $p = 0.0419$).

2.3. Growth Stimulation Bioassay

Seventy-one fungal isolates were tested for stimulation of maize growth (Table 2). The average number of conidia per maize seed was $2.43 \times 10^6 \pm 1.99 \times 10^5$. There was a significant difference in the average number of conidia per maize seed between isolates of different genera ($\chi^2(1) = 5.2406$, $p = 0.\ 0.0221$) and between different habitats or origin of the isolate ($\chi^2(4) = 21.219$, $p = 0.0002$). Isolates that originated from maize fields had a higher number of conidia per maize seed ($p \leq 0.05$) as opposed to isolates from meadows, soil, blueberry field, or insects. Furthermore, isolates of the genus *Metarhizium* had a higher number of conidia per maize seed as opposed to genus *Beauveria*.

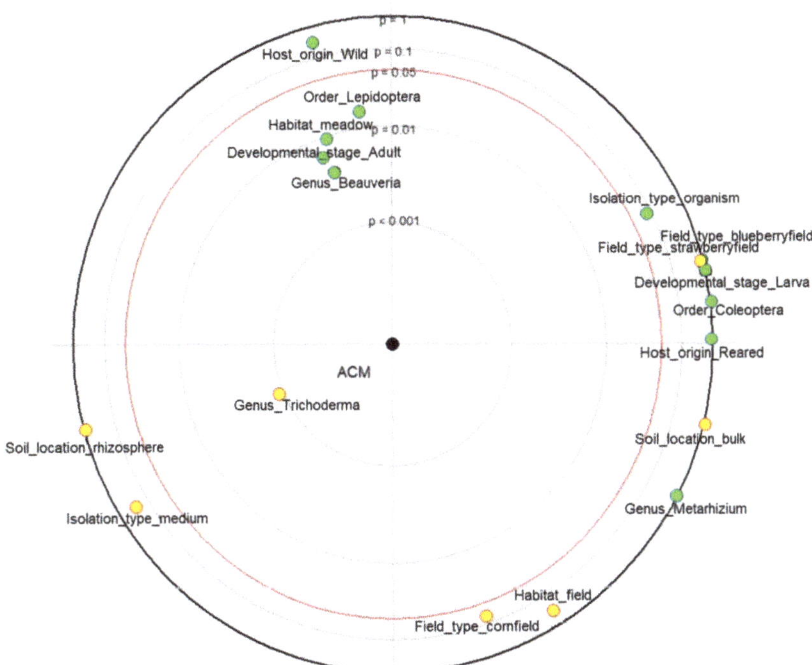

Figure 1. Correlation circle of predictor variables toward mortality rate (ACM). Green dots indicate positive correlation with the dependent variable and yellow dots indicate negative correlation with the dependent variable. Predictor variables within the red circle are significantly correlated with the dependent variable ($p < 0.05$).

Table 2. Growth stimulating effects of maize treated with selected fungal isolates and grown in twice autoclaved substrate for 21 days. Data presented are the mean values ± SE (n = 15 for Chapalu variety (3 replicates with 5 seeds each) and n = 30 for Belokranjka variety (3 replicates with 10 seeds each). Green fill indicates isolates with significant growth promoting properties, red fill indicates isolates with growth inhibitory properties and grey fill indicates a significant difference from negative control (sterile 0.1% Tween 80), $p \leq 0.05$. Striped horizontal lines separate individual experiments.

Isolate	Taxon	Maize Variety	Emergence Success [n]	Emergence Time [d]	Root Length [cm]	Shoot Length [cm]	Total Plant Length [cm]	Plant Dry Weight [g]
Control		Chapalu	**4.67 ± 0.33**	4.33 ± 0.33	22.57 ± 1.15	32.87 ± 1.15	55.43 ± 2.45	0.30 ± 0.04
1154	*MB*	Chapalu	4.67 ± 0.33	4.13 ± 0.13	21.67 ± 1.63	30.90 ± 0.78	52.57 ± 2.28	0.29 ± 0.02
1868	*MB*	Chapalu	4.67 ± 0.33	4.33 ± 0.33	21.50 ± 0.15	29.60 ± 1.19	51.09 ± 1.21	0.28 ± 0.02
2121	*BB*	Chapalu	5.00 ± 0.00	4.40 ± 0.23	21.50 ± 1.10	32.27 ± 0.74	53.75 ± 1.73	0.30 ± 0.01
Control		Chapalu	**5.00 ± 0.00**	**4.27 ± 0.07**	**20.97 ± 0.45**	**29.97 ± 0.13**	**50.95 ± 0.37**	**0.27 ± 0.02**
2631	*MR*	Chapalu	4.67 ± 0.33	4.38 ± 0.32	21.23 ± 0.47	26.37 ± 0.43	47.58 ± 0.88	0.24 ± 0.01
2632	*MR*	Chapalu	4.33 ± 0.33	4.40 ± 0.31	21.40 ± 1.51	27.97 ± 0.65	49.37 ± 1.69	0.26 ± 0.01
2245	*MR*	Chapalu	5.00 ± 0.00	4.53 ± 0.07	18.80 ± 0.56	27.23 ± 1.43	46.04 ± 1.62	0.22 ± 0.02
2246	*MR*	Chapalu	4.67 ± 0.33	4.17 ± 0.17	18.50 ± 0.40	28.10 ± 0.81	46.56 ± 1.19	0.23 ± 0.01
Control		Chapalu	**5.00 ± 0.00**	**4.93 ± 0.07**	**21.23 ± 0.64**	**31.47 ± 0.83**	**52.70 ± 1.02**	**0.27 ± 0.01**
2215	*MR*	Chapalu	4.67 ± 0.33	4.43 ± 0.03	21.55 ± 0.75	29.30 ± 0.82	50.85 ± 0.08	0.25 ± 0.01
2216	*MR*	Chapalu	4.67 ± 0.33	4.57 ± 0.23	22.39 ± 0.55	29.82 ± 0.32	52.21 ± 0.73	0.25 ± 0.01
2299	*BB*	Chapalu	4.33 ± 0.33	4.40 ± 0.31	21.53 ± 0.48	26.43 ± 1.41	47.96 ± 1.81	0.21 ± 0.01
2300	*BB*	Chapalu	4.67 ± 0.33	4.40 ± 0.23	22.25 ± 1.02	27.40 ± 1.09	49.64 ± 2.10	0.22 ± 0.01
Control		Chapalu	**5.00 ± 0.00**	**4.13 ± 0.07**	**18.60 ± 0.95**	**29.57 ± 0.78**	**48.16 ± 1.71**	**0.25 ± 0.01**
2635	*MR*	Chapalu	4.67 ± 0.33	4.37 ± 0.09	21.40 ± 1.14	28.53 ± 0.96	49.94 ± 1.57	0.23 ± 0.03
2637	*MR*	Chapalu	4.67 ± 0.33	4.62 ± 0.50	19.93 ± 0.46	28.67 ± 1.08	48.59 ± 1.51	0.23 ± 0.02
2641	*MR*	Chapalu	5.00 ± 0.00	4.40 ± 0.31	19.57 ± 0.71	28.27 ± 1.07	47.81 ± 0.70	0.22 ± 0.02
Control		Chapalu	**4.67 ± 0.33**	**4.00 ± 0.00**	**16.30 ± 0.47**	**30.30 ± 1.30**	**46.61 ± 1.78**	**0.22 ± 0.01**
2697	*ND*	Chapalu	5.00 ± 0.00	4.07 ± 0.07	17.90 ± 0.23	27.27 ± 2.83	45.20 ± 2.70	0.23 ± 0.02

Table 2. *Cont.*

Isolate	Taxon	Maize Variety	Emergence Success [n]	Emergence Time [d]	Root Length [cm]	Shoot Length [cm]	Total Plant Length [cm]	Plant Dry Weight [g]
2698	*MR*	Chapalu	4.33 ± 0.67	4.44 ± 0.08	19.13 ± 0.99	27.60 ± 1.80	46.73 ± 2.17	0.21 ± 0.01
2699	*MR*	Chapalu	5.00 ± 0.00	4.20 ± 0.00	18.07 ± 0.61	31.27 ± 0.67	49.35 ± 1.20	0.22 ± 0.01
2700	*MR*	Chapalu	4.33 ± 0.33	4.53 ± 0.15	19.17 ± 0.24	33.30 ± 1.65	52.46 ± 1.83	0.26 ± 0.01
Control		Chapalu	**5.00 ± 0.00**	**4.07 ± 0.07**	**20.07 ± 1.28**	**30.33 ± 1.21**	**50.43 ± 2.28**	**0.22 ± 0.03**
2298	*BB*	Chapalu	4.33 ± 0.33	4.85 ± 0.52	20.80 ± 1.12	30.30 ± 0.66	51.09 ± 1.80	0.26 ± 0.02
2243	*MR*	Chapalu	5.00 ± 0.00	4.93 ± 0.13	15.60 ± 1.21	27.00 ± 0.95	42.59 ± 1.96	0.21 ± 0.01
2636	*MR*	Chapalu	5.00 ± 0.00	4.80 ± 0.23	16.77 ± 0.70	28.47 ± 0.38	45.26 ± 0.91	0.20 ± 0.02
2642	*MR*	Chapalu	4.33 ± 0.33	4.50 ± 0.25	17.20 ± 0.85	34.53 ± 0.32	51.72 ± 0.73	0.25 ± 0.01
Control		Chapalu	**4.67 ± 0.33**	**4.13 ± 0.07**	**20.70 ± 0.31**	**32.93 ± 1.88**	**53.61 ± 1.80**	**0.31 ± 0.05**
2148	*MR*	Chapalu	5.00 ± 0.00	4.27 ± 0.18	21.37 ± 0.45	35.77 ± 0.47	57.15 ± 0.50	0.32 ± 0.02
2151	*MR*	Chapalu	4.67 ± 0.33	4.57 ± 0.12	19.47 ± 0.58	36.77 ± 0.78	56.22 ± 0.50	0.32 ± 0.02
2152	*MR*	Chapalu	4.33 ± 0.33	4.28 ± 0.17	21.50 ± 0.64	35.67 ± 2.11	57.19 ± 2.76	0.31 ± 0.03
2701	*MR*	Chapalu	5.00 ± 0.00	4.47 ± 0.07	20.57 ± 0.62	33.60 ± 0.55	54.15 ± 0.38	0.33 ± 0.01
Control		Chapalu	**4.67 ± 0.33**	**4.20 ± 0.12**	**20.27 ± 0.26**	**34.70 ± 0.59**	**54.98 ± 0.83**	**0.32 ± 0.01**
2703	*MB*	Chapalu	4.33 ± 0.33	4.25 ± 0.25	24.43 ± 1.56	35.00 ± 1.59	59.44 ± 0.11	0.36 ± 0.01
2009	*MR*	Chapalu	5.00 ± 0.00	4.47 ± 0.18	20.70 ± 0.46	33.97 ± 2.72	54.65 ± 2.84	0.31 ± 0.04
2010	*MG*	Chapalu	5.00 ± 0.00	4.13 ± 0.13	20.40 ± 1.68	31.53 ± 1.82	51.94 ± 3.47	0.27 ± 0.04
2011	*MR*	Chapalu	5.00 ± 0.00	4.53 ± 0.13	20.93 ± 0.94	30.83 ± 1.43	51.81 ± 2.37	0.23 ± 0.00
Control		Chapalu	**4.67 ± 0.33**	**4.43 ± 0.07**	**13.03 ± 0.03**	**15.90 ± 1.46**	**28.91 ± 1.44**	**0.15 ± 0.01**
2239	*MR*	Chapalu	5.00 ± 0.00	4.59 ± 0.21	14.63 ± 1.76	18.90 ± 0.20	33.50 ± 1.56	0.19 ± 0.01
2704	*BB*	Chapalu	3.33 ± 0.33	5.00 ± 0.25	11.10 ± 0.50	18.00 ± 1.06	29.11 ± 1.30	0.17 ± 0.00
2247	*MG*	Chapalu	3.33 ± 0.33	4.81 ± 0.13	15.33 ± 1.06	18.80 ± 0.65	34.13 ± 1.61	0.19 ± 0.01
2752	*TA*	Chapalu	3.00 ± 0.58	5.29 ± 0.14	13.93 ± 0.73	18.40 ± 1.34	32.33 ± 2.03	0.19 ± 0.03
2815	*TB*	Chapalu	3.67 ± 0.33	4.71 ± 0.22	13.33 ± 1.48	18.20 ± 1.04	31.53 ± 2.35	0.20 ± 0.02
2878	*TH*	Chapalu	3.33 ± 0.33	4.44 ± 0.34	14.70 ± 1.51	18.73 ± 0.50	33.39 ± 1.62	0.17 ± 0.01
2882	*TA*	Chapalu	3.33 ± 0.33	4.84 ± 0.30	14.27 ± 1.08	20.00 ± 0.60	34.32 ± 1.11	0.21 ± 0.01
2883	*TG*	Chapalu	3.33 ± 0.33	4.41 ± 0.12	14.67 ± 1.21	19.47 ± 0.65	34.13 ± 1.83	0.17 ± 0.03
Control		Belokranjka	**9.00 ± 0.58**	**5.19 ± 0.05**	**23.87 ± 1.58**	**32.73 ± 0.83**	**56.58 ± 2.41**	**0.31 ± 0.03**
2686	*MR*	Belokranjka	10.00 ± 0.00	5.07 ± 0.20	24.50 ± 3.10	30.97 ± 0.58	55.48 ± 2.55	0.30 ± 0.01
2687	*MR*	Belokranjka	10.00 ± 0.00	5.33 ± 0.07	24.30 ± 2.31	30.70 ± 0.23	54.96 ± 2.22	0.30 ± 0.01
2690	*MB*	Belokranjka	9.33 ± 0.33	5.14 ± 0.07	23.23 ± 0.43	29.10 ± 0.40	52.35 ± 0.21	0.30 ± 0.01
2692	*MR*	Belokranjka	10.00 ± 0.00	5.20 ± 0.06	24.13 ± 2.70	30.90 ± 1.06	55.01 ± 2.82	0.29 ± 0.03
Control		Belokranjka	**9.67 ± 0.33**	**5.41 ± 0.11**	**25.17 ± 0.59**	**32.80 ± 1.36**	**57.95 ± 1.96**	**0.30 ± 0.02**
2152	*MR*	Belokranjka	10.00 ± 0.00	5.43 ± 0.07	28.00 ± 2.11	33.47 ± 0.23	61.46 ± 2.21	0.29 ± 0.01
Control		Belokranjka	**9.33 ± 0.33**	**5.11 ± 0.00**	**28.53 ± 0.87**	**26.87 ± 0.93**	**55.39 ± 0.82**	**0.27 ± 0.01**
2146	*MR*	Belokranjka	9.33 ± 0.33	5.22 ± 0.16	26.33 ± 1.87	30.00 ± 0.35	56.36 ± 1.75	0.35 ± 0.02
2147	*MR*	Belokranjka	10.00 ± 0.00	4.90 ± 0.21	24.83 ± 1.42	26.03 ± 0.43	50.84 ± 1.51	0.25 ± 0.02
2251	*MR*	Belokranjka	9.33 ± 0.33	5.31 ± 0.22	28.33 ± 3.80	28.00 ± 0.78	56.27 ± 4.42	0.27 ± 0.03
2789	*MR*	Belokranjka	9.67 ± 0.33	5.07 ± 0.14	27.23 ± 1.95	26.10 ± 1.01	53.37 ± 2.55	0.28 ± 0.02
2793	*MR*	Belokranjka	9.33 ± 0.67	4.93 ± 0.09	26.27 ± 2.03	27.47 ± 1.83	53.71 ± 1.97	0.30 ± 0.00
2794	*MR*	Belokranjka	8.67 ± 0.88	4.85 ± 0.03	29.03 ± 2.22	29.43 ± 0.50	58.47 ± 2.68	0.35 ± 0.03
2795	*MR*	Belokranjka	9.67 ± 0.33	5.27 ± 0.18	24.60 ± 2.01	29.10 ± 1.46	53.69 ± 0.85	0.27 ± 0.01
Control		Belokranjka	**9.67 ± 0.33**	**5.11 ± 0.21**	**19.60 ± 1.91**	**25.23 ± 1.12**	**44.81 ± 2.82**	**0.25 ± 0.01**
2645	*MR*	Belokranjka	9.67 ± 0.33	5.00 ± 0.06	23.57 ± 0.52	26.23 ± 1.36	49.76 ± 1.90	0.34 ± 0.06
2691	*MR*	Belokranjka	9.00 ± 0.00	5.26 ± 0.04	24.40 ± 3.21	25.50 ± 1.19	49.91 ± 4.36	0.30 ± 0.01
2693	*MR*	Belokranjka	10.00 ± 0.00	5.07 ± 0.09	23.93 ± 0.91	26.27 ± 0.78	50.19 ± 1.61	0.32 ± 0.04
2790	*MR*	Belokranjka	9.67 ± 0.33	5.34 ± 0.18	21.50 ± 0.49	27.37 ± 1.92	48.86 ± 2.42	0.31 ± 0.02
Control		Belokranjka	**8.33 ± 0.33**	**5.47 ± 0.16**	**20.63 ± 1.77**	**23.40 ± 1.08**	**44.05 ± 2.27**	**0.31 ± 0.03**
2634	*MR*	Belokranjka	9.33 ± 0.33	5.56 ± 0.20	21.20 ± 1.10	25.40 ± 0.71	46.57 ± 1.06	0.34 ± 0.06
2214	*MR*	Belokranjka	9.33 ± 0.33	5.25 ± 0.04	23.57 ± 0.73	27.07 ± 0.87	50.63 ± 0.22	0.33 ± 0.03
2243	*MR*	Belokranjka	9.00 ± 0.58	5.23 ± 0.07	23.90 ± 1.99	28.37 ± 1.06	52.25 ± 2.88	0.36 ± 0.00
2702	*MR*	Belokranjka	9.00 ± 0.00	5.07 ± 0.04	22.10 ± 2.80	27.07 ± 1.39	49.18 ± 3.64	0.32 ± 0.00
2791	*MR*	Belokranjka	9.33 ± 0.33	5.10 ± 0.10	25.33 ± 4.53	27.10 ± 1.61	52.42 ± 6.14	0.32 ± 0.03
Control		Belokranjka	**9.00 ± 0.58**	**4.91 ± 0.21**	**24.20 ± 0.38**	**28.30 ± 0.51**	**52.52 ± 0.21**	**0.31 ± 0.01**
2250	*MG*	Belokranjka	8.33 ± 0.88	5.13 ± 0.07	24.20 ± 2.08	26.77 ± 0.42	50.98 ± 1.81	0.26 ± 0.02
2685	*MR*	Belokranjka	9.00 ± 0.58	4.99 ± 0.11	22.50 ± 0.25	29.97 ± 1.39	52.45 ± 1.59	0.30 ± 0.02
Control		Belokranjka	**8.67 ± 0.33**	**5.23 ± 0.17**	**24.10 ± 0.75**	**26.63 ± 0.12**	**50.75 ± 0.67**	**0.29 ± 0.02**
2640	*MR*	Belokranjka	9.67 ± 0.33	5.17 ± 0.09	23.37 ± 2.00	26.10 ± 0.29	49.47 ± 1.89	0.28 ± 0.02
2694	*MR*	Belokranjka	9.67 ± 0.33	5.28 ± 0.32	24.87 ± 1.07	27.90 ± 0.66	52.77 ± 1.05	0.26 ± 0.00
2695	*MR*	Belokranjka	9.33 ± 0.33	5.11 ± 0.06	24.40 ± 2.94	26.57 ± 0.90	50.96 ± 3.77	0.28 ± 0.02
2788	*MR*	Belokranjka	8.67 ± 0.67	5.15 ± 0.27	23.87 ± 2.12	28.20 ± 1.33	52.07 ± 2.48	0.32 ± 0.02
2792	*MR*	Belokranjka	9.33 ± 0.33	5.57 ± 0.36	25.20 ± 3.54	28.10 ± 0.96	53.27 ± 4.46	0.29 ± 0.01

Table 2. *Cont.*

Isolate	Taxon	Maize Variety	Emergence Success [n]	Emergence Time [d]	Root Length [cm]	Shoot Length [cm]	Total Plant Length [cm]	Plant Dry Weight [g]
2796	*MR*	Belokranjka	8.67 ± 1.33	4.98 ± 0.11	21.90 ± 1.20	27.70 ± 2.06	49.62 ± 0.86	0.32 ± 0.02
Control		Belokranjka	**8.67 ± 0.88**	**4.67 ± 0.13**	**22.93 ± 1.71**	**28.00 ± 1.13**	**50.96 ± 2.47**	**0.29 ± 0.02**
2688	*MR*	Belokranjka	9.00 ± 0.58	5.10 ± 0.23	26.80 ± 0.68	28.97 ± 0.37	55.76 ± 0.48	0.36 ± 0.03
Control		Belokranjka	**8.33 ± 0.67**	**4.52 ± 0.24**	**15.53 ± 1.49**	**21.43 ± 0.88**	**36.97 ± 2.00**	**0.15 ± 0.01**
2154	*MR*	Belokranjka	7.67 ± 0.33	5.09 ± 0.25	14.90 ± 0.92	22.43 ± 0.78	37.35 ± 1.07	0.14 ± 0.01
Control		Belokranjka	**9.67 ± 0.33**	**4.14 ± 0.10**	**22.97 ± 0.84**	**26.53 ± 0.55**	**49.54 ± 1.18**	**0.23 ± 0.01**
2150	*MR*	Belokranjka	10.00 ± 0.00	4.47 ± 0.37	23.13 ± 1.39	27.07 ± 1.20	50.23 ± 2.56	0.26 ± 0.01
2240	*MR*	Belokranjka	9.67 ± 0.33	4.28 ± 0.09	24.07 ± 0.77	27.53 ± 1.78	51.60 ± 2.41	0.25 ± 0.02

Note: *MB*: *Metarhizium brunneum*; *MR*: *Metarhizium robertsii*; *MG*: *Metarhizium guizhouense*; *BB*: *Beauveria bassiana*; *TA*: *Trichoderma atroviride*; *TB*: *Trichoderma brevicompactum*; *TG*: *Trichoderma gamsii*; *TH*: *Trichoderma harzianum*; *ND*: No data.

In Chapalu variety, there was no significant effect of tested isolates on emergence success and total plant length (root + shoot length). ANOVA showed a significant prolongation of emergence time with two *M. robertsii* isolates (2698 and 2700) and one *T. atroviride* (2752). Root length was significantly reduced by two *M. robertsii* isolates (2243 and 2636). Shoot length was significantly reduced by *B. bassiana* (2299) but increased by *T. atroviride* (2882) and *T. gamsii* (2883). Plant dry weight was significantly reduced by two *B. bassiana* isolates (2299 and 2300) and *M. robertsii* (2011).

In Belokranjka variety, there was no significant effect of tested isolates on emergence success, shoot length, and total plant length (root + shoot length). ANOVA showed a significant prolongation of emergence time with *M. robertsii* (2154). Root length was significantly increased by three *M. robertsii* isolates (2691, 2693, and 2688) and plant dry weight was significantly increased by two *M. robertsii* isolates (2146 and 2794).

2.4. Enhancement of Nutrient Utilization by Fungi in Maize

Thirty fungal isolates were further tested for growth stimulation of maize (Chapalu variety only) in sand with or without fertilizers (General Hydroponics, Flora Series®) (Table 3). In the absence of fertilizer, there was no significant effect of the tested isolates on emergence success and total plant length. However, ANOVA showed a significant prolongation of emergence time with *M. guizhouense* (2010) and a significant reduction in root length with *B. bassiana* (2299) and *M. robertsii* (2148). Shoot length was significantly reduced by *M. robertsii* (2642) but increased by *M. robertsii* (2011). Plant dry weight was significantly increased by *M. robertsii* (2216) in unfertilized sand.

There was no significant effect of tested isolates on emergence success, root length, total plant length, and dry weight in the presence of fertilizer. ANOVA showed a significant prolongation of emergence time of maize treated with *M. brunneum* (2703), *M. robertsii* (2009), and *M. guizhouense* (2010). Shoot length was significantly increased with *M. robertsii* (2011) in fertilized sand.

When all data were analyzed together, fertilization caused an average increase in shoot length of 32.3% ($t(74) = 13.13$, $p = < 0.0001$) and an increase in total plant length of 13.2% ($t(74) = 6.47$, $p = < 0.0001$) compared to unfertilized plants. Fertilizer itself had no significant effect of on emergence success, emergence time, root length and plant dry weight.

Two-way ANOVA was performed to determine if fungal isolates altered the growth parameters of maize in fertilized vs. unfertilized sand compared to untreated maize. Dunnett's multiple comparison test showed a prolonged emergence time in fertilized sand when treated with *B. bassiana* (2009). The isolate *M. guizhouense* (2010) prolonged emergence time in fertilized sand as well as in unfertilized sand. *M. robertsii* (2631) significantly increased root length in unfertilized sand and *M. robertsii* (2632) in fertilized sand. On the other hand, *B. bassiana* (2299) and *M. robertsii* (2148) significantly reduced root length in unfertilized sand. Two isolates of *M. brunneum* (1868 and 1154) and *B. bassiana* (2121) increased dry weight in fertilized sand.

Table 3. Growth stimulating effects of selected fungal isolates on maize (Chapalu variety) grown in sand with/without fertilizer for 21 days. Data presented are the mean values ± SE (n = 15 (3 replicates with 5 seeds each)). Green fill indicates isolates with significant growth promoting properties, red fill indicates isolates with growth inhibitory properties and grey fill indicates a significant difference from negative control (sterile 0.1% Tween 80), $p \leq 0.05$.

		Emergence Success [n]		Emergence Time [d]		Root Length [cm]		Shoot Length [cm]		Total Plant Length [cm]		Plant Dry Weight [g]	
Isolate	**Taxon**	**No Fertilizer**	**Fertilizer**	**No Fertilizer**	**Fertilizer**	**No Fertilizer**	**Fertilizer**	**No Fertilizer**	**Fertilizer**	**No Fertilizer**	**Fertilizer**	**No Fertilizer**	**Fertilizer**
Control		**5.00 ± 0.00**	**4.33 ± 0.67**	**4.25 ± 0.00**	**4.42 ± 0.21**	**18.90 ± 0.00**	**16.27 ± 0.58**	**14.10 ± 0.00**	**16.77 ± 1.30**	**32.93 ± 0.00**	**33.05 ± 1.72**	**0.21 ± 0.00**	**0.27 ± 0.02**
1154	*MB*	5.00 ± 0.00	4.67 ± 0.33	4.60 ± 0.12	4.83 ± 0.34	16.23 ± 1.42	14.13 ± 0.74	12.93 ± 0.72	16.40 ± 0.46	29.11 ± 1.18	30.53 ± 1.00	0.25 ± 0.02	0.42 ± 0.04
1868	*MB*	4.67 ± 0.33	5.00 ± 0.00	4.33 ± 0.29	4.53 ± 0.07	17.90 ± 0.70	16.73 ± 0.72	13.50 ± 0.06	16.77 ± 0.81	31.38 ± 0.77	33.49 ± 0.40	0.24 ± 0.01	0.38 ± 0.05
2121	*BB*	5.00 ± 0.00	4.67 ± 0.33	4.72 ± 0.17	5.02 ± 0.13	15.53 ± 0.87	15.27 ± 0.23	15.07 ± 0.74	18.40 ± 0.59	30.65 ± 1.24	33.68 ± 0.76	0.30 ± 0.00	0.41 ± 0.09
Control		**5.00 ± 0.00**	**4.67 ± 0.33**	**4.67 ± 0.13**	**5.17 ± 0.20**	**14.63 ± 0.90**	**15.43 ± 0.92**	**13.40 ± 2.15**	**18.70 ± 1.36**	**28.04 ± 2.92**	**34.15 ± 2.29**	**0.28 ± 0.03**	**0.30 ± 0.02**
2631	*MR*	4.67 ± 0.33	5.00 ± 0.00	4.63 ± 0.09	4.40 ± 0.12	20.03 ± 0.96	18.77 ± 0.19	13.80 ± 0.65	17.47 ± 0.50	33.84 ± 1.20	36.23 ± 0.57	0.30 ± 0.01	0.31 ± 0.04
2632	*MR*	4.33 ± 0.67	5.00 ± 0.00	4.42 ± 0.14	4.53 ± 0.07	17.47 ± 0.99	19.67 ± 1.62	12.87 ± 0.43	17.50 ± 0.86	30.38 ± 1.32	37.17 ± 1.98	0.30 ± 0.03	0.27 ± 0.03
2245	*MR*	4.67 ± 0.33	5.00 ± 0.00	5.37 ± 0.63	4.53 ± 0.24	18.63 ± 0.58	19.33 ± 1.43	14.07 ± 0.73	17.83 ± 0.47	32.72 ± 0.99	37.16 ± 1.35	0.23 ± 0.02	0.27 ± 0.02
2246	*MR*	5.00 ± 0.00	4.67 ± 0.33	4.67 ± 0.24	5.15 ± 0.18	18.13 ± 0.64	19.33 ± 1.83	12.37 ± 0.37	18.67 ± 0.76	30.49 ± 1.00	37.99 ± 2.43	0.26 ± 0.03	0.30 ± 0.02
Control		**5.00 ± 0.00**	**5.00 ± 0.00**	**5.53 ± 0.37**	**5.47 ± 0.29**	**18.93 ± 0.92**	**17.95 ± 1.23**	**13.34 ± 1.26**	**18.80 ± 0.96**	**32.27 ± 1.92**	**36.75 ± 1.70**	**0.20 ± 0.02**	**0.23 ± 0.02**
2215	*MR*	5.00 ± 0.00	5.00 ± 0.00	5.87 ± 0.35	5.80 ± 0.12	17.15 ± 1.35	19.25 ± 1.35	13.19 ± 1.10	17.18 ± 0.33	30.33 ± 2.06	36.43 ± 1.40	0.19 ± 0.00	0.21 ± 0.01
2216	*MR*	4.67 ± 0.33	4.00 ± 0.58	5.68 ± 0.28	5.27 ± 0.27	18.92 ± 1.85	17.62 ± 1.43	13.50 ± 0.14	16.97 ± 0.97	32.42 ± 1.85	34.59 ± 1.36	0.25 ± 0.01	0.21 ± 0.01
2299	*BB*	4.67 ± 0.33	5.00 ± 0.00	5.93 ± 0.58	5.87 ± 0.29	12.94 ± 1.10	15.99 ± 1.95	14.98 ± 0.26	18.75 ± 0.36	27.92 ± 1.27	34.73 ± 1.60	0.18 ± 0.01	0.19 ± 0.01
2300	*BB*	4.67 ± 0.33	4.67 ± 0.33	5.18 ± 0.32	5.98 ± 0.21	19.94 ± 1.20	14.48 ± 0.74	15.33 ± 0.42	19.26 ± 0.22	35.26 ± 1.58	33.74 ± 0.58	0.23 ± 0.00	0.25 ± 0.02
Control		**5.00 ± 0.00**	**4.33 ± 0.67**	**5.27 ± 0.29**	**5.22 ± 0.51**	**19.67 ± 1.56**	**17.50 ± 1.53**	**13.97 ± 0.58**	**18.40 ± 0.57**	**33.63 ± 0.98**	**35.90 ± 1.97**	**0.19 ± 0.01**	**0.20 ± 0.02**
2635	*MR*	4.67 ± 0.33	4.67 ± 0.33	5.97 ± 0.27	6.13 ± 0.77	18.20 ± 0.90	16.30 ± 1.40	13.47 ± 1.22	19.30 ± 1.05	31.68 ± 1.18	35.62 ± 2.23	0.18 ± 0.01	0.21 ± 0.00
2637	*MR*	4.33 ± 0.33	4.67 ± 0.33	5.68 ± 0.09	5.65 ± 0.13	17.00 ± 0.75	19.60 ± 1.04	14.23 ± 0.38	19.63 ± 0.26	31.25 ± 1.14	39.23 ± 1.25	0.17 ± 0.00	0.22 ± 0.03
2641	*MR*	5.00 ± 0.00	5.00 ± 0.00	4.93 ± 0.07	5.07 ± 0.13	17.40 ± 0.61	18.73 ± 2.03	14.03 ± 0.42	20.07 ± 0.61	31.42 ± 0.61	38.77 ± 2.58	0.20 ± 0.01	0.21 ± 0.01
Control		**5.00 ± 0.00**	**4.67 ± 0.33**	**4.93 ± 0.07**	**5.07 ± 0.07**	**19.73 ± 1.36**	**18.27 ± 1.92**	**15.90 ± 0.23**	**19.33 ± 0.44**	**35.61 ± 1.28**	**37.59 ± 2.35**	**0.42 ± 0.04**	**0.41 ± 0.02**
2697	*ND*	5.00 ± 0.00	5.00 ± 0.00	4.40 ± 0.12	4.87 ± 0.24	19.03 ± 1.68	16.77 ± 1.93	14.30 ± 1.21	20.00 ± 0.96	33.34 ± 2.80	36.75 ± 2.64	0.40 ± 0.03	0.46 ± 0.04
2698	*MR*	4.67 ± 0.33	5.00 ± 0.00	4.80 ± 0.12	5.20 ± 0.40	17.10 ± 1.12	16.50 ± 1.65	15.10 ± 0.45	18.17 ± 0.90	32.20 ± 1.39	34.69 ± 2.31	0.37 ± 0.03	0.45 ± 0.04
2699	*MR*	4.67 ± 0.33	5.00 ± 0.00	5.42 ± 0.42	5.20 ± 0.23	19.87 ± 2.63	21.47 ± 2.86	14.67 ± 1.28	21.13 ± 0.78	34.51 ± 3.89	42.59 ± 2.23	0.37 ± 0.04	0.36 ± 0.02
2700	*MR*	4.67 ± 0.33	5.00 ± 0.00	5.08 ± 0.08	4.67 ± 0.24	19.43 ± 1.93	20.33 ± 0.83	15.33 ± 0.41	19.27 ± 0.48	34.73 ± 2.28	39.63 ± 0.99	0.34 ± 0.03	0.39 ± 0.01
Control		**5.00 ± 0.00**	**4.67 ± 0.33**	**5.33 ± 0.35**	**5.50 ± 0.06**	**21.07 ± 1.69**	**19.00 ± 1.30**	**15.67 ± 1.17**	**21.63 ± 0.85**	**36.75 ± 2.84**	**40.65 ± 0.84**	**0.39 ± 0.02**	**0.28 ± 0.07**
2298	*BB*	4.67 ± 0.33	4.67 ± 0.33	5.57 ± 0.12	4.92 ± 0.14	21.03 ± 1.07	22.33 ± 1.32	16.60 ± 0.61	21.63 ± 0.81	37.63 ± 1.58	43.97 ± 1.46	0.45 ± 0.03	0.44 ± 0.03
2243	*MR*	5.00 ± 0.00	5.00 ± 0.00	5.40 ± 0.23	5.20 ± 0.23	20.80 ± 1.22	21.27 ± 0.89	16.10 ± 0.30	21.27 ± 0.72	36.89 ± 1.52	42.51 ± 1.38	0.39 ± 0.04	0.39 ± 0.08
2636	*MR*	5.00 ± 0.00	5.00 ± 0.00	5.60 ± 0.35	5.13 ± 0.24	19.90 ± 1.35	20.87 ± 1.01	14.83 ± 1.03	20.20 ± 1.54	34.77 ± 2.35	41.04 ± 2.56	0.33 ± 0.04	0.37 ± 0.03
2642	*MR*	5.00 ± 0.00	4.67 ± 0.33	5.60 ± 0.12	5.20 ± 0.12	19.57 ± 1.12	21.07 ± 1.16	13.27 ± 1.03	18.93 ± 1.13	32.83 ± 2.10	39.99 ± 2.30	0.36 ± 0.01	0.32 ± 0.08
Control		**5.00 ± 0.00**	**5.00 ± 0.00**	**5.33 ± 0.47**	**4.87 ± 0.07**	**21.87 ± 0.84**	**18.77 ± 1.53**	**16.97 ± 0.38**	**21.03 ± 1.65**	**38.89 ± 1.14**	**39.79 ± 1.96**	**0.19 ± 0.01**	**0.21 ± 0.01**
2148	*MR*	5.00 ± 0.00	5.00 ± 0.00	4.40 ± 0.23	4.87 ± 0.07	17.67 ± 1.02	18.20 ± 1.51	14.63 ± 1.04	19.53 ± 1.02	32.28 ± 1.21	37.74 ± 1.92	0.20 ± 0.01	0.24 ± 0.01
2151	*MR*	4.67 ± 0.33	4.67 ± 0.33	4.83 ± 0.28	4.83 ± 0.34	20.80 ± 2.21	20.07 ± 0.70	15.87 ± 0.34	22.20 ± 0.42	36.65 ± 2.56	42.27 ± 0.38	0.21 ± 0.02	0.21 ± 0.00
2152	*MR*	4.33 ± 0.33	5.00 ± 0.00	5.15 ± 0.69	4.87 ± 0.13	20.50 ± 0.85	18.90 ± 0.87	16.07 ± 1.55	21.40 ± 0.85	36.59 ± 1.96	40.31 ± 0.08	0.19 ± 0.00	0.21 ± 0.02
2701	*MR*	5.00 ± 0.00	4.33 ± 0.33	4.80 ± 0.12	5.43 ± 0.70	20.90 ± 0.78	18.40 ± 0.87	15.50 ± 0.95	21.33 ± 0.38	36.37 ± 1.19	39.72 ± 1.23	0.20 ± 0.01	0.23 ± 0.00

Table 3. *Cont.*

		Emergence Success [n]		Emergence Time [d]		Root Length [cm]		Shoot Length [cm]		Total Plant Length [cm]		Plant Dry Weight [g]	
Isolate	**Taxon**	**No Fertilizer**	**Fertilizer**	**No Fertilizer**	**Fertilizer**	**No Fertilizer**	**Fertilizer**	**No Fertilizer**	**Fertilizer**	**No Fertilizer**	**Fertilizer**	**No Fertilizer**	**Fertilizer**
Control		**4.33 ± 0.33**	**5.00 ± 0.00**	**4.90 ± 0.21**	**4.73 ± 0.24**	**19.47 ± 2.25**	**19.13 ± 0.58**	**14.83 ± 1.39**	**20.30 ± 1.30**	**34.27 ± 3.62**	**39.43 ± 1.83**	**0.21 ± 0.01**	**0.22 ± 0.00**
2703	*MB*	5.00 ± 0.00	5.00 ± 0.00	5.53 ± 0.27	5.87 ± 0.64	19.10 ± 1.66	21.13 ± 0.52	15.90 ± 0.32	18.40 ± 0.35	34.95 ± 1.35	39.51 ± 0.65	0.20 ± 0.00	0.23 ± 0.01
2009	*MR*	5.00 ± 0.00	3.67 ± 0.67	6.00 ± 0.53	5.87 ± 0.59	17.57 ± 0.62	17.97 ± 1.08	15.13 ± 1.82	22.87 ± 1.67	32.71 ± 2.32	40.83 ± 2.67	0.21 ± 0.02	0.23 ± 0.01
2010	*MG*	5.00 ± 0.00	5.00 ± 0.00	7.40 ± 0.90	6.20 ± 0.35	19.80 ± 1.59	16.37 ± 0.47	15.40 ± 1.56	21.93 ± 1.03	35.21 ± 0.57	38.30 ± 1.49	0.21 ± 0.00	0.21 ± 0.02
2011	*MR*	4.33 ± 0.33	4.67 ± 0.33	5.22 ± 0.46	5.60 ± 0.35	19.53 ± 0.78	19.40 ± 0.06	19.13 ± 0.42	22.97 ± 0.60	38.66 ± 1.16	42.40 ± 0.54	0.22 ± 0.02	0.23 ± 0.01

Note: *MB*: *Metarhizium brunneum*; *MR*: *Metarhizium robertsii*; *MG*: *Metarhizium guizhouense*; *BB*: *Beauveria bassiana*; *TA*: *Trichoderma atroviride*; *TB*: *Trichoderma brevicompactum*; *TG*: *Trichoderma gamsii*; *TH*: *Trichoderma harzianum*; ND: No data.

Fungal treatment had a significant effect on emergence time in two out of eight experiments, on root length in three out of eight experiments, on shoot length in four out of eight experiments, and on plant dry weight in one out of eight experiments. However, no significant effect of fungi on emergence success was observed.

Fertilization significantly affected emergence time and plant dry weight in one out of eight experiments, root length in two out of eight experiments, and in all experiments the presence of fertilizer significantly affected shoot length. No significant effect of fertilizer on emergence success was observed.

3. Discussion

A total of 71 fungal isolates were obtained mainly from soil samples, by using the *Galleria–Tenebrio* bait method, but also using selective media and mycosed insects found in different agroecosystems. Overall, the most frequently isolated representatives were *M. robertsii* (i.e., 75% of isolates). Therefore, it is possible that the isolation techniques favor this species. However, Sharma et al. [28] also used the *Galleria-Tenebrio* bait method, where twice as many *B. bassiana* than *M. robertsii* were isolated. Moreover, Medo and Cagáň [29] used the *Galleria* bait method to isolate fungi, but the predominant species was *B. bassiana* and no *M. robertsii* was isolated.

In the present study, only *M. robertsii* and *M. guizhouense* were isolated with *Tenebrio* as bait, while with *Galleria* as bait approximately half of the isolates were *B. bassiana* and the other half belonged to the genus *Metarhizium*. There are some reports where *B. bassiana* was recovered more frequently when *Galleria* was used as bait, while *Tenebrio* bait resulted in more frequent isolation of *Metarhizium* species [28,30,31]. Therefore, to obtain more representative and less biased information about the entomopathogenic fungal community in an agroecosystem, it is recommended to increase the number of arthropod species used as bait.

A total of 71 isolates were tested for virulence against *T. molitor* and maize growth stimulation. The isolates differed significantly in their degree of virulence. The most virulent isolates were those obtained from lepidopteran insect hosts and from mycosed wild adult coleopterans. One would expect higher virulence from isolates derived from *T. molitor* baits, which is the same species as the model insect used in our bioassays, but this was not the case in our study. The positive controls (Actara, Force) and the commercial bioinsecticides (Mycotal, Met52) showed very low mortality after 14 days: Mycotal, Force, and Met52 around 10% ACM or less, and Actara less than 50% ACM. Mycotal and Met52 are both biopesticides primarily intended for whitefly and thrips control (Met52 also fungus gnats and mites), but have also been tested for Coleoptera [32,33]. Force and Actara are, among others, used to control coleopteran pests. Although mortality was low 14 days after Actara treatments, it is worth noting that mealworms were ecologically dead (i.e., insects were lethargic, spasming, no longer feeding) a few days after treatment.

Although our analysis suggests a higher correlation of *B. bassiana* with mortality, one cannot conclude that one species is more pathogenic than the others. The level of virulence often varies within species and even clades, as shown by the phylogenetic analyses of Medo et al. [34] and Lopes et al. [35]. Moreover, the seven most virulent isolates in our study belong to three different species. The origin of soil samples and their chemical and physical properties can have significant effects on the presence, abundance, and pathogenicity of insect-associated entomopathogens. For example, the infection rate of pupae of the Mediterranean fruit fly, *Ceratitis capitata* (Wiedemann) was higher with fungi isolated from soils with a sandy texture and high organic matter content [36] and in soils with a water potential of −0.1 MPa [37].

Our study also suggests a stronger association of *B. bassiana* with meadows. Higher abundance and diversity of *B. bassiana* in more semi-natural habitats and less physically disturbed soils has also been observed in other studies and is likely the result of many biotic and abiotic factors, such as increased humidity, reduced ultra-violet radiation, and temperature, reduced agricultural activities (e.g., tillage or fungicide use), higher insect diversity, etc. [38,39].

Trichoderma species are important biological control agents due to their antagonistic properties against various pathogenic fungi. Some species are capable of colonizing plants, including maize, in addition to increasing photosynthetic rate [40], root and shoot growth, plant biomass [41], and enhancing the immune system of plants [42]. However, there are also a few reports on the entomopathogenic properties of *Trichoderma*, where direct damage to insect pests has been observed. *Trichoderma viride* Pers. derived chitinases have effectively degraded the chitinous vital structures of *Bombyx mori* (Linnaeus, 1758) larvae [26], *Trichoderma koningiopsis* Samuels, Carm. Suárez and H.C.Evans have shown significant entomotoxicity against *Delia radicum* (L.) pupae in the soil environment, and *T. atroviride* against *D. radicum* eggs in in vitro tests [25], while *T. harzianum* caused up to 80% larval mortality against the Egyptian cotton leafworm *Spodoptera littoralis* (Boisduval, 1833) [23] and up to 100% mortality of *T. molitor* larvae [24]. In contrast, the *Trichoderma* isolates tested in this study showed little or no pathogenicity against *T. molitor*. In terms of stimulation of maize growth, shoot length was significantly increased by 25.8% by *T. atroviride* (2882) and 22.5% by *T. gamsii* (2883). However, significant prolongation of emergence time was observed with *T. atroviride* (2752). Ousley et al. [43] also reported no significant growth-promoting or even inhibitory properties of *Trichoderma*, especially in relation to germination rate. *Metarhizium robertsii* isolates (2698, 2700 and 2154) also significantly prolonged the emergence time. Razinger et al. [44] and Kuzhuppillymyal-Prabhakarankutty et al. [45] also reported a lower germination rate of maize seed; this could be a consequence of the method by which the conidia were applied to the seeds, namely by using carboxymethyl cellulose or methylcellulose. In our case, the maize seeds were soaked in a suspension of fungal conidia using only 0.1% Tween 80 to overcome the difficulties with the hydrophobic properties of the conidia of the fungal species under study and to allow adequate adhesion of the conidia to the maize kernels. Therefore, the method of conidia attachment to the seeds used may not be the (only) reason for the inhibition of germination and emergence; more likely the reason lies in the fungi tested. It should be noted that entomopathogenicity may have evolved later, especially within the Clavicipitaceae, meaning that their ancestors used plants or plant debris as a food source [46]. This could explain the inhibitory effect of entomopathogenic fungi, as their metabolites, i.e., destruxins, might also be toxic for plants [47].

In general, there are very few studies observing the emergence speed of plant seeds treated with entomopathogens [48,49]. The focus of most research is more prone to study germination rate rather than emergence time. However, rapid and reliable emergence is of particular importance to maize seed growers, especially in temperate regions, where maize is usually planted in spring in soil with suboptimal temperatures for emergence [50]. Rapid emergence also shortens the time plants are exposed to (soil) pests and reduces weed infestation [51,52].

Metarhizium and *Beauveria* species as typical entomopathogens were also tested for their growth stimulation properties to maize. *Beauveria bassiana* isolate (2299) significantly reduced shoot length and plant dry weight (isolates 2299 and 2300) in the variety Chapalu. Rivas-Franco et al. [53] also noticed a reduction in root and shoot dry weight in maize seeds treated with *B. bassiana*. However, Kuzhuppillymyal-Prabhakarankutty et al. [45] observed higher plant dry and fresh weight as well as better performance of coated maize exposed to drought. In addition, Russo et al. [54] detected positive effects on all yield characteristics, seed germination, and measured growth parameters when maize was inoculated by a leaf spraying technique. Tall and Meyling [55] reported increased root and shoot biomass in maize treated with *B. bassiana* and grown in nutrient-rich soil. However, when nutrient availability was low, they observed reduced plant growth compared to the control, which may indicate that fungi act as potential resource sink. Our results showed no significant growth stimulation of maize treated with *B. bassiana* growing in sand with added fertilizers. However, in the absence of fertilizers, *B. bassiana* (2299) and *M. robertsii* (2148) showed a significant decrease in root length and *M. robertsii* (2642) also showed a decrease in shoot length, which could indicate the uptake of nutrients by the fungi in an environment where resources are scarce.

Metarhizium robertsii isolates significantly reduced root length (isolates 2243 and 2636) and plant dry weight (isolate 2011) in the Chapalu variety. In contrast, other isolates of the same species significantly increased root length (isolates 2691, 2693, and 2688) and plant dry weight (isolates 2146 and 2794) in the Belokranjka variety. The effects of coating maize seeds with *Metarhizium* are often beneficial. Razinger et al. [44] reported a significant increase in fresh weight of maize by coating seeds with *M. robertsii*, but no effect on plant length, whereas colonized maize plants of Ahmad et al. [1] were greater in length and shoot biomass. Kabaluk and Ericsson [56] treated maize seeds with *Metarhizium anisopliae* (Metschn.) Sorokīn conidia, which resulted in increased stand density and plant fresh weight in a wireworm-infected field. However, their laboratory experiments showed that treating maize with 3.8×10^8 conidia per seed actually reduced seed germination and root growth, indicating the possibility of a potential limit of conidia per seed at which seed viability is not at risk.

4. Materials and Methods

4.1. Isolation of Fungi

Entomopathogenic fungi were isolated either from naturally present sporulating insect cadavers, or from soil samples using *Galleria mellonella* (Linnaeus, 1758) and *T. molitor* larvae as bait [57], or from serially diluted soil suspensions plated on semi-selective media as described by Cooke [58] and Williams et al. [59]. In the latter two cases, soil samples were obtained from maize fields (mainly bulk soil) or from Karst extensive hay meadows, accommodating a high diversity of Poaceae species (soil from the Poaceae rhizosphere). Sampling sites and host/medium characteristics are summarized in Table 1. A Nikon (SMZ800, Nikon Corp., Melville, NY, USA) binocular was used to identify sporulating structures formed by fungi on cadavers that were placed in droplets of sterile water to generate spore suspensions. Aliquots of the suspension were moved over the surface of potato dextrose agar supplemented with bacteria suppressing antibiotics (streptomycin and penicillin) to generate single spore cultures. The isolates obtained were identified on the basis of morphological characters seen on the insect cadavers or in pure culture or through DNA barcoding according to Razinger et al. [60]. In brief, molecular barcode sequences of the intron-rich part of the elongation factor 1-alpha (*tef*) were obtained by adopting the strategies described by Bischoff et al. [61] but using the EF2 primer of O'Donnell et al. [62]. The 50 μL reaction mixture for PCR consisted of 5 μL of Taq PCR buffer with (NH4)2SO4 (Fermentas, Waltham, MA, USA), 2 mM MgCl2, 0.2 mM dNTP (Promega, Madison, WI, USA), 0.5 mM of each of the primers, 1 unit of native Taq polymerase (Fermentas, Waltham, MA, USA) and 1 μL of genomic DNA. In PCR, we used an initial denaturation step at 94 °C for 3 min, 5 cycles of 94 °C for 60 s (denaturation), 56 °C for 45 s (annealing), 72 °C for 60 s (elongation), and 35 cycles as described before but with an annealing temperature of 53 °C, and a final extension at 72 °C for 8 min. Sequencing reactions were performed at the Macrogen Europe sequencing facility (Amsterdam, The Netherlands) in both directions by using the same primers as used in PCR. The data were inspected and edited with the aid of the software program BioEdit v7.2.0 [63]. Representative sequences were deposited at NCBI database.

4.2. Fungal Virulence toward Tenebrio molitor

Single-dose virulence testing was performed on larvae of mealworms *T. molitor*, reared at the Agricultural Institute of Slovenia. Fungal strains were subcultured on Potato Dextrose agar (PDA; Biolife, Italy) and incubated in an incubation chamber (IPP 500, Memmert) at 22 °C for 14 days or longer to obtain the required amount of sporulating structures. Spores were washed-off by pipetting approximately 10–15 mL of sterile 0.1% Tween 80 solution onto the top of cultures and scraping colonies with a Drigalski spatel. The obtained suspensions were collected into sterile 50 mL Falcon tubes. Haemocytometer (Bürker-Türk, BRAND GMBH + CO. KG, Wertheim, Germany) counting was used to adjust obtained suspensions to a concentration of 1×10^8 conidia ml^{-1} [44]. The viability of conidia was determined by counting germinated conidia after 24 h of incubation of the diluted suspension sample.

Thirty larvae per strain were immersed in 1×10^8 mL^{-1} conidial suspension for 15 s, with a slight stirring. Two commercial insecticides were used as positive controls: 0.1% tap-water dilution of Actara 25 WG (Syngenta, Switzerland; active ingredient Thiamethoxam, 25% *w*/*w*) and 0.1% Force 1.5G (Syngenta, Switzerland; a.i. Tefluthrin, 0.15% *w*/*w*). In addition, two commercial bioinsecticides were used as reference biocontrol agents: 0.1% Mycotal (Koppert, Netherlands; a.i. *Lecanicillium muscarium* (Petch) Zare and W.Gams Ve6) and 1% Met52 EC (Novozymes, France; a.i. *M. brunneum* strain F52). Sterile 0.1% Tween 80 was used as a negative control. Mealworms were afterwards transferred into a petri dish (each strain to a separate Petri dish) and allowed to dry under a laminar flow hood for 20–30 min. Each mealworm was placed in its own well in a six-well plate with a few pieces of oatmeal as food. Five replicates of six-well plates were made per strain (n = 30 per strain). Treated mealworms were kept in a loosely closed cardboard box in an incubation chamber for 2 weeks set to 75% r.h., 21 °C and 14:10 h (light:dark) regime. The number of dead or immobile larvae was checked every 3 days. Dead larvae were incubated at room temperature on water agar to confirm infection by the fungi. For further information on the virulence bioassay see Supplementary Materials Figures S1 and S2.

4.3. *Maize Growth Biostimulation Tests*

4.3.1. Maize Seed Treatment

The fungal suspensions for the growth stimulation trials were prepared as described above. Maize seeds were soaked in the suspension or in sterile 0.1% Tween 80 (control treatment) and placed on an orbital shaker for 1 h and 15 min at 200 RPM. The seeds were then placed on filter paper and dried in a laminar flow hood for 1 h.

For each experiment, the number of conidia of each fungal strain per maize seed was evaluated. Three ml of 0.1% Tween 80 was added to 10 inoculated maize seeds in a Falcon tube (Deltalab, Barcelona, Spain) and vortexed for 10 s at 3000 rpm. The Falcon tube was left on an orbital shaker for 30 min at 600 rpm and afterwards vortexed again for 10 s at 3000 rpm. The number of conidia was determined using a hemocytometer [64].

4.3.2. Growth Stimulation Bioassay

Two maize varieties, namely Chapalu (Saatzucht Gleisdorf, Austria) and Belokranjka (Organic farm Župnca, Slovenia), were used for the growth stimulation assays. The experiments with Chapalu variety were conducted with 5 seeds and 3 replicates and with Belokranjka variety with 10 seeds and 3 replicates. Seventy-one fungal isolates were tested for potential growth stimulation of maize in (i) twice autoclaved commercial planting substrate (Potgrond H, Klasmann, Germany).

Coated seeds were planted in 12 L plastic pots containing the substrates and kept in an incubation chamber at 22 °C/20 °C day/night temperature with a photoperiod of 14:10 h (light:dark) and 70–75% r.h. The number of emerged sprouts was counted every day until the end of seedling emergence. Three weeks after planting, growth parameters such as root length, shoot length and plant dry weight were measured on the harvested maize plants. For obtaining the dry weight, the substrate was carefully washed from the roots and all plants from one pot were placed in a paper bag, dried at 60 °C for 48 h, and weighed (BP301S, Sartorius).

4.3.3. Fungal Nutrient Utilization Enhancement in Maize

Thirty fungal isolates were further tested for their potential enhancement of nutrient utilization in Chapalu variety only. Tests were performed in (ii) non-autoclaved sand and (iii) non-autoclaved sand with mineral fertilizers FloraMicro:FloraGro:FloraBlooom

(General Hydroponics, Flora Series®, Europe) added to the sand on the 7th and 14th day of the experiment in the ratio FloraMicro:FloraGro:FloraBlooom = 2:1:1 mL per 3.79 L of water on day 7 and 4:5:1 mL per 3.79 L of water on day 14.

Coated seeds were planted in 0.25 L plastic pots with fertilized or unfertilized sand. The growth conditions and evaluation parameters were the same as in the 'Growth stimulation bioassay'.

4.4. Data Analysis

The time-based larval mortality was analyzed using Kaplan–Meier survival analysis and its significance was analyzed using the log-rank (Mantel–Cox) test. When multiple survival curves were compared, the significance threshold was corrected using the Bonferroni method [65]. Survival analysis and calculation of median lethal time (LT50) were performed using GraphPad Prism 5.00 (GraphPad Software, Inc., La Jolla, CA, USA). Additionally, Abbott's corrected mortality (ACM) was calculated to eliminate the effect of natural or unexplained mortality of the negative control group [66].

Focused principal component analysis (FPCA) was implemented for a more accurate interpretation of correlation of predictor variables, in our case fungal origin, habitat characteristics, and isolation method, toward mortality rate (ACM) using the packages "psy" [67] and "dummies" [68] in R 3.6.1 [69]. Selected parameters were as follows: genus of the isolates, habitat type (field vs. meadow), soil sample location (bulk vs. rhizosphere), field type, isolation method/type (insect host vs. selective medium), insect host order, and their origin (wild vs. reared) and developmental stage (adult vs. larva). The significance of the analysis was tested using the non-parametric Kruskal–Wallis test, followed by Dunn's post hoc test, where the p-value was adjusted using the Benjamini–Hochberg (BH) correction. Normally distributed data were tested using the one-way ANOVA, followed by a post hoc Tukey HSD test. For this purpose, packages "dplyr" [70] and "rstatix" [71] were used.

All growth stimulation data were subjected to one-way ANOVA followed by a Bonferroni–Holm multiple comparisons test. For experiments where fertilizer was one of the parameters, also two-way ANOVA, followed by Dunnett's multiple comparison test, was used in order to compare the effect of substrate (fertilized vs. unfertilized) and fungal isolates on the growth parameters of Chapalu maize. The analyses were carried out using GraphPad Prism software.

5. Conclusions

The aim of this study was to find the ideal fungal isolate that would combine two important characteristics of entomopathogenic and biostimulative fungi, namely the ability to infect insect pests and promote plant growth, and to test whether fungal virulence depends on the source of the isolate(s). The isolates *M. brunneum* (1154) and *B. bassiana* (2121) showed the highest mortality (100%) against *T. molitor*. High virulence was observed in isolates from wild adult mycosed insects, meadow habitats, and Lepidopteran hosts, but due to the uneven distribution of samples, we cannot draw any conclusive inferences. *Trichoderma atroviride* (2882) and *T. gamsii* (2883) showed the greatest promotion of plant growth, followed by two *M. robertsii* isolates (2693 and 2794). Even though we did not find the super fungus, *M. robertsii* (2693) came closest to meet our requirements. Maize seeds inoculated with this isolate showed a positive effect on all measured growth and emergence parameters while causing the death of 73% of *T. molitor* larvae. Therefore, it would be beneficial to test this isolate in a tri-trophic system that also includes a pest organism, e.g., wireworms, to determine its potential effect on maize stand density and/or yield increase.

Supplementary Materials: The following are available online at https://www.mdpi.com/article/10.3390/plants10112498/s1, Figure S1: Most virulent fungal isolates in pure culture (left) and on mycosed *Tenebrio molitor* larvae and adults (right). Isolates 1154 and 1868 are representatives of *Metarhizium brunneum*, isolates 2251 and 2637 representatives of *Metarhizium robertsii* and 2300 and 2121 are representatives of *Beauveria bassiana*. Figure S2: Virulence bioassay. Spore suspension in

50 mL Falcon tubes (left). After 15 s immersion in spore suspension, mealworm larvae were placed in a six-well plate (right).

Author Contributions: Conceptualization, J.R.; methodology, E.P., J.R. and J.L.; formal analysis, E.P. and J.R.; investigation, E.P., J.R. and J.L.; resources, E.P., J.R. and J.L.; data curation, E.P. and J.R.; writing—original draft preparation, E.P.; writing—review and editing, J.R.; visualization, E.P.; supervision, J.R.; funding acquisition, J.R. All authors have read and agreed to the published version of the manuscript.

Funding: This research was funded by the Slovenian Research Agency (ARRS), grant number 100-18-0401 to E.P. and grant no. P4-0072 (Agrobiodiversity program) and by the European Union's Horizon 2020 research and innovation program under grant agreement no. 817946 (Excalibur project) and no. 771367 (ECOBREED project).

Institutional Review Board Statement: Not applicable.

Data Availability Statement: Not applicable.

Acknowledgments: We would like to thank to Hans-Josef Schroers, Agricultural Institute of Slovenia, for isolating and identifying studied fungal isolates as well as commenting the manuscript and to Marko Mechora for technical assistance.

Conflicts of Interest: The authors declare no conflict of interest.

References

1. Ahmad, I.; Jiménez-Gasco, M.D.M.; Luthe, D.S.; Shakeel, S.N.; Barbercheck, M.E. Endophytic *Metarhizium robertsii* promotes maize growth, suppresses insect growth, and alters plant defense gene expression. *Biol. Control* **2020**, *114*, 104167. [CrossRef]
2. Barra-Bucarei, L.; González, M.G.; Iglesias, A.F.; Aguayo, G.S.; Peñalosa, M.G.; Vera, P.V. *Beauveria bassiana* multifunction as an endophyte: Growth promotion and biologic control of *Trialeurodes vaporariorum*, (Westwood) (Hemiptera: Aleyrodidae) in tomato. *Insects* **2020**, *11*, 591. [CrossRef] [PubMed]
3. Reinbacher, L.; Bacher, S.; Knecht, F.; Schweizer, C.; Sostizzo, T.; Grabenweger, G. Preventive field application of *Metarhizium brunneum* in cover crops for wireworm control. *Crop Prot.* **2021**, *150*, 105811. [CrossRef]
4. Jaronski, S.T.; Jackson, M.A. Mass production of entomopathogenic Hypocreales. In *Manual of Techniques in Invertebrate Pathology*, 2nd ed.; Lacey, L.A., Ed.; Academic Press: London, UK, 2012; pp. 225–284.
5. Vega, F.E.; Goettel, M.S.; Blackwell, M.; Chandler, D.; Jackson, M.A.; Keller, S.; Koike, M.; Maniania, N.K.; Monzón, A.; Ownley, B.H.; et al. Fungal entomopathogens: New insights on their ecology. *Fungal Ecol.* **2009**, *2*, 149–159. [CrossRef]
6. Fisher, J.J.; Rehner, S.A.; Bruck, D.J. Diversity of rhizosphere associated entomopathogenic fungi of perennial herbs, shrubs and coniferous trees. *J. Invertebr. Pathol.* **2011**, *106*, 289–295. [CrossRef]
7. Bamisile, B.S.; Dash, C.K.; Akutse, K.S.; Keppanan, R.; Afolabi, O.G.; Hussain, M.; Qasim, M.; Wang, L. Prospects of endophytic fungal entomopathogens as biocontrol and plant growth promoting agents: An insight on how artificial inoculation methods affect endophytic colonization of host plants. *Microbiol. Res.* **2018**, *217*, 34–50. [CrossRef]
8. Biere, A.; Bennett, A.E. Three-way interactions between plants, microbes and insects. *Funct. Ecol.* **2013**, *27*, 567–573. [CrossRef]
9. Gruden, K.; Lidoy, J.; Petek, M.; Podpečan, V.; Flors, V.; Papadopoulou, K.K.; Pappas, M.L.; Martinez-Medina, A.; Bejarano, E.; Biere, A.; et al. Ménage à Trois: Unraveling the Mechanisms Regulating Plant-Microbe-Arthropod Interactions. *Trends Plant Sci.* **2020**, *25*, 1215–1226. [CrossRef]
10. Lacey, L.A.; Grzywacz, D.; Shapiro-Ilan, D.I.; Frutos, R.; Brownbridge, M.; Goettel, M.S. Insect pathogens as biological control agents: Back to the future. *J. Invertebr. Pathol.* **2015**, *132*, 1–41. [CrossRef]
11. Rocha, I.; Ma, Y.; Souza-Alonso, P.; Vosátka, M.; Freitas, H.; Oliveira, R.S. Seed Coating: A Tool for Delivering Beneficial Microbes to Agricultural Crops. *Front. Plant Sci.* **2019**, *10*, 1357. [CrossRef]
12. Waliwitiya, R.; Isman, M.B.; Vernon, R.S.; Riseman, A. Insecticidal Activity of Selected Monoterpenoids and Rosemary Oil to *Agriotes obscurus* (Coleoptera: Elateridae). *J. Econ. Entomol.* **2005**, *98*, 1560–1565. [CrossRef]
13. Furlan, L. IPM thresholds for *Agriotes* wireworm species in maize in Southern Europe. *J. Pest Sci.* **2014**, *87*, 609–617. [CrossRef]
14. Canassa, F.; Tall, S.; Moral, R.A.; Lara, I.A.R.D.; Delalibera, I.; Meyling, N.V. Effects of bean seed treatment by the entomopathogenic fungi *Metarhizium robertsii* and *Beauveria bassiana* on plant growth, spider mite populations and behavior of predatory mites. *Biol. Control* **2019**, *132*, 199–208. [CrossRef]
15. Qin, X.; Zhao, X.; Huang, S.; Deng, J.; Li, X.; Luo, Z.; Zhang, Y. Pest management via endophytic colonization of tobacco seedlings by the insect fungal pathogen *Beauveria bassiana*. *Pest Manag. Sci.* **2021**, *77*, 2007–2018. [CrossRef]
16. Biswas, C.; Dey, P.; Satpathy, S.; Satya, P.; Mahapatra, B.S. Endophytic colonization of white jute (*Corchorus capsularis*) plants by different *Beauveria bassiana* strains for managing stem weevil (*Apion corchori*). *Phytoparasitica* **2013**, *41*, 17–21. [CrossRef]
17. Faeth, S.H.; Fagan, W.F. Fungal endophytes: Common host plant symbionts but uncommon mutualists. *Integr. Comp. Biol.* **2002**, *42*, 360–368. [CrossRef]

18. Saikkonen, K.; Wäli, P.; Helander, M.; Faeth, S.H. Evolution of endophyte-plant symbioses. *Trends Plant Sci.* **2004**, *9*, 275–280. [CrossRef]
19. Leger, R.J.S. Studies on adaptations of *Metarhizium anisopliae* to life in the soil. *J. Invertebr. Pathol.* **2008**, *98*, 271–276. [CrossRef]
20. Bruck, D.J. Fungal entomopathogens in the rhizosphere. *BioControl* **2010**, *55*, 103–112. [CrossRef]
21. Harman, G.E.; Howell, C.R.; Viterbo, A.; Chet, I.; Lorito, M. *Trichoderma* species—Opportunistic, avirulent plant symbionts. *Nat. Rev. Microbiol.* **2004**, *2*, 43–56. [CrossRef]
22. Atanasova, L.; Crom, S.L.; Gruber, S.; Coulpier, F.; Seidl-Seiboth, V.; Kubicek, C.P.; Druzhinina, I.S. Comparative transcriptomics reveals different strategies of *Trichoderma* mycoparasitism. *BMC Genom.* **2013**, *14*, 121. [CrossRef]
23. Ahmed, A.M.; El-Katatny, M.H. Entomopathogenic fungi as biopesticides against the Egyptian cotton leaf worm, Spodoptera littoralis: Between biocontrol-promise and immune-limitation. *J. Egypt. Soc. Toxicol.* **2007**, *37*, 39–51.
24. Shakeri, J.; Foster, H.A. Proteolytic activity and antibiotic production by *Trichoderma harzianum* in relation to pathogenicity to insects. *Enzym. Microb. Technol.* **2007**, *40*, 961–968. [CrossRef]
25. Razinger, J.; Lutz, M.; Schroers, H.J.; Urek, G.; Grunder, J. Evaluation of insect associated and plant growth promoting fungi in the control of cabbage root flies. *J. Econ. Entomol.* **2014**, *107*, 1348–1354. [CrossRef]
26. Berini, F.; Caccia, S.; Franzetti, E.; Congiu, T.; Marinelli, F.; Casartelli, M.; Tettamanti, G. Effects of *Trichoderma* viride chitinases on the peritrophic matrix of Lepidoptera. *Pest Manag. Sci.* **2016**, *72*, 980–989. [CrossRef]
27. Dhinaut, J.; Balourdet, A.; Teixeira, M.; Chogne, M.; Moret, Y. A dietary carotenoid reduces immunopathology and enhances longevity through an immune depressive effect in an insect model. *Sci. Rep.* **2017**, *7*, 12429. [CrossRef]
28. Sharma, L.; Oliveira, I.; Torres, L.; Marques, G. Entomopathogenic fungi in Portuguese vineyards soils: Suggesting a '*Galleria-Tenebrio*-bait method' as bait-insects *Galleria* and *Tenebrio* significantly underestimate the respective recoveries of *Metarhizium (robertsii)* and *Beauveria (bassiana)*. *MycoKeys* **2018**, *38*, 1–23. [CrossRef]
29. Medo, J.; Cagáň, L'. Factors affecting the occurrence of entomopathogenic fungi in soils of Slovakia as revealed using two methods. *Biol. Control* **2011**, *59*, 200–208. [CrossRef]
30. Hughes, W.O.H.; Thomsen, L.; Eilenberg, J.; Boomsma, J.J. Diversity of entomopathogenic fungi near leaf-cutting ant nests in a neotropical forest, with particular reference to *Metarhizium anisopliae* var. *anisopliae*. *J. Invertebr. Pathol.* **2004**, *85*, 46–53. [CrossRef]
31. Esparza Mora, M.A.; Costa Rouws, J.R.; Fraga, M.E. Occurrence of entomopathogenic fungi in atlantic forest soils. *Microbiol. Discov.* **2016**, *4*, 1. [CrossRef]
32. Ibrahim, R.A. Laboratory Evaluation of Entomopathogenic Fungi, Commercial Formulations, against the Rhinoceros Beetle, *Oryctes agamemnon arabicus* (Coleoptera: Scarabaeidae). *Egypt. J. Biol. Pest Control* **2017**, *27*, 49–55.
33. Clifton, E.H.; Jaronski, S.T.; Hajek, A.E. Virulence of Commercialized Fungal Entomopathogens Against Asian Longhorned Beetle (Coleoptera: Cerambycidae). *J. Insect Sci.* **2020**, *20*, 1–6. [CrossRef] [PubMed]
34. Medo, J.; Medová, J.; Michalko, J.; Cagáň, L'. Variability in virulence of Beauveria spp. soil isolates against *Ostrinia nubilalis*. *J. Appl. Entomol.* **2021**, *145*, 92–103. [CrossRef]
35. Lopes, R.B.; Mesquita, A.L.M.; Tigano, M.S.; Souza, D.A.; Martins, I.; Faria, M. Diversity of indigenous Beauveria and *Metarhizium* spp. in a commercial banana field and their virulence toward *Cosmopolites sordidus* (Coleoptera: Curculionidae). *Fungal Ecol.* **2013**, *6*, 356–364. [CrossRef]
36. Hallouti, A.; Ait Hamza, M.; Zahidi, A.; Ait Hammou, R.; Bouharroud, R.; Ait Ben Aoumar, A.; Boubaker, H. Diversity of entomopathogenic fungi associated with Mediterranean fruit fly *(Ceratitis capitata* (Diptera: Tephritidae)) in Moroccan Argan forests and nearby area: Impact of soil factors on their distribution. *BMC Ecol.* **2020**, *20*, 64. [CrossRef] [PubMed]
37. Quesada-Moraga, E.; Ruiz-GarcíA, A.; Santiago-Álvarez, C. Laboratory evaluation of entomopathogenic fungi Beauveria bassiana and Metarhizium anisopliae against puparia and adults of Ceratitis capitata (Diptera: Tephritidae). *J. Econ. Entomol.* **2006**, *99*, 1955–1966. [CrossRef]
38. Meyling, N.V.; Lübeck, M.; Buckley, E.P.; Eilenberg, J.; Rehner, S.A. Community composition, host range and genetic structure of the fungal entomopathogen *Beauveria* in adjoining agricultural and seminatural habitats. *Mol. Ecol.* **2009**, *18*, 1282–1293. [CrossRef]
39. Goble, T.A.; Dames, J.F.; Hill, M.P.; Moore, S.D. The effects of farming system, habitat type and bait type on the isolation of entomopathogenic fungi from citrus soils in the Eastern Cape Province, South Africa. *BioControl* **2010**, *55*, 399–412. [CrossRef]
40. Vargas, W.A.; Mandawe, J.C.; Kenerley, C.M. Plant-derived sucrose is a key element in the symbiotic association between *Trichoderma* virens and maize plants. *Plant Physiol.* **2009**, *151*, 792–808. [CrossRef]
41. Saravanakumar, K.; Li, Y.; Yu, C.; Wang, Q.Q.; Wang, M.; Sun, J.; Gao, J.X.; Chen, J. Effect of *Trichoderma harzianum* on maize rhizosphere microbiome and biocontrol of *Fusarium* Stalk rot. *Sci. Rep.* **2017**, *7*, 1771. [CrossRef]
42. Xu, Y.; Zhang, J.; Shao, J.; Feng, H.; Zhang, R.; Shen, Q. Extracellular proteins of *Trichoderma guizhouense* elicit an immune response in maize (*Zea mays*) plants. *Plant Soil* **2020**, *449*, 133–149. [CrossRef]
43. Ousley, M.A.; Lynch, J.M.; Whipps, J.M. Effect of *Trichoderma* on plant growth: A balance between inhibition and growth promotion. *Microb. Ecol.* **1993**, *26*, 277–285. [CrossRef] [PubMed]
44. Razinger, J.; Praprotnik, E.; Schroers, H.J. Bioaugmentation of Entomopathogenic Fungi for Sustainable *Agriotes* Larvae (Wireworms) Management in Maize. *Front. Plant Sci.* **2020**, *11*, 535005. [CrossRef] [PubMed]

45. Kuzhuppillymyal-Prabhakarankutty, L.; Tamez-Guerra, P.; Gomez-Flores, R.; Rodriguez-Padilla, M.C.; Ek-Ramos, M.J. Endophytic *Beauveria bassiana* promotes drought tolerance and early flowering in corn. *World J. Microbiol. Biotechnol.* **2020**, *36*, 47. [CrossRef] [PubMed]
46. Spatafora, J.W.; Sung, G.H.; Sung, J.M.; Hywel-Jones, N.L.; White, J.F. Phylogenetic evidence for an animal pathogen origin of ergot and the grass endophytes. *Mol. Ecol.* **2007**, *16*, 1701–1711. [CrossRef]
47. Wang, B.; Kang, Q.; Lu, Y.; Bai, L.; Wang, C. Unveiling the biosynthetic puzzle of destruxins in *Metarhizium* species. *Proc. Natl. Acad. Sci. USA* **2012**, *109*, 1287–1292. [CrossRef]
48. Machado, D.F.M.; Tavares, A.P.; Lopes, S.J.; da Silva, A.C.F. *Trichoderma* spp. na emergência e crescimento de mudas de cambará (*Gochnatia polymorpha* (Less.) Cabrera). *Rev. Árvore* **2015**, *39*, 167–176. [CrossRef]
49. Campos, B.F.; Araújo, A.J.C.; Felsemburgh, C.A.; Vieira, T.A.; Lustosa, D.C. *Trichoderma* contributes to the germination and seedling development of açaí palm. *Agriculture* **2020**, *10*, 456. [CrossRef]
50. Eagles, H.A.; Brooking, I.R. Populations of maize with more rapid and reliable seedling emergence than cornbelt dents at low temperatures. *Euphytica* **1981**, *30*, 755–763. [CrossRef]
51. Knezevic, S.Z.; Weise, S.F.; Swanton, C.J. Interference of Redroot Pigweed (*Amaranthus retroflexus*) in Corn (*Zea mays*). *Weed Sci.* **1994**, *42*, 568–573. [CrossRef]
52. Chiduza, C.; Dube, E. Maize production challenges in high biomass input smallholder farmer conservation agriculture systems: A practical research experience from South Africa. *Afr. Crop Sci. Soc.* **2013**, *11*, 23–27.
53. Rivas-Franco, F.; Hampton, J.G.; Morán-Diez, M.E.; Narciso, J.; Rostás, M.; Wessman, P.; Jackson, T.A.; Glare, T.R. Effect of coating maize seed with entomopathogenic fungi on plant growth and resistance against *Fusarium graminearum* and *Costelytra giveni*. *Biocontrol Sci. Technol.* **2019**, *29*, 877–900. [CrossRef]
54. Russo, M.L.; Scorsetti, A.C.; Vianna, M.F.; Cabello, M.N.; Ferreri, N.; Pelizza, S. Endophytic Effects of *Beauveria bassiana* on Corn (*Zea mays*) and Its Herbivore, *Rachiplusia nu* (Lepidoptera: Noctuidae). *Insects* **2019**, *10*, 110. [CrossRef]
55. Tall, S.; Meyling, N.V. Probiotics for Plants? Growth Promotion by the Entomopathogenic Fungus *Beauveria bassiana* Depends on Nutrient Availability. *Microb. Ecol.* **2018**, *76*, 1002–1008. [CrossRef]
56. Kabaluk, J.T.; Ericsson, J.D. *Metarhizium anisopliae* seed treatment increases yield of field corn when applied for wireworm control. *Agron. J.* **2007**, *99*, 1377–1381. [CrossRef]
57. Zimmermann, G. The '*Galleria* bait method' for detection of entomopathogenic fungi in soil. *J. Appl. Entomol.* **1986**, *102*, 213–215. [CrossRef]
58. Cooke, B.W. Fungi in Polluted Water and Sewage: II. Isolation Technique. *Sewage Ind. Waste.* **1954**, *26*, 661–674.
59. Williams, J.; Clarkson, J.M.; Mills, P.R.; Cooper, R.M. A selective medium for quantitative reisolation of *Trichoderma harzianum* from *Agaricus bisporus* compost. *Appl. Environ. Microbiol.* **2003**, *69*, 4190–4191. [CrossRef]
60. Razinger, J.; Schroers, H.J.; Urek, G. Virulence of *Metarhizium brunneum* to field collected *Agriotes* spp. Wireworms. *J. Agric. Sci. Technol.* **2018**, *20*, 309–320.
61. Bischoff, J.F.; Rehner, S.A.; Humber, R.A. *Metarhizium frigidum* sp. nov.: A cryptic species of *M. anisopliae* and a member of the *M. flavoviride* complex. *Mycologia* **2006**, *98*, 737–745. [CrossRef]
62. O'Donnell, K.; Kistlerr, H.C.; Cigelnik, E.; Ploetz, R.C. Multiple evolutionary origins of the fungus causing panama disease of banana: Concordant evidence from nuclear and mitochondrial gene genealogies. *Proc. Natl. Acad. Sci. USA* **1998**, *95*, 2044–2049. [CrossRef]
63. Hall, T. BioEdit: A user-friendly biological sequence alignment editor and analysis program for Windows 95/98/NT. *Nucleic Acids Symp. Ser.* **1999**, *41*, 95–98.
64. Bastidas, Ó. Cell Counting with Neubauer Chamber—Basic Hemocytometer Usage. Available online: https://moam.info/cell-counting-with-neubauer-chamber-celeromics_597734c21723dde38bd38984.html (accessed on 20 September 2021).
65. Panevska, A.; Hodnik, V.; Skočaj, M.; Novak, M.; Modic, Š.; Pavlic, I.; Podržaj, S.; Zarić, M.; Resnik, N.; Maček, P.; et al. Pore-forming protein complexes from *Pleurotus* mushrooms kill western corn rootworm and Colorado potato beetle through targeting membrane ceramide phosphoethanolamine. *Sci. Rep.* **2019**, *9*, 5073. [CrossRef]
66. Abbott, W.S. A method of computing the effectiveness of an insecticide. *J. Econ. Entomol.* **1925**, *18*, 265–267. [CrossRef]
67. Falissard, B. psy: Various Procedures Used in Psychometry. R Package Version 1.1. 2012. Available online: https://CRAN.R-project.org/package=psy (accessed on 28 September 2021).
68. Brown, C. dummies: Create Dummy/Indicator Variables Flexibly and Efficiently. R Package Version 1.5.6. 2012. Available online: https://CRAN.R-project.org/package=dummies (accessed on 26 October 2021).
69. R Core Team. *R: A Language and Environment for Statistical Computing*; R Foundation for Statistical Computing: Vienna, Austria, 2019; Available online: https://www.R-project.org/ (accessed on 8 October 2021).
70. Wickham, H.; François, R.; Henry, L.; Müller, K. dplyr: A Grammar of Data Manipulation. R Package Version 1.0.5. 2021. Available online: https://CRAN.R-project.org/package=dplyr (accessed on 28 September 2021).
71. Kassambara, A. rstatix: Pipe-Friendly Framework for Basic Statistical Tests. R Package Version 0.7.0. 2021. Available online: https://CRAN.R-project.org/package=rstatix (accessed on 28 September 2021).

MDPI

Article

Changes in Major Phenolic Compounds of Seeds, Skins, and Pulps from Various *Vitis* spp. and the Effect of Powdery and Downy Mildew Diseases on Their Levels in Grape Leaves

Arif Atak [1,*], Zekiye Göksel [2] and Yusuf Yılmaz [3]

1 Department of Viticulture, Atatürk Horticultural Central Research Institute, Yalova 77102, Turkey
2 Department of Food Quality, Atatürk Horticultural Central Research Institute, Yalova 77102, Turkey; zekiyegoksel@gmail.com
3 Department of Food Engineering, Faculty of Engineering and Architecture, Burdur Mehmet Akif Ersoy University, Burdur 15030, Turkey; yilmaz4yusuf@yahoo.com
* Correspondence: atakarif@gmail.com; Tel.: +90-505-480-4130

Abstract: The main purpose of this study is to determine the contents of 3 major phenolic compounds (gallic acid, catechin, and epicatechin) in 22 different grape cultivars/hybrids obtained from 2 different breeding programs. Additionally, changes in these phenolic components in the grape leaves of some resistant/tolerant species were determined in relation to powdery and downy mildew diseases in viticulture. The skin, pulp, and seeds of grape berries were analysed over two years, while changes in the phenolic contents of grape leaves were determined before and after these diseases for two years. The major phenolic contents of new hybrids/cultivars were compared with those of popular cultivars in different parts of the grapes, and significant differences in phenolic contents were found among hybrids/cultivars and different grape parts. Variations in the contents of phenolics in grape seeds, skins, and pulp were high, but seeds contained higher levels of these phenolics than pulp and skin. Analyses of the relationship between two viticultural diseases and phenolic changes in resistant/tolerant cultivars in relation with the susceptible "Italia" cultivar revealed that an increase in the content of the phenolic compounds was found after powdery mildew disease. Hybrids/cultivars with high phenolic contents are recommended to develop new superior cultivars, which are resistant to grape fungal diseases, in breeding programs.

Keywords: gallic acid; catechin; epicatechin; fungal diseases; grape berry; hybrids

Citation: Atak, A.; Göksel, Z.; Yılmaz, Y. Changes in Major Phenolic Compounds of Seeds, Skins, and Pulps from Various *Vitis* spp. and the Effect of Powdery and Downy Mildew Diseases on Their Levels in Grape Leaves. *Plants* **2021**, *10*, 2554. https://doi.org/10.3390/plants10122554

Academic Editors: Špela Mechora and Dragana Šunjka

Received: 25 October 2021
Accepted: 17 November 2021
Published: 23 November 2021

1. Introduction

Grapes are one of the richest sources of phenolics among fruits, and many of them are renowned for their therapeutic or health-promoting properties, making grapes an important fruit for human health [1]. Phenolic compounds are an extensive family of numerous natural bioactive compounds with health benefits [2] and constitute one of the most common and widespread groups of substances in plants [3]. Most of the phenolic compounds in plants are important for pigmentation, plant reproduction activities, juice and wine production and flavour formation, and as substrates for enzymatic browning, while playing an important role in the resistance of plants against diseases [4]. The concentration of these compounds in different parts of plants may vary, and stress factors, to which plants are exposed, may change their contents in grape berries [5]. Although the phenolic profiles of grapes depend on various factors, such as cultivar, maturity [6], genetic diversity [7], viticulture practices [8], soil characteristics [9], environmental stress and vine health status [9], the composition of phenolic compounds in grape berries strongly depends on cultivars [10–12]. The distribution of phenolic compounds in grape berries seems irregular. About 64% of the total of free phenolic compounds are found in seeds, 30% in skins, and 6% in pulps, and phenolic compounds in seeds, skin, and pulp are represented by flavan-3-ols, flavonols, and hydroxycinnamic acids, respectively [13,14]. Grape seeds and skins are important sources of phytochemicals, such as gallic acid, catechin, and epicatechin, and suitable raw materials for the production of antioxidative dietary supplements [15,16].

Secondary metabolites in plants may act as a part of defence mechanism against herbivores, microbes, viruses, and competing plants, and signal compounds to attract pollinating or seed-dispersing animals, as well as protection of the plant from ultraviolet radiation and oxidants [17,18]. While plant phenolics as secondary metabolites may inhibit the growth of insects [19], their resistance role against fungi is more dynamic than their role against insects. Infection, wounding or herbivory, in plants may induce the production of several classes of secondary metabolites. Genetic variation in the speed and extent of such induction may account for the difference between resistant and susceptible cultivars. When plants are exposed to diseases, a rapid accumulation of phenolic compounds are usually observed as a part of their defence mechanism in the infected part first, and this slows down the development of pathogens [20]. Researchers working on this topic have previously found a correlation between increased host resistance and high phenolic compound content [21,22]. A variety of factors contribute to the ability of plants to resist attacks by different pathogenic microorganisms, and the level of phenolics in different parts of grapes may differ. Concentration of phenolics may increase significantly after the infection of fungal diseases [23,24].

Downy mildew (*P. viticola*) and powdery mildew (*U. necator*) are the most common destructive grapevine diseases that occur worldwide, particularly in warm and humid climates [25]. According to recent global surveys, researchers and agricultural professionals in the main grape growing regions have considered these diseases as the most harmful for grape production [26,27]. Although these diseases damage almost all the above-ground organs of vine, they cause great damage, especially in grape berries. With different breeding studies carried out in the world, new grape cultivars resistant to these diseases are developed continually [28,29]. Such studies are carried out by two different exclusive research institutes, Tekirdağ Viticulture Research Institute (TVRI) and Yalova Atatürk Horticultural Central Research Institutes (YAHCRI), in Turkey for many years [29]. The cultivars and hybrids used in this study have been developed under the breeding programs of these institutes.

Flavonoids are a group of the most abundant biologically active phytonutrients among polyphenolic compounds present in grapes, and they represent a large family of secondary metabolites, with nearly 6000 structures identified in plants. Catechin and epicatechin are among the most common flavonoids found in grapes [30,31]. Gallic acid is a trihydroxybenzoic acid and classified as a phenolic acid, which may have various therapeutic properties, including antioxidant, anti-cancer, anti-inflammatory, antifungal, and antiviral activities [32]. Gallic acid, catechin, and epicatechin have been studied by many researchers, especially their concentrations in different parts of grape berries from various cultivars because of their health-beneficial properties [33,34].

Monitoring the major phenolics of grape berries and response of grape plants to viticultural diseases are critical for the selection of superior cultivars/hybrids in the breeding programs of viticultural studies. Therefore, the contents of gallic acid, epicatechin, and catechin in three different berry parts (pulp, skin, and seed) of grape cultivars/hybrids from the breeding programs of TVRI and YAHCRI were monitored for two years in the first part of the study. In the second part, cultivars/hybrids that were resistant/tolerant to powdery and downy mildew diseases at different rates were selected and the contents of gallic acid, epicatechin, and catechin in grape leaves were evaluated before and after the disease. Finally, the relationship between the contents of phenolics and the diseases of cultivars/hybrids were determined.

2. Materials and Methods

2.1. Plant Materials

In the first part of the study, the contents of 3 different phenolic compounds present in 3 different parts of grape berries (skin, pulp, and seed) were determined in 22 cultivars/hybrids grafted on Kober 5 BB rootstocks over two years. The properties of grape cultivars/hybrids used in this first part are given in Table 1.

Table 1. Main characteristics and origin of grape cultivars/hybrids used in this study.

Cultivar/Hybrid	Species	Origin of Material	Berry Colour	Special Flavour	Seed Status
Isabella	Interspecies (*V. labrusca* × *V. vinifera*)	Common Cultivar	Black	Foxy	Seeded
Özer Karası		TVRI *	Black	No	Seeded
BX1-166		TVRI	Yellow/Green	No	Seeded
FX1-1		TVRI	Yellow/Green	No	Seeded
FX1-10		YAHCRI	Yellow/Green	No	Seeded
Alphonse Lavallée	*V. vinifera*	Common Cultivar	Black	No	Seeded
Muscat Hamburg		Common Cultivar	Black	Muscat	Seeded
Yalova Misketi		YAHCRI **	Black	Muscat	Seeded
Trakya İlkeren		TVRI	Blue/Black	No	Seeded
Bilecik İrikarası		Common Cultivar	Black	No	Seeded
İsmetbey		YAHCRI	Black	No	Seeded
KXP-10		TVRI	Black	No	Seeded
Tekirdağ Çekdsz.		TVRI	Dark Red/Purple	No	Seedless
Reçel Üzümü		TVRI	Red/Black	No	Seedless
Güz Gülü		TVRI	Rose	No	Seedless
Pembe 77		YAHCRI	Dark Pink	No	Seeded
Uslu		YAHCRI	Rose	No	Seeded
83/1		YAHCRI	Rose	No	Seeded
85/1		YAHCRI	Yellow/Green	Muscat	Seeded
53/1		YAHCRI	Yellow/Green	No	Seeded
86/1		YAHCRI	Yellow/Green	Muscat	Seeded
130/1		YAHCRI	Yellow/Green	No	Seedless
Italia		Common Cultivar	Yellow/Green	Muscat	Seeded

* TVRI, Tekirdağ Viticulture Research Institute (Tekirdağ, Turkey); ** YAHCRI, Yalova Atatürk Horticultural Central Research Institutes (Yalova, Turkey).

Standard grape cultivars were grown in the Marmara region, northwest Turkey, with an exception of Isabella, a local grapevine cultivar in the Black Sea region in northern Turkey. It is grown especially in humid areas because of its resistance to fungal diseases. Other cultivars/hybrids were selected from the disease-resistant cultivation breeding program of TVRI (Tekirdağ, Turkey). The fresh berry samples were harvested from 22 different cultivars between the third week of August and end of September during two consecutive growing seasons. Soluble solid levels ranged from 25.3 to 30.4° Brix during the harvest. Plants were grown at the ACHRI vineyards in Yalova, Turkey.

In the second part of the study, 10 grape cultivars/hybrids were inoculated with powdery and downy mildew diseases artificially in a greenhouse over 2 years. *V. vinifera* "Italia" was used as control because of its sensitivity to the diseases. Related analyses were conducted in the Food Technology Laboratory at YAHCRI. Disease inoculations were carried out in a greenhouse using 2-year-old potted vines grown in 5 L pots filled with a soil mixture (1/3 garden soil, 1/3 peat moss, and 1/3 compost). Optimum climatic conditions were ensured for the development of both diseases in the greenhouse.

2.2. Inoculation of Vines and Evaluation of Fungal Diseases

All cultivars/hybrids were planted in pots and cultivated in the greenhouse for downy and powdery mildew inoculations. Inoculation was applied according to Wang et al. [35] and Boso et al. [36]. Fungal conidia were collected from infected leaves from the YAHCRI vineyards. For the propagation of powdery mildew, inoculum vines were sprayed with a suspension of sporangia (40,000 sporangia per mL of distilled water) on the abaxial leaf side and plants were completely covered with polyethylene covers overnight. Next day, polyethylene covers were removed, and incubation lasted 5–6 days at 25 °C. This procedure was repeated after a week.

The vine leaves were inoculated with a conidial suspension at 2×10^5 conidia mL^{-1} by spraying the upper surface of the leaves for downy mildew inoculation. The inoculated leaves were immediately covered with thin plastic for 6 h. Fogging was applied for a limited period in order to stimulate the formation of the diseases at a desired level. Both disease inoculations were taken place independently in two separate compartments in the greenhouse.

Depending on the vigour of vines, four-to-six young leaves from the shoot tip were selected from each vine, and were observed for powdery mildew at different times during June–August. The severity of infections on leaves was determined based on the percentage of disease spots observed on the entire leaf area [37], according to the procedure described in Table 2. Disease severity was scored 3 weeks after inoculation. Since plants in a greenhouse usually develop faster than those in an open field, disease inoculation and scoring were done earlier. All leaves of each plant were observed at different times during May–August for a downy mildew disease. The infection severity on leaves was determined based on the percentage of disease spots observed on the entire leaf area (Table 2). Scoring was done after 6 weeks of inoculation.

Table 2. Scoring scale used for downy and powdery mildew diseases in grape leaves (from 1: very low, to 9: very high).

Scale	Level	Symptom/Reaction		
		Powdery Mildew	**Downy Mildew**	**Host Response**
1	Very low	Tiny spots or no symptoms; neither visible sporulation nor mycelium	Tiny necrotic spots or no symptoms; neither sporulation nor mycelium	Extremely Resistant
3	Low	Limited patches < 2 cm diameter; limited sporulation and mycelium; the presence of *Uncinula* is only indicated by a slight curling of the blade	Small patches < 1 cm in diameter; little sporulation or mycelium	Resistant
5	Medium	Patches usually limited with a diameter of 2–5 cm	Little patches 1–2 cm diameter; more or less strong sporulation; irregular formation of mycelium	Tolerant
7	High	Vast patches; some limited; strong sporulation and abundant mycelium	Vast patches; strong sporulation and abundant mycelium; leaf drop later than below	Susceptible
9	Very high	Very vast unlimited patches or totally attached leaf blades; strong sporulation and abundant mycelium	Vast patches or totally attached leaf blades; strong sporulation and dense mycelium; very early leaf drop	Extremely Susceptible

2.3. Collection and Preparation of Samples

Samples in the first part of the study were taken as follows: the clusters of cultivar/hybrids were checked every week for the ideal harvest time, and those that reached the ideal Brix ratio were harvested. At the time of harvest, 2–3 kg samples were collected from different vines and different parts of vines, representing each cultivar. A bunch of grapes was brought to the laboratory immediately after harvest. The seed, skin, and pulp parts of the grapes were carefully separated manually. Each part was divided into three equal parts, representing replicates. Seeds were partially dried at 45 °C for 4 h in a convection oven (Memmert UN110, Nurnberg, Germany) in order to facilitate the grinding process. The pulp and skin parts were frozen at −20 °C until analysis. The partially dried seeds were ground with a coffee grinder (Bosch, MKM 6000, Istanbul, Turkey), whereas the pulp or skin parts were chopped with a blender immediately after thawing. Analyses were performed in duplicates within a month.

Grape cultivars/hybrids in the second part of the study were kept in separate sections in the greenhouse with 3 replications and at least 3 vines in each repeat. Samples were collected from all replicates before and after diseases. The first 6 leaves of the grape cultivars/hybrids from the shoot tip (before and 3 weeks after disease inoculation) were used for analyses. The first 6 leaves of each of the 3 vines (healthy and infected vines) were used, and the leaves were separated carefully from their petioles and washed in pure water. Each part was divided into three equal parts, representing replicates. Clean leaves were dried for 48 h at room temperature under dark. Dried leaves were ground at a high speed for 60 s by a grinder, and then 2 g of leaf powder was added to 40 mL of methanol and shaken for 60 min at 60 °C in a water bath. The samples were centrifuged for 10 min at 7000 rpm (5810 Eppendorf AG, Hamburg, Germany). The supernatants were collected in amber bottles and stored at −20 °C until analysis. Three extracts were obtained for each grape cultivar/hybrid: the first week of June (healthy), end of June (downy mildew), and end of July (powdery mildew).

2.4. Determination of Major Phenolic Compounds

The contents of gallic acid, catechin, and epicatechin in the leaf extracts of grape cultivars/hybrids were determined by using a chromatographic unit (HP1100 System HPLC, Agilent Technologies Inc., Palo Alto, CA, USA), according to the method of Katalinic et al. [38] with some modifications.

The separations were conducted at room temperature in an Agilent Eclipse XDB-C18 column (4.6 × 250 mm, particle size 5 µm), protected by a guard column. The compounds were detected with an HP 1100 series ultraviolet (UV) Diode Array Detector (DAD). The mobile phase included A: 2.0% acetic acid in distilled water, and B: acetonitrile. The column was eluted at 1.0 mL/min under a linear gradient from 5% mobile phase B to 75% over 20 min, to 100% over 5 min, isocratic for 5 min, to 25% over 5 min, and to 5% over 5 min. The injection volume was 20 µL for each sample. Phenolic compounds were identified according to the retention times of the available pure compounds and the UV–Vis data obtained from authentic standards and/or published in previous studies [39]. Gallic acid, catechin, and epicatechin (Sigma-Aldrich) were quantified at 280 nm. Their contents in grape seeds, skins, and pulps and leaf samples were expressed as milligrams per weight (mg 100 g^{-1}). The results of these phenolic compounds were the averages of triplicates.

2.5. Statistical Analysis

An analysis of variance (ANOVA) was used to determine significant differences among means. The data were presented as arithmetic means of three replications ± standard deviations, which represented the means of two consecutive years. For each parameter, the LSD (the least significant difference) was used to determine the level of significant differences for all accessions. Differences at $p < 0.05$ were considered significant. The correlation coefficients (R) of a parametric Pearson's test was used to evaluate covariance relationships among variables. The statistical analyses were performed using JMP 15.0 software [40].

3. Results

3.1. Evaluation of Fungal Diseases

The resistance level of the cultivars/hybrids against downy and powdery mildew diseases was monitored for two years after the artificial inoculation of vine, and results are given in Table 3. Results indicated that the cultivar "Isabella" was very resistant to downy mildew disease, while the cultivar "Özer Karası" and the hybrid of FX1-1 exhibited resistance to this disease (with a score of 3). Other five cultivars/hybrids determined as tolerant to downy mildew diseases.

Table 3. Scores of grape cultivars/hybrids after artificial inoculation of downy and powdery mildew diseases (1: High Resistance; 3: Resistance; 5: Tolerance; 7: Susceptible; 9: High Susceptible).

Cultivar/Hybrid	Species	Downy Mildew	Powdery Mildew
Isabella	Interspecies (*V. vinifera* × *V. labrusca*)	1	1
Özer Karası		3	3
FX1-1		3	3
FX1-10		5	1
BX1-166		5	5
KXP-10	*V. vinifera*	5	5
86/1		5	7
Güz Gülü		5	5
Italia		5	7

In terms of a powdery mildew disease, results, which are similar to the scores for downy mildew diseases, were obtained in general (Table 3). The cultivar "Italia", known to be susceptible to powdery mildew diseases, was used as a control cultivar, and it was found "susceptible" to powdery mildew diseases in our study. In addition, the 86/1 hybrid had a score of 7, indicating that it was "susceptible" to powdery mildew diseases. Of the other cultivars/hybrids, "Isabella" and "FX1-10" had "a high resistance", while "FX1-1" and "Özer Karası" were resistant. "Güzgülü", "BX1-166", and "KXP-10" showed "tolerance" to powdery mildew disease.

3.2. Determination of Phenolic Compounds in Grape Leaves

The gallic acid, catechin, and epicatechin contents of grape leaves from different cultivars/hybrids were monitored for two years before and after downy and powdery mildew diseases, and results are given in Table 4. In addition, Pearson's correlation coefficients among the contents of phenolic compounds in grape leaves before and after diseases are given in Table 5, representing the averages of the two harvesting seasons.

Table 4. Contents of major phenolic compounds before and after downy/powdery mildew diseases in grape leaves.

Cultivars/ Hybrids	Catechin * (mg 100 g^{-1})					
	Before Disease		After Downy Mildew		After Powdery Mildew	
	1st Year	2nd Year	1st Year	2nd Year	1st Year	2nd Year
Isabella	0.09 ± 0.02 [c] **	15.63 ± 0.07 [b]	Nd ***	1.60 ± 0.06 [e]	0.10 ± 0.00 [c]	1.14 ± 0.03 [d]
Özer Karası	0.23 ± 0.02 [b]	54.50 ± 2.40 [a]	Nd	3.83 ± 0.01 [a]	Nd	3.67 ± 4.34 [cd]
Fx1-1	0.21 ± 0.02 [b]	0.23 ± 0.00 [f]	0.09 ± 0.02 [c]	2.38 ± 0.06 [d]	Nd	1.18 ± 0.08 [d]
Fx1-10	0.06 ± 0.01 [c]	5.63 ± 0.02 [e]	0.25 ± 0.10 [b]	0.24 ± 0.04 [g]	0.13 ± 0.00 [c]	4.83 ± 0.44 [bc]
BX1-166	0.29 ± 0.02 [a]	9.12 ± 0.31 [d]	0.09 ± 0.01 [c]	3.52 ± 0.06 [b]	2.62 ± 0.18 [c]	8.81 ± 0. 24 [a]
KXP-10	0.21 ± 0.03 [b]	53.50 ± 0.42 [a]	0.87 ± 0.10 [a]	1.49 ± 0.00 [e]	45.83 ± 4.35 [a]	3.86 ± 0.06 [cd]
86/1	Nd	12.77 ± 0.07 [c]	Nd	0.17 ± 0.00 [g]	Nd	1.60 ± 0.03 [cd]
Güzgülü	Nd	8.38 ± 0.06 [d]	Nd	2.79 ± 0.07 [c]	Nd	4.12 ± 0.06 [b–d]
Italia	Nd	0.09 ± 0.00 [f]	0.05 ± 0.00 [c]	0.87 ± 0.15 [f]	13.67 ± 0.29 [b]	7.27 ± 0.21 [b]
	Epicatechin (mg 100 g^{-1})					
Isabella	Nd	0.71 ± 0.04 [e]	Nd	0.30 ± 0.00 [i]	42.52 ± 6.04 [e]	1.24 ± 0.02 [cd]
Özer Karası	Nd	12.20 ± 0.42 [b]	Nd	5.43 ± 0.18 [d]	127.81 ± 14.15 [c]	Nd
Fx1-1	13.64 ± 3.14 [c]	17.30 ± 0.31 [a]	48.70 ± 5.12 [c]	2.50 ± 0.11 [e]	Nd	4.18 ± 0.11 [ab]
Fx1-10	27.83 ± 3.68 [a]	0.88 ± 0.01 [e]	Nd	1.01 ± 0.01 [h]	Nd	1.79 ± 0.00 [a]
BX1-166	Nd	4.70 ± 0.06 [d]	Nd	15.37 ± 0.30 [a]	20.92 ± 1.71 [f]	0.68 ± 0.01 [de]
KXP-10	Nd	11.25 ± 1.03 [c]	86.66 ± 9.19 [a]	7.66 ± 0.25 [b]	101.01 ± 2.41 [d]	Nd
86/1	Nd	1.17 ± 0.07 [e]	Nd	1.50 ± 0.07 [g]	427.00 ± 8.49 [a]	4.55 ± 5.24 [a]
Güzgülü	7.58 ± 1.17 [d]	1.14 ± 0.05 [e]	63.09 ± 1.66 [b]	6.40 ± 0.11 [c]	7.53 ± 0.20 [fg]	Nd
Italia	18.69 ± 0.69 [b]	0.62 ± 0.01 [e]	70.29 ± 4.84 [b]	1.94 ± 0.02 [f]	176.27 ± 1.66 [b]	0.14 ± 0.00 [e]

Table 4. *Cont.*

	Gallic Acid (mg 100 g^{-1})					
Isabella	1.24 ± 0.10 ^{e}	2.33 ± 0.04 ^{c}	0.51 ± 0.07 ^{fg}	2.02 ± 0.03 ^{a}	2.68 ± 0. 43 ^{cd}	3.52 ± 0.03 ^{e}
Özer Karası	1.82 ± 0.11 ^{a}	2.81 ± 0.01 ^{b}	0.28 ± 0.01 ^{g}	0.24 ± 0.06 ^{g}	2.57 ± 0.29 ^{cd}	3.26 ± 0.08 ^{f}
Fx1-1	1.56 ± 0.13 ^{bc}	1.28 ± 0.11 ^{g}	2.53 ± 0.35 ^{a}	1.05 ± 0.07 ^{c}	1.51 ± 0.03 ^{d}	2.61 ± 0.01 ^{h}
Fx1-10	1.50 ± 0.09 ^{cd}	1.44 ± 0.06 ^{f}	0.82 ± 0.09 $^{c–e}$	0.69 ± 0.13 ^{e}	1.86 ± 0.19 ^{cd}	2.77 ± 0.14 ^{g}
BX1-166	1.31 ± 0.09 ^{de}	3.55 ± 0.07 ^{a}	0.92 ± 0.02 ^{cd}	0.90 ± 0.14 ^{d}	1.91 ± 0.20 ^{cd}	2.48 ± 0.11 ^{i}
KXP-10	0.96 ± 0.09 ^{f}	0.55 ± 0.07 ^{h}	0.97 ± 0.01 ^{c}	0.88 ± 0.11 ^{d}	4.91 ± 0.04 ^{ab}	4.98 ± 0.11 ^{b}
86/1	1.79 ± 0.02 ^{ab}	3.62 ± 0.03 ^{a}	0.64 ± 0.04 $^{d–f}$	0.65 ± 0.07 ^{e}	4.20 ± 0.36 ^{ab}	6.51 ± 0.01 ^{a}
Güzgülü	1.63 ± 0.12 $^{a–c}$	1.86 ± 0.08 ^{d}	1.50 ± 0.03 ^{b}	1.39 ± 0.13^{b}	4.06 ± 1.58 ^{ab}	4.53 ± 0.04 ^{d}
Italia	1.52 ± 0.07 ^{cd}	1.58 ± 0.04 ^{e}	0.61 ± 0.05 ^{ef}	0.51 ± 0.02 ^{f}	3.02 ± 0.09 ^{bc}	4.78 ± 0.11 ^{c}

* For each phenolic compound within a column, different superscripts across the table indicate significant differences at $p \leq 0.05$. ** All means are expressed as means ± standard deviation (n = 3). *** Nd = not detected.

Table 5. Pearson's correlation coefficients (R) among major phenolic compounds before (BD) and after downy mildew (DM) or powdery mildew diseases (PM) in grape leaves (n = 9).

Variable, Disease Condition	By Variable, Disease Condition	Correlation Coefficient	Lower 95%	Upper 95%	Significance Probability	Sign of Correlation
Gallic acid, DM	Gallic acid, BD	−0.3561	−0.8251	0.4034	0.3468	
Gallic acid, PM	Gallic acid, BD	−0.0298	−0.6805	0.6471	0.9393	
	Gallic acid, DM	−0.2248	−0.7735	0.5164	0.5608	
Catechin, BD	Gallic acid, BD	−0.1289	−0.7305	0.5853	0.7410	
	Gallic acid, DM	−0.4373	−0.8535	0.3197	0.2392	
	Gallic acid, PM	0.2791	−0.4726	0.7957	0.4671	
Catechin, DM	Gallic acid, BD	0.0781	−0.6181	0.7056	0.8416	
	Gallic acid, DM	0.1181	−0.5925	0.7253	0.7621	
	Gallic acid, PM	−0.3554	−0.8248	0.4041	0.3480	
	Catechin, BD	0.4408	−0.3158	0.8547	0.2351	
Catechin, PM	Gallic acid, BD	−0.6395	−0.9150	0.0428	0.0637	
	Gallic acid, DM	−0.1769	−0.7526	0.5521	0.6488	
	Gallic acid, PM	0.4257	−0.3323	0.8496	0.2532	
	Catechin, BD	0.4882	−0.2603	0.8702	0.1824	
	Catechin, DM	0.0616	−0.6282	0.6972	0.8749	
Epicatechin, BD	Gallic acid, BD	−0.5155	−0.8788	0.2259	0.1554	
	Gallic acid, DM	0.1934	−0.5401	0.7599	0.6180	
	Gallic acid, PM	−0.4900	−0.8708	0.2581	0.1806	
	Catechin, BD	−0.2703	−0.7922	0.4800	0.4818	
	Catechin, DM	−0.1225	−0.7274	0.5896	0.7536	
	Catechin, PM	−0.0092	−0.6692	0.6589	0.9812	

Table 5. *Cont.*

Variable, Disease Condition	By Variable, Disease Condition	Correlation Coefficient	Lower 95%	Upper 95%	Significance Probability	Sign of Correlation
Epicatechin, DM	Gallic acid, BD	−0.6992	−0.9310	−0.0654	0.0361 *	
	Gallic acid, DM	0.2901	−0.4632	0.8001	0.4488	
	Gallic acid, PM	0.3828	−0.3772	0.8347	0.3092	
	Gallic acid, BD	0.0931	−0.6087	0.7131	0.8118	
	Catechin, DM	0.1435	−0.5755	0.7373	0.7127	
	Catechin, PM	0.7056	0.0781	0.9327	0.0337*	
	Epicatechin, BD	0.2261	−0.5154	0.7740	0.5585	
Epicatechin, PM	Gallic acid, BD	0.4776	−0.2732	0.8668	0.1935	
	Gallic acid, DM	−0.5198	−0.8801	0.2204	0.1515	
	Gallic acid, PM	0.6913	0.0502	0.9289	0.0392 *	
	Catechin, BD	0.1081	−0.5990	0.7205	0.7819	
	Catechin, DM	−0.4882	−0.8702	0.2604	0.1825	
	Catechin, PM	0.0093	−0.6589	0.6693	0.9811	
	Epicatechin, BD	−0.4049	−0.8425	0.3545	0.2797	
	Epicatechin, DM	−0.1681	−0.7486	0.5584	0.6655	

* indicates that the correlation coefficient is statistically significant at $p \leq 0.05$.

The catechin contents of grape leaves for different grape cultivars/hybrids were compared in Table 4 within each harvesting season, and they seemed to fluctuate over the two harvesting seasons, although a statistical comparison was not included in the study. In the first year, the catechin contents remained low during the healthy period of the first harvesting season before mildew diseases; they increased especially after powdery mildew infections. It was noticeable that the catechin contents of leaves before diseases in the second year were generally higher than those in the first year. The values in Table 4 indicate that the effect of diseases on the catechin contents of leaves was dependent on the grape cultivar/hybrid. In terms of the numerical results in Table 4, although the catechin content of grapes leaves for the "Italia" cultivar, highly susceptible to diseases, increased after diseases, the leaves of the disease-resistant cultivars "Isabella" and "Özer Karası" had a low content of catechin in general. These results show that the catechin content in grape leaves may be influenced by mildew diseases differently depending on the susceptibility of grape vine to diseases. Among all cultivars/hybrids, the grape leaves of "KXP-10" and "Özer Karası" (disease resistant) with dark berries had a catechin content of about 54 mg 100 g^{-1} in the healthy period before diseases (Table 4). In terms of correlations within catechin itself, the highest, but insignificant correlation (0.49) was found between the values obtained before disease (Catechin, BD) and after powdery mildew disease (Catechin, PM) ($p > 0.05$).

The epicatechin contents of grape leaves increased numerically for almost all cultivars/hybrids in the 1st year of harvesting, regardless of mildew disease conditions in plants. In the 2nd year, this trend was unclear. Although seasonal changes in phenolics or the effect of diseases on major phenolics were not compared statistically in this study, this result could be attributed to the fact that the epicatechin response of vine might be highly influenced by the harvesting season (most likely climatic conditions in a season) and the severity of mildew diseases as well as the variety of grapes. An increase in the epicatechin contents was observed for the leaves of the "Italia" cultivar, which is more susceptible to mildew diseases than others, after diseases. The epicatechin contents of grape leaves from the disease-resistant cultivars "Isabella" and "Özer Karası" seemed to increase especially after powdery mildew disease. In terms of the averages of the two harvesting years, the catechin content of grape leaves increased for the cultivars/hybrids, which was much more noticeable after powdery mildew. Among the grape cultivars/hybrids studied, the epicatechin content of the "86/1" hybrid with yellow flesh colour and intense Muscat flavour after powdery mildew infection in the 1st year of harvesting was 427 mg 100 g^{-1}. After downy mildew infection, the leaves of the "KXP-10" hybrid in dark fruit colour had an epicatechin content of about 87 mg 100 g^{-1}.

The results of the correlation study within epicatechin itself show that the highest, but insignificant correlation (R = −0.40) was found before diseases (Epicatechin, BD) and after powdery mildew disease (Epicatechin, PM) ($p > 0.05$). When we look at the correlation between other phenolic compounds and epicatechin, the epicatechin content of grape leaves after downy mildew disease (Epicatechin, DM) was significantly and positively correlated with the catechin content of leaves after powdery mildew (Catechin, PM) (R = 0.71), but negatively correlated with the gallic acid content before disease (Gallic acid, BD) (R = −0.70) ($p < 0.05$). Moreover, the epicatechin content of grape leaves after powdery mildew disease (Epicatechin, PM) was positively correlated with the gallic acid content after powdery mildew disease (Gallic acid, PM) (R = 0.69) ($p < 0.05$).

In general, small increases in the gallic acid contents of grape leaves were found numerically in the 2nd harvesting year after powdery mildew infection for all cultivars/hybrids. Furthermore, the gallic acid contents of grape leaves seemed to decrease after downy mildew infections in comparison to the healthy period. The reduction in the gallic acid contents of grape leaves before the disease and after downy mildew infection in the 2nd harvesting year was similar for all cultivars/hybrids. Among all cultivars/hybrids studied, the gallic acid contents of grape leaves after powdery mildew disease were 6.51 mg 100 g^{-1} for the "86/1" hybrid with yellow berry colour and intense Muscat flavour and 4.98 mg 100 g^{-1} for the "KXP-10" hybrid with dark berry colour in the 2nd year of harvesting. Especially after powdery mildew, the grape leaves from the cultivars/hybrids with intense Muscat flavour and dark coloured berries tended to have higher gallic acid contents than those with less flavoured and light coloured berries.

The highest, but insignificant correlation (R = −0.37) was found between the gallic acid contents of grapes leaves before disease values (Gallic acid, BD) and downy mildew infection (Gallic acid, DM) ($p > 0.05$). Negative and significant correlation coefficients in Table 5 indicate that the gallic acid contents of healthy vines decreases after powdery mildew infection or vice versa. As reported in the previous paragraph, the gallic acid content of grape leaves after powdery mildew disease (Gallic acid, PM) was positively correlated with the epicatechin content after powdery mildew disease (Epicatechin, PM) with R = 0.69 ($p < 0.05$).

3.3. *Determination of Phenolic Compounds in Different Parts of the Grape Berry*

The contents of catechin, epicatechin, and gallic acid in the skins, pulps, and seeds of grape berries from 22 cultivars/hybrids for two years are given in Tables 6–8.

Especially the seed parts of grape berries from the cultivars/hybrids studied contained generally high amounts of catechin in comparison to their skin or pulp parts. The catechin contents of grape pulps ranged from 0.50 to 0.81 mg 100 g^{-1} in the 1st year and from 0.54 to 0.76 mg 100 g^{-1} in the 2nd year, while the catechin contents of grape skins ranged from 0.03 to 4.59 mg 100 g^{-1} in the 1st year and from 0.03 to 3.22 mg 100 g^{-1} in the 2nd year. Moreover, grape seeds had a catechin content ranging from 48.17 to 494.91 mg 100 g^{-1} in the 1st year and from 44.95 to 494.95 mg 100 g^{-1} in the 2nd year. The highest catechin content in seeds was detected for the "KXP-10" hybrid grapes as 492 mg 100 g^{-1} in the 1st year and 495 mg 100 g^{-1} in the 2nd year of harvesting ($p < 0.05$). Additionally, the second highest catechin content was determined for the seeds of the "85/1" hybrid, which has the yellow coloured berries and intense Muscat flavour (about 370 mg 100 g^{-1}) ($p < 0.05$). The pulp and skin parts of the grape cultivars/hybrids studied had a low content of catechin. In terms of the catechin content of skins and pulps, a high content was determined in cultivars/hybrids with dark coloured berry, such as "Isabella" and "Trakya İlkeren", or light coloured berries, such as "85/1" and "86/1", containing an intense Muscat flavour. Catechin could not be detected in the skins of two cultivars/hybrids ("Güz Gülü" and "130/1") in both years, probably due to its very low content (Table 6).

Table 6. Catechin contents (mg 100 g^{-1}) of pulps, skins, and seeds of grape cultivars/hybrids used in the study in two growing seasons.

Cultivar/ Hybrid	Pulp *		Skin		Seed	
	1st Year	2nd Year	1st Year	2nd Year	1st Year	2nd Year
Isabella	0.81 ± 0.03 ^{a} **	0.76 ± 0.01 ^{a}	1.86 ± 0.07 ^{e}	1.88 ± 0.01 ^{d}	48.17 ± 2.99 ^{m}	44.95 ± 1.55 ^{l}
Alphonse L.	0.70 ± 0.03 ^{cd}	0.68 ± 0.02 $^{b–d}$	3.07 ± 0.05 ^{c}	3.02 ± 0.22 ^{ab}	103.18 ± 2.70 ^{k}	100.80 ± 5.60 ^{ij}
Muscat Ham.	0.72 ± 0.01 ^{bc}	0.69 ± 0.02 ^{bc}	1.21 ± 0.02 ^{h}	1.20 ± 0.10 ^{f}	176.05 ± 2.85 ^{f}	176.55 ± 2.25 ^{ef}
Yalova Misketi	0.69 ± 0.11$^{c–e}$	0.65 ± 0.00 $^{d–g}$	0.69 ± 0.51 ^{j}	0.11 ± 0.01 ^{j}	262.11 ± 6.20 ^{c}	263.60 ± 3.20 ^{c}
Trakya İlkeren	0.65 ± 0.02 $^{c–g}$	0.68 ± 0.01 $^{b–d}$	3.28 ± 0.06 ^{b}	3.22 ± 0.36 ^{a}	187.45 ± 1.96 ^{e}	187.10 ± 10.80 ^{de}
Bilecik İrikarası	0.69 ± 0.06 $^{c–e}$	0.66 ± 0.02 $^{c–f}$	0.76 ± 0.02 ^{j}	0.74 ± 0.06 ^{h}	117.85 ± 1.84 ^{j}	118.00 ± 3.90 ^{i}
İsmetbey	0.63 ± 0.02 $^{d–g}$	0.64 ± 0.01 $^{e–g}$	0.04 ± 0.00 ^{l}	0.04 ± 0.00 ^{j}	130.86 ± 1.73 ^{i}	135.00 ± 3.20 ^{h}
KXP-10	0.68 ± 0.06 $^{c–f}$	0.69 ± 0.00 ^{bc}	2.55 ± 0.06 ^{d}	2.49 ± 0.44 ^{c}	491.91 ± 2.76 ^{a}	494.95 ± 18.05 ^{a}
Özer Karası	0.66 ± 0.08 $^{c–g}$	0.65 ± 0.04 $^{d–g}$	1.73 ± 0.05 ^{ef}	1.73 ± 0.01 ^{de}	163.73 ± 2.02 ^{g}	163.50 ± 17.80 ^{fg}
Tekirdağ Çekirdeksizi	0.69 ± 0.05 $^{c–e}$	0.65 ± 0.06 $^{d–g}$	0.48 ± 0.04 ^{k}	0.44 ± 0.02 ^{i}	Seedless	Seedless
Reçel Üzümü	0.79 ± 0.08 ^{ab}	0.71 ± 0.02 ^{b}	0.47 ± 0.04 ^{k}	0.44 ± 0.01 ^{i}	Seedless	Seedless
Güz Gülü	0.60 ± 0.02 $^{f–h}$	0.66 ± 0.01 $^{d–g}$	Nd ***	Nd	Seedless	Seedless
Pembe 77	0.62 ± 0.03 $^{d–g}$	0.66 ± 0.01 $^{d–g}$	1.67 ± 0.06 ^{fg}	1.60 ± 0.03 ^{e}	151.56 ± 6.80 ^{h}	154.10 ± 19.70 ^{g}
Uslu	0.60 ± 0.05 $^{f–h}$	0.66 ± 0.01 $^{d–g}$	2.58 ± 0.04 ^{d}	2.61 ± 0.05 ^{c}	191.78 ± 1.17 ^{de}	191.93 ± 0.88 ^{d}
83/1	0.65 ± 0.05 $^{c–g}$	0.67 ± 0.01 $^{c–e}$	0.77 ± 0.02 ^{j}	0.77 ± 0.11 ^{gh}	197.16 ± 1.23 ^{d}	198.35 ± 1.85 ^{d}
FX1-1	0.52 ± 0.04 ^{hi}	0.58 ± 0.02 ^{h}	1.15 ± 0.07 ^{hi}	1.10 ± 0.05 ^{f}	83.42 ± 2.97 ^{l}	82.65 ± 2.04 ^{k}
85/1	0.66 ± 0.03 $^{c–g}$	0.67 ± 0.02 $^{c–e}$	4.59 ± 0.05 ^{a}	2.87 ± 0.06 ^{b}	370.58 ± 7.06 ^{b}	367.70 ± 2.50 ^{b}
BX1-166	0.50 ± 0.03 ^{i}	0.54 ± 0.05 ^{i}	1.00 ± 0.02 ^{i}	0.98 ± 0.03 ^{fg}	113.59 ± 1.78 ^{j}	109.00 ± 6.86 ^{ij}
53/1	0.61 ± 0.02 $^{e–g}$	0.67 ± 0.02 ^{cd}	0.03 ± 0.01 ^{l}	0.03 ± 0.00 ^{j}	117.26 ± 2.67 ^{j}	171.80 ± 3.60 ^{f}
86/1	0.68 ± 0.06 $^{c–f}$	0.63 ± 0.00 ^{fg}	1.53 ± 0.10 ^{g}	3.20 ± 0.10 ^{a}	170.89 ± 1.46 ^{f}	170.25 ± 2.95 ^{f}
FX1-10	0.59 ± 0.03 ^{gh}	0.63 ± 0.02 ^{g}	1.08 ± 0.05 ^{hi}	1.12 ± 0.10 ^{f}	100.84 ± 0.91 ^{k}	101.33 ± 2.19 ^{j}
130/1	0.67 ± 0.02 $^{c–g}$	0.70 ± 0.02 ^{bc}	Nd	Nd	Seedless	Seedless

* Different superscripts within a column indicate significant differences at $p \leq 0.05$. ** All means are expressed as means ± standard deviations (n = 3). *** Nd = not detected.

The seeds of the grape cultivars/hybrids studied contained more epicatechin than their skins or pulps for the two harvesting years. The epicatechin contents of grape pulps ranged from 0.01 to 1.06 mg 100 g^{-1} in the 1st year and from 0.01 to 1.23 mg 100 g^{-1} in the 2nd year. The epicatechin contents of grape skins ranged from 0.14 to 1.79 mg 100 g^{-1} in the 1st year and from 0.10 to 1.79 mg 100 g^{-1} in the 2nd year. Morevoer, the epicatechin contents of the grape seeds ranged from 20.65 to 106.91 mg 100 g^{-1} in the 1st year and from 20.50 to 107.20 mg 100 g^{-1} in the 2nd year. Among the cultivars/hybrids, the highest epicatechin content was determined in the seeds of "86/1" hybrid grapes with yellow-coloured fruits and an intense Muscat flavour (107 mg 100 g^{-1}), which was followed by the seeds of the "KXP-10" hybrid grapes with black berry colour in both years (about 100 mg 100 g^{-1}) ($p < 0.05$). The epicatechin content of the seeds of "Isabella" and "Özer Karası", two dark berry cultivars, was about 21 mg 100 g^{-1}, and their epicatechin content was the lowest ($p < 0.05$). The high levels of epicatechin in pulps and skins were obtained from cultivars with dark berry colour, such as the "Bilecik İrikarası" and "İsmetbey" cultivars. The skins or pulps of the cultivar "Pembe 77" grapes with a dark pink berry colour had a relatively high epicatechin content. The epicatechin contents in the pulps or skins of some grape cultivars/hybrids could not be detected in some of the harvesting years, possibly due to the existence of their very low levels (Table 7).

Table 7. Epicatechin contents (mg 100 g^{-1}) of pulps, skins, and seeds of grape cultivars/hybrids used in the study in two growing seasons.

Cultivar/ Hybrid	Pulp *		Skin		Seed	
	1st Year	2nd Year	1st Year	2nd Year	1st Year	2nd Year
Isabella	0.05 ± 0.00 $^{e–h}$ **	Nd ***	Nd	Nd	20.65 ± 0.17 ^{n}	20.70 ± 0.90 ^{j}
Alphonse L.	0.24 ± 0.01 $^{b–d}$	0.23 ± 0.02 ^{d}	1.22 ± 0.01 ^{d}	1.21 ± 0.03 ^{c}	44.72 ± 0.87 ^{jk}	44.65 ± 1.4 ^{h}
Muscat Ham.	0.12 ± 0.01 $^{e–g}$	0.12 ± 0.01 ^{fg}	0.28 ± 0.03 ^{j}	0.26 ± 0.04 ^{f}	52.48 ± 0.71 ^{i}	51.70 ± 0.70 ^{g}
Yalova Misketi	0.01 ± 0.00 ^{gh}	0.01 ± 0.00 ^{i}	Nd	Nd	72.74 ± 0.81 ^{d}	73.50 ± 1.10 ^{d}
Trakya İlkeren	0.16 ± 0.00 $^{c–e}$	0.16 ± 0.02 ^{ef}	1.39 ± 0.02 ^{b}	1.37 ± 0.01 ^{b}	56.13 ± 3.24 ^{h}	58.60 ± 1.20 ^{f}
Bilecik İ.K.	0.02 ± 0.00 ^{gh}	0.01 ± 0.00 ^{i}	1.79 ± 0.02 ^{a}	1.79 ± 0.08 ^{a}	43.01 ± 1.32 ^{kl}	42.50 ± 3.70 ^{h}
İsmetbey	1.06 ± 0.03 ^{a}	1.03 ± 0.07 ^{b}	Nd	Nd	33.88 ± 1.62 ^{m}	35.65 ± 0.85 ^{i}
KXP-10	0.03 ± 0.00 $^{f–h}$	0.91 ± 0.01 ^{c}	0.91 ± 0.01 ^{f}	0.91 ± 0.01 ^{d}	99.45 ± 1.91 ^{b}	99.65 ± 0.55 ^{b}
Özer Karası	0.29 ± 0.02 ^{b}	Nd	Nd	Nd	20.95 ± 0.81 ^{n}	20.50 ± 2.30 ^{j}
Tekirdağ Ç.	0.14 ± 0.02 $^{d–f}$	0.14 ± 0.02 ^{ef}	0.10 ± 0.00 ^{m}	0.10 ± 0.02 ^{g}	Seedless	Seedless
Reçel Üzümü	0.02 ± 0.00 ^{gh}	0.01 ± 0.00 ^{i}	Nd	Nd	Seedless	Seedless
Güz Gülü	Nd	Nd	Nd	Nd	Seedless	Seedless
Pembe 77	0.02 ± 0.00 ^{gh}	1.23 ± 0.11 ^{a}	1.25 ± 0.02 ^{c}	1.23 ± 0.11 ^{c}	79.52 ± 0.46 ^{c}	79.30 ± 9.22 ^{c}
Uslu	Nd	Nd	Nd	Nd	66.44 ± 2.17 ^{e}	66.80 ± 1.90 ^{e}
83/1	Nd	Nd	0.67 ± 0.02 ^{g}	0.65 ± 0.00 ^{e}	46.17 ± 2.59 ^{j}	45.85 ± 0.35 ^{h}
FX1-1	0.26 ± 0.18 ^{bc}	0.24 ± 0.06 ^{d}	0.22 ± 0.03 ^{k}	Nd	62.46 ± 1.70 ^{g}	59.25 ± 3.53 ^{f}
85/1	0.23 ± 0.01 $^{b–d}$	0.23 ± 0.06 ^{d}	1.09 ± 0.03 ^{e}	Nd	63.42 ± 0.50 ^{fg}	62.90 ± 0.30 ^{ef}
BX1-166	0.08 ± 0.01 $^{e–h}$	0.06 ± 0.02 ^{hi}	0.32 ± 0.03 ^{i}	Nd	36.11 ± 0.56 ^{m}	34.14 ± 1.92 ^{i}
53/1	0.31 ± 0.22 ^{b}	0.19 ± 0.01 ^{de}	Nd	Nd	40.73 ± 1.15 ^{l}	45.40 ± 3.90 ^{h}
86/1	Nd	Nd	0.56 ± 0.03 ^{h}	Nd	106.91 ± 1.15 ^{a}	107.20 ± 6.10 ^{a}
FX1-10	0.11 ± 0.08 $^{e–g}$	0.08 ± 0.04 ^{gh}	0.14 ± 0.02 ^{l}	0.11 ± 0.03 ^{g}	65.80 ± 1.42 ^{ef}	61.15 ± 2.02 ^{f}
130/1	Nd	Nd	Nd	Nd	Seedless	Seedless

* Different superscripts within a column indicate significant differences at $p \leq 0.05$. ** All means are expressed as means ± standard deviations (n = 3). *** Nd = not detected.

Based on the results of grape pulps, the gallic acid contents ranged from 0.57 to 0.76 in the 1st year and from 0.56 to 0.76 mg 100 g^{-1}in the 2nd year. The gallic acid contents of the grape skins ranged from 0.59 to 1.47 mg 100 g^{-1}in the 1st year and from 0.58 to 1.43 mg 100 g^{-1}in the 2nd year. Moreover, the gallic acid contents of the grape seeds ranged from 1.35 to 6.88 mg 100 g^{-1}in the 1st year and from 1.26 to 6.88 mg 100 g^{-1} in the 2nd year. Among different grape parts, grape seeds had the highest content of gallic acid in the manner of the other two phenolic compounds. Furthermore, compared to the catechin and epicatechin levels, the content of gallic acid in seeds was somewhat limited. The seeds of the "İsmetbey" cultivar grapes with black berry colour contained the highest gallic acid content (about 6.9 mg 100 g^{-1}) in the 1st and 2nd harvesting years ($p < 0.05$) while its difference from the seeds of "FX1-10" was found insignificant in the 2nd year ($p > 0.05$). Moreover, in the 1st year, the second highest gallic acid content was determined in the seeds of the "FX1-10" hybrid grapes with yellow berry flesh colour. In the pulps and skins, the gallic acid contents were generally found high for the cultivars/hybrids with dark berry colour, such as "Isabella", "83/1", "Pembe 77", and "Reçel Üzümü", and for the hybrid of "86/1" with an intense Muscat flavour (Table 8).

Table 8. Gallic acid contents of pulps, skins, and seeds of grape cultivars/hybrids used in the study in two growing seasons (mg 100 g^{-1}).

Cultivar/ Hybrid	Pulp *		Skin		Seed	
	1st Year	2nd Year	1st Year	2nd Year	1st Year	2nd Year
Isabella	0.76 ± 0.01 ^{a} **	0.76 ± 0.01 ^{a}	0.59 ± 0.03 ^{d}	0.59 ± 0.01 ^{k}	1.35 ± 0.10 ^{i}	1.26 ± 0.05 ^{k}
Alphonse L.	0.69 ± 0.02 $^{c–e}$	0.67 ± 0.01 $^{c–e}$	0.85 ± 0.03 ^{c}	0.82 ± 0.01 ^{d}	4.50 ± 0.31 ^{d}	4.16 ± 0.20 ^{ef}
Muscat Ham.	0.68 ± 0.01 $^{c–f}$	0.69 ± 0.00 ^{c}	0.72 ± 0.01 ^{cd}	0.72 ± 0.01 ^{ef}	3.93 ± 0.10 ^{e}	3.95 ± 0.05 ^{fg}
Yalova Misketi	0.65 ± 0.01 $^{h–j}$	0.65 ± 0.00 ^{fg}	1.20 ± 0.18 ^{b}	0.98 ± 0.05 ^{c}	3.71 ± 0.06 ^{e}	3.72 ± 0.03 ^{g}
Trakya İlkeren	0.68 ± 0.03 $^{c–f}$	0.68 ± 0.01 ^{cd}	0.68 ± 0.03 ^{cd}	0.67 ± 0.00 $^{f–i}$	5.27 ± 0.64 ^{c}	5.47 ± 0.27 ^{b}
Bilecik İ.K.	0.66 ± 0.01 $^{g–i}$	0.66 ± 0.01 $^{d–f}$	0.63 ± 0.02 ^{d}	0.63 ± 0.02 $^{h–k}$	1.58 ± 0.13 ^{i}	1.47 ± 0.05 ^{k}
İsmetbey	0.64 ± 0.02 $^{i–j}$	0.64 ± 0.01 ^{gh}	0.69 ± 0.04 ^{cd}	0.69 ± 0.00 ^{fg}	2.17 ± 0.10 ^{h}	4.73 ± 0.08 ^{d}
KXP-10	0.70 ± 0.02 ^{bc}	0.69 ± 0.00 ^{bc}	0.75 ± 0.03 ^{cd}	0.76 ± 0.07 ^{e}	6.88 ± 0.04 ^{a}	6.88 ± 0.22 ^{a}
Özer Karası	0.65 ± 0.02 $^{h–j}$	0.65 ± 0.04 $^{c–g}$	0.65 ± 0.04 ^{d}	0.65 ± 0.01 $^{g–j}$	3.08 ± 0.04 ^{f}	3.04 ± 0.26 ^{hi}
Tekirdağ Ç.	0.64 ± 0.01 $^{i–j}$	0.65 ± 0.06 ^{fg}	0.59 ± 0.02 ^{d}	0.59 ± 0.02 ^{k}	Seedless	Seedless
Reçel Üzümü	0.72 ± 0.00 ^{b}	0.72 ± 0.01 ^{b}	0.63 ± 0.01 ^{d}	0.62 ± 0.00 $^{i–k}$	Seedless	Seedless
Güz Gülü	0.66 ± 0.01 $^{g–i}$	0.66 ± 0.01 $^{d–f}$	0.64 ± 0.01 ^{d}	0.63 ± 0.01 $^{h–k}$	Seedless	Seedless
Pembe 77	0.66 ± 0.01 $^{g–i}$	0.66 ± 0.01 $^{d–f}$	1.32 ± 0.03 ^{ab}	1.31 ± 0.05 ^{b}	2.92 ± 0.06 ^{fg}	2.98 ± 0.31 ^{hi}
Uslu	0.66 ± 0.02 $^{g–i}$	0.66 ± 0.01 $^{d–f}$	0.73 ± 0.03 ^{cd}	0.72 ± 0.01 ^{ef}	2.67 ± 0.15 ^{g}	2.54 ± 0.03 ^{j}
83/1	0.68 ± 0.03 $^{c–f}$	0.67 ± 0.01 $^{c–e}$	1.47 ± 0.07 ^{a}	1.43 ± 0.11 ^{a}	3.21 ± 0.02 ^{f}	3.21 ± 0.02 ^{h}
FX1-1	0.57 ± 0.03 ^{l}	0.56 ± 0.03 ^{i}	0.61 ± 0.02 ^{d}	0.61 ± 0.01 ^{jk}	4.40 ± 0.38 ^{d}	4.25 ± 0.27 ^{e}
85/1	0.67 ± 0.02 $^{e–h}$	0.67 ± 0.02 $^{c–e}$	1.42 ± 0.49 ^{a}	0.68 ± 0.01 $^{f–h}$	3.86 ± 0.18 ^{e}	3.93 ± 0.02 ^{fg}
BX1-166	0.59 ± 0.02 ^{l}	0.60 ± 0.01 ^{h}	0.59 ± 0.03 ^{d}	0.58 ± 0.04 ^{k}	2.88 ± 0.05 ^{fg}	2.90 ± 0.08 ^{i}
53/1	0.68 ± 0.01 $^{c–f}$	0.68 ± 0.00 ^{cd}	0.63 ± 0.01 ^{d}	0.62 ± 0.02 $^{i–k}$	2.27 ± 0.10 ^{h}	5.01 ± 0.09 ^{c}
86/1	0.63 ± 0.02 ^{jk}	0.63 ± 0.00 ^{gh}	1.44 ± 0.11 ^{a}	0.82 ± 0.01 ^{d}	3.74 ± 0.05 ^{e}	3.77 ± 0.04 ^{g}
FX1-10	0.62 ± 0.01 ^{k}	0.63 ± 0.01 ^{gh}	0.64 ± 0.03 ^{d}	0.66 ± 0.02 $^{g–j}$	6.35 ± 0.44 ^{b}	6.63 ± 0.11 ^{ab}
130/1	0.70 ± 0.02 ^{bc}	0.69 ± 0.00 ^{c}	0.68 ± 0.01 ^{cd}	0.65 ± 0.00 $^{g–j}$	Seedless	Seedless

* Different superscripts within a column indicate significant differences at $p \leq 0.05$. ** All means are expressed as means ± standard deviation (n = 3).

Correlations among major phenolic components in three different parts of the grape berries were determined over the averages for two harvesting seasons. A positive and highly significant correlation coefficient (R = 0.95) was found between the catechin content of pulps (Catechin, Pulp) and the gallic acid of skins (Gallic acid, Skin) ($p < 0.05$). In addition, significant correlations were found for the phenolic components of grape seeds. While the correlation between gallic acid (Gallic acid, Seed) and the epicatechin contents of seeds (Epicatechin, Seed) (R = 0.77) was high and significant ($p < 0.05$), the catechin content of seeds (Catechin, Seed) was positively correlated with the epicatechin contents of seeds (Epicatechin, Seed) (R = 0.74) ($p < 0.05$). Moreover, there was a positive correlation coefficient between the catechin (Catechin, Seed) and gallic acid contents of seeds (Gallic acid, Seed) (R = 0.70) ($p < 0.05$). This correlation analysis showed that the contents of almost all phenolic compounds in grape seeds increased more than those in the other two parts of grape berries (Table 9).

Table 9. Pearson's correlation coefficients (R) among major phenolic compounds in different parts of grapes (seeds, skins, and pulps) (n = 22).

Variable (Compound, Part)	By Variable (Compound, Part)	Correlation Coefficient	Lower 95%	Upper 95%	Significance Probability	Sign of Correlation
Gallic acid (Skin)	Gallic acid (Pulp)	−0.0155	−0.4343	0.4088	0.9454	
Catechin (Pulp)	Gallic acid (Pulp)	0.0866	−0.3477	0.4903	0.7015	
	Gallic acid (Skin)	0.9538	0.8901	0.9809	<0.0001 *	

Table 9. *Cont.*

Variable (Compound, Part)	By Variable (Compound, Part)	Correlation Coefficient	Lower 95%	Upper 95%	Significance Probability	Sign of Correlation
Catechin (Skin)	Gallic acid (Pulp)	0.1320	−0.3067	0.5244	0.5583	
	Gallic acid (Skin)	0.2235	−0.2187	0.5896	0.3173	
	Catechin (Pulp)	0.3376	−0.0980	0.6646	0.1244	
Epicatechin (Pulp)	Gallic acid (Pulp)	−0.1293	−0.5224	0.3092	0.5664	
	Gallic acid (Skin)	0.0713	−0.3611	0.4786	0.7524	
	Catechin (Pulp)	−0.0011	−0.4225	0.4207	0.9960	
	Catechin (Skin)	0.0003	−0.4213	0.4219	0.9988	
Epicatechin (Skin)	Gallic acid (Pulp)	0.1002	−0.3356	0.5007	0.6572	
	Gallic acid (Skin)	0.2852	−0.1551	0.6309	0.1983	
	Catechin (Pulp)	0.2420	−0.2000	0.6022	0.2778	
	Catechin (Skin)	0.4435	0.0269	0.7288	0.0387 *	
	Epicatechin (Pulp)	0.1315	−0.3071	0.5241	0.5597	
Gallic acid (Seed)	Gallic acid (Pulp)	−0.2728	−0.6228	0.1682	0.2194	
	Gallic acid (Skin)	0.2028	−0.2393	0.5752	0.3655	
	Catechin (Pulp)	0.1714	−0.2697	0.5531	0.4456	
	Catechin (Skin)	0.5039	0.1045	0.7633	0.0168 *	
	Epicatechin (Pulp)	0.3185	−0.1191	0.6525	0.1486	
	Epicatechin (Skin)	0.2882	−0.1519	0.6329	0.1934	
Epicatechin (Seed)	Gallic acid (Pulp)	−0.2651	−0.6176	0.1762	0.2332	
	Gallic acid (Skin)	0.5137	0.1175	0.7688	0.0145 *	
	Catechin (Pulp)	0.5203	0.1264	0.7724	0.0131 *	
	Catechin (Skin)	0.5358	0.1475	0.7810	0.0102 *	
	Epicatechin (Pulp)	0.2019	−0.2402	0.5746	0.3677	
	Epicatechin (Skin)	0.3505	−0.0834	0.6727	0.1098	
	Gallic acid (Seed)	0.7746	0.5243	0.9017	<0.0001 *	
Catechin (Seed)	Gallic acid (Pulp)	0.0129	−0.4109	0.4322	0.9545	
	Gallic acid (Skin)	0.4117	−0.0120	0.7101	0.0570	
	Catechin (Pulp)	0.4531	0.0389	0.7344	0.0342 *	
	Catechin (Skin)	0.5494	0.1664	0.7884	0.0081 *	
	Epicatechin (Pulp)	0.2565	−0.1851	0.6120	0.2491	
	Epicatechin (Skin)	0.3047	−0.1342	0.6436	0.1680	
	Gallic acid (Seed)	0.6983	0.3921	0.8652	0.0003 *	
	Epicatechin (Seed)	0.7388	0.4607	0.8848	<0.0001 *	

* indicates that the correlation coefficient is statistically significant at $p \leq 0.05$.

4. Discussion

In this study, changes in the catechin, epicatechin, and gallic acid contents of grape leaves after two important fungal diseases and differences in the contents of these compounds in skins, pulps, and skins of grape berries for two harvesting seasons were determined. After downy mildew and powdery mildew diseases, differences in the contents of these phenolic compounds were observed among grape leaves depending on the cultivar/hybrid, harvesting year, and the type of phenolic compounds.

Although there may be a variety of factors influencing the concentration of phenolic compounds in grapes, their concentration is highly dependent on the cultivar types and even species. There may be differences in the phenolic contents of grapes among different cultivars, as well as within the fruits of the same cultivar grown in different regions [41] Several factors, such as the type of cultivar, processing techniques, viticultural practices, geographical region, and climatic conditions, may significantly influence the phenolic compositions of grapes [42–46]. In a study on the contents of some phenolic compounds in different grape cultivars, Eyduran et al. [47] reported that the quantity of phenolic compounds could vary depending on the type of cultivars. Doshi et al. [48] also reported that the concentrations of rutin and quercetin hydrates representing the flavanols group of phenolics might change in different organs of grapevine, and sometimes it could be barely detected.

In species, such as *Vitis labrusca*, *Vitis riparia*, and *Vitis rupestris* hybrids, differences in the chemical structures of phenolic components have been reported in the literature. The contents of phenolic compounds may differ in *V. vinifera* species [49], and for this reason, in most hybrid cultivars that are tolerant or resistant to grapevine diseases, the composition of these components can flactuate considerably. In our study, it was noteworthy that the contents of some phenolic components were higher in the disease-resistant varieties, such as "Özer Karası" and "Isabella", which are cross-bred between species, especially in their skins and pulps.

Yaman et al. [50] reported that resveratrol levels in two different grape cultivars might be dependent on the vegetation time, cultivar, and region. In another study on changes in the total phenolic contents and seven phenolic compounds (gallic acid, catechin, catechol, chlorogenic acid, o-coumaric acid, rutin, and quercetin) of the shoot tips from "Cardinal" and "Uslu" grape cultivars collected in different months, Baydar [51] reported that the concentration of these phenolic compounds varied depending on the type of cultivar and harvesting month. The flavonol and anthocyanin contents of five red fungus-resistant grape cultivars ("Frontenac", "Maréchal Foch", "Marquette", "Sabrevois", and "St. Croix") were characterised from berry (skin, seed, and free-run must) to wine to evaluate varietal differences and relationships between the berry and wine composition by Gagne et al. [52], and they reported that the principal component analysis of berry composition showed significant differences among the cultivars. In our study, we also found that the contents of phenolic compounds varied depending on the type of grape cultivar/hybrid.

In the present study, there were significant increases in the contents of catechin, epicatechin, and gallic acid in 22 different cultivars/hybrids, especially after powdery mildew disease. Using two wine grape cultivars ("Cabernet Sauvignon" and "Sauvignon Blanc") and a table grape ("Thompson Seedless"), Taware et al. [53] determined the total phenolic contents and some phenolic compounds in the leaves, berries, and wines from healthy and powdery mildew-infected grapes, and they reported higher phenolic contents in the leaves of wine grapes compared to "Thompson Seedless". Moreover, they reported that this disease significantly altered the phenolic profile of the leaves, berries, and wines while the foliar infection resulted in the accumulation of phenolic compounds in leaves and reduction in berries and wines because of cluster disease infection. In a study by Romero-Perez et al. [54], concentrations of phenolic compounds increased considerably in grape berries infected by powdery mildew disease in comparison to healthy grape berries. Santos et al. [55] compared the contents of phenolic compounds and trans-resveratrol in different berry parts from *V. vinifera* and *V. labrusca* cultivars/genotypes, and reported that

V. labrusca cultivars/genotypes contained more phenolic compounds and trans-resveratrol than *V. vinifera* cultivars. Dani et al. [56] found a high level of phenolic compounds in the leaves of a *V. labrusca* cultivar, and reported that these compounds reduced the vine damage from lipids and proteins significantly. This result could explain how and why *V. labrusca* or interspecies cultivars with high contents of phenolic compounds in our study had significantly minimum damages after the inoculation of diseases. Rebello et al. [57] determined the contents of some phenolic components in the skins, pulps, and seeds of the "BRS Violeta" cultivar, a hybrid grape, and reported a very high content of phenolic components in skins, most probably because of the very thick skin of this cultivar (46% of grape weight).

Coklar [58] investigated the phenolic profile of whole berry, skin, and seeds of the local "Ekşikara" (*V. vinifera*) cultivar and determined the effect of harvesting year and altitude of the vineyard location. Anthocyanins, resveratrol, rutin, and isorhamnetin-3-glucoside were mainly found in skins, while monomers and dimers of flavan-3-ols were detected mainly in seeds. In the study, altitude had a drastic effect on phenolic compounds in whole berry, skins, and seeds, and very high amounts of catechin and epicatechin, especially in the seeds. The quantities of gallic acid, epicatechin, and catechin varied depending on harvesting year. Our results were in good agreement with this study.

Based on the chemical structure, phenolic compounds are mostly classified into flavonoid and non-flavonoids. Flavonoids are found mainly in grape seeds and skins while proanthocyanidins in grapes are present mainly in berry skins and seeds. Grape seed proanthocyanidins comprise only (+)-catechin, (−)-epicatechin, and procyanidins, whereas grape skin proanthocyanidins comprise both prodelphinidins and procyanidins [16,59]. Procyanidins are dimers resulting from the union of monomeric units of flavanols (+)-catechin. Among grape cultivars, there are differences in procyanidin concentrations, but their profile remains mostly unchanged. Prodelphinidins are only present in grape skin and their monomers are catechin, epicatechin, gallocatechin, and epigallocatechin units. Proanthocyanidins (procyanidins and prodelphinidins) are the major phenolic compounds in grape seeds and skins, and about 60–70% of total polyphenols are stored in seeds [16,59]. Similarly, in our study, monomeric phenolic compounds, such as catechin and epicatechin, were found at a very high concentration in grape seeds in comparison to grape skins.

Mulero et al. [60] reported that the skin, pulp, and seeds of grapes contain an enormous amount of different phenolic compounds, while Rodriguez-Montealegre et al. [61] found that the phenolic composition of grapes and different grape parts may depend on multiple factors, including climate, ripeness, berry size, grapevine cultivar, and viticulture practices. Our results also indicated significant differences in the contents of phenolic compounds in different berry parts of grapes from various cultivars/hybrids. Moreover, the catechin and epicatechin contents of seeds were much higher than those of pulps and skins. Most phenolic compounds are located in different parts of grape berries, and it would be beneficial to re-evaluate the processing techniques for many food products so that the phenolic components migrate or are incorporated into these products. Many phenolic components originating from grape skins and seeds could be health-beneficial especially during processing for food products. Phenolic compounds are related to not only human health, but also the fight of plants against diseases, and their composition may vary under the effect of different stress factors for plants.

5. Conclusions

The contents of major phenolic compounds, such as catechin, epicatechin, and gallic acid, in grape leaves increased especially after powdery mildew disease, but this increase seemed to be independent from the cultivar being disease resistant/tolerant. In order to understand the phenolic response of vine plants against downy and powdery mildew diseases, more comprehensive studies are needed. In addition, registration studies for novel grape cultivars should be accelerated so that candidate grape hybrids/cultivars, such as "86/1", "85/1", and "KXP-10", which contained high levels of major phenolic compounds, could be used in the grape juice industry. Besides these hybrids, (interspecies) cultivars, such as "Isabella" and "Özer Karası", with a high content of phenolics could also be used as parents in future breeding studies planning to develop new grape cultivars with a rich phenolic content.

Author Contributions: Methodology, A.A., Z.G. and Y.Y.; data curation, A.A. and Z.G.; writing–original draft preparation, A.A. and Z.G.; conceptualization, A.A.; investigation, Z.G. and A.A.; resources, A.A.; supervision, Y.Y.; project administration, Y.Y.; writing—review and editing, A.A. and Y.Y.; software, A.A. and Z.G.; formal analysis, A.A. and Z.G.; validation, Z.G. and A.A. All authors have read and agreed to the published version of the manuscript.

Funding: This research was funded by the Scientific and Technological Research Council of Turkey (TÜBİTAK, project number 113O641) and General Directorate of Agricultural Research and Policies (TAGEM).

Institutional Review Board Statement: Not applicable.

Informed Consent Statement: Not applicable.

Data Availability Statement: The authors will make the results available if requested.

Acknowledgments: The authors appreciate TÜBİTAK and TAGEM for their financial support. Moreover, we would like to thank the researchers in the "Viticulture" and "Food Technology" Departments of Yalova Atatürk Horticultural Central Research Institute for their assistance.

Conflicts of Interest: The authors declare no conflict of interest.

References

1. Georgiev, V.; Ananga, A.; Tsolova, V. Recent advances and uses of grape flavonoids as nutraceuticals. *Nutrients* **2014**, *6*, 391–415. [CrossRef]
2. Hornedo-Ortega, R.; González-Centeno, M.R.; Chira, K.; Jourdes, M.; Teissedre, P.L. Phenolic compounds of grapes and wines: Key compounds and implications in sensory perception. In *Chemistry and Biochemistry of Winemaking, Wine Stabilization and Aging*; Cosme, F., Nunes, F.M., Filipe-Ribeiro, L., Eds.; IntechOpen: London, UK, 2021. [CrossRef]
3. Whiting, D.A. Natural phenolic compounds 1900–2000: A bird's eye view of a century's chemistry. *Nat. Prod. Rep.* **2001**, *18*, 583–606. [CrossRef]
4. Oksana, S.; Brestic, M.; Rai, M.; Shao, H.B. Plant phenolic compounds for food, pharmaceutical and cosmetics production. *J. Med. Plants Res.* **2012**, *6*, 2526–2539. [CrossRef]
5. Cheynier, V. Phenolic compounds: From plants to foods. *Phytochem. Rev.* **2012**, *11*, 153–177. [CrossRef]
6. Fanzone, M.; Zamora, F.; Jofré, E.; Assof, F.; Peña-Neira, Á. Phenolic composition of Malbec grape skins and seeds from Valle de Uco (Mendoza, Argentina) during ripening. Effect of cluster thinning. *J. Agric. Food Chem.* **2011**, *59*, 6120–6136. [CrossRef] [PubMed]
7. Bustamante, L.; Sáez, V.; Hinrichsen, P.; Castro, M.H.; Vergara, C.; Von Baer, D.; Mardones, C. Differences in *Vvufgt* and *VvmybA1* gene expression levels and phenolic composition in table grape (*Vitis vinifera* L.) 'Red Globe' and its somaclonal variant 'Pink Globe'. *J. Agric. Food Chem.* **2017**, *65*, 2793–2804. [CrossRef]
8. De Pascali, S.A.; Coletta, A.; Del-Coco, L.; Basile, T.; Gambacorta, G.; Fanizzi, F.P. Viticultural practice and winemaking effects on metabolic profile of Negroamaro. *Food Chem.* **2014**, *161*, 112–119. [CrossRef]
9. Cheng, G.; Fa, J.Q.; Xi, Z.M.; Zhang, Z.W. Research on the quality of the wine grapes in corridor area of China. *Food Sci. Technol.* **2015**, *35*, 38–44. [CrossRef]
10. Rusjan, D.; Veberič, R.; Mikulič-Petkovšek, M. The response of phenolic compounds in grapes of the variety 'Chardonnay' (*Vitis vinifera* L.) to the infection by phytoplasma Bois noir. *Eur. J. Plant Pathol.* **2012**, *133*, 965–974. [CrossRef]
11. Yang, J.; Xiao, Y. Grape phytochemicals and associated health benefits. *Crit. Rev. Food Sci.* **2013**, *53*, 1202–1225. [CrossRef]
12. Aubert, C.; Chalot, G. Chemical composition, bioactive compounds, and volatiles of six table grape varieties (*Vitis vinifera* L.). *Food Chem.* **2018**, *240*, 524–533. [CrossRef]
13. Teixeira, A.; Eiras-Dias, J.; Castellarin, S.D.; Gerós, H. Berry phenolics of grapevine under challenging environments. *Int. J. Mol. Sci.* **2013**, *14*, 18711–18739. [CrossRef]
14. Yilmaz, Y.; Göksel, Z.; Erdoğan, S.S.; Öztürk, A.; Atak, A.; Özer, C. Antioxidant activity and phenolic content of seed, skin and pulp parts of 22 grape (*Vitis vinifera* L.) cultivars (4 common and 18 registered or candidate for registration). *J. Food Process Pres.* **2014**, *39*, 1682–1691. [CrossRef]
15. Yilmaz, Y.; Toledo, R.T. Major flavonoids in grape seeds and skins: Antioxidant capacity of catechin, epicatechin, and gallic acid. *J. Agric. Food Chem.* **2004**, *52*, 255–260. [CrossRef]

16. Cosme, F.; Pinto, T.; Vilela, A. Phenolic compounds and antioxidant activity in grape juices: A chemical and sensory view. *Beverages* **2018**, *4*, 22. [CrossRef]
17. Swain, T. Secondary compounds as protective agents. *Annu. Rev. Plant Physiol.* **1977**, *28*, 479–501. [CrossRef]
18. Kutchan, T.M. Ecological arsenal and developmental dispatcher. The paradigm of secondary metabolism. *Plant Physiol.* **2001**, *125*, 58–60. [CrossRef] [PubMed]
19. Grayer, R.J.; Kimmins, F.M.; Padgham, D.E.; Harborne, J.B.; Ranga Rao, D.V. Condensed tannins levels and resistance of groundnuts against *Aphis craccivora*. *Phytochemistry* **1992**, *31*, 3795–3800. [CrossRef]
20. Bennett, R.N.; Wallsgrove, R.M. Secondary metabolites in plant defence mechanisms. *New Phytol.* **1994**, *127*, 617–633. [CrossRef]
21. Orlando, R.; Magro, P.; Rugini, E. Pectic enzymes as a selective pressure tool for in vitro recovery of strawberry plants with fungal disease resistance. *Plant Cell Rep.* **1997**, *16*, 272–276. [CrossRef] [PubMed]
22. Atak, A.; Göksel, Z.; Çelik, H. Relations between downy/powdery mildew diseases and some phenolic compounds in *Vitis* spp. *Turk. J. Agric. For.* **2017**, *41*, 69–81. [CrossRef]
23. Baydar, N.G.; Babalık, Z.; Türk, F.H.; Çetin, E.S. Phenolic composition and antioxidant activities of wines and extracts of some grape varieties grown in Turkey. *J. Agric. Sci.* **2011**, *17*, 67–76. [CrossRef]
24. Mazid, M.; Khan, T.A.; Mohammad, M. Secondary metabolites in defence mechanism of plants. *Biol. Med.* **2011**, *3*, 232–249.
25. Capriotti, L.; Baraldi, E.; Mezzetti, B.; Limera, C.; Sabbadini, S. Biotechnological approaches: Gene overexpression, gene silencing, and genome editing to control fungal and oomycete diseases in grapevine. *Int. J. Mol. Sci.* **2020**, *21*, 5701. [CrossRef]
26. Bois, B.; Zito, S.; Calonnec, A. Climate vs grapevine pests and diseases worldwide: The first results of a global survey. *OENO One* **2017**, *51*, 133–139. [CrossRef]
27. Martínez-Bracero, M.; Alcázar, P.; Velasco-Jiménez, M.J.; Galán, C. Fungal spores affecting vineyards in Montilla-Moriles caddleSouthern Spain. *Eur. J. Plant Pathol.* **2019**, *153*, 1–13. [CrossRef]
28. Caddle-Davidson, L. Variation within and between *Vitis* spp. for foliar resistance to the downy mildew pathogen *Plasmopara viticola*. *Plant Dis.* **2008**, *92*, 1577–1584. [CrossRef] [PubMed]
29. Söylemezoğlu, G.; Atak, A.; Boz, Y.; Ünal, A.; Sağlam, M. Viticulture in Turkey. *Chronica Hortic.* **2016**, *56*, 27–32.
30. Ali, K.; Maltese, F.; Choi, Y.; Verpoorte, R. Metabolic constituents of grapevine and grape-derived products. *Phytochem. Rev.* **2010**, *9*, 357–378. [CrossRef] [PubMed]
31. Nassiri-Asl, M.; Hosseinzadeh, H. Review of the pharmacological effects of *Vitis vinifera* (grape) and its bioactive compounds. *Phytother. Res.* **2009**, *23*, 1197–1204. [CrossRef]
32. Rather, S.A.; Sarumathi, A.; Anbu, S.; Saravanan, N. Gallic acid protects against immobilization stress-induced changes in wistar rats. *J. Stress Physiol. Biochem.* **2013**, *9*, 136–147.
33. Applequist, W.L.; Johnson, H.; Rottinghaus, G. (+)-Catechin, (−)-epicatechin, and gallic acid content of seeds of hybrid grapes hardy in Missouri. *Am. J. Enol. Vitic.* **2008**, *59*, 98–102.
34. Güler, A.; Candemir, A. Total phenolic and flavonoid contents, phenolic compositions and color properties of fresh grape leaves. *Turk. J. Agric. Nat. Sci.* **2014**, *24*, 778–782.
35. Wang, Y.; Li, Y.; He, P.; Chen, J.; Lamikanra, O.; Lu, J. Evaluation of foliar resistance to *Uncinula necator* in Chinese wild *Vitis* species. *Vitis* **1995**, *34*, 159–164.
36. Boso, S.; Martinez, M.C.; Unger, S.; Kassemeyer, H.H. Evaluation of foliar resistance to downy mildew in different cv. Albariño clones. *Vitis* **2006**, *45*, 23–27.
37. IPGRI; UPOV; OIV. *Descriptors for Grapevines (Vitis spp.)*; International Plant Genetic Resources Institute: Rome, Italy; International Union for the Protection of New Varieties of Plants: Geneva, Switzerland; Office International de la Vigne et du Vin: Paris, France, 1997.
38. Katalinic, V.; Mozina, S.S.; Generalic, I.; Skroza, D.; Ljubenkov, I.; Klancnik, A. Phenolic profile, antioxidant capacity, and antimicrobial activity of leaf extracts from six *Vitis vinifera* L. varieties. *Int. J. Food Prop.* **2013**, *16*, 45–60. [CrossRef]
39. Castillo-Muñoz, N.; Fernández-González, M.; Gómez-Alonso, S.; García-Romero, E.; Hermosín-Gutiérrez, I. Red-color related phenolic composition of Garnacha Tintorera (*Vitis vinifera* L.) grapes and red wines. *J. Agric. Food Chem.* **2009**, *57*, 7883–7891. [CrossRef]
40. SAS Institute. *JMP Statistical Discovery Software*; JMP 15.0 Edition of programme; SAS Institute Inc.: Cary, NC, USA, 2020.
41. Portu, J.; López, R.; Santamaría, P.; Garde-Cerdán, T. Methyl jasmonate treatment to increase grape and wine phenolic content in Tempranillo and Graciano varieties during two growing seasons. *Sci. Hortic.* **2018**, *240*, 378–386. [CrossRef]
42. Dani, C.; Oliboni, L.S.; Vanderlinde, R.; Bonatto, D.; Salvador, M.; Henriques, J.A.P. Phenolic content and antioxidant activities of white and purple juices manufactured with organically- or conventionally-produced grapes. *Food Chem. Toxicol.* **2007**, *45*, 2574–2580. [CrossRef]
43. Fuleki, T.; Ricardo-da-Silva, J.M. Effects of cultivar and processing method on the contents of catechins and procyanidins in grape juice. *J. Agric. Food Chem.* **2003**, *51*, 640–646. [CrossRef]
44. Leblanc, M.R.; Johnson, C.E.; Wilson, P.W. Influence of pressing method on juice stilbene content in Muscadine and Bunch Grapes. *J. Food Sci.* **2008**, *73*, H58–H62. [CrossRef] [PubMed]
45. Natividade, M.M.P.; Corrêa, L.C.; Souza, S.V.C.; Pereira, G.E.; Lima, L.C.O. Simultaneous analysis of 25 phenolic compounds in grape juice for HPLC: Method validation and characterization of São Francisco Valley samples. *Microchem. J.* **2013**, *110*, 665–674. [CrossRef]

46. Talcott, S.T.; Lee, J.H. Ellagic acid and flavonoid antioxidant content of Muscadine wine and juice. *J. Agric. Food Chem.* **2002**, *50*, 3186–3192. [CrossRef]
47. Eyduran, S.P.; Akin, M.; Ercişli, S.; Eyduran, E.; Maghradze, D. Sugars, organic acids, and phenolic compounds of ancient grape cultivars (*Vitis vinifera* L.) from Igdır province of Eastern Turkey. *Biol. Res.* **2015**, *48*, 2–8. [CrossRef]
48. Doshi, P.; Adsule, P.; Banerjee, K.; Oulkar, D. Phenolic compounds, antioxidant activity and insulinotropic effect of extracts prepared from grape (*Vitis vinifera* L.) by products. *J. Food Sci. Technol.* **2015**, *52*, 181–190. [CrossRef]
49. Rubilar, M.; Pinelo, M.; Shene, C.; Sineiro, J.; Nunez, M.J. Separation. HPLC-MS identification of phenolic antioxidants from agricultural residues: Almond hulls and grape pomace. *J. Agric. Food Chem.* **2007**, *55*, 10101–10109. [CrossRef] [PubMed]
50. Yaman, Ü.R.; Adıgüzel, B.Ç.; Yücel, U.; Çetinkaya, N. Effect of vegetation time and climatic conditions on trans-resveratrol concentrations in Cabernet Sauvignon and Merlot wines from different regions in Turkey. *S. Afr. J. Enol. Vitic.* **2016**, *37*, 85–92. [CrossRef]
51. Baydar, N.G. Phenolic composition of grapevine shoot tips collected in different months and their effects on the explant browning. *Biotechnol. Biotechnol. Equip.* **2006**, *20*, 41–46. [CrossRef]
52. Gagne, M.P.; Angers, P.; Pedneault, K. Phenolic compounds profile of berries and wines from five fungus-resistant grape varieties. *Ann. Food Process Preserv.* **2016**, *1*, 1003.
53. Taware, P.B.; Kondıram, N.D.; Dasharath, P.O.; Sangram, H.P.; Banerjee, K. Phenolic alterations in grape leaves berries and wines due to foliar and cluster powdery mildew infections. *Int. J. Pharma Bio Sci.* **2010**, *1*, 1–14.
54. Romero-Perez, A.I.; Lamuela-Raventos, M.R.; Andres-Lacueva, C.; Torre-Boronat, d.l.M.C. Method for the quantitative extraction of resveratrol and piceid isomers in grape berry skins. Effect of powdery mildew on the stilbene content. *J. Agric. Food Chem.* **2001**, *49*, 210–215. [CrossRef] [PubMed]
55. Santos, L.P.; Morais, D.R.; Souza, N.E.; Cottica, S.M.; Boroski, M.; Visentainer, J.V. Phenolic compounds and fatty acids in different parts of *Vitis labrusca* and *V. vinifera* grapes. *Food Res. Int.* **2011**, *44*, 1414–1418. [CrossRef]
56. Dani, C.; Oliboni, L.S.; Agostini, F.; Funchal, C.; Serafini, L.; Henriques, J.A.; Salvador, M. Phenolic content of grapevine leaves (Vitis labrusca var. Bordo) and its neuroprotective effect against peroxide damage. *Toxicol. In Vitro* **2010**, *24*, 148–153. [CrossRef] [PubMed]
57. Rebello, L.P.G.; Lago-Vanzela, E.S.; Barcia, M.T.; Ramos, A.M.; Stringheta, P.C.; Da-Silva, R.; Castillo-Munoz, N.; Gómez-Alonso, S.; Hermosín-Gutiérrez, I. Phenolic composition of the berry parts of hybrid grape cultivar BRS Violeta (BRS Rubea × IAC 1398-21) using HPLC-DAD-ESI-MS/MS. *Food Res. Int.* **2013**, *54*, 354–366. [CrossRef]
58. Coklar, H. Antioxidant capacity and phenolic profile of berry, seed, and skin of Ekşikara (*Vitis vinifera* L) grape: Influence of harvest year and altitude. *Int. J. Food Prop.* **2017**, *20*, 2071–2087. [CrossRef]
59. Auger, C.; Teissedre, P.L.; Gerain, P.; Lequeux, N.; Bornet, A.; Serisier, S.; Besançon, P.; Caporiccio, B.; Cristol, J.P.; Rouanet, J.M. Dietary wine phenolics catechin, quercetin, and resveratrol efficiently protect hypercholesterolemic hamsters against aortic fatty streak accumulation. *J. Agric. Food Chem.* **2005**, *53*, 2015–2021. [CrossRef]
60. Mulero, J.; Pardo, F.; Zaffrilla, P. Antioxidant activity and phenolic composition of organic and conventional grapes and wines. *J. Food Compost. Anal.* **2010**, *23*, 569–574. [CrossRef]
61. Rodriguez-Montealegre, R.; Romero Peces, R.; Chacon Vozmediano, J.L.; Martinez Gascuena, J.; Garcia Romero, E. Phenolic compounds in skins and seeds of ten grape *Vitis vinifera* grown in a warm climate. *J. Food Compost. Anal.* **2006**, *19*, 687–693. [CrossRef]

MDPI

Article

Extracts from Environmental Strains of *Pseudomonas* spp. Effectively Control Fungal Plant Diseases

Valentina Librizzi [1,†], Antonino Malacrinò [1,*,†], Maria Giulia Li Destri Nicosia [1], Nataly Barger [2], Tal Luzzatto-Knaan [2], Sonia Pangallo [1], Giovanni E. Agosteo [1] and Leonardo Schena [1]

1 Department AGRARIA, University of Reggio Calabria, 89122 Reggio Calabria, Italy; librizzi.valentina@libero.it (V.L.); giulia.lidestri@unirc.it (M.G.L.D.N.); sonia.pangallo@unirc.it (S.P.); geagosteo@unirc.it (G.E.A.); lschena@unirc.it (L.S.)

2 Department of Marine Biology, Leon H. Charney School of Marine Sciences, University of Haifa, Haifa 3498838, Israel; nbarger@univ.haifa.ac.il (N.B.); tluzzatto@univ.haifa.ac.il (T.L.-K.)

* Correspondence: antonino.malacrino@unirc.it

† These authors contributed equally to this work.

Abstract: The use of synthetic chemical products in agriculture is causing severe damage to the environment and human health, but agrochemicals are still widely used to protect our crops. To counteract this trend, we have been looking for alternative strategies to control plant diseases without causing harm to the environment or damage to our health. However, these alternatives are still far from completely replacing chemical products. Microorganisms have been widely known as a biological tool to control plant diseases, but their use is still limited due to the high variability in their efficacy, together with issues in product registration. However, the metabolites produced by these microorganisms can represent a novel tool for the environment-friendly management of plant diseases, while reducing the issues mentioned above. In this study, we explore the soil microbial diversity in natural systems to look for microorganisms with the potential to be used in pre- and post-harvest protection against fungal plant pathogens. Using a simple workflow, we isolated 22 bacterial strains that were tested both in vitro and in vivo for their ability to counteract the growth of common plant pathogens. The three best isolates, identified as members of the bacterial genus *Pseudomonas*, were used to produce a series of alcoholic extracts, which were then tested for their action against plant pathogens in simulated real-world applications. Results show that extracts from these isolates have an exceptional biocontrol activity and can be successfully used to control plant pathogens in operational setups. Thus, this study shows that the environmental microbiome is an important source of microorganisms producing metabolites that might provide an alternative strategy to synthetic chemical products.

Keywords: *Penicillium*; *Botrytis*; *Colletotrichum*; *Alternaria*; *Monilinia*; post-harvest diseases

Citation: Librizzi, V.; Malacrinò, A.; Li Destri Nicosia, M.G.; Barger, N.; Luzzatto-Knaan, T.; Pangallo, S.; Agosteo, G.E.; Schena, L. Extracts from Environmental Strains of *Pseudomonas* spp. Effectively Control Fungal Plant Diseases. *Plants* **2022**, *11*, 436. https://doi.org/10.3390/plants11030436

Academic Editors: Špela Mechora and Dragana Šunjka

Received: 19 January 2022
Accepted: 3 February 2022
Published: 5 February 2022

1. Introduction

Evidence that a lot of the chemical products used in agriculture are harmful to the environment and human health has been around for several decades [1–3]. This has generated a huge response from consumers, farmers, scientists, and policy makers, which has promoted the use of agricultural products obtained with low or no chemical inputs; at the same time, this has fostered research on alternative strategies to prevent damage from pests and pathogens [4]. Microorganisms have been found to be potential competitors of chemical pesticides. However, microbial-based products still struggle to penetrate the market and replace synthetic molecules, mainly because of the high variability of their action, availability, and persistence, together with issues in product registration [5–8].

Microorganisms have a long story of being used to control pre-harvest and post-harvest pathogens [9–11]. These microbes belong to several taxonomical groups [12] and exploit a wide variety of mechanisms to contrast the development of fungal plant pathogens, including the production of metabolites, enzymes, and siderophores; the competition for nutrients and space; and the modulation of plant physiology (e.g., the induction of systemic resistance) [13]. Several potential biocontrol agents have been studied, and a few have been commercialized [14], but their marketability is still insufficient because of their low efficiency and reliability compared to chemical products [11]. *Bacillus*, *Pseudomonas*, and *Paenibacillus* are all bacterial genera that include strains with biocontrol activity [7,15,16], and they are commonly found to be associated with plants and soil. Although microbial biocontrol agents are still not competitive on the market, recent studies on plant and soil microbiomes show that the diversity of these microbial communities is incredibly high [17,18]. Thus, there is still a wide potential pool of microorganisms in natural environments that might offer new opportunities for the biocontrol of pre- and post-harvest diseases.

Screenings for biocontrol organisms have been mainly performed on plant tissues or the rhizosphere of diseased plants, resulting in isolates with high capacity for contrasting plant pathogens [7]. Biocontrol organisms can be isolated from the environment and screened in mass to select those with antagonistic action against the target pathogens. In this framework, the extreme diversity of the environmental microbiome [17] might represent a powerful source of microorganisms with the potential to be used in pre- and post-harvest protection. Within these complex microbial communities, the competition between species and strains is very high, and some microbes can diversify to reduce the fitness of competing species (e.g., antibiotics, fungicides), so they increase their competitive ability and occupy new niches. These microbial strains and the metabolites they produce thus have a high potential to be isolated and used to counteract the growth of pathogens that damage agricultural products. Here, we test the hypothesis that the environmental microbiome is a source of microorganisms that can be isolated, selected, cultivated, and used as a source of active metabolites to control fungal plant diseases in real-world applications.

2. Materials and Methods

2.1. Study Overview

In this study, we exploited the soil microbial diversity in natural environments in order to discover novel bacterial isolates as a potential source of bioactive compounds that can be used in plant protection (Figure 1). We started by collecting soil from different environments with low anthropic impact, creating a microbial wash from each of these soils and plating each microbial wash together with a suspension of spores of *Penicillium digitatum*. This allowed us to select 22 bacterial isolates that showed antifungal activity. All these isolates were then tested for their antifugal activity both in vitro (dual-culture assay) and in vivo (co-inoculating fruits together with a pathogen) against 5 different fungal pathogens (Table 1). Subsequently, we selected the 3 isolates that showed the best results in both assays, and we taxonomically identified them as species of the bacterial genus *Pseudomonas*. To better investigate the molecules with antifungal action, we produced an alcoholic extract from each bacterial isolate, characterized by using metabolomics and tested for: (i) efficacy to prevent diseases caused by fungal pathogens, (ii) induction of systemic resistance, (iii) curative activity of early-stage infections, (iv) efficacy in field trials. Data analysis was performed using R 4.1.0 [19] with the packages *lme4* [20] and *car* [21]. The modeling strategy is detailed for each trial described below. The package *emmeans* [22] was used to extract post hoc contrasts.

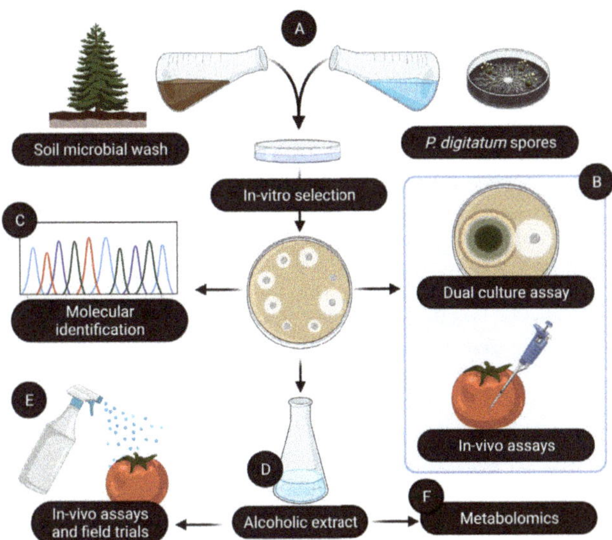

Figure 1. Workflow used in this study. (**A**) A microbial wash from environmental samples was mixed with a solution containing a high density of spores of *Penicillium digitatum*. (**B**) Isolates were screened for their efficacy using a wide set of in vitro and in vivo trials and (**C**) identified as *Pseudomonas* spp. by sequencing a portion of the 16S rRNA gene. From the most promising bacterial isolates, we produced a set of alcoholic extracts (**D**) that were then tested (**E**) for their efficacy in vivo and in setups simulating real-world applications, and (**F**) using an untargeted metabolomics approach, we attempted an annotation of the molecules likely to have antifungal activity. Created with BioRender.com.

Table 1. Summary of the host–pathogen combinations used for each trial. Tests were performed using alcoholic extracts from bacterial cultures, except the experiment marked with "*", where we used live bacterial cultures.

Trial	**Fruit**	**Variety**	**Pathogen**
Preventive antifungal activity of live bacteria *	Apple Grape Olive Tangerine Tomato Tomato	Golden delicious Italia Ottobratica Avana Datterino Datterino	*Penicillium expansum* *Botrytis cinerea* *Colletotrichum acutatum* *Penicillium digitatum* *Alternaria alternata* *Botrytis cinerea*
Preventive antifungal activity of extracts	Apricot Tangerine Tomato	Tsunami Avana Datterino	*Monilinia fructicola* *Penicillium digitatum* *Botrytis cinerea*
Induction of resistance	Apricot Grape Tomato	Tsunami Italia Datterino	*Monilinia fructicola* *Botrytis cinerea* *Botrytis cinerea*
Curative effects	Tangerine	Avana	*Penicillium digitatum*
Post-harvest disease control on olives	Olive	Ottobratica	*Colletotrichum acutatum*
Post-harvest disease control on cherries	Sweet cherry	Ferrovia, Giorgia	*Monilinia fructicola*
Field trial	Grape	Gaglioppo	*Uncinula necator*

2.2. Isolation of Potential Biocontrol Agents

Potential biocontrol agents were isolated from soil collected in five different forests located in southern Italy (Table S1). At each location, we collected soil from the top 5 cm below the litter, bulking together 10 subsamples of ~100 g from 10 randomly selected spots within a radius of 50 m. The soil was then sieved through a 2 mm mesh to eliminate large debris (e.g., leaves, roots, stones) and stored at 4 °C until further processing.

We selected potential biocontrol agents by growing together a microbial mix obtained from soil and a high-density suspension of spores of *P. digitatum* (10^6 spores/mL). The microbial mix was obtained from each location (Table S1) by mixing 10g of soil with 100 mL of water using a magnetic stirrer for ~20 min. Each soil was then serially diluted to a 1:1000 ratio with sterile water. This process was repeated three times for each location, and the three subsamples were bulked together before plate inoculation. The spore suspension of *P. digitatum* was obtained by harvesting conidia from isolates grown on PDA (Potato Dextrose Agar) plates at 20 °C for 7–10 days. Conidia were collected with a spatula, suspended in sterile distilled water, filtered through a double layer of sterile gauze, and vortexed for 1 min to ensure uniform mixing. Conidia from this stock solution were counted using a Thoma cell and then diluted to a density of 10^6 spores/mL. A total of 100 µL of each soil microbial mix was then co-inoculated with 100 µL of *P. digitatum* spore suspension on PDA, evenly distributed with a sterile spatula and incubated at 25 °C for 7 days. The plates were periodically observed to identify bacteria producing inhibition halos, yielding 125 bacterial colonies that were then isolated on Nutrient Broth Agar (NBA). These isolates were grown on NBA plates for 7 days at 24 °C, and then roughly screened to select isolates with different morphological features (e.g., different colony shape, color, and size). This yielded 22 bacterial isolates that were then used for the experiments below.

2.3. In Vitro Antifungal Activity

The 22 bacterial isolates were first tested to estimate their capacity to inhibit in vitro the growth of three common plant pathogens: *Botrytis cinerea*, *Alternaria alternata*, and *Phytophtora palmivora*. Tests were run using a dual-culture assay [7], growing together combinations of each bacterial isolate with one of the three pathogens, and then estimating the growth reduction of the fungal pathogen. Each PDA Petri dish was inoculated with ~0.16 cm^2 agar plug with fungal mycelium (obtained from a pure culture of the pathogen) placed on one side of the plate. On the opposite side of the plate, a single bacterial isolate was inoculated in a single spot by transferring cells from the pure cultures using a sterile needle. Each pathogen-bacterial isolate combination was replicated 3 times, and a set of 3 plates for each pathogen without the bacterial isolate served as control. Plates were incubated at 22 °C for 5 days in the dark, and the radial growth of the pathogen was then measured for each plate. The inhibition of fungal growth was then estimated as percentage reduction compared to control plates. Data were fit to a linear model, including *isolate ID*, *pathogen ID*, and their interaction as fixed factors.

2.4. Preventive Antifungal Activity of Live Bacteria

The same 22 bacterial isolates were tested for their ability to contrast fungal infection on fruits. For this assay, we used six different fruit–pathogen combinations (Table 1). Fruit surface was sterilized by immersion in a 2% sodium hypochlorite solution for 2 min, washed with tap water, air-dried, and fixed onto polypropylene panels using a double-sided tape [23]. Fruits were kept 1–2 cm apart to avoid nesting, and wounded to a uniform and standard depth of 3 mm using a nail with diameter of 1 mm. Wounds were inoculated by applying 10 µL of bacterial suspension or water (negative control). The bacterial inoculum was prepared by growing each isolate in Erlenmeyer flasks with 5 mL of Luria–Bertani (LB) broth on a rotary shaker at 100 rpm, for 48 h at 25 °C. The cells were harvested by centrifugation (1100× *g* for 15 min), rinsed twice with water, and resuspended in 5 mL of distilled water. All bacterial suspensions were diluted at a 1:10 ratio before inoculation. After two hours from bacterial inoculation, the same wounds were inoculated with 10 µL of fungal conidial suspension obtained from each fungal species, as reported above for *P. digitatum*, and diluted with distilled water in order to have 5×10^4 conidia/mL (*P. digitatum*) or 10^5 conidia/mL (*P. expansum*, *A. alternata*, *Colletotrichum acutatum*, and *B. cinerea*). These concentrations were chosen according to preliminary trials. A pomegranate peel extract (PGE) at a concentration of 6 mg/L was used as a positive control [23,24]. Tests were performed on 15 fruits (apple, tomato) or 30 fruits (grape, olive, tangerine) for each treatment.

After the inoculations, fruits were maintained at room temperature (22 ± 2 °C) in plastic boxes containing wet paper to ensure high relative humidity, and the presence or absence of fungal growth was recorded after 7 days. Within each host–pathogen combination, disease incidence for groups treated with a bacterial isolate was compared to the negative control by fitting a generalized linear model, using *isolate ID* as the fixed factor and specifying a binomial error distribution. According to the results from these assays (see below), all the following trials focused only on isolates B01, B05, and B09.

2.5. Preventive Antifungal Activity of Extracts

Alcoholic extracts from isolates B01, B05, and B09 were prepared to test the efficacy of metabolites produced by these bacteria in protecting harvested fruits from fungal diseases. Each bacterial isolate was grown in an Erlenmeyer Flask containing 20 mL of Nutrient Broth (NB) for 2 days at 28 °C on a rotary shaker. Then, bacterial suspensions were transferred into 500 mL glass bottles containing 80 mL of absolute ethanol. Bottles were vigorously shaken for 5 min and then kept overnight on a rotary shaker. The extract was aliquoted in 1.5 mL tubes and transferred to a rotary evaporator overnight. Once dry, the pellet (approximately 0.02 g) was transferred to a single tube and suspended in 66.7 mL of absolute ethanol and then stored at −20 °C until use.

We tested the efficacy of these extracts to prevent fungal disease in fruits artificially inoculated with pathogens using three host–pathogen combinations (Table 1): apricot—*M. fructicola*, tangerine—*P. digitatum*, and tomato—*B. cinerea*. The fruits selected had a uniform size, and they were surface-sterilized and wounded, as described above. Each wound received 10 µL of bacterial extract diluted with sterile distilled water, with final concentrations at 1.5, 3, 6, 12, 24, and 36 mg/L. The fruits treated with a 10 µL water solution containing 12 mL/L of ethanol (similar to the ethanol concentration inoculated in fruits treated with 36 mg/L of extract) were used as a negative control. Once dry (∼2 h), the wounds were inoculated with a 10 µL spore suspension containing 5×10^4 (*P. digitatum*) or 5×10^5 (*B. cinerea* and *M. fructicola*) conidia/mL. Tests were run on 45 (apricots), 54 (tangerine), or 180 (tomato) fruits per group. The fruits were incubated at room temperature (22 ± 2 °C), in plastic boxes containing wet paper to ensure high relative humidity, and after 7 days, they were scored for the presence or absence of fungal growth. Data were then fit to a generalized linear model using the *isolate ID* and *dose* (and their interaction) as fixed factors and specifying a binomial error distribution.

We used an untargeted metabolomics approach to characterize the composition of the alcoholic extract of the three isolates B01, B05, and B09. Analyses were conducted according to Luzzatto-Knaan et al. [25]. Briefly, liquid chromatography was carried out on a UPLC ultimate 3000 dionex system with a C-18 column (Phenomenex 1.7 µm C18 50 × 2.1 mm) in the following conditions: A- ACN: 0.1% FA; B- H20: 0.1% FA. Flow: 0.5 mL/min; 0 min: 90% B, 0.5 min: 90% B, 3 min: 50% B, 8 min: 1% B, 11 min: 1% B, 11.5 min: 90% B, 12.5 min: 90% B. Mass spectrometry measurements were carried out on a Bruker Maxis impact QTQF system in an ESI positive mode. The method used was tune positive MS/MS, with the following MS conditions: Ionization mode: ESI positive; Capillary: 4000 V; Corona: 4000 nA; Nebulizer: 2 Bar; Dry Gas: 5 L/min; Dry temp: 200; Vaporizer Temp: 450; Active: 5; Exclude after: 4; Release after: 0.5 min; Absolute Threshold: 213 cts; Relative Threshold: 0%; Spectra rate: 1 Hz; Precursor Ion list: exclude. For MS2, Auto MS/MS transitions were used. LC-MS raw data files were converted to the mzXML format by Compass

DataAnalysis 4.2 software (Bruker Daltonics). Feature detection was performed by GNPS molecular networking based on MS/MS spectra.

2.6. Induction of Resistance

We tested the efficacy of extracts from the B01, B05, and B09 bacterial isolates in inducing resistance in fruits artificially inoculated with pathogens. For this test, we used three host–pathogen combinations (Table 1): apricot—*M. fructicola*, grape—*B. cinerea*, and tomato—*B. cinerea*. The fruits selected had a uniform size, and they were surface sterilized, as described above. In this case, each fruit was wounded twice at a distance of ~2 cm. We first inoculated the bacterial extract into one wound at a concentration of 1.5, 12, and 36 mg/L. Then, after 24 h, the fungal pathogen was inoculated into the other wound by applying 10 µL of a suspension containing 5×10^5 conidia/mL. In this way, we were able to test whether the bacterial extracts could induce resistance mechanisms in fruits, reducing their susceptibility to fungal pathogens. A group of fruits inoculated with a water solution containing 12 mL/L of ethanol served as a negative control. Fruits (45 per group) were incubated at room temperature (22 ± 2 °C), in plastic boxes containing wet paper to ensure high relative humidity, and after 7 days, they were scored for the presence/absence of fungal growth. Data were then fit to a generalized linear model, using *isolate ID* and *dose* (and their interaction) as fixed factors and specifying a binomial error distribution.

2.7. Curative Effects

We tested the curative effect (control of pre-existing infections) of extracts from the B01, B05, and B09 bacterial isolates. Tangerine fruits were selected to have a uniform size, surface-sterilized as described above, wounded, and inoculated with 10 µL of a suspension containing 5×10^4 conidia/mL of *P. digitatum*. After 24 h from inoculating the pathogen, we inoculated 10 µL of each extract at the concentration of 1.5, 12, and 36 mg/L. This time frame was considered enough for the germination of the conidia of *P. digitatum* and the starting of the infection process. A group of fruits inoculated with a water solution containing 12 mL/L of ethanol served as a negative control. Fruits (45 per group) were incubated at room temperature (22 ± 2 °C), in plastic boxes containing wet paper to ensure high relative humidity, and after 7 days, they were scored for the presence/absence of fungal growth. Data were then fit to a generalized linear model, using the *isolate ID* and *dose* (and their interaction) as fixed factors and specifying a binomial error distribution.

2.8. Control of Post-Harvest Rots on Olives

After selecting and assessing the biocontrol potential of our bacterial extracts, we tested their efficacy in situations close to real-world applications. In this first trial, we tested whether extracts from the isolates B01, B05, and B09 were able to control post-harvest fruit decay caused by *Colletotrichum* spp. on olive fruits (Table 1). The fruits were collected in a commercial orchard located within the Gioia Tauro plain (southern Italy) where *Colletotrichum* species (mainly *C. acutatum s. str.* and *C. godetiae*) are endemic, selected to be uniform in size and ripeness, avoiding fruits with lesions and any symptoms of fungal disease. Fruits were then divided into 4 groups (~1500 fruits each) and treated with one of each of the three isolates or water as a negative control. The olives were left to dry for ~2 h at room temperature, transferred to plastic boxes, and incubated at room temperature (22 ± 2 °C). The boxes contained wet paper to ensure high relative humidity. After 7 days, we scored each fruit for the presence/absence of fungal rots, and we scored the decay severity using an empirical scale, according to the amount of fruit surface showing symptoms: 0 (no rot), 1 (<25%), 2 (25–50%), 3 (50–75%), 4 (>75%). This allowed us to calculate McKinney's index [26], which considers the disease incidence (presence/absence) together with its severity. Data were then fit to a linear model, using *treatment* as a fixed factor.

2.9. Control of Post-Harvest Rots on Sweet Cherries

In this second trial, we tested the efficacy of the alcoholic extracts of the B09 bacterial isolate to control post-harvest diseases (mainly *M. fructicola*) on sweet cherries. We selected only this bacterial isolate because it is the one consistently showing the best results across all the previous trials. For this experiment, we collected two cherry varieties (Ferrovia and Giorgia) in a commercial orchard located in Puglia (southern Italy). Fruits with lesions or disease symptoms were discarded. The fruits (50 per group) were dipped into a solution containing the alcoholic extract from isolate B09 at different concentrations (6, 12, and 24 mg/L) or tap water (negative control), dried at room temperature for 2 h on blotting paper, placed in plastic trays, covered with plastic sheet, and stored at 2 ± 1 °C. Incidence and severity of decay were evaluated after 14 days of cold storage, using the same empirical scale described above. Data were then fit to a linear model, using *treatment* as fixed factor.

2.10. Control of Powdery Mildew in a Commercial Grapevine Orchard

In this third trial, we tested the efficacy of the alcoholic extract from isolate B09 in controlling powdery mildew (*Uncinula necator*) in grapevine commercial orchards. Field trials were conducted in Cirò (southern Italy), in a 10-year-old vineyard cultivated with the variety Gaglioppo (Table 1) and managed using an organic farming approach. The trial consisted of three blocks, each divided into 4 groups, which were treated using: (i) the alcoholic extract from isolate B09 at a concentration of 24 mg/L; (ii) a pomegranate peel extract (PGE) at a concentration of 6 mg/L as a positive control [24,27]; (iii) chemical products as normally used for the rest of the vineyard as second positive control (see below); (iv) no treatment (negative control). Treatments were applied following the farm schedule. This trial was performed in 2021, when the severity of powdery mildew was rather low, so the farm scheduled only two treatments on 6 June (copper-based Coprantol Hi Bio 2.0, Syngenta) and on 5 July (sulfur-based Tjovit Jet, Syngenta). All the treatments above were applied on the same days. The incidence and severity of the powdery mildew were scored on 3 July and 29 July on leaves and bunches. The leaves were divided into mature and young in order to differentiate those already present at the time of the treatment from those that developed later (and did not directly receive the treatment). We evaluated symptoms on 72 bunches and 90 young and mature leaves for each treatment, scoring their severity on an empirical scale, according to the amount of fruit/leaf surface showing symptoms: 0 (no damage), 1 (<20%), 2 (20–40%), 3 (40–60%), 4 (>60%). In addition, we tested whether our treatments would influence the sugar content of grapes. The data were gathered on 31 August after collecting 48 berries per replicate (144 for each treatment) and measuring the total soluble solids (°Brix) with a refractometer. Data were then fit to a linear model, using *treatment* as a fixed factor.

3. Results

3.1. In Vitro and In Vivo Antifungal Activity

First, we tested the efficacy of the 22 bacterial isolates in reducing the growth of three plant pathogens (*B. cinerea*, *A. alternata*, and *P. palmivora*) in vitro and their efficacy on reducing post-harvest fruit disease in six different host–pathogen combinations (Table 1).

The in vitro assays suggest that each isolate has a different efficacy in reducing the growth of fungal pathogens, and this depends on the identity of the pathogen itself (isolate × pathogen interaction $F = 98.63$; $df = 42, 132$; $p < 0.001$; Figure S1). While no bacterial isolate was able to efficiently inhibit all three pathogens, isolates B02, B03, B09, and B17 were able to reduce the growth of both *B. cinerea* and *P. palmivora* by >40%, while isolates B13 and B21 were able to reduce the growth of both *B. cinerea* and *A. alternata* by >40% (Figure S1).

In the in vivo assays, regardless the host–pathogen combination (Table 1), all the fruits in the negative control group (ethanol) developed the fungal disease, while no fruit in the positive control group (pomegranate peel extract, PGE) showed symptoms of decay. Thus, we decided to test the efficacy of each bacterial isolate against the negative control, within each host–pathogen combination (Table S2). Given that all the fruits from the negative control group developed decay, differences would represent the efficacy of the isolate in controlling the fungal pathogen. The isolates numbered from B12 to B22 were not able to significantly reduce the incidence of decay in any host–pathogen combination (Table S2); therefore, we decided to focus on the isolates numbered from B01 to B11. All these isolates were able to significantly reduce decay in olives inoculated with *Colletotrichum acutatum s. str.* (Table S2). Most of the isolates were able to control *B. cinerea* in tomato (except B01, B04, B10, and B11) and grape (except B06, B07, and B11), as well as *P. digitatum* in tangerine (except B07, B08, and B11). On the other hand, just a few isolates were able to effectively reduce decay by *P. expansum* in apple (B05, B09), and symptoms by *A. alternata* in tomato (B01, B02, B05, B09). In general, isolates B05 and B09 were able to significantly reduce decay in all host–pathogen combinations, and were used for the following tests together with isolate B01. As reported above and in the Supplementary results, isolates B01, B05, and B09 were identified as *Pseudomonas* spp.

3.2. Preventive Antifungal Activity of Extracts

Given the efficacy shown by isolates B01, B05, and B09 in protecting fruits from fungal pathogens, we then tested the hypothesis that their action was mainly driven by the metabolites they produce. Thus, we prepared an alcoholic extract from each isolate and tested their efficacy in preventing the development of fungal decay in tangerine (*P. digitatum*), tomato (*B. cinerea*), and apricot (*M. fructicola*).

In tangerines, we observed the effect of the treatment (B01, B05, B09, or control; $\chi^2 = 207.16$; df = 3; $p < 0.001$), but no effect was driven by dose or the treatment x dose interaction ($p > 0.05$). Indeed, while all fruits in the control group showed symptoms of decay, no fruits treated with the B01 or B09 extract showed any sign of decay, and only 3.71% of fruits treated with the B05 extract showed symptoms. Similarly, we found an effect driven by the treatment ($\chi^2 = 207.16$; df = 3; $p < 0.001$) when testing our extracts on apricots against *M. fructicola*, but no effect was driven by dose or the treatment x dose interaction ($p > 0.05$). Moreover, in this case, all control fruits showed signs of decay, while only 4.44% of fruits treated with isolate B01 showed the disease by *M. fructicola*; no fruits treated with the B05 and B09 extracts showed any symptoms. When testing the efficacy of the alcoholic extracts on tomatoes against *B. cinerea*, we found an effect driven by the treatment ($\chi^2 = 406.95$; df = 3; $p < 0.001$) but no effect driven by dose or the treatment x dose interaction ($p > 0.05$). Additionally, in this case, all untreated fruits developed fungal decay, while all three isolates were able to reduce the incidence of the fungal disease to 15.5% (B09), 17.22% (B05), and 25% (B01), with no differences between extracts, as suggested by the pairwise contrasts ($p > 0.05$).

3.3. Induction of Resistance

We also tested the extracts from the B01, B05, and B09 isolates for their ability to induce resistance against post-harvest pathogens by inoculating extracts and pathogens into two different wounds on each fruit. Results show that all three isolates were able to fully protect the apricots against *M. fructicola* at any dose, while all control fruits developed fungal decay ($\chi^2 = 202.44$; df = 3; $p < 0.001$). Similarly, the B01 and B09 extracts fully protected the tomatoes against *B. cinerea* regardless the dose, while 22.2% of the fruits treated with the B05 extract showed symptoms, and all the control fruits developed decay.

3.4. Curative Effects

In addition, we tested the curative effects of our extracts by inoculating tangerine fruits with spores of *P. digitatum* and treating them with the extracts after 24 h, allowing time for the spores to germinate and begin to infect the fruit. Furthermore, in this case, the results suggest an effect driven by treatment ($\chi^2 = 159.48$; df = 4; $p < 0.001$), with all negative control fruits showing symptoms, all positive control fruits (PGE) not showing any symptoms, only 17.7% of fruits treated with B05 and B09 extracts showing decay, and 26.6% of those treated with the B01 extract showing signs of infection.

3.5. Metabolomics

Given the efficacy of the extracts from all three isolates (B01, B05, and B09), we used an untargeted metabolomics approach to try to identify the molecules that might be responsible of their antifungal activity. By removing compounds identified in the blank samples and in the bacterial growth medium, we found 337 molecules associated with our bacterial extracts. The majority of these molecules did not find a match in the database, while we found phenylpropanoids and polyketides (3.56%), benzenoids (2.97%), organoheterocyclic compounds (1.78%), organic nitrogen compounds (0.89%), lipids and lipid-like molecules (0.3%), and organophosphorous compounds (0.3%). All the extracts shared 71 compounds, of which 11 were putatively annotated as: epitestosterone-like, tolcapone-like, avobenzone-like, pheniramine N-Oxide-like, neoeriocitrin-like, carbamazepine-like, N-succinylmexiletine-like, altenusin-like, phenazine-1-carboxylic acid-like, sphingosine-like, and velutin-like. Extracts from isolate B01 were characterized by a unique signature of 18 compounds, of which none was accurately identified. Similarly, 22 metabolites were uniquely identified in extracts from isolate B05, but we were not able to identify any of them. Extracts from B09 show a unique blend of 36 metabolites, of which we were able to identify only one as cryptotanshinone-like.

3.6. Control of Post-Harvest Rots on Olives

After testing the extracts from the B01, B05, and B09 isolates in controlled conditions, we ran three trials to simulate their use in simulated operative conditions. The first trial assessed whether the three extracts could help in controlling post-harvest rots on olives caused by *Colletotrichum spp*. The control group resulted in 100% of fruits showing decay, while those treated with PGE resulted in no fruits showing any symptoms. Analyses suggest an effect driven by the treatment (F = 200.25, df = 4, 10, $p < 0.001$), and post hoc contrasts suggest differences between the three extracts and both the control and PGE groups ($p < 0.001$), but with no differences between the extracts ($p > 0.05$).

3.7. Control of Post-Harvest Rots on Sweet Cherries

Given the high performance of extracts from isolate B09, we focused this and the next trial only on this isolate. Moreover, in this case, we compared its performance to a control group and a treatment with PGE. Results (Figure 2) show an effect driven by the treatment for both the varieties Ferrovia ($\chi^2 = 402.89$; df = 2; $p < 0.001$) and Giorgia ($\chi^2 = 232.3$; df = 2; $p < 0.001$). In both cases, post hoc contrasts show similar results: the control group had the highest incidence and severity of diseases, followed by the B09 extract, and PGE.

3.8. Control of Powdery Mildew in a Commercial Grapevine Orchard

As the last trial, we tested the efficacy of the extract from isolate B09 in controlling powdery mildew in field conditions, comparing it with PGE, a chemical treatment, and an untreated control. In this case, we tested the effects on disease incidence and severity in young leaves, old leaves, and fruits at two time points. The results (Figure 3) suggest no effect of any treatment on young leaves (F = 3.07; df = 3, 8 ; $p = 0.09$), while all treatments reduced the disease symptoms in old leaves compared to control (F = 7.94; df = 3, 8; $p = 0.008$), but no differences were recorded between treatments ($p > 0.05$). During the first sampling on fruits, we did not observe any effect driven by the treatment (F = 1.84; df = 3, 8; $p = 0.21$), while on the second sampling, we observed a treatment effect (F = 8.13; df = 3, 8; $p = 0.008$) mainly driven by the reduction in disease severity in the PGE group compared to the control ($p = 0.006$). Finally, we also tested whether the treatments impact the amount of sugars in fruits at harvest, and we did not find any significant effect (F = 1.14; df = 3, 8; $p = 0.38$).

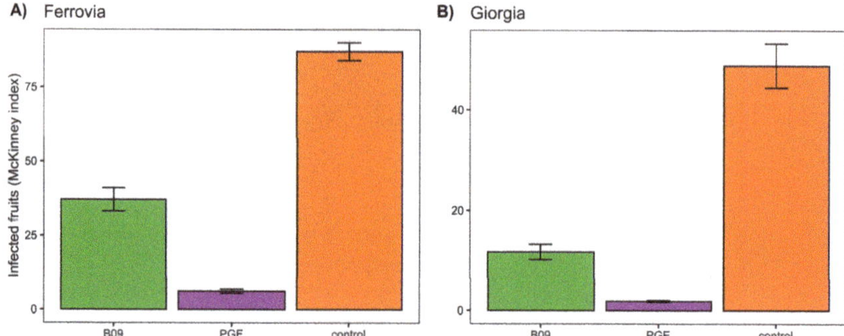

Figure 2. Efficacy of alcoholic extracts from bacterial isolates B01, B05, and B09 in reducing the incidence and severity (both accounted for using McKinney's index) of fungal rot (mainly caused by *Monilinia fructicola*) on two varieties of sweet cherries: (**A**) Ferrovia and (**B**) Giorgia. The extract is compared to a plant extract with strong antifungal activity (pomegranate peel extract, PGE) and an untreated control group.

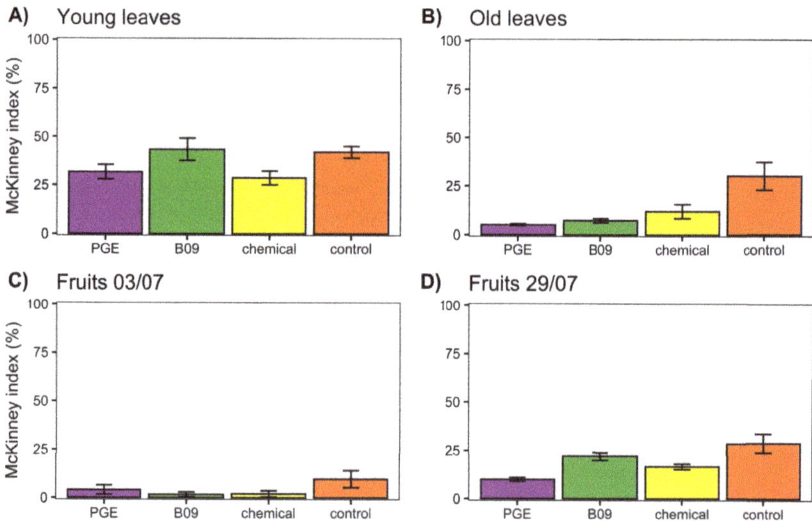

Figure 3. Efficacy of an alcoholic extract from the bacterial isolate B09 in reducing the incidence and severity (both accounted for using McKinney's index) of the damage caused by powdery mildew on grapes in field conditions. The extract is compared to a plant extract with strong antifungal activity (pomegranate peel extract, PGE), the chemical treatments that are normally performed in the farm, and an untreated control group.

4. Discussion

In this study, we isolated microorganisms from the environment and selected a group with a high potential for the biological control of pre- and post-harvest pathogens. In addition, we found that alcoholic extracts from bacterial isolates have good performance against fungal pathogens, also when tested under conditions simulating real-world applications.

As a first step, we used a simple method of selecting microorganisms from the environment, which yielded 125 bacterial isolates. These were further reduced to 22 isolates, which were then tested for their in vitro and in vivo activity to contrast the growth of fungal plant pathogens. The in vitro tests showed that all the 22 isolates were able to significantly reduce the growth of at least one of the three fungal pathogens we used (*A. alternata*, *B. cinerea*, *P. palmivora*), and the outcome strongly varied according to the isolate–pathogen combination. On the other hand, the in vivo tests showed that roughly half of our isolates were able to significantly reduce the incidence of decay, although only two isolates were able to consistently prevent fungal growth in all six trials. These isolates (B05 and B09) were further selected for more focused bioassays together with an additional isolate (B01). All three isolates were identified to belong to the genus *Pseudomonas*, although we were not able to identify them at the species level. This bacterial genus hosts a wide diversity of species with different ecological and functional roles [28,29], and they can be pathogens or beneficial associates to plants [29,30]. Strains of *Pseudomonas* have been previously studied for their potential in controlling plant pathogens [31]. Given the high diversity of *Pseudomonas* strains in soils [32] and their known biocontrol activity [33], it is not surprising that the strains with the highest performance belong to this genus. Indeed, previous studies reported strains of *Pseudomonas* with in vitro antifungal activity against *A. alternata* [34], *B. cinerea* [35], and *P. palmivora* [36]. In addition, members of the genus *Pseudomonas* have been reported to reduce the incidence of fungal disease in agricultural products during the post-harvest phase. For example, strains of *Pseudomonas* isolated from disease-suppressive composts successfully protected blueberry fruits against *A. alternata* and *B. cinerea* [37]. In other studies, isolates of *P. fluorescens* were able to inhibit *B. cinerea* in apples [38], or *Pseudomonas synxantha* showed the ability to control rots by *M. fructicola* and *M. fructigena* in peaches [39], together with many other examples for citrus [40], Chinese cherry [41], strawberry [42], and grapes [43].

In vitro tests showed that fungal growth was severely reduced around the bacterial colonies, suggesting that antifungal activity was performed by the metabolites produced by the bacterial isolates. Thus, to test this idea, we prepared an alcoholic extract from each of the three bacterial strains, and we tested the extracts for their performance in preventing the development of fungal pathogens, curing fungal diseases at early stage, and inducing resistance to fungal pathogen in fruits. Results show that extracts from our isolates were able to completely prevent decay by *P. digitatum* in tangerine and by *M. fructicola* in apricot, with a maximum disease incidence of ∼4%, while in tomato, the incidence of rots by *B. cinerea* reached a maximum of 25%. Similarly, the extracts from the same isolates were able to induce systemic resistance in apricots and tomatoes, and fully prevent the growth of the inoculated pathogens (*M. fructicola* and *B. cinerea*, respectively). In addition, the same isolates were able to cure early-stage infections (∼24 h from exposure) in tangerine fruits inoculated with *P. digitatum*, with a maximum disease incidence of ∼27%. This demonstrates that the *Pseudomonas* strains we isolated are able to produce metabolites that can act both directly and indirectly against fungal plant pathogens. The production of bioactive compounds from different strains of *Pseudomonas* has been previously reported [38,42,44–46]. Given the high bioactivity of the extracts from our isolates against fungal pathogens, we used untargeted metabolomics to attempt to identify the metabolites responsible of their efficacy. Indeed, we identified a pool of 71 metabolites common to all three bacterial isolates, but also a pool of unique metabolites produced by each of the three strains. Although we were only able to identify a few compounds, we found the presence of altenusin-like and phenazine-1-carboxylic acid-like molecules, both previously reported to have antifungal activity [47,48].

Our results pose interesting perspectives on using metabolites extracted from these bacteria as a tool for controlling post-harvest fungal pathogens. To follow on this idea, we set up three experiments where we simulated the use of our extracts in situations close to the actual needs of pre- and post-harvest operators. In the first trials, we tested the efficacy of the extracts against rots caused by *Colletotrichum* sp. on olives, and the results showed that they were able to reduce the incidence of the disease from 100% (control) to about 25%. Given the highest performance of the isolate B09 in most of the preliminary assays, we focused the other two field trials only on this isolate. In the second trial, we evaluated the efficacy of the extracts from isolate B09 in protecting sweet cherries from post-harvest decay mainly caused by *M. fructicola*. Results suggest a reduction of disease incidence and severity from 100% (control) down to 15–30%, according to the variety. As the last field trial, we tested the extract from isolate B09 directly in a vineyard, investigating its potential in controlling damages caused by powdery mildew in fruits and leaves of grapes. Likewise, in this case, extracts from our isolate proved to have high potential, and they effectively controlled powdery mildew on old leaves (those that directly received the sprayed formula), with results similar to the chemical control procedures normally performed by farmers. Only a few previous studies tested the efficacy of *Pseudomonas* isolates or their metabolites in controlling fungal plant pathogens in field conditions. For example, Wang et al. [49] showed that the pre-harvest application of *P. fluorescens* significantly reduced decay caused by *Penicillium* spp. Post-harvest fumigations with volatile organic compounds produced by *P. fluorescens* resulted in the control of disease caused by *Penicillium* spp. in oranges [40], and by *B. cinerea* in grapes [43]. This suggests that a detailed chemical characterization of the extracts used in this study might unveil the single or multiple molecules responsible for their bioactivity, opening new possibilities for the control of plant diseases. This also poses interesting perspectives on understanding the underlying molecular processes that regulate the production of these metabolites, and on identifying markers that can help in the selection of bacterial strains able to produce higher amounts, and thus likely to show higher performance as biocontrol agents.

In addition to our tests on *Pseudomonas* strains isolated from soil, for this paper, we often used the pomegranate peel extract (PGE) as a positive control. This product has been extensively tested by our group [23,24,27,50], and the results showed above further strengthen the possibility of using this extract as an alternative tool for plant disease control, perhaps in combination with other techniques such as the bacterial extracts used in this study.

Agriculture is at a turning point. Consumers are aware of the damages that chemical treatments cause to themselves and the environment, and this creates an increasingly higher amount of requests for produce that has been grown using sustainable approaches. Our study shows that by exploring the soil biodiversity, it is possible to select bacterial strains that produce metabolites with a high potential in the pre- and post-harvest control of fungal diseases. This type of screening can help us to find microbial metabolites that can finally become competitive with synthetic molecules and put an end to the use of chemical pesticides.

Supplementary Materials: The following supporting information can be downloaded at: https://www.mdpi.com/article/10.3390/plants11030436/s1, Figure S1: Inhibition of fungal growth by each microbial isolate in the dual-culture assay, for each isolate–pathogen combination; Table S1: Details about the sampling sites, the number of bacteria isolated, and those that were selected to be tested in this study; Table S2: Results from testing the decay incidence in fruits treated with each of the bacterial isolates against the negative control, for each host–pathogen combination; Supplementary results: Molecular identification of selected isolates. Figure S2: Phylogenetic tree comparing the 16S rRNA of isolates B01, B05, and B09 with a collection of sequences available on SILVA (Quast et al., 2012).

Author Contributions: Conceptualization, M.G.L.D.N. and L.S.; methodology, V.L., G.E.A., and L.S.; experiments, V.L., S.P., N.B., and T.L.-K.; data analysis, A.M.; writing—original draft preparation, A.M. and L.S.; writing—review and editing, all coauthors; funding acquisition, G.E.A. and L.S. All authors have read and agreed to the published version of the manuscript.

Funding: This research was funded by The Italian Ministry of Education, University, and Research (MIUR) with "PON Ricerca e competitività 2007–2013": (i) "Modelli sostenibili e nuove tecnologie per la valorizzazione delle olive e dell'olio extravergine di oliva prodotto in Calabria (PON03PE_00090_02); and (ii) Modelli sostenibili e nuove tecnologie per la valorizzazione delle filiere vegetali mediterranee (PON03PE_00090_03).

Conflicts of Interest: The authors declare no conflicts of interest.

References

1. Carson, R. *Silent Spring*; Technical Report; Reverside Press: Cambridge, MA, USA, 1962.
2. Elahi, E.; Weijun, C.; Zhang, H.; Nazeer, M. Agricultural intensification and damages to human health in relation to agrochemicals: Application of artificial intelligence. *Land Use Policy* **2019**, *83*, 461–474. [CrossRef]
3. Sellare, J.; Meemken, E.M.; Qaim, M. Fairtrade, agrochemical input use, and effects on human health and the environment. *Ecol. Econ.* **2020**, *176*, 106718. [CrossRef]
4. Möhring, N.; Ingold, K.; Kudsk, P.; Martin-Laurent, F.; Niggli, U.; Siegrist, M.; Studer, B.; Walter, A.; Finger, R. Pathways for advancing pesticide policies. *Nat. Food* **2020**, *1*, 535–540. [CrossRef]
5. Isman, M.B. Botanical insecticides: For richer, for poorer. *Pest Manag. Sci.* **2008**, *64*, 8–11. [CrossRef] [PubMed]
6. Wisniewski, M.; Droby, S.; Norelli, J.; Liu, J.; Schena, L. Alternative management technologies for postharvest disease control: The journey from simplicity to complexity. *Postharvest Biol. Technol.* **2016**, *122*, 3–10. [CrossRef]
7. Raymaekers, K.; Ponet, L.; Holtappels, D.; Berckmans, B.; Cammue, B.P. Screening for novel biocontrol agents applicable in plant disease management–a review. *Biol. Control* **2020**, *144*, 104240. [CrossRef]
8. ALKahtani, M.D.; Fouda, A.; Attia, K.A.; Al-Otaibi, F.; Eid, A.M.; Ewais, E.E.D.; Hijri, M.; St-Arnaud, M.; Hassan, S.E.D.; Khan, N.; et al. Isolation and characterization of plant growth promoting endophytic bacteria from desert plants and their application as bioinoculants for sustainable agriculture. *Agronomy* **2020**, *10*, 1325. [CrossRef]
9. Janisiewicz, W. Biocontrol of postharvest diseases of apples with antagonist mixtures. *Phytopathology* **1988**, *78*, 194–198. [CrossRef]
10. Droby, S.; Wisniewski, M.; Macarisin, D.; Wilson, C. Twenty years of postharvest biocontrol research: Is it time for a new paradigm? *Postharvest Biol. Technol.* **2009**, *52*, 137–145. [CrossRef]
11. Morales-Cedeno, L.R.; del Carmen Orozco-Mosqueda, M.; Loeza-Lara, P.D.; Parra-Cota, F.I.; de Los Santos-Villalobos, S.; Santoyo, G. Plant growth-promoting bacterial endophytes as biocontrol agents of pre-and post-harvest diseases: Fundamentals, methods of application and future perspectives. *Microbiol. Res.* **2021**, *242*, 126612. [CrossRef]
12. Naseri, B.; Younesi, H. Beneficial microbes in biocontrol of root rots in bean crops: A meta-analysis (1990–2020). *Physiol. Mol. Plant Pathol.* **2021**, *116*, 101712. [CrossRef]
13. Legein, M.; Smets, W.; Vandenheuvel, D.; Eilers, T.; Muyshondt, B.; Prinsen, E.; Samson, R.; Lebeer, S. Modes of action of microbial biocontrol in the phyllosphere. *Front. Microbiol.* **2020**, *11*, 1619. [CrossRef]
14. Zhang, H.; Mahunu, G.K.; Castoria, R.; Yang, Q.; Apaliya, M.T. Recent developments in the enhancement of some postharvest biocontrol agents with unconventional chemicals compounds. *Trends Food Sci. Technol.* **2018**, *78*, 180–187. [CrossRef]
15. Ab Rahman, S.F.S.; Singh, E.; Pieterse, C.M.; Schenk, P.M. Emerging microbial biocontrol strategies for plant pathogens. *Plant Sci.* **2018**, *267*, 102–111. [CrossRef]
16. Hafez, Y.M.; Attia, K.A.; Kamel, S.; Alamery, S.F.; El-Gendy, S.; Al-Doss, A.A.; Mehiar, F.; Ghazy, A.I.; Ibrahim, E.I.; Abdelaal, K.A. Bacillus subtilis as a bio-agent combined with nano molecules can control powdery mildew disease through histochemical and physiobiochemical changes in cucumber plants. *Physiol. Mol. Plant Pathol.* **2020**, *111*, 101489. [CrossRef]
17. Fierer, N. Embracing the unknown: Disentangling the complexities of the soil microbiome. *Nat. Rev. Microbiol.* **2017**, *15*, 579–590. [CrossRef] [PubMed]
18. Trivedi, P.; Leach, J.E.; Tringe, S.G.; Sa, T.; Singh, B.K. Plant–microbiome interactions: From community assembly to plant health. *Nat. Rev. Microbiol.* **2020**, *18*, 607–621. [CrossRef]
19. R Core Team. *R: A Language and Environment for Statistical Computing*; R Foundation for Statistical Computing: Vienna, Austria, 2020.
20. Bates, D.M. *lme4: Mixed-Effects Modeling with R*; 2010.
21. Fox, J.; Weisberg, S. *An R Companion to Applied Regression*, 3rd ed.; Sage: Thousand Oaks, CA, USA, 2019.
22. Lenth, R.V. *Emmeans: Estimated Marginal Means, Aka Least-Squares Means*; R Package Version 1.6.2-1; 2021.
23. Li Destri Nicosia, M.G.; Pangallo, S.; Raphael, G.; Romeo, F.V.; Strano, M.C.; Rapisarda, P.; Droby, S.; Schena, L. Control of postharvest fungal rots on citrus fruit and sweet cherries using a pomegranate peel extract. *Postharvest Biol. Technol.* **2016**, *114*, 54–61. [CrossRef]
24. Belgacem, I.; Li Destri Nicosia, M.G.; Pangallo, S.; Abdelfattah, A.; Benuzzi, M.; Agosteo, G.E.; Schena, L. Pomegranate Peel Extracts as Safe Natural Treatments to Control Plant Diseases and Increase the Shelf-Life and Safety of Fresh Fruits and Vegetables. *Plants* **2021**, *10*, 453. [CrossRef]
25. Luzzatto-Knaan, T.; Melnik, A.V.; Dorrestein, P.C. Mass spectrometry uncovers the role of surfactin as an interspecies recruitment factor. *ACS Chem. Biol.* **2019**, *14*, 459–467. [CrossRef]
26. McKinney, H. Influence of soil temperature and moisture on infection of wheat seedlings. by *Helminthosporium sativum*. *J. Agric. Res.* **1923**, *26*, 195.
27. Pangallo, S.; Li Destri Nicosia, M.G.; Scibetta, S.; Strano, M.C.; Cacciola, S.O.; Belgacem, I.; Agosteo, G.E.; Schena, L. Preharvest and Postharvest Applications of a Pomegranate Peel Extract to Control Citrus Fruit Decay During Storage and Shelf Life. *Plant Dis.* **2021**, *105*, 1013–1018. [CrossRef] [PubMed]
28. Spiers, A.J.; Buckling, A.; Rainey, P.B. The causes of *Pseudomonas* Divers. *Microbiol.* **2000**, *146*, 2345–2350. [CrossRef] [PubMed]
29. Loper, J.E.; Hassan, K.A.; Mavrodi, D.V.; Davis, E.W.; Lim, C.K.; Shaffer, B.T.; Elbourne, L.D.; Stockwell, V.O.; Hartney, S.L.; Breakwell, K.; et al. Comparative genomics of plant-associated *Pseudomonas* spp.: Insights Divers. Inherit. Trait. Involv. Multitrophic Interact. *PLoS Genet.* **2012**, *8*, e1002784. [CrossRef]

30. Hu, J.; Wei, Z.; Weidner, S.; Friman, V.P.; Xu, Y.C.; Shen, Q.R.; Jousset, A. Probiotic *Pseudomonas* Communities Enhanc. Plant Growth Nutr. Assim. Via Divers.-Mediat. Ecosyst. Funct. *Soil Biol. Biochem.* **2017**, *113*, 122–129. [CrossRef]
31. Höfte, M.; Altier, N. Fluorescent pseudomonads as biocontrol agents for sustainable agricultural systems. *Res. Microbiol.* **2010**, *161*, 464–471. [CrossRef]
32. Li, L.; Al-Soud, W.A.; Bergmark, L.; Riber, L.; Hansen, L.H.; Magid, J.; Sørensen, S.J. Investigating the diversity of *Pseudomonas* spp. Soil Using Cult. Depend. Indep. Tech. *Curr. Microbiol.* **2013**, *67*, 423–430. [CrossRef]
33. McSpadden Gardener, B.B. Diversity and ecology of biocontrol *Pseudomonas* spp. *Agric. Syst. Phytopathol.* **2007**, *97*, 221–226. [CrossRef]
34. Trivedi, P.; Pandey, A.; Palni, L.M.S. In vitro evaluation of antagonistic properties of *Pseudomonas Corrugata*. *Microbiol. Res.* **2008**, *163*, 329–336. [CrossRef]
35. Barka, E.A.; Gognies, S.; Nowak, J.; Audran, J.C.; Belarbi, A. Inhibitory effect of endophyte bacteria on *Botrytis Cinerea* Its Influ. Promot. Grapevine Growth. *Biol. Control* **2002**, *24*, 135–142. [CrossRef]
36. Alsultan, W.; Vadamalai, G.; Khairulmazmi, A.; Saud, H.M.; Al-Sadi, A.M.; Rashed, O.; Jaaffar, A.K.M.; Nasehi, A. Isolation, identification and characterization of endophytic bacteria antagonistic to *Phytophthora Palmivora* Causing Black Pod Cocoa Malaysia. *Eur. J. Plant Pathol.* **2019**, *155*, 1077–1091. [CrossRef]
37. Kurniawan, O.; Wilson, K.; Mohamed, R.; Avis, T.J. *Bacillus Pseudomonas* spp. Provid. Antifung. Act. Gray Mold *Alternaria* Rot Blueberry Fruit. *Biol. Control* **2018**, *126*, 136–141. [CrossRef]
38. Wallace, R.L.; Hirkala, D.L.; Nelson, L.M. Mechanisms of action of three isolates of *Pseudomonas fluorescens* active against postharvest grey mold decay of apple during commercial storage. *Biol. Control.* **2018**, *117*, 13–20. [CrossRef]
39. Aiello, D.; Restuccia, C.; Stefani, E.; Vitale, A.; Cirvilleri, G. Postharvest biocontrol ability of *Pseudomonas synxantha* against *Monilinia fructicola* and *Monilinia fructigena* on stone fruit. *Postharvest Biol. Technol.* **2019**, *149*, 83–89. [CrossRef]
40. Wang, Z.; Mei, X.; Du, M.; Chen, K.; Jiang, M.; Wang, K.; Zalán, Z.; Kan, J. Potential modes of action of *Pseudomonas fluorescens* ZX during biocontrol of blue mold decay on postharvest citrus. *J. Sci. Food Agric.* **2020**, *100*, 744–754. [CrossRef] [PubMed]
41. Wang, C.; Wang, Y.; Wang, L.; Fan, W.; Zhang, X.; Chen, X.; Wang, M.; Wang, J. Biocontrol potential of volatile organic compounds from *Pseudomonas chlororaphis* ZL3 against postharvest gray mold caused by *Botrytis cinerea* on Chinese cherry. *Biol. Control.* **2021**, *159*, 104613. [CrossRef]
42. Wafaa M, H.; Mostafa, A.E.S. Production and optimization of *Pseudomonas fluorescens* biomass and metabolites for biocontrol of strawberry grey mould. *Am. J. Plant Sci.* **2012**, *2012*.
43. Zhong, T.; Wang, Z.; Zhang, M.; Wei, X.; Kan, J.; Zalán, Z.; Wang, K.; Du, M. Volatile organic compounds produced by *Pseudomonas fluorescens* ZX as potential biological fumigants against gray mold on postharvest grapes. *Biol. Control* **2021**, *163*, 104754. [CrossRef]
44. Dowling, D.N.; O'Gara, F. Metabolites of *Pseudomonas* involved in the biocontrol of plant disease. *Trends Biotechnol.* **1994**, *12*, 133–141. [CrossRef]
45. Ligon, J.M.; Hill, D.S.; Hammer, P.E.; Torkewitz, N.R.; Hofmann, D.; Kempf, H.J.; Pée, K.H.v. Natural products with antifungal activity from *Pseudomonas* biocontrol bacteria. *Pest Manag. Sci. Former. Pestic. Sci.* **2000**, *56*, 688–695. [CrossRef]
46. Zhang, Y.; Li, T.; Liu, Y.; Li, X.; Zhang, C.; Feng, Z.; Peng, X.; Li, Z.; Qin, S.; Xing, K. Volatile organic compounds produced by *Pseudomonas chlororaphis* subsp. *aureofaciens* SPS-41 as biological fumigants to control *Ceratocystis fimbriata* in postharvest sweet potatoes. *J. Agric. Food Chem.* **2019**, *67*, 3702–3710. [CrossRef] [PubMed]
47. Johann, S.; Rosa, L.H.; Rosa, C.A.; Perez, P.; Cisalpino, P.S.; Zani, C.L.; Cota, B.B. Antifungal activity of altenusin isolated from the endophytic fungus Alternaria sp. against the pathogenic fungus Paracoccidioides brasiliensis. *Rev. Iberoam. Micol.* **2012**, *29*, 205–209. [CrossRef] [PubMed]
48. Puopolo, G.; Masi, M.; Raio, A.; Andolfi, A.; Zoina, A.; Cimmino, A.; Evidente, A. Insights on the susceptibility of plant pathogenic fungi to phenazine-1-carboxylic acid and its chemical derivatives. *Nat. Prod. Res.* **2013**, *27*, 956–966. [CrossRef] [PubMed]
49. Wang, Z.; Zhong, T.; Chen, X.; Xiang, X.; Du, M.; Zalán, Z.; Kan, J. Multiple pre-harvest applications of antagonist *Pseudomonas fluorescens* ZX induce resistance against blue and green molds in postharvest citrus fruit. *LWT* **2022**, *155*, 112922. [CrossRef]
50. Pangallo, S.; Li Destri Nicosia, M.G.; Agosteo, G.E.; Schena, L. Control of olive anthracnose and leaf spot disease by bloom treatments with a pomegranate peel extract. *J. Saudi Soc. Agric. Sci.* **2021**, in press. [CrossRef]

MDPI

Article

Evaluation of the Impact of Different Management Methods on *Tetranychus urticae* (Acari: Tetranychidae) and Their Predators in Citrus Orchards

Amine Assouguem [1,2,*], Mohammed Kara [3], Hamza Mechchate [4,*], Fahd A. Al-Mekhlafi [5], Fahd Nasr [6], Abdellah Farah [2] and Abderahim Lazraq [1]

1 Laboratory of Functional Ecology and Environment, Faculty of Sciences and Technology, Sidi Mohamed Ben Abdellah University, Fez 30000, Morocco; lazraqab@gmail.com
2 Laboratory of Applied Organic Chemistry, Faculty of Sciences and Technology, Sidi Mohamed Ben Abdellah University, Fez 30000, Morocco; farah.abdellah1@gmail.com
3 Laboratory of Biotechnology and Conservation and Valorization of Natural Resources (LBCVRN) (Ex LBPRN), Sidi Mohamed Ben Abdellah University, Fez 30000, Morocco; mohammed.kara@usmba.ac.ma
4 Laboratory of Inorganic Chemistry, Department of Chemistry, University of Helsinki, P.O. Box 55, FI-00014 Helsinki, Finland
5 Department of Zoology, College of Science, King Saud University, Riyadh 11451, Saudi Arabia; falmekhlafi@ksu.edu.sa
6 Department of Pharmacognosy, College of Pharmacy, King Saud University, Riyadh 11451, Saudi Arabia; fahdnasr74@gmail.com
* Correspondence: assougam@gmail.com (A.A.); hamza.mechchate@helsinki.fi (H.M.)

Citation: Assouguem, A.; Kara, M.; Mechchate, H.; Al-Mekhlafi, F.A.; Nasr, F.; Farah, A.; Lazraq, A. Evaluation of the Impact of Different Management Methods on *Tetranychus urticae* (Acari: Tetranychidae) and Their Predators in Citrus Orchards. *Plants* **2022**, *11*, 623. https://doi.org/10.3390/plants11050623

Academic Editors: Špela Mechora, Dragana Šunjka and Fabrizio Araniti

Received: 8 January 2022
Accepted: 21 February 2022
Published: 25 February 2022

Publisher's Note: MDPI stays neutral with regard to jurisdictional claims in published maps and institutional affiliations.

Abstract: To evaluate the effectiveness of eco-friendly treatments based on detergents classified as non-hazardous and black soap on the pest *Tetranychus urticae* Koch 1836, and their predators (*Euseius stipulatus* Athias-Henriot, 1960, *Typhlodromus* sp., *Phytoseiulus persimilis* Athias-Henriot, 1957), different treatments were applied to citrus orchards planted with Valencia late (Orange) in the Mechraa Belksiri region of Morocco (T0 = control experiment; T1 = spirodiclofen 0.5 L/Ha; T2 = 125 L/Ha (5%) of black soap; T3 = detergent; 4 L/Ha of Oni product + 2 L/Ha of Tide product). The results obtained during the whole monitoring period indicated that the three treatments used, namely spirodiclofen, black soap, and detergents, ensured a reduction in the rate of population of the pest *T. urticae* compared to the untreated plot. In the untreated plot, the average was 45.01 A± 4.90 mobile forms, while the plot treated with spirodiclofen it was only 21.10 C ± 2.71, the black soap 31.49 B ± 3.35, and in the plot treated with detergents, the average was similar to that obtained by spirodiclofen (22.90 C ± 2.18). On the predators (*E. stipulatus*, *P. persimilis*, and *Typhlodropmus* sp.), the black soap and the treatment with detergents were less harmful compared to the chemical spirodiclofen.

Keywords: *Tetranychus urticae*; *Euseius stipulatus*; *Typhlodromus* sp.; *Phytoseiulus persimilis*; citrus orchard; monitoring; predators; pest; treatments

1. Introduction

Citrus is one of the world's major fruit crops and is grown in over 100 countries [1]. It belongs to the Rutaceae family, with 140 genera and 1300 species, including fundamental groups like orange, lemons, mandarin, and pummelos. They are cultivated in tropical and subtropical areas [2]. The total worldwide production of citrus is 139.80 million tons [3].

In Morocco, thanks to the Ministry of Agriculture, citrus orchards have reached 130,000 ha with a total annual production of over 2.2 million tons [4]. However, the yield at the national level is still low compared to that achieved by other European and American countries [5]. The Ministry of Agriculture is currently trying to remedy constraints that are hampering the increase of citrus production, including lack of manpower, maturity of orchards, and biotic and abiotic constraints [6,7]. The negative impact caused by pests reduces the quantity and quality of production. Indeed, spider mites, aphids, medflies, and diaspine scales are pests of primary economic importance [8,9]. When conditions are favorable and in the absence of adequate methods of control, significant damage is often observed on the fruits, twigs, leaves, and young shoots of citrus [10]. *Tetranychus urticae* Koch, 1836 is an economically major pest to agricultural crops and ornamental plants throughout the world [11], being found in Europe, Asia, Africa, the Caribbean islands, and North America [12]. It is known for developing a resistance to pesticides [13], and causes significant damage to agricultural crops, such as defoliation, leaf yellowing, and leaf burning [14,15]. *Euseius stipulatus* Athias-Henriot, 1960 (Acari: Phytoseiidae), *Phytoseiulus persimilis* Athias-Henriot, 1957 (Acari: Phytoseiidae), and *Typhlodromus* sp. (Acari: Phytoseiidae) are the main predators of *T. urticae* in many regions of the world [16,17]. Spirodiclofen (Envidor 240 SC, Bayer, SA) is an acaricide of the keto-enol family. It acts by contact and ingestion as an inhibitor of lipid biosynthesis (LBI) [18]. It has a powerful ovicidal action, controls all juvenile stages, and significantly reduces fertility in females [19].

Black soap, brown in color, is biodegradable, non-polluting, and an excellent insecticide [20]. This product is effective against insects like mealy bugs, aphids, whiteflies, thrips, and spider mites. Through simple contact, it blocks the respiratory pores [2,21]. It does not produce toxic residues and does not affect natural predators. These products are approved by organic agriculture (EEC regulation 2092/91). To minimize the negative impact of these pests on citrus production, the use of chemical applications is often the solution adopted by growers [22,23]. Spirodiclofen often gives good results in the control of spider mites, namely, the biological parameters, including developmental time, survival rate, and in particular, the fecundity of *T. urticae* [24,25]. The misuse of this active ingredient often has harmful impacts on the agroecosystems [26,27]. The development of new eco-friendly approaches is a necessity. In this context, our study aims to evaluate the impact of spirodiclofen, black soap, and a mixture of two detergents (Oni and Tide products) on the pest *T. urticae* and its predators (*E. stipulatus*, *Typhlodromus* sp., and *P. persimilis*).

2. Materials and Methods

2.1. Study Area

This study was performed in Mechra Bel Ksiri, located north of Oued Sebou in the Gharb region, at an altitude of 300–500 m above sea level (Figure 1) [28]. The climate is temperate to hot. This region is well known for the production of citrus fruits, cereals, and vegetable crops due to the suitable properties of the climate and soil [29]. The Gharb region is characterized by a Mediterranean climate with annual precipitation ranging between 480 and 600 mm/year, and an average temperature of 27 °C in summer and 13 °C in winter [30].

2.2. Sampling Design

A Valencia late (*Citrus sinensis*) orchard with a high spider mite infestation was selected to study the impact of spirodiclofen, black soap, and a mixture of two detergents (Oni product and Tide product) on *Tetranychus urticae* populations and their predators. Spirodiclofen was used because of its effectiveness against a wide range of biting/sucking pests, including spider mites. The orchard covers 4 ha of Valencia late trees. The orchard was divided into 4 plots of 1 ha (Figure 2), and each plot was treated with a specific dose of the treatment: (i) T0 treated with water only (as a control experiment), (ii) T1 = spirodiclofen 0.5 L/Ha, which is the recommended dose for treating citrus mites, (iii) T2 = 125 L/Ha (5%) of black soap, and (iv) T3 = detergent 4 L/Ha of Oni product (sodium C14–17 alkyl sec sulfonate, sodium C12–13 pareth sulfate) and 2 L/Ha of Tide product (sodium C10–16 alkylbenzene sulfonate, sodium borate, and propylene glycol). These two detergents are often used by local farmers to control pests and minimize the use of pesticides. They are not classified as hazardous, according to the European directive 99/45/EC on dangerous preparations [31]. The treatments were applied with the Teyme Eolo sprayer (Teyme Tecnologia Agricola, Girona, Spain), with turbulent nozzles 12 mm in diameter, delivering 1.55 L/min at 20 bar pressure. The towed sprayer delivers 2500 L of spray liquid per hectare, at a rate of 6 L of spray solution for each tree (Figure 2). To evaluate the effect of each product used, a block of 10 trees was selected and monitored weekly. Ten leaves were collected from each tree, from different directions (north, east, south, and west) and at different heights of the tree (10 replicates were performed independently) [32–34]. We left two untreated lines (12 m) between the different treated plots to avoid overlapping treatments. The total number of predators (*P. persimilis*, *Typhlodromus* sp., and *E. stipulatis*) and phytophagous mite (*T. urticae*) found on the 10 leaves of each plot was recorded [35]. All mites, except eggs and mites in quiescence, were counted on both sides of each leaf with a professional eye loop 10×. Then, to confirm the number of mites on each leaf, the collected leaves were transferred directly into polyethylene bags referenced to the laboratory for observation under a binocular microscope. Inspections were conducted 3 days after the treatment: 12 April (week 1), 19 April (week 2), 26 April (week 3), 3 May (week 4), 10 May (week 5), 17 May (week 6), 24 May (week 7), and 1 June (week 8).

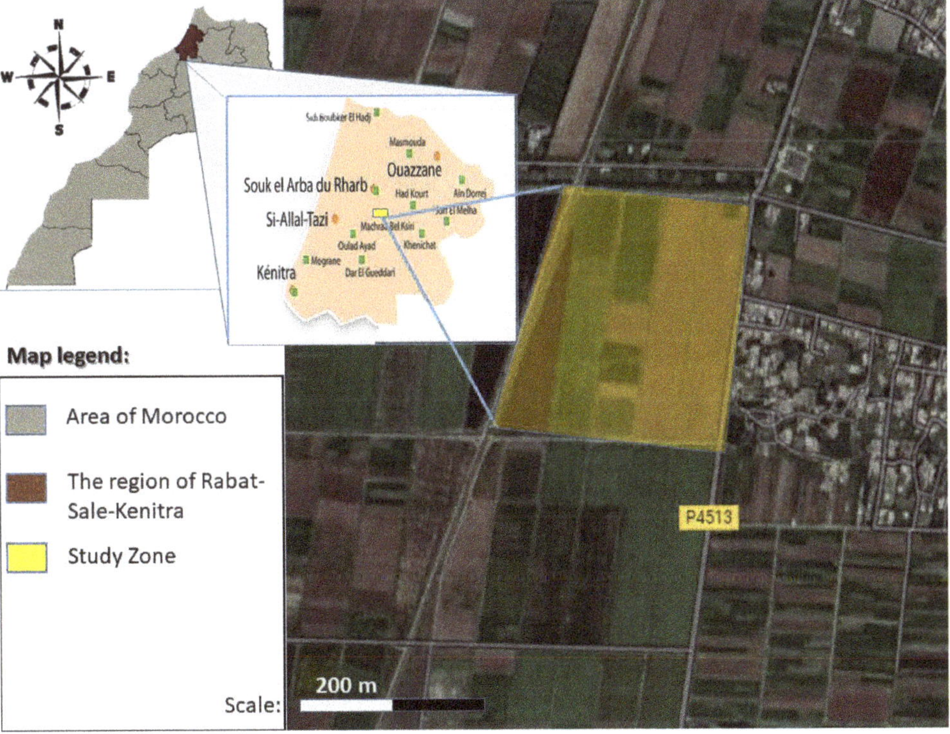

Figure 1. Location of experimental orchards.

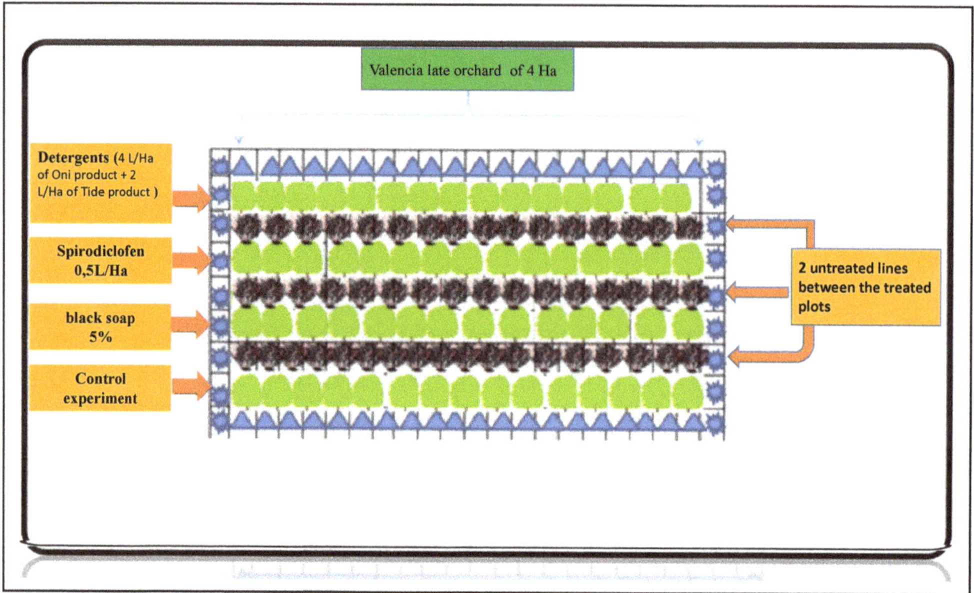

Figure 2. The different types of treatments used in the citrus orchard.

2.3. Statistics

Statistics of the data were performed using Minitab software, version 1.1.19, Minitab, Sydney, NSW, Australia. The results were given as percentage and mean ± SD. In order to evaluate the effectiveness of the treatments used on the pests and predators, we calculated the average of mobile forms found of *T. urticae* in (T0 = control experiment; T1 = plot treated with spirodiclofen; T2 = plot treated with black soap; T3 = plot treated with a mixture of two detergents). In parallel, we evaluated the impact of the treatments on natural enemies. We calculated the averages of each predator (*P. persimilis*, *Typhlodromus* sp. and *E. stipulatis*) in T0, T1, T2, and T3 [36]. We tested for normality and homogeneity of variance for all variables with the Kolmogorov–Smirnov test. The impact of treatments, monitoring dates, and their interactions was compared using the general linear model (GLM) univariate, followed by a post hoc Tukey test at $p < 0.05$. The principal component analyses (PCA) were accomplished using Minitab 19.1 software (Minitab, State College, PA, USA) to elucidate the relationship between the different mites studied and the treatments tested.

3. Results

3.1. Leaf Occupancy Rate by Mites in the Different Plots

During the whole monitoring period, on the untreated plot, the occupancy of the inspected leaves by the pest *Tetranychus urticae* was $n = 3601$ mobile forms (54%); moreover, the predator *Typhlodromus* sp. presented the most important proportion with 18% ($n = 1212$), followed by *P. persimilis* with 15% ($n = 1032$) and *E. stipulatus* with 13% ($n = 838$).

In the plot treated with spirodiclofen, the abundance of these mites was low compared to the untreated plot. The pest *T. urticae* presented 46% with 1688 mobile forms. The predator *P. persimilis* presented 22% ($n = 805$), while *E. stipulatus* and *Typhlodromus* sp. presented the same proportion of 16% ($n = 595$ and $n = 571$, respectively).

In the plot managed with black soap, we recorded 2519 mobile forms (50%) for the pest *T. urticae*, 19% ($n = 948$) for the predator *P. persimilis*, 17% ($n = 832$) for *Typhlodromus* sp., and only 14% ($n = 728$) for *E. stipulatus*.

In the plot treated with detergents, we recorded 42% (n = 1832) of *T. urticae*, 23% (n = 1022) of *P. persimilis*, 18% (n = 809) of *Typhlodromus* sp., and 17% (n = 766) of *E. stipulatus* (Figure 3).

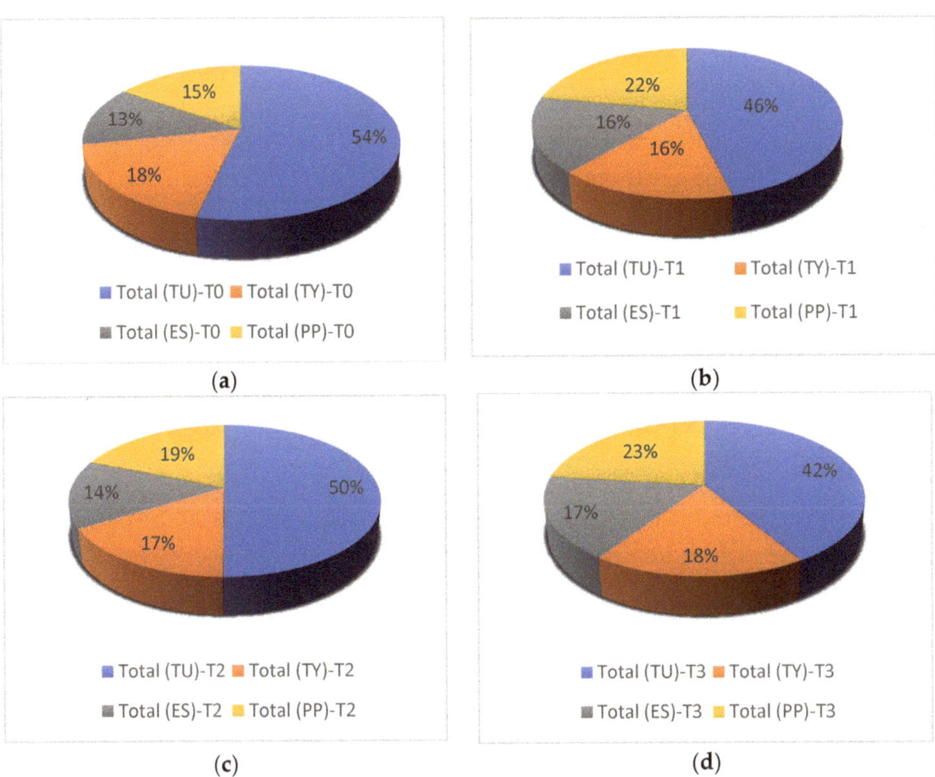

Figure 3. The proportion of the different mites studied (the phytophagous mite *Tetranychus urticae* (TU) and their predators *Euseius stipulatus* (ES), *Typhlodromus* sp. (TY), and *Pytoseiulus persimilis* (PP)) according to the treatments used (T0 = control experiment; T1 = spirodiclofen 0.5 L/Ha; T2 = 125 L/Ha (5%) of black soap; T3 = 4 L/Ha of Oni product + 2 L/Ha of Tide product). (**a**) Control experiment; (**b**) Spirodiclofen 0.5 L/Ha; (**c**) Black soap 125 L/Ha; (**d**) T3 = Mixture of two detergents.

3.2. The Influence of Treatments on the Different Mites Studied

Regarding the effect of the different treatments used on the mites studied, the treatment with detergents and spirodiclofen showed the highest efficiency on the populations of the pest *T. urticae* (Figure 4), while on the predators (*E. stipulatus*, *P. persimilis*, and *Typhlodropmus* sp.), the black soap and the treatment with detergents were less harmful compared to the chemical product (spirodiclofen). Table 1 confirms these results; spirodiclofen treatment significantly reduced *T. urticae* populations with a mean of 21.10 C ± 2.71 compared to the control (45.01 A ± 4.90). The same result was observed with the detergent treatment, which reduced the *T. urticae* levels to an average of 22.90 C ± 2.18. Black soap provided a significant reduction for the mean of this pest (31.49 B ± 3.35) compared to the control; however, this treatment was less effective on *T. urticae* compared to the treatment with spirodiclofen and detergents.

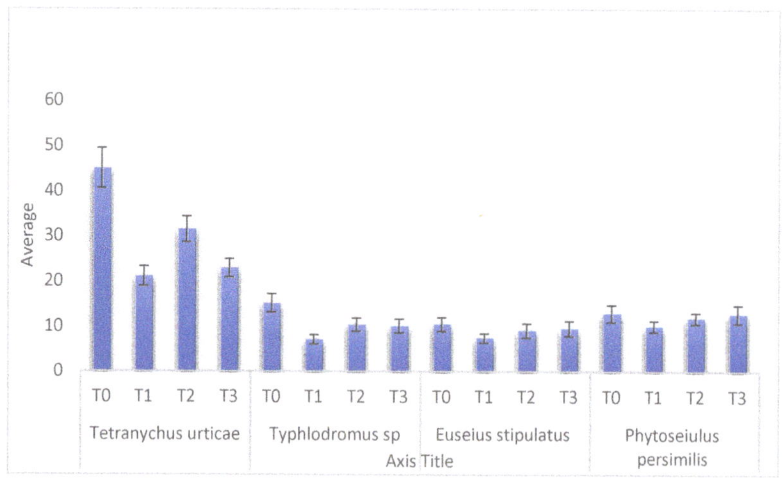

Figure 4. Comparison between the means of different mites according to the treatments (T0 = control experiment; T1 = spirodiclofen 0.5 L/Ha; T2 = 125 L/Ha (5%) of black soap; T3 = detergents; 4 L/Ha of Oni product + 2 L/Ha of Tide product).

Table 1. The impact of different treatments used on the mites studied (TU: *T. urticae*; ES: *E. stipulatus*; TY: *Typhlodromus* sp.; PP: *P. persimilis*) in citrus orchards. Values in the same column with different superscripts are significantly different ($p < 0.05$).

	TU	TY	ES	PP
Control experiment (T0)	$45.01^{A} \pm 4.90$	$15.15^{A} \pm 1.83$	$10.47^{A} \pm 1.63$	$16.02^{A} \pm 2.12$
Spirodiclofen 0.5 L/Ha (T1)	$21.10^{C} \pm 2.71$	$7.13^{C} \pm 1.28$	$7.44^{B} \pm 1.37$	$10.06^{B} \pm 1.15$
Black soap 5% (T2)	$31.49^{B} \pm 3.35$	$10.40^{B} \pm 2.05$	$9.10^{AB} \pm 1.25$	$11.85^{AB} \pm 1.11$
Detergent (T3)	$22.90^{C} \pm 2.18$	$10.11^{B} \pm 2.14$	$9.57^{AB} \pm 1.47$	$14.68^{A} \pm 1.87$

The highest average of *Typhlodromus* sp. was in the water treatment, with an average of 15.15 A ± 1.83 and the treatment with black soap (T3) and detergents (T4) were less harmful on the predator *Typhlodromus* sp. with the means of 10.40 B ± 2.05 for the plot treated with black soap and 10.11 B ± 2.14 for the one sprayed with detergents, while the treatment with spirodiclofen (T2) had the most harmful impact on the population of *Typhlodromus* sp., with an average of 7.13 C ± 1.28.

For the predator *E. stipulates*, the average was 10.47 A ± 1.63 in the plot treated with water (T0), while it was 9.10 AB ± 1.25 in the plot treated with black soap (T3), and 9.57 AB ± 1.47 in the plot treated with detergents (T4). The average of this population was low in the plot treated with the chemical product spirodiclofen, with a mean of 7.44 B ± 1.37.

On *P. persimilis* populations, the treatment with detergents was the least harmful, with an average (14.68 A ± 1.87) nearly similar to that for the plot treated only with water (16.02 A ± 2.12). With black soap, the average was 11.85 AB ± 1.11, whereas the plot treated with spirodiclofen was 10.06 B ± 1.15.

In Figure 5, the eigenvalues of the first two principal components explain 92.7 % of the variation in the data. The first principal component explains 86.5% of the total variance. The variables with the highest association with the first principal component (CP1) are *T. urticae* (0.50), *E. stipulatus* (0.50), and *P. persimilis* (0.50), while the variable *Typhlodromus* sp. (−0.75) is negatively associated with the second principal component (CP2). However, the projection of the scoring and the contribution diagram can visually show a positive contribution of the four mites studied (*T. urticae, P. persimilis, Typhlodromus* sp. and *E. stipulatis*) on the first main axis in positive correlation with T0 water treatment, while spirodiclofen (T2), detergents (T4), and black soap (T3) treatment were correlated negatively.

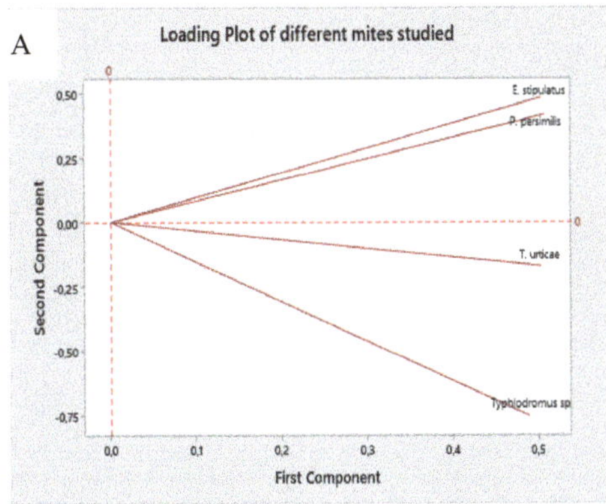

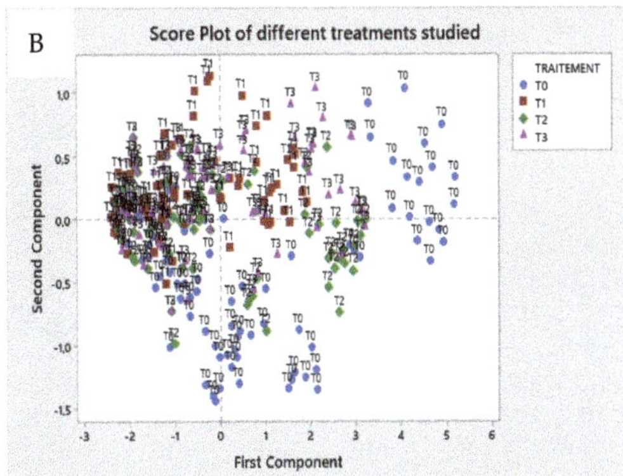

Figure 5. Principal component analysis (PCA) of the different mites studied (*T. urticae; E. stipulatus; Typhlodromus* sp.; *P. persimilis*) according to the different treatments (T0 = control experiment; T1 = spirodiclofen 0.5 L/Ha; T2 = 125 L/Ha (5%) of black soap; T3 = detergents; 4 L/Ha of Oni product + 2 L/Ha of Tide product) (**A**); double projection diagram for the two components (**B**).

3.3. *Fluctuation of Mites According to Treatments and Follow-Up Dates*

According to the results in Table 2, we can conclude that the interaction between the treatment used and the monitoring dates has a significant effect on the variation of the means of the different mites studied. During the first week, three days after the spraying of the treatments, we observed a significant decrease in the rate of *T. urticae*; we had an average of 6.30 F ± 1.16 for the spirodiclofen, 8.20 E ± 1.69 for the black soap (Table 3), and 07.00 F ± 1.49 for the detergent treatment. In contrast, in the untreated plot T0, and during the same week (W1), this average almost doubled 14.30 E ± 2.00. Despite the increase in temperature from 27 to 34 °C from week 1 to week 5 (Table 3), the rate of *T. urticae* had a slight increase during the first five weeks after treatment with spirodiclofen compared to the average reached in the untreated plot (Figure 6). In week 5, these averages were 16.30 D ± 1.41 in the plot treated with spirodiclofen and 36.00 C ± 2.35 in the untreated plot. Beyond that, this pest had a significant increase during the last three weeks to reach an average of 53.60 A ± 4.77 in the plot treated with spirodiclofen (T1), and 86.60 A ± 7.06 in the untreated plot during week 8, where the temperature reached 39 °C.

Table 2. General linear model (GLM) of analysis of variance for the mean density of *T. urticae* (**a**), *Typhlodromus* sp. (**b**), *P. persimilis* (**c**), and *E. stipulatus* (**d**) according to the treatments, monitoring dates, and their interactions.

		(a) *T. urticae*			
Source	**DF**	**Adj SS**	**Adj MS**	**F-Value**	***p*-Value**
Monitoring dates	7	135.717	19,388.1	1579.57	0.000 *
Treatments	3	28.572	9523.9	775.92	0.000 *
M. dates * Treatments	21	18.865	898.4	73.19	0.000 *
Error	288	3535	12.3		
Total	319	186.689			
		(b) *Typhlodromus* sp.			
Source	**DF**	**Adj SS**	**Adj MS**	**F-Value**	***p*-Value**
Monitoring dates	7	7966.9	1138.13	211.42	0.000 *
Treatments	3	2634.3	878.11	163.12	0.000 *

Table 2. *Cont.*

M. dates * Treatments	21	579.6	27.60	5.13	0.000 *
Error	288	1550.4	5.38		
Total	319	12,731.2			
		(c) *P. persimilis*			
Source	**DF**	**Adj SS**	**Adj MS**	**F-Value**	***p*-Value**
Monitoring dates	7	30,048	4292.56	473.97	0.000 *
Treatments	3	1745	581.70	64.23	0.000 *
M. dates * Treatments		3012	143.44	15.84	0.000 *
Error	288	2608	9.06		
Total	319	37,413			
		(d) *E. stipulatus*			
Source	**DF**	**Adj SS**	**Adj MS**	**F-Value**	***p*-Value**
Monitoring dates	7	8996.3	1285.19	304.76	0.000 *
Treatments	3	389.7	129.90	30.80	0.000 *
M. dates * Treatments		409.6	19.50	4.62	0.000 *
Error	288	1214.5	4.22		
Total	319	11,010.1			

* Statistically significant *p*-values ($p < 0.05$).

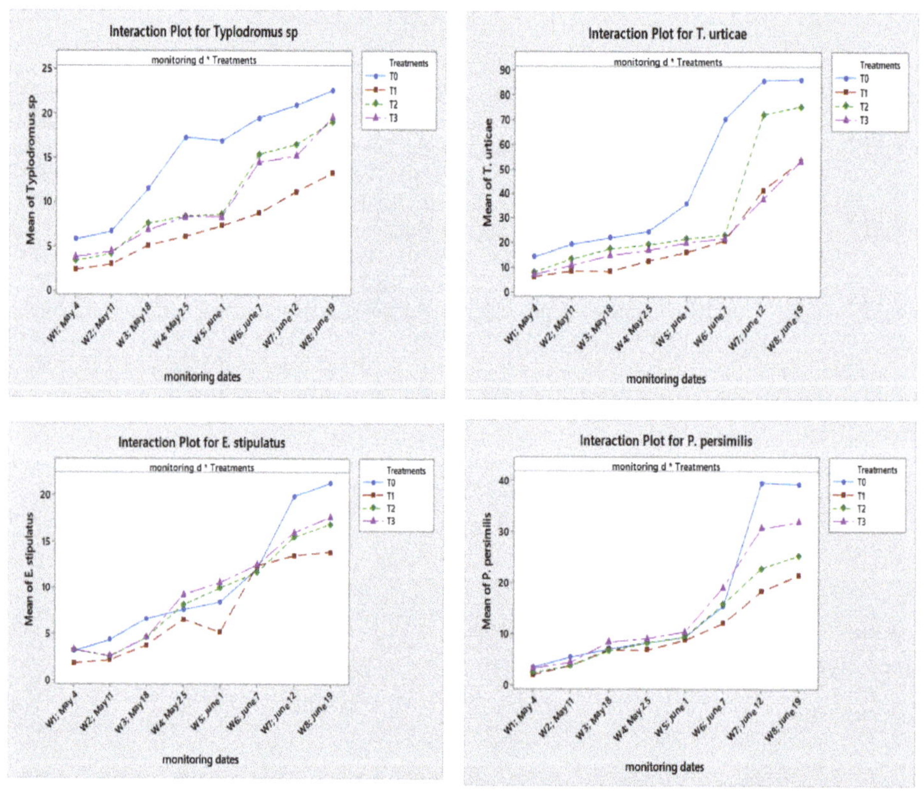

Figure 6. Variation of the means of *T. urticae* and their predators (ES: *E. stipulatus*; TY: *Typhlodromus* sp.; PP: *P. persimilis*) in the four plots (T0 = control experiment; T1 = spirodiclofen 0.5 L/Ha; T2 = 125 L/Ha (5%) of black soap; T3 = detergent; 4 L/Ha of Oni product + 2 L/Ha of Tide product) according to the interaction between different monitoring dates and treatments (d = dates).

Table 3. Fluctuation of the different species studied (TU: *T. urticae*; ES: *E. stipulatus*; TY: *Typhlodromus* sp.; PP: *P. persimilis*) in the four treated plots (T0 = control experiment; T1 = spirodiclofen 0.5 L/Ha; T2 = 125 L/Ha (5%) of black soap; T3 = detergents; 4 L/Ha of Oni product + 2 L/Ha of Tide product) according to the different monitoring dates (W: week).

	TU-TO	TU-T1	TU-T2	TU-T3	TY-T0	TY-T1	TY-T2	TY-T3	ES-T0	ES-T1	ES-T2	ES-T3	PP-T0	PP-T1	PP-T2	PP-T3
W1	14.30 ^{E} ± 2.00	6.30 ^{F} ± 1.16	8.20 ^{E} ± 1.69	07.00 ^{F} ± 1.49	5.80 ^{D} ± 1.33	2.40 ^{F} ± 0.63	3.40 ^{D} ± 1.57	3.80 ^{D} ± 1.39	3.20 ^{D} ± 1.81	1.90 ^{D} ± 0.94	3.30 ^{D} ± 1.25	3.30 ^{D} ± 1.25	3.50 ^{D} ± 1.90	2.00 ^{D} ± 1.16	2.40 ^{E} ± 1.02	3.20 ^{D} ± 1.41
W2	19.50 ^{DE} ± 1.51	8.60 ^{F} ± 1.71	13.50 ^{D} ± 2.17	10.80 ^{F} ± 2.39	6.70 ^{D} ± 1.14	3.00 ^{EF} ± 1.04	4.20 ^{D} ± 1.24	4.40 ^{D} ± 1.24	4.40 ^{CD} ± 1.64	2.20 ^{D} ± 1.31	2.60 ^{D} ± 1.07	2.60 ^{D} ± 1.07	5.50 ^{CD} ± 1.44	3.80 ^{D} ± 1.16	3.80 ^{DE} ± 1.34	4.40 ^{D} ± 1.22
W3	22.10 ^{D} ± 2.13	8.50 ^{F} ± 1.35	17.70 ^{CD} ± 4.08	14.90 ^{E} ± 4.33	11.50 ^{C} ± 1.43	5.10 ^{DE} ± 1.37	7.60 ^{C} ± 1.53	6.80 ^{CD} ± 1.56	6.70 ^{CD} ± 1.54	3.80 ^{CD} ± 1.44	4.60 ^{D} ± 1.95	4.60 ^{D} ± 1.53	7.10 ^{CD} ± 1.93	6.90 ^{C} ± 1.53	6.70 ^{CD} ± 1.37	8.40 ^{C} ± 1.43
W4	24.80 ^{D} ± 1.47	12.80 ^{E} ± 1.22	19.30 ^{BC} ± 2.22	17.0 ^{DE} ± 2.87	17.30 ^{B} ± 1.87	6.10 ^{CD} ± 1.10	8.40 ^{C} ± 1.43	8.30 ^{C} ± 1.47	7.70 ^{C} ± 1.70	6.60 ^{BC} ± 1.67	8.20 ^{C} ± 1.22	9.30 ^{C} ± 1.82	8.30 ^{C} ± 1.40	6.90 ^{B} ± 1.89	8.30 ^{C} ± 1.35	9.10 ^{C} ± 1.33
W5	36.00 ^{C} ± 2.35	16.30 ^{D} ± 1.41	21.80 ^{BC} ± 3.15	20.0 ^{CD} ± 2.44	16.90 ^{B} ± 1.79	7.30 ^{CD} ± 1.41	8.60 ^{C} ± 1.49	8.30 ^{C} ± 1.30	8.50 ^{BC} ± 1.87	5.20 ^{B} ± 1.41	10.00 ^{BC} ± 1.29	10.60 ^{BC} ± 1.71	9.40 ^{C} ± 1.59	8.80 ^{B} ± 1.91	9.40 ^{C} ± 1.67	10.40 ^{C} ± 1.4
W6	70.60 ^{B} ± 4.09	21.00 ^{C} ± 1.33	23.40 ^{B} ± 1.43	21.90 ^{C} ± 1.79	19.50 ^{A} ± 1.84	8.70 ^{BC} ± 1.45	15.40 ^{B} ± 2.14	14.50 ^{B} ± 1.94	12.10 ^{B} ± 1.54	12.40 ^{A} ± 1.26	11.70 ^{B} ± 1.82	12.50 ^{B} ± 1.78	15.50 ^{B} ± 2.04	12.10 ^{A} ± 2.01	15.90 ^{B} ± 2.04	19.20 ^{B} ± 1.7
W7	86.20 ^{A} ± 6.42	41.70 ^{B} ± 3.71	72.30 ^{A} ± 5.29	38.00 ^{B} ± 1.29	20.90 ^{A} ± 2.84	11.20 ^{AB} ± 2.44	16.50 ^{AB} ± 2.84	15.20 ^{B} ± 2.35	19.90 ^{A} ± 2.19	13.50 ^{A} ± 1.78	15.50 ^{A} ± 1.54	16.00 ^{A} ± 1.24	39.60 ^{A} ± 2.94	18.40 ^{A} ± 2.56	22.90 ^{A} ± 2.91	30.70 ^{A} ± 2.4
W8	86.60 ^{A} ± 7.06	53.60 ^{A} ± 4.77	75.70 ^{A} ± 4.69	53.60 ^{A} ± 3.77	22.60 ^{A} ± 2.77	13.30 ^{A} ± 1.17	19.10 ^{A} ± 1.27	19.60 ^{A} ± 2.55	21.30 ^{A} ± 2.75	13.90 ^{A} ± 1.83	16.90 ^{A} ± 1.96	17.70 ^{A} ± 1.57	39.30 ^{A} ± 2.87	21.60 ^{A} ± 2.04	25.90 ^{A} ± 2.87	32.0 ^{A} ± 2.97

Values in the same column with different superscripts are significantly different ($p < 0.05$).

For the treatment based on detergents, we obtained similar results to those achieved with spirodiclofen. These products ensured a significant reduction of the pest *T. urticae* during the eight-week monitoring period. Despite the increase in temperature from 27 °C to 38 °C (Table 4), the treatment with black soap ensured a significant decrease in the rates of *T. urticae* during the first six weeks after application compared to the untreated plot (Figure 6). During the sixth week, these averages were 23.40 B ± 1.43 on the plot sprayed with black soap and 70.60 B ± 4.09 on the untreated plot. The number of this pest increased during the last 2 weeks of monitoring (weeks 7 and 8) to reach 75.70 A ± 4.69 in week 8, almost the same as the average reached on the untreated plot (86.60 A ± 7.06).

Table 4. Temperatures recorded during the monitoring period.

Follow-Up Date	4 May	11 May	18 May	25 May	1 June	7 June	12 June	19 June
Temperature	27	28	30	32	34	38	37	39

During the whole monitoring period, Figure 6 indicates that the three treatments used caused a decrease in the means of *T. urticae* from 12 April (week 1) to 19 May (week 8) compared to the untreated plot; the effectiveness of spirodiclofen was important compared to the control by black soap and detergents.

Regarding the interaction between treatments and monitoring dates (Figure 6), all three treatments used had a significant impact on the abundance of *Typhlodromus* sp. over time. This influence was greater in the spirodiclofen than the detergent and black soap treatments.

For *Typhlodromus* sp., this difference was less important during the first five weeks after spraying the treatments. On the date of 18 May, these averages were 7.30 CD ± 1.41 in T1 (spirodiclofen), 8.60 C ± 1.49 in T2 (black soap), and 8.30 C ± 1. 30 in T3 (detergents). Then, there was a significant difference between the treatment with spirodiclofen and the treatments with black soap and detergents from the sixth week until the last week. The averages were 8.70 BC ± 1.45 in T1, 15.40 B ± 2.14 in T2, and 14.50 B ± 1.94 in T3.

During the first five weeks, the detergent and black soap treatments showed less harmful impacts compared to spirodiclofen on the *E. stipulatus* population. The averages at this time (1 June) were 5.20 B ± 1.41 in T1 (spirodiclofen), 10.00 BC ± 1.29 in T2 (black soap), and 10.60 BC ± 1.71 in T3 (detergents).

For *P. persimilis,* the detergent treatment showed almost similar results as in the control, and showed the least harmful impact, in the last three weeks of the follow-up (from 7 to 19 June). In the plot treated with detergents, we had an average of 32.0 A ± 2.97 for this species; in the untreated plot, we had 39.30 A ± 2.87; in the plot treated with black soap, the average was 25.90 A ± 2.87; and the average was 21.60 A ± 2.04 in the one treated with spirodiclofen.

4. Discussion

The biodiversity of phytoseiid mites nationally and internationally has started to decline catastrophically, as well as the increase in resistance developed by the pests against acaricides, allowing unusual increases in the population of the latter. This work aims to compare the impact of detergents, black soap, and spirodiclofen on the pest *T. urticae* on the one hand and on phytoseiid mites on the other hand.

The results collected during the study period showed that the three treatments used, namely, spirodiclofen, black soap, and detergents, ensured a decrease in the population rate of the pest *T. urticae* compared to the untreated plot. In the untreated plot, the average was 45.01 A ± 4. 90 mobile forms, while the plot treated with spirodiclofen was only 21.10 C ± 2.71, the black soap gave 31.49 B ± 3.35, and average of the plot treated with detergents was similar to that obtained by spirodiclofen at 22.90 C ± 2.18. The same results were reported by [26,37]: the application of spirodiclofen ensured the decrease of the survival rate and fecundity of *T. urticae*. In another study, the efficacy of spirodiclofen and other acaricides against *T. urticae* on greenhouse strawberries was tested at different times after application. At seven days, spirodiclofen showed significant efficacy at 86.6%. This efficacy reached 90% on day 10 and started to decrease slightly after that period [38]. In addition, black soap also showed significant efficacy against several insects such as *Bemisia tabaci* Gennadius, 1889, *Aphis craccivora* Koch, 1854 and *parlatoria ziziphi* Lucas, 1853 [39,40]. The pesticidal activity of the agricultural detergents SU-120 and Tecsa® Fruta was evaluated on *T. urticae* and *Myzus persicae* Sulzer, 1776. Several concentrations of SU-120 were applied to nymphs and adults of *T. urticae*. Mortality was measured in both cases at 24 h. On *T. urticae*, both detergents caused mortality greater than 70% [41–43].

This study revealed that all three treatments used had an impact on the predators studied compared to the control, with the influence of spirodiclofen being greater than that of the detergents and black soap treatments. The effect of spirodiclofen was studied on *Neoseiulus californicus* McGregor, 1954, a potential predator of *T. urticae*, and the results indicated that the adverse consequences on the population growth parameters were significant [44]. In a small field plot test on spirodiclofen-treated apple trees, its use in a conventional chemical control program affected populations of *Forficula auricularia* Linnaeus, 1758 (Dermaptera), which is an important generalist predator in orchards, regulating populations of several damaging pest species, while the number of earwigs increased in the orchards with the compatible spray program (Integrated Pest Management) [45]. Many studies confirm that black soap and detergents can replace conventional chemical treatments and constitute an important tool in integrated pest management because of their low phytotoxicity [43,46].

The results revealed that there was an impact of the treatments used on the three predators compared to the control, and this influence was greater with the spirodiclofen than with the detergents and black soap treatments. Despite the benefits of pesticides, such as the rapidity of action in the reduction of number of pests and their easy use when compared to natural extracts from plants [16], chemical control has many limitations; the examination of the action spectrum of the active components used throughout the world reveals that 46% of acaricides and 72% of insecticides are globally toxic towards auxiliary arthropods and public health [47]. Concerning the impact of the treatments according to the dates on *T. urticae*, we noticed that the three treatments ensured a decrease in the abundance of this pest. On the other hand, the efficacy of spirodiclofen was superior to the black soap and detergents; this can be explained by the high toxicity of the pesticide [22], as it can be remedied by monthly repetitive spraying with black soap or/and detergents, which remain less toxic for the plant, the predators, and the agroecosystem in general [39]. Botanical pesticides from the Lamiaceae, the Asteraceae, the Myrtaceae, and the Apiaceae taxons can also be used as a complementary alternative in the control of *T. urticae* [48]. We can explain the increase of the number of pests during the last two weeks of follow-up by the increase in the temperature, which remains a paramount parameter for population fluctuation [49,50].

5. Conclusions

The following conclusions can be drawn from this work: the black soap and detergent products can replace conventional chemical treatments and are an important tool in integrated pest management, provided they are used correctly. The repetitive spraying of eco-friendly products can ensure the reduction of the pest rates in the long term. Finally, the release of natural enemies can be ensured after four weeks of the treatment with these products, since their persistence does not exceed five weeks, according to our results.

Author Contributions: Conceptualization, A.A. and A.L.; methodology, A.A.; investigation, M.K.; data curation, H.M.; writing—original draft preparation, A.A.; writing, review and editing, F.A.A.-M. and F.N.; supervision, A.L. and A.F. All authors have read and agreed to the published version of the manuscript.

Funding: This research was funded by Researchers Supporting Project number (RSP-2021/112), King Saud University, Riyadh, Saudi Arabia.

Institutional Review Board Statement: Not applicable.

Informed Consent Statement: Not applicable.

Data Availability Statement: Data available upon request.

Acknowledgments: The authors are grateful for Researchers Supporting Project number (RSP-2021/112), King Saud University, Riyadh, Saudi Arabia. Open access funding provided by University of Helsinki.

Conflicts of Interest: The authors declare no conflict of interest.

References

1. Citrus Fruit Fresh and Processed Statistical Bulletin 2020 | Policy Support and Governance | Food and Agriculture Organization of the United Nations. Available online: https://www.fao.org/policy-support/tools-and-publications/resources-details/fr/c/1439010/ (accessed on 28 January 2022).
2. Bora, H.; Kamle, M.; Mahato, D.K.; Tiwari, P.; Kumar, P. Citrus essential oils (CEOs) and their applications in food: An overview. *Plants* **2020**, *9*, 357. [CrossRef] [PubMed]
3. Tahir, R. Impact of Foliar Application of Zn on Growth Yield and Quality Production of Citrus: A Review. *Indian J. Pure Appl. Biosci.* **2020**, *8*, 529–534. [CrossRef]
4. Afechtal, M.; Djelouah, K.; Cocuzza, G.; D'Onghia, A.M. Large Scale Survey of Citrus Tristeza Virus (CTV) and Its Aphid Vectors in Morocco. *Acta Hortic.* **2015**, *1065*, 753–757. [CrossRef]
5. Mazih, A. Status of citrus IPM in the southern mediterranean basin Morocco, North Africa. *Acta Hortic.* **2015**, *1065*, 1097–1104. [CrossRef]
6. Rachid, E.; Ahmed, M. Current status and future prospects of ceratitis capitata wiedemann (Diptera: Tephritidae) control in Morocco. *J. Entomol.* **2018**, *15*, 47–55. [CrossRef]
7. Assouguem, A.; Kara, M.; Benmessaoud, S.; Mechchate, H.; El Jabboury, Z.; Farah, A.; Lazraq, A. Ceratitis capitata Wiedemann (Diptera:Tephritidae) in Moroccan Citrus orchards: Chemical and Sustainable Control Methods and Priorities for Future Research. *Trop. J. Nat. Prod. Res.* **2021**, *5*, 1703–1708.
8. Assouguem, A.; Kara, M.; Mansouri, I.; Imtara, H.; Alzain, M.N.; Mechchate, H.; Conte, R.; Squalli, W.; Farah, A.; Lazraq, A. Evaluation of the effectiveness of spirotetramat on the diaspine scale parlatoria pergandii in citrus orchards. *Agronomy* **2021**, *11*, 1562. [CrossRef]
9. Smaili, M.C.; El Ghadraoui, L.; Gaboun, F.; Benkirane, R.; Blenzar, A. Impact of some alternative methods to chemical control in controlling aphids (Hemiptera: Sternorrhyncha) and their side effects on natural enemies on young Moroccan citrus groves. *Phytoparasitica* **2014**, *42*, 421–436. [CrossRef]
10. Smaili, M.C. Current Pest Status and the Integrated. Available online: https://scholar.google.com/scholar?hl=fr&as_sdt=0%2C5&q=Smaili+MC+%282017%29+Current+pest+status+and+the+integrated+pest+management+strategy+in+the+citrus+groves+in+Morocco.+In%3A+Abstracts+of+the+IOBC+citrus+working+group+meeting+on+citrus+pests%2C+disea (accessed on 10 July 2021).
11. Xue, W.; Snoeck, S.; Njiru, C.; Inak, E.; Dermauw, W.; Van Leeuwen, T. Geographical distribution and molecular insights into abamectin and milbemectin cross-resistance in European field populations of Tetranychus urticae. *Pest Manag. Sci.* **2020**, *76*, 2569–2581. [CrossRef]
12. Migeon, A.; Nouguier, E.; Dorkeld, F.; Cbgp, U.M.R.; Ird, I.; Montpellier, C.; Cedex, M. Spider Mites Web: A comprehensive database for the Tetranychidae Alain. *Trends Acarol.* **2010**, 557–560. [CrossRef]
13. Grbić, M.; Van Leeuwen, T.; Clark, R.M.; Rombauts, S.; Rouzé, P.; Grbić, V.; Osborne, E.J.; Dermauw, W.; Ngoc, P.C.T.; Ortego, F.; et al. The genome of Tetranychus urticae reveals herbivorous pest adaptations. *Nature* **2011**, *479*, 487–492. [CrossRef] [PubMed]
14. Fakhour, S. L' acarien jaune, Tetranychus urticae Koch. (Acari: Tetranychidae) sur Maïs fourrager au Maroc: Impact et stratégies de lutte. In Proceedings of the Deuxième Colloque sur les Acariens des Cultures, Montpellier, France, 24–25 October 2005.
15. Nachman, G.; Zemek, R. Interactions in a tritrophic acarine predator-prey metapopulation system V: Within-plant dynamics of Phytoseiulus persimilis and Tetranychus urticae (Acari: Phytoseiidae, Tetranychidae). *Exp. Appl. Acarol.* **2003**, *29*, 35–68. [CrossRef] [PubMed]
16. Attia, S.; Grissa, K.L.; Lognay, G.; Bitume, E.; Hance, T.; Mailleux, A.C. A review of the major biological approaches to control the worldwide pest Tetranychus urticae (Acari: Tetranychidae) with special reference to natural pesticides: Biological approaches to control Tetranychus urticae. *J. Pest Sci.* **2013**, *86*, 361–386. [CrossRef]
17. Smaili, M.C.; Boutaleb-Joutei, A.; Blenzar, A. Beneficial insect community of Moroccan citrus groves: Assessment of their potential to enhance biocontrol services. *Egypt. J. Biol. Pest Control* **2020**, *30*, 47. [CrossRef]
18. De Maeyer, L.; Geerinck, R. The multiple target use of spirodiclofen (Envidor 240 SC) in IPM pomefruit in Belgium. *Commun. Agric. Appl. Biol. Sci.* **2009**, *74*, 225–232.

19. Ramdani, C.; El Fakhouri, K.; Sbaghi, M.; Bouharroud, R.; Boulamtat, R.; Aasfar, A.; Mesfioui, A.; Bouhssini, M. El Chemical Composition and Insecticidal Potential of Six Essential Oils from Morocco against Dactylopius opuntiae. *Insects* **2021**, *12*, 1007. [CrossRef]
20. Daniel, E.C.; Wyss, C.L. Applications de soufre en automne: Une nouvelle manière de lutter contre lériophyide à galles du poirier. *Rev. Suisse Vitic. Arboric. Hortic.* **2004**, *36*, 199–203.
21. Buitenhuis, R.; Brownbridge, M.; Brommit, A.; Saito, T.; Murphy, G. How to start with a clean crop: Biopesticide dips reduce populations of Bemisia tabaci (Hemiptera: Aleyrodidae) on greenhouse poinsettia propagative cuttings. *Insects* **2016**, *7*, 48. [CrossRef]
22. Attia, S.; Grissa, K.L.; Zeineb, G.G.; Mailleux, A.C.; Lognay, G.; Hance, T. Assessment of the acaricidal activity of several plant extracts on the phytophagous mite Tetranychus urticae (Tetranychidae) in Tunisian citrus orchards. *Bull. Soc. R. Belg. D'Entomol.* **2011**, *147*, 71–79.
23. Van Leeuwen, T.; Tirry, L.; Yamamoto, A.; Nauen, R.; Dermauw, W. The economic importance of acaricides in the control of phytophagous mites and an update on recent acaricide mode of action research. *Pestic. Biochem. Physiol.* **2015**, *121*, 12–21. [CrossRef]
24. Saber, M.; Ahmadi, Z.; Mahdavinia, G. Sublethal effects of spirodiclofen, abamectin and pyridaben on life-history traits and life-table parameters of two-spotted spider mite, Tetranychus urticae (Acari: Tetranychidae). *Exp. Appl. Acarol.* **2018**, *75*, 55–67. [CrossRef] [PubMed]
25. Bostanian, N.J.; Trudeau, M.; Lasnier, J. Management of the two-spotted spider mite, Tetranychus urticae [Acari: Tetranychidae] in eggplant fields. *Phytoprotection* **2003**, *84*, 1–8. [CrossRef]
26. Hu, J.; Wang, C.; Wang, J.; You, Y.; Chen, F. Monitoring of resistance to spirodiclofen and five other acaricides in Panonychus citri collected from Chinese citrus orchards. *Pest Manag. Sci.* **2010**, *66*, 1025–1030. [CrossRef]
27. Faria, J.M.S.; Rodrigues, A.M. Essential Oils as Potential Biopesticides in the Control of the Genus *Meloidogyne*: A Review. *Biol. Life Sci. Forum* **2021**, *3*, 26. [CrossRef]
28. Assouguem, A.; Kara, M.; Mechchate, H.; Alzain, M.N.; Noman, O.M.; Imtara, H.; Hano, C.; Ibrahim, M.N.; Benmessaoud, S.; Farah, A.; et al. Evaluation of the Effect of Different Concentrations of Spirotetramat on the Diaspine Scale Parlatoria ziziphi in Citrus Orchards. *Agronomy* **2021**, *11*, 1840. [CrossRef]
29. Benyahia, H.; Talha, A.; Fadli, A.; Chetto, O.; Omari, F.E.; Beniken, L. Performance of 'Valencia late' sweet orange (Citrus sinensis) on different rootstocks in the gharb region (Northwestern Morocco). *Annu. Res. Rev. Biol.* **2017**, *20*, 1–11. [CrossRef]
30. Hassouni, T.; Belghyti, D. Distribution of gastrointestinal helminths in chicken farms in the Gharb region—Morocco. *Parasitol. Res.* **2006**, *99*, 181–183. [CrossRef]
31. Eur-lex EUR-Lex—31999L0045—EN—EUR-Lex. Available online: https://eur-lex.europa.eu/legal-content/FR/ALL/?uri=celex:31999L0045 (accessed on 30 November 2021).
32. Bouharroud, R. Acariens clés des Agrumes au Maroc, Agriculture du Maghreb. Available online: http://www.agri-mag.com/2017/06/acariens-cles-des-agrumes-au-maroc/ (accessed on 18 April 2021).
33. Martínez-Ferrer, M.T.; Jacas, J.A.; Ripollés-Moles, J.L.; Aucejo-Romero, S. Approaches for sampling the twospotted spider mite (Acari: Tetranychidae) on clementines in Spain. *J. Econ. Entomol.* **2006**, *99*, 1490–1499. [CrossRef]
34. Marullo, R.; Bonsignore, C.P.; Vono, G. Thrips: A review of sampling methods in relation to their habitats. *Bull. Insectol.* **2021**, *74*, 241–251.
35. Benziane, T.; Abbassi, M.; Bihi, T. Evaluation de deux méthodes de lutte intégrée contre les ravageurs en vergers d'agrumes. *J. Appl. Entomol.* **2003**, *127*, 51–63. [CrossRef]
36. Haddad, D.N.; Sadoudi, F.G.-M. Management of a main citrus pest black parlatoria scale Parlatoria ziziphi (lucas) (hemiptera: Diaspididae) in the mediterranean basin. *SciFed Virol. Res. J.* **2018**, *1*, 184–199.
37. Van Pottelberge, S.; Khajehali, J.; Van Leeuwen, T.; Tirry, L. Effects of spirodiclofen on reproduction in a susceptible and resistant strain of Tetranychus urticae (Acari: Tetranychidae). *Exp. Appl. Acarol.* **2009**, *47*, 301–309. [CrossRef] [PubMed]
38. Wang, Z.; Cang, T.; Wu, S.; Wang, X.; Qi, P.; Wang, X.; Zhao, X. Screening for suitable chemical acaricides against two-spotted spider mites, Tetranychus urticae, on greenhouse strawberries in China. *Ecotoxicol. Environ. Saf.* **2018**, *163*, 63–68. [CrossRef] [PubMed]
39. Egho, E.; Emosairue, S. Evaluation of native soap (local black soap) for the control of major insect pests and yield of cowpea (*Vigna unguiculata* (L) WALP in ASABA, Southern Nigeria. *Agric. Biol. J. N. Am.* **2010**, *1*, 938–945. [CrossRef]
40. ABD-RABOU, S. Evaluation of Aphytis melinus De Bach (Hymenoptera: Chalcidoidea: Aphelinidae) in Citrus Orchards as a Biocontrol Agent of Black Scale, Parlatoria ziziphi (Lucas) (Hemiptera: Coccoidea: Diaspididae) in Egypt. *J. Biol. Control* **2009**, *23*, 37–41. [CrossRef]
41. Gill, H.K.; Goyal, G. (Eds.) *Integrated Pest Management (IPM): Environmentally Sound Pest Management*; IntechOpen: London, UK, 2016. Available online: https://www.intechopen.com/books/5252 (accessed on 28 January 2022).
42. Curkovic, T.; Araya, J.; Canales, C.; Medina, Á. Evaluation of two agriculture detergents as control alternatives for green peach aphid and twospotted spidermite, two pests affecting peach orchards in Chile. *Acta Hortic.* **2006**, *713*, 405–407. [CrossRef]
43. Curkovic, T. Detergents and Soaps as Tools for IPM in Agriculture. In *Integrated Pest Management (IPM): Environmentally Sound Pest Management*; IntechOpen: London, UK, 2016. Available online: https://www.intechopen.com/chapters/51590 (accessed on 28 January 2022).

44. Sarbaz, S.; Goldasteh, S.; Zamani, A.A.; Solymannejadiyan, E.; Vafaei Shoushtari, R. Side effects of spiromesifen and spirodiclofen on life table parameters of the predatory mite, Neoseiulus californicus McGregor (Acari: Phytoseiidae). *Int. J. Acarol.* **2017**, *43*, 380–386. [CrossRef]
45. Fountain, M.T.; Harris, A.L. Non-target consequences of insecticides used in apple and pear orchards on *Forficula auricularia* L. (Dermaptera: Forficulidae). *Biol. Control* **2015**, *91*, 27–33. [CrossRef]
46. Bajwa, U.; Singh Sandhu, K. Effect of handling and processing on pesticide residues in food-a review. *J. Food Sci. Technol.* **2014**, *51*, 201–220. [CrossRef]
47. Van Leeuwen, T.; Vontas, J.; Tsagkarakou, A.; Dermauw, W.; Tirry, L. Acaricide resistance mechanisms in the two-spotted spider mite Tetranychus urticae and other important Acari: A review. *Insect Biochem. Mol. Biol.* **2010**, *40*, 563–572. [CrossRef]
48. Rincón, R.A.; Rodríguez, D.; Coy-Barrera, E. Botanicals against Tetranychus urticae Koch under Laboratory Conditions: A Survey of Alternatives for Controlling Pest Mites. *Plants* **2019**, *8*, 272. [CrossRef]
49. Bayu, M.S.Y.I.; Ullah, M.S.; Takano, Y.; Gotoh, T. Impact of constant versus fluctuating temperatures on the development and life history parameters of Tetranychus urticae (Acari: Tetranychidae). *Exp. Appl. Acarol.* **2017**, *72*, 205–227. [CrossRef]
50. Abdallah, A.M.; Ismail, M.S.M.; AboGhalia, A.H.; Soliman, M.F.M. Factors affecting population dynamics of Tetranychus urticae and its predators on three economic plants in Ismailia, Egypt. *Int. J. Trop. Insect Sci.* **2019**, *39*, 115–124. [CrossRef]

Article

Field Margin Plants Support Natural Enemies in Sub-Saharan Africa Smallholder Common Bean Farming Systems

Baltazar J. Ndakidemi [1,*], Ernest R. Mbega [1], Patrick A. Ndakidemi [1], Steven R. Belmain [2], Sarah E. J. Arnold [1,2], Victoria C. Woolley [2] and Philip C. Stevenson [2,3]

1 Department of Sustainable Agriculture, The Nelson Mandela African Institution of Science and Technology, School of Life Sciences and Bioengineering, Arusha P.O. Box 447, Tanzania; ernest.mbega@nm-aist.ac.tz (E.R.M.); patrick.ndakidemi@nm-aist.ac.tz (P.A.N.); s.e.j.arnold@greenwich.ac.uk (S.E.J.A.)
2 Agriculture, Health and Environment Department, Faculty of Engineering & Science, Natural Resources Institute, Medway Campus, University of Greenwich, Chatham Maritime, Kent ME4 4TB, UK; s.r.belmain@greenwich.ac.uk (S.R.B.); v.woolley291@gmail.com (V.C.W.); p.stevenson@kew.org (P.C.S.)
3 Royal Botanic Gardens, Kew, Richmond, Surrey TW9 3AE, UK
* Correspondence: noebien@gmail.com

Abstract: Flower-rich field margins provide habitats and food resources for natural enemies of pests (NEs), but their potential, particularly in the tropics and on smallholder farms, is poorly understood. We surveyed field margins for plant-NE interactions in bean fields. NEs most often interacted with *Bidens pilosa* (15.4% of all interactions) and *Euphorbia heterophylla* (11.3% of all interactions). In cage trials with an aphid-infested bean plant and a single flowering margin plant, the survival of *Aphidius colemani*, the most abundant parasitoid NE in bean fields, was greater in the presence of *Euphorbia heterophylla* than *Bidens pilosa*, *Tagetes minuta*, and *Hyptis suaveolens*. UV-fluorescent dye was applied to flowers of specific field margin plant species and NE sampled from within the bean crop and field margins using sweep-netting and pan-traps respectively. Captured insects were examined for the presence of the dye, indicative of a prior visit to the margin. Lady beetles and assassin bugs were most abundant in plots with *B. pilosa* margins; hoverflies with *T. minuta* and *Parthenium hysterophorus* margins; and lacewings with *T. minuta* and *B. pilosa* margins. Overall, NE benefitted from field margin plants, and those possessing extra floral nectaries had an added advantage. Field margin plants need careful selection to ensure benefits to different NE groups.

Keywords: aphid; conservation biocontrol; parasitoid; *Aphidius colemani*; floral resource plant; field margin

Citation: Ndakidemi, B.J.; Mbega, E.R.; Ndakidemi, P.A.; Belmain, S.R.; Arnold, S.E.J.; Woolley, V.C.; Stevenson, P.C. Field Margin Plants Support Natural Enemies in Sub-Saharan Africa Smallholder Common Bean Farming Systems. *Plants* **2022**, *11*, 898. https://doi.org/10.3390/plants11070898

Academic Editors: Špela Mechora and Dragana Šunjka

Received: 3 March 2022
Accepted: 23 March 2022
Published: 28 March 2022

1. Introduction

The common bean (*Phaseolus vulgaris*) is important for food security for millions of people in sub-Saharan Africa (SSA). It is a major source of dietary protein, carbohydrates, and minerals [1], and contributes to soil fertility by fixing nitrogen with yield benefits for subsequent crops [2–5]. However, common bean production is constrained by several factors, including insect pests [6–9]. Synthetic pesticides are the primary technology used to manage bean pests, but this has adverse effects on human health, contributes to biodiversity loss, and leads to the resurgence of secondary pests [8,10]. More natural pest regulating approaches are required, such as conservation biological control, but this has not been adequately addressed in sub-Saharan Africa [11,12].

For example, natural enemies (NEs) that predate or parasitize insect pests can be a key component of sustainable pest management [13–15]. NEs benefit from non-crop plants in agricultural systems through the provision of shelter, nectar, and pollen for effective biological control [11,16,17]. It is possible to optimize the pest management contribution of NEs by managing field margin plants. However, to maximize this benefit, it is necessary to understand the specific advantages of each flowering plant to NE to improve natural pest regulation (NPR) and increase crop productivity. For instance, some flowers are tubular and lack nectaries, and these are not suitable for NEs [18–22].

Most studies focus on NE and the floral resource requirements in a specific context; either the benefit of the plants on NE in a controlled environment (e.g., cage), or whether the NE interact in the field with the margin plants and enhance NPR [11,12,16,17]. However, combining this information could help to better understand which plants will be most valuable in specific approaches supporting conservation biological control.

Field margin plants differ in their attractiveness to NEs [23], and few studies have been conducted on the contribution of these plants to supporting NEs of bean pests. Thus, we intended to test the following hypotheses: (1) field margin plants influence assemblage natural enemies on beans; (2) specific field margin plants influence differently the fecundity and survival of NEs on beans; and (3) NEs of bean pests use specific margin plants before migrating into the crop to provide NPR services.

2. Results

2.1. Interactions between Natural Enemies and Field Margin Plants

This field trial surveyed the interactions of the field margin plant species with NE in bean fields. When observing NE-plant interactions on transect walks on bean fields, most insect groups investigated had interactions with most species of plant investigated. Overall, 5597 NE-plant interactions were observed, and the greatest number of interactions were recorded with the margin plant *Bidens pilosa* (861 interactions), followed by *Euphorbia heterophylla* (631). Parasitoid wasps were the NE group with the greatest number of observed interactions with field margin plants (724 interactions) (Figure 1).

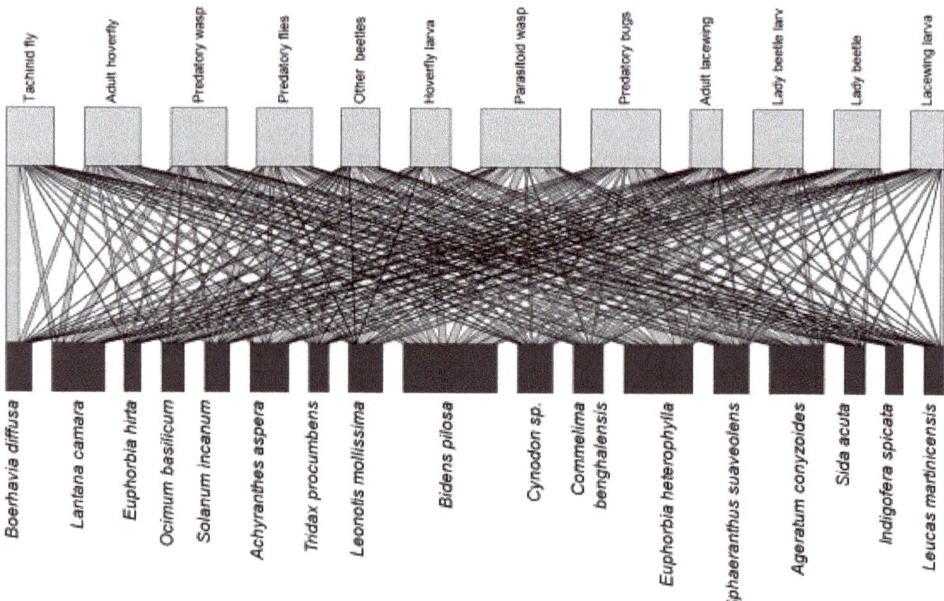

Figure 1. Interactions of natural enemies with field margin plants observed during transect walks in margins of bean fields over 12 months through visual observation, with *Bidens pilosa* having a high number of interactions with natural enemies compared to other field margin plants. The lower row shows plant species present, and the upper row shows the natural enemy guilds; the width of the linking bars indicates the frequency of the interactions observed.

2.2. Effects of Flowering Plant Resources on Parasitism and Survival of Aphidius colemani

In the controlled environment (cages), we investigated the potential of the key field margin plants to support the survival of the *Aphis fabae* parasitoid *Aphidius colemani* and optimize parasitism. When aphid-infested bean plants were caged with *A. colemani* parasitoids and a flowering margin companion plant, the survival of *A. colemani* was enhanced on all plant species compared to the negative control, demonstrating that nectar from all species tested supported NE survival. However, the effect differed significantly among the different flower treatments ($p < 0.001$; Figure 2). The plant that supported significantly improved survival of *A. colemani* compared to other plants, as well as the positive control, was *E. heterophylla*. There was no significant difference ($F_{4,16} = 1.126$, $p = 0.381$) between the number of mummies produced by *A. colemani* given access to any floral resource plants or the positive control (Figure 3).

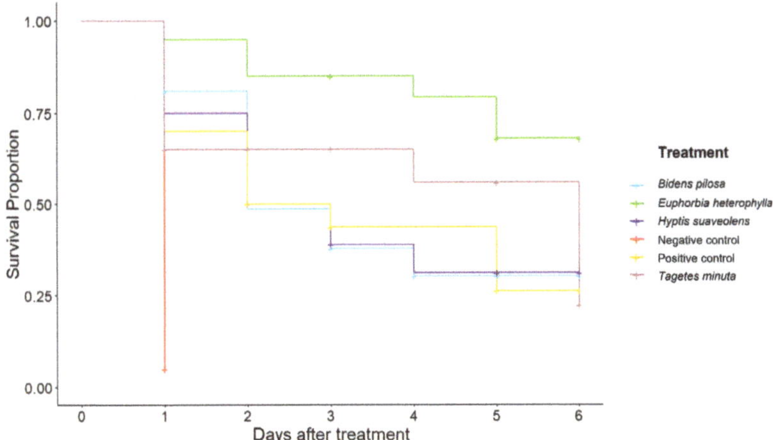

Figure 2. Survival of *Aphidiuscolemani* when provided different field margin plant species, sugar water (positive control), or only water (negative control). A '+' represents a censored individual.

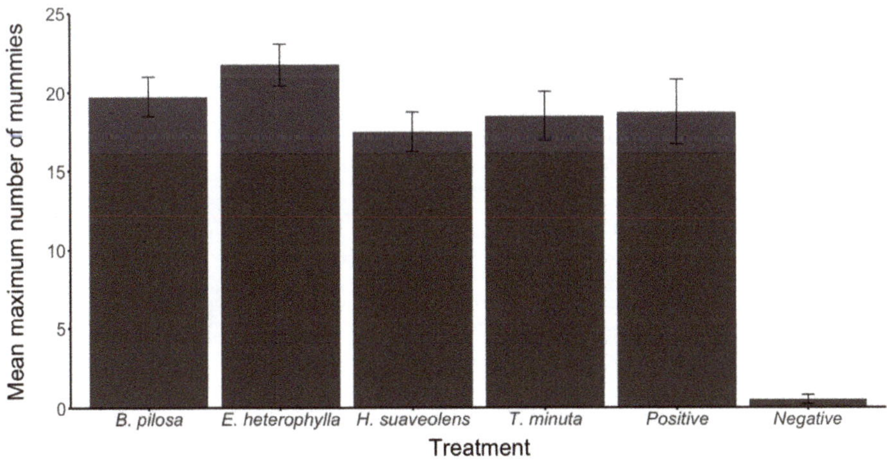

Figure 3. The mean number of *Aphis fabae* mummies produced per cage containing four females and two males of *Aphidius colemani*. Treatments: *B. pilosa-Bidens pilosa*; *E. heterophylla-Euphorbia heterophylla*; *H. suaveolens-Hyptis suaveolens*; *T. minuta-Tagetes minuta*; *Positive*-Positive control; *Negative*-Negative control.

2.3. Effect of Different Field Margin Plants in Supporting Natural Enemies in Bean Fields Fluorescent Dye

On a station trial, we cultivated bean plots surrounded by one of three field margin plant species, and treated the flowers with a fluorescent dye. We then captured insects in the crop (via sweep netting), and examined them for the presence of the dye, which identified the insects that had interacted with the field margin before moving into the crop, as well as pan-trapping within the margins. Lady beetles, hoverflies, assassin bugs, and lacewings differed in the number of fluorescent-labelled individuals captured the crop, according to the surrounding plant species (Table 1). Lady beetles, lacewings, and assassin bugs with fluorescent dye were particularly numerous in the *B. pilosa*-edged plots (and for lacewings, *Tagetes minuta*), indicating that these species regularly used *B. pilosa* before moving into the crop. Conversely, hoverflies were more numerous in plots surrounded by dye-marked *Parthenium hysterophorus* and *T. minuta*. Plots with *P. hysterophrous* margins were much less frequently used by lady beetles, lacewings, and assassin bugs (Table 1).

Table 1. Mean ± (SEM) numbers of natural enemies labelled with UV fluorescent powder within bean crops surrounded by different field margin plants. Different letters indicate significant differences between treatments within natural enemy groups. Significant differences were calculated using a GLM with Poisson distribution, followed by pairwise comparisons and a Holm multiple comparisons test.

Treatment	Mean Number of Natural Enemies (±SEM)					
	Lady Beetle	Hoverfly	Assassin Bug	Lacewing	Parasitoid Wasp	Long-Legged Fly
Biden pilosa	9.50 ± 2.02 a	2.50 ± 0.65 b	5.25 ± 0.95 a	6.50 ± 1.04 a	2.00 ± 0.82 a	1.50 ± 0.87 a
Control (no plant)	2.50 ± 1.04 b	5.50 ± 2.26 ab	0.75 ± 0.48 b	2.00 ± 0.41 b	1.75 ± 0.63 a	1.75 ± 0.48 a
Parthenium hysterophorus	4.25 ± 0.85 b	7.25 ± 0.48 a	2.25 ± 0.48 b	2.00 ± 0.82 b	2.25 ± 0.75 a	1.75 ± 0.48 a
Tagetes minuta	6.25 ± 1.03 ab	8.25 ± 0.85 a	0.75 ± 0.48 b	7.00 ± 1.47 a	3.25 ± 1.32 a	2.75 ± 0.48 a

Values followed by the same letters (a and b) within the column are not significantly different ($p < 0.05$).

The number of natural enemies caught in pan traps in field margins also varied significantly depending on which field margin plant was present (Table 2), following similar patterns to the crop. Again, *B. pilosa* plots favored lady beetles, assassin bugs, lacewings, and parasitoid wasps within the margins, while *T. minuta* and *P. hysterophorus* favored hoverflies in the margins (Table 2).

Table 2. The mean ± (SEM) number of natural enemies caught in pan traps in field plots with different field margin plants. Different letters indicate significant differences between treatments within natural enemy groups. Significant differences were calculated using a GLM with Poisson distribution, followed by pairwise comparisons and a Holm multiple comparisons test.

Treatment	Mean Number of Natural Enemies (±SEM)					
	Lady Beetle	Hoverfly	Assassin Bug	Lacewing	Parasitoid Wasp	Long-Legged Fly
Control (no plant)	1.83 (± 0.63) a	1.92 (± 0.54) a	1.25 (± 0.70) a	1.50 (± 0.86) a	1.75 (± 0.49) a	1.75 (± 0.63) a
Bidens pilosa	5.92 (± 1.05) b	2.67 (± 0.77) ab	3.75 (± 0.35) b	3.50 (± 0.42) b	5.31 (± 1.53) b	1.67 (± 0.45) a
Parthenium hysterophorus	2.17 (± 1.95) a	4.50 (± 0.82) b	2.42 (± 0.78) ab	2.42 (± 0.86) ab	3.23 (± 0.93) b	1.08 (± 0.34) a
Tagetes minuta	3.33 (± 0.88) a	4.58 (± 0.83) b	3.42 (± 1.23) b	3.58 (± 0.93) b	4.58 (± 1.32) b	0.75 (± 0.22) a

Values followed by the same letters (a and b) within the column are not significantly different ($p < 0.05$).

The data for the field and laboratory trials are found in Supplementary Materials (Table S1).

3. Discussion

The composition of plants for a field margin that effectively supports natural enemies (NEs) may require different plant communities than for pollinators, though some plants may provide nectar and pollen to both natural enemies and pollinators [24]. The species that support NEs most effectively in East Africa are still poorly understood. Some common field margin species, such as *E. heterophylla* (Euphorbiaceae), *P. hyterophorus*, *B. pilosa*, *T. minuta* (Asteraceae), and *H. suaveolens* (Laminaceae), are invasive to SSA. However, their potential has been explored for pest control, pollination, and medicinal activities [11,16,25–27]. Our study aimed to determine which field margin plant species in SSA were beneficial to Nes in smallholder bean farms. The transect walk showed that Nes interact with multiple field margin plant species, although certain species had a higher number of interactions (*B. pilosa* and *E. heterophylla*). NEs depend on pollen and nectar from the plants, and plants provide alternative hosts in the absence of crops [28–35]. Similar results were found in a recent study by Arnold et al. [16] which showed that *B. pilosa* and *Euphorbia* sp. were preferred by natural enemies and pollinators in SSA. The use of *B. pilosa* and *E. heterophylla* by NEs could indicate that they provide valuable food resources or habitat. The observed interactions of NEs with *E. heterophylla* concur with the study by Patt [36] showing that this species provided nectar for lady beetles (*Coelophora inequalis*, *Cryptolaemus montrouzieri*, and *Harmonia axyridis*). Similarly, *B. pilosa* is effective in attracting populations of lady beetles (*Cycloneda sanguinea*) and hoverflies (*Pseudodoros* sp.) [37,38]. In addition, chemical cues from *B. pilosa* play a role in attracting natural enemies [39].

In station trials, we found that plots surrounded by *B. pilosa* margins were used frequently by lady beetles, parasitoids, and assassin bugs; *T. minuta* field margins were associated with catches of hoverflies, assassin bugs, lacewings, and parasitoids; and *P. hysterophorus* only with higher numbers of hoverflies and parasitoids. Furthermore, NEs caught inside the field crops with fluorescent dye indicated that the insects visited the flowers (possibly consuming nectar and/or pollen) before moving into the crop where they can provide pest control benefits. Relatively few long-legged flies and parasitoids were captured with the fluorescent dye, but this could be due to their small size rather than a lack of interaction with field margin species. Previous studies have shown the importance of *B. pilosa*, *T. minuta*, and *P. hysterophorus* in supporting NEs [27,37–40]. Floral resources from non-crop habitats are expected to support NPR by NEs [40–46]. Thus, selecting suitable plants for NEs is an important component of agricultural landscaping, as some plants will be better at supporting NEs. For instance, providing adult hoverflies with floral resources can enhance biological control by their larvae [47,48]. Moreover, pollen from some plants is superior to others in enhancing the performance of NEs [49], and this might explain why, in our data, one plant species, *E. heterophylla*, supported *A. colemani* better than the positive control.

Significantly more parasitoids were recorded in plots with field margin plants compared to the control (without field margin plants). In the transect walks, parasitoids were also the NE group with the most plant interactions. This could suggest that parasitoids are a NE group for which field margin plants are particularly important, providing carbohydrates, amino acids, and vitamins in nectar that enhance their pest controlling activities and optimize their metabolism [29,30,35]. This was demonstrated by using a cage trial experiment, which showed that all plants, i.e., *B. pilosa*, *H. suaveolens*, *T. minuta*, and *E. heterophylla*, resulted in improved *A. colemani* survival, and showed similar results to the positive control, which provided an in-cage carbohydrate food supply. Survival of parasitoids on *E. heterophylla* was greater than even the positive control and all other plants, suggesting that this species provided a greater nutritional benefit to *A. colemani*. Our results concur with similar studies, showing that access to flowers prolongs the lifespan and increases parasitism by *A. colemani*. For instance, studies on *A. colemani* and *Diadegma insulare* have shown improved performance compared with controls both in the field and when caged with flowering plant species such as *Fagopyrum esculentum*, *Conium maculatum*, *Photinia* × *fraseri*, *Brassica kaber*, *Barbarea vulgaris*, *Salvia apiana*, *Ligustrum japonicum*, *Lantana camara*, *Eriogonum fasciculatum*, *Daucus carota*, and *Thlaspi arvense* [50–53].

One of the reasons that *B. pilosa*, *H. suaveolens* and *E. heterophylla* supported *A. colemani* survival during the cage trial and that *B. pilosa* and *E. heterophylla* had high numbers of interactions with NEs in field trials could be the presence of extrafloral nectaries on these species [36,54–56]. Extrafloral nectaries are easily accessible, and the nectar composition differs from floral nectar and may be secreted differently [57]. Extrafloral nectar sugars are typically more concentrated than floral nectar, and normally it is present in larger volumes and secreted for a longer period [49,57]. Plants with extrafloral nectaries can be particularly important for NEs, as well as attractive to parasitoids [58], because these insects have mouthparts that are not suited to feeding on floral corollas; hence, they depend on plants with extrafloral or otherwise exposed flower nectaries [19,59,60]. Indeed, *E. heterophylla* produces extrafloral nectar right up to fruit maturation, possibly to provide food resources to attract natural enemies of seed and fruit pests [36].

Although *T. minuta* does not have extrafloral nectaries, it supported *A. colemani* survival in cage trials and plots, while *T. minuta* field margins had greater numbers of hoverflies, assassin bugs, lacewings, and parasitoids compared to control plots. This concurs with previous reports that *T. minuta* supports NEs. *T. minuta* increased the longevity of the egg parasitoid *Trichogramma minutum*, which enhanced the parasitism of the *Grapholita molesta* eggs [40]. Other species of *Tagetes*, including *T. erecta*, increased the longevity of *Cyrtorhinus lividipennis*, a NE of rice brown planthopper (*Nilaparvata lugens*) [61]. The extrafloral resources from the field margin plants could have additional benefits to NEs, and support biological control. Incorporating those field margins with extrafloral resources could bring positive effects for pest control in bean fields.

4. Materials and Methods

4.1. Interactions between Natural Enemies and Field Margin Plants

This field trial was carried out at Kwa Sadala Village in Hai District, Kilimanjaro region (3° 10′ 0″ S, 37° 10′ 0″ E). A total of eight sites with a high diversity margin and eight sites with a poor margin were visited. The high and poor diversity fields were determined by measuring the plant diversity in each farm before the selection of the sites. A transect walk along one margin, for the length of the field, was performed, and the visual observation of the NEs visits to the specific plant flowers was recorded. The sampling was conducted monthly during a year, and this coincided with specific bean crop development stages (1,5,9,14, etc., weeks after bean emergence).

4.2. Effect of Flowering Plant Resources on Parasitism and Survival of A. colemani

Aphidius colemani adults were obtained from *Aphis fabae* mummies collected from bean fields at Kwa Sadala in Hai District, Kilimanjaro region. *A. colemani* was selected for the cage trial because it has been reported as a primary parasitoid of *A. fabae* in SSA [12]. Moreover, this species is commercially produced for the biological control of many aphid species [62,63].

They were reared on potted bean plants infested with *A. fabae* in a wooden netted cage 30 × 30 × 60 cm. The plants were watered every three days. The *A. fabae* colonies were established from insects collected from farmers' fields at Kwa Sadala village, the location of the field trials.

Bean seeds were grown in pots, then after five weeks, they were infested with 60 *A. fabae* (nymphs and apterous adults) [64]. The seeds from four field margin weeds (*Tagetes minuta*, *Hyptis suaveolens*, *Euphorbia heterophylla*, and *Bidens pilosa*) were germinated in pots before being planted out in fields. The experiment consisted of six treatments (*T. minuta*, *H. suaveolens*, *E. heterophylla*, *B. pilosa*, positive control, and negative control), and each treatment was replicated four times. Each cage contained one of these treatments with a potted bean plant infested with *A. fabae*. The positive control contained 10% sugar solution (glucose) as often as it was needed [65] and a potted bean plant infested with *A. fabae*, while the negative control had only a potted bean plant infested with *A. fabae*. Plants were watered every three days.

After leaving the aphids in the cage for 24 h to acclimatize, four female parasitoids and two male parasitoids were introduced to each cage. For the first seven days, the number of live parasitoids and mummies was counted daily to determine the survival of the first generation. Following this, the counting was performed three times a week. The number of parasitoids that emerged from mummies was recorded. The experiment was carried out for one entire lifecycle of parasitoids (approximately one month). The parasitoids and aphids were maintained under controlled conditions, with an average temperature of 25–27 °C, 66–68% R.H., and under natural lighting.

4.3. Effect of Different Field Margin Plants in Supporting Natural Enemies in Bean Fields

A field experiment was carried out at Kwa Sadala Village in Hai District, Kilimanjaro region, to monitor NE movement between field margins and crops. In total, three plant species introduced as above; *T. minuta*, *B. pilosa*, and *P. hysterophorus* were cultivated as the field margin. We initially selected *B. pilosa* and *E. heterophylla* due to the high number of interactions in the transect walk experiment. However, *E. heterophylla* failed to develop in the field, and thus other plants (*T. minuta*, *B. pilosa* and *P. hysterophorus*) were selected for subsequent field studies, as they occur frequently in SSA, have previously reported associations with beneficial insects [11,16,25,27], and are therefore straightforward for smallholders to acquire [26,66].

The experimental layout was composed of four treatment plots containing common bean (*Phaseolus vulgaris*) with four replications, 0.5 m field margin plants surrounded by three plots in each replication, each plot with the specific field margin plant, and the remaining plot was the control without a field margin. The plots measured 15 × 15 m, and the distance between plots was 15 m (Figure 4). During the flowering period of beans in the fifth week, powdered UV fluorescent dye (Baker Ross Ltd., Harlow, UK) was applied in the field margin plant flowers using a soft paintbrush. After 24 h, natural enemies were collected inside the field using sweep nets, then examined using a UV torch (365 nm; UVGear, Surrey, UK) to detect any fluorescent dye. This allowed the identification of insects that had visited the different field margin plants before being caught in the crop fields as an indication of the potential value of different species to different NEs. Pan traps were also used to collect NEs in the field margins [11]: two pan traps were placed in the field margin of each plot and natural enemies were sampled for three months (April, May, and June), coinciding with bean development stages. The collections were preserved in 70% ethanol for further identification.

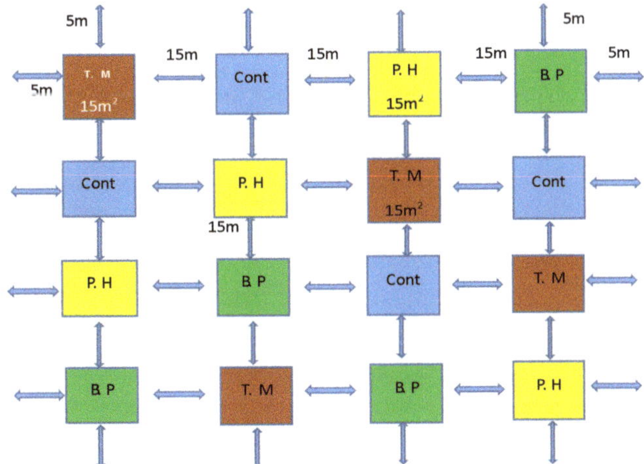

Figure 4. Field margin plants surrounding bean experimental plots for the fluorescent dye trial: T.M-*Tagetes minuta*; Cont-Control; P.H-*Parthenium hysterophorus*; B.P-*Biden pilosa*).

4.4. Statistical Analysis

The interactions between NEs and different field margin plants were plotted using the 'bipartite' package [67] in R (RStudio Version 1.2.1335). Insects caught in field margins using pan traps were grouped into functional categories of NEs; catch distributions were checked for normality using a Shapiro–Wilk test. To assess the effect of field margin plants on the number of natural enemies caught, a histogram was plotted to assess the distribution of the data. Following this, GLM with Poisson distribution (RStudio Version 1.2.1335) was selected, as the data were not normally distributed. The fit of models was assessed with Chi-squared goodness of fit test, and all were found to follow this distribution. The month of sampling and field margin plant species were included in the GLM as covariates without interactions. Following this, pairwise comparisons were performed with the Holm multiple comparisons test in the 'emmeans' package [68].

The number of *A. colemani* surviving in cage trials was analyzed over six days using the Kaplan–Meier estimator of survival in R (RStudio Version 1.2.1335; [69]. For this analysis, surviving individuals were censored at the end of the experiment, and individuals were censored if it was not possible to monitor their survival for the duration of the experiment (e.g., for escaped individuals). Pairwise comparisons between treatments were then performed using a log-rank test with Benjamini–Hochberg correction. To assess the parasitism of *A. colemani*, the number of mummies in each cage was analyzed using ANOVA. Prior to the analysis, the negative control was removed due to lack of variance, and the normality of the remaining data was assessed using the Shapiro–Wilk test [50].

To analyze the number of insects labelled with the fluorescent dye and those captured in the field margin, GLM assuming Poisson distribution with a log link was used, followed by pairwise comparisons and a Holm multiple comparisons test. the normality was assessed using the Shapiro–Wilk test.

5. Conclusions

Flowering plants provide food and shelter for NEs and can promote natural pest regulation in crops. Our study highlights the potential of field margin plants in supporting populations of NEs in smallholder farms, and shows that conservation biological control could be used to promote NEs in these agro-ecosystems. Certain plant species appear to be preferred by different NE groups and provide different benefits. In transect walks, the highest number of Nes were observed interacting with *B. pilosa* and *E. heterophylla*. In addition, *B. pilosa*, *T. minuta*, and *P. hysterophorus* supported different groups of natural enemies when planted as a field margin. NE groups were shown to interact with flowers of these field margin plants, suggesting that they are supported by the provision of nectar and pollen. This is corroborated by cage trials where *B. pilosa*, *E. heterophylla*, *H. suaveolens*, and *T. minuta* enhanced the survival of *A. colemani*, most likely through the provision of nectar [27,37,39,40,70]. However, it is important to consider the wider implications of using these plants in conservation biological control, for example, *P. hysterophorus* is toxic [71,72], and other plants may be invasive to the area and present a challenge as weeds. Some field margin plant species might also provide food and shelter for specific pests, and therefore it is crucial to study the biology of the host plants and how they interact with pests.

Supplementary Materials: The following supporting information can be downloaded at: https://www.mdpi.com/article/10.3390/plants11070898/s1, Table S1: Experimental data.

Author Contributions: Conceptualization, B.J.N., E.R.M., P.A.N., S.R.B., S.E.J.A., V.C.W. and P.C.S.; methodology, B.J.N., E.R.M., P.A.N., S.R.B., S.E.J.A., V.C.W. and P.C.S.; formal analysis, B.J.N., S.R.B., V.C.W.; investigation, B.J.N.; resources, B.J.N., E.R.M., P.A.N., S.R.B., S.E.J.A., V.C.W. and P.C.S.; data curation, B.J.N., S.R.B., V.C.W.; writing—original draft preparation, B.J.N.; writing—review and editing, B.J.N., E.R.M., P.A.N., S.R.B., S.E.J.A., V.C.W. and P.C.S.; visualization, B.J.N., E.R.M., P.A.N., S.R.B., S.E.J.A., V.C.W. and P.C.S.; supervision, E.R.M., P.A.N., S.R.B., S.E.J.A., V.C.W. and P.CS.; project administration, E.R.M., P.A.N., S.R.B., S.E.J.A. and P.C.S.; funding acquisition, S.R.B., S.E.J.A. and P.C.S.; All authors have read and agreed to the published version of the manuscript.

Funding: The funding for this work was received from the UKRI GCRF project: Natural Pest Regulation in Orphan Crop Legumes in Africa (NaPROCLA) to PCS, grant number BB/R020361/1 and the World Bank through The Africa Centre for Research, Agricultural Advancement, Teaching Excellence and Sustainability (CREATES) at the Nelson Mandela African Institution of Science and Technology (NM-AIST).

Institutional Review Board Statement: Not applicable.

Informed Consent Statement: Not applicable.

Data Availability Statement: Data is contained in Supplementary Materials.

Acknowledgments: The authors are grateful to the smallholder farmers of Kwa Sadala village in Kilimanjaro for permitting sampling on their farms and supporting this project, the area's agricultural officers, field assistants, village leaders, field laborers, and NM-AIST technicians for contributing to various activities in different stages of the project.

Conflicts of Interest: The authors declare that they have no conflict of interest.

References

1. Temba, M.C.; Njobeh, P.B.; Adebo, O.A.; Olugbile, A.O.; Kayitesi, E. The role of compositing cereals with legumes to alleviate protein-energy malnutrition in Africa. *Int. J. Food Sci.* **2016**, *51*, 543–554. [CrossRef]
2. Franke, A.C.; Laberge, G.; Oyewole, B.D.; Schulz, S. A comparison between legume technologies and fallow, and their effects on maize and soil traits, in two distinct environments of the West African savannah. *Nutr. Cycl. Agroecosyst.* **2008**, *82*, 117–135. [CrossRef]
3. Franke, A.C.; Van den Brand, G.J.; Vanlauwe, B.; Giller, K.E. Sustainable intensification through rotations with grain legumes in Sub-Saharan Africa: A review. *Agric. Ecosyst. Environ.* **2018**, *261*, 172–185. [CrossRef] [PubMed]
4. Kamanga, B.C.G.; Waddington, S.R.; Robertson, M.J.; Giller, K.E. Risk analysis of maize-legume crop combinations with smallholder farmers varying in resource endowment in central Malawi. *Exp. Agric.* **2010**, *46*, 1. [CrossRef]
5. Ojiem, J.O.; Franke, A.C.; Vanlauwe, B.; De Ridder, N.; Giller, K.E. Benefits of legume–maize rotations: Assessing the impact of diversity on the productivity of smallholders in Western Kenya. *Field Crops Res.* **2014**, *168*, 75–85. [CrossRef]
6. Abate, T.; van Huis, A.; Ampofo, J.K.O. Pest management strategies in traditional agriculture: An African perspective. *Annu. Rev. Entomol.* **2000**, *45*, 631–659. [CrossRef] [PubMed]
7. Allen, D.J. *Pests, Diseases, and Nutritional Disorders of the Common Bean in Africa: A Field Guide (No. 260)*; CIAT: Palmira, Colombia, 1996; p. 131.
8. Otieno, M.; Steffan-Dewenter, I.; Potts, S.G.; Kinuthia, W.; Kasina, M.J.; Garratt, P.D. Enhancing legume crop pollination and natural pest regulation for improved food security in changing African landscapes. *Glob. Food Sec.* **2020**, *26*, 100394. [CrossRef]
9. Wortmann, C.S. *Atlas of Common Bean (Phaseolus vulgaris L.) Production in Africa (No. 297)*; CIAT: Palmira, Colombia, 1998.
10. Belmain, S.R.; Neal, G.; Ray, D.E.; Golob, P. Insecticidal and vertebrate toxicity associated with ethnobotanicals used as post-harvest protectants in Ghana. *Food Chem. Toxicol.* **2001**, *39*, 287–291. [CrossRef]
11. Mkenda, P.A.; Ndakidemi, P.A.; Stevenson, P.C.; Sarah, S.E.; Belmain, S.R.; Chidege, M.; Gurr, G.M. Field margin vegetation in tropical African bean systems harbours diverse natural enemies for biological pest control in adjacent crops. *Sustainability* **2019**, *11*, 6399. [CrossRef]
12. Mkenda, P.A.; Ndakidemi, P.A.; Stevenson, P.C.; Arnold, S.E.J.; Chidege, M.; Gurr, G.M.; Belmain, S.R. Characterization of hymenopteran parasitoids of *Aphis fabae* in an African smallholder bean farming system through sequencing of coi 'mini-barcodes'. *Insects* **2019**, *10*, 331. [CrossRef]
13. Bale, J.S.; Van Lenteren, J.C.; Bigler, F. Biological control and sustainable food production. *Philos. Trans. R. Soc. Lond. Biol. Sci.* **2008**, *363*, 761–776. [CrossRef] [PubMed]
14. Gurr, G.M.; Reynolds, O.L.; Johnson, A.C.; Desneux, N.; Zalucki, M.P.; Furlong, M.J.; Li, Z.Y.; Akutse, K.S.; Chen, J.H.; Gao, X.W.; et al. Landscape ecology and expanding range of biocontrol agent taxa enhance prospects for diamondback moth management. A review. *Agron. Sustain. Dev.* **2018**, *38*, 1–16. [CrossRef]
15. Sampaio, M.V.; Bueno, V.H.P.; Silveira, L.C.P.; Auad, A.M. *Biological control of insect pests in the tropics: In Tropical Biology and Conservation Management*; Del Claro, K., Oliveira, P.S., Rico-Gray, V., Eds.; Eolss Publishers: Oxford, UK, 2009; pp. 28–70.
16. Arnold, S.E.J.; Elisante, F.; Mkenda, P.A.; Tembo, Y.L.B.; Ndakidemi, P.A.; Gurr, G.M.; Darbyshire, I.A.; Belmain, S.R.; Stevenson, P.C. Beneficial insects are associated with botanically rich margins with trees on small farms. *Sci. Rep.* **2021**, *11*, 1–11. [CrossRef]
17. Kishinevsky, M.; Keasar, T.; Bar-Massada, A. Parasitoid abundance on plants: Effects of host abundance, plant species, and plant flowering state. *Arthropod-Plant Int.* **2017**, *11*, 155–161. [CrossRef]
18. Géneau, C.E.; Wäckers, F.L.; Luka, H.; Daniel, C.; Balmer, O. Selective flowers to enhance biological control of cabbage pests by parasitoids. *Basic Appl. Ecol.* **2012**, *13*, 85–93. [CrossRef]
19. Patt, J.M.; Hamilton, G.; Lashomb, J.H. Impact of strip-insectary intercropping with flowers on conservation biological control of the Colorado potato beetle. *Adv. Hortic. Sci.* **1997**, *11*, 175–181.

20. Sivinski, J.; Wahl, D.; Holler, T.; Al Dobai, S.; Sivinski, R. Conserving natural enemies with flowering plants: Estimating floral attractiveness to parasitic Hymenoptera and attraction's relationship to flower and plant morphology. *Biol. Control* **2011**, *58*, 208–214. [CrossRef]
21. Tschumi, M.; Albrecht, M.; Entling, M.H.; Jacot, K. Targeted flower strips effectively promote natural enemies of aphids. *IOBCwprs Bull.* **2014**, *100*, 131–135. [CrossRef]
22. Wäckers, F.L.; van Reijn, P.C.J. Pick and mix: Selecting flowering plants to meet the requirements of target biological control insects. In *Biodiversity and Insect Pests: Key Issues for Sustainable Management*; Gurr, G.M., Wratten, S.D., Snyder, W.E., Eds.; John Wiley & Sons Ltd.: Chichester, UK, 2012; pp. 139–165.
23. Fitzgerald, J.D.; Solomon, M.G. Can flowering plants enhance numbers of beneficial arthropods in U.K. apple and -pear orchards? *Biocontrol Sci. Technol.* **2004**, *14*, 291–300. [CrossRef]
24. Carrié, R.J.; George, D.R.; Wäckers, F.L. Selection of floral resources to optimise conservation of agriculturally-functional insect groups. *J. Insect Conserv.* **2012**, *16*, 635–640. [CrossRef]
25. Adda, C.; Atachi, P.; Hell, K.; Tamò, M. Potential use of the bushmint, *Hyptis suaveolens*, for the control of infestation by the pink stalk borer, *Sesamia calamistis* on maize in southern Benin, West Africa. *J. Insect Sci.* **2011**, *11*, 33. [CrossRef]
26. Hillocks, R.J. The potential benefits of weeds with reference to smallholder agriculture in Africa. *J. Integr. Pest Manag.* **1998**, *3*, 155–167. [CrossRef]
27. Ojija, F.; Arnold, S.E.J.; Treydte, A.C. Impacts of alien invasive *Parthenium hysterophorus* on flower visitation by insects to co—Flowering plants. *Arthropod Plant Interact.* **2019**, *13*, 719–734. [CrossRef]
28. Bianchi, F.J.J.A.; van der Werf, W. Model evaluation of the function of prey in non-crop habitats for biological control by ladybeetles in agricultural landscapes. *Ecol. Model.* **2004**, *171*, 177–193. [CrossRef]
29. Gurr, G.M.; Wratten, S.D.; Landis, D.A.; You, M. Habitat management to suppress pest populations: Progress and prospects. *Annu. Rev. Entomol.* **2017**, *62*, 91–109. [CrossRef]
30. Jonsson, M.; Wratten, S.D.; Landis, D.A.; Gurr, G.M. Recent advances in conservation biological control of arthropods by arthropods. *Biol. Control* **2008**, *45*, 172–175. [CrossRef]
31. Landis, D.A.; Wratten, S.D.; Gurr, G.M. Habitat management to conserve natural enemies of arthropod pests in agriculture. *Annu. Rev. Entomol.* **2000**, *45*, 175–201. [CrossRef]
32. Lundgren, J.G. Nutritional aspects of non-prey foods in the life histories of predaceous Coccinellidae. *Biol. Control.* **2009**, *51*, 294–305. [CrossRef]
33. Mitsunaga, T.; Shimoda, T.; Yano, E. Influence of food supply on longevity and parasitization ability of a larval endoparasitoid, *Cotesia plutellae* (Hymenoptera: Braconidae). *Appl. Entomol. Zool.* **2004**, *39*, 691–697. [CrossRef]
34. Tylianakis, J.M.; Didham, R.K.; Wratten, S.D. Improved fitness of aphid parasitoids receiving resource subsidies. *Ecology* **2004**, *85*, 658–666. [CrossRef]
35. Wäckers, F.L.; van Rijn, P.C.J.; Bruin, J. (Eds.) *Plant-Provided Food for Carnivorous Insects: A Protective Mutualism and Its Applications*; Cambridge University Press: Cambridge, UK, 2005.
36. Patt, J.M. Occurrence of coccinellids that prey on *Diaphorina citri* (Hemiptera: Liviidae) on *Euphorbia heterophylla* (Euphorbiaceae) and *Chamaecrista fasciculata* (Fabaceae) in a South Florida Residential Area. *Fla. Entomol.* **2018**, *101*, 131–134. [CrossRef]
37. Fonseca, M.M.; Lima, E.; Lemos, F.; Venzon, M.; Janssen, A. Non-crop plant to attract and conserve an aphid predator (Coleoptera: Coccinellidae) in tomato. *Biol. Control.* **2017**, *115*, 129–134. [CrossRef]
38. Grombone-Guaratini, M.T.; Solferini, V.N.; Semir, J. Reproductive biology in species of *Bidens*, L. (Asteraceae). *Sci. Agric.* **2004**, *61*, 185–189. [CrossRef]
39. Quispe, R.; Mazón, M.; Rodríguez-Berrío, A. Do refuge plants favour natural pest control in maize crops? *Insects* **2017**, *8*, 71. [CrossRef]
40. Shearer, P.W.; Atanassov, A. Impact of peach extrafloral nectar on key biological characteristics of *Trichogramma minutum* (Hymenoptera: Trichogrammatidae). *J. Econ. Entomol.* **2004**, *97*, 789–792. [CrossRef]
41. Dufour, R. *Farmscaping to Enhance Biological Control*; Appropriate Technology Transfer for Rural Areas: Fayetteville, AR, USA, 2000; p. 40.
42. Gurr, G.M.; Wratten, S.D.; Altieri, M.A. Ecological engineering: A new direction for agricultural pest management. *AFBM J.* **2004**, *1*, 28–35. [CrossRef]
43. Heimoana, V.; Pilkington, L.J.; Raman, A.; Mitchell, A.; Nicol, H.I.; Johnson, A.C.; Gurr, G.M. Integrating spatially explicit molecular and ecological methods to explore the significance of non-crop vegetation to predators of brassica pests. *Agric. Ecosyst. Environ.* **2017**, *239*, 12–19. [CrossRef]
44. Jervis, M.A.; Lee, J.C.; Heimpel, G.E. Use of behavioural and life history studies to understand the effects of habitat manipulation. In *Advances in Habitat Manipulation for Arthropods*; Gurr, G.M., Wratten, S.D., Altieri, M.A., Eds.; Ecological Engineering for Pest, Management; CSIRO Publishing: Melbourne, Australia, 2004; pp. 65–100.
45. Mitsunaga, T.; Mukawa, S.; Shimoda, T.; Suzuki, Y. The influence of food supply on the parasitoid against *Plutella xylostella* L. (Lepidoptera: Yponomeutidae) on the longevity and fecundity of the pea leafminer, *Chromatomyia horticola* (Goureau) (Diptera: Agromyzidae). *Appl. Entomol. Zool.* **2006**, *41*, 277–285. [CrossRef]
46. Rahat, S.; Gurr, G.M.; Wratten, S.D.; Mo, J.; Neeson, R. Effect of plant nectars on adult longevity of the stinkbug parasitoid, *Trissolcus basalis*. *Int. J. Pest Manag.* **2005**, *51*, 321–324. [CrossRef]

47. Hogg, B.N.; Bugg, R.L.; Daane, K.M. The attractiveness of common insectary and harvestable floral resources to beneficial insects. *Biol. Control* **2011**, *56*, 76–84. [CrossRef]
48. Sutherland, J.P.; Sullivan, M.S.; Poppy, G.M. Distribution and abundance of aphidophagous hoverflies (Diptera: Syrphidae) in wildflower patches and field margin habitats. *Agric. For. Entomol.* **2001**, *3*, 57–64. [CrossRef]
49. Lu, Z.X.; Zhu, P.Y.; Gurr, G.M.; Zheng, X.S.; Read, D.M.; Heong, K.L.; Yang, Y.J.; Xu, H.X. Mechanisms for flowering plants to benefit arthropod natural enemies of insect pests: Prospects for enhanced use in agriculture. *Insect Sci.* **2014**, *21*, 1–12. [CrossRef] [PubMed]
50. Charles, J.J.; Paine, T.D. Fitness effects of food resources on the polyphagous aphid parasitoid, *Aphidius colemani* Viereck (Hymenoptera: Braconidae: Aphidiinae). *PLoS ONE* **2016**, *11*, e0147551. [CrossRef]
51. Idris, A.B.; Grafius, E. Wildflowers as nectar sources for *Diadegma insulare* (Hymenoptera: Ichneumonidae), a parasitoid of diamondback moth (Lepidoptera: Yponomeutidae). *Environ. Entomol.* **1995**, *24*, 1726–1735. [CrossRef]
52. Lee, J.C.; Heimpel, G.E. Floral resources impact the longevity and oviposition rate of a parasitoid in the field. *J. Anim. Ecol.* **2008**, *77*, 565–572. [CrossRef] [PubMed]
53. Munir, S.; Dosdall, L.M.; Keddie, A. Selective effects of floral food sources and honey on life-history traits of a pest–parasitoid system. *Entomol. Exp. Appl.* **2018**, *166*, 500–507. [CrossRef]
54. Díaz-Castelazo, C.; Rico-Gray, V.; Oliveira, P.S.; Cuautle, M. Extrafloral nectary-mediated ant-plant interactions in the coastal vegetation of Veracruz, Mexico: Richness, occurrence, seasonality, and ant foraging patterns. *Ecoscience* **2004**, *11*, 472–481. [CrossRef]
55. Hernandez, L.M.; Otero, J.T.; Manzano, M.R. Biological control of the greenhouse whitefly by *Amitus fuscipennis*: Understanding the role of extrafloral nectaries from crop and noncrop vegetation. *Biol. Control* **2013**, *67*, 227–234. [CrossRef]
56. Koptur, S.; William, P.; Olive, Z. Ants and plants with extrafloral nectaries in fire successional habitats on Andros (Bahamas). *Fla. Entomol.* **2010**, *93*, 89–99. [CrossRef]
57. Mizell, R.F. *Many Plants Have Extrafloral Nectaries Helpful to Beneficials*; University of Florida Cooperative Extension Service, Institute of Food and Agricultural Sciences: EDIS, Florida, FL, USA, 2001. Available online: http://edis.ifas.ufl.edu/in175 (accessed on 2 March 2022).
58. Patt, J.M.; Rohrig, E. Laboratory evaluations of the foraging success of *Tamarixia radiata* (Hymenoptera: Eulophidae) on flowers and extrafloral nectaries: Potential use of nectar plants for conservation biological control of Asian citrus psyllid (Hemiptera: Liviidae). *Fla. Entomol.* **2017**, *100*, 149–156. [CrossRef]
59. González-Teuber, M.; Heil, M. Nectar chemistry is tailored for both attractions of mutualists and protection from exploiters. *Plant Signal. Behav.* **2009**, *4*, 809–813. [CrossRef]
60. Wäckers, F.L.; Van Rijn, P.C.; Heimpel, G.E. Honeydew as a food source for natural enemies: Making the best of a bad meal? *Biol. Control.* **2008**, *45*, 176–184. [CrossRef]
61. Zhu, P.; Lu, Z.; Heong, K.; Chen, G.; Zheng, X.; Xu, H.; Yang, Y.; Nicol, H.I.; Gurr, G.M. Selection of nectar plants for use in ecological engineering to promote biological control of rice pests by the predatory bug, *Cyrtorhinus lividipennis*, (Heteroptera: Miridae). *PLoS ONE* **2014**, *9*, e0108669. [CrossRef]
62. Fernandez, C.; Nentwig, W. Quality control of the parasitoid *Aphidius colemani* (Hym., Aphidiidae) used for biological control in greenhouses. *J. Appl. Entomol.* **1997**, *121*, 447–456. [CrossRef]
63. Starý, P. Field establishment of *Aphidius colemani* Vier.(Hym., Braconidae, Aphidiinae) in the Czech Republic. *J. Appl. Entomol.* **2002**, *126*, 405–408. [CrossRef]
64. Adisu, B.; Starý, P.; Freier, B.; Büttner, C. Aphidius colemani Vier.(Hymenoptera, Braconidae, Aphidiinae) detected in cereal fields in Germany. *Anz. Schädlingskunde J. Pest Sci.* **2002**, *75*, 89–94. [CrossRef]
65. Dannon, E.A.; Tamò, M.; Van Huis, A.; Dicke, M. Effects of volatiles from *Maruca vitrata* larvae and caterpillar-infested flowers of their host plant *Vigna unguiculata* on the foraging behavior of the parasitoid *Apanteles taragamae*. *J. Chem. Ecol.* **2010**, *36*, 1083–10991. [CrossRef]
66. Yogesh, W.H.; Deepali, G.M.; Sagar, B.P. Larvicidal effects of *Parthenium hysterophorus* against polyphagous pests *Spodoptera litura* and *Spodoptera littoralis*. *Indian J. Appl. Res.* **2019**, *2*, 18–19.
67. Dormann, C.F.; Fruend, J.; Bluethgen, N.; Gruber, B. Indices, graphs and null models: Analyzing bipartite ecological networks. *Open J. Ecol.* **2009**, *2*, 7–24. [CrossRef]
68. Length, R. Emmeans: Estimated Marginal Means, Aka Least-Squares Means. R Package Version 1.4.1. 2019. Available online: https://CRAN.R-project.org/package=emmeans (accessed on 2 March 2022).
69. Kassambara, A.; Kosinski, M.; Biecek, P.; Fabian, S. Package 'survminer'. Drawing Survival Curves Using 'ggplot2' (R Package Version 0.4.9). 2017. Available online: https://CRAN.R-project.org/package=survminer (accessed on 2 March 2022).
70. Koptur, S. Nectar as fuel for plant protectors. In *Plant-Provided Food for Carnivorous Insects: A Protective Mutualism and Its Application*; Cambridge University Press: Cambridge, UK, 2005; pp. 75–108.
71. Patel, S. Harmful and beneficial aspects of *Parthenium hysterophorus*: An update. *3 Biotech* **2011**, *1*, 1–9. [CrossRef]
72. Ramos, A.; Rivero, R.; Visozo, A.; Piloto, J.; García, A. Parthenin, a sesquiterpene lactone of *Parthenium hysterophorus* L. is a high toxicity clastogen. *Mutat. Res. Genet. Toxicol. Environ. Mutagen.* **2002**, *514*, 19–27. [CrossRef]

MDPI

Article

Effect of Irrigation Dose on Powdery Mildew Incidence and Root Biomass of Sessile Oaks (*Quercus petraea* (Matt.) Liebl.)

Winicjusz Kasprzyk [1], Marlena Baranowska [2,*], Robert Korzeniewicz [2], Jolanta Behnke-Borowczyk [2] and Wojciech Kowalkowski [2]

1 The State Forests National Forest Holding, Forest District Jawor, Myśliborska 3 Str., 59-400 Jawor, Poland; seeman6@wp.pl

2 Faculty of Forestry and Wood Technology, Poznan University of Life Sciences, Wojska Polskiego 28 Str., 60-637 Poznań, Poland; robert.korzeniewicz@up.poznan.pl (R.K.); jbehnke@up.poznan.pl (J.B.-B.); wojciech.kowalkowski@up.poznan.pl (W.K.)

* Correspondence: marlenab@up.poznan.pl

Abstract: The sessile oak is one of the most significant forest tree species in Europe. This species is vulnerable to various stresses, among which drought and powdery mildew have been the most serious threats. The aim of this study was to determine the influence of irrigation levels (overhead sprinklers) on the damage caused by powdery mildew to *Quercus petraea* growing in a nursery setting. Four irrigation rates were used: 100%, 75%, 50% and 25% of the full rate. The area of the leaves was measured and the ratio between the dry mass of the roots and the dry mass of the entire plant was calculated after the growing season in years' 2015 and 2016. Limiting the total amount of water provided to a level between 53.6 mm $\times$ m^{-2} and 83.6 mm $\times$ m^{-2}, particularly in the months when total precipitation was low (VII and VIII 2015), a supplemental irrigation rate between 3 and 9 mm $\times$ m^{-2} resulted in a lower severity of oak powdery mildew on leaves and lead to a favorable allocation of the biomass of the sessile oak seedlings to the root system. The severity of infection on oak leaf blades was lower when irrigation rates were reduced. The greatest mean degree of infestation in 2015 was noted in the 100% irrigation rate (14.6%), 75% (6.25%), 50% (4.35%) and 25% (5.47%). In 2016, there was no significant difference between the mean area of leaves infected by powdery mildew depending on the applied irrigation rate. The shoot-root biomass rate showed greater variation under limited irrigation rates. Controlling the irrigation rate can become an effective component of integrated protection strategies against this pathogen.

Keywords: *Erysiphe alphitoides*; the shoot-biomass amount; Compu Eye, Leaf & Symptom Area

Citation: Kasprzyk, W.; Baranowska, M.; Korzeniewicz, R.; Behnke-Borowczyk, J.; Kowalkowski, W. Effect of Irrigation Dose on Powdery Mildew Incidence and Root Biomass of Sessile Oaks (*Quercus petraea* (Matt.) Liebl.). *Plants* **2022**, *12*, 1248. https://doi.org/10.3390/plants11091248

Academic Editors: Špela Mechora and Dragana Šunjka

Received: 4 April 2022
Accepted: 28 April 2022
Published: 5 May 2022

1. Introduction

The sessile oak (*Quercus petraea* (Mattuschka) Liebl.) is one of the most significant economic and ecologic forest tree species in Europe. This species is vulnerable to various abiotic stresses, among which drought has been considered to be the most serious threat [1]. Apart from the negative influence of abiotic factors on oak trees, there are also additional unfavorable biotic factors that affect the trees [2]. Powdery mildew caused by *Erysiphe alphitoides* (Griffon and Maubl.) U. Braun and S. Takam. is one of the most dangerous diseases of oaks. *Erysiphe alphitoides* is a non-indigenous fungal species from North America that was first reported in Europe in 1907 [3] and was first studied in Poland in 1909 [4].

Other species of *Erysiphe* spp. that occur on oak leaves include: *E. abbreviata* (Peck) U. Braun and S. Takam., *E. calocladophora* (G.F. Atk.) U. Braun and S. Takam., *E. extensa* (Cooke and Peck) U. Braun and S. Takam., *E. gracilis* R.Y. Zheng and G.Q. Chen, *E. hypophylla* (Nevod.) U. Braun and Cunningt, *E. quercicola* S. Takam. and U. Braun and *E. polygoni* D. C. [5–7]. The use of molecular biology techniques allowed us to state that they can occur in Europe in four haplotypes of the pathogen. Three of them are from *E. alphitoides*

(100% ITS compliance), *E. hypophylla* (99.4% ITS agreement) and *Phyllactinia guttata* (Wallr.) Lév. (97.64% ITS agreement). A fourth, common in Europe, has 100% ITS compatibility with *E. quercicola* found in *Quercus* species in Asia and several tropical mildew species including *Oidium heveae* B.A. Steinm [8]. Sucharzewska [9] reported that in Poland, there are two species of powdery mildew: *E. alphitoides* and *E. hypophylla*. *Erysiphe hypophylla* is more numerous in the western part of Poland. These results are also confirmed by Behnke-Borowczyk and Baranowska-Wasilewska [10] and Roszak et al. [11], because during the study of powdery mildew affecting oaks from western Poland, they confirmed the presence of only *E. alphitoides*.

Erysiphe alphitoides causes a disease of the leaves and affects the non-lignified tissues of oak trees of a range of plants including *Quercus robur* L., *Q. conferta* (Kit.) Vuk., *Q. pubescens* Willd., *Q. cerris* L. and *Q. pyrenaica* Willd. [12]. It has been identified also on the leaves of *Fagus sylvatica* L., *Castanea sativa* Mill., other host plants such as on the *Eucalyptus gunnii* Hook. f. London [13]. Powdery mildew is an obligate parasite that attacks oaks of all age classes [8,14], resulting in exponentially increasing damage, particularly in young seedlings in the early phases of ontogeny [15,16], but only young developing leaves are susceptible of the disease [17]. Increased *E. alphitoides* prevalence in spring is associated with higher oak mildew severity in autumn [18]. Annually occurring infections can lead to necrosis and eventually death of young trees [19]. Non-lignification of the affected shoots may enable recurring infections, which occur in summer and concern mainly fresh and regenerating leaves [2,20]. Oak powdery mildew reduces the uptake of CO_2 (net), hinders stomatal conductance and lowers the nitrogen content of leaves [15]. Direct exposure to sun light as well as a warmer and dryer microclimate (e.g., conditions in gaps of a tree stand) both enhance spread of oak powdery mildew [21,22]. Selochnik et al. [23] mentioned an optimum disease development for mean temperature in June around 20 °C. Research pointed to the detrimental effects of rain on powdery mildew fungi, by washing off spores and damaging mycelium at the leaf surface [24]. Sivapalan [25] showed that *E. alphitoides* for which conidia were able to germinate as well in water as on oak leaves, the ability of the fungus to establish a parasitic relationship with the host strongly decreased with the duration in water.

Due to the economic significance of oaks, powdery mildew is of interest to scientists. Research has focused, for example, on understanding the influence of abiotic factors and climate change on promoting spread of the pathogen [12,15,26,27], as well as on the possibilities of exploiting biological methods to reduce the occurrence of oak powdery mildew [28,29]. Lately, it has been reported that the failure of natural regeneration of pedunculate oak trees in Europe attributable to infestation by oak powdery mildew is linked to factors such as animal pressure, changes in land use and the diminishing levels of ground water [30]. However, an interesting example is given by Demeter et al. [31], who reported the failure of natural regeneration of sessile oaks, despite the lack of animal pressure, sufficient accessibility to water and no changes in land use, in a region where prior to the appearance of oak powdery mildew, there was successful widespread natural regeneration of oak trees [32]. Every year, the native tree species range of tree stands in Europe, including oak trees, increases. It is associated with a more extensive culture of oak trees in nurseries, and this, in turn, results in a greater risk of outbreaks of powdery mildew [11]. In the nurseries, which are producing seedlings of trees, several irrigation and methods are available and used. Irrigation can reduce leaching, time of work, use of fertilizer and foliage diseases. Subirrigation systems are commonly used by nursery plant producers. Drip irrigation has been used in areas where water is scarce. Drip irrigation uses up to 70% less water as compared to a flood irrigation system. However, it is expensive compared to other irrigation methods [33]. Drip irrigation was effective in reducing pathogens such as *Colletotrichum acutatum* (J. H. Simmonds) that cause anthroids in strawberry production [34]. Overhead irrigation is the most common system in outdoor nursery areas. Overhead irrigation is water inefficient (as much as 80%). Major irrigation methods chosen for field nursery irrigation system are overhead sprinklers and microirrigation. The choice of these methods, again, depends on crop, water type and quality of plants [33].

In this context, it is of particular interest to determine whether standard practice in nurseries, i.e., irrigation, can be modified to limit the occurrence of powdery mildew in sessile oaks. Any modifications to irrigation schedules should be implemented in a manner that does not distort the normal shoot-root ratio and will not lead to lasting physiological changes in plants. Hence, the aim of the current study was to determine the influence of the irrigation rate provided for the sessile oaks (in a field nursery) on the damage caused by powdery mildew. It is hypothesized that different irrigation rates are likely to impact the severity of the sessile oak leaf infestation caused by powdery mildew. Specifically, we predict that under a lower irrigation rate (i) the severity of infection of the leaf blade of the oak will be reduced and (ii) the biomass ratio of shoots and roots will be greater.

2. Results

Analysis of the variability in soil moisture content in 2015 showed statistically significant differences between variants, blocks and variability over time. The 25% variant was clearly less wet than the other variants on the first two measurement dates. As a result of the deepening drought and the increase in temperatures in the third measuring period, a reduction in humidity was also observed in the 50 and 75% variants (Table 1). A consistent reduction in the general level of soil moisture by weight was also observed, and deepened with time. The results of measurements of moisture content by weight and volume carried out in 2016 did not show statistically significant differences. The correlation analysis showed the existence of a statistically significant relationship between the weight and volume moisture, which was the basis for the calculations that allowed derivation of the regression equation (Table 1).

Table 1. Average values of soil moisture by weight (x, confidence intervals (CI), standard deviations (SD) and coefficients of variation (V) for each variants of experiment in 2015 and in 2016. The same letters (a, b) next to the means mean that the means do not differ significantly (Tukey's test; $\alpha = 0.05$).

Date of Measurement	Variant	x	CI −95.0%	CI +95.0%	SD	V
6 July 2015	25%	16.80 (a)	13.36	20.24	6.45	38.39%
	50%	19.87 (b)	17.09	22.64	5.21	26.22%
	75%	19.45 (b)	16.84	22.05	4.88	25.09%
	100%	20.51 (b)	17.26	23.76	6.10	29.74%
28 July 2015	25%	16.80 (a)	12.62	20.97	7.83	46.60%
	50%	20.40 (b)	17.51	23.28	5.41	26.51%
	75%	19.08 (b)	16.91	21.25	4.07	21.33%
	100%	22.03 (b)	18.42	25.65	6.78	30.78%
6 August 2015	25%	15.72 (a)	12.24	19.20	6.53	41.54%
	50%	17.91 (ab)	15.55	20.28	4.44	24.79%
	75%	16.94 (ab)	14.65	19.22	4.28	25.27%
	100%	19.22 (b)	16.16	22.28	5.74	29.86%
2016	25%	18.22 (a)	5.28	31.16	8.13	44.62%
	50%	15.74 (a)	12.16	19.31	2.25	14.29%
	75%	19.17 (a)	8.23	30.11	6.87	35.84%
	100%	18.07 (a)	7.57	28.56	6.60	36.52%

ANOVA revealed that depending on the adopted irrigation rate in 2015, the area of oak leaves affected by powdery mildew differed significantly ($F = 11.71$; $p < 0.001$). In 2016, there was no statistically significant difference between the mean area of leaves infected by powdery mildew depending on the applied irrigation rate ($F = 2.04$; $p = 0.115$) (Figure 1a,b). The greatest mean degree of infestation in 2015 was noted in the case of 100% irrigation rate (14.6%) and with $\alpha = 0.05$, differed significantly from the irrigation rate (75–6.25%;

50–4.35%; 25–5.47%; Figures 1 and 2). In 2016, the mean area of leaves infected, regardless of the irrigation variant, fluctuated between 9.33% and 12.54%.

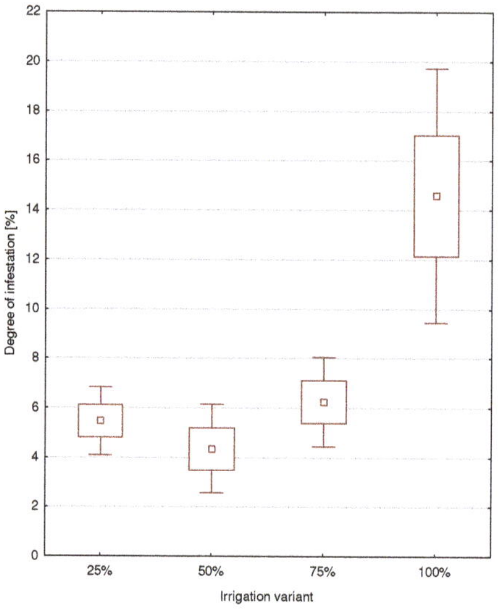

(a)

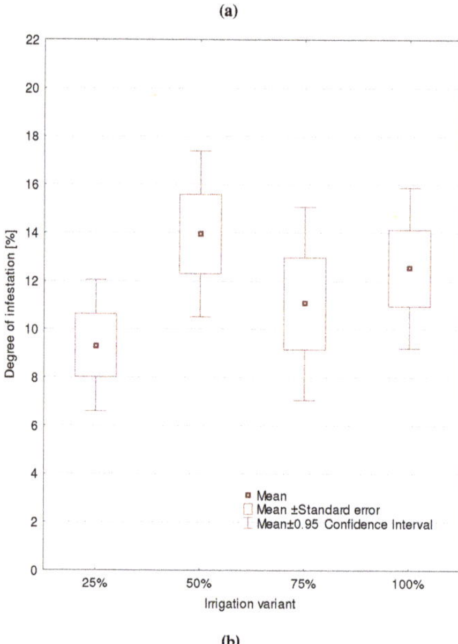

(b)

Figure 1. The mean degree of infested leaves of the sessile oak in 2015 (**a**) and in 2016 (**b**) for each variant including standard errors and confidence intervals.

Figure 2. An example of scans of oak leaves taken from plants growing in block A in 2015, infected with powdery mildew of oak, made with a scanner with a resolution of 300 dpi. (**a**) Full dose (100%), (**b**) 75% of the dose, (**c**) 50% of the dose, (**d**) 25% of the dose.

The ratio of the dry root mass to dry seedling mass of the sessile oak (F = 6.85; p = 0.0004) differed significantly between the different irrigation rates in 2015, and also in

2016 (F = 3.67; $p = 0.016$). The greatest mean value of the ratio in 2015 was observed in the case of oaks which had been irrigated with 50% of the irrigation rate (0.64), and the second highest value occurred at 75% of the irrigation rate (0.65). These results differed significantly from the mean for 100% of the irrigation rate (0.54). There were no statistically significant differences in the mean ratio of the dry root mass to dry seedling mass in the case of oaks irrigated with 25% of the irrigation rate (0.59) and the other means (Figure 3a,b). However, in 2016, the highest value of this ratio was observed in the oaks irrigated with 25% of the irrigation rate (0.54) and this differed significantly from the value for the oaks irrigated with 75% of the irrigation rate (0.46). The mean ratio did not differ from others in the case of the oaks irrigated with 50% (0.5) and 100% of the irrigation rate (100%) (Figure 3a,b).

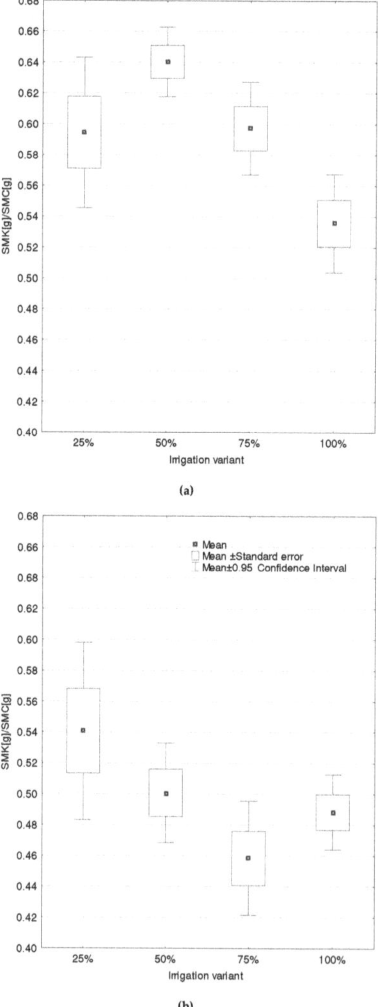

Figure 3. Mean rate of dry root mass (SMK) to dry plant mass (SMC) of the sessile oak in years' 2015 (**a**) and 2016 (**b**) for each variant including standard errors and confidence intervals.

3. Discussion

The results reported herein show that it is possible to influence the degree of powdery mildew infestation, which occurs on the leaves of the sessile oak, by controlling irrigation. The experiment confirms Sharma et al. [35] findings that higher irrigation rate is conducive to the development of the pathogen, but the differences were significant only during the first year. Sharma et al. [35] carried out research to evaluate the effect of irrigation schedules on the incidence of powdery mildew disease (*E. graminis tritici* E. Marchal) in wheat (*Triticum aestivum* L. emend. Fiori and Paol.).

A combination of favorable temperatures and optimal humidity are vital for the survival and transmission of oak powdery mildew [16]. Pap et al. [16] claim that other elements which contribute to the effectiveness of powdery mildew infections are either excess or lack of nutrients and water in the soil, as well as access to space and light. Optimal conditions contribute to the acceleration of ontogenesis, which decreases the severity of powdery mildew on leaves and, at the same time, limits the negative outcomes of oak powdery mildew infection [16]. The optimum for growth of oak powdery mildew is at 25 °C and 96% relative humidity [16,36]. Light had a significant impact on the growth of mycelium of oak powdery mildew that reached the maximum value in full light [16]. Moreover, it has been shown that the more wetted the leaves, the greater the impact of powdery mildew on the plant [37]. The mycelium of oak powdery mildew appears to be hydrophilic [38]. Despite the existing difference in the content of nutrients in the soil, no differences in powdery mildew of oak were noted between the same variants of the experiment, which were located in different blocks of the experiment. In view of all the observations listed above, controlling the irrigation rate would appear to be of key significance in the protection of oak trees against oak powdery mildew in nurseries, and this conclusion is supported by the results of the experiment in 2015.

Water is crucial for the growth and development of plants. A long-lasting period with a lack of sufficient water inevitably leads to dehydration and can lead to the death of a plant. It can also increase susceptibility to infectious diseases [39]. In this context, scientists have been intrigued by the influence of oak powdery mildew on changes in stomatal conductance and other consequences of infection by this fungus [40,41]. Oak powdery mildew has been shown to increase transpiration of the infected leaves. Another element which evoked interest was the impact of the pathogen, which affects only leaves, on water consumption of the entire plant [15]. However, the precise irrigation rate that would impede the development of the disease in nurseries, without disturbing growth and development of seedlings, has not been determined. Taking into account the larger amounts of water in 2016, which came from precipitation and lower air temperatures, it can be concluded that the excess of water may contribute to the intensification of infection from powdery oak mildew.

In 2015 and 2016, weather conditions may have influenced the existence of differences between soil moisture, the amount of leaf infestation by powdery mildew of oak and the dry mass of oak roots. The year 2015 was drier compared to 2016, which confirms the existence of differences in soil moisture in 2015 and no differences in 2016. Lower powdery mildew infection of oak may result from the presence of higher current air temperatures in 2015 and lower rainfall. A similar relationship was observed by Markovic et al. [42], who, at a lower air temperature of 17–21 °C and air humidity of 85–100%, recorded a very high rate of seedling infection.

All powdery mildew species can germinate and get infected in the absence of water [43]. Sucharzewska [9] informs that the greatest degree of powdery mildew infection was noted at times when the growing season had been very warm and dry. According to Marçais et al. [18] it is also possible that a microclimate close to the ground, with higher humidity, may provide good conditions for *E. quercicola* development. The results of the current study indicate that oak powdery mildew infected trees which were irrigated by a range of irrigation rates; however, the degree of leaf infection was less severe under lower irrigation rates, but only in years when total precipitation was lower. Thus, it is possible to decrease the infection of the oak trees in nurseries by limiting and controlling the irrigation rate. As a result, it is advisable to use irrigation rates in the range between 53.6 mm/m^2 and 83.6 mm/m^2 in the July–August period.

The mycelium of oak powdery mildew mainly covers leaves [8]; hence, it is possible to use the leaves covered by fungal hyphae in order to assess the severity of the infection [14]. Oak powdery mildews infect young developing leaves in spring and summer [19]. In Europe, the white mycelium of oak powdery mildew on leaves and shoots often appears in early spring [15]. The secondary infections occur during the summer and concern mainly the young foliage regenerating after insect defoliation [2]. While analyzing the severity of infection on oak leaf blades in post-flood trees stands (leaves in full isolation), concluded that the maximal infection by powdery mildew did not exceed 22%. Our results concur because even with the highest irrigation rate, the infection did not exceed 22%. These findings indicate that the higher irrigation rates in nurseries increase in degree of the leaf blade infection by *Erysiphe* spp.

It is worth emphasizing that numerous studies on oak powdery mildew have focused on the interactions between the pathogen and pedunculate oak trees [15,16,37,44–46], and there are fewer publications on the interactions of this fungus and sessile oak trees [18,19,47,48], most likely because sessile oaks occur less frequently (cover smaller areas) [48]. Nevertheless, the latter species is a valuable component of deciduous tree stands, and both oak species can freely interbreed [49]. Given that the sessile oak is more resilient to droughts than the pedunculate oak [50], it is probable that the species will gain in significance in the future, particularly in view of climate change. Hence, it is important to devote more attention to the study of the interactions between stress factors, such as oak powdery mildew, and the sessile oak.

The variation of irrigation rate that was applied in the current work influenced also the rate of development of the root-mass. In the dry areas, seedlings increased water consumption and invested more in root system development. This later allowed the trees to draw water more effectively from deeper layers of the soil, where greater reservoirs of water reside especially in dry seasons [51]. That is why using lower irrigation rates stimulated oaks to invest more in development of the root biomass. Similar dependence has been reported by Gieger and Thomas [52], Thomas and Gausling [53] and Sustani et al. [54]. However, Broadmeadow [55] noticed a clear influence of optimal water conditions on the shaping of unfavorable shoot-root rate in the sessile oak caused by the excessive increase of the aerial component of the trees. The relationship between the shoot of Scots pine and the shoot of ash and root biomass was affected by irrigation. These plants tended towards increased shoot allocation. However, in oak, there were large differences in the shoot-root relationship. These are difficult to explain in terms of irrigated effects, which is also confirmed by reports by Broadmeadow et al. [55].

4. Materials and Methods

4.1. Experiment Design

The experiment focused on sessile oaks which were grown from seeds collected from a seed-tree stand (stand age–159 years) in subdivision 315d in Legnica Forest District (Poland, WGS 84–51.3016 E:16.0833). The experiment was established in the Muchów Forest Nursery in subdivision 194f in Jawor Forest District (Poland, WGS 84–51.0076 E:16.0187)-open field nursery.

The seeds were planted by hand in an open field nursery (0.606 kg acorns/m^2) on the 6 May 2015 in 4 rows, and were then covered with sand and protected with nonwoven fabric (spacing between seeds was 5 cm and between rows was 33.3 cm, Figure 4d). During the experiment (in 2015 and in 2016), no protective measures were undertaken. A randomized complete block design, with 4 blocks, was used in the experiment. Each plot was 2 m long and 1.5 m wide (about 200 oaks grew on each plot; Figure 4a–d.). Four irrigations rates were used: (1) a fully-calculated rate–100% (control group–12 mm/m^2; reference evapotranspiration-ETo), (2) 75% of the full rate (9 mm/m^2), (3) 50% of the full rate (6 mm/m^2), (4) 25% of the full rate (3 mm/m^2) (Figure 4c). Density of the trees in each experimental unit was uniform. Plants were surface irrigation-overhead sprinklers method (Figure 4a,b). The atmospheric conditions were the same in all places of the experiment (Table 2 and Figure 5).Variations of the irrigation rate were established based on the document: "Guidelines for irrigating forest nurseries on open areas" [56]. In order to achieve initial stable and balanced growth in the period following sowing of the acorns, the irrigation rate on all plots was identical and as advised [56]. A necessary irrigation rate for the so-called 2nd period of irrigation of a one-year-old material was determined on the basis of Smorowski's method described in the Guidelines document [56].

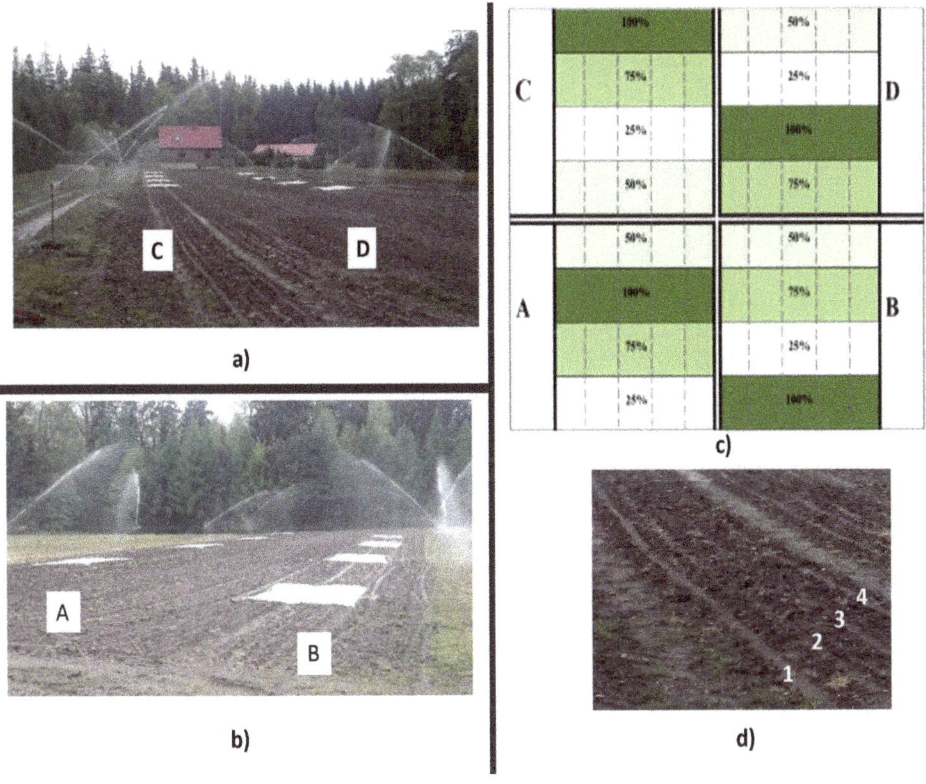

Figure 4. The experimental set-up divided into variants of the experiment in Muchów Forest Nursery; (**a**) Block C and D and working sprinklers, (**b**) block A and B and working sprinklers, (**c**) experiment design; letters stand for block names; 100%-full rate (control), 75% of the rate, 50% of the rate and 25% of the rate; the dashed line marks the rows where the oak was planted; the desired variant of the experiment is marked with a separate color, (**d**) rows in which acorns were planted; the numbers (1–4) are rows.

Table 2. Total quantity of precipitation and irrigation rate for each variant and irrigation dates in years' 2015 and 2016, VII—July, VIII—August—summer in Poland.

Variant	Total Amount of Precipitation [mm] VII 2015	Dose of Irrigation [mm]				Total Amount of Precipitation [mm] VIII 2015	Dose of Irrigation [mm]	Total Amount of Precipitation [mm] VII 2016	Dose of Irrigation [mm]		Total Amount of Precipitation [mm] VIII 2016	Dose of Irrigation [mm]	
		6 July 2015	19 July 2015	24 July 2015	29 July 2015		5 August 2015		13 July 2016	27 July 2016		16 August 2016	31 August 2016
25%	34.2	3.0	0.0	3.0	3.0	7.4	3.0	132	3.5	3.5	28.2	3.5	3.5
50%		6.0	3.0	6.0	6.0		6.0		7	7		7	7
75%		9.0	6.0	9.0	9.0		9.0		10.5	10.5		10.5	10.5
100%		12.0	9.0	12.0	12.0		12.0		14	14		14	14

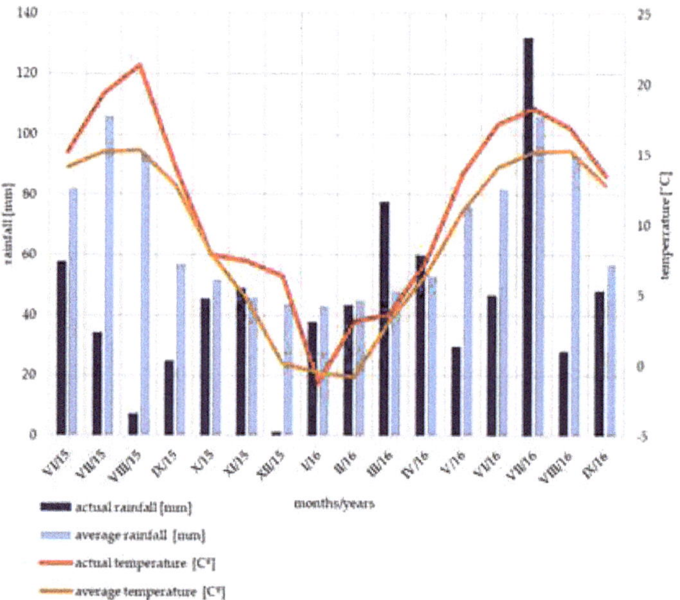

Figure 5. The average rainfall [mm] and temperature [°C] and actual rainfall [mm] and temperature [°C] in Muchów Forest Nursery, Poland (I—January; II—February, III—March etc.).

The rate of easily accessible water in the soil was verified at 7.7% of the soil volume. Average daily water consumption for evapotranspiration (similar to ETo) was assumed to be 2.3 mm. High annual precipitation exceeding 610 mm [56]. The preferable depth for dampening the soil was established at 15 cm in both years of the experiment. The net irrigation dose (d) was calculated according to the formula:

$$d = 0.1 \times Wd \times h \text{ [mm]},$$

where Wd—water content readily available in % of soil volume; h—the desired depth of soil moistening. The gross irrigation dose (d) was calculated according to the formula:

$$D = d/k_e \text{ [mm]},$$

where d/k_e—technical efficiency factor of irrigation (0.85) [23]. The frequency of watering (T) was calculated by the formula:

$$T = d/E,$$

where d—the net irrigation dose, E—daily water consumption for evapotranspiration [23].

The calculated net dose of irrigation was 11.55 mm (13.58 mm gross); however, the calculated frequency for sprinkling irrigation was 5 days. In order to determine the sprinkler flow rate per 1 h, a pluviometer from a Davis Vantage Pro weather station was used, which was positioned centrally on the experimental fields (before seeding). The achieved mean value of the flow rate per 1 sprinkler was 3.5 mm $\times$ m^{-2} $\times$ h^{-1}. RainBird EWH 14070 sprinklers were used in the experiment (Figure 4a,b). The adopted methods permitted careful control of the irrigation rate for each variant by controlling sprinkler run time. The sprinkler was turned off after a set period during which a precise irrigation rate for the particular variant was delivered.

Precipitation reaching 3 mm, which occurred between each artificial irrigation, was also taken into consideration. Plants in the 100% dose variant were watered between 5:00 a.m. and 9:00 a.m., 75% from 5:00 a.m. to 8:00 a.m., 50% from 5:00 a.m. to 7:00 a.m. and 25% from 5:00 a.m. to 6:00 a.m. If precipitation occurred, then the next irrigation rate was rectified according to the extent of the precipitation [56]. In cases when the sum of precipitation in the period from the earlier irrigation reached the approximated value to 1 full irrigation rate, then the date of the next irrigation was postponed by another 5 days. This situation only happened twice: on 10 July 2015 and 19 July 2015, when above 3 mm of rain fell the day before. The total quantity of water from the sprinklers and from the precipitation is presented in Table 1. Meteorological data during the time of the research are given in Figure 4. Average data are derived from the closest meteorological station in Wojcieszów Górny (Poland, WGS 84–50.9516 E:15.9214.), while the local data come from the Davis Vantage Pro meteorological station, located in the Muchów Forest Nursery. We have not noticed the influence of the seasons on the development of powdery mildew of oak.

4.2. Soil Moisture

Chemical analyses of soil were commissioned to the Seed Testing Station of the National Forests Research and Implementation Centre in Bedoń (report no 7 of 25 March 2014—internal data, unpublished), where in two soil samples provided for analyses the following parameters were determined: pH in KCl using the electrochemical technique, contents of phosphorus and potassium according to Egner-Rhiem by inductively coupled plasma atomic emission spectrometry [57], magnesium content according to Schachtschabel by inductively coupled plasma atomic emission spectrometry [58], nitrogen content by the direct method using the TruSpec CHNS apparatus [57] organic carbon content by the modified Tiurin method [57], while contents of $N\text{-}NO_3$ and $N\text{-}NH_4$ by the electrochemical method following extraction in 0.03 N acetic acid [57]. The results of soil analyses are given in Table 3.

Table 3. Results of chemical analyses of soils.

Parameters	Blocks	
	A and B	C and D
pH in H_2O	5.8	6.3
pH in KCl	4.6	5.5
N [%]	0.16	0.30
C [%]	2.65	6.58
C/N	16.30	22.00
P_2O_5 [mg/100 g]	5.78	9.9
Ca [mg/100 g]	113	326
K [mg/100 g]	9.4	26.7
Mg [mg/100 g]	4.8	15
Na [mg/100 g]	0.32	0.4

In order to confirm or exclude the influence of the irrigation rate on the soil moisture content by weight and volume, three measurements of soil moisture content by the weight method were carried out on each experimental plot in 2015 and one measurement of soil moisture content with simultaneous measurement of volumetric moisture in 2016. Moisture measurements were made after 4 h from the end of irrigation. In 2015, 64 samples were taken with an Egner cane at a depth of between 10 and 15 cm. Before drying, each sample was thoroughly mixed and a cohesive sample, weighing 50 g, was separated from it by pouring the soil into a ceramic container with a constant weight placed on a weighing pan. Then, the samples were dried at a temperature of 105 °C until a constant weight was obtained, which lasted about 4 h. After completion of drying, the soil samples together with the containers were weighed to an accuracy of 0.001 g.

The weight moisture content of each soil sample was calculated according to formula:

$$W = \frac{\text{mmt} - \text{mst}}{\text{mst} - \text{mt}} * 100 \text{ [mm]},$$

where mt—weight of the container mmt—weight of the vessel with moist soil mst—weight of the vessel with dry soil [59]. The HH2 Meter was used to measure volumetric moisture of the soil in 2016 by Delta T with the ECHO EC-5 sensor by Decagon Devices, which allows to obtain a result in percentage by volume. Volumetric moisture measurements were performed in the places where soil samples were taken to determine the weight moisture content, thus facilitating determination of the relationship between both types of moisture and enabling the derivation of the regression equation.

The data were analyzed statistically implementing a two-factor model without interactions. the data expressed as percentages were subjected to the Bliss transformation.

Additionally, for soil moisture by weight, in 2015, an analysis of variance was performed in a system with repeated measurements in order to determine changes in soil moisture over time. The sphericality of the data was confirmed by the Mauchley test. For weight and volume moisture, in 2016, a linear correlation analysis was performed yielding a correlation coefficient R, determination coefficient R^2 and estimation of the standard error of the assessment, as well as the coefficients of the regression equation. Statistical analyses were performed using the Statistica 6.0 software (StatSoft) [60].

4.3. *The Weighing of Dry Mass Seedlings and Calculate Leaves Area*

The weighing of dry mass seedlings was conducted after the growing season in years' 2015 and 2016 (in September-late summer). The leaves, which were collected in order to measure the area of the leaves, were collected before digging out the plants. Leaves were taken randomly from each part of the plant (apical, median, basal). Five trees from each variant of the experiment and from each block in each year of the experiment were extracted (in total 160). From each tree, the leaves were collected for further research (in total 160 pieces). The leaves were scanned in order to calculate their area and, in order to do that, a scanner with a resolution of 300 dpi was used, together with Compu Eye, Leaf & Symptom Area software, which was used to calculate the area of the leaves [61]. The infestation of leaf blades by powdery mildew was expressed as the rate of the affected part (showing etiological symptoms) to the non-infected part (without etiological symptoms) expressed in % (degree of infestation—the severity of the disease, Figure 6). The dry mass [g] of particular organs was weighed after the samples were dried in a thermostatic cabinet (POL-EKO-Aparatura, typ ST 1200 B60 photoperiod) at the temperature of 60 °C. The ratio between dry mass of roots and dry mass of the entire plant was established [g × g^{-1}]. Measurements were made on an annual basis due to the invasiveness of the applied methodology, which was dependent on seedlings being extracted from their beds.

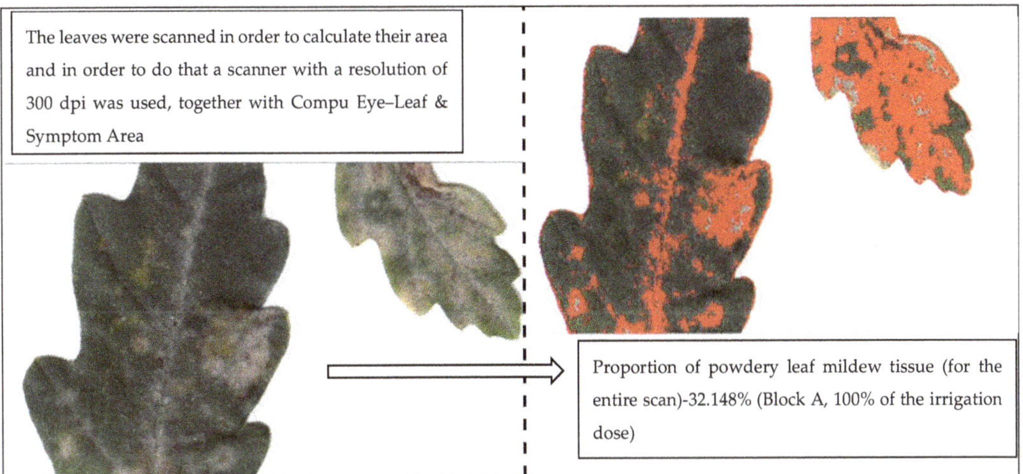

Figure 6. Determine the degree of infestation of scanned leaves with use Compu Eye, Leaf & Symptom Area.

Based on previous research by Behnke-Borowczyk and Baranowska-Wasilewska (2017) and Roszak et al. (2019) it was considered that powdery mildew of oak caused by *E. alphitoides* will dominate in the Muchów Forest Nursery.

4.4. Statistical Analysis

The Shapiro–Wilk test was used to assess whether data conformed to the Gaussian distribution. In the cases when a normal distribution could not be confirmed, the data were Box-Cox transformation, and after meeting the assumptions of homogeneity of variance, the Levene's test was applied. When the null hypothesis was rejected (implying no differences between mean values), Tukey's range test for equal sample size was implemented post-hoc.

5. Conclusions

Variation in the rate of irrigation influenced the severity of oak powdery mildew infections on the leaves of sessile oak trees.

The shoot-root biomass rate was greater under limited irrigation rates.

Limiting the total rate of water in July and August (summer) to a level between 53.6 mm × m^{-2} and 83.6 mm × m^{-2}, particularly in the months when total precipitation is low, and with a supplemental irrigation rate of between 3 and 9 mm × m^{-2} results in lower severity of oak powdery mildew on leaves. Concurrently, it leads to a favorable allocation of the biomass of the sessile oak seedlings to the root system, which lasts for at least 2 years.

Given the assumptions of the Green Deal, and the implementation of regulations concerning plant protection products, i.e., pesticides [62], it is crucial to conduct further research into the significance of irrigation rates in regulating, and particularly limiting, the severity of oak powdery mildew. It is plausible that controlling the irrigation rate can become an element of the integrated protection method against the pathogen.

Controlling spraying doses (reducing irrigation) can be an effective way to reduce the occurrence of powdery oak on seedlings produced in tunnels and greenhouses; therefore, research should be continued, especially under these conditions.

Author Contributions: Conceptualization, W.K. (Winicjusz Kasprzyk) and W.K. (Wojciech Kowalkowski); methodology, W.K. (Winicjusz Kasprzyk) and W.K. (Wojciech Kowalkowski); resources, W.K. (Winicjusz Kasprzyk); writing—original draft preparation, M.B. and W.K. (Winicjusz Kasprzyk); writing—review and editing, M.B., R.K. and J.B.-B.; visualization, R.K., M.B; supervision, J.B.-B., W.K. (Winicjusz Kasprzyk), W.K. (Wojciech Kowalkowski) and M.B. All authors have read and agreed to the published version of the manuscript.

Funding: This research was funded by the General Directorate of the State Forests: Develop nursery irrigation guidelines and irrigation strategies using the regulated water deficit and maintaining the research potential of the Department of Silviculture.

Institutional Review Board Statement: Not applicable.

Informed Consent Statement: Not applicable.

Data Availability Statement: No new data were created or analyzed in this study. Data sharing is not applicable to this article.

Conflicts of Interest: The authors declare no conflict of interest.

References

1. Eaton, E.; Caudullo, G.; Oliveira, S.; de Rigo, D. *Quercus robur* and *Quercus petraea* in Europe: Distribution, habitat, usage and threats. *Eur. Atlas For. Tree Species* **2016**, 160–163.
2. Thomas, F.M.; Blank, R.; Hartmann, G. Abiotic and biotic factors and their interactions as causes of oak decline in Central Europe. *For. Pathol.* **2002**, *32*, 277–307. [CrossRef]
3. Foex, M.E. L'invasion Des Chenes d'Europe Par Le Blanc Ou Oidium. *Rev. Eaux. For.* **1941**, *79*, 338–349.
4. G.C.H.; Lange, M. Nordic Journal of Botany. *Mycologia* **1982**, *74*, 354. [CrossRef]
5. Braun, U.; Takamatsu, S. Preparation of a Color Atlas of Hyphomycetous Mycosphaerellaceae Associated with Fruit and Forest Trees Using Classical and DNA Barcoding Methods in North of Iran View Project. *Schlechtendalia* **2000**, *4*, 1–33.
6. Braun, U.; Cunnington, J.; Brielmaier-Liebetanz, U.; Ale-Agha, N.; Heluta, V.P. Miscellaneous Notes on Some Powdery Mildew Fungi. *Schlechtendalia* **2013**, *10*, 91–95.
7. Lee, H.B.; Kim, C.J.; Mun, H.Y.; Lee, K.-H. First Report of Erysiphe quercicola Causing Powdery Mildew on Ubame Oak in Korea. *Plant Dis.* **2011**, *95*, 77. [CrossRef]
8. Mougou, A.; Dutech, C.; Desprez-Loustau, M.-L. New insights into the identity and origin of the causal agent of oak powdery mildew in Europe. *For. Pathol.* **2008**, *38*, 275–287. [CrossRef]
9. Sucharzewska, E. The development of *Erysiphe alphitoides* and *E. hypophylla* in the urban environment. *Acta Mycol.* **2013**, *44*, 109–123. [CrossRef]
10. Behnke-Borowczyk, J.; Baranowska-Wasilewska, M. Population of oak powdery mildew in western Poland. *Acta Sci. Pol. Silvarum Colendarum Ratio Ind. Lignaria* **2015**, *16*, 227–232. [CrossRef]
11. Roszak, R.; Baranowska, M.; Belka, M.; Behnke-Borowczyk, J. Zastosowanie Technik Biologii Molekularnej Do Detekcji *Erysiphe alphitoides* (Griffon & Maubl.) U. Braun and S. Takam. w Organach Roślinnych. *Sylwan* **2019**, *163*, 1–6. [CrossRef]
12. Desprez-Loustau, M.-L.; Feau, N.; Mougou-Hamdane, A.; Dutech, C. Interspecific and intraspecific diversity in oak powdery mildews in Europe: Coevolution history and adaptation to their hosts. *Mycoscience* **2011**, *52*, 165–173. [CrossRef]
13. Łukowski, A.; Adamska, T.; Wagrodzka, O.; Witkowski, R.; Zaluska, M.; Calka, P.; Milczarek, A.; Banaszczak, P. Ocena Zagrożenia Drzewostanów Sosnowych (*Pinus* sp.) Przez Foliofagi Zimujące w Ściółce Leśnej w Arboretum SGGW w Rogowie. *For. Lett.* **2013**, *104*, 59–66.
14. Szewczyk, W.; Kuźmiński, R.; Mańka, M.; Kwaśna, H.; Łakomy, P.; Baranowska-Wasilewska, M.; Behnke-Borowczyk, J. Occurrence of *Erysiphe alphitoides* in oak stands affected by flood disaster. *For. Res. Pap.* **2015**, *76*, 73–77. [CrossRef]
15. Hajji, M.; Dreyer, E.; Marçais, B. Impact of *Erysiphe alphitoides* on transpiration and photosynthesis in *Quercus robur* leaves. *Eur. J. Plant Pathol.* **2009**, *125*, 63–72. [CrossRef]
16. Pap, P.; Rankovic, B.; Masirevic, S. Effect of temperature, relative humidity and light on conidial germination of oak powdery mildew (*Microsphaera alphitoides* Griff. et Maubl.) under controlled conditions. *Arch. Biol. Sci.* **2013**, *65*, 1069–1077. [CrossRef]
17. Gorb, E.V.; Gorb, S.N. Powdery Mildew Fungus Erysiphe Alphitoides Turns Oak Leaf Surface to the Highly Hydrophobic State. In *Biologically Inspired Systems*; Springer: Berlin/Heidelberg, Germany, 2021; pp. 171–185.

18. Marçais, B.; Piou, D.; Dezette, D.; Desprez-Loustau, M.-L. Can Oak Powdery Mildew Severity be Explained by Indirect Effects of Climate on the Composition of the Erysiphe Pathogenic Complex? *Phytopathology* **2017**, *107*, 570–579. [CrossRef]
19. Edwards, M.; Ayres, P. Seasonal changes in resistance of *Quercus petraea* (sessile oak) leaves to Microsphaera Alphitoides. *Trans. Br. Mycol. Soc.* **1982**, *78*, 569–571. [CrossRef]
20. Marçais, B.; Bréda, N. Role of an opportunistic pathogen in the decline of stressed oak trees. *J. Ecol.* **2006**, *94*, 1214–1223. [CrossRef]
21. Lonsdale, D. Review of oak mildew, with particular reference to mature and veteran trees in Britain. *Arboric. J.* **2015**, *37*, 61–84. [CrossRef]
22. Newsham, B.K.K.; Oxborough, K.; White, R.; Greenslade, P.D.; McLeod, A.R. UV-B radiation constrains the photosynthesis of *Quercus robur* through impacts on the abundance of *Microsphaera alphitoides*. *For. Pathol.* **2000**, *30*, 265–275. [CrossRef]
23. Selochnik, N.N.; Kondrashova, N.K. Effects of Powdery Mildew and Fungicides on Oak Seedlings. *Les. Khozyalstvo* **1996**, *6*, 51–53.
24. Sivapalan, A. Effects of impacting rain drops on the growth and development of powdery mildew fungi. *Plant Pathol.* **1993**, *42*, 256–263. [CrossRef]
25. Sivapalan, A. Effects of water on germination of powdery mildew conidia. *Mycol. Res.* **1993**, *97*, 71–76. [CrossRef]
26. Grzebyta, J.; Karolewski, P.; Zytkowiak, R.; Giertych, M.J.; Werner, A.; Zadworny, M.; Oleksyn, J. Effects of Elevated Temperature and Fluorine Pollution on Relations between the Pedunculate Oak (*Quercus robur*) and Oak Powdery Mildew (*Microsphaera alphitoides*). *Dendrobiology* **2005**, *53*, 27–33.
27. Marçais, B.; Desprez-Loustau, M.-L. European oak powdery mildew: Impact on trees, effects of environmental factors, and potential effects of climate change. *Ann. For. Sci.* **2014**, *71*, 633–642. [CrossRef]
28. Kiss, L. Natural occurrence of ampelomyces intracellular mycoparasites in mycelia of powdery mildew fungi. *N. Phytol.* **1998**, *140*, 709–714. [CrossRef]
29. Kiss, L. A review of fungal antagonists of powdery mildews and their potential as biocontrol agents. *Pest Manag. Sci.* **2003**, *59*, 475–483. [CrossRef]
30. Demeter, L.; Bede-Fazekas, Á.; Molnár, Z.; Csicsek, G.; Ortmann-Ajkai, A.; Varga, A.; Molnár, Á.; Horváth, F. The legacy of management approaches and abandonment on old-growth attributes in hardwood floodplain forests in the Pannonian Ecoregion. *Forstwiss. Centralblatt* **2020**, *139*, 595–610. [CrossRef]
31. Demeter, L.; Molnár, Á.P.; Öllerer, K.; Csóka, G.; Kiš, A.; Vadász, C.; Horváth, F.; Molnár, Z. Rethinking the natural regeneration failure of pedunculate oak: The pathogen mildew hypothesis. *Biol. Conserv.* **2021**, *253*, 108928. [CrossRef]
32. Fekete, L. Bereg Vármegye Erdőtenyésztési Viszonyairól. *Erdészeti Lapok* **1890**, *29*, 94–121.
33. Paudel, K.P.; Pandit, M.; Hinson, R. Irrigation water sources and irrigation application methods used by U.S. plant nursery producers. *Water Resour. Res.* **2016**, *52*, 698–712. [CrossRef]
34. Daugovish, O.; Bolda, M.; Kaur, S.; Mochizuki, M.J.; Marcum, D.; Epstein, L. Drip Irrigation in California Strawberry Nurseries to Reduce the Incidence of Colletotrichum Acutatum in Fruit Production. *HortScience* **2012**, *47*, 368–373. [CrossRef]
35. Sharma, A.; Babu, K.S. Effect of planting options and irrigation schedules on development of powdery mildew and yield of wheat in the North Western plains of India. *Crop Prot.* **2004**, *23*, 249–253. [CrossRef]
36. Hewitt, H. Conidial germination in *Microsphaera alphitoides*. *Trans. Br. Mycol. Soc.* **1974**, *63*, 587–589. [CrossRef]
37. Klamerus-Iwan, A.; Witek, W. Variability in the Wettability and Water Storage Capacity of Common Oak Leaves (*Quercus robur* L.). *Water* **2018**, *10*, 695. [CrossRef]
38. Gorb, E.V.; Hosoda, N.; Miksch, C.; Gorb, S.N. Slippery pores: Anti-adhesive effect of nanoporous substrates on the beetle attachment system. *J. R. Soc. Interface* **2010**, *7*, 1571–1579. [CrossRef]
39. Pérez-Pastor, A.; Ruiz-Sánchez, M.C.; Conesa, M.R. Drought Stress Effect on Woody Tree Yield. In *Water Stress and Crop Plants: A Sustainable Approach*; John Wiley & Sons, Ltd.: London, UK, 2016; pp. 356–374, ISBN 978-1-119-05445-0.
40. Niederleitner, S.; Knoppik, D. Effects of the cherry leaf spot pathogen *Blumeriella jaapiion* gas exchange before and after expression of symptoms on cherry leaves. *Physiol. Mol. Plant Pathol.* **1997**, *51*, 145–153. [CrossRef]
41. Pinkard, E.A.; Mohammed, C.L. Photosynthesis of Eucalyptus globulus with Mycosphaerella leaf disease. *N. Phytol.* **2006**, *170*, 119–127. [CrossRef]
42. Markovic, M.; Stajic, S.; Markovic, N.; Hadrovic, S. Оригинални Научни Рад Efficiency of Biological Protection in Nursery Production. *ШУМАРСТВО For.* **2020**, *3–4*, 73–82.
43. Severoglu, Z.; Ozyigit, I.I. Powdery Mildew Disease in Some Natural and Exotic Plants of Istanbul, Turkey. *Pak. J. Bot.* **2012**, *44*, 387–393.
44. Vaštag, E.E.; Kastori, R.R.; Orlović, S.S.; Bojović, M.M.; Kesić, L.A.; Pap, P.L.; Stojnić, S.M. Effects of oak powdery mildew (*Erysiphe alphitoides* [Griffon and Maubl.] U. Braun and S. Takam.) on photosynthesis of pedunculate oak (*Quercus robur* L.). *Zb. Matic Srp. Prir. Nauk.* **2019**, 43–56. [CrossRef]
45. Copolovici, L.; Väärtnõu, F.; Estrada, M.P. Niinemets, Ülo Oak powdery mildew (*Erysiphe alphitoides*)-induced volatile emissions scale with the degree of infection in Quercus robur. *Tree Physiol.* **2014**, *34*, 1399–1410. [CrossRef] [PubMed]
46. Brüggemann, N.; Schnitzler, J.-P. Influence of Powdery Mildew (*Microsphaera alphitoides*) on Isoprene Biosynthesis and Emission of Pedunculate Oak (*Quercus robur* L.). *Leaves. J. Appl. Bot.* **2001**, *75*, 91–96.
47. Jactel, H.; Petit, J.N.; Desprez-Loustau, M.-L.; Delzon, S.; Piou, D.; Battisti, A.; Koricheva, J. Drought effects on damage by forest insects and pathogens: A meta-analysis. *Glob. Chang. Biol.* **2012**, *18*, 267–276. [CrossRef]

48. Ducousso, A.; Bordacs, S. EUFORGEN Technical Guidelines for Genetic Conservation and Use for Pedunculate and Sessile Oaks (*Quercus robur* and *Q. petraea*). *Int. Plant Genet. Res. Inst.* **2004**, 1–6.
49. Curtu, A.L.; Gailing, O.; Finkeldey, R. Evidence for hybridization and introgression within a species-rich oak (*Quercus* spp.) community. *BMC Evol. Biol.* **2007**, *7*, 218-15. [CrossRef]
50. Savill, P. Quercus, L. In *The Silviculture of Trees Used in British Forestry*; CABI, Cambridge University Press: Cambridge, UK, 2019; pp. 256–276.
51. Engelbrecht, B.M.J.; Kursar, T.A.; Tyree, M.T. Drought effects on seedling survival in a tropical moist forest. *Trees* **2005**, *19*, 312–321. [CrossRef]
52. Gieger, T.; Thomas, F.M. Differential response of two Central-European oak species to single and combined stress factors. *Trees* **2005**, *19*, 607–618. [CrossRef]
53. Thomas, F.M.; Gausling, T. Morphological and physiological responses of oak seedlings (*Quercus petraea* and *Q. robur*) to moderate drought. *Ann. For. Sci.* **2000**, *57*, 325–333. [CrossRef]
54. Sustani, F.B.; Jalali, S.; Sohrabi, H.; Shirvani, A. Biomass allocation of chestnut oak (*Quercus castaneifolia* C.A. Mey) seedlings: Effects of provenance and light gradient. *J. For. Sci.* **2014**, *60*, 443–450. [CrossRef]
55. Broadmeadow, M.S.J.; Jackson, S.B. Growth responses of *Quercus petraea*, *Fraxinus excelsior* and *Pinus sylvestris* to elevated carbon dioxide, ozone and water supply. *N. Phytol.* **2000**, *146*, 437–451. [CrossRef]
56. Dominiewska, A. *Wytyczne Nawadniania Szkółek Leśnych na Powierzchniach Otwartych*; Centrum Informacyjne La-sów Państwowych: Warszawa, Poland, 2002; p. 64.
57. Ostrowska, A.; Porębska, G.; Borzyszkowski, J.; Król, H.; Gawliński, S. *Właściwości Gleb Leśnych i Metody Ich Oznaczania*; Instytut Ochrony Środowiska: Warszawa, Poland, 2001; p. 108.
58. Kabała, C.; Karczewska, A. *Metodyka Analiz Laboratoryjnych Gleb I Roślin*, 8th ed.; Uniwersytet Przyrodniczy we Wrocławiu Instytut Nauk o Glebie i Ochrony Środowiska: Wrocław, Poland, 2019; p. 48.
59. Myślińska, E. *Laboratoryjne Badania Gruntów*; Wydawnictwo Naukowe PWN: Warszawa, Poland, 1998; p. 278.
60. Rabiej, M. *Statystyka z Programem Statistica*; Wydawnictwo Helion: Gliwice, Poland, 2012; p. 344.
61. Bakr, E.M. A new software for measuring leaf area, and area damaged by *Tetranychus urticae* Koch. *J. Appl. Èntomol.* **2005**, *129*, 173–175. [CrossRef]
62. EUROPEAN COMMISSION EUR-Lex Access to European Union Law. Available online: https://eur-lex.europa.eu/legal-content/PL/TXT/PDF/?uri=CELEX:52019DC0640&from=EN (accessed on 1 January 2022).

MDPI

Article

Isolation, Identification, and Biocontrol Potential of Root Fungal Endophytes Associated with Solanaceous Plants against Potato Late Blight (*Phytophthora infestans*)

Abbas El-Hasan *, Grace Ngatia, Tobias I. Link and Ralf T. Voegele

Department of Phytopathology, Institute of Phytomedicine, Faculty of Agricultural Sciences, University of Hohenheim, Otto-Sander-Str. 5, D-70599 Stuttgart, Germany; grace.ngatia@uni-hohenheim.de (G.N.); tobias.link@uni-hohenheim.de (T.I.L.); ralf.voegele@uni-hohenheim.de (R.T.V.)

* Correspondence: aelhasan@uni-hohenheim.de; Tel.: +49-711-459-22392

Abstract: Late blight of potato caused by *Phytophthora infestans* is one of the most damaging diseases affecting potato production worldwide. We screened 357 root fungal endophytes isolated from four solanaceous plant species obtained from Kenya regarding their in vitro antagonistic activity against the potato late blight pathogen and evaluated their performance in planta. Preliminary in vitro tests revealed that 46 of these isolates showed potential activity against the pathogen. Based on their ITS-sequences, 37 out of 46 endophytes were identified to species level, three isolates were connected to higher taxa (phylum or genus), while two remained unidentified. Confrontation assays, as well as assays for volatile or diffusible organic compounds, resulted in the selection of three endophytes (KB1S1-4, KA2S1-42, and KB2S2-15) with a pronounced inhibitory activity against *P. infestans*. All three isolates produce volatile organic compounds that inhibit mycelial growth of *P. infestans* by up to 48.9%. The addition of 5% extracts obtained from KB2S2-15 or KA2S1-42 to *P. infestans* sporangia entirely suppressed their germination. A slightly lower inhibition (69%) was achieved using extract from KB1S1-4. Moreover, late blight symptoms and the mycelial growth of *P. infestans* were completely suppressed when leaflets were pre-treated with a 5% extract from these endophytes. This might suggest the implementation of such biocontrol candidates or their fungicidal compounds in late blight control strategies.

Keywords: antifungal endophytes; secondary metabolites; biological control; volatile compounds; anti-oomycete; *Phytophthora infestans*; sporangia germination

Citation: El-Hasan, A.; Ngatia, G.; Link, T.I.; Voegele, R.T. Isolation, Identification, and Biocontrol Potential of Root Fungal Endophytes Associated with Solanaceous Plants against Potato Late Blight (*Phytophthora infestans*). *Plants* **2022**, *11*, 1605. https://doi.org/10.3390/plants11121605

Academic Editors: Špela Mechora and Dragana Šunjka

Received: 1 June 2022
Accepted: 14 June 2022
Published: 18 June 2022

1. Introduction

Since its earliest epidemic outbreak in the 1840s, late blight of potato incited by *Phytophthora infestans* (Mont.) de Bary has been the most severe biotic constraint threatening potato production worldwide. Under favorable environmental conditions and host susceptibility, the heterothallic fungal-like oomycete can massively produce aerially dispersible sporangia that can either directly germinate or release large numbers of water motile zoospores able to completely destroy not only potato, but also tomato crops within a few days [1–3]. The global economic impact associated with late blight on potato in terms of losses incurred due to damage and costs of disease control is conservatively estimated at more than 6 billion USD annually, with losses being heavier in developing rather than developed countries [1,4]. In Sub-Saharan Africa, losses attributed to late blight range from 30–75% [5]. However, when the disease is initiated early in the cropping season losses might increase up to 100% [6].

Over several decades, fighting late blight disease has largely been based on the use of chemicals [7–10]. Recently, ecologists have succeeded in raising public concern and awareness about the impacts of agrochemicals on the environment [11]. From this perspective, increased public and scientific desire has been elevated to develop alternative environmental-friendly control strategies that might be less harmful to both the consumer and the ecosystem [12–15]. Introducing beneficial microorganisms as biological control agents (BCAs) might represent a sustainable and reliable solution to replace chemical fungicides in late blight management. Nonpathogenic endophytic microorganisms have been shown to have potential as BCAs in many agricultural systems [16,17]. Endophytic microorganisms may exert their biocontrol activities by producing antimicrobial compounds that suppress plant pathogens or by inducing defense reactions within the plants [18,19]. Moreover, endophytes can also bestow other beneficial properties to their hosts (e.g., nitrogen fixation and/or phytohormone production), which may lead to a reduced use of agrochemicals and maintenance of biodiversity in plant-associated communities [20]. Several attempts regarding the activities of BCAs against the potato late blight pathogen have been reported [21,22]. In 1997, Ng and Webster [23] demonstrated that treating potato foliage with crude extracts of *Xenorhabdus bovienii*, led to a development of late blight symptoms in only 4% of the pathogen-treated plants compared to the untreated control [23]. Daayf et al. [24] established that *Bacillus subtilis* and *Rahnella aquatilis* restricted the growth of *P. infestans* in vitro, reporting inhibitions up to 81%. They also showed that *Serratia plymuthica* could inhibit the growth of the pathogen by >75% on detached potato leaves. In a potted plant experiment, a commercial preparation of *B. subtilis* applied as a foliar spray immediately after *P. infestans* inoculation was effective against late blight, suppressing the disease to below 40% [25]. Loliam et al. [26] demonstrated that *Streptomyces rubrolavendulae* could increase the survival of tomato and chili seedlings from 51.42% to 88.57% and 34.10% to 76.71%, respectively, in *P. infestans*-infested soils. Fungal endophytes isolated from *Espeletia* spp., including *Trichoderma asperellum*, *Aureobasidium pullulans*, *Nigrospora oryzae*, *Chaetomium globosum*, and *Penicillium commune* completely inhibited growth of *P. infestans* in vitro, while *Paecilomyces sinensis* and *Pestalotiopsis disseminate* recorded inhibitions of >60% [27]. Moreover, Gupta [28] also showed that lesion areas caused by *P. infestans* on detached potato leaves were significantly reduced after treatment with spore suspensions of *Trichoderma viride*, *Penicillium viridcatum*, *Myrothecium verrucaria*, or *Trichoderma harzianum* compared to the untreated control. Under controlled conditions, *Trichoderma atroviride* was observed to reduce late blight disease severity by 27% relative to the untreated control in potato plants [29].

Solanaceous plants providing diverse niches for endophytic associations have been shown to harbor a diversity of endophytic fungal species in their leaves, stems, and roots that enhance growth and suppress plant pathogens [14,30,31]. Kim et al. [12] showed that fermentation broths of *F. oxysporum* EF119, isolated from roots of red pepper, controlled late blight by >90% compared to the control in intact tomato seedlings grown under controlled conditions. Andrade-Linares et al. [14] observed that tomato plants colonized by dark septate endophytes isolated from tomato roots recorded enhanced shoot biomass during early stages of vegetative growth. Recently, de Vries et al. [32] have shown that a root endophyte, *Phoma eupatorii*, could suppress mycelial growth of a broad spectrum of *P. infestans* isolates in vitro and also protect tomato plants through the production of anti-oomycete compounds in planta.

Despite several attempts to characterize endophytes successfully controlling late blight disease on potatoes, their effects often did not deliver consistent disease suppression comparable to their chemical counterparts [2–4,6]. Due to this lack of consistency of biocontrol activity, research efforts in terms of discovering new bioactive endophytes preferably with multiple modes of action should be accelerated. Hence, the objectives of the present study were (i) to isolate root endophytic fungi associated with four solanaceous plants obtained from diverse regions in Kenya, (ii) to characterize these endophytes according to their phylogeny, (iii) to evaluate their inhibitory effects against *P. infestans* in co-culture, (iv) to explore the modes of action of the most successful fungal endophytes by studying the suppressive effects of their diffusible and volatile metabolites and, (v) to validate the efficacy of these endophytes in combating *P. infestans* in planta.

2. Results

2.1. Isolation of Endophytic Fungi from Roots of Solanaceous Plants

A total of 357 isolates of fungal endophytes were obtained from roots of potato, tomato, bell pepper, and nightshade from Nyandarua, Kiambu and Kilifi regions in Kenya. The highest number of endophytes were isolated from nightshade (30.5%), followed by potato (25.5%), tomato (23%), and bell pepper (21%) (Figure 1). Among all endophytes, 112 (31.4%) were identified as *Fusarium* spp., 104 (29.1%) readily sporulated on PDA, while 141, accounting for 39.5%, formed no spores on this medium.

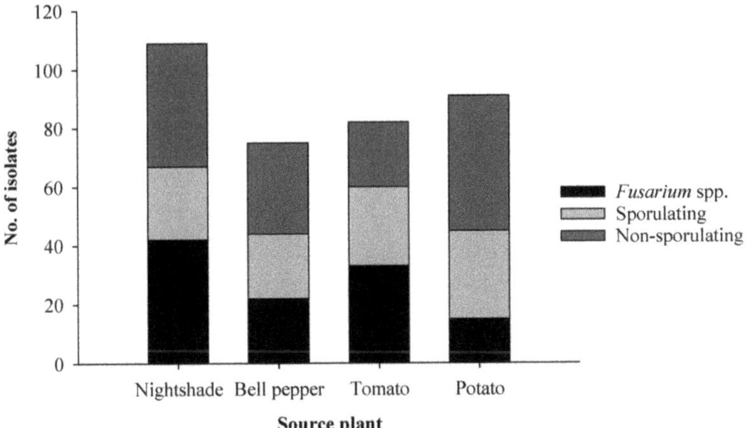

Figure 1. Distribution of the total fungal root endophytes found in four solanaceous host plants and classified into three categories: fusarial, sporulating, and non-sporulating endophytes.

2.2. Screening of Endophytes for Anti-Oomycete Activity

The results obtained from the primary high throughput screening assay showed that 64 of the total isolates (n = 357) had potential activity against *P. infestans*. However, four of these proved fastidious and ceased to grow in culture while 14 were morphologically identified as *Fusarium* spp. and excluded from subsequent analyses. Among the remaining 46 potentially active isolates, 63% were obtained from Kilifi (Figure 2), with those isolated from bell pepper in this region accounting for 32.6% of the total number of potential antagonists. Other isolates from Kilifi showing potential activity against *P. infestans* were obtained from tomato (17.4%) and nightshade (13%). All antagonistic isolates from Kiambu were from nightshade, representing 15.2% of the selected potential antagonists while those from Nyandarua were isolated from potato and nightshade accounting for 17.4% and 4.3%, respectively. Generally, the number of antagonists isolated per plant species was 15 isolates each for bell pepper and nightshade and eight isolates each for tomato and potato. Interestingly, only nightshade plants were accompanied by potential antagonists from all three regions.

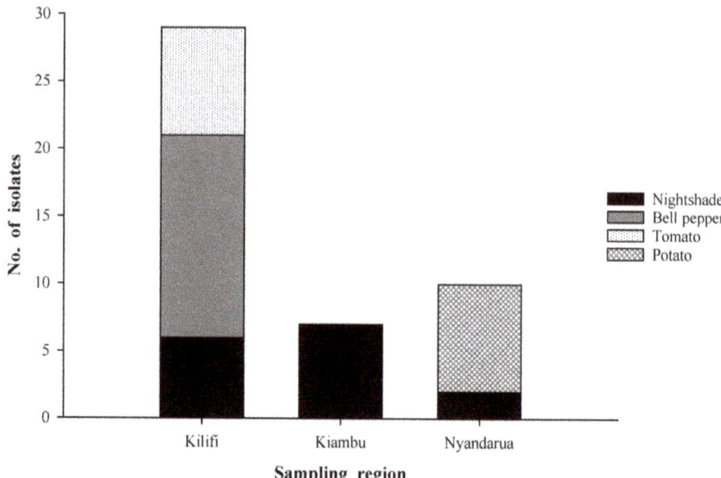

Figure 2. Distribution of root fungal endophytes with potential activity against *P. infestans* in relation to host plant and sampling region.

2.3. Characterization of the Fungal Endophytes

Based on their ITS sequences, 46 endophytic fungal isolates screened in the dual culture were characterized. The isolates were considered conspecific to species on the NCBI database when their ITS sequences (ITS1-5.8S-ITS4) matched those of the reference with an identity of ≥99% [33]. Based on this criterion, 37 of the 46 sequences were identified to the species level, three isolates were affiliated to higher taxa (phylum or order), while two matched unidentified fungi (Table 1).

The similarities of the remaining four endophytes (NP3S4-63, KA2S1-42, KB2S2-16, and KB2S2-15) did not meet the threshold and showed associations to higher taxa (genus, family, and order). The 37 isolates belonged to 18 species within the 13 genera including: *Albifimbria, Aspergillus, Myrothecium, Cylindobasidium, Epicoccum, Macrophomina, Penicillium, Plectosphaerella, Purpureocillium, Pyrenochaeta, Rhizoctonia, Mucor*, and *Colletotrichum*.

The 46 endophytic isolates and their closest BLAST entries form eight distinctive clades, which agrees nicely with their taxonomic identity (Figure 3). Three of the clades represent the two fungal phyla Zygomycota and Basidiomycota, which contain one and three fungal species, respectively. The Zygomycete *Mucor moelleri* was found to belong to the order Mucorales, while the Basidiomycetes could be placed in two orders, Cantharellales (*Rhizoctonia solani*) and Agaricales (*Cylindrobasidium evolvens*), according to Crous et al. [34]. More than 90% of the endophytic isolates were Ascomycetes distributed into three classes, namely Dothideomycetes, Sordariomycetes, and Eurotiomycetes, which fell into four clades (Figure 3). Dothideomycetes and Eurotiomycetes form the bulk of the identified Ascomycetes representing 45.2% and 35.7%, respectively. All Eurotiomycetes identified belong to the order Eurotiales, while the identified Dothiodeomycetes comprise the orders Pleosporales and Botryosphaeriales [34]. Two isolates (KB1S1-4 and KB2S2-15) cluster with the Dothideomycetes, however, they could not be placed in any of the two orders with absolute certainty. Identified Sordariomycetes form the minority of identified Ascomycetes (19%) and are grouped into two orders, Hypocreales and Glomerellales (Figure 3). Unknown isolate KB1S4-9 clustered with Sordariomycetes and was inferred to belong to the Glomerellales as it grouped with species within this order [34]. The closest BLAST match for isolate KB2S2-15 formed the eighth clade (Figure 3).

Table 1. Molecular identity of endophytic fungi from four solanaceous plants.

Isolates	Accession Numbers	Host Plant	Best BLAST Match (Accession Number)	Identity (%)
KB1S2-7	MG214581	*Capsicum annum*	*Macrophomina phaseolina* (KT862032)	99
NA2S2-45	MG214587	*Solanum nigrum*	*Mucor moelleri* (KP900321)	99
KA1S1-37	MG214583	*S. nigrum*	*Macrophomina phaseolina* (FJ395243)	100
OA3S1-49	MG214585	*S. nigrum*	*Macrophomina phaseolina* (KT768130)	100
KA1S4-40	MG214584	*S. nigrum*	*Macrophomina phaseolina* (KT768130)	100
OA3S1-51	MG214604	*S. nigrum*	*Rhizoctonia solani* (JQ616871)	100
NP2S2-61	MG214605	*S. tuberosum*	*Rhizoctonia solani* (KT783526)	99
KA1S1-34	MG214568	*S. nigrum*	*Aspergillus aculeatus* (KJ862074)	100
KB1S1-3	MG214573	*C. annum*	*Aspergillus* sp. (EF669604)	99
KT1S1-20	MG214569	*Lycopersicon esculentum*	*Aspergillus flavipes* (KC426997)	100
OA1S1-46	MG214592	*S. nigrum*	*Penicillium simplicissimum* (KM396382)	99
KB1S4-8	MG214560	*C. annum*	*Albifimbria terrestris* (KU845884)	99
KB2S2-17	MG214582	*C. annum*	*Macrophomina phaseolina* (JN672592)	100
KB2S4-18	MG214565	*C. annum*	Ascomycota sp. (KT240142)	100
KB1S4-10	MG214561	*C. annum*	*Albifimbria terrestris* (KU845884)	99
KT1S1-21	MG214571	*L. esculentum*	*Aspergillus ochraceopetaliformis* (JQ647894)	100
KT1S1-24	MG214575	*L. esculentum*	*Aspergillus terreus* (KP793450)	99
KT2S1-27	MG214576	*L. esculentum*	*Aspergillus terreus* (KX011595)	100
KB1S4-13	MG214590	*C. annum*	*Penicillium oxalicum* (KU743897)	100
KB3S4-19	MG214579	*C. annum*	*Cylindrobasidium evolvens* (KT201654)	100
KT3S4-33	MG214567	*L. esculentum*	*Aspergillus aculeatus* (KM278131)	100
KT1S1-22	MG214574	*L. esculentum*	*Aspergillus terreus* (KT778597)	100
OA3S1-52	MG214598	*S. nigrum*	*Pyrenochaeta lycopersici* (AB695298)	99
KB2S2-16	MG214586	*C. annum*	*Massarinaceae* sp. (JF502440)	97
KB1S4-9	MG214563	*C. annum*	Fungal endophyte isolate (KP335438)	99
NP1S4-59	MG214578	*S. tuberosum*	*Colletotrichum coccodes* (JX294026)	99
NA2S2-44	MG214591	*S. nigrum*	*Penicillium simplicissimum* (KM396382)	99
KA1S2-39	MG214580	*S. nigrum*	*Epicoccum nigrum* (KT276982)	100
OA2S1-48	MG214595	*S. nigrum*	*Plectosphaerella oligotrophica* (KX446769)	99
KB1S4-11	MG214588	*C. annum*	*Myrothecium cinctum* (DQ135998)	100
OA3S1-50	MG214594	*S. nigrum*	*Plectosphaerella cucumerina* (KT826571)	100
KB1S4-12	MG214570	*C. annum*	*Aspergillus foveolatus* (AB249010)	100
KT1S1-23	MG214572	*L. esculentum*	*Aspergillus ochraceopetaliformis* (JQ647894)	100
KA1S1-35	MG214577	*S. nigrum*	*Aspergillus terreus* (KM491895)	99
NP1S5-64	MG214602	*S. tuberosum*	*Pyrenochaeta lycopersici* (AY649593)	99
NP2S1-55	MG214599	*S. tuberosum*	*Pyrenochaeta lycopersici* (AY649594)	100
NP2S1-60	MG214603	*S. tuberosum*	Uncultured *Pyrenochaeta* (JQ247357)	100
KB1S1-1	MG214596	*C. annum*	*Purpureocillium lilacinum* (KT224843)	100
NP3S4-63	MG214593	*S. tuberosum*	*Periconia* sp. (KP269005)	97
KA2S1-42	MG214566	*S. nigrum*	Uncultured *Pleosporales* (GU909731)	94
NP2S1 56	MG214600	*S. tuberosum*	*Pyrenochaeta lycopersici* (AY649594)	100
NP3S2-62	MG214601	*S. tuberosum*	*Pyrenochaeta lycopersici* (AY649594)	100
OA1S1-47	MG214597	*S. nigrum*	*Pyrenochaeta lycopersici* (KF494161)	100
KB1S1-4	MG214562	*C. annum*	Uncultured fungus (FJ528715)	99
KT2S2-29	MG214589	*L. esculentum*	*Penicillium citrinum* (KT385733)	100
KB2S2-15	MG214564	*C. annum*	Uncultured ectomycorrhiza (JX043219)	91

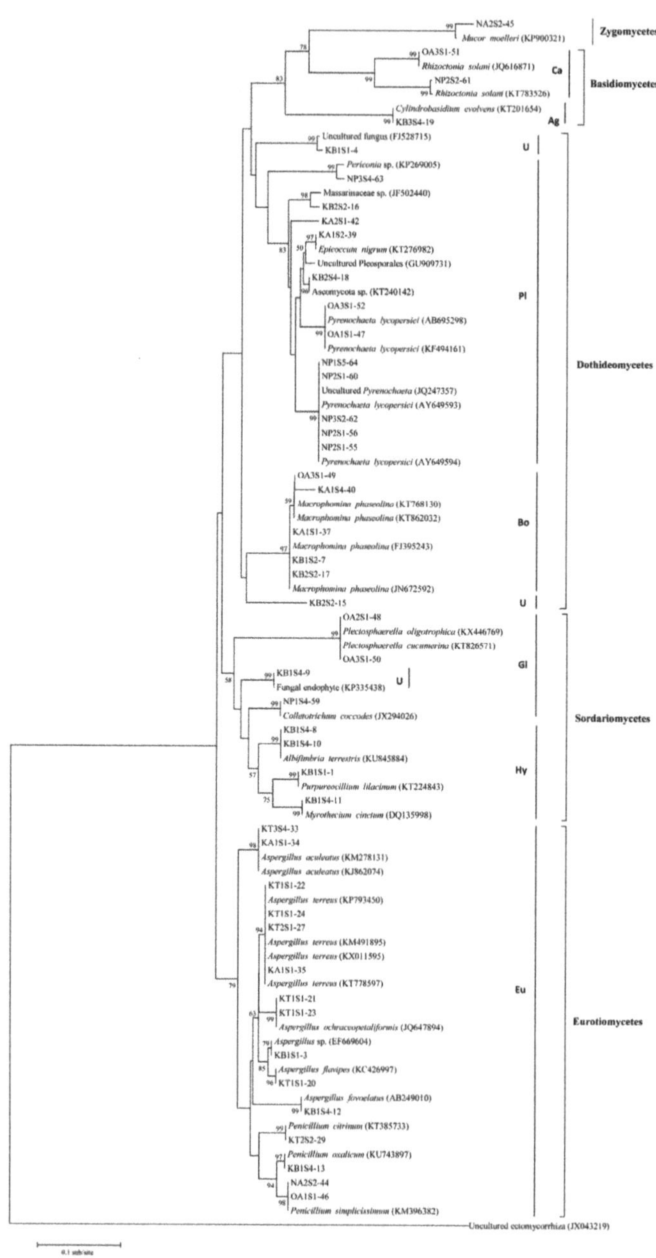

Figure 3. Phylogram of 46 rDNA ITS sequences of root endophytic fungi and their closest BLAST matches based on neighbor-joining analysis. Bootstrap values of >50% are shown at branching points. Ca: Cantharellales; Ag: Agaricales; U: Unknown; Pl: Pleosporales; Bo: Botryosphaeriales; Gl: Glomerellales; Hy: Hypocreales; and Eu: Eurotiales.

A deeper look at the distribution of fungal taxa with potential activity against *P. infestans* in relation to the source host plant showed that tomato accommodated two genera, namely

Aspergillus (87.5%) and *Penicillium* (12.5%), while four genera, with *Pyrenochaeta* being the most abundant (62.5%), were found in potato (Figure 4).

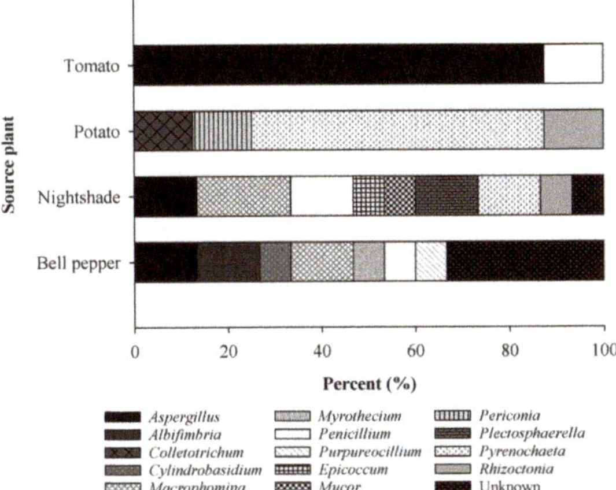

Figure 4. Comparative distribution of taxa with antagonistic activity against *P. infestans* from four solanaceous plant species.

On the other hand, nightshade and bell pepper harbored more diverse fungi giving rise to eight and seven known genera, respectively. Both host plants also harbored unknown species, with the proportion being greater in bell pepper (33.3%) than in nightshade (6.7%).

2.4. In Vitro Activity of Endophytic Fungi against Mycelial Growth of P. infestans

In confrontation assays, mycelial growth retardation of *P. infestans* in the presence of one of the 46 fungal endophytic isolates varied significantly (Table 2). *T. harzianum* along with two endophytes (KB1S2-7 and KA1S1-34) suppressed mycelial growth of the pathogen by 84.5%, 78.2%, and 76.5%, respectively. The other endophytes, however, were either only moderate (KB1S4-10 and KT1S1-21), or slight (KT2S2-29 and KB2S2-15) growth inhibitors. Macroscopic observations of the interaction zones showed that some endophytes completely overgrew the pathogen. Members of these endophytes including *T. harzianum*, *Mucor moelleri*, and *Macrophomina phaseolina* recorded the highest inhibition (70.4–84.5%) (Table 2). *Albifimbria terrestris* and *Penicillium simplicissimum* partially overgrew pathogen colonies giving moderate growth inhibition (51.8–56.3%). In other cases, the growth of both endophyte and pathogen stopped once their colonies came into contact. Mycelial growth inhibition in this group was dependent on the growth rate of the endophytes. Fungal endophytes within the genera *Aspergillus*, *Albifimbria*, *Macrophomina*, and *Cylindrobasidium* showed this type of interaction. Interestingly, other endophytes within the genera *Aspergillus*, *Penicillium*, *Purpureocillium*, and *Pyrenochaeta* created inhibition zones with the pathogen. These endophytes showed slight to moderate inhibition percentages (13.3–46.9%) and the inhibition areas created between endophyte and pathogen varied markedly (Table 2). Three unidentified isolates (KB1S1-4, KA2S1-42, and KB2S2-15) formed significantly ($\alpha = 0.05$) larger inhibition zones measuring 18.8, 18.3, and 14.8 mm, respectively.

Table 2. Mycelial growth inhibition of *P. infestans* and inhibition zones caused by endophytic fungi (arranged in descending order in terms of their mycelial growth inhibition) in dual culture.

Isolate	Highest BLAST Affinities	Mycelial Growth Inhibition (%) *	Inhibition Zone (mm) *
T16	*Trichoderma harzianum*	84.5 [a]	-
KB1S2-7	*Macrophomina phaseolina*	78.8 [ab]	-
KA1S1-34	*Aspergillus aculeatus*	76.7 [ab]	-
NA2S2-45	*Mucor moelleri*	73.8 [ab]	-
KB2S2-17	*Macrophomina phaseolina*	72.0 [ab]	-
KA1S1-37	*Macrophomina phaseolina*	70.4 [ab]	-
OA3S1-49	*Macrophomina phaseolina*	70.2 [ab]	-
KA1S4-40	*Macrophomina phaseolina*	69.2 [ab]	-
OA3S1-51	*Rhizoctonia solani*	69.2 [ab]	-
NP2S2-61	*Rhizoctonia solani*	68.3 [bc]	-
KB1S1-3	*Aspergillus* sp.	68.2 [bc]	-
OA1S1-46	*Penicillium simplicissimum*	56.3 [dc]	-
KB2S4-18	Ascomycota sp.	54.1 [de]	-
KT1S1-23	*Aspergillus ochraceopetaliformis*	53.5 [de]	9.3 [cd]
KA1S1-35	*Aspergillus terreus*	52.0 [d–f]	7.0 [c–e]
KB1S4-10	*Albifimbria terrestris*	51.8 [d–f]	-
KT1S1-21	*Aspergillus ochraceopetaliformis*	49.5 [d–g]	-
KT1S1-24	*Aspergillus terreus*	47.0 [d–h]	-
NP1S4-59	*Colletotrichum coccodes*	46.9 [d–h]	-
KT2S1-27	*Aspergillus terreus*	46.5 [d–h]	-
NA2S2-44	*Penicillium simplicissimum*	45.0 [d–i]	-
NP1S5-64	*Pyrenochaeta lycopersici*	43.9 [d–j]	10.0 [cd]
KB1S4-13	*Penicillium oxalicum*	43.5 [d–j]	-
NP2S1-55	*Pyrenochaeta lycopersici*	42.9 [e–k]	7.6 [c–e]
KB1S4-8	*Albifimbria terrestris*	41.5 [e–k]	-
NP2S1-60	Uncultured *Pyrenochaeta*	39.4 [f–l]	17.0 [a]
KB1S1-1	*Purpureocillium lilacinum*	39.0 [f–l]	6.5 [de]
KA1S2-39	*Epicoccum nigrum*	39.0 [f–l]	-
NP3S4-63	*Periconia* sp.	38.4 [g–m]	11.0 [bc]
KB3S4-19	*Cylindrobasidium evolvens*	37.3 [g–m]	-
KA2S1-42	Uncultured Pleosporales	37.0 [g–n]	18.3 [a]
NP2S1-56	*Pyrenochaeta lycopersici*	36.5 [g–n]	11.3 [bc]
KT3S4-33	*Aspergillus aculeatus*	36.4 [g–o]	-
NP3S2-62	*Pyrenochaeta lycopersici*	34.5 [h–o]	7.3 [c–e]
KT1S1-20	*Aspergillus flavipes*	32.2 [i–p]	-
OA1S1-47	*Pyrenochaeta lycopersici*	31.4 [j–q]	8.0 [c–e]
KT1S1-22	*Aspergillus terreus*	31.1 [j–q]	-
OA2S1-48	*Plectosphaerella oligotrophica*	30.5 [k–q]	-
KB1S4-11	*Myrothecium cinctum*	28.1 [l–r]	-
OA3S1-50	*Plectosphaerella cucumerina*	26.1 [m–r]	-
OA3S1-52	*Pyrenochaeta lycopersici*	24.7 [n–r]	-
KB1S1-4	Uncultured fungus	24.2 [o–r]	18.8 [a]
KB1S4-12	*Aspergillus foveolatus*	22.3 [p–s]	-
KB2S2-16	*Massarinaceae* sp.	20.1 [q–s]	-
KB1S4-9	Fungal endophyte isolate	20.0 [q–s]	-
KT2S2-29	*Penicillium citrinum*	18.9 [rs]	4.8 [e]
KB2S2-15	Uncultured ectomycorrhiza	13.3 [s]	14.8 [ab]

* Means with the same letter are not significantly different at $\alpha = 0.05$.

2.5. Potential of Endophytes Secreting Volatile Organic Compounds Active against *P. infestans*

Based on the results obtained from dual culture experiments, six endophytic fungal isolates (NA2S2-45, KB2S2-17, KB1S1-4, KA2S1-42, KB2S2-15, and KB2S4-8) were selected and subjected to further characterization regarding their ability to produce volatile organic compounds (VOCs) with anti-oomycete activity. All endophytes were found to produce VOCs with varying activities against mycelial growth of *P. infestans* (Figure 5).

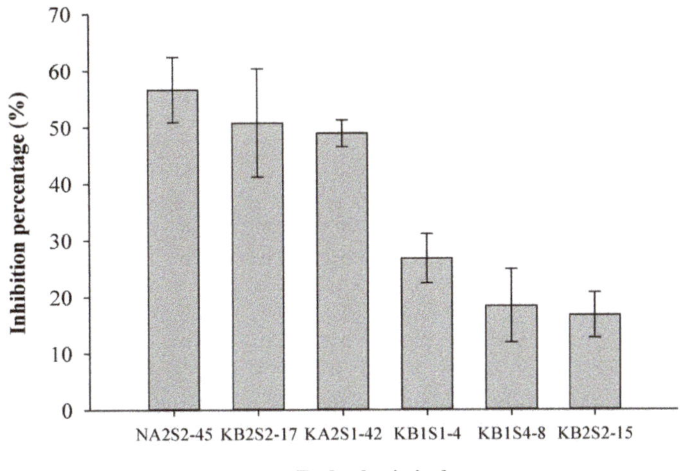

Figure 5. Suppressive effects (%) of the VOCs secreted by the endophytic fungi against mycelial growth of *P. infestans*. Error bars represent the standard error of means (n = 8).

M. moelleri (NA2S2-45), *M. phaseolina* (KB2S2-17), and an unidentified Pleosporales species (KA2S1-42) retarded the growth of the pathogen to a similar extent, recording inhibition percentages of 56.7%, 50.8%, and 48.9%, respectively. Lower suppression levels were observed in the case of *Alibifimbria terrestris* KB1S4-8 (18.4%) and two unidentified fungi, KB1S1-4 (26.8%) and KB2S2-15 (16.7%).

2.6. *Effect of Crude Extracts from Selected Endophytes on Sporangial Germination of P. infestans*

To assess the potential of crude extracts (CEs) obtained from selected fungal isolates against *P. infestans*, sporangia were allowed to germinate in the presence of 5% of individual CEs and germination was evaluated after 16 h. Sporangia germination in 5% acetone (solvent control) recorded ≥90% and differed only insignificantly from that of the water control. Sporangia reacted differentially to the individual CEs applied (Table 3).

Table 3. Effect of crude extracts from the endophytes on sporangial germination and germ tube elongation of *P. infestans*.

CEs Source	Inhibition (%) [1]	
	Sporangial Germination [2]	Germ tube Growth [3]
KB2S2-15	100.00 [a]	100.00 [a]
KA2S1-42	100.00 [a]	100.00 [a]
KB1S1-4	69.38 [b]	82.73 [b]
NA2S2-45	30.04 [c]	−1.92 [e]
Acetone (solvent) control	0.00 [d]	0.00 [e]

(1) Inhibition percentages calculated relative to the solvent control. Means with the same letter within each column are not significantly different at $\alpha = 0.05$, ($p < 0.0001$), n (2) = 1600, n (3) = 160.

While the crude extract (5%) from NA2S2-45 significantly retarded sporangia germination (30%), no suppressive effect on germ tube growth was detected. A similar dosage of the crude extract from KB1S1-4 suppressed sporangial germination by approximately 70% and diminished germ tube elongation by >80% compared to the solvent control. The amendment of *P. infestans* sporangia with CEs obtained from two endophytic fungi (KB2S2-15 and KA2S1-42) entirely suppressed their germination (Table 3). As a result, CEs from KB2S2-15, KA2S1-42, and KB1S1-4 showed suppressive effects against sporangia germination and germ tube development and hence were selected for further in vivo investigations.

2.7. *In Vivo Activity of the Crude Extracts against P. infestans*

To validate the in vitro results, the effect of CEs on leaf blight development on detached potato leaves was investigated. CEs were incorporated into sporangial suspension to give a dosage of (5%, w:v). This mixture was inoculated on the abaxial surface of detached leaflets on either side of the midrib. The results obtained from two independent experiments revealed that detached leaflets treated with either acetone or CEs in the absence of the pathogen did not show any phytotoxic damage (Figure 6b,d–f).

Figure 6. Detached potato leaflets treated or not with crude extracts of the endophytes and inoculated (+*Pi*, uppercase latters) or not (−*Pi*, lowercase letters) with sporangial suspension of *P. infestans*. (**A**,**a**): water control, (**B**,**b**): solvent control, (**C**,**c**): Infinito®, (**D**,**d**): KB1S1-4, (**E**,**e**): KB2S1-15, and (**F**,**f**): KA2S1-42.

Under a high relative humidity, necrotic lesions covered with white fluffy hyphae and sporangia of *P. infestans* developed on the inoculated leaflets treated only with water (control) within 7 d (Figure 6A). Similarly, no decrease in lesion development or mycelial colonization was detected on inoculated leaflets treated with 5% acetone compared with the water control treatment (Figure 6B). In contrast, late blight symptoms and mycelial growth were completely suppressed when the systemic broad-spectrum fungicide Infinito® was applied (Figure 6C). Interestingly, the application of 5% CEs of the endophytes KB1S1-4,

KA2S1-42, and KB2S2-15 yielded neither visible lesions nor hyphal growth of *P. infestans* on the inoculated leaflets (Figure 6D–F).

3. Discussion

The screening and identification of microorganisms with antagonistic properties against soilborne pathogens are indispensable first steps in the search for potential biocontrol agents [27,33]. In the current study, root endophytic fungi were isolated from four solanaceous plant species obtained from three regions in Kenya with diverse climatic conditions and soil properties (Table 1). This might increase the heterogeneity and number of fungal species obtained, and hence the possibility of identifying endophytes with unusual adaptation strategies and potential bioactivity against *P. infestans* [27]. The sampling regions were a few to several hundred kilometers apart (1000 km maximum distance), differed in elevation (2532 m maximum difference), climate, and physio-chemical soil properties (Table S1).

In addition to soil and climatic factors, host plant species also influence the composition of root microbiomes, with roots being shown to harbor more endophytic biodiversity than the rest of the plant organs [35,36]. These factors may have contributed to the realization of 357 endophytic fungal isolates from Kilifi, Nyandarua, and Kiambu from the roots of four solanaceous plant species sampled, namely *S. tuberosum*, *L. esculentum*, *S. nigrum*, and *C. annuum*. However, only a limited proportion of approximately 13% of the isolates showed potential activity against *P. infestans*. Interestingly, 63% of the endophytes were obtained from Kilifi (Figure 2), a non-potato growing region, implying that spatial separation could be a predominant factor limiting pathogen–antagonist interactions. Similar findings of the occurrence of unique endophytic fungal species from Kilifi were reported by Bogner et al. [37]. These authors attributed their findings to the hot and humid climate experienced in the coastal region.

The type of host plant may also have been an influential factor in the availability of antagonists. Bell pepper and nightshade each gave rise to the largest number of antagonistic endophytes (Figure 2), which was consistent with the finding of Kim et al. [12], that red pepper roots harbored endophytic fungi with potent activity against *P. infestans*. The endophytes from nightshade and bell pepper were significantly more diverse than those from other host plants (Figure 4). In these two plant species, unidentified endophytes were also captured, all isolated from Kilifi. The diversity of antagonists was lower in potato and least in tomato (Figure 4).

Although challenges in the use of sequence information on the NCBI database for molecular identification of fungi have been cited, it remains a useful tool in the identification of species, especially for non-sporulating endophytes [27,38]. In the current study, there was considerable agreement between the phylogenetic relationships of the endophytes and their corresponding closest BLAST matches, as demonstrated by the high bootstrap support at the terminal nodes (Table 1 and Figure 3). The exceptions were isolates KB2S2-15 and KA2S1-42, for which BLAST matches with only little similarity were found that did not cluster in the phylogeny. These endophytes were non-sporulating on different nutrient media, hampering morphological characterization. Similarly, isolates KB1S1-4, KB1S4-9, KB2S2-16, and KB2S4-18 were also non-sporulating and their identity remained concealed, as their closest BLAST matches were unknown fungal species. The isolates KB1S1-3, NP2S1-60, and NP3S4-63 could not be identified to the species level, but molecular and morphological traits established their genera as *Aspergillus*, *Pyrenochaeta*, and *Periconia*, respectively. On a broader perspective, most of the antagonistic endophytes characterized in this study were Ascomycetes, while the occurrence of Basidiomycetes and Zygomycetes was limited to one or a few isolates. The Ascomycota is the largest and most diverse fungal phylum, and its members are usually the dominant endophytes in plants in relation to other fungal groups [39–42].

By observing the interactions of 46 potential fungal endophytes in dual culture with *P. infestans*, some antagonists rapidly grew over the pathogen colony resulting in the highest inhibition percentages. Apart from *M. moelleri*, this group comprises mainly of members of possible potato pathogens. While the pathogenicity of these endophytes has not been yet verified, *M. moelleri* has been recently indicated as a potential antagonist against different tomato pathogens including *Athelia rolfsii* and *Colletotrichum gloeosporiodes* [43]. *T. harzianum* T16 also showed a similar interaction with *P. infestans*. Previous reports of the activity of *Trichoderma* spp. against *P. infestans* and other pathogens have demonstrated their capacity for complete growth over pathogen colonies in dual culture, mycoparasitism, and antibiosis through the production of active compounds such as 6-pentyl-α-pyrone, viridiofungin A, harzianolide, and harzianic acid [44–47]. In exclusion of putative potato pathogens, *Aspergillus* spp., *Penicillium* spp., *Albifimbria terrestris*, and *Cylindrobasidium evolvens* were among the antagonistic fungal endophytes identified. Members of *Aspergillus* and *Penicillium* have been shown to have endophytic lifestyles in plants as well as exhibiting bioactivity against plant pathogens (e.g., *Botrytis cinerea, Trichothecium roseum, Sclerotinia sclerotiorum, Fusarium oxysporum*, and *Rhizoctonia solani*) with *A. niger*, *A. terreus*, and *P. citrinum* being some of the well-studied antagonists [33,39,42,48,49]. In addition, a protein derived from *Aspergillus giganteus* was shown to have activity against *P. infestans* in vitro [50]. The Basidiomycete *C. evolvens*, as a saprophyte capable of infecting wounded stem and root tissue, is associated with decay in woody plants [51,52]. In this study, *C. evolvens* was isolated as an endophyte from the roots of *C. annuum* and showed a moderate antagonistic activity against *P. infestans* in dual culture (Table 2). Likewise, *A. terrestris* was isolated from the roots of *C. annuum* and is morphologically similar to *A. verrucaria*, whose basionym was *Myrothecium verrucaria*, a known plant pathogen that has been formulated into bioherbicides or nematicides [53–55]. Interestingly, some endophytes exerted their antagonism by antibiosis as indicated by the inhibition zones. In a comparable study, CEs of *Aureobasidium pullulans* isolated from *Espeletia* spp., a native Andean plant, showed in vitro activity against *P. infestans* [27].

In addition, endophytic fungi are also capable of producing VOCs. VOCs from *M. moelleri*, *M. phaseolina*, and an unknown endophyte KA2S1-42 showed a higher mycelial growth inhibition of *P. infestans* than those released by KB1S1-4, *A. terrestris*, and KB2S2-15 (Figure 5). Previous reports have shown that *Mucor* spp. are able to produce ethanol while *M. phaseolina* has been associated with the production of several fatty acid methyl esters with an inhibitory activity against *Sclerotinia sclerotiorum* [42,56]. On the other hand, *A. verrucaria* excreted bioactive compounds, including antibacterial cyclopeptides that showed herbicidal activities as well [57,58]. Moreover, the *A. verrucaria* isolate SYE-1 exhibited a wide antifungal activity against *B. cinerea*, *Lasiodiplodia theobromae*, and *Elsinoë ampelina* on grapevine [59]. *Myrothecium inundatum*, a close relative of *A. terrestris*, has been implicated in the production of a variety of hydrocarbon VOCs, or their derivatives, potent against *Pythium ultimum* and *S. sclerotiorum* [60].

VOCs have been indicated to have a greater coverage of soil and organic substrata than diffusible organic compounds (DOCs), a characteristic attributed to their gaseous state [61]. In the current study, three unknown antagonists KA2S1-42, KA1S1-4, and KB2S2-15 showed the capability of producing VOCs as well as DOCs that affect *P. infestans*. In addition, *M. moelleri* showed a high inhibition of *P. infestans* mycelium both in dual culture and through the production of VOCs. These findings led to the selection of the antagonistic endophytes KA1S1-4, KB2S2-15, KA2S1-42, and *M. moelleri* for further testing while putative pathogens were left out of subsequent studies.

Consistent with dual culture findings, CEs from the antagonistic endophytes KB1S1-4, KB2S2-15, and KA2S1-42, showed a suppressive activity against sporangia germination and the germ tube growth of *P. infestans*. While CEs from KB1S1-4 at a concentration of 5% exhibited considerable suppressive effects against sporangial germination and germ tube elongation of the pathogen, CEs from KB2S2-15 and KA2S1-42 entirely blocked sporangia germination (Table 3). In a similar set up, Linkies et al. [62] found that upon exposure to ethyl-acetate extracts from *Chaetomium cochliodes* and *C. elatum* at a dosage of 30%, sporangia germination of P. infestans was considerably inhibited. However, this dosage is much higher (six-fold) than that used in our study.

On the other hand, although CEs from *M. moelleri* (NA2S2-45) showed a relatively weak activity against sporangia germination, it did not inhibit germ tube development, suggesting that the anti-*Phytophthora* metabolites produced by this fungus are rather confined to the gaseous phase, which could not be extracted from the culture filtrates in concentrations sufficient to affect the germination of *P. infestans* sporangia. Several reports agree with the ability of endophytic, rhizospheric, or phyllospheric fungi in producing diffusible compounds with anti-oomycete activity [12,27,45,63,64]. Tellenbach et al. [65] reported that a strain of *Phialocephala europaea* was able to produce metabolites containing sclerin and sclerotinin with activity against *Phytophthora citricola*. Our earlier studies showed that strain T23 of *T. harzianum* (recently re-identified as *T. asperellum*) secreted viridiofungin A with a strong inhibitory effect against sporangia germination of *P. infestans* [45]. In the current study, although the biologically active metabolites produced by KB1S1-4, KB2S2-15, and KA2S1-42 have not been identified, it was evident that these endophytes produce potent compounds with a strong activity against *P. infestans*.

In the detached leaflet assay, the application of CEs from the endophytes KB1S1-4, KB2S2-15, and KA2S1-42 entirely protected inoculated leaflets and suppressed the hyphal growth of *P. infestans* (Figure 6). Likewise, Bae et al. [66] found that CEs from *Trichoderma atroviride* showed inhibitory activities against *Phytophthora sojae*, *P. capsici*, and *P. melonis* and induced defense reactions in the detached leaves of pepper and tomato plants. Furthermore, Kim et al. [12] reported a substantial in vivo activity of *Fusarium oxysporum* strain EF119 against tomato late blight. Similar results were also reported by Chandrakala et al. [64], where culture filtrates from *T. virens* and *T. viride* proved to be effective against sporangia germination of *P. infestans* and impeded the establishment of late blight on potato.

Overall, we report preliminary evidence for targeting the late blight pathogen by particular fungal endophytes both in vitro and in vivo. Based on their promising in vitro performance against the late blight pathogen, three endophytes (KA1S1-4, KB2S2-15, and KA2S1-42) were selected from a total of 357 isolates. These endophytic isolates proved to be very active in protecting potato leaflets from *P. infestans*. This biological activity is associated with DOCs and VOCs with anti-oomycete properties. However, the purification and elucidation of the chemical structures of these bioactive molecules have not been accomplished in this study. Therefore, it is still a rich field for future investigations, particularly in terms of clarifying modes of action of the metabolites involved to determine whether they target single or multiple pathways in *P. infestans*. This will shed light on the possibility of combining them to improve their efficacy and hence both the reliability and durability of the biocontrol.

4. Materials and Methods

4.1. Sampling Regions and Collection of Plant Materials

Root samples from four solanaceous plant species including potato (*Solanum tuberosum* L.), tomato (*Lycopersicon esculentum* L. Mill.), bell pepper (*Capsicum annuum* L.), and African nightshade (*Solanum nigrum* L.), were collected from three sampling regions in Kenya (Nyandarua, Kilifi, and Kiambu) with variable soil types. From each region, root systems of twelve apparently healthy plants were obtained along with representative soil samples. Immediately after collection, intact roots were washed thoroughly under running tap water to remove soil and adhering debris, air dried, and stored at 4 °C until further processing. In order to determine their physio-chemical properties, soil samples were analyzed at the Soil Science laboratories of Jomo Kenyatta University of Agriculture and Technology (Juja, Kenya) (Table S1). pH and electrical conductivity (EC) were determined in a soil: water mixture (2:5, w:v), while total soil organic carbon was measured using the Walkley–Black rapid titration method [67].

4.2. Isolation of Fungal Root Endophytes and P. infestans

Root segments of 3–4 cm from primary and secondary roots of each plant were surface sterilized by immersion in 70% alcohol for 1 min and immediately transferred into 2.5% sodium hypochlorite (Carl Roth, Karlsruhe, Germany) for 5 min. Root segments were rinsed three times with sterile distilled water for 5 min and blotted between sterile paper towels to remove excess moisture. To ascertain the success of surface sterilization, sterilized root segments were briefly imprinted on PDA (Figure S1). Subsequently, the ends of the root segments were trimmed off and further divided into three fragments of approx. 1 cm length, which were placed on PDA amended with 0.05 g/L chloramphenicol (Merck KGaA, Darmstadt, Germany). Finally, plates were sealed with Parafilm and incubated at 22 ± 1 °C in the dark. Only endophytic fungi emerging from successfully surface sterilized roots (Figure S2) were subcultured onto fresh PDA. Isolates were purified through either single-spore isolation or hyphal tip transfer for sporulating and non-sporulating colonies, respectively. Pathogens (e.g., *Fusarium* spp.) were excluded from further tests.

P. infestans was isolated from single lesions of potato leaves obtained from a late blight susceptible variety (Duke of York) growing under field conditions in Hohenheim, Stuttgart, Germany. To induce fresh sporulation, blighted leaves were placed in a humid chamber at 20 °C for 48 h. Colonized leaf segments were placed between two surface sterilized tuber slices (5 mm thickness) and incubated for 7 d at 20 °C in the dark (Figure S3). Mycelia emerging from the upper side of a slice were transferred to unclarified V8-based agar medium containing 200 mL V8 juice, 2 g $CaCO_3$, 0.05 g β-sitosterol, 0.05 g ampicillin, 0.05 g vancomycin, 0.01 g pentachloronitrobenzene (PCNB), and 15 g agar in 800 mL deionized water. A pure culture of the pathogen generated from the tip of a single hypha was maintained on corn meal agar (CMA, 17 g/L; Sigma-Aldrich, Munich, Germany) and reactivated after every 3–4 transfers on CMA by inoculation on sterile potato leaflets.

4.3. Primary Screening of Endophytes for Anti-Oomycete Activity against P. infestans

A high throughput confrontation assay was established to screen 357 potential antagonistic endophytes against *P. infestans*. Owing to the slow growth of *P. infestans*, 5 mm Ø agar plugs from the border of an actively growing colony were inoculated on the center of 20% V8 agar plates and incubated at 20 °C for 72 h in darkness. Subsequently, agar plugs (5 mm Ø) from four different endophytic fungi were placed equidistant, 5 mm from the edge of the Petri plates pre-inoculated with the pathogen (Figure S4). Control treatments were set up in a similar way in the absence of endophytes. Isolates observed to retard the growth of *P. infestans* compared to the control were selected for further testing and purified to generate axenic cultures.

4.4. DNA Extraction, PCR Conditions, and Sequencing

Genomic DNA was extracted from mycelia of root endophytic fungi selected from the screening experiment using the method described by Liu et al. [68]. PCR was conducted in a 40 µL reaction mixture using ITS1 (5′-TCCGTAGGTGAACCTGCGG-3′) and ITS4 (5′-TCCTCCGCTTATTGATATGC-3′) primers [69] to amplify the internal transcribed spacer regions (ITS1-5.8S-ITS2). A single PCR reaction was comprised of 8 µL of Phusion® HF buffer (5×), 0.8 µL of dNTPs (10 mM), 1 µL of each of the forward and reverse primers (10 µM), 0.4 µL of Phusion® polymerase (2 U/µL), and 27.8 µL of ultra-pure water. The PCR program started with an initial denaturation step at 98 °C for 30 s, followed by 35 cycles at 98 °C for 10 s, 54 °C for 20 s, and 72 °C for 35 s and a final extension step at 72 °C for 10 min. Success of amplification was ascertained on 1% agarose gels visualized with 0.05% ethidium bromide with the aid of a gel documentation system (Quantum 1100 PEQLAB, VWR, Darmstadt, Germany). Amplified PCR products were purified using Innu PREP PCR-pure kit (Analytik Jena AG, Jena, Germany) and their concentration adjusted to meet requirements for Sanger sequencing by Source Bioscience Company (Berlin, Germany). Sequences from single reads with the forward primer (ITS1) were trimmed and edited with GENtle v 2.0 and compared to those deposited at the NCBI database. Endophytic fungi whose sequences showed a similarity of >99% to database entries were considered identical to the reference fungi [33], and their taxonomic classification carried out using the MYCOBANK Database [34]. Evolutionary history was deduced using the Neighbour-Joining method as described by Saitou and Nei [70] with a bootstrap test of 1000 replicates to determine the percentage when replicate trees of related taxa clustered together [71]. The maximum composite likelihood method [72] was used to calculate the evolutionary distances, used to infer the phylogenetic tree. Finally, sequence data were submitted to GenBank.

4.5. Establishment the Anti-Oomycete Activity of the Endophytes

To validate the inhibitory effect of screened fungal endophytes against *P. infestans*, a dual culture assay was performed. The assay was similar to that described under 4.3 except that the pathogen and each unique endophytic isolate were inoculated at 7 cm distance. *P. infestans* was also inoculated on 20% V8 agar plates 72 h prior the antagonistic isolates. Three replicates were prepared for each isolate and plates inoculated only with the pathogen served as control. *Trichoderma harzianum* (T16) shown to have antagonistic properties against several plant pathogens [45,46] was used as a positive control, while pathogen colonies grown in the absence of the endophytes served as negative control. The experiment was laid out in a randomized complete block design. The initial (72 h post inoculation; hpi) and the final (12 d post inoculation; dpi) colony diameter of the pathogen were measured. The percentage of mycelial growth inhibition was calculated using the formula ascribed by Edgington et al. [73]. In addition, the inhibition zones created between the pathogen and the tested endophytic fungal colonies were recorded.

4.6. Impact of Volatile Organic Compounds Extracted from the Endophytes on Mycelial Growth of P. infestans

The ability of selected endophytic fungi to produce volatile organic compounds (VOCs) was assessed following the modified procedure of El-Hasan et al. [74]. Briefly, agar plugs (5 mm Ø) bearing *P. infestans* mycelium were placed on the center of V8 agar plates and incubated for 72 h. Subsequently, the lid was replaced with an upside-down agar plate inoculated with an agar plug colonized by an endophyte. Both Petri dishes were separated with a sterile cellophane sheet and held together with Parafilm. Positive and negative controls consisted of *T. harzianum* and sterile PDA plugs, respectively. Mycelial growth inhibition (in %) was calculated as described above.

4.7. Suppressive Activity of Crude Extracts of Root Endophytic Fungi against Sporangial Germination of P. infestans

Putative endophytic fungi that formed inhibition zones in dual culture and/or produced active VOCs were tested for their ability to produce diffusible organic compounds with anti-oomycete activity. To this end, 1 L of 20% V8 broth was inoculated with ten agar plugs (5 mm Ø) of an actively growing endophyte culture. The cultures were incubated on an orbital shaker at 125 rpm and 20 °C in the dark for 12 d. Control treatments, where the endophytes were absent, were set up. Crude extracts (CEs) were prepared from culture filtrates according to El-Hasan et al. [45]. The resulting CE residues were re-dissolved in 2 mL acetone.

Sporangia of *P. infestans* were produced on detached potato leaflets. For this purpose, an agar plug (3 × 3 mm) was placed on the abaxial side of a detached leaflet surface sterilized with 2.5% sodium hypochlorite. A drop of sterile deionized water was introduced on the interface and the inoculated leaflets were placed in a humid chamber made by inverting the Petri dish containing water agar over lids containing sterile moist filter paper. Cultures were incubated at 20/18 °C (light/dark) for 7 d in a growth chamber with a 16 h light cycle. Sporangia were harvested by briefly vortexing infected leaflets in a sterile centrifuge tube containing V8 liquid medium. Mycelia and leaf fragments were trapped using double layers of sterile muslin cloth and the resulting sporangial suspension (5×10^4 sporangia/mL) was used in subsequent experiments.

To determine the effect of CEs on sporangia germination, 25 µL of each extract were combined with an equal volume of water in a sterile 1.5 mL microcentrifuge tube. The tubes were aseptically left open for three hours to allow solvent evaporation. Subsequently, 475 µL of the sporangia suspension were added to the CE solution. Cultures were incubated at 20 °C in the dark to allow sporangia germination, the experiment was terminated after 16 h by the addition of 100 µL lactophenol blue. Acetone was used as solvent control. The percentage of germinated sporangia and the length of germ tubes were determined using a light microscope (Axioskop 2, Carl Zeiss Microscopy GmbH, Göttingen, Germany). Images were taken with an AxioCam MRm digital camera (Carl Zeiss) and measurements determined with the corresponding AxioVision software (SE64 Release 4.8.3 SP1; Carl Zeiss).

4.8. Activity of the Crude Extracts against P. infestans on Detached Potato Leaflets

For the in vivo bioassay, leaflets were harvested from the fourth to sixth fully expanded leaves obtained from potato plants (var. Duke of York) grown under greenhouse conditions. CEs that showed activity against germination of *P. infestans* sporangia were tested for their activity in vivo against the pathogen on detached leaflets. Acetone was evaporated from the crude extracts as described above, and each extract reconstituted to give a concentration of 5% in a sporangial suspension. Two droplets of 50 µL each were inoculated on the abaxial surface of detached leaflets on either side of the midrib. Control treatments were set up in a similar way and comprised water, 5% acetone, and a conventional fungicide (Infinito®; Bayer CropScience, Langenfeld, Germany). Leaflets were placed individually in a humid chamber and incubated for 7 d at 20/18 °C (light/dark) with a 16 h photoperiod. All treatments were set up in four replicates. The assay was conducted twice.

4.9. Data Analysis

If not mentioned elsewhere, experiments were repeated at least twice in triplicate in a completely randomized design. All statistical analyses were performed using SAS software (SAS Institute Inc.). The MIXED and GLIMMIX procedures were applied to determine whether there were treatment effects in the experiments conducted within this study. Data were assessed and transformed when necessary to ensure they met model assumptions. Pairwise comparisons among treatments were carried out using Tukey test ($\alpha = 0.05$).

Supplementary Materials: The following supporting information can be downloaded at: https://www.mdpi.com/article/10.3390/plants11121605/s1, Table S1. Characteristics of sampling regions; Figure S1. Isolation of fungal root endophytes. (a) Endophytic fungi emerging from cultured roots, (b) Clean short press plate, (c) Contaminated short press plate; Figure S2. Isolation of fungal root endophytes from solanaceous plants, (A) Endophytic fungi arising from cultured roots, (B) Pure fungal colonies; Figure S3. Isolation of *P. infestans* (a) *P. infestans* mycelia growing on the surface of an inoculated potato tuber slice (b) Pure colony of *P. infestans* on 20% V8 agar (c) Mycelia of *P. infestans* on potato leaflets inoculated with the pathogen; Figure S4. Dual antagonistic assay of fungal endophytes and *P. infestans*. (a) Co-culture of *P. infestans* (center colony) with four root endophytes to screen for activity against the pathogen's mycelia (b) Setup of dual culture assay on the introduction of an endophyte 72 h after *P. infestans* (left colony) inoculation (c) Schematic representation of the measurement (R2) used in calculating percentage of inhibition.

Author Contributions: Conceptualization, A.E.-H., G.N. and R.T.V.; methodology, A.E.-H. and G.N.; validation, G.N. and A.E.-H.; formal analysis, A.E.-H., G.N. and T.I.L.; investigation, G.N.; resources, A.E.-H. and G.N.; data curation, G.N. and A.E.-H.; writing—original draft preparation, A.E.-H. and G.N.; writing—review and editing, A.E.-H., T.I.L. and R.T.V.; visualization, G.N. and A.E.-H.; supervision, A.E.-H. and R.T.V.; project administration, R.T.V.; funding acquisition, G.N. and R.T.V. All authors have read and agreed to the published version of the manuscript.

Funding: This research was funded by the German Academic Exchange Service (DAAD), grant number A/12/97759, and Food Security Center (FSC), University of Hohenheim, no grant number.

Institutional Review Board Statement: Not applicable.

Informed Consent Statement: Not applicable.

Data Availability Statement: The data presented in this study are all contained within the figures and tables.

Acknowledgments: The kind assistance of Heike Popovitsch, Bianka Maiwald and Sybille Berger of the Institute of Phytomedicine, University of Hohenheim is greatly acknowledged.

Conflicts of Interest: The authors declare no conflict of interest.

References

1. Haverkort, A.J.; Struik, P.C.; Visser, R.G.F.; Jacobsen, E. Applied biotechnology to combat late blight in potato caused by *Phytophthora Infestans*. *Potato Res.* **2009**, *52*, 249–264. [CrossRef]
2. Kromann, P.; Miethbauer, T.; Ortiz, O.; Forbes, G.A. Review of potato biotic constraints and experiences with integrated pest management interventions. In *Integrated Pest Management*; Pimental, D., Peshin, R., Eds.; Springer: Dordrecht, The Netherlands, 2014; pp. 245–268.
3. Ivanov, A.A.; Ukladov, E.O.; Golubeva, T.S. *Phytophthora infestans*: An Overview of Methods and Attempts to Combat Late Blight. *J. Fungi* **2021**, *7*, 1071. [CrossRef] [PubMed]
4. Fry, W.E.; Birch, P.R.J.; Judelson, H.S.; Grünwald, N.J.; Danies, G.; Everts, K.L.; Gevens, A.J.; Gugino, B.K.; Johnson, D.A.; Johnson, S.B.; et al. Five reasons to consider *Phytophthora infestans* a reemerging pathogen. *Phytopathology* **2015**, *105*, 966–981. [CrossRef] [PubMed]
5. Olanya, O.M.; Adipala, E.; Hakiza, J.J.; Kedera, J.C.; Ojiambo, P.; Mukalazi, J.M.; Forbes, G.; Nelson, R. Epidemiology and population dynamics of *Phytophthora infestans* in sub Saharan Africa. *Afr. Crop Sci. J.* **2001**, *9*, 185–194. [CrossRef]
6. Chepsergon, J.; Motaung, T.E.; Bellieny-Rabelo, D.; Moleleki, L.N. Organize, Don't Agonize: Strategic Success of *Phytophthora* Species. *Microorganisms* **2020**, *8*, 917. [CrossRef]
7. Leach, S.S. Effects of copper and copper fungicide soil residues on *Phytophthora infestans*. *Am. Potato J.* **1966**, *43*, 431–438. [CrossRef]
8. Bruck, R.I.; Fry, W.E.; Apple, A.E. Effect of metalaxyl, an acylalanine fungicide, on developmental stages of *Phytophthora infestans*. *Phytopathology* **1980**, *70*, 597–601. [CrossRef]
9. Wale, S.; Platt, H.W.; Cattlin, N.D. Introduction. In *Diseases, Pests and Disorders of Potatoes: A Colour Handbook*; Wale, S.H., Platt, W., Cattlin, N.D., Eds.; Manson Publishing: London, UK, 2008; pp. 6–8.
10. Cohen, Y.; Rubin, A.E.; Galperin, M. Effective control of two genotypes of *Phytophthora infestans* in the field by three oxathiapiprolin fungicidal mixtures. *PLoS ONE* **2021**, *16*, e0258280. [CrossRef]
11. Rani, L.; Thapa, K.; Kanojia, N.; Sharma, N.; Singh, S.; Grewal, A.; Srivastav, A.L.; Kaushal, J. An extensive review on the consequences of chemical pesticides on human health and environment. *J. Clean. Prod.* **2021**, *283*, 124657. [CrossRef]
12. Kim, H.-Y.; Choi, G.J.; Lee, H.B.; Lee, S.-W.; Lim, H.K.; Jang, K.S.; Son, S.W.; Lee, S.O.; Cho, K.Y.; Sung, N.D.; et al. Some fungal endophytes from vegetable crops and their anti-oomycete activities against tomato late blight. *Lett. Appl. Microbiol.* **2007**, *44*, 332–337. [CrossRef]
13. Portz, D.; Koch, E.; Slusarenko, A.J. Effects of garlic (*Allium sativum*) juice containing allicin on *Phytophthora infestans* and downy mildew of cucumber caused by *Pseudoperonospora cubensis*. *Eur. J. Plant Pathol.* **2008**, *122*, 197–206. [CrossRef]
14. Andrade-Linares, D.R.; Grosch, R.; Restrepo, S.; Krumbein, A.; Franken, P. Effects of dark septate endophytes on tomato plant performance. *Mycorrhiza* **2011**, *21*, 413–422. [CrossRef] [PubMed]
15. Hashemi, M.; Tabet, D.; Sandroni, M.; Benavent-Celma, C.; Seematti, J.; Andersen, C.; Grenville-Briggs, L. The hunt for sustainable biocontrol of oomycete plant pathogens, a case study of *Phytophthora infestans*. *Fungal Biol. Rev.* **2022**, *40*, 53–69. [CrossRef]
16. Miller, S.L. Functional diversity in fungi. *Can. J. Bot.* **1995**, *73*, 50–57. [CrossRef]
17. Berg, G.; Krechel, A.; Lottmann, J.; Faltin, F.; Ulrich, A.; Hallmann, J.; Grosch, R. Endophytes: A new source for multi-target Biological Control Agents? *IOBC/WPRS Bull.* **2004**, *27*, 161–165.
18. van Loon, L.C.; Bakker, P.A.H.M.; Pieterse, C.M.J. Systemic resistance induced by rhizospere bacteria. *Annu. Rev. Phytopathol.* **1998**, *36*, 453–483. [CrossRef]
19. Kloepper, J.W.; Ryu, C.-M. Bacterial endophytes as elicitors of induced systemic resistance. In *Microbial Root Endophytes, SOIL Biology*; Schulz, B., Boyle, C., Sieber, T.N., Eds.; Springer Science + Business Media Verlag: Berlin, Germany, 2006; Volume 9, pp. 33–52.
20. Feng, Y.; Shen, D.; Song, W. Rice endophyte Pantoea agglomerans YS19 promotes host plant growth and affects allocations of host photosynthates. *J. Appl. Microbiol.* **2006**, *100*, 938–945. [CrossRef]
21. Mizubuti, E.S.G.; Júnior, L.V.; Gregory, F.A. Management of late blight with alternative products. *Pest. Technol.* **2007**, *1*, 106–116.
22. De Vrieze, M.; Germanier, F.; Vuille, N.; Weisskopf, L. Combining Different Potato-Associated Pseudomonas Strains for Improved Biocontrol of *Phytophthora infestans*. *Front. Microbiol.* **2018**, *9*, 2573. [CrossRef]
23. Ng, K.K.; Webster, J.M. Oomycete activity of Xenorhabdus bovienii. *Can. J. Plant Pathol.* **1997**, *2*, 125–132. [CrossRef]
24. Daayf, F.; Adam, L.; Fernando, W.G.D. Comparative screening of bacteria for biological control of potato late blight (strain US-8), using in-vitro, detached-leaves, and whole-plant testing systems. *Can. J. Plant Pathol.* **2003**, *25*, 276–284. [CrossRef]

25. Stephan, D.; Schmitt, A.; Carvalho, S.M.; Seddon, B.; Koch, E. Evaluation of biocontrol preparations and plant extracts for the control of *Phytophthora infestans* on potato leaves. *Eur. J. Plant Pathol.* **2005**, *112*, 235–246. [CrossRef]
26. Loliam, B.; Morinaga, T.; Chaiyanan, S. Biocontrol of *Phytophthora infestans*, fungal pathogen of seedling damping off disease in economic plant nursery. *Psyche* **2012**, *2012*, 324317. [CrossRef]
27. Miles, L.A.; Lopera, C.A.; González, S.; Cepero de García, M.C.; Franco, A.E.; Restrepo, S. Exploring the biocontrol potential of fungal endophytes from an Andean Colombian Paramo ecosystem. *BioControl* **2012**, *57*, 697–710. [CrossRef]
28. Gupta, J. Efficacy of biocontrol agents against *Phytophthora infestans* on potato. *Int. J. Eng. Sci.* **2016**, *6*, 2249–2252.
29. Al-Mughrabi, K.I. Biological control of *Phytophthora infestans* of potatoes using *Trichoderma atroviride*. *Pest. Technol.* **2008**, *2*, 104–108.
30. Andrade-Linares, D.R.; Grosch, R.; Franken, P.; Rexer, K.-H.; Kost, G.; Restrepo, S.; de Garcia, M.C.C.; Maximova, E. Colonization of roots of cultivated *Solanum lycopersicum* by dark septate and other ascomycetous endophytes. *Mycologia* **2011**, *103*, 710–721. [CrossRef]
31. Kannan, K.P.; Muthumary, J. Comparative analysis of endophytic mycobiota in different tissues of medicinal plants. *Afr. J. Microbiol. Res.* **2012**, *6*, 4219–4225.
32. de Vries, S.; von Dahlen, J.K.; Schnake, A.; Ginschel, S.; Schulz, B.; Rose, L.E. Broad-spectrum inhibition of *Phytophthora infestans* by root endophytes. *FEMS Microbiol. Ecol.* **2018**, *94*, fiy037. [CrossRef]
33. Kusari, P.; Kusari, S.; Spiteller, M.; Kayser, O. Endophytic fungi harbored in *Cannabis sativa* L.: Diversity and potential as biocontrol agents against host plant-specific phytopathogens. *Fungal Divers.* **2012**, *60*, 137–151. [CrossRef]
34. Crous, P.W.; Gams, W.; Stalpers, J.A.; Robert, V.; Stegehuis, G. MycoBank: An online initiative to launch mycology into the 21st century. *Stud. Mycol.* **2004**, *50*, 19–22.
35. Lv, Y.; Zhang, F.; Chen, J.; Cui, J.; Xing, Y.; Li, X.; Guo, S. Diversity and antimicrobial activity of endophytic fungi associated with the alpine plant *Saussurea involucrata*. *Biol. Pharm. Bull.* **2010**, *33*, 1300–1306. [CrossRef] [PubMed]
36. Bokati, D.; Herrera, J.; Poudel, R. Soil influences colonization of root-associated fungal endophyte communities of maize, wheat and their progenitors. *J. Mycol.* **2016**, *2016*, 8062073. [CrossRef]
37. Bogner, C.W.; Kariuki, G.M.; Elashry, A.; Sichtermann, G.; Buch, A.; Mishra, B.; Thines, M.; Grundler, F.M.W.; Schouten, A. Fungal root endophytes of tomato from Kenya and their nematode biocontrol potential. *Mycol. Prog.* **2016**, *15*, 30. [CrossRef]
38. Arnold, A.E.; Lutzoni, F. Diversity and host range of foliar fungal endophytes: Are tropical leaves biodiversity hotspots? *Ecology* **2007**, *88*, 541–549. [CrossRef]
39. Vitorino, L.C.; Silva, F.G.; Soares, M.A.; Souchie, E.L.; Lima, W.C. The isolation and characterization of endophytic microorganisms from *Hyptis marrubioides* Epling roots. *Afr. J. Biotechnol.* **2012**, *11*, 12766–12772.
40. Karst, J.; Piculell, B.; Brigham, C.; Booth, M.; Hoeksema, J.D. Fungal communities in soils along a vegetative ecotone. *Mycologia* **2013**, *105*, 61–70. [CrossRef]
41. Lau, M.K.; Arnold, A.E.; Johnson, N.C. Factors influencing communities of foliar fungal endophytes in riparian woody plants. *Fungal Ecol.* **2013**, *6*, 365–378. [CrossRef]
42. Chowdhary, K.; Kaushik, N. Fungal endophyte diversity and bioactivity in the Indian medicinal plant *Ocimum sanctum*. *PLoS ONE* **2015**, *10*, e0141444. [CrossRef]
43. Nartey, L.; Pu, Q.; Zhu, W.; Zhang, S.; Li, J.; Yao, Y.; Hu, X. Antagonistic and plant growth promotion effects of *Mucor moelleri*, a potential biocontrol agent. *Microbiol. Res.* **2022**, *255*, 126922. [CrossRef]
44. El-Hasan, A.; Walker, F.; Schöne, J.; Buchenauer, H. Antagonistic effect of 6-pentyl-alpha-pyrone produced by *Trichoderma harzianum* toward *Fusarium moniliforme*. *J. Plant Dis. Prot.* **2007**, *114*, 62–68. [CrossRef]
45. El-Hasan, A.; Walker, F.; Schöne, J.; Buchenauer, H. Detection of viridiofungin A and other antifungal metabolites excreted by *Trichoderma harzianum* active against different plant pathogens. *Eur. J. Plant Pathol.* **2009**, *124*, 457–470. [CrossRef]
46. El-Hasan, A.; Walker, F.; Klaiber, I.; Schöne, J.; Pfannstiel, J.; Voegele, R.T. New Approaches to Manage Asian Soybean Rust (*Phakopsora pachyrhizi*) Using *Trichoderma* spp. or Their Antifungal Secondary Metabolites. *Metabolites* **2022**, *12*, 507. [CrossRef]
47. Matroudi, S.; Zamani, M.R.; Motallebi, M. Antagonistic effects of three species of *Trichoderma* sp. on *Sclerotinia sclerotiorum*, the causal agent of canola stem rot. *Egypt. J. Biol.* **2009**, *11*, 37–44.
48. Tripathi, A.; Sharma, N.; Tripathi, N. Biological control of plant diseases: An overview and the *Trichoderma* system as biocontrol agents. In *Management of Fungal Plant Pathogens*; Arya, A., Perelló, A.E., Eds.; CAB International: Oxfordshire, UK, 2010; pp. 110–121.
49. Massimo, N.C.; Devan, M.M.N.; Arendt, K.R.; Wilch, M.H.; Riddle, J.M.; Furr, S.H.; Steen, C.; U'Ren, J.M.; Sandberg, D.C.; Arnold, E. Fungal endophytes in above-ground tissues of desert plants: Infrequent in culture, but highly diverse and distinctive symbionts. *Microb. Ecol.* **2015**, *70*, 61–76. [CrossRef]
50. Vila, L.; Lacadena, V.; Fontanet, P.; del Pozo, M.A.; Segundo, B.S. A protein from the mold *Aspergillus giganteus* is a potent inhibitor of fungal plant pathogens. *Mol. Plant-Microbe Interact.* **2001**, *14*, 1327–1331. [CrossRef]
51. Vasiliauskas, R.; Stenlid, J. Fungi inhabiting stems of *Picea abies* in a managed stand in Lithuania. *For. Ecol. Manag.* **1998**, *109*, 119–126. [CrossRef]
52. Vasiliauskas, R.; Stenlid, J. Population structure and genetic variation in *Cylindrobasidium evolvens*. *Mycol. Res.* **1998**, *102*, 1453–1458. [CrossRef]

53. Warrior, P.; Rehberger, L.A.; Beach, M.; Grau, P.A.; Kirfman, G.W.; Conley, J.M. Commercial development and introduction of DiTeraTM, a new nematicide. *Pest. Manag. Sci.* **1999**, *59*, 376–379. [CrossRef]
54. Hoagland, R.E.; Boyette, D.C.; Abbas, H.K. *Myrothecium verrucaria* isolates and formulations as bioherbicide agents for kudzu. *Biocontrol Sci. Technol.* **2007**, *17*, 721–731. [CrossRef]
55. Lombard, L.; Houbraken, J.; Decock, C.; Samson, R.A.; Meijer, M.; Réblová, M.; Groenewald, J.Z.; Crous, P.W. Generic hyper-diversity in Stachybotriaceae. *Persoonia* **2016**, *36*, 156–246. [CrossRef] [PubMed]
56. Kurakov, A.V.; Lavrent, R.B.; Nechitailo, T.Y.; Golyshin, P.N.; Zvyagintsev, D.G. Diversity of facultatively anaerobic microscopic mycelial fungi in soils. *Microbiology* **2008**, *77*, 90–98. [CrossRef]
57. Zou, X.; Niu, S.; Ren, J.; Li, E.; Liu, X.; Che, Y. Verrucamides A–D, antibacterial cyclopeptides from *Myrothecium verrucaria*. *J. Nat. Prod.* **2011**, *74*, 1111–1116. [CrossRef] [PubMed]
58. Walker, H.L.; Tilley, A.M. Evaluation of an isolate of *Myrothecium verrucaria* from sicklepod (*Senna obtusifolia*) as a potential mycoherbicide agent. *Biol. Control* **1997**, *10*, 104–112. [CrossRef]
59. Li, Z.; Chang, P.; Gao, L.; Wang, X. The endophytic fungus *Albifimbria verrucaria* from wild grape as an antagonist of *Botrytis cinerea* and other grape pathogens. *Phytopathology* **2019**, *110*, 843–850. [CrossRef]
60. Banerjee, D.; Strobel, G.A.; Booth, E.; Geary, B.; Sears, J.; Spakowicz, D.; Busse, S. An endophytic *Myrothecium inundatum* producing volatile organic compounds. *Mycosphere* **2010**, *1*, 229–240.
61. Boddy, L. Interactions between fungi and other microbes. In *The Fungi*, 3rd ed.; Watkinson, S.C., Boddy, L., Money, N.P., Eds.; Academic Press: London, UK, 2016; pp. 337–360.
62. Linkies, A.; Jacob, S.; Zink, P.; Maschemer, M.; Maier, W.; Koch, E. Characterization of cultural traits and fungicidal activity of strains belonging to the fungal genus *Chaetomium*. *J. Appl. Microbiol.* **2021**, *131*, 375–391. [CrossRef]
63. Bae, S.J.; Mohanta, T.K.; Chung, J.Y.; Ryu, M.; Park, G.; Shim, S.; Hong, S.-B.; Seo, H.; Bae, D.-W.; Bae, I.; et al. *Trichoderma* metabolites as biological control agents against *Phytophthora pathogens*. *Biol. Control* **2016**, *92*, 128–138. [CrossRef]
64. Chandrakala, A.; Chadrashekar, S.C.; Jyothi, G.; Ravikumar, B.M. Effect of cell-free culture filtrates of bio-control agents on the spore germination and infection by *Phytophthora infestans* causing late blight of potato. *Glob. J. Biol. Agric. Health Sci.* **2012**, *1*, 40–45.
65. Tellenbach, C.; Sumarah, M.W.; Grünig, C.R.; Miller, J.D. Inhibition of *Phytophthora* species by secondary metabolites produced by the dark septate endophyte *Phialocephala europaea*. *Fungal Ecol.* **2013**, *6*, 12–18. [CrossRef]
66. Bae, H.; Roberts, D.P.; Lim, H.-S.; Strem, M.D.; Park, S.-C.; Ryu, C.-M.; Melnick, R.L.; Bailey, B.A. Endophytic *Trichoderma* isolates from tropical environments delay disease onset and induce resistance against *Phytophthora capsici* in hot pepper using multiple mechanisms. *Mol. Plant-Microbe Interact.* **2011**, *24*, 336–351. [CrossRef] [PubMed]
67. Schulte, E.E.; Hoskins, B. Recommended soil organic matter tests. In *Recommended Soil Testing Procedures for the Northeastern United States, The Northeast Coordinating Committee for Soil Testing (NEC 1812), Northeastern Regional Publication Cooperative Bulletin No. 493*, 3rd ed.; Agricultural Experiment Station, Univ. of Delaware: Newark, NY, USA, 2011; pp. 63–74.
68. Liu, D.; Coloe, S.; Baird, R.; Pedersen, J. Rapid mini-preparation of fungal DNA for PCR. *J. Clin. Microbiol.* **2000**, *38*, 471. [CrossRef] [PubMed]
69. White, T.J.; Bruns, T.; Lee, S.; Taylor, J. PCR protocols: A guide to methods and applications. In *Amplification and Direct Sequencing of Fungal Ribosomal DNA Genes for Phylogenetics*; Innis, M.A., Gelfand, D.H., Sninsky, J.J., White, T.J., Eds.; Academic Press Inc.: San Diego, CA, USA, 1990; pp. 315–322.
70. Saitou, N.; Nei, M. The neighbor-joining method: A new method for reconstructing phylogenetic trees. *Mol. Biol. Evol.* **1987**, *4*, 406–425.
71. Felsenstein, J. Confidence limits on phylogenies: An approach using the bootstrap. *Evolution* **1985**, *39*, 783–791. [CrossRef]
72. Tamura, K.; Nei, M.; Kumar, S. Prospects for inferring very large phylogenies by using the neighbor-joining method. *Proc. Natl. Acad. Sci. USA* **2004**, *101*, 11030–11035. [CrossRef]
73. Edgington, L.V.; Khew, K.L.; Barron, G.L. Fungitoxic spectrum of benzimidazole compounds. *Phytopathology* **1971**, *61*, 42–44. [CrossRef]
74. El-Hasan, A.; Schöne, J.; Höglinger, B.; Walker, F.; Voegele, R.T. Assessment of the antifungal activity of selected biocontrol agents and their secondary metabolites against *Fusarium graminearum*. *Eur. J. Plant Pathol.* **2018**, *150*, 91–103. [CrossRef]

MDPI

Article

Efficacy of Biorational Products for Managing Diseases of Tomato in Greenhouse Production

Luis Fernando Esquivel-Cervantes [1], Bertha Tlapal-Bolaños [1,*], Juan Manuel Tovar-Pedraza [2,*], Oscar Pérez-Hernández [3], Santos Gerardo Leyva-Mir [1] and Moisés Camacho-Tapia [4]

1 Departamento de Parasitología Agrícola, Universidad Autónoma Chapingo, Texcoco 56230, Mexico; luis-esqcer1@hotmail.com (L.F.E.-C.); lsantos@correo.chapingo.mx (S.G.L.-M.)
2 Laboratorio de Fitopatología, Coordinación Regional Culiacán, Centro de Investigación en Alimentación y Desarrollo, Culiacán 80110, Mexico
3 Department of Plant Sciences and Plant Pathology, Montana State University, Bozeman, MT 59717, USA; oscar.perezhernandez@montana.edu
4 Laboratorio Nacional de Investigación y Servicio Agroalimentario y Forestal, Universidad Autónoma Chapingo, Texcoco 56230, Mexico; moises.camachotapia@gmail.com
* Correspondence: btlapalb@chapingo.mx (B.T.-B.); juan.tovar@ciad.mx (J.M.T.-P.)

Abstract: Gray mold (*Botrytis cinerea*), late blight (*Phytophthora infestans*), powdery mildew (*Leveillula taurica*), pith necrosis (*Pseudomonas corrugata*), and bacterial canker (*Clavibacter michiganensis*) are major diseases that affect tomato (*Solanum lycopersicum* L.) in greenhouse production in Mexico. Management of these diseases depends heavily on chemical control, with up to 24 fungicide applications required in a single season to control fungal diseases, thus ensuring a harvestable crop. While disease chemical control is a mainstay practice in the region, its frequent use increases the production costs, likelihood of pathogen-resistance development, and negative environmental impact. Due to this, there is a need for alternative practices that minimize such effects and increase profits for tomato growers. The aim of this study is to evaluate the effect of biorational products in the control of these diseases in greenhouse production. Four different treatments, including soil application of *Bacillus* spp. or *B. subtilis* and foliar application of *Reynoutria sachalinensis*, *Melaleuca alternifolia*, harpin αβ proteins, or bee honey were evaluated and compared to a conventional foliar management program (control) in a commercial production greenhouse in Central Mexico in 2016 and 2017. Disease incidence was measured at periodic intervals for six months and used to calculate the area under the disease progress curve (AUDPC). Overall, the analysis of the AUDPC showed that all treatments were more effective than the conventional program in controlling most of the examined diseases. The tested products were effective in reducing the intensity of powdery mildew and gray mold, but not that of bacterial canker, late blight, and pith necrosis. Application of these products constitutes a disease management alternative that represents cost-saving to tomato growers of about 2500 U.S. dollars per production cycle ha^{-1}, in addition to having less negative impact on the environment. The products tested in this study have the potential to be incorporated in an integrated program for management of the examined diseases in tomato in this region.

Keywords: *Solanum lycopersicum*; integrated disease management; plant extracts; defense inducers; incidence; AUDPC

Citation: Esquivel-Cervantes, L.F.; Tlapal-Bolaños, B.; Tovar-Pedraza, J.M.; Pérez-Hernández, O.; Leyva-Mir, S.G.; Camacho-Tapia, M. Efficacy of Biorational Products for Managing Diseases of Tomato in Greenhouse Production. *Plants* **2022**, *11*, 1638. https://doi.org/10.3390/plants11131638

Academic Editors: Špela Mechora and Dragana Šunjka

Received: 21 May 2022
Accepted: 18 June 2022
Published: 21 June 2022

1. Introduction

Tomato (*Solanum lycopersicum* L.) is one of the most widely cultivated plants in the world and is appreciated for its high nutritional value [1]. Mexico is the ninth largest producer, with 3.4 million tons annually, and the largest exporter of tomato worldwide [2]. Out of the total production of tomato in Mexico, 40% is conducted in greenhouses. A major constraint in greenhouse tomato production is diseases, several of which occur each season. Among the most frequent diseases are those caused by fungi, oomycetes, and bacteria; damage is reported in pre- and postemergence, transplanting, and crop development [3,4]. The most concerning diseases in this region are caused by oomycetes such as *Pythium aphanidermatum* and *Phytophthora capsici*, and by fungi such as *Rhizoctonia solani*, *Fusarium oxysporum*, *F. solani*, *Verticillium dahliae*, *Macrophomina phaseolina*, *Botrytis cinerea*, *Sclerotinia sclerotiorum*, and *Sclerotium rolfsii*. During crop development, it is common to find *Phytophthora infestans*, *Botrytis cinerea*, *Alternaria solani*, *Sclerotinia sclerotiorum*, *Phoma lycopersici*, *Septoria lycopersici*, and *Leveillula taurica*. Bacterial diseases caused by the species *Clavibacter michiganensis*, *Pseudomonas corrugata*, *Pseudomonas syringae* pv. *tomato*, *Xanthomonas vesicatoria*, and *Pectobacterium carotovorum* [5] also affect the crop in the growing season. All of these diseases cause damage to roots and above-ground parts, including fruit, and inflict high economic losses [5], mainly due to their cost of control. Virtually, the only way to control these diseases is with the use of chemical products. Many growers in this region carry out up to 24 applications in a growing season to ensure a harvestable crop. The repeated fungicide applications elevate production costs, leaving growers with small margins that render the operations unsustainable. In addition, the numerous applications have a negative impact on the environment. This latter aspect has become more concerning due to the public pressure and demand for healthy tomato fruit. In view of that, there is an urgent need for alternative sustainable strategies to manage these diseases. A relatively new approach to disease control in the region is the use of biorational products. Biorational products are low-environmental-impact products for use in agriculture [6–8]. They include biopesticides (fungi or bacteria), botanicals (plant extracts, oils), minerals, as well as products for crop stress management [6,7]. Several products are of commercial use at present, for example, a large number of bacterial strains have been isolated and identified as biocontrol agents against tomato diseases [8]. Among them, the genus *Bacillus* is widely distributed and recognized as an effective biocontrol agent of crop diseases [9]. *Bacillus*-based biological control agents (BCAs) have great potential in integrated pest management (IPM) systems; however, most research has focused on BCAs as alternatives to synthetic chemical fungicides or bactericides and not as part of an integrated disease management system [10]. The main mechanisms of action of *Bacillus* spp. are the excretion of antibiotics, toxins, siderophores, and lytic enzymes as well as the induction of systemic resistance [9]. Currently, various commercial formulations have strains of the genus *Bacillus* as an active ingredient, due to their colonization capacity, easy reproduction, and high persistence associated with the formation of endospores, the last being a characteristic of special interest as it allows them to survive under abiotic stress conditions, facilitating their production and storage for long periods of time [11]. The efficacy and performance of these products in commercial tomato production are unknown.

The aim of this study is to evaluate the efficacy of biorational products in controlling diseases in greenhouse tomato production. The ultimate goal is to be able to incorporate these tactics into an integrated disease management system in commercial tomato production.

2. Results

In 2016, the lowest incidence values of diseases caused by *B. cinerea* and *L. taurica* were observed with soil-applied *Bacillus* spp. (Figure 1), while the lowest incidence values for late blight and pith necrosis occurred in the treatments with *B. subtilis* (Figure 2). However, no difference was observed in the incidence of bacterial canker between treatments based on *Bacillus* spp. and *B. subtilis*. In 2017, the lowest incidence of the disease caused by *B. cinerea* was observed in tomato plants treated with *B. subtilis*; however, for the other diseases, the lowest incidence values occurred in plants treated with soil-applied *Bacillus* spp. (Figures 3 and 4).

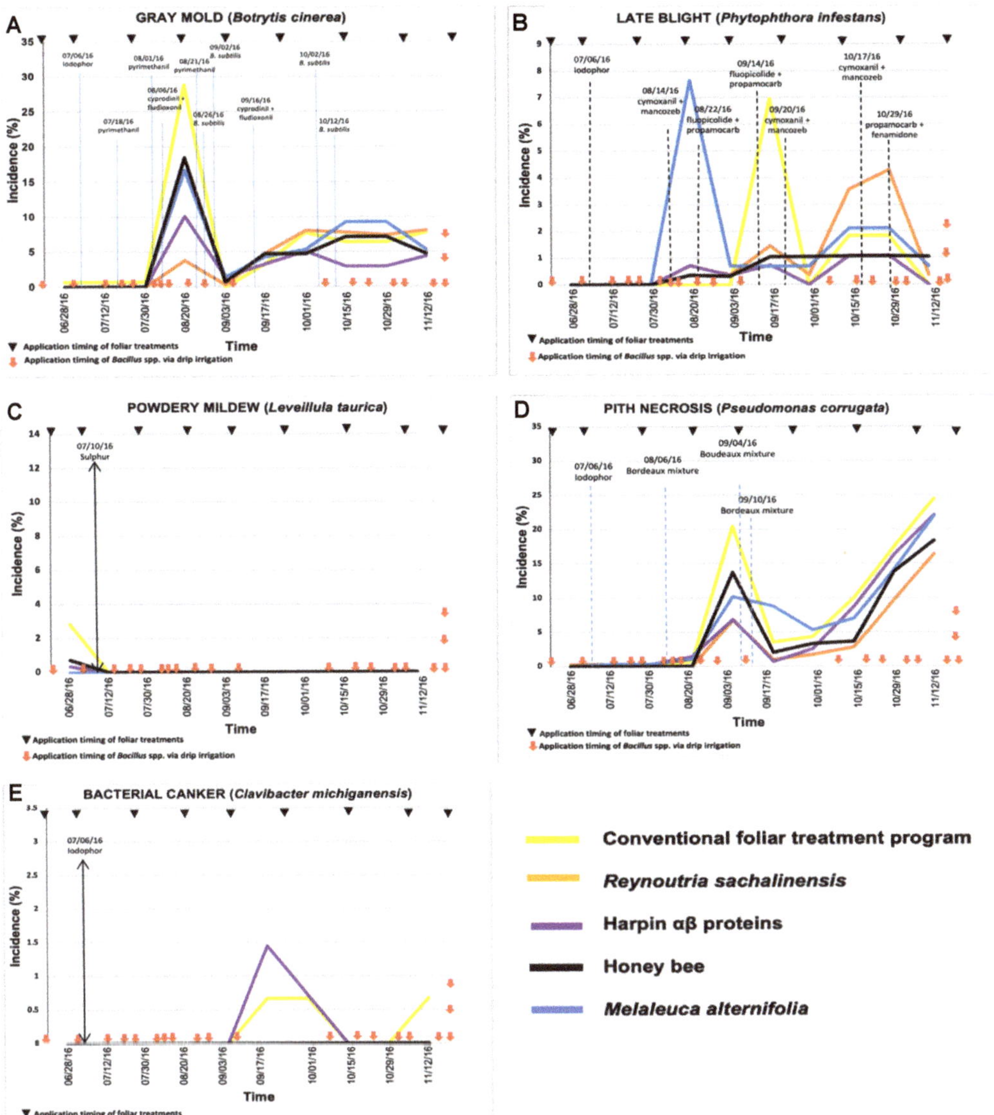

Figure 1. Temporal progress of five major tomato diseases—(**A**) gray mold (*Botrytis cinerea*), (**B**) late blight (*Phytophthora infestans*), (**C**) powdery mildew (*Leveillula taurica*), (**D**) pith necrosis (*Pseudomonas corrugata*), and (**E**) bacterial canker (*Clavibacter michiganensis*)—under the effect of different treatments consisting of soil application of *Bacillus* spp. via drip irrigation and foliar application of biorational products via spraying in greenhouse tomato production in Mexico in 2016. Dotted vertical lines in graphs indicate the approximate timing of application of fungicides in the conventional management program or in any other treatment due to low intensity tolerance to disease. Incidence values are the average of three replications.

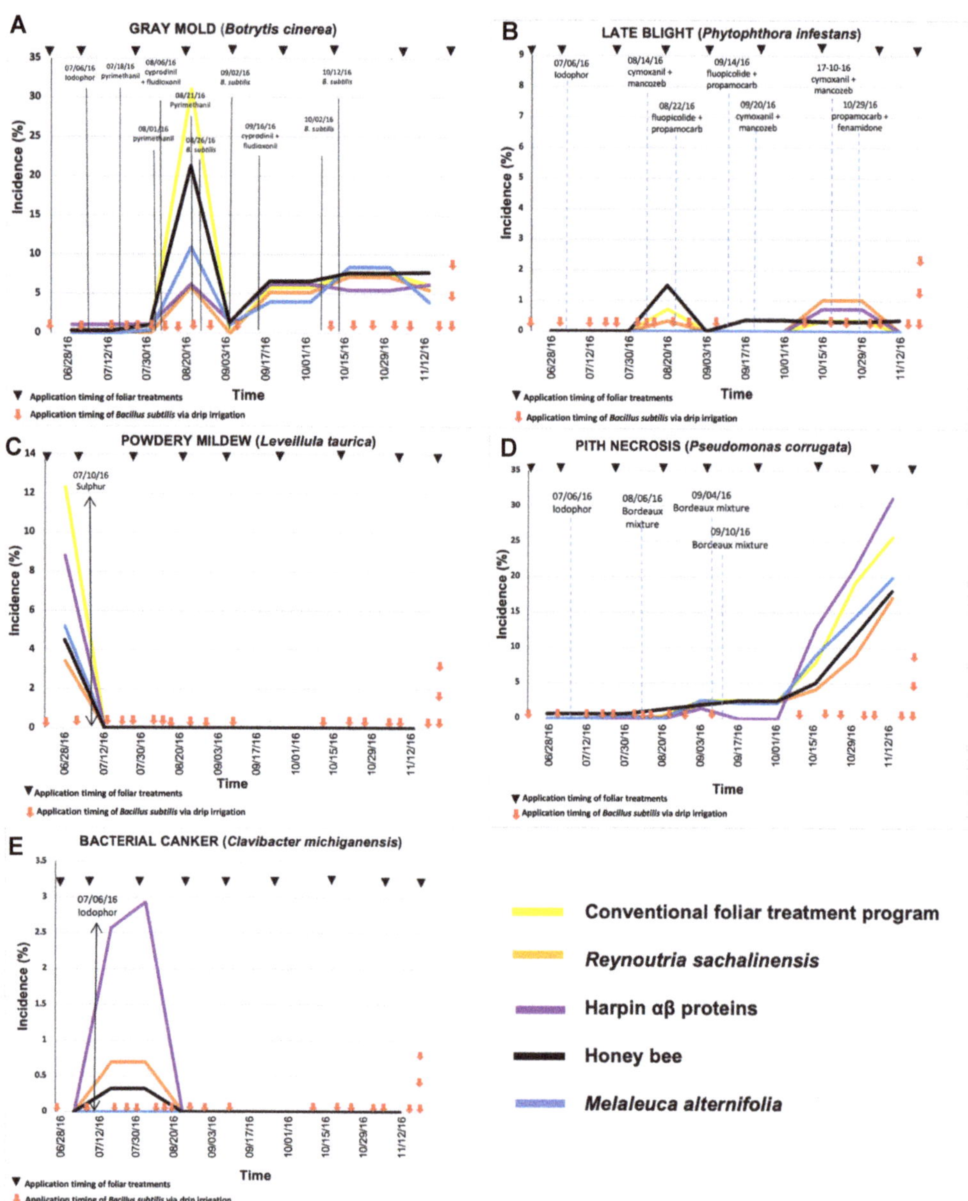

Figure 2. Temporal progress of major tomato diseases—(**A**) gray mold (*Botrytis cinerea*), (**B**) late blight (*Phytophthora infestans*), (**C**) powdery mildew (*Leveillula taurica*), (**D**) pith necrosis (*Pseudomonas corrugata*), and (**E**) bacterial canker (*Clavibacter michiganensis*)—in five different treatment combinations consisting of soil application of *Bacillus subtilis* via drip irrigation and foliar application of biorational products in tomato greenhouse production in Experiment 2 in 2016. Dotted vertical lines in graphs indicate the approximate timing of application of fungicides in the conventional management program or in any other treatment due to low-intensity tolerance to disease. Incidence values are the average of three replications.

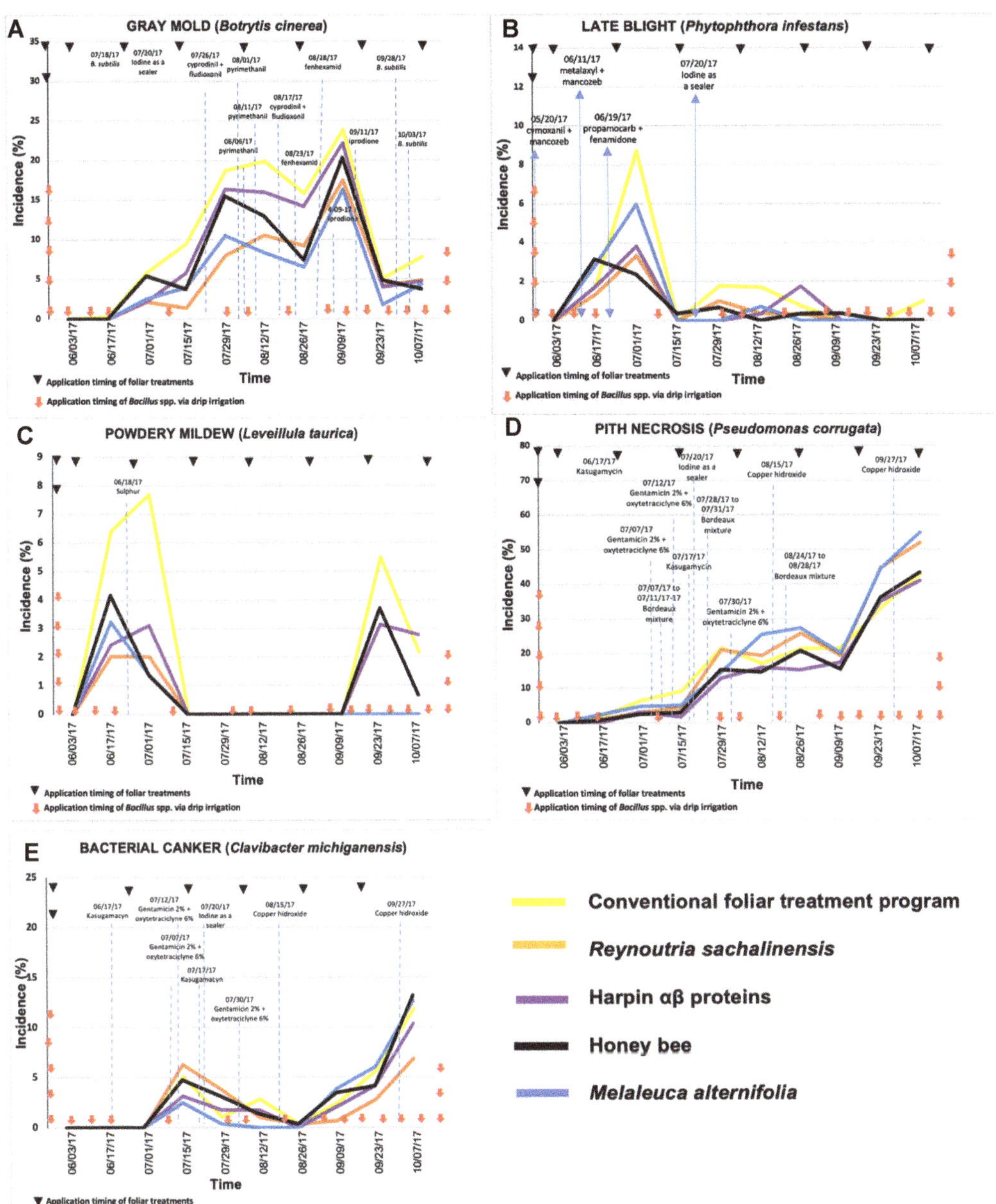

Figure 3. Temporal progress of major tomato diseases—(**A**) gray mold (*Botrytis cinerea*), (**B**) late blight (*Phytophthora infestans*), (**C**) powdery mildew (*Leveillula taurica*), (**D**) pith necrosis (*Pseudomonas corrugata*), and (**E**) bacterial canker (*Clavibacter michiganensis*)—in five different treatment combinations consisting of soil application of *Bacillus* spp. Via drip irrigation and foliar application of biorational products in tomato greenhouse production in Experiment 1 in 2017. Dotted vertical lines in graphs indicate the approximate timing of application of fungicides in the conventional management program or in any other treatment due to low-intensity tolerance to disease. Incidence values are the average of three replications.

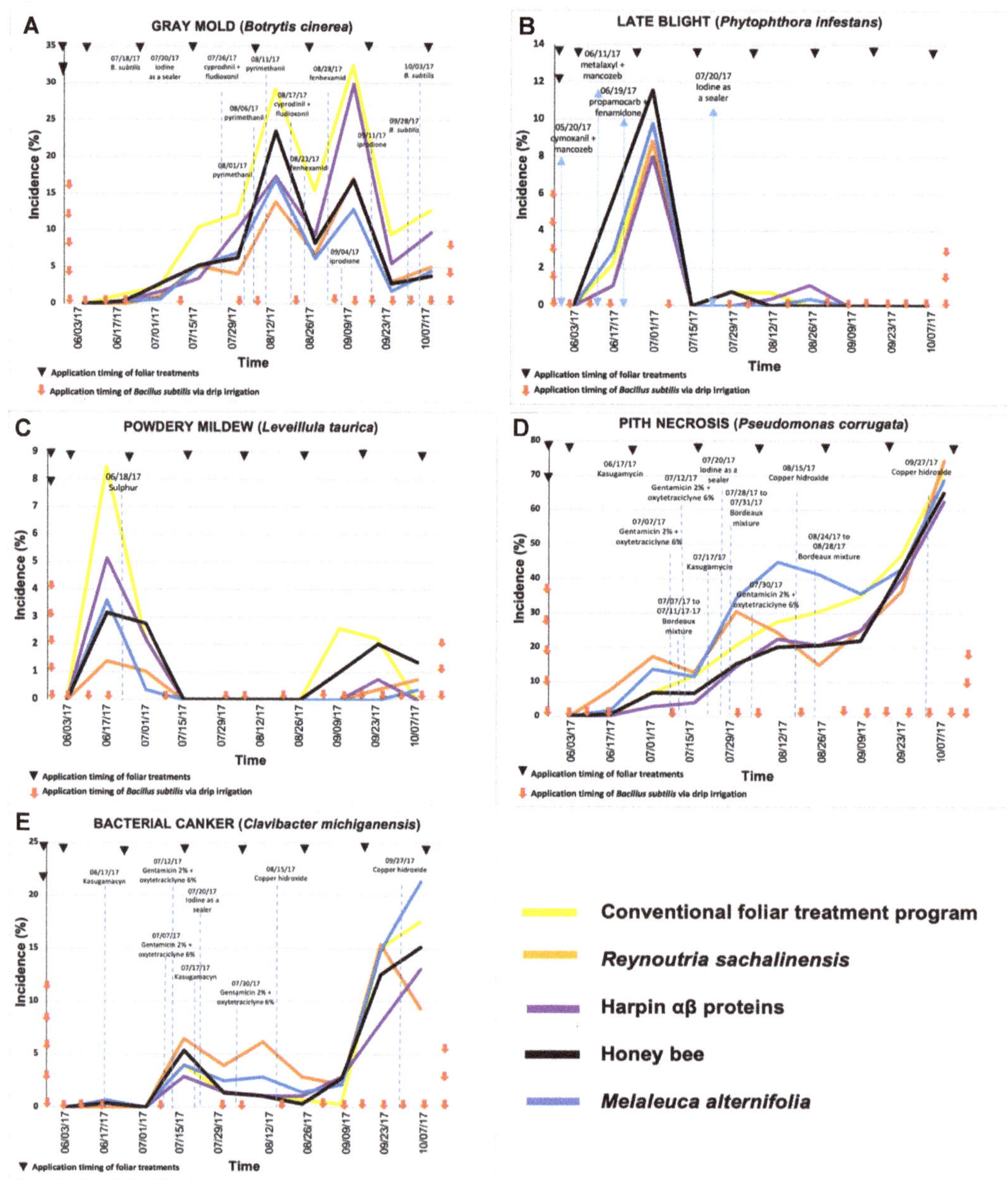

Figure 4. Temporal progress of major tomato diseases—(**A**) gray mold (*Botrytis cinerea*), (**B**) late blight (*Phytophthora infestans*), (**C**) powdery mildew (*Leveillula taurica*), (**D**) pith necrosis (*Pseudomonas corrugata*), and (**E**) bacterial canker (*Clavibacter michiganensis*)—in five different treatment combinations consisting of soil application of *Bacillus subtilis* via drip irrigation and foliar application of biorational products in tomato greenhouse production in Experiment 2 in 2017. Dotted vertical lines in graphs indicate the approximate timing of application of fungicides in the conventional management program or in any other treatment due to low-intensity tolerance to disease. Incidence values are the average of three replications.

2.1. Treatment Effect in 2016: Experiment 1

Gray mold. The tomato plants that received the biorational foliar treatments resulted in numerically low AUDCP compared to the conventional treatment program. Treatment with harpin proteins and *R. sachalinensis* exhibited the highest level of disease control, as indicated by the low AUDPC (Table 1). However, the difference was not significant (Table 1). All treatments indicated a lower incidence of disease than the base-conventional treatment. In terms of incidence, the end-of-season incidence of gray mold showed a significant difference among treatments on August ($P = 0.0025$). The treatments with the lowest incidence of gray mold in this year were those including *R. sachalinensis* and harpin αβ proteins with 3.77 and 10.11%, respectively, while the highest incidence occurred in the nontreated plants with 20.82%. In the rest of the growing season, the difference was not significant ($P > 0.05$) and only a numerical difference was observed, with the best treatment being that of harpin proteins, with incidence below 5% (Figure 1A). Peaks of incidence were more evident through midseason with disease going up and down (Figure 1).

Table 1. Effect of soil-application of *Bacillus* spp. and foliar application of different biorational products in the control of major tomato diseases in greenhouse in Mexico in 2016 and 2017.

Treatment	AUDPC [a] Gray Mold		AUDPC Late Blight		AUDPC Powdery Mildew		AUDPC Pith Necrosis		AUDPC Bacterial Canker	
	2016	2017	2016	2017	2016	2017	2016	2017	2016	2017
T1 [c]	59.24 [ab]	119.67 [a]	10.62 [a]	15.16 [a]	1.43 [a]	20.65 [a]	68.90 [a]	153.03 [a]	3.03 [ab]	23.26 [a]
T2 [d]	36.04 [bc]	53.65 [c]	10.94 [a]	6.35 [b]	0.18 [a]	4.05 [b]	30.65 [b]	164.30 [a]	0.0 [b]	18.31 [a]
T3 [e]	27.80 [c]	82.93 [b]	3.98 [a]	7.64 [ab]	0.17 [a]	10.09 [ab]	48.43 [ab]	121.60 [a]	6.52 [a]	18.07 [a]
T4 [f]	48.80 [ab]	53.41 [c]	14.31 [a]	9.51 [ab]	0.0 [a]	4.60 [b]	58.39 [ab]	171.64 [a]	0.0 [b]	19.07 [a]
T5 [g]	45.03 [abc]	67.8 [bc]	5.36 [a]	7.25 [ab]	0.35 [a]	9.57 [ab]	45.52 [ab]	129.71 [a]	0.0 [b]	23.74 [a]

[a] AUDPC—Area under the disease progress curve. [b] Mean values followed by different letters in the column indicate significant differences among treatments according to a Duncan test ($P < 0.05$). Value in each cell is from three replications. [c] T1—Soil application of *Bacillus* spp. + conventional foliar applications. [d] T2—Soil application of *Bacillus* spp. and foliar application of extracts of *R. sachalinensis*. [e] T3—Soil application of *Bacillus* spp. via drip irrigation + foliar application of αβ harpins. [f] T4—Soil application of *Bacillus* spp. via drip irrigation + foliar application of *Melalueca alternifolia*. [g] T5—Soil application of *Bacillus* spp. via drip irrigation + foliar application of bee honey.

Late blight. Incidence of late blight in the experiment with *Bacillus* spp. (Figure 1B) did not show a significant statistical difference on the dates between treatments, but it did show a numerical difference, where the highest incidence was in August for the treatment with *M. alternifolia*, in September for the control, and from October onwards in the treatment with *R. sachalinensis*, while the lowest incidence during these months occurred in the treatments with harpin αβ proteins and bee honey. The AUDPC values in the experiment with *Bacillus* spp. did not exhibit a statistical difference, but a numerical difference (Table 1), where the lowest was in the treatments with harpin proteins and bee honey with 3.98 and 5.36, respectively. The highest value was not recorded in the nontreated control (10.62), but in the treatment with *M. alternifolia* (14.31).

Powdery mildew. As the season progressed, incidence of powdery mildew in the experiment with *Bacillus* spp. (Figure 1C) showed no statistical difference and only presented incidence at the beginning of the growing season, which was in June in the nontreated control with 2.8% against *M. alternifolia* with 0%. The AUDPC values for powdery mildew (Table 1) showed no significant difference either, where the highest value being 1.43 for the nontreated plants and the lowest (0.0) for the treatment with *M. alternifolia*.

Pith necrosis. Incidence of pith necrosis (Figure 1D) in the experiment with *Bacillus* spp. showed a statistical difference between treatments on 17 September ($P = 0.013$) and 1 October ($P = 0.034$). The treatments with the lowest incidence on 17 September were bee honey (0.72%), harpin proteins (0.99%), *R. sachalinensis* (2.01%), and the nontreated control (3.36%), while the most affected was *M. alternifolia* (8.83%). In October, the best was *R. sachalinensis* with 1.7% and the most damaged was *M. alternifolia*. The AUDPC values for

Bacillus spp. did not show statistical differences, only numerical (Table 1), where the best treatment was *R. sachalinensis* (30.65) followed by bee honey (45.52).

Bacterial canker. The temporal progress of bacterial canker in the experiment with *Bacillus* spp. (Figure 1E) had an incidence of less than 1.5% in all treatments and there were only outbreaks of this disease in the treatment based on harpin proteins; however, they were so low that they did not represent a statistical difference with the other treatments. The AUDPC values for bacterial canker in the *Bacillus* spp. experiment showed no differences (Table 1).

2.2. Treatment Effect in 2016: Experiment 2

The incidence of gray mold in the experiment with *B. subtilis* (Figure 2A) presented a statistical difference and corresponded to the same date ($P = 0.0020$) as in the experiment with *Bacillus* spp., where the highest incidence was reported in the nontreated plants with 31.08% and the lowest in the treatments with *R. sachalinensis* (5.8%) and harpin proteins (6.24%). The most favorable conditions for *B. cinerea* occurred in August in the two trials; so, conventional fungicides were applied preventively and curatively when the incidence exceeded 5%. In the AUDPC values for gray mold, there was no statistical difference, as the treatment with *R. sachalinensis* reached 33.81, which was the lowest value, followed by the treatments with harpin proteins and *M. alternifolia*, with 37.08 and 39.17, respectively (Table 2).

Table 2. Effect of combined use of *Bacillus subtilis* applied to the drip irrigation and foliar sprays of biorational products for controlling tomato diseases under greenhouse conditions in 2016 and 2017.

Treatment	AUDPC [a] Gray Mold		AUDPC Late Blight		AUDPC Powdery Mildew		AUDPC Pith Necrosis		AUDPC Bacterial Canker	
	2016	2017	2016	2017	2016	2017	2016	2017	2016	2017
T1 [c]	61.20 [ab]	103.03 [a]	1.45 [a]	12.5 [a]	6.16 [a]	15.49 [a]	48.05 [a]	215.93 [a]	0.00 [a]	31.55 [a]
T2 [d]	33.81 [b]	55.34 [bc]	2.41 [a]	9.89 [a]	1.71 [b]	3.16 [b]	28.92 [a]	207.08 [a]	3.12 [a]	41.85 [a]
T3 [e]	37.08 [ab]	83.11 [ab]	1.47 [a]	10.54 [a]	4.40 [ab]	8.06 [ab]	51.21 [a]	161.81 [a]	10.6 [a]	24.39 [a]
T4 [f]	39.17 [ab]	52.17 [c]	0.00 [a]	13.04 [a]	2.59 [b]	4.16 [b]	40.74 [a]	261.23 [a]	0.00 [a]	39.31 [a]
T5 [g]	56.53 [ab]	72.15 [bc]	3.08 [a]	18.01 [a]	2.25 [b]	9.60 [ab]	35.24 [a]	167.89 [a]	2.75 [a]	31.38 [a]

[a] AUDPC—Area under the disease progress curve. [b] Mean values followed by different letters in the column indicate significant differences among treatments according to Duncan test ($P < 0.05$). [c] T1—Soil application of *Bacillus subtilis* + conventional foliar applications. [d] T2—Soil application of *Bacillus subtilis* and foliar application of extracts of *R. sachalinensis*. [e] T3—Soil application of *Bacillus subtilis* via drip irrigation + foliar application of αβ harpins. [f] T4—Soil application of *Bacillus subtilis* via drip irrigation + foliar application of *Melalueca alternifolia*. [g] T5—Soil application of *Bacillus subtilis* via drip irrigation + foliar application of bee honey.

As the season progressed, disease incidence of late blight caused by *P. infestans* in the experiment with *B. subtilis* (Figure 2B) only exhibited a numerical difference, where the highest incidence was in August with bee honey with 1.49% and the best treatments were *M. alternifolia* and harpin proteins with 0%; in October, the highest incidence was observed in the treatment with *R. sachalinensis*, while the lowest incidence in the whole growing season was in *M. alternifolia* with 0%. For the experiment with *B. subtilis*, there was also no difference among treatments with respect to the AUDPC values for late blight since this presented very low levels (0 to 3.5) in all treatments (Table 2).

The incidence of powdery mildew (Figure 2C) showed a difference on June 28 (0.04%), between the control (12.31%), *R. sachalinensis* (3.43%), bee honey (4.51%), and *M. alternifolia* (5.19%). In addition, for the AUDPC data (Table 2), there was a statistical difference (0.046), and the best treatments for powdery mildew were *R. sachalinensis*, bee honey, and *M. alternifolia* with values of 1.71, 2.25, and 2.59, respectively

The incidence of pith necrosis did not differ among the treatments evaluated and the disease remained below 5% (Figure 2D). However, it was in October that the incidence increased, being the lowest in the treatment with *R. sachalinensis* (17.15%), whereas the highest levels were found in the plants treated with harpin proteins (31.21%). The development of the disease caused by *P. corrugata* according to the temporal dynamics was favored from October, where the incidence increased without decreasing again; this damage is related to the age of the plant in the last months of production. In the case of the AUDPC values (Table 2), there was no statistical difference, but there was a numerical difference, and it was observed that all treatments had a similar effect; however, the lowest AUDPC values were observed in the treatments with *R. sachalinensis* (28.92) and bee honey (35.24), while the highest AUDPC value was in the treatment with harpin proteins (51.21).

The bacterial canker in the experiment with *B. subtilis* (Figure 2E) had an incidence of less than 3% in all treatments and there were only outbreaks of this disease in the treatment with harpin proteins; however, they were so low that they did not represent a statistical difference with the other treatments. The AUDPC data for bacterial canker in the *B. subtilis* experiment (Table 2) also reported no differences among treatments.

2.3. Treatment Effect in 2017: Experiment 1

Disease incidence of gray mold in the experiment with *Bacillus* spp. (Figure 3A) showed a significant statistical difference on 12 August (0.022), 26 August (0.019), 9 September (0.042), and 23 September (0.009). The treatments with the lowest incidence on August 12 were *R. sachalinensis* (13.90%), *M. alternifolia* (17.02%), and harpin proteins (17.38%). Noteworthily, for the dates of 26 August as well as September 9 and 23, the best treatment was with *M. alternifolia*. On the other hand, the AUDPC values for gray mold indicated a statistical difference (0.0074), where it was observed that the best foliar treatments were the plant extracts based on *M. alternifolia* and *R. sachalinensis* (Table 1).

The incidence of late blight in the experiment with *Bacillus* spp. (Figure 3B) showed no significant statistical difference in dates among treatments but it did show a numerical difference. The highest values were in July in the control with 8.71% and the lowest in bee honey (2.37%), whereas in the rest of the growing season the values were below 2% in all treatments; so, the numerical difference was minimal. The AUDPC data for late blight (Table 1) did not indicate statistical differences and only had numerical differences, where the best treatment was *R. sachalinensis*.

The incidence of powdery mildew in the experiment with *Bacillus* spp. (Figure 3C) showed a statistical difference on 1 July ($P = 0.0072$), among treatments, where the best were *M. alternifolia*, bee honey, *R. sachalinensis*, and harpin $\alpha\beta$ proteins. In the rest of 2017, the levels decreased to 0%, and it was in September when an increase was again observed in the control (5.47%) against 0% in the treatment with *R. sachalinensis*. The AUDPC values for powdery mildew reported not statistical but numerical differences (Table 1), where the best treatments were *R. sachalinensis* and *M. alternifolia*.

The disease incidence of pith necrosis in the experiment with *Bacillus* spp. (Figure 3D) showed a statistical difference on 29 July ($P = 0.0054$) and 7 October ($P = 0.0212$). The best treatment for the control of pith necrosis was with the treatment based on harpin proteins, since it recorded the lowest values throughout the growing season. The AUDPC data for pith necrosis did not present statistical differences and all treatments had the same effect (Table 1), but the lowest AUDPC was for the treatment with harpin proteins.

The incidence of bacterial canker (Figure 3E) showed no statistical difference between dates in the treatments, but it did show a numerical difference. Throughout the growing season, there was an increase in mid-July in the treatment with *R. sachalinensis* and the lowest incidence was in the treatment with *M. alternifolia*. The AUDPC data for bacterial canker also showed no differences among treatments (Table 1).

2.4. Treatment Effect in 2017: Experiment 2

In 2017, the experiment with *B. subtilis* for gray mold control (Figure 4A) showed a statistical difference on 29 July ($P = 0.046$) and 26 August ($P = 0.024$), among treatments. The treatment with the lowest incidence was that of 29 July with *R. sachalinensis* (7.97%); however, for 26 August, the best was with *M. alternifolia* (6.52%). The most favorable conditions for the disease caused by *B. cinerea* were from July to September according to the temporal dynamics. For the AUDPC values of gray mold (Table 2), statistical differences were also shown (P = 0.035) and the best treatments were *M. alternifolia* and *R. sachalinensis*.

In the experiment with *B. subtilis* for late blight control, no statistical difference was found among treatments, only numerical, and the highest incidence occurred in July in the treatment with bee honey (11.53%) and the lowest in the treatment with harpin αβ proteins (8.0%) compared with the untreated control (8.80%). In the remainder of 2017, the values were <1%, so there was practically no presence of this disease. Favorable conditions for the development of *P. infestans* were observed in June and July 2017, similar to the conditions that occurred in 2016 (Figure 4B). For the experiment with *B. subtilis*, there was also no significant statistical difference in the AUDPC data (Table 2), since practically all treatments had very similar levels; even so, among these levels, the least affected corresponded to the treatment with *R. sachalinensis*.

In the experiment with *B. subtilis*, there was no statistical difference in the foliar treatments for powdery mildew management (Figure 4C), only numerical, and it was the treatment with *R. sachalinensis* that presented the lowest incidence of the disease. Favorable conditions for the development of *L. taurica* were observed in June 2017 and at the end of September, where environmental conditions were favorable and marked the moment to make the decision to apply the necessary prevention measures. In the experiment with *B. subtilis*, there was a statistical difference (P = 0.048) in the AUDPC values (Table 2) and the best treatments were *R. sachalinensis* and *M. alternifolia*.

The temporal progress of pith necrosis in the experiment with *B. subtilis* presented a statistical difference on 15 July (P = 0.0139), where the best treatment was with harpin proteins, while in the rest of the growing season, there was no difference among treatments, but the lowest incidence was observed in plants treated with harpin proteins (Figure 4D). The AUDPC data for pith necrosis (Table 2) did not present statistical differences and all treatments had the same effect, but the lowest AUDPC value was for the treatment with harpin αβ proteins.

In the experiment with *B. subtilis*, there was a statistical difference in the incidence values for bacterial canker on 23 September (P = 0.0257). According to the temporal progress, the favorable conditions for the development of *C. michiganensis* were observed from July to September (Figure 4E), while the AUDPC values did not show significant differences among treatments (Table 2).

3. Discussion

Tomato diseases with the greatest impact in the 2016 and 2017 growing seasons were gray mold, late blight, powdery mildew, pith necrosis, and bacterial canker. The results from the present study showed that *Bacillus* spp. (PHC Colonize®) and *B. subtilis* (Fungifree AB®) applied as a soil treatment prevented the development of fungal diseases in the root of tomato plants, which are common in the area where the study was conducted. Other studies have reported that *Bacillus* species are a good alternative for the integrated management of soilborne tomato diseases, such as Rhizoctonia or Pythium damping-off, and Fusarium wilt [12,13].

Our study showed that disease incidence caused by *B. cinerea* was significantly reduced with the application of foliar treatments based on extracts of *R. sachalinensis* and *M. alternifolia* in both growing seasons. Previous studies have shown that *M. alternifolia* oil destroys the cell wall and changes the composition of the cell membrane of *B. cinerea* hyphae, increasing its permeability and allowing the release of cellular material [14]; whereas, in the case of *R. sachalinensis*, it has been found that this extract significantly reduced the percentage of *B. cinerea* colonization on strawberry leaves [15]. In addition, the volatile phase of the extracts is known to be more toxic and effective than the contact phase for the control of plant pathogenic fungi [14,16].

According to the incidence of late blight caused by the oomycete *P. infestans*, it was observed that the treatment with harpin αβ proteins presented the lowest values, which has also been reported in previous studies reporting the reduction of the disease caused by

P. infestans in tomato plants by foliar sprays of harpin proteins [17]. On the other hand, plant extracts based on *R. sachalinensis* and *M. alternifolia* did not satisfactorily reduce the disease caused by *P. infestans*. In this regard, Seidl Johnson et al. [18] found that *R. sachalinensis* extract was ineffective in controlling the disease caused by different isolates of *P. infestans* in tomato. In contrast, Reuveni et al. [19] reported that foliar sprays of *M. alternifolia* at 0.5–1% were effective in prophylactically reducing the disease caused by *P. infestans* under greenhouse conditions.

In relation to the development of powdery mildew caused by *L. taurica*, this disease was favored at the beginning of the growing season; so, preventive fungicide applications should be made prior to an outbreak occurring. In the present study, it was determined that foliar sprays of *M. alternifolia* and *R. sachalinensis* extracts are effective for the management of powdery mildew in tomato plants. Similar results were observed in other studies, where *M. alternifolia* extract was reported to have a good effect in the control of powdery mildew in different crops, including tomato [19] and cucumber [20]. Similarly, the effect of *R. sachalinensis* extract in the control of powdery mildew in tomato has been verified [21,22], as well as in the control of powdery mildew in cucumber [23,24], where foliar sprays significantly reduced conidial germination and disease severity due to the direct effect on the fungus and the induction of plant defense responses through the formation of callose papillae, hydrogen peroxide accumulation, and induction of the salicylic acid-dependent pathway [25].

In the case of pith necrosis, caused by *Pseudomonas corrugata*, the products evaluated in this study did not control the disease. Tomato pith necrosis appears to develop when there are low night temperatures, high nitrogen levels, and high humidity [1]. According to the temporal progress, favorable conditions for the development of the disease were observed from July and continued until October; therefore, prior to this time is when the necessary preventive measures should be taken, such as applications with copper-based products, local treatment with Bordeaux mixture, and bactericide applications. In addition, it has been recommended to avoid high doses of nitrogen fertilization and regularly sanitize tools such as clippers and pruning shears [1].

Regarding *C. michiganensis*, this plant pathogenic bacterium did not cause catastrophic damage and the disease was kept under control throughout the growing season, which could be related to the application of *Bacillus*. In this regard, Abo-Elyousr et al. [26] recorded that *B. subtilis* and *B. amyloliquefaciens* can successfully decrease bacterial canker disease in tomato plants by producing antibiotics, siderophores, and lytic enzymes. In 2017, the development of bacterial canker was significantly higher compared with that in 2016, which is attributed to the fact that the pathogen was possibly in the seed, given that there are reports indicating that the onset of this disease or primary infection results from internally or superficially infested seeds [27], which leads to carrying out preventive measures, such as those used here, to avoid damage from this or any other problem. On the other hand, it was observed that the application of harpin $\alpha\beta$ proteins had no control effect on bacterial canker, coinciding with the findings reported by Ustun et al. [28] in Turkey. Similarly, Obradovic et al. [29] reported that harpin proteins did not induce effective defense responses of tomato plants against the bacterium *Xanthomonas vesicatoria*.

It should be noted that although there were no statistical differences in all the combinations of *Bacillus* with the foliar treatments in this study, there was a numerical difference, which showed important results since the number of diseased plants was reduced. In addition, the lowest values were in the interactions where the *Bacillus* spp.-based product was used.

In summary, there is convincing evidence that the soil and foliar application of the six biorational products tested in this study are effective in controlling potato late blight, gray mold, powdery mildew, pith necrosis, and bacterial canker in greenhouse tomato production compared to the conventional treatment used by growers in this region. The findings in this study contribute to the development of integrated disease management strategies, which include cultural practices aimed at reducing inoculum or avoiding conditions that predispose crops to the development of the disease; the application of biological products; and, ultimately, the use of chemical measures. This implementation in the tomato crop allowed combining natural inducers (bee honey and harpin αβ proteins), plant extracts (*R. sachalinensis* and *M. alternifolia*), beneficial microorganisms (*Bacillus* spp.), and cultural and chemical alternatives, with the last element only being applied when necessary. The products tested in this study can be integrated into this management system; in addition to saving on chemical applications and reducing the negative impact on the environment, they allowed opening up a panorama for integrated crop management (ICM) that can be further refined for a more sustainable use of resources.

4. Materials and Methods

4.1. Description of the Experiments

Two experiments were conducted in 2016 (and repeated in 2017) in a low-technology 1000-m^2 greenhouse located in the municipality of Coatepec Harinas, State of Mexico, Mexico. The greenhouse space was divided into two sections of ~500 m^2 to establish the two experiments. In each experiment, six biorational commercial products (Table 3) were tested in a number of treatment combinations (described in Table 4), each including soil application of either of two *Bacillus*-containing commercial products as a base treatment. The six products that were tested in this study were selected because of their reported effect on diseases in other cropping systems and because of their affordability for growers.

Table 3. Biorational products tested for control of tomato diseases in the greenhouse trials.

Product; Manufacturer	Active Ingredient	Rate
PHC Colonize®, PHC	*Bacillus* spp.	2.0 kg/ha
Fungifree AB®, FMC	*Bacillus subtilis*	2.0 kg/ha
Regalia Maxx®, FMC	Extract of *Reynoutria sachalinensis*	1.25 L/ha
Messenger Gold®, PHC	Harpin αβ proteins	150 g/ha
Timorex Gold®, Syngenta	Extract of *Melaleuca alternifolia*	5 mL/L
Bee honey	Bee honey	1 mL/L

Table 4. Description of the treatment combinations tested in this study for control of major tomato diseases in greenhouse production in 2016 and 2017.

Treatment [a]	Description
1	Soil application of *Bacillus* spp. via drip irrigation + foliar application (conventional) of grower's treatments (Control)
2	Soil application of *Bacillus* spp. via drip irrigation + foliar application of *Reynoutria sachalinensis* at 20-day intervals
3	Soil application of *Bacillus* spp. via drip irrigation + foliar application of αβ harpins at 20-day intervals
4	Soil application of *Bacillus* spp. via drip irrigation + foliar application of *Melalueca alternifolia* at 20-day intervals
5	Soil application of *Bacillus* spp. via drip irrigation + foliar application of bee honey at 20-day intervals

[a] Corresponds to Experiment 1. For Experiment 2, the base treatment *Bacillus* spp. was changed to *B. subtilis*, but the foliar treatments remained unchanged.

The soil in this greenhouse had an initial pH of 4.8, EC of 0.56, and organic matter content of 5.4. The soil is not fumigated and has a history of occurrence of the major tomato diseases. In addition, the area of study is an area where these diseases are prevalent and expected to occur each year. For the case of potato late blight and gray mold, one or two applications of fungicides were necessary to ensure a harvestable crop. Weed control during both growing seasons was carried out manually.

4.2. Experiment 1

In this experiment, a *Bacillus*-spp.-based product (PHC Colonize®) was applied to the soil in one of the 500 m^2 greenhouse sections through a drip irrigation system. Eight days later (28 June 2016), four week-old tomato seedlings cv. Aguamiel (Vilmorin®, Paris, France) were transplanted to marked rows inside the greenhouse section. Row spacing was 1.20 m and interplant spacing was 25 cm. The treatments were disposed in a randomized complete block design with three replications. The experimental units consisted of two 10 m-long, 1.20 m-wide rows. Thereafter, application of *Bacillus* spp. to the soil started eight days after transplant and continued for the four first months and then were switched to 15-day intervals for the last two months of the study. Foliar applications also started eight days after transplant but continued at 20-day intervals until the end of the study. Applications were carried out with a backpack sprayer when the plants were <60 cm-tall and with a stretcher power sprayer when the plants were >60 cm-tall. The equipment was calibrated at each application and the water pH was adjusted to 5.5 to 6 and amended with an adherent.

4.3. Experiment 2

This experiment was carried out in the other 500-m^2 greenhouse section. The experiment design and size of experimental units remained the same as in experiment 1. However, the soil treatment in the irrigation system was changed to *Bacillus subtilis* instead of *Bacillus* spp. The foliar treatments were the same as in the previous experiment. Both experiments were repeated in 2017.

4.4. Statistical Analyses

The incidence of the five major fungi, oomycetes, and bacteria diseases that occurred on the crop during six months of study was evaluated on each experimental unit at 15-day-intervals. Incidence was calculated with the following formula:

$$\boldsymbol{PI} = \frac{n}{N} \times 100$$

where $\boldsymbol{PI}$ = Percentage of incidence of the disease, n = number of seedlings or plants with symptoms of the disease, and N = total number of seedlings or plants in the experimental unit.

With the incidence data (temporal incidence percentages), the area under the disease progress curve (AUDPC) was calculated for each disease using the following formula [30]:

$$\text{AUDPC} = \sum_{i=l}^{n-1} \left(\frac{y_i + y_{i+l}}{2}\right)(t_{i+l} - t_i)$$

where n is the number of assessments, i is the time point of observation, y_i is the disease incidence, and t_i is the number of days from the start of the experiment to the measurement date.

The value of the AUDPC for each disease was analyzed with a one-way ANOVA using PROC GLM procedure on SAS version 9.4 (SAS Institute, Cary, NC, USA). The assumption of normality and homogeneity of variance was satisfied in the two trials. Differences between treatments within the trials were further compared by a Duncan test at $P \leq 0.05$ level. The AUDPC data from each experiment were analyzed separately because the species of *Bacillus* used in the conventional treatment (control) in experiment 1 differed from that in experiment 2. Moreover, given that each experiment was repeated once only, it made more sense to keep the year effect as fixed rather than random.

Author Contributions: Conceptualization: L.F.E.-C., B.T.-B. and J.M.T.-P.; formal analysis: L.F.E.-C., B.T.-B. and M.C.-T.; investigation: L.F.E.-C., S.G.L.-M. and J.M.T.-P.; writing—original-draft preparation: L.F.E.-C., S.G.L.-M., O.P.-H., B.T.-B. and J.M.T.-P.; writing—review and editing: S.G.L.-M., B.T.-B., O.P.-H. and J.M.T.-P.; supervision: M.C.-T., S.G.L.-M., B.T.-B. and J.M.T.-P. All authors have read and agreed to the published version of the manuscript.

Funding: This research received no external funding.

Institutional Review Board Statement: Not applicable.

Informed Consent Statement: Not applicable.

Data Availability Statement: Not applicable.

Acknowledgments: The authors wish to thank Universidad Autónoma Chapingo for administrative support of this research project.

Conflicts of Interest: The authors declare no conflict of interest.

References

1. Koike, S.T.; Gladders, P.; Paulus, A.O. *Vegetable Diseases: A Colour Handbook*; Manson Publishing: London, UK, 2007.
2. SIAP. Servicio de Información Agroalimentaria y Pesquera. Panorama Agroalimentario. 2020. Available online: https://nube.siap.gob.mx/gobmx_publicaciones_siap/pag/2020/Atlas-Agroalimentario-2020 (accessed on 10 December 2021).
3. Blancard, D. *Tomato Diseases: Identification, Biology and Control*, 2nd ed.; CRC Press: Boca Raton, FL, USA, 2013.
4. Vallad, G.E.; Messelink, G.; Smith, H.A. Crop Protection: Pest and Disease Management. In *Tomatoes*, 2nd ed.; Euvelink, E., Ed.; CAB International: Boston, MA, USA, 2018; pp. 207–257.
5. Jones, J.B.; Zitter, T.A.; Momol, T.M.; Miller, S.A. *Compendium of Tomato Diseases and Pests*, 2nd ed.; APS Press: St. Paul, MN, USA, 2014.
6. Shrestha, S.; Hausbeck, M.K. Evaluation of geranium cultivars and biorational products to control botrytis blight in the greenhouse. *Plant Health Prog.* **2021**, *22*, 474–482. [CrossRef]
7. Akanmu, A.O.; Babalola, O.O.; Venturi, V.; Ayilara, M.S.; Adeleke, B.S.; Amoo, A.E.; Sobowale, A.A.; Fadiji, A.E.; Glick, B.R. Plant Disease Management: Leveraging on the Plant-Microbe-Soil Interface in the Biorational Use of Organic Amendments. *Front. Plant Sci.* **2021**, *12*, 700507. [CrossRef] [PubMed]
8. Singh, V.K.; Singh, A.K.; Kumar, A. Disease management of tomato through PGPB: Current trends and future perspective. *3 Biotech* **2017**, *7*, 255. [CrossRef] [PubMed]
9. Miljaković, D.; Marinković, J.; Balesěević-Tubić, S. The significance of *Bacillus* spp. in disease suppression and growth promotion of field and vegetable crops. *Microorganisms* **2020**, *8*, 1037. [CrossRef] [PubMed]
10. Jacobsen, B.J.; Zidack, N.K.; Larson, B.J. The role of *Bacillus*-based biological control agents in integrated pest management systems: Plant diseases. *Phytopathology* **2014**, *94*, 1272–1275. [CrossRef] [PubMed]
11. Villarreal-Delgado, M.F.; Villa-Rodríguez, E.D.; Cira-Chávez, L.A.; Estrada-Alvarado, M.I.; Parra-Cota, F.I.; De los Santos-Villalobos, S. The genus *Bacillus* as a biological control agent and its implications in the agricultural biosecurity. *Rev. Mex. Fitopatol.* **2017**, *36*, 95–130. [CrossRef]
12. Mandal, A.K.; Maurya, P.K.; Dutta, S.; Chattopadhyay, A. Effective management of major tomato diseases in the Gangetic Plains of Eastern India through integrated approach. *Agric. Res. Technol.* **2017**, *10*, 109–117. [CrossRef]
13. Samaras, A.; Roumeliotis, E.; Ntasiou, P.; Karaoglanidis, G. *Bacillus subtilis* MBI600 promotes growth of tomato plants and induces systemic resistance contributing to the control of soilborne pathogens. *Plants* **2021**, *10*, 1113. [CrossRef]
14. Shao, X.; Cheng, S.; Wang, H.; Yu, D.; Mungai, C. The possible mechanism of antifungal action of tea tree oil on *Botrytis cinerea*. *J. Appl. Microbiol.* **2013**, *114*, 1642–1649. [CrossRef]
15. Oliveira, M.S.; Cordova, L.G.; Marin, M.V.; Peres, N.A. Fungicide dip treatments for management of *Botrytis cinerea* infection on strawberry transplants. *Plant Health Prog.* **2018**, *19*, 279–283. [CrossRef]
16. Soylu, E.M.; Kurt, S.; Soylu, S. *In vitro* and *in vivo* antifungal activities of the essential oils of various plants against tomato grey mould disease agent *Botrytis cinerea*. *Int. J. Food Microbiol.* **2010**, *143*, 183–189. [CrossRef]
17. Fontanilla, M.; Montes, M.; De Prado, R. Effects of the foliar-applied protein "Harpin (Ea)" (messenger) on tomatoes infected with *Phytophthora infestans*. *Commun. Agric. Appl. Biol. Sci.* **2005**, *70*, 41–45. [PubMed]
18. Seidl Johnson, A.C.; Jordan, S.A.; Gevens, A.J. Efficacy of organic and conventional fungicides and impact of application timing on control of tomato late blight caused by US-22, US-23, and US-24 isolates of *Phytophthora infestans*. *Plant Dis.* **2015**, *99*, 641–647. [CrossRef] [PubMed]
19. Reuveni, M.; Neifeld, D.; Dayan, D.; Kotzer, Y. BM-608-A novel organic product based on essential tea tree oil for the control of fungal diseases in tomato. *Acta Hortic.* **2008**, *808*, 129–132. [CrossRef]
20. Reuveni, M.; Sanches, E.; Barbier, M. Curative and suppressive activities of essential tea tree oil against fungal plant pathogens. *Agronomy* **2020**, *10*, 609. [CrossRef]
21. Konstantinidou-Doltsinis, S.; Markellou, E.; Kasselaki, A.M.; Fanouraki, M.N.; Koumaki, C.M.; Schmitt, A.; Liopa-Tsakalidis, A.; Malathrakis, N.E. Efficacy of Milsana, a formulated plant extract from *Reynoutria sachalinensis*, against powdery mildew of tomato (*Leveillula taurica*). *Biocontrol* **2006**, *51*, 375–392. [CrossRef]
22. Baysal-Gurel, F.; Miller, S.A. Management of powdery mildew in greenhouse tomato production with biorational products and fungicides. In Proceedings of the IV International Symposium on Tomato Diseases, Orlando, FL, USA, 24–27 June 2013; Volume 1069, pp. 179–184. [CrossRef]

23. Petsikos-Panayotarou, N.; Schmitt, A.; Markellou, E.; Kalamarakis, A.E.; Tzempelikou, K.; Siranidou, E.; Konstantinidou-Doltsinis, S. Management of cucumber powdery mildew by new formulations of *Reynoutria sachalinensis* (F. Schmidt) Nakai extract. *J. Plant Dis. Prot.* **2002**, *109*, 478–490.
24. Bokshi, A.; Jobling, J.; McConchie, R. A single application of Milsana® followed by Bion® assists in the control of powdery mildew in cucumber and helps overcome yield losses. *J. Hortic. Sci. Biotechnol.* **2008**, *83*, 701–706. [CrossRef]
25. Margaritopoulou, T.; Toufexi, E.; Kizis, D.; Balayiannis, G.; Anagnostopoulos, C.; Theocharis, A.; Rempelos, L.; Troyanos, Y.; Leifert, C.; Markellou, E. *Reynoutria sachalinensis* extract elicits SA-dependent defense responses in courgette genotypes against powdery mildew caused by *Podosphaera xanthii*. *Sci. Rep.* **2020**, *10*, 3354. [CrossRef]
26. Abo-Elyousr, K.A.M.; Khalil Bagy, H.M.M.; Hashem, M.; Alamri, S.A.M.; Mostafam, Y.S. Biological control of the tomato wilt caused by *Clavibacter michiganensis* subsp. *michiganensis* using formulated plant growth-promoting bacteria. *Egypt. J. Biol. Pest Control* **2019**, *29*, 54. [CrossRef]
27. Sen, Y.; van der Wolf, J.; Visser, R.G.F.; van Heusden, S. Bacterial canker of tomato: Current knowledge of detection, management, resistance, and interactions. *Plant Dis.* **2019**, *99*, 4–13. [CrossRef] [PubMed]
28. Ustun, N.; Ulutas, E.; Yasarakinci, N.; Kilic, T. Efficacy of some plant activators on bacterial canker of tomato in Aegean Region of Turkey. *Acta Hortic.* **2009**, *808*, 405–408. [CrossRef]
29. Obradovic, A.; Jones, J.B.; Momol, M.T.; Olson, S.M.; Jackson, L.E.; Balogh, B.; Guven, K.; Iriarte, F.B. Integration of biological control agents and systemic acquired resistance inducers against bacterial spot on tomato. *Plant Dis.* **2005**, *89*, 712–716. [CrossRef] [PubMed]
30. Madden, L.V.; Hughes, G.; van den Bosch, F. *The Study of Plant Disease Epidemics*; APS Press: St. Paul, MN, USA, 2007.

MDPI

Article

The Effect of *Thymus vulgaris* L. Hydrolate Solutions on the Seed Germination, Seedling Length, and Oxidative Stress of Some Cultivated and Weed Species

Bojan Konstantinović [1], Milena Popov [1,*], Nataša Samardžić [1], Milica Aćimović [2], Jovana Šućur Elez [1], Tijana Stojanović [1], Marina Crnković [1] and Miloš Rajković [2]

[1] Department of Phytomedicine and Environmental Protection, Faculty of Agriculture, University of Novi Sad, 21000 Novi Sad, Serbia; bojan.konstantinovic@polj.edu.rs (B.K.); natasa.samardzic@polj.uns.ac.rs (N.S.); jovana.sucur@polj.uns.ac.rs (J.Š.E.); tijana.stojanovic@polj.edu.rs (T.S.); marina.crnkovic@polj.uns.ac.rs (M.C.)

[2] Institute of Field and Vegetable Crops Novi Sad, 21101 Novi Sad, Serbia; milica.acimovic@ifvcns.ns.ac.rs (M.A.); rajkovicmilos@gmail.com (M.R.)

* Correspondence: milena.popov@polj.edu.rs

Citation: Konstantinović, B.; Popov, M.; Samardžić, N.; Aćimović, M.; Šućur Elez, J.; Stojanović, T.; Crnković, M.; Rajković, M. The Effect of *Thymus vulgaris* L. Hydrolate Solutions on the Seed Germination, Seedling Length, and Oxidative Stress of Some Cultivated and Weed Species. *Plants* **2022**, *11*, 1782. https://doi.org/10.3390/plants11131782

Academic Editors: Špela Mechora and Dragana Šunjka

Received: 25 May 2022
Accepted: 1 July 2022
Published: 5 July 2022

Abstract: The aim of this study was to determine the effect of the hydrolates obtained as the by-products of the *Thymus vulgaris* essential oil steam distillation process. The bioassays, which were undertaken in order to determine the effect on germination and initial growth of seedlings of some cultivated and weed species, were performed under controlled conditions with different concentrations of the hydrolates. Seeds of *Glycine max*, *Helianthus annuus*, *Zea mays*, *Triticum aestivum*, *Daucus carota* subsp. *sativus*, *Allium cepa*, *Medicago sativa*, and *Trifolium repens*, and six weed species—*Amaranthus retroflexus*, *Chenopodium album*, *Portulaca oleracea*, *Echinochloa crus-galli*, *Sorghum halepense*, and *Solanum nigrum*—were treated with 10, 20, 50, and 100% *T. vulgaris* hydrolate solution. The obtained results showed that the *T. vulgaris* hydrolate had the least negative effect on the germination of cultivated species, such as soybean, sunflower and maize, whereas clover and alfalfa were the most sensitive. By comparison, all the tested weed species expressed high susceptibility. It can be concluded that the *T. vulgaris* hydrolate has an herbicidal effect, in addition to its potential as a biopesticide in terms of integrated weed management.

Keywords: oxidative stress; germination; inhibition; biopesticide

1. Introduction

The hydrolates or hydrosols are the by-products of the essential oil steam distillation process. They consist of condensate water and small amounts of essential oil compounds, which are mostly soluble in water [1]. Hydrosols have been recently noticed for their biological properties, including antifungal, antibacterial and antioxidant activity [2,3]. Today, hydrolates are widely used in the pharmaceutical, food, and cosmetic industries, and in aromatherapy [4]. Their production is easy and cheap, and they appear to be less toxic to human health than essential oils [2].

Thymus vulgaris L. is a perennial flowering aromatic plant that has been used globally for many centuries for medicinal and culinary purposes [5]. Essential oils and lipophilic substances are abundant in this plant [6]. According to [7], *T. vulgaris* extracts are rich in the aromatic compounds thymol, carvacrol, p-cymene, and γ-terpinene. The chemical composition of an essential oil depends on the harvesting time of the plant material, the phenological phase of the plant, the growing area of the plants, and other parameters [8]. Thus, the results of chemical composition studies of *T. vulgaris* essential oil and hydrolates vary. For example, it was determined that the major component of *T. vulgaris* essential oil is thymol (36.1%), whereas, in the case of hydrosol, thymol (98.1%) and carvacrol (1.9%) were dominant [9]. According to [8], the major component of the essential oil obtained before flowering is thymol (46.74%), whereas [10] discovered that *T. vulgaris* essential oil contained 68.1% thymol and 3.5% carvacrol.

Thymol and carvacrol have antioxidant properties [11,12]. In the fields of human and veterinary medicine, and in plant protection, the antibacterial and antifungal activity of thymol and carvacrol [13–15], the nematocidal effect of *T. vulgaris* essential oil [16], and the antiviral properties of the extracts [17–19], essential oil [20–22], and hydrolates [23] have been reported. With regard to its antibacterial properties, *T. vulgaris* hydrolate has practical applications in the food industry for the microbiological safety of fresh-cut tomatoes and cucumbers [24], carrots and apples [25], and lettuce [26].

In recent years, in order to reduce the use of pesticides in the control of diseases, pests, and weeds in agriculture, great efforts have been made to identify new biopesticides that are as effective as synthetic chemicals. Thymol and carvacrol are examples of bioactive compounds that may have the potential to become an integral part of agricultural practice as biopesticides. These compounds are potentially useful in controlling the emission of odors and pathogens in swine waste [27]. Due to its fast degradation rates in tropical soil and water, thymol is considered to have a low environmental risk in terms of tropical crop production [28]. According to [29], thymol and carvacrol showed high inhibition at low concentrations against weed seeds, such as those of red-root amaranth (*Amaranthus retroflexus* L.), wild radish (*Raphanus raphanistrum* L.), and wild mustard (*Sinapis arvensis* L.). In the case of *Sorghum bicolor* L. seeds, carvacrol applied in a concentration of 3 mmol L^{-1} showed higher efficacy than the tested commercial herbicide (2,4-D), leading to the inhibition of the germination rate (~40%) and the germination speed (~56%) [30]. Furthermore, it was concluded that, among the studied essential oils, thymol can be used as an environmentally friendly root-repellent agent instead of synthetic herbicides [31].

The aim of this study was to test the different *T. vulgaris* hydrolate solution (THS) concentrations against the seed germination and seedling lengths of some of the most important weed species that occur in Serbia: red-root amaranth (*Amaranthus retroflexus* L.), common lambsquarters (*Chenopodium album* L.), common purslane (*Portulaca oleracea* L.), cockspur grass (*Echinochloa crus-galli* (L.) P. Beauv.), Johnson grass (*Sorghum halepense* (L.) Pers.), and black nightshade (*Solanum nigrum* L.). The other aim was to study the phytotoxic potential of the hydrolate towards the most common field and vegetable crop seeds: soybean (*Glycine max* (L.) Merr.), sunflower (*Helianthus annuus* L.), maize (*Zea mays* L.), wheat (*Triticum aestivum* L.), carrot (*Daucus carota* subsp. *sativus* (Hoffm.) Schübl. & Martens), onion (*Allium cepa* L.), alfalfa (*Medicago sativa* L.), and white clover (*Trifolium repens* L.).

2. Results

2.1. Seed Germination

2.1.1. Germination Percentage (GP)

Cultivated species: The highest germination percentage was registered in sunflower seeds (ranging from 73 to 96%), whereas the carrot seeds expressed the lowest germination percentage, ranging from 14 to 0% (Table 1).

Weed species: The tested hydrolate, depending on the hydrolate solution percentage, led to a reduction in germination ranging from 53 to 100%, with the exception of the black nightshade, in which 10 and 20% hydrolate solutions led to a reduction of ≤5%; by comparison, in the case of 50 and 100% THS, the reduction was 86 and 100%, respectively (Table 2).

Table 1. The obtained results for the calculated germination indices in the case of the tested cultivated species treated with *T. vulgaris* hydrolate solutions.

Tested Plant	Concentration	GP (%)	CVG (% Day^{-1})	GI (−)	t_{50} (Time)	GRI (%/Day)
soybean	control	93	23.48	627	2.78	33.94
	10%	74	30.58	572	2.48	31.41
	20%	71	28.18	529	2.48	27.82
	50%	66	30.56	510	2.39	27.21
	100%	60	24.00	410	2.77	19.21
maize	control	97	22.35	633	4.15	27.45
	10%	80	19.66	473	4.80	19.36
	20%	82	18.34	455	4.56	18.13
	50%	65	18.36	361	4.69	15.81
	100%	21	23.33	141	3.25	8.71
onion	control	85	23.94	580	3.69	26.29
	10%	75	19.74	445	4.07	16.42
	20%	33	17.94	179	4.56	6.30
	50%	21	17.95	114	4.59	4.07
	100%	10	13.51	36	7.00	1.42
sunflower	control	89	33.21	711	1.81	44.89
	10%	96	34.78	780	1.77	49.72
	20%	88	19.43	515	4.18	18.13
	50%	83	21.01	518	3.81	18.42
	100%	73	25.70	519	2.68	23.79
alfalfa	control	83	35.78	681	2.17	36.08
	10%	85	32.69	675	2.18	35.73
	20%	76	23.03	506	2.76	24.99
clover	control	93	37.65	776	2.29	41.29
	10%	95	38.00	791	2.28	42.34
	20%	90	23.87	613	2.90	30.06
carrot	control	39	16.46	192	4.85	7.09
	10%	14	11.76	35	8.20	1.70
	20%	1	12.50	3	7.50	0.12
wheat	control	100	43.10	868	1.73	52.47
	10%	89	18.39	495	4.36	18.22
	20%	66	22.53	433	3.65	17.26
	50%	9	20.93	56	4.17	2.14

GP—germination percentage (ISTA, 2015); CVG—coefficient of velocity of germination (Jones and Sanders, 1987); GI—germination index (Bench et al., 1991); t_{50}—median germination time (Farooq et al., 2005); GRI—germination rate index (Esechie, 1994 after modification).

Table 2. The obtained results for the calculated germination indices in the case of the tested weed species treated with *T. vulgaris* hydrolate solutions.

Tested Plant	Concentration	GP (%)	CVG (% Day^{-1})	GI (−)	t_{50} (Time)	GRI (%/Day)
red-root amaranth	control	83	24.78	578	2.56	30.41
	10%	26	14.86	111	6.29	4.03
cockspur	control	22	19.47	129	3.92	4.60
	10%	20	15.15	88	6.33	3.30
	20%	9	15.25	40	5.75	1.47
	50%	4	17.39	21	5.33	0.71
common purslane	control	100	38.46	840	1.88	51.41
	10%	46	22.77	304	3.88	13.86

Table 2. *Cont.*

Tested Plant	Concentration	GP (%)	CVG (% Day^{-1})	GI (–)	t_{50} (Time)	GRI (%/Day)
Johnson grass	control	23	13.30	80	7.50	3.35
	10%	17	12.06	46	7.90	2.10
	20%	6	11.76	15	8.00	0.71
	50%	1	11.11	2	8.50	0.11
common lambsquarters	control	51	20.00	306	4.19	12.95
	10%	47	18.58	264	4.43	9.91
	20%	1	14.29	4	6.50	0.14
black nightshade	control	84	13.25	290	7.68	12.16
	10%	100	14.14	393	6.47	15.13
	20%	95	14.80	403	6.34	15.27
	50%	14	11.57	33	8.25	1.64

GP—germination percentage (ISTA, 2015); CVG—coefficient of velocity of germination (Jones and Sanders, 1987); GI—germination index (Bench et al., 1991); t_{50}—median germination time (Farooq et al., 2005); GRI—germination rate index (Esechie, 1994 after modification).

2.1.2. Coefficient of Velocity of Germination (CVG)

Cultivated species: The highest CVG values for the applied hydrolate solutions (from 10 to 100%) were noted in the following order: clover; soybean; soybean; sunflower (in the range from 25.70 to 38.00% day^{-1}). The lowest CVG values for 10 and 20% THS were observed in carrot (11.76 and 12.50% day^{-1}, respectively), whereas, for 50 and 100% hydrolate solutions, they were recorded for onion (17.95 and 13.51% day^{-1}, respectively) (Table 1).

Weed species: The highest CVG value when 10% THS was applied was noted for common purslane seeds (22.77% day^{-1}), whereas, in the case of 20 and 50% hydrolate solution, it was observed in cockspur (15.25 and 17.39% day^{-1}, respectively). Conversely, Johnson grass expressed the lowest CVG values (in the range from 11.11 to 12.06% day^{-1}) when 10, 20, and 50% hydrolate solutions were applied (Table 2).

2.1.3. Germination Index (GI) and Germination Rate Index (GRI)

Cultivated species: Not taking into account the complete germination reduction, clover had the highest GI and GRI values for 10 and 20% hydrolate solutions (791/42.34 and 613/30.06% day^{-1}, respectively), whereas sunflower expressed the highest values when 50 and 100% THS was applied (518/18.42 and 519/23.79% day^{-1}, respectively). The lowest GI and GRI values for 10 and 20% hydrolate solutions were noted for carrot (35/1.70 and 3/0.12% day^{-1}, respectively), whereas, in the case of 50 and 100% THS, the lowest values were observed for wheat (56/2.14% day^{-1}) and onion (36/1.42% day^{-1}) (Table 1).

Weed species: Not taking into account the complete germination reduction, the highest GI and GRI values were observed for common purslane (840/51.41 and 304/13.86% day^{-1}, respectively), whereas the lowest values were noted for Johnson grass (in the range from 2/0.11 to 46/2.10% day^{-1}) (Table 2).

2.1.4. Median Germination Time (t_{50})

Cultivated species: Not taking into account the complete germination reduction, the highest t_{50} values for the applied hydrolate solutions (from 10 to 100%) were noted in the following order: carrot; carrot; maize; onion (in the range from 4.69 to 8.20 days); whereas the lowest values were observed as follows: sunflower; soybean; soybean; sunflower (ranging from 1.77 to 2.68 days) (Table 1).

Weed species: The highest t_{50} values were noted for Johnson grass (in the range from 7.90 to 8.50), whereas the lowest values were observed in the case of common purslane (10% hydrolate solution: –3.88) and cockspur (20 and 50%: 5.75 and 5.33) (Table 2).

The obtained results for the calculated germination indices (GP, CVG, GI, t_{50}, and GRI) are shown in Tables 1 and 2.

2.1.5. Shoot and Root Length

The examination of the hydrolate effect on the initial growth of the studied field and vegetable seedlings showed that the sunflower seeds expressed a stimulative effect in terms of shoot and root seedling length when treated with the lower hydrolate concentrations (10%), whereas, in the case of the other tested plants, a statistically significant reduction, proportional to the increase in hydrolate concentration, was observed.

The TH reduced the seedling growth in all of the tested plant species, with the exception of black nightshade, which, like in the case of sunflower, exhibited shoot and root growth stimulation when treated with 10 and 20% hydrolate solution. However, 50% THS inhibited the growth of black nightshade seedlings (Tables 3 and 4).

Table 3. The shoot and root seedling length of the field and vegetable crop species treated with 10, 20, 50, and 100% *T. vulgaris* hydrolate solutions.

Tested Plant	Variable	Shoot Length (mm)	Root Length (mm)
onion	control	13.70 ± 10.15 [a]	7.65 ± 6.40 [a]
	10%	6.59 ± 5.69 [b]	2.86 ± 2.80 [b]
	20%	0.85 ± 1.98 [c]	0.59 ± 1.44 [c]
	50%	1.08 ± 1.64 [c]	0.60 ± 0.98 [c]
	100%	0.17 ± 0.53 [c]	0.17 ± 0.53 [c]
maize	control	25.26 ± 12.87 [a]	69.29 ± 25.76 [a]
	10%	19.49 ± 17.00 [b]	54.31 ± 49.06 [b]
	20%	13.80 ± 14.75 [c]	29.66 ± 28.76 [c]
	50%	10.78 ± 14.81 [c]	20.09 ± 27.81 [d]
	100%	4.87 ± 10.05 [d]	5.53 ± 13.48 [e]
sunflower	control	34.41 ± 21.64 [a]	50.01 ± 38.16 [b]
	10%	38.00 ± 19.68 [a]	64.51 ± 46.27 [a]
	20%	29.31 ± 23.66 [b]	40.90 ± 41.06 [b]
	50%	9.47 ± 7.91 [c]	13.18 ± 10.70 [c]
	100%	7.32 ± 6.54 [c]	12.26 ± 11.13 [c]
soybean	control	27.87 ± 13.03 [a]	13.92 ± 8.65 [b]
	10%	14.76 ± 11.18 [b]	19.91 ± 19.83 [a]
	20%	12.13 ± 9.80 [b]	18.56 ± 18.65 [a]
	50%	8.94 ± 7.37 [c]	11.70 ± 12.40 [b]
	100%	7.70 ± 7.41 [c]	6.28 ± 6.61 [c]
wheat	control	114.12 ± 19.17 [a]	114.11 ± 19.14 [a]
	10%	30.28 ± 26.61 [b]	40.36 ± 34.36 [b]
	20%	22.47 ± 26.56 [c]	22.92 ± 26.09 [c]
	50%	3.72 ± 13.82 [d]	2.63 ± 9.69 [d]
carrot	control	3.65 ± 7.38 [a]	4.32 ± 6.95 [a]
	10%	0.37 ± 1.07 [b]	1.65 ± 4.54 [b]
	20%	0.04 ± 0.40 [b]	0.04 ± 0.40 [c]
white clover	control	12.74 ± 4.99 [a]	31.85 ± 16.72 [a]
	10%	9.66 ± 3.47 [b]	25.64 ± 10.86 [b]
	20%	6.99 ± 3.26 [c]	17.17 ± 10.18 [c]
alfalfa	control	12.08 ± 7.71 [a]	28.51 ± 19.45 [a]
	10%	10.35 ± 5.35 [b]	27.29 ± 17.64 [a]
	20%	7.50 ± 5.44 [c]	12.46 ± 11.92 [b]

The data are mean values ± standard error. [a–d] Values without the same superscripts within each column differ significantly ($p < 0.05$).

Table 4. The shoot and root seedling length of the weed species treated with 10, 20, 50, and 100% *T. vulgaris* hydrolate solutions.

Tested Plant	Variable	Shoot Length (mm)	Root Length (mm)
common purslane	control	10.49 ± 2.55 [a]	19.76 ± 4.53 [a]
	10%	1.54 ± 1.78 [b]	1.71 ± 2.01 [b]
red-root amaranth	control	15.61 ± 8.32 [a]	14.43 ± 7.72 [a]
	10%	1.81 ± 3.29 [b]	1.96 ± 3.41 [b]
common lambsquarters	control	4.12 ± 5.63 [a]	7.65 ± 12.77 [a]
	10%	2.07 ± 3.21 [b]	1.89 ± 2.46 [b]
	20%	-	0.05 ± 0.50 [c]
cockspur grass	control	8.90 ± 17.75 [a]	10.70 ± 23.20 [a]
	10%	6.13 ± 13.74 [ab]	4.70 ± 11.07 [bc]
	20%	3.08 ± 10.33 [bc]	1.47 ± 5.88 [b]
Johnson grass	control	10.48 ± 21.54 [a]	10.38 ± 22.58 [a]
	10%	2.38 ± 9.94 [bc]	1.78 ± 7.33 [bc]
	20%	5.52 ± 15.09 [b]	4.81 ± 13.25 [b]
	50%	0.42 ± 2.96 [c]	-
black nightshade	control	2.57 ± 5.48 [c]	9.69 ± 19.86 [c]
	10%	10.96 ± 3.80 [a]	43.88 ± 13.68 [a]
	20%	8.77 ± 4.60 [b]	19.64 ± 15.25 [b]
	50%	1.22 ± 4.17 [d]	2.97 ± 9.20 [d]

The data are mean values ± standard error. [a–d] Values without the same superscripts within each column differ significantly ($p < 0.05$).

The effect of the *T. vulgaris* hydrolate solutions on the germination rate and seedling length can be observed in Figure 1, where maize, white clover, and red-root amaranth are shown as the examples.

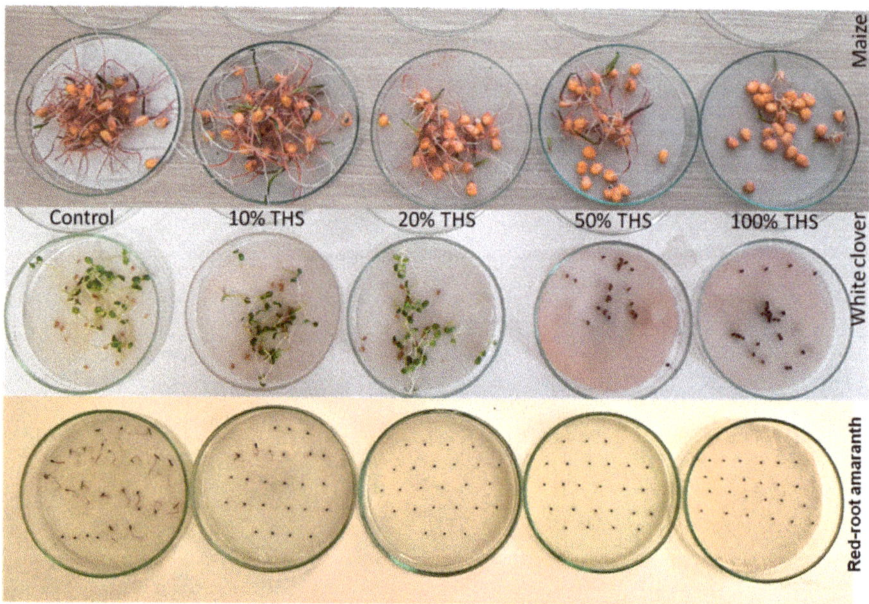

Figure 1. The effect of the applied *T. vulgaris* hydrolate solutions on maize and white clover.

2.2. Biochemical Parameters

2.2.1. MDA Content

The accumulation of the malondialdehyde (MDA) was notably higher in alfalfa seedlings after the treatment with 20% TH solution (115.74 nmol MDA/g FW). In the common lambsquarters seedlings, the amount of MDA after the treatment with 10% TH solution was 165.73 nmol MDA/g FW, whereas, in the case of the cockspur grass seedlings after treatment with the 20% hydrolate solution, it was 83.96 nmol MDA/g FW. The accumulation of the MDA indicates that 10 and 20% hydrolate extracts have a negative effect on these three species by inducing oxidative stress, along with the induction of the lipid peroxidation process. By comparison, the red-root amaranth seedlings showed higher oxidative stress in the control, whereas, in the case of Johnson grass, the highest MDA content was noted in the seedlings treated with 20% TH solution (74.51 nmol MDA/g FW). The results of the tested seedlings of the remaining species showed no statistically significant differences in terms of the lipid peroxidation intensity.

2.2.2. SOD Activity

A significant decrease in SOD activity was detected in the case of maize (116.40 U/g FW) and soybean (5.69 U/g FW) seedlings when treated with the highest THS concentration, and in sunflower (33.85 U/g FW) and white clover (2.47 U/g FW) seedlings when treated with 20% THS. An increase in SOD activity was noted in the seedlings of wheat (all the tested concentrations, ranging between 122.96 and 125.39 U/g FW), carrot (10% hydrolate, 150.95 U/g FW), cockspur grass (20% hydrolate, 313.76 U/g FW), black nightshade (all the tested concentrations, with the highest in the case of 10% hydrolate, 236.92 U/g FW), and Johnson grass (all the tested concentrations, with the highest in the case of 20% hydrolate, 337.81 U/g FW). The alfalfa seedlings showed the highest SOD activity when treated with 20% hydrolate (282.83 U/g FW) and the lowest in the case of 10% hydrolate (175.61U/g FW). The results of the tested seedlings of the remaining species did not show any statistically significant differences in terms of the SOD activity (Tables 5 and 6).

Table 5. Biochemical analysis of the treated field and vegetable crop species 14 days after the treatment.

Tested Plant	Variable	LP nmol/g FW	SOD U/g FW	nmol $O_2^{\cdot-}$/g FW
wheat	control	69.40 ± 16.60 [a]	119.09 ± 0.81 [a]	58.01 ± 4.06 [a]
	10%	67.91 ± 8.23 [a]	124.46 ± 0.38 [b]	89.84 ± 25.86 [a]
	20%	48.09 ± 4.86 [a]	122.96 ± 0.50 [b]	252.08 ± 63.13 [ab]
	50%	55.27 ± 8.72 [a]	125.39 ± 1.31 [b]	319.63 ± 103.44 [b]
maize	control	62.00 ± 1.90 [a]	120.44 ± 0.14 [d]	15.83 ± 3.88 [a]
	10%	38.66 ± 4.75 [a]	118.91 ± 0.21 [bc]	34.25 ± 9.08 [bc]
	20%	69.25 ± 27.45 [a]	119.53 ± 0.42 [c]	19.66 ± 3.71 [ab]
	50%	63.19 ± 23.16 [a]	118.35 ± 0.07 [b]	46.38 ± 1.21 [c]
	100%	49.51 ± 8.59 [a]	116.40 ± 0.00 [a]	5.11 ± 0.63 [a]
sunflower	control	52.20 ± 5.65 [a]	82.66 ± 0.57 [c]	57.63 ± 6.15 [a]
	10%	42.03 ± 10.17 [a]	50.60 ± 16.59 [ab]	115.60 ± 7.12 [b]
	20%	43.00 ± 6.69 [a]	33.85 ± 11.59 [a]	56.08 ± 16.80 [a]
	50%	41.13 ± 4.83 [a]	67.20 ± 5.48 [bc]	76.26 ± 18.75 [ab]
	100%	37.24 ± 6.66 [a]	71.64 ± 0.26 [bc]	99.39 ± 5.96 [b]
soybean	control	95.20 ± 4.59 [a]	107.96 ± 0.86 [c]	22.38 ± 1.15 [ab]
	10%	96.92 ± 8.93 [a]	24.91 ± 0.99 [b]	30.09 ± 1.71 [bc]
	20%	97.15 ± 16.08 [a]	23.01 ± 4.43 [b]	18.21 ± 2.40 [a]
	50%	86.53 ± 8.75 [a]	10.68 ± 1.05 [a]	33.75 ± 5.81 [c]
	100%	89.89 ± 24.37 [a]	5.69 ± 2.94 [a]	19.71 ± 2.77 [ab]

Table 5. *Cont.*

Tested Plant	Variable	LP nmol/g FW	SOD U/g FW	nmol $O_2^{\cdot-}$/g FW
onion	control	36.87 ± 5.63 [a]	111.58 ± 2.46 [a]	90.44 ± 42.56 [a]
	10%	37.92 ± 8.46 [a]	97.41 ± 2.74 [a]	290.53 ± 25.37 [a]
	20%	35.17 ± 5.62 [a]	116.34 ± 14.87 [a]	817.26 ± 117.71 [b]
carrot	control	18.03 ± 2.20 [a]	99.19 ± 6.91 [a]	60.83 ± 14.08 [a]
	10%	18.54 ± 0.81 [a]	150.95 ± 14.96 [b]	165.92 ± 33.20 [b]
white clover	control	43.73 ± 8.47 [a]	23.46 ± 2.68 [c]	46.48 ± 6.47 [b]
	10%	38.65 ± 3.78 [a]	8.72 ± 0.95 [b]	25.45 ± 2.64 [a]
	20%	60.90 ± 10.18 [a]	2.47 ± 0.63 [a]	30.07 ± 4.56 [ab]
alfalfa	control	97.48 ± 20.25 [ab]	188.94 ± 1.74 [b]	17.77 ± 7.08 [a]
	10%	51.35 ± 11.12 [a]	175.61 ± 0.78 [a]	24.64 ± 7.01 [a]
	20%	115.74 ± 19.05 [b]	282.83 ± 1.20 [c]	49.91 ± 6.47 [b]

The data are mean values ± standard error. [a–d] Values without the same superscripts within each column differ significantly ($p < 0.05$).

Table 6. Biochemical analysis of the treated weed species 14 days after the treatment.

Tested Plant	Variable	LP nmol/g FW	SOD U/g FW	nmol $O_2^{\cdot-}$/g FW
common lambsquarters	control	18.48 ± 2.38 [a]	79.52 ± 2.91 [a]	154.49 ± 8.15 [a]
	10%	165.73 ± 0.67 [b]	334.59 ± 95.64 [a]	3366.46 ± 412.48 [b]
red-root amaranth	control	43.88 ± 2.81 [b]	80.56 ± 3.06 [a]	77.48 ± 1.87 [a]
	10%	31.41 ± 2.42 [a]	101.83 ± 10.51 [a]	82.48 ± 19.03 [a]
cockspur grass	control	37.81 ± 1.41 [a]	142.56 ± 7.65 [a]	151.52 ± 45.71 [a]
	10%	32.90 ± 0.20 [a]	126.47 ± 8.79 [a]	256.70 ± 89.31 [a]
	20%	83.96 ± 3.75 [b]	313.76 ± 9.55 [b]	536.88 ± 20.80 [b]
black nightshade	control	48.92 ± 4.32 [a]	84.67 ± 2.40 [a]	305.66 ± 126.34 [a]
	10%	48.46 ± 9.95 [a]	236.92 ± 58.78 [b]	1105.43 ± 9.45 [a]
	20%	30.48 ± 2.91 [a]	179.82 ± 3.76 [ab]	644.68 ± 460.14 [a]
	50%	35.23 ± 6.61 [a]	100.55 ± 2.79 [a]	420.61 ± 85.40 [a]
Johnson grass	control	37.40 ± 1.77 [ab]	77.59 ± 2.43 [a]	224.38 ± 14.53 [a]
	10%	33.36 ± 0.12 [a]	166.76 ± 57.79 [ab]	241.64 ± 46.46 [a]
	20%	74.51 ± 2.82 [c]	337.81 ± 31.68 [c]	731.04 ± 290.72 [a]
	50%	43.63 ± 2.98 [b]	228.08 ± 3.93 [b]	338.85 ± 8.69 [a]

The data are mean values ± standard error. [a–c] Values without the same superscripts within each column differ significantly ($p < 0.05$).

2.2.3. $O_2^{\cdot-}$/Radicals

Without taking into account common purslane, for which there was not enough plant material to carry out the analysis, the total quantity of removed superoxide anion radicals was proportionally higher with the increase in the TH concentrations in the seedlings of: wheat (319.63 nmol $O_2^{\cdot-}$//g FW), onion (817.26 nmol $O_2^{\cdot-}$//g FW), carrot (165.92 nmol $O_2^{\cdot-}$//g FW), alfalfa (49.91 nmol $O_2^{\cdot-}$//g FW), common lambsquarters (3366.46 nmol $O_2^{\cdot-}$//g FW), and cockspur grass (536.88 nmol $O_2^{\cdot-}$//g FW). The lowest quantities of removed superoxide anion radicals were noted in the seedlings of sunflower (56.08 nmol $O_2^{\cdot-}$//g FW) and soybean (18.21 nmol $O_2^{\cdot-}$//g FW) treated with 20% hydrolate, in the seedlings of white clover (25.45 nmol $O_2^{\cdot-}$//g FW) treated with 10% hydrolate, and in those of maize (5.11 nmol $O_2^{\cdot-}$//g FW) treated with 100% hydrolate. The results of the tested seedlings of the remaining species showed no statistically significant differences in terms of the quantity of removed superoxide anion radicals.

3. Discussion

The most important components of TH are the phenols carvacrol and thymol. Carvacrol and thymol have the potential to be used as bioherbicides and may help to reduce the use of synthetic herbicides and minimize damage to biodiversity and human health [32]. To date, these compounds have been associated with a phytotoxic effect on many weed and cultivated species. It was found that oregano (*Origanum acutidens* (Hand.-Mazz.) Ietsw.) essential oil and the phenols carvacrol and thymol had a phytotoxic effect on the seed germination and plant growth of red-root amaranth, common lambsquarters, and curly dock (*Rumex crispus* L.) [33]. The pepper-rosmarin (*Lippia sidoides* Cham.) essential oil, which contains thymol as its main component (84.90%), presented negative allelopathic effects on garden lettuce (*Lactuca sativa* L.) culture [34]. The phytotoxic activity of the *T. vulgaris* plant has been proven for various types of extracts and oils [35,36] and in soil experiments [37].

In this experiment, it was found that high concentrations of TH limited the seed germination of some weed species, such as common lambsquarters, red-root amaranth, and common purslane, whereas no negative effect was found on the germination of some field and vegetable crops. Seed germination efficiency depends on the seed's size and weight; thus, large seeds contain more nutrients and are usually capable of faster germination and growth than small seeds [38]. Moreover, seedlings of large-seeded species have higher survival rates than small-seeded species [39]. This may partly explain the good germination of sunflower, soybean, and maize seeds, even after treatment with higher concentrations of the hydrolates, whereas, in small seeds of common lambsquarters, red-root amaranth, and common purslane, germination was completely absent after these treatments. In addition, germination of the mentioned weeds was significantly influenced by the concentrations of thymol and carvacrol. By investigating the sensitivity of common lambsquarters and red-root amaranth seeds to thymol and carvacrol, it was found that the germination was completely absent when 10 mg thymol and 9.8 mg carvacrol were applied [33]. Other studies concluded that ormadere (*Tanacetum chiliophyllum* var. *chiliophyllum* (Fisch. & Mey.) Sch. Bip.) essential oil, which is also rich in borneol, also inhibits the germination of these two weed species [40]. In a previous study [41], it was confirmed that carvacrol and thymol are the main compounds that induce the total inhibitory effect against seed germination of common purslane and cockspur grass. It was also found that carvacrol completely inhibits the germination of common purslane seeds [42]. According to [43], thymol has the greatest impact on cockspur grass, reducing its seed germination and shoot growth; this was confirmed in our research, in which 50 and 100% of TH rich in thymol completely inhibited the germination of this weed species, and 20% hydrolate reduced the seedlings' length. In studies of the effect of oregano (*Origanum vulgare* L.) essential oil on Johnson grass, it was noted that carvacrol-rich (73.7%) essential oil inhibited Johnson grass seeds' germination (52.7%) [44]. Our research confirms the hypothesis that carvacrol, in addition to thymol, plays a significant role in inhibiting the germination of this weed. Regarding black nightshade, a previous study proved that conehead thyme (*Thymbra capitata* (L.) Cav.) essential oil (which has carvacrol as its main compound) blocked black nightshade germination and seedling growth at 0.5 μL/mL [42]. Although our research showed that lower concentrations of TH stimulate germination and seedlings' growth, higher concentrations (50 and 100%) led to their reduction.

Maize and sunflower showed the greatest resistance to the *T. vulgaris* hydrolates. The highest concentration of the applied hydrolate did not completely inhibit the germination of the maize seeds, although it was confirmed that thymol and carvacrol had an inhibitory effect on maize seeds' germination [41] and growth [45].

According to [46], borneol, carvacrol, and thymol significantly inhibited the garden pepperwort (*Lepidium sativum* L.) radicle length by 10^{-4} M, and borneol reduced the garden radish (*Raphanus sativus* L.) radicle length by the same amount. When the effect of *T. vulgaris* essential oil on some crops and weeds was examined, it was found that cockspur grass inhibited the radicle and seedling length, whereas, in the case of garden radish, bell pepper (*Capsicum annuum* L.), and garden lettuce, seed germination was completely absent [47]. Treatment with the individual components (thymol, carvacrol, and borneol) showed that thymol stimulated the germination of garden radish seeds, but thymol and carvacrol completely inhibited the seed germination of bell pepper, garden lettuce, common lambsquarters, and common purslane. Moreover, thymol inhibited the cockspur grass radicle and seedling length. In our research, TH inhibited the germination and seedling growth of cockspur grass, common purslane, and common lambsquarters, and those of the other tested weed species, which is in accordance with the results of the above-mentioned studies.

Usually, the major components of hydrosol are the same as those present in the essential oils [3]. For the three main components of the studied THS (thymol, borneol, and carvacrol), the phytotoxic effects on some weed and cultivated species were determined.

Biochemical analysis showed that the tested crops expressed different sensitivity to THS, with alfalfa being the most sensitive. By comparison, a statistically significant increase in MDA accumulation was recorded in the tested weeds: common lambsquarters, cockspur grass, and Johnson grass. This means that the THS provoked stress in these plants and strongly affected the lipid peroxidation [48].

4. Materials and Methods

4.1. Tested Plants

In this experiment, the seeds of the field and vegetable crops, i.e., soybean, sunflower, maize, wheat, carrot, onion, alfalfa, and white clover, and of the weeds, i.e., red-root amaranth, common lambsquarters, common purslane, cockspur grass, Johnson grass, and black nightshade, were used. The field and vegetable crop seeds were obtained from The Institute of Field and Vegetable crops in Novi Sad, and the weed seeds were collected from several localities during 2019 and 2020, and confirmed and deposited at the Herbarium of The Department of Plant and Environmental Protection, Faculty of Agriculture, University of Novi Sad.

4.2. Hydrolate

Thymus vulgaris cv. "N19" plants were grown at The Institute of Field and Vegetable crops, Novi Sad (experimental field in Bački Petrovac). Steam distillation was performed in a small-scale distillation unit according to [49]. After two hours (according to the requirements of the European Pharmacopoeia), the essential oil was separated from the aqueous layer and the hydrolate was purified using filter paper and stored in the refrigerator at 8 °C during the whole experiment. Simultaneous distillation–extraction using the Likens–Nickerson apparatus was performed to isolate the volatile compounds, which were further analyzed by GC-FID and GC-MS, according to [50]. The main volatile compound in *T. vulgaris* hydrolate was thymol with 73.6%, followed by borneol (7.1%), carvacrol (4.4%), linalool (2.8%), terpinen-4-ol (2.8%), and 1-octen-3-ol (2.5%). The other 25 compounds were present in percentage shares of less than 1.0% (Figure 2).

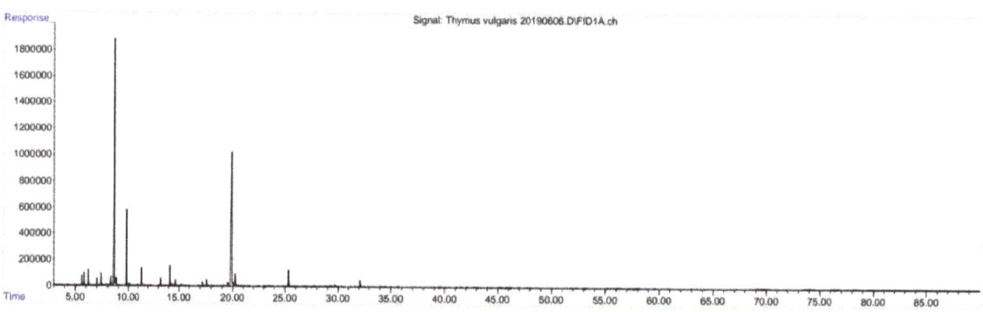

Figure 2. GC-FID chromatogram of *T. vulgaris* hydrolate.

The TH was diluted with distilled water in order to make 10, 20, 50, and 100% hydrolate solution, which was undertaken shortly before setting the experiment.

4.3. Seed Germination and Seedlings Length

After sterilization, which was performed according to [51], 25 seeds of each plant species were placed in a Petri dish (Ø 12 mm) on filter paper soaked with 10 mL of the particular hydrolate solution, i.e., distilled water in the case of the controls, in four replicates. The seeds of the tested weeds, and of the field and vegetable crops, were kept in a climate chamber at 22/20 °C during a 12 h photoperiod and at 60 ± 2% humidity for 10 days. The exceptions were Johnson grass and black nightshade, for which the seeds were kept at a higher temperature (30/26 °C). The seed germination rate and seedling length were recorded at the same time once a day during the whole experiment. The influence of TH on the tested seeds was determined by measuring the seedlings' shoot and root length [52], and by recording the number of germinated seeds on the last day of the experiment.

In order to better understand the obtained results considering the germination of the tested seeds of the cultivated and weed species, several germination indices were calculated:

- Germination percentage (GP) [%] [53] represents the final germination percentage of the seed population and is calculated according to Equation (1):

$$GP = \frac{Ng}{Nt} \times 10 \quad (1)$$

  where Ng is the number of the germinated seeds and Nt is the total number of the seeds.

- Coefficient of velocity of germination (CVG) [% day^{-1}] [54] represents the time required in order to reach the final germination percentage, and is calculated by Equation (2):

$$CVG = \frac{\Sigma_{i=1}^{k} Ni}{\Sigma_{i=1}^{k} Ni \times Ti} \times 100 \quad (2)$$

  where Ti is the time from the start of the experiment to the ith interval; Ni is the number of the seeds germinated in the ith interval (the number corresponding to the ith interval, not the accumulated number); and k is the total number of the intervals.

- Germination index (GI) [%] [55] reflects the germination speed; thus, a higher GI value indicates a faster germination rate. It is calculated according to Equation (3):

$$GI = (10 \times n_1) + (9 \times n_2) + \cdots + (1 \times n_{10}) \quad (3)$$

  where $n_1, n_2 \ldots n_{10}$ represent the number of germinated seeds on the 1st, 2nd, and subsequent days until the 10th day; 10, 9 ... 1 are the weights that are given to the number of germinated seeds on the 1st, 2nd, and subsequent days until the 10th day.

- Median germination time (t_{50}) [time] [56] represents the time required in order to reach the 50% of the final germination and is calculated by Equation (4):

$$t_{50} = \frac{Ti + \left(\frac{N}{2} - Ni\right) \times (Tj - Ti)}{Nj - Ni} \quad (4)$$

  where N is the final number of germinated seeds; Ni and Nj are the total number of seeds germinated in adjacent counts at time Ti and Tj, when $Ni < \frac{N}{2} < Nj$.

- Germination rate index (GRI) (% day^{-1}) [57] (after a modification) reflects the germination speed without distinguishing between the days with higher or lower germination since the percentage is evenly spread across the time frame. It is calculated according to Equation (5):

$$\text{GRI} = \frac{G_1}{1} + \frac{G_2}{2} + \cdots + \frac{G_x}{x} \quad (5)$$

where G_1 is the germination percentage × 100 on the 1st day after sowing; G_2 is the germination percentage × 100 on the 2nd day after sowing, etc., and G_x is the germination percentage × 100 on the xth day after sowing.

The seedlings of the tested plants were collected at the end of the experiment for the biochemical analysis.

4.4. Biochemical Analysis of the Tested Plants

For the determination of the biochemical parameters, 2 g of the fresh plant material (leaf) treated with TH, and the controls (untreated plants), were homogenized in phosphate buffer (10 mL; 0.1 M, pH 7.0). After centrifugation, the supernatants were used for the biochemical analyses. The biochemical parameters were determined spectrophotometrically using an UV/VIS spectrophotometer (Thermo Scientific Evolution 220 (Waltham, MA, USA)).

Superoxide dismutase (SOD) (EC 1.15.1.1) activity was determined according to the method of [58] with minor modifications. One unit of SOD activity was defined as the quantity of enzymes required to inhibit photochemical reduction of nitro blue tetrazolium (NBT) chloride by 50%. The SOD activity was expressed in U/g of the fresh weight (FW). The quantity of removed superoxide anion radicals was determined by the method of [59]. The total quantity of removed superoxide anion radicals ($O_2^{\cdot-}$) is reported in nmol $O_2^{\cdot-}$ per g of the fresh weight (nmol $O_2^{\cdot-}$/g FW). The content of malondialdehyde (MDA), which is the end product of the lipid peroxidation process, was measured at 532 nm using the thiobarbituric acid (TBA) test [58]. The total quantity of TBA-reactive substances is reported in nmol of the MDA equivalents per g of the fresh weight (nmol MDA/g FW).

4.5. Statistical Analysis

The values of the biochemical parameters are expressed as the mean ± standard error of the mean, and were tested by ANOVA followed by the comparison of the means by Duncan's multiple range test ($p < 0.05$). The data were analyzed using TIBC STATISTICA version 14.

5. Conclusions

The studied THS showed an inhibitory effect on the weed species, i.e., red-root amaranth, common lambsquarters, common purslane, cockspur grass, Johnson grass, and black nightshade.

Regardless of the applied THS concentration, sunflower seeds showed the highest germination rates, whereas red-root amaranth seeds showed the highest sensitivity. The lower hydrolate concentrations (10%) had a stimulative effect in terms of sunflower shoot and root seedlings' length, whereas, in the case of the other tested plants, a statistically significant reduction was noted.

The main compounds present in TH (thymol, borneol, and carvacrol) were able to inhibit the seed germination and seedling growth of several weeds, and exhibited a less phytotoxic effect on some of the tested field and vegetable crops. The tested hydrosol represents a potential source of an alternative and environmentally acceptable weed management compound for selective weed control.

Author Contributions: Conceptualization, M.P.; methodology, M.P., J.Š.E. and M.A.; software, T.S.; validation, N.S.; formal analysis, M.C. and M.A.; investigation, M.P. and T.S.; resources, M.R.; data curation, M.P. and T.S.; writing—original draft preparation, M.P. and T.S.; writing—review and editing, M.P. and T.S.; visualization, M.C.; supervision, B.K.; funding acquisition, B.K. All authors have read and agreed to the published version of the manuscript.

Funding: This research received no external funding.

Institutional Review Board Statement: Not applicable.

Informed Consent Statement: Not applicable.

Data Availability Statement: The data is contained within this manuscript. All data, tables and figures in this manuscript are original.

Conflicts of Interest: The authors declare no conflict of interest.

References

1. Aćimović, M.; Tešević, V.; Smiljanić, K.; Cvetković, M.; Stanković, J.; Kiprovski, B.; Sikora, V. Hydrolates–by-products of essential oil distillation: Chemical composition, biological activity and potential uses. *Adv. Technol.* **2020**, *9*, 54–70. [CrossRef]
2. D'Amato, S.; Serio, A.; Chavez López, C.; Paparella, A. Hydrosols: Biological activity and potential as antimicrobials for food applications. *Food Control* **2018**, *86*, 126–137. [CrossRef]
3. Andola, H.C.; Purohit, V.K.; Chauhan, R.S.; Arunachalam, K. Standardize quality standards for aromatic hydrosols. *Med. Plants* **2014**, *6*, 161–162. [CrossRef]
4. Labadie, C.; Cerutti, C.; Carlin, F. Fate and control of pathogenic and spoilage micro-organisms in orange blossom (*Citrus aurantium*), and rose flower (*Rosa centifolia*) hydrosols. *J. Appl. Microbiol.* **2016**, *121*, 1568–1579. [CrossRef] [PubMed]
5. Kuete, V. Thymus vulgaris. In *Medicinal Spices and Vegetables from Africa–Therapeutic Potential against Metabolic, Inflammatory, Infectious and Systemic Diseases*; Kuete, V., Ed.; Academic Press: Cambridge, MA, USA; Elsevier: Amsterdam, The Netherlands, 2017; p. 599. [CrossRef]
6. Nabavi, S.M.; Marchese, A.; Izadi, M.; Curti, V.; Daglia, M.; Nabavi, S.F. Plants belonging to the genus *Thymus* as antibacterial agents: From farm to pharmacy. *Food Chem.* **2015**, *173*, 339–347. [CrossRef]
7. Kowalczyk, A.; Przychodna, M.; Sopata, S.; Bodalska, A.; Fecka, I. Thymol and thyme essential oil-new insights into selected therapeutic applications. *Molecules* **2020**, *9*, 4125. [CrossRef]
8. Moisa, C.; Lupitu, A.; Pop, G.; Chambre, D.; Copolovici, L.; Cioca, G.; Bungau, S.G.; Copolovici, D.M. Variation of the chemical composition of *Thymus vulgaris* essential oils by phenological stage. *Rev. Chim.* **2019**, *70*, 633–637. [CrossRef]
9. Hay, Y.O.; Abril-Sierra, M.A.; Sequeda-Castañeda, L.G.; Bonnafous, C.; Raynaud, C. Evaluation of combinations of essential oils and essential oils with hydrosols on antimicrobial and antioxidant activities. *J. Pharm. Pharmacogn. Res.* **2018**, *6*, 216–230.
10. Rota, M.; Herrera, A.; Martínez, R.M.; Sotomayor, J.A.; Jordán, M.J. Antimicrobial activity and chemical composition of *Thymus vulgaris*, *Thymus zygis* and *Thymus hyemalis* essential oils. *Food Control* **2008**, *19*, 681–687. [CrossRef]
11. Yildiz, S.; Turan, S.; Kiralan, M.; Ramadan, M.F. Antioxidant properties of thymol, carvacrol, and thymoquinone and its efficiencies on the stabilization of refined and stripped corn oils. *J. Food Meas. Charact.* **2021**, *15*, 621–632. [CrossRef]
12. Beena; Kumar, D.; Rawat, D.S. Synthesis and antioxidant activity of thymol and carvacrol based Schiff bases. *Bioorg. Med. Chem. Lett.* **2013**, *23*, 641–645. [CrossRef] [PubMed]
13. Guarda, A.; Rubilar, J.F.; Miltz, J.; Galotto, M.J. The antimicrobial activity of microencapsulated thymol and carvacrol. *Int. J. Food Microbiol.* **2011**, *146*, 144–150. [CrossRef] [PubMed]
14. Gavaric, N.; Smole Mozina, S.; Kladar, N.; Bozin, B. Chemical Profile, Antioxidant and Antibacterial Activity of Thyme and Oregano Essential Oils, Thymol and Carvacrol and Their Possible Synergism. *J. Essent. Oil-Bear. Plants* **2015**, *18*, 1013–1021. [CrossRef]
15. Brito, C.; Hansen, H.; Espinoza, L.; Faúndez, M.; Olea, A.F.; Pino, S.; Díaz, K. Assessing the control of postharvest gray mold disease on tomato fruit using mixtures of essential oils and their respective hydrolates. *Plants* **2021**, *10*, 1719. [CrossRef] [PubMed]
16. Andrés, M.F.; González-Coloma, A.; Muñoz, R.; De la Peña, F.; Fernando, J.L.; Burillo, J. Nematicidal potential of hydrolates from the semi industrial vapor-pressure extraction of Spanish aromatic plants. *Environ. Sci. Pollut. Res.* **2018**, *25*, 29834–29840. [CrossRef]
17. Sardari, S.; Mobaien, A.; Ghassemifard, L.; Kamali, K.; Khavasi, N. Therapeutic effect of thyme (*Thymus vulgaris*) essential oil on patients with COVID19: A randomized clinical trial. *J. Adv. Med. Biomed. Res.* **2021**, *29*, 83–91. [CrossRef]
18. Rezatofighi, S.E.; Seydabadi, A.; Seyyed Nejad, S.M. Evaluating the efficacy of *Achillea millefolium* and *Thymus vulgaris* extracts against newcastle disease virus in *Ovo*. *Jundishapur J. Microbiol.* **2014**, *7*, e9016. [CrossRef]
19. Lelešius, R.; Karpovaitė, A.; Mickienė, R.; Drevinskas, T.; Tiso, N.; Ragažinskienė, O.; Kubilienė, L.; Maruška, A.; Šalomskas, A. In vitro antiviral activity of fifteen plant extracts against avian infectious bronchitis virus. *BMC Vet. Res.* **2019**, *29*, 178. [CrossRef]
20. Vimalanathan, S.; Hudson, J. Anti-influenza virus activity of essential oils and vapors. *Am. J. Essent.* **2014**, *2*, 47–53.
21. Feriotto, G.; Marchetti, N.; Costa, V.; Beninati, S.; Tagliati, F.; Mischiati, C. Chemical composition of essential oils from *Thymus vulgaris*, *Cymbopogon citratus*, and *Rosmarinus officinalis*, and their effects on the HIV-1 Tat protein function. *Chem. Biodivers.* **2018**, *15*, e1700436. [CrossRef]
22. Catella, C.; Camero, M.; Lucente, S.; Fracchiolla, G.; Sblano, S.; Tempesta, M.; Martella, V.; Buonavoglia, C.; Lanave, G. Virucidal and antiviral effects of *Thymus vulgaris* essential oil on feline coronavirus. *Res. Vet. Sci.* **2021**, *137*, 44–47. [CrossRef] [PubMed]
23. Kaewprom, K.; Chen, Y.H.; Lin, C.F.; Chiou, M.T.; Lin, C.N. Antiviral activity of *Thymus vulgaris* and *Nepeta cataria* hydrosols against porcine reproductive and respiratory syndrome virus. *Thai J. Vet. Med.* **2017**, *47*, 25–33.

24. Sağdiç, O.; Ozturk, I.; Tornuk, F. Inactivation of non-toxigenic and toxigenic *Escherichia coli* O157:H7 inoculated on minimally processed tomatoes and cucumbers: Utilization of hydrosols of *Lamiaceae* spices as natural food sanitizers. *Food Control* **2013**, *30*, 7–14. [CrossRef]
25. Tornuk, F.; Cankurt, H.; Ozturk, I.; Sagdic, O.; Bayram, O.; Yetim, H. Efficacy of various plant hydrosols as natural food sanitizers in reducing Escherichia coli O157:H7 and *Salmonella typhimurium* on fresh cut carrots and apples. *Int. J. Food Microbiol.* **2011**, *148*, 30–35. [CrossRef] [PubMed]
26. Ozturk, I.; Tornuk, F.; Caliskan-Aydogan, O.; Durak, M.Z.; Sagdic, O. Decontamination of iceberg lettuce by some plant hydrosols. *LWT* **2016**, *74*, 48–54. [CrossRef]
27. Varel, V.H. Carvacrol and thymol reduce swine waste odor and pathogens: Stability of oils. *Curr. Microbiol.* **2002**, *44*, 38–43. [CrossRef]
28. Liu, B.; Chen, B.; Zhang, J.; Wang, P.; Feng, G. The environmental fate of thymol, a novel botanical pesticide, in tropical agricultural soil and water. *Toxicol. Environ. Chem.* **2017**, *99*, 223–232. [CrossRef]
29. Azirak, S.; Karaman, S. Allelopathic effect of some essential oils and components on germination of weed species. *Acta Agric. Scand.-B Soil Plant Sci.* **2008**, *58*, 88–92. [CrossRef]
30. de Assis Alves, T.; Fontes Pinheiro, P.; Praça-Fontes, M.M.; Andrade-Vieira, L.F.; Barelo Corrêa, K.; de Assis Alves, T.; da Cruz, F.A.; Lacerda Júnior, V.; Ferreira, A.; Bastos Soares, T.C. Toxicity of thymol, carvacrol and their respective phenoxyacetic acids in *Lactuca sativa* and *Sorghum bicolor*. *Ind. Crops Prod.* **2018**, *114*, 59–67. [CrossRef]
31. Balboul, T.; Ophir, A.; Dotan, A. Essential oils as natural root-repellent herbicides for drip irrigation systems. *Polym. Adv. Technol.* **2022**, 1–10. [CrossRef]
32. Pinheiro, P.F.; Costa, A.V.; de Assis Alves, T.; Galter, I.N.; Pinheiro, C.A.; Pereira, A.F.; Oliveira, C.M.; Fontes, M.M. Phytotoxicity and cytotoxicity of essential oil from leaves of *Plectranthus amboinicus*, carvacrol, and thymol in plant bioassays. *J. Agric. Food Chem.* **2015**, *63*, 8981–8990. [CrossRef] [PubMed]
33. Kordali, S.; Cakir, A.; Ozer, H.; Cakmakci, R.; Kesdek, M.; Mete, E. Antifungal, phytotoxic and insecticidal properties of essential oil isolated from Turkish *Origanum acutidens* and its three components, carvacrol, thymol and p-cymene. *Bioresour. Technol.* **2008**, *99*, 8788–8795. [CrossRef] [PubMed]
34. Marco, C.A.; Teixeira, E.; Simplício, A.; Oliveira, C.; Costa, J.; Feitosa, J. Chemical composition and allelopathyc activity of essential oil of *Lippia sidoides Cham. Chil. J. Agric. Res.* **2021**, *72*, 157–160. [CrossRef]
35. Hemada, M.; El-Darier, S. Comparative study on composition and biological activity of essential oils of two *Thymus* species grown in Egypt. *Am.-Eurasian J. Agric. Environ. Sci.* **2011**, *11*, 647–654.
36. Arouiee, H.; Ghasemimohsenabad, S.; Azizi, M.; Neamati, S.H. Allelopathic effects of some medicinal plants extracts on seed germination and growth of common weeds in Mashhad Area. In Proceedings of the 8th International Symposium on Biocontrol and Biotechnology, Pattaya, Thailand, 4–6 October 2010.
37. Linhart, Y.B.; Gauthier, P.; Keefover-Ring, K.; Thompson, J.D. Variable phytotoxic effects of *Thymus vulgaris* (*Lamiaceae*) terpenes on associated species. *Int. J. Plant Sci.* **2015**, *176*, 20–30. [CrossRef]
38. Sadeghi, H.; Khazaei, F.; Sheidaei, S.; Yari, L. Effect of seed size on seed germination behavior of safflower (*Carthamus tinctorius* L.). *J. Agric. Biol. Sci.* **2011**, *6*, 5–8.
39. Moles, A.T.; Westoby, M. Seed size and plant strategy across the whole life cycle. *Oikos* **2006**, *113*, 91–105. [CrossRef]
40. Salamci, E.; Kordali, S.; Kotan, R.; Cakir, A.; Kaya, Y. Chemical compositions, antimicrobial and herbicidal effects of essential oils isolated from Turkish *Tanacetum aucheranum* and *Tanacetum chiliophyllum* var. *chiliophyllum*. *Biochem. Syst. Ecol.* **2007**, *35*, 569–581. [CrossRef]
41. Ibáñez, M.D.; Blázquez, M.A. Phytotoxicity of essential oils on selected weeds: Potential hazard on food crops. *Plants* **2018**, *7*, 79. [CrossRef]
42. Verdeguer, M.; Torres-Pagan, N.; Muñoz, M.; Jouini, A.; García-Plasencia, S.; Chinchilla, P.; Berbegal, M.; Salamone, A.; Agnello, S.; Carrubba, A.; et al. Herbicidal activity of *Thymbra capitata* (L.) Cav. essential oil. *Molecules* **2020**, *25*, 2832. [CrossRef]
43. Saad, M.; Gouda, N.; Abdelgaleil, S. Bioherbicidal activity of terpenes and phenylpropenes against *Echinochloa crus-galli*. *J. Environ. Sci. Health-B Pestic. Food Contam. Agric. Wastes* **2019**, *54*, 954–963. [CrossRef] [PubMed]
44. Matković, A.; Marković, T.; Vrbničanin, S.; Sarić-Krsmanović, M.; Božić, D. Chemical composition and in vitro herbicidal activity of five essential oils on seeds of Johnson grass (*Sorghum halepense* L. Pers.). *Lek. Sirovine* **2018**, *38*, 44–50. [CrossRef]
45. Zunino, M.P.; Zygadlo, J.A. Effect of monoterpenes on lipid oxidation in maize. *Planta* **2004**, *219*, 303–309. [CrossRef]
46. De Martino, L.; Mancini, E.; Rolim de Almeida, L.F.; De Feo, V. The antigerminative activity of twenty-seven monoterpenes. *Molecules* **2010**, *15*, 6630–6637. [CrossRef] [PubMed]
47. Angelini, L.G.; Carpanese, G.; Cioni, P.L.; Morelli, I.; Macchia, M.; Flamini, G. Essential oils from Mediterranean *Lamiaceae* as weed germination inhibitors. *J. Agric. Food Chem.* **2003**, *51*, 6158–6164. [CrossRef] [PubMed]
48. Šućur, J.; Konstantinović, B.; Crnković, M.; Bursić, V.; Samardžić, N.; Malenčić, Đ.; Prvulović, D.; Popov, M.; Vuković, G. Chemical composition of *Ambrosia trifida* L. and its allelopathic influence on crops. *Plants* **2021**, *10*, 2222. [CrossRef]
49. Aćimović, M.; Cvetković, M.; Stanković Jeremić, J.; Pezo, L.; Varga, A.; Čabarkapa, I.; Kiprovski, B. Biological activity and profiling of *Salvia sclarea* essential oil obtained by steam and hydrodistillation extraction methods via chemometrics tools. *Flavour Fragr. J.* **2022**, *37*, 20–32. [CrossRef]

50. Aćimović, M.; Lončar, B.; Stanković Jeremić, J.; Cvetković, M.; Pezo, L.; Pezo, M.; Todosijević, M.; Tešević, V. Weather conditions influence on lavandin essential oil and hydrolate quality. *Horticulturae* **2022**, *8*, 281. [CrossRef]
51. Voll, E.; Voll, C.E.; Filho, R.V. Allelopathic effects of aconitic acid on wild poinsettia (*Euphorbia heterophy* L. La) and morningglory (*Ipomoea grandifo* Lia). *J. Environ. Sci. Health-B* **2005**, *40*, 69–75. [CrossRef]
52. Marinov-Serafimov, P.; Dimitrova, T.S.; Golubinova, I.; Ilieva, A. Study of suitability of some solutions in allelopathic researches. *Herbologia* **2007**, *8*, 1–10.
53. ISTA-International Seed Testing Association. International rules for seed testing. *Seed Sci. Technol.* **1993**, *21*, 142–168.
54. Jones, K.W.; Sanders, D. The influence of soaking paper seed in water or potassium salt solutions on germination at three temperatures. *J. Seed Technol.* **1987**, *11*, 97–102. [CrossRef]
55. Bench, A.R.; Fenner, M.; Edwards, P. Changes in germinability, ABA content and ABA embryonic sensitivity in developing seeds of *Sorghum bicolor* (L.) Moench induced by water stress during grain filling. *New Phytol.* **1991**, *118*, 339–347. [CrossRef] [PubMed]
56. Farooq, M.; Basra, S.M.A.; Ahmad, N.; Hafeez, K. Thermal hardening: A new seed vigor enhancement tool in rice. *J. Integr. Plant Biol.* **2005**, *47*, 187–193. [CrossRef]
57. Esechie, H. Interaction of salinity and temperature on the germination of sorghum. *J. Agron. Crop Sci.* **1994**, *172*, 194–199. [CrossRef]
58. Mandal, S.; Mitra, A.; Mallick, N. Biochemical characterization of oxidative burst during interaction between *Solanum lycopersicum* and *Fusarium oxysporum* f. sp. *lycopersici*. *Physiol. Mol. Plant Pathol.* **2008**, *72*, 56–61. [CrossRef]
59. Misra, H.P.; Fridovich, I. The role of superoxide anion in the autoxidation of epinephrine and a simple assay for superoxide dismutase. *J. Biol. Chem.* **1972**, *247*, 3170–3175. [CrossRef]

Article

Application of an Enzymatic Hydrolysed L-α-Amino Acid Based Biostimulant to Improve Sunflower Tolerance to Imazamox

Eloy Navarro-León [1,*], Elisabet Borda [2], Cándido Marín [2], Nuria Sierras [2], Begoña Blasco [1] and Juan M. Ruiz [1]

[1] Department of Plant Physiology, Faculty of Sciences, University of Granada, 18071 Granada, Spain
[2] R & D Plant Health, Bioiberica S.A.U., 08389 Barcelona, Spain
* Correspondence: enleon@ugr.es; Tel.: +34-958-243255

Abstract: Herbicides, commonly used in agriculture to control weeds, often cause negative effects on crops. Safeners are applied to reduce the damage to crops without affecting the effectiveness of herbicides against weeds. Plant biostimulants have the potential to increase tolerance to a series of abiotic stresses, but very limited information exists about their effects on herbicide-stressed plants. This study aims to verify whether the application of a potential safener such as Terra-Sorb®, an L-α-amino acid-based biostimulant, reduces the phytotoxicity of an Imazamox-based herbicide and to elucidate which tolerance mechanisms are induced. Sunflower plants were treated with Pulsar® 40 (4% Imazamox) both alone and in combination with Terra-Sorb®. Plants treated with the herbicide in combination with Terra-Sorb® showed higher growth, increased acetolactate synthase (ALS) activity, and amino acid concentration with respect to the plants treated with Imazamox alone. Moreover, the biostimulant protected photosynthetic activity and reduced oxidative stress. This protective effect could be due to the glutathione S-transferase (GST) induction and antioxidant systems dependent on glutathione (GSH). However, no effect of the biostimulant application was observed regarding phenolic compound phenylalanine ammonium-lyase (PAL) activity. Therefore, this study opens the perspective of using Terra-Sorb® in protecting sunflower plants against an imazamox-based herbicide effect.

Keywords: amino acids; biostimulant; glutathione; herbicide; safener; sunflower

Citation: Navarro-León, E.; Borda, E.; Marín, C.; Sierras, N.; Blasco, B.; Ruiz, J.M. Application of an Enzymatic Hydrolysed L-α-Amino Acid Based Biostimulant to Improve Sunflower Tolerance to Imazamox. *Plants* **2022**, *11*, 2761. https://doi.org/10.3390/plants11202761

Academic Editors: Špela Mechora and Dragana Šunjka

Received: 14 September 2022
Accepted: 17 October 2022
Published: 19 October 2022

1. Introduction

Weeds are included among the main biotic limitations that affect optimal crop productivity. Weeds establish a competitive relationship with the main crop to obtain water and nutrients essential for the growth and development of plants. This competition may reduce the yield and quality of the crops by more than 50% [1,2]. Herbicide application is currently the most widely used and efficient agricultural practice to restrict the proliferation of weeds. Herbicides are biologically active organic compounds that alter the basic physiological processes of plants, causing their death. For instance, the mechanism of action of the Imazamox [(5-(methoxymethyl)-2-(4-methyl-5-oxo-4-propan-2-yl-1H-imidazol-2-yl) pyridine-3-carboxylic acid)] herbicide is the inhibition of acetolactate synthase (ALS) enzymes. ALS is necessary for the synthesis of essential branched-chain amino acids, such as valine, leucine, and isoleucine, necessary for plant growth [3,4].

Herbicides must meet two antagonistic criteria: on the one hand, they must allow for the efficient control of a wide range of weed species and, on the other hand, be harmless or less phytotoxic for the crops they are applied to [5]. The most characteristic visible symptoms of phytotoxicity that appear a few days after the application of herbicides on plants are an accumulation of anthocyanins, chlorosis, and necrotic spots on the leaves [6]. Different studies have shown that herbicides restrict plant growth and development through increased oxidative stress and decreased antioxidant and photosynthetic activities. Thus, a wide range of herbicide types causes oxidative stress-inducing the accumulation of reactive oxygen species (ROS) such as H_2O_2 and O_2^- [7,8]. This phenomenon was observed in plants treated with ALS-inhibiting herbicides such as mesosulfuron- methyl, although there are a lack of studies about the effect of imazamox herbicides on oxidative stress [9].

Concerning photosynthesis, the analysis of chlorophyll (Chl) *a* fluorescence is a powerful tool for studying this process and the effects of stress on it [10]. The photosynthetic process is initiated when light is absorbed by the pigments of the membrane antenna complexes. Part of the absorbed energy is transferred as excitation energy and trapped by the reaction centers (RCs), where it is useful for photosynthetic reactions, and the other part is dissipated as heat and low-energy light (fluorescence) [11]. In Chl *a* fluorescence analysis, a dark-adapted leaf with the photosystem II (PSII) electron-acceptor pools completely oxidized was illuminated with a high-intensity light source. This illumination produced a rapid polyphasic rise of fluorescence until the electron acceptors in the photosystems were reduced. This polyphasic curve of fluorescence presented several phases (O, J, I, P). Derived from this curve, multiple parameters indicative of the photosystem functioning were available [10]. Several studies showed that the analysis of Chl *a* fluorescence is very useful for estimating the damage caused by Imazamox-based herbicides, although they do not directly affect the photosynthesis process [12,13].

Certain compounds called protectors or "safeners" are usually applied together with herbicides to reduce phytotoxicity. Safeners are applied to reduce damage to crops without affecting the effectiveness of herbicides against weeds [5]. Currently, the use of herbicides associated with "safeners" represents roughly 30% of the value of herbicide sales [14]. Safeners induce the resistance responses of plants to herbicide toxicity through the following mechanisms: retention on the leaf surface to avoid herbicide drift, decreased absorption and translocation, desensitization of the herbicide target protein, and detoxification or herbicide metabolism in plants [6]. This latter mechanism of herbicide resistance involves the performance of a series of physiological processes called non-target site resistance (NTSR). Two important plant mechanisms of the NTSR are glutathione (GSH) homeostasis and phenolic compounds [15].

GSH is a tripeptide made up of three amino acids: glutamic acid, cysteine, and glycine. The synthesis and availability of these amino acids are essential for reducing the toxic effects of herbicides. Thus, GSH, together with the action of the enzyme glutathione S-transferase (GST), produces the herbicide conjugation with GSH, forming a complex that later facilitates the degradation of the herbicide into non-toxic compounds for plants [16]. Furthermore, GSH acts directly in the detoxification of toxic radicals produced by herbicides in plants, such as H_2O_2 and methylglyoxal, through the induction of GSH-peroxidase enzymes and the glyoxalase system [17,18].

In addition, the application of phenolics in safeners produces an improvement in sensitive plants to the resistance of herbicides [19,20]. Salicylic acid is a phenolic compound that acts as a hormone and induces the antioxidant response of plants, which could explain its protective effect. Furthermore, the concentration of phenolic compounds increases very significantly in plants resistant to the Imazamox herbicide, enhancing the plant antioxidant response. Hence, the direct use of salicylic acid and other polyphenols as safeners, or the use of compounds that stimulate the synthesis of these compounds from secondary metabolism, could be very useful in herbicide tolerance [21]. Despite the effectiveness and low cost of most commercial safeners, some are toxic to different organisms in aquatic and terrestrial ecosystems [22]. Therefore, the search for new herbicide protective compounds that are efficient and more environmentally friendly is imperative and currently represents a new direction of research in herbicide development.

The use of biostimulants as a means for increasing the tolerance of plants when exposed to different stress factors, including herbicides, could be very positive [16,23]. Thus, the growth of plants that are under stress conditions such as drought, nutrient shortage, and other factors, including herbicides, could be improved by biostimulant application [24–27]. Surprisingly, very limited information exists about the effects of L-α-amino acid-based biostimulants on herbicide-stressed plants [28].

The R&D department of Bioiberica, S.A. company, developed an L-α-amino acid-based biostimulant called Terra-Sorb®. It was obtained through an exclusive enzymatic hydrolysis technology (Enzyneer®) of animal-origin proteins, and it contains all of the biologically active amino acids in plants. Other studies observed that the application of this biostimulant improved the regulation of Rubisco, tolerance to cold stress, and nitrogen deficiency [29–31].

The first objective of this study is to verify whether the combined application of an Imazamox-based herbicide with the biostimulant Terra-Sorb® reduces the phytotoxic effect compared to treatment with the herbicide application alone. This herbicide-biostimulant combination was chosen given the negative effects of the herbicide on the amino acid synthesis and the significant amino acid content of the biostimulant. In addition, through the analysis of ALS activity, AAs and phenolics profiles, GSH metabolism, and stress indicators, this study tries to elucidate which mechanisms of resistance to herbicides are induced by the application of the foliar Bioiberica biostimulant (Terra-Sorb®).

2. Results and Discussion

Herbicide application is currently the most widely used and efficient agricultural practice to restrict weed proliferation. Herbicides are biologically active organic compounds that alter the basic physiological processes, causing plant death [5]. However, different studies showed that herbicides have also restricted the growth and development of plants for agronomic interests. One of the most obvious symptoms of herbicide phytotoxicity is biomass loss, which has been well characterized in numerous species and is, therefore, one of the most widely used indices to define the degree of resistance or sensitivity to this stress type [5,6]. Therefore, biomass production and RGR were measured as indicators of the protective effect of the Terra-Sorb® biostimulant against herbicide application. The application of the herbicide Pulsar_40 to sunflower plants caused wilting symptoms, turgor loss, the appearance of chlorotic spots on leaves, and less growth between leaf nodes (Figure 1). However, the application of the Terra-Sorb® biostimulant effectively mitigated these negative symptoms (Figure 1), and plants showed 52% more shoot biomass and almost six times more RGR compared to plants treated only with the herbicide (Figure 2).

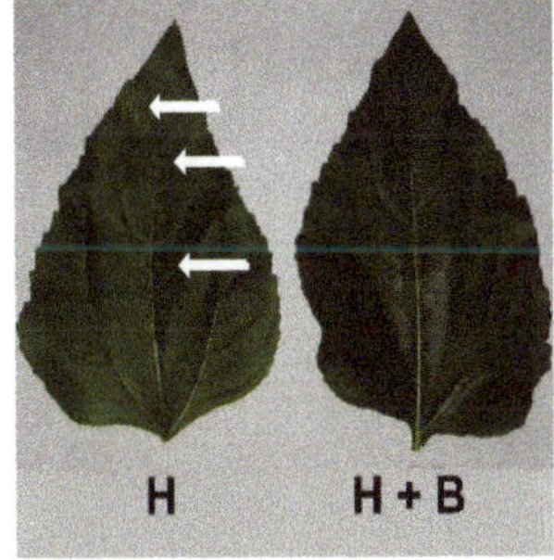

Figure 1. Sunflower plants and leaves 4-days after the application of the treatments (H: Pulsar_40; H + B: Pulsar_40 + Terra-Sorb®). Arrows on sunflower leaves indicate chlorotic spots on plants treated with the herbicide (H).

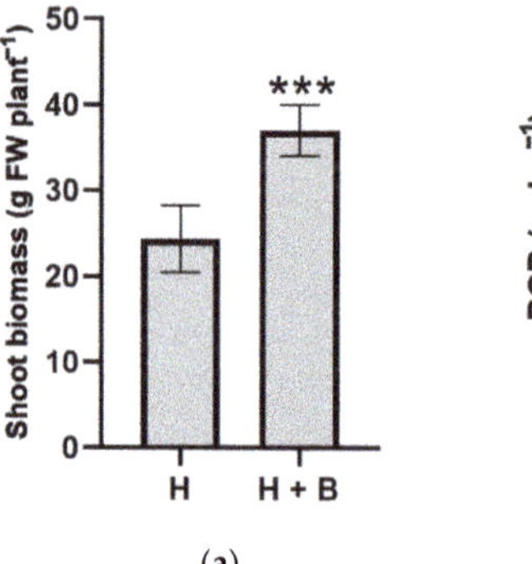

(a)

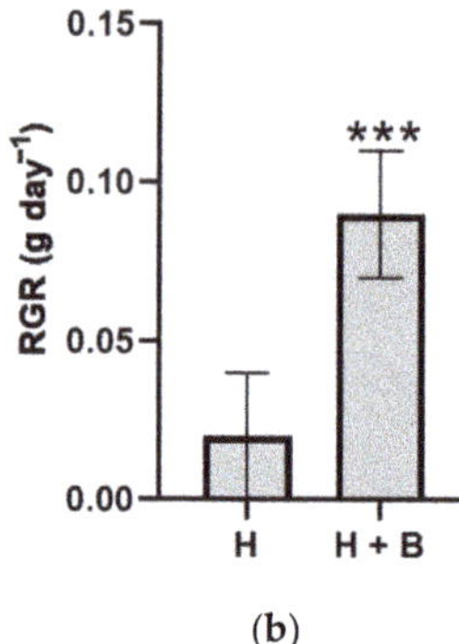

(b)

Figure 2. Biomass production (**a**) and relative growth rate (RGR) (**b**) of shoot of sunflower plants subjected to the treatments (H: Pulsar 40; H + B: Pulsar 40 + Terra-Sorb®). Values are expressed as means ± standard error (n = 9). The level of significance was $p < 0.001$ (***).

The analysis of ALS enzyme activity is the main indicator of the phytotoxic effect of imidazolinone-type herbicides such as Imazamox (Pulsar 40) when used in sunflower plants [3,4]. In our experiment, plants supplied with the Terra-Sorb® biostimulant and the herbicide registered greater ALS enzyme activities compared to plants treated only with the Pulsar 40 herbicide (Figure 3). This result indicates the protective effect of ALS activity by Terra Sorb® foliar against the herbicide.

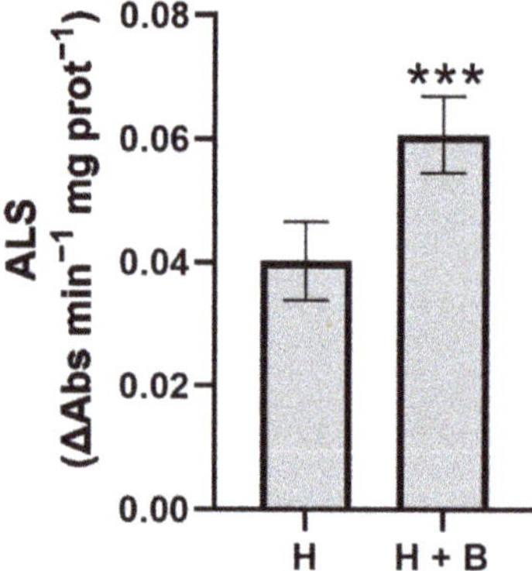

Figure 3. ALS activity in sunflower plants subjected to the treatments (H: Pulsar-40; H + B: Pulsar-40 + Terra Sorb® foliar). Values are expressed as means ± standard error (n = 9). The level of significance was represented by $p < 0.001$ (***).

We analyzed the aminogram profile of the sunflower plants to relate the ALS activity to the amino acids profile of the sunflower plants. Thus, the lower ALS enzyme activity (Figure 3) might reduce concentrations of leucine, isoleucine, and valine in plants treated with the herbicide, as observed in other studies [32,33]. Otherwise, the sunflower plants treated with the biostimulant Terra-Sorb® together with the herbicide showed higher concentrations of these amino acids, which suggests the protective effect of this biostimulant to avoid the reduction of amino acid synthesis (Table 1). This effect could have a great contribution to the best growth observed in the sunflower plants of the combined treatment (Figures 1 and 2). In addition, plants supplied with Terra-Sorb® and the herbicide presented higher levels of alanine and aspartate, which are precursors of leucine, isoleucine, and valine [34]. Likewise, the concentrations of other amino acids, such as glutamate, glycine, serine, and proline, were higher in plants treated with the combined treatment compared to plants treated only with the herbicide (Table 1). The increase in these amino acids might be essential to the protective effect that this compound exerts against the herbicide.

Table 1. Aminogram (μg g^{-1} DW) in sunflower plants subjected to the treatments.

	H	H + B	*p*-Value
Alanine	145.69 ± 2.30	210.92 ± 3.33	***
Arginine	37.26 ± 0.59	36.13 ± 0.57	NS
Asparagine	ND	ND	
Aspartate	237.33 ± 2.08	306.49 ± 4.85	***
Glutamate	381.49 ± 6.03	430.49 ± 6.81	***
Glutamine	ND	ND	
Glycine	169.06 ± 2.67	195.13 ± 3.09	***
Histidine	11.72 ± 0.22	11.49 ± 0.21	NS
Isoleucine	460.64 ± 7.28	751.21 ± 11.88	***
Leucine	821.21 ± 12.98	1113.13 ± 17.60	***
Lysine	7.23 ± 0.11	7.30 ± 0.12	NS
Phenylalanine	479.71 ± 7.58	485.64 ± 7.68	NS
Proline	43.20 ± 0.68	62.93 ± 1.00	***
Serine	635.08 ± 10.04	691.70 ± 10.94	***
Threonine	42.86 ± 0.68	42.63 ± 0.67	NS
Tryptophan	355.27 ± 5.62	361.70 ± 5.72	NS
Tyrosine	512.92 ± 8.11	514.62 ± 8.14	NS
Valine	1008.64 ± 15.95	1475.79 ± 23.33	***
Methionine	75.51 ± 1.19	75.88 ± 1.20	NS
Cysteine	7.39 ± 0.13	7.23 ± 0.12	NS

H: Pulsar-40; H + B: Pulsar-40 + Terra-Sorb®. Values are means ± standard error (n = 9). The levels of significance were represented as NS ($p > 0.05$) and *** ($p < 0.001$). ND—Non detected.

In addition to biomass production and ALS activity, the use of different photosynthetic parameters is commonly used as an indicator of the phytotoxic effects of many herbicide types [9,13,35,36]. Photosynthetic activity analysis through fluorescence is a quick, non-invasive method to check the physiological state of plants. This technique is based on the induction of fluorescence emission to a dark-adapted leaf. The fluorescence may be represented over time with a curve showing O, J, I, and P phases [10], as observed in Figure 4. In the present study, sunflower plants treated with the combined treatment showed lower fluorescence values during all the OJIP phases of fluorescence emission, lower V_J and DI_o/RC, and higher S_m values in comparison to plants treated with the herbicide alone (Figures 4 and 5). These results suggest that a lower energy dissipation, and therefore better use of light energy, is produced in plants supplied with the biostimulant under herbicide toxic conditions. Other OJIP parameters also indicated a positive influence of the Terra-Sorb® application. Thus, the greater RC/ABS and γ_{RC} point led to a greater availability of RC and energy capture. In addition, the higher N, ψ_o, Φ_{Eo} indicated a better quinone A reduction and electron transport capacities. Finally, the great increments in PI_{ABS} and PI_{total} showed an overall improvement in photosynthetic performance (Figure 5). In their study, Balabanova et al. [12] = observed a diminution in these parameters as a response to Imazamox in sunflower plants but did not observe a positive effect in the biostimulant application together with the herbicide compared to the herbicide alone. These results suggest different responses of fluorescence parameters in response to different biostimulants.

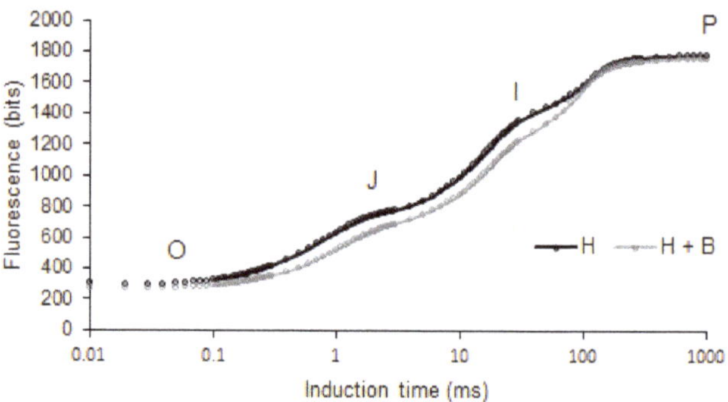

Figure 4. Native fluorescence induction curves of sunflower plants subjected to the treatments (H: Pulsar-40; H + B: Pulsar-40 + Terra Sorb® foliar). Represented values are means from nine data (*n* = 9).

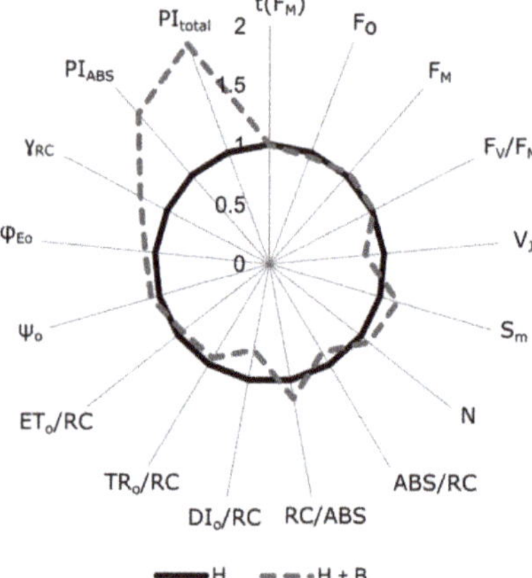

Figure 5. Radar plot showing OJIP-test parameters derived from Chl *a* fluorescence induction in sunflower plants subjected to H + B treatment: Pulsar-40 + Terra Sorb® foliar. Data normalization were made to enable the representation of all parameters on the same scale. The dotted gray line values represent the relative increase or decrease in plants of the H + B treatment with respect to plants of the H treatment (black line).

The application of herbicides usually produces an induction of oxidative stress. Indeed, the concentrations of indicators such as MDA and ROS tend to increase in plants affected by herbicide phytotoxicity [9]. Under these stress conditions, the application of biostimulants could reduce cellular oxidative damage and, therefore, the peroxidation of membrane lipids by regulating the antioxidant defense and reducing ROS levels in plants [37]. Accordingly, in the present experiment, sunflower plants treated with the Terra-Sorb® biostimulant showed lower MDA, O_2^- and H_2O_2 values, which confirmed the protective effect of the Terra-Sorb® in reducing the phytotoxic effect of the Pulsar-40 herbicide on sunflower plants (Table 2).

Table 2. Stress indicators in sunflower plants subjected to the treatments.

	MDA (μg g^{-1} FW)	O_2^- (μg g^{-1} FW)	H_2O_2 (μg g^{-1} FW)
H	9.68 ± 0.49	1.64 ± 0.35	12.64 ± 1.47
H + B	8.41 ± 0.35	1.08 ± 0.27	8.33 ± 1.81
p-value	***	*	***

H: Pulsar-40; H + B: Pulsar 40 +Terra-Sorb®. Values are means ± standard error (n = 9). The levels of significance were represented as * ($p < 0.05$) and *** ($p < 0.001$).

One of the most important mechanisms in non-target herbicide resistance involves the action of the GSH tripeptide and the GST enzyme [38]. The combined action of the GSH-GST system produces the GSH–herbicide complex, from which the herbicide undergoes a series of transformations, thus reducing its phytotoxic effects [38–40]. In the present experiment, sunflower plants treated with the herbicide plus biostimulant presented greater GSH concentrations and GST activities than plants supplied with the herbicide alone. The greater GSH levels could be favored by the higher presence of glutamic, glycine, and serine (Table 1). This last amino acid produces cysteine through the enzyme serine acetyltransferase, which, together with glutamic and glycine, is one of the three amino acids that make up the GSH tripeptide [41]. Briefly, the induction of the GSH-GST in plants treated with the Terra-Sorb® and Pulsar-40 herbicide could be defined as an action mechanism that would explain the protective biostimulant effect inducing the phytotransformation and/or degradation of the herbicide, and therefore reducing its intracellular concentration and its phytotoxicity in plants.

There are a lot of studies which focus on the involvement of GSH acting directly in the detoxification of toxic radicals produced by the herbicide application on plants, such as ROS and the methylglyoxal compound. This detoxification process is crucial in the so-called NTSR and is carried out by enzymes such as GPX and glyoxalases (Gly I and Gly II) [17,18]. In our experiment, plants supplied with the combined treatment showed higher GPX and Gly activities compared to the plants treated with the herbicide alone (Table 3). The induction of these enzymes in sunflower plants by the application of the Terra-Sorb® could contribute to the observed minimum values of the oxidative stress indicators (Table 2).

Table 3. GSH homeostasis in sunflower plants subjected to the treatments.

	GSH (mg g^{-1} FW)	GST (ΔAbs min^{-1} mg $prot^{-1}$)	GPX (ΔAbs min^{-1} mg $prot^{-1}$)	Gly I (ΔAbs min^{-1} mg $prot^{-1}$)	Gly II (ΔAbs min^{-1} mg $prot^{-1}$)
H	0.14 ± 0.02	0.32 ± 0.07	0.04 ± 0.01	0.06 ± 0.01	0.05 ± 0.01
H + B	0.27 ± 0.05	1.04 ± 0.27	0.07 ± 0.01	0.15 ± 0.03	0.10 ± 0.02
p-value	***	**	***	***	***

H: Pulsar_40; H + B: Pulsar_40 +Terra-Sorb®. Values are means ± standard error (n = 9). The levels of significance were represented as ** ($p < 0.01$) and *** ($p < 0.001$).

Finally, regarding ROS detoxification, it should be noted that higher proline levels were detected in plants supplied with the biostimulant (Table 1). Proline is a very reliable indicator of resistance against abiotic stress since it has an osmoprotective, osmoregulatory, and antioxidant role against the generation of ROS [42,43]. The increase in proline concentration in the plants subjected to the combined treatment could explain the decrease in ROS levels, especially H_2O_2 (Table 2).

The phenolics profile was analyzed because these compounds constitute another herbicide resistance mechanism that enhances the antioxidant response of plants [21]. Moreover, the application of a phenol compound-like salicylic acid as a "safener" improves herbicide resistance in sensitive plants because it induces antioxidant plant responses [19,20]. These results have led to the proposal of the direct use of salicylic acid and other polyphenols as "safeners" or the use of compounds that stimulate their synthesis from secondary metabolism [20,21]. Additionally, we analyzed the phenylalanine ammonium-lyase (PAL) enzyme activity, which is key in the synthesis of phenolic compounds [44]. Nevertheless, neither the PAL activity nor the polyphenolic compounds which were detected varied significantly between the applied treatments (Figure 6 and Table 4). Considering that neither the PAL activity (Figure 5) nor the concentration of salicylic acid nor the rest of the phenolic compounds (Table 4) were modified with the application of Terra-Sorb®, we can rule out this physiological process as part of the mode of action that could be involved in the protective effect of the biostimulant.

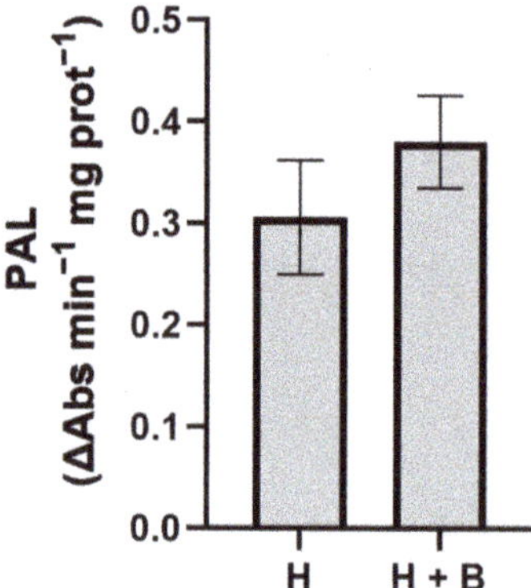

Figure 6. PAL activity in sunflower plants subjected to the treatments (H: Pulsar-40; H + B: Pulsar-40 + Terra-Sorb®). Values are expressed as means ± standard error (n = 9).

Table 4. Phenolic compound profile (mg g^{-1} DW) in sunflower plants subjected to the treatments.

	H	H + B	p-Value
5-CQA (chlorogenic acid)	3.17 ± 0.10	3.17 ± 0.13	NS
5-CQA-Isomer	4.03 ± 0.05	4.03 ± 0.05	NS
Di-caffeoylquinic acid	10.56 ± 2.10	10.56 ± 2.02	NS
SA (µg g^{-1} DW)	4.03 ± 0.06	4.07 ± 0.06	NS

H: Pulsar_40; H + B: Pulsar_40 + Terra-Sorb®. Values are means ± standard error (n = 9). The level of significance is represented as NS ($p > 0.05$).

3. Materials and Methods

3.1. Plant Material and Experimental Design

Sunflower seeds, *Helianthus annuus* var. Neoma (first generation Clearfield) were sown in individual pots (13 cm in upper diameter, 10 cm in lower diameter, 12.5 cm in height, and a volume of 2 L) filled with peat. A total of 48 plants were used for the experiment. These plants were grouped in 6 plastic trays (8 plants per tray). The sunflower plants grew in a culture chamber in the Department of Plant Physiology of the University of Granada under controlled conditions with relative humidity 60–80%, temperature 25 °C/15 °C (day/night), and 16 h/8 h photoperiod with a PPFD (photosynthetic photon-flux density) of 350 µmol^{-2} s^{-1} (measured with an SB quantum 190 sensor, LI-COR Inc., Lincoln, NE, USA). The light source was provided by fluorescent tubes (Philips Master TL-D 58W/840 REFLEX, Holland; 400–700 nm). Plants were watered with tap water until the first true leaves emerged. After, a complete Hogland-type nutrient solution was applied to the plants as irrigation composed of: 4 mM KNO_3, 2 mM $Ca(NO_3)_2$, 2 mM $MgSO_4$, 1 mM KH_2PO_4, 1 mM NaH_2PO_4, 2 µM $MnCl_2$, 1 µM $ZnSO_4$, 0.25 µM $CuSO_4$, 0.1 µM Na_2MoO_4, 125 µM Fe-EDDHA, and 50 µM H_3BO_3 (pH 5.8). To avoid salt accumulation, the nutrient solution in the trays was removed, the trays were thoroughly cleaned using distilled water, and a new nutritive solution was added. This process was repeated every three days.

Treatments were applied 47 days after sowing and when the sunflower plants presented 6 fully expanded mature leaves. A foliar application with a volume of 0.02 mL/cm^2 was applied to the plants of the two different treatments:

Herbicide treatment (H): Pulsar_40 (4 mL/L) + Dash_HC (2 mL/L).

Herbicides + Biostimulant (HB): Pulsar_40 (4 mL/L) + Dash_HC (2 mL/L) + Terra Sorb® foliar (8 mL/L).

Pulsar_40 herbicide (BASF Española, S.L., Barcelona, Spain) contains Imazamox at 4% weight/volume (*w/v*). The different treatments were diluted using distilled water to a final volume of 100 mL at the concentrations described above. Dash_HC (BASF Española, S.L., Barcelona) was added as a surfactant, which is composed of methyl oleate/palmitate 37.5%, fat alcohol polyalkoxylate phosphate 22.5%, and oleic acid 5%.

The foliar treatments were applied to plants in 3 trays randomly distributed in the cultivation chamber. A total of 24 plants received each of the treatments. Two samplings of the plant materials (leaves) were carried out. The first sampling was on the treatments application date in which only the fresh weight (FW) of the shoots was measured. Four days after the application of the treatments, the fluorescence of chlorophyll *a* was analyzed, and subsequently, the second sampling of the plant material was carried out. The plant material was washed and subsequently dried on filter paper to obtain the FW. From the FW data of both samplings, the relative growth rate (RGR) was calculated. A part of the plant material from the second sampling was subjected to quick freezing using liquid nitrogen and subsequently transferred to the lab for the determination of biochemical parameters. The remaining plant material was lyophilized to determine the profile of phenolic compounds. For each analysis, 3 subsamples were obtained from the frozen or lyophilized material. Then, from each of the sub-samples, the analysis was repeated in triplicate.

3.2. Relative Growth Rate (RGR)

The RGR was calculated using the increase in FW of the plants from the moment of the application of treatments to the moment of sampling using the equation:

$$RGR = (\ln FW_f - \ln FW_i)/(T_f - T_i)$$

T is the time (number of days), and the subscripts indicate the first (i) and last sampling (f) [45].

3.3. Determination of Enzymatic Activities

The assay method for the acetolactate synthase (ALS) enzyme was carried out following the method described by Malkawi et al. [46], in which the acetolactate reaction product was measured spectrophotometrically at an absorbance of 525 nm.

To determine the enzyme activities related to GSH homeostasis [glutathione S-transferase, glutathione peroxidase (GPX), glyoxalase I (Gly I), and glyoxalase II (Gly II)], the enzyme extracts were obtained by grinding 0.5 g of leaves in 5 mL of 50 mM phosphate buffer (pH 7.0) containing 100 mM KCl, 5 mM mercaptoethanol, and 10% glycerol. The homogenate was centrifuged at 21,500× *g* for 10 min, and the supernatant was used to determine the enzymatic activities [47].

Glutathione S-transferase (GST) activity was measured according to Hasanuzzaman et al. [47], where the reaction mixture contained 1.5 mM GSH, 1 mM 1-chloro-2,4-dinitrobenzene (CDNB), and enzyme extract. The increase in absorbance at 340 nm was recorded. GPX activity was measured as described by Nahar et al. [48]. The absorbance was measured at 412 nm in a maximum time of 5 min. Gly I activity was analyzed according to the method of Hasanuzzaman et al. [47] and modified as follows: after adding methylglyoxal, the reaction started, and the increase in the absorbance at 240 nm was measured. Gly II activity was analyzed as described by Hasanuzzaman et al. [47]. The reaction was started by adding S-lactoylglutathione, and the increase in the absorbance at 412 nm was recorded spectrophotometrically.

The phenylalanine ammonium lyase (PAL) activity was analyzed according to Rivero et al. [49] by a method which measured the production of cinnamic acid at 290 nm.

3.4. Aminogram Analysis

The soluble amino acids were extracted following the method of Bieleski and Turner [50] modified as follows: 0.1 g of fresh leaves were homogenized in 1 mL of MCW (methanol:chloroform:water, 12:5:1). An amount of 50 μL of L-2 aminobutyric acid was added as an internal standard. The mixture was centrifuged at 2300× g for 10 min. An amount of 700 μL of Milli-Q water and 1.2 mL of chloroform were added to the resulting supernatant and incubated for 24 h at 4 °C. Then, the aqueous phase was obtained, which was lyophilized, and the resulting dry extract was diluted with 0.1 M HCl. The instrumental analysis of soluble amino acids was carried out using the Pre-column AccQ Tag Ultra Derivatization Kit (Waters, Milford, MA, USA). Derivatization was carried out according to the manufacturer's protocol. For derivatization, 60 μL of borate buffer was added to 10 μL of the sample, 10 μL of 0.1 N NaOH, and 20 μL of reconstituted AccQ Tag Ultra reagent. LC fluorescence analysis was performed on the Waters Acquity® UHPLC system equipped with the Acquity fluorescence detector. UHPLC separation was performed on the AccQ Tag Ultra column (2.1 × 100 mm, 1.7 μm) from Waters. The flow rate was 0.7 mL min^{-1}, and the column temperature was kept at 55 °C. The injection volume was 1 μL, and detection was established at an excitation wavelength of 266 nm and an emission wavelength of 473 nm. The solvent system consisted of two eluents: 1:20 dilution of concentrated AccQ Tag Ultra eluent A and AccQ Tag Ultra eluent B.

3.5. Chl *a* Fluorescence

The analysis was performed on intact plants. For each measurement, a small area of the leaf was kept in darkness using a special clip. Chl *a* fluorescence kinetics was determined using the Handy PEA Chlorophyll Fluorimeter (Hansatech Ltd., King's Lynn, Norfolk, UK); OJIP phases were induced by red light (650 nm) with a light intensity of 3000 μmol photons m^{-2} s^{-1}. One measurement was made on nine fully developed leaves from nine different plants from each treatment (n = 9). The analyzed leaves were in the middle position of the plant. This analysis was performed at midday The phases of the OJIP fluorescence were analyzed by the JIP test and the following parameters were obtained: time to reach fluorescence maximum ($t(F_M)$), fluorescence origin (F_O), fluorescence maximum (F_M), variable fluorescence (F_V), maximum quantum yield of primary PSII photochemistry (F_V/F_M), fluorescence at J-step (2 ms) (V_J), standardized area above the fluorescence curve (S_m), the number of quinone A (Q_A) redox turnovers until F_M was reached (N), the apparent antenna size of active photosystem II (PSII) reaction center (RC) (ABS/RC), proportion of active RC (RC/ABS), dissipated energy flux per RC at t = 0 (DI_o/RC), trapping flux leading to Q_A reduction per RC (TR_o/RC), electron transport flux per RC at t = 0 (ET_o/RC), the efficiency that a trapped exciton can move an electron further than Q_A into the electron transport chain (ψ_o), quantum yield for electron transport from Q_A to plastoquinone (φ_{Eo}), the probability that the PSII Chl molecule functions as RC (γ_{RC}), performance index of electron flux from PSII-based to intersystem acceptors (PI_{ABS}), and performance index of electron flux to the final photosystem I (PSI) electron acceptors (PI_{total}) [51]. The results of these parameters are shown in Figure 5. The represented data were obtained by dividing the mean of H + B treatment plants by the mean of H treatment plants for each parameter to normalize the data and enable the comparison of parameters of different scales.

3.6. Determination of the Concentration of Oxidative Indicators (Malondialdehyde (MDA), H_2O_2, and O_2^-)

MDA concentration was determined according to the method described by Fu and Huang [52]. Fresh materials were extracted with TBA + TCA, and after extraction, the absorbance was recorded at 532 nm and 600 nm to the correct turbidity. H_2O_2 concentration was measured colorimetrically according to Junglee et al. [53] based on the reaction with KI and reading absorbance at 350 nm. For O_2^- determination, the method described by Xiao et al. [54] was followed. The method was based on the reaction of the sample extract with hydroxylamine, sulfanilic acid, and α-1-naphthylamine, and the color intensity was measured at 530 nm.

3.7. Determination of GSH Concentration

The determination of GSH concentration was carried out following the method of Law et al. [55]. This method is based on the specificity of the enzyme GSH reductase for oxidized glutathione. Finally, the samples were read at 412 nm against a GSH standard curve.

3.8. Determination of the Phenolic Compound Profile

Dry lyophilized leaves (50 mg) were extracted with 1 mL of methanol 70% *v*/*v* in a vortex for 1 min; it was then heated to 70 °C for 30 min in a heat bath, stirring every 5 min using a vortex and centrifuged at 12,000× *g* for 10 min at 4 °C. The supernatant was collected, and the methanol was completely removed using a rotary steam. The dried material obtained was redissolved in 1 mL of ultrapure water and filtered through a 0.22-micron Millex-HV13 filter (Millipore, Billerica, MA, USA). Phenolic compounds were determined using a high-performance ion-exchange liquid chromatography method which separated the phenolic compounds according to the procedure of Moreno et al. [56]. First, the separated phenolic compounds were identified from the extracted samples following their MS^2-[MH] fragments in HPLC-DAD-ESI-MSN, carried out on a Luna C18 100A column (250 × 4.6 mm, 5 microns in particle size; Phenomenex, Macclesfield, UK). For mobile phases A and B, respectively, water was used: formic acid (99:1, *v*/*v*) and acetonitrile A and B, with a flow rate of 800 µL/min. The linear gradient started with 1% solvent B, reaching 17% solvent B in 15 min to 17 min, 25% at 22, 35% at 30, and 50% at 35, which was kept in isocratic mode up to 45 min. Chromatograms were recorded at 330 nm. HPLC-DAD-ESI analysis was carried out on an Agilent 1200 HPLC (Agilent Technologies, Waldbronn, Germany) coupled to a serial mass detector.

3.9. Statistical Analysis

The results were statistically evaluated using an analysis of variance, and simple ANOVA with a 95% confidence interval. For all parameters analyzed, the mean and standard error were calculated from nine data ($n = 9$). Differences between treatment means were compared using Fisher's least significant differences (LSD) test at a 95% probability level. The significance levels were expressed as: * $p < 0.05$; ** $p < 0.01$; *** $p < 0.001$; NS—not significant.

4. Conclusions

In conclusion, the application of the Terra-Sorb® biostimulant shows a protective effect against stress due to the imazamox-based Pulsar-40 herbicide. Thus, the combined treatment incremented plant growth and ALS enzymatic activity and maintained leucine, isoleucine, and valine concentrations. In addition, the application of the biostimulant combined with the imazamox herbicide protected photosynthetic activity and significantly reduced oxidative stress in the sunflower plants. This protective effect could be based on the induction of the GSH-GST and antioxidant enzymatic systems. In addition, a higher accumulation of proline could substantially reduce the radical and toxic reactive compounds produced by herbicide application.

Author Contributions: Conceptualization, E.B., C.M., N.S. and J.M.R.; methodology, E.N.-L. and B.B.; validation, B.B.; formal analysis, E.N.-L. and B.B.; data curation, E.N.-L. and B.B.; writing—original draft preparation, E.N.-L. and J.M.R.; writing—review and editing, J.M.R., B.B., E.B., C.M. and N.S.; project administration, J.M.R., E.B., C.M. and N.S. All authors have read and agreed to the published version of the manuscript.

Funding: This research was funded by the PAI program (Plan Andaluz de Investigación, Grupo de Investigación AGR282).

Institutional Review Board Statement: Not applicable.

Informed Consent Statement: Not applicable.

Data Availability Statement: Not applicable.

Conflicts of Interest: The authors declare no conflict of interest.

References

1. Shennan, C. Biotic Interactions, Ecological Knowledge and Agriculture. *Philos. Trans. R. Soc. B Biol. Sci.* **2008**, *363*, 717–739. [CrossRef] [PubMed]
2. Munsif, F.; Shah, S.; Samuel, A.L.; Amin, M.; Shan, A.; Ahmad, A.; Ali, Z. Integration of Weed Control Methods with Seed Rates for Improving Wheat Yield. *Pak. J. Weed Sci. Res.* **2014**, *20*, 155–165.
3. Pang, S.S.; Guddat, L.W.; Duggleby, R.G. Molecular Basis of Sulfonylurea Herbicide Inhibition of Acetohydroxyacid Synthase. *J. Biol. Chem.* **2003**, *278*, 7639–7644. [CrossRef] [PubMed]
4. Tan, S.; Evans, R.R.; Dahmer, M.L.; Singh, B.K.; Shaner, D.L. Imidazolinone-Tolerant Crops: History, Current Status and Future. *Pest Manag. Sci.* **2005**, *61*, 246–257. [CrossRef]
5. Rosinger, C. Herbicide Safeners: An Overview. *Jul.-Kühn-Arch.* **2014**, *443*, 516–525.
6. Powles, S.B.; Yu, Q. Evolution in Action: Plants Resistant to Herbicides. *Annu. Rev. Plant Biol.* **2010**, *61*, 317–347. [CrossRef]
7. Tripthi, D.K.; Varma, R.K.; Singh, S.; Sachan, M.; Guerriero, G.; Kushwaha, B.K.; Bhardwaj, S.; Ramawat, N.; Sharma, S.; Singh, V.P.; et al. Silicon Tackles Butachlor Toxicity in Rice Seedlings by Regulating Anatomical Characteristics, Ascorbate-Glutathione Cycle, Proline Metabolism and Levels of Nutrients. *Sci. Rep.* **2020**, *10*, 14078. [CrossRef]
8. Gomes, M.P.; le Manac'h, S.G.; Hénault-Ethier, L.; Labrecque, M.; Lucotte, M.; Juneau, P. Glyphosate-Dependent Inhibition of Photosynthesis in Willow. *Front. Plant Sci.* **2017**, *8*, 207. [CrossRef]
9. Zhao, N.; Yan, Y.; Luo, Y.; Zou, N.; Liu, W.; Wang, J. Unravelling Mesosulfuron-Methyl Phytotoxicity and Metabolism-Based Herbicide Resistance in Alopecurus Aequalis: Insight into Regulatory Mechanisms Using Proteomics. *Sci. Total Environ.* **2019**, *670*, 486–497. [CrossRef]
10. Strasser, R.J.; Tsimilli-Michael, M.; Srivastava, A. Analysis of the Chlorophyll a Fluorescence Transient. In *Chlorophyll a Fluorescence*; Springer: Dordrecht, The Netherlands, 2004; pp. 321–362.
11. Maxwell, K.; Johnson, G.N. Chlorophyll Fluorescence—A Practical Guide. *J. Exp. Bot.* **2000**, *51*, 659–668. [CrossRef]
12. Balabanova, D.A.; Paunov, M.; Goltsev, V.; Cuypers, A.; Vangronsveld, J.; Vassilev, A. Photosynthetic Performance of the Imidazolinone Resistant Sunflower Exposed to Single and Combined Treatment by the Herbicide Imazamox and an Amino Acid Extract. *Front. Plant Sci.* **2016**, *7*, 1559. [CrossRef] [PubMed]
13. Weber, J.F.; Kunz, C.; Peteinatos, G.G.; Santel, H.-J.; Gerhards, R. Utilization of Chlorophyll Fluorescence Imaging Technology to Detect Plant Injury by Herbicides in Sugar Beet and Soybean. *Weed Technol.* **2017**, *31*, 523–535. [CrossRef]
14. Duhoux, A.; Pernin, F.; Desserre, D.; Délye, C. Herbicide Safeners Decrease Sensitivity to Herbicides Inhibiting Acetolactate-Synthase and Likely Activate Non-Target-Site-Based Resistance Pathways in the Major Grass Weed Lolium sp. (Rye-Grass). *Front. Plant Sci.* **2017**, *8*, 1310. [CrossRef]
15. Délye, C.; Gardin, J.A.C.; Boucansaud, K.; Chauvel, B.; Petit, C. Non-Target-Site-Based Resistance Should Be the Centre of Attention for Herbicide Resistance Research: Alopecurus Myosuroides as an Illustration. *Weed Res.* **2011**, *51*, 433–437. [CrossRef]
16. Neshev, N.; Balabanova, D.; Yanev, M.; Mitkov, A. Is the Plant Biostimulant Application Ameliorative for Herbicide-Damaged Sunflower Hybrids? *Ind. Crops Prod.* **2022**, *182*, 114926. [CrossRef]
17. Hasanuzzaman, M.; Nahar, K.; Alam, M.M.; Bhuyan, M.H.M.B.; Oku, H.; Fujita, M. Exogenous Nitric Oxide Pretreatment Protects *Brassica napus* L. Seedlings from Paraquat Toxicity through the Modulation of Antioxidant Defense and Glyoxalase Systems. *Plant Physiol. Biochem.* **2018**, *126*, 173–186. [CrossRef] [PubMed]
18. Hernández Estévez, I.; Rodríguez Hernández, M. Plant Glutathione S-Transferases: An Overview. *Plant Gene* **2020**, *23*, 100233. [CrossRef]
19. Fayez, K.; Ali, E. Impact of Glyphosate Herbicide and Salicylic Acid on Seed Germination, Cell Structure and Physiological Activities of Faba Bean (*Vicia faba* L.) Plant. *Annu. Res. Rev. Biol.* **2017**, *17*, 1–15. [CrossRef]
20. Deng, X.; Zheng, W.; Zhou, X.; Bai, L. The Effect of Salicylic Acid and 20 Substituted Molecules on Alleviating Metolachlor Herbicide Injury in Rice (*Oryza sativa*). *Agronomy* **2020**, *10*, 317. [CrossRef]
21. Arda, H.; Kaya, A.; Alyuruk, G. Physiological and Genetic Effects of Imazamox Treatment on Imidazolinone-Sensitive and Resistant Sunflower Hybrids. *Water Air Soil Pollut.* **2020**, *231*, 118. [CrossRef]
22. Sivey, J.D.; Lehmler, H.-J.; Salice, C.J.; Ricko, A.N.; Cwiertny, D.M. Environmental Fate and Effects of Dichloroacetamide Herbicide Safeners: "Inert" yet Biologically Active Agrochemical Ingredients. *Environ. Sci. Technol. Lett.* **2015**, *2*, 260–269. [CrossRef]
23. Brown, P.; Saa, S. Biostimulants in Agriculture. *Front. Plant Sci.* **2015**, *6*, 671. [CrossRef]

24. Taha, R.S.; Alharby, H.F.; Bamagoos, A.A.; Medani, R.A.; Rady, M.M. Elevating Tolerance of Drought Stress in Ocimum Basilicum Using Pollen Grains Extract; a Natural Biostimulant by Regulation of Plant Performance and Antioxidant Defense System. *S. Afr. J. Bot.* **2020**, *128*, 42–53. [CrossRef]
25. ur Rehman, H.; Alharby, H.F.; Alzahrani, Y.; Rady, M.M. Magnesium and Organic Biostimulant Integrative Application Induces Physiological and Biochemical Changes in Sunflower Plants and Its Harvested Progeny on Sandy Soil. *Plant Physiol. Biochem.* **2018**, *126*, 97–105. [CrossRef]
26. Alharby, H.; Alzahrani, Y.; Rady, M. Seeds Pretreatment with Zeatins or Maize Grain-Derived Organic Biostimulant Improved Hormonal Contents, Polyamine Gene Expression, and Salinity and Drought. *Int. J. Agric. Biol.* **2020**, *24*, 714–724.
27. Desoky, E.-S.M.; Elrys, A.S.; Mansour, E.; Eid, R.S.M.; Selem, E.; Rady, M.M.; Ali, E.F.; Mersal, G.A.M.; Semida, W.M. Application of Biostimulants Promotes Growth and Productivity by Fortifying the Antioxidant Machinery and Suppressing Oxidative Stress in Faba Bean under Various Abiotic Stresses. *Sci. Hortic.* **2021**, *288*, 110340. [CrossRef]
28. Calvo, P.; Nelson, L.; Kloepper, J.W. Agricultural Uses of Plant Biostimulants. *Plant Soil* **2014**, *383*, 3–41. [CrossRef]
29. Martínez-Esteso, M.; Vilella-Antón, M.; Sellés-Marchart, S.; Martínez-Márquez, A.; Botta-Català, A.; Piñol-Dastis, R.; Bru-Martínez, R. A DIGE Proteomic Analysis of Wheat Flag Leaf Treated with TERRA-SORB®Foliar, a Free Amino Acid High Content Biostimulant. *J. Integr. OMICS* **2016**, *6*, 188. [CrossRef]
30. Gavelienė, V.; Pakalniškytė, L.; Novickienė, L.; Balčiauskas, L. Effect of Biostimulants on Cold Resistance and Productivity Formation in Winter Rapeseed and Winter Wheat. *Ir. J. Agric. Food Res.* **2018**, *57*, 71–83. [CrossRef]
31. Navarro-León, E.; López-Moreno, F.J.; Borda, E.; Marín, C.; Sierras, N.; Blasco, B.; Ruiz, J.M. Effect of l -amino Acid-based Biostimulants on Nitrogen Use Efficiency (NUE) in Lettuce Plants. *J. Sci. Food Agric.* **2022**. [CrossRef]
32. Fernández-Escalada, M.; Zulet-González, A.; Gil-Monreal, M.; Royuela, M.; Zabalza, A. Physiological Performance of Glyphosate and Imazamox Mixtures on Amaranthus Palmeri Sensitive and Resistant to Glyphosate. *Sci. Rep.* **2019**, *9*, 18225. [CrossRef]
33. Gil-Monreal, M.; Royuela, M.; Zabalza, A. Hypoxic Treatment Decreases the Physiological Action of the Herbicide Imazamox on Pisum Sativum Roots. *Plants* **2020**, *9*, 981. [CrossRef] [PubMed]
34. Batista-Silva, W.; Heinemann, B.; Rugen, N.; Nunes-Nesi, A.; Araújo, W.; Braun, H.; Hildebrandt, T. The Role of Amino Acid Metabolism during Abiotic Stress Release. *Plant Cell Environ.* **2019**, *42*, 1630–1644. [CrossRef] [PubMed]
35. Christensen, M.G.; Teicher, H.B.; Streibig, J.C. Linking Fluorescence Induction Curve and Biomass in Herbicide Screening. *Pest Manag. Sci.* **2003**, *59*, 1303–1310. [CrossRef]
36. Linn, A.I.; Mink, R.; Peteinatos, G.G.; Gerhards, R. In-field Classification of Herbicide-resistant *Papaver rhoeas* and *Stellaria media* Using an Imaging Sensor of the Maximum Quantum Efficiency of Photosystem II. *Weed Res.* **2019**, *59*, 357–366. [CrossRef]
37. Zulfiqar, F.; Casadesús, A.; Brockman, H.; Munné-Bosch, S. An Overview of Plant-Based Natural Biostimulants for Sustainable Horticulture with a Particular Focus on Moringa Leaf Extracts. *Plant Sci.* **2020**, *295*, 110194. [CrossRef]
38. Balabanova, D.; Remans, T.; Vassilev, A.; Cuypers, A.; Vangronsveld, J. Possible Involvement of Glutathione S-Transferases in Imazamox Detoxification in an Imidazolinone-Resistant Sunflower Hybrid. *J. Plant Physiol.* **2018**, *221*, 62–65. [CrossRef]
39. Abu-Qare, A.W.; Duncan, H.J. Herbicide Safeners: Uses, Limitations, Metabolism, and Mechanisms of Action. *Chemosphere* **2002**, *48*, 965–974. [CrossRef]
40. DeRidder, B.P.; Dixon, D.P.; Beussman, D.J.; Edwards, R.; Goldsbrough, P.B. Induction of Glutathione *S* -Transferases in Arabidopsis by Herbicide Safeners. *Plant Physiol.* **2002**, *130*, 1497–1505. [CrossRef]
41. Li, Q.; Gao, Y.; Yang, A. Sulfur Homeostasis in Plants. *Int. J. Mol. Sci.* **2020**, *21*, 8926. [CrossRef]
42. Sánchez-Rodríguez, E.; Rubio-Wilhelmi, M.; Cervilla, L.M.; Blasco, B.; Rios, J.J.; Rosales, M.A.; Romero, L.; Ruiz, J.M. Genotypic Differences in Some Physiological Parameters Symptomatic for Oxidative Stress under Moderate Drought in Tomato Plants. *Plant Sci.* **2010**, *178*, 30–40. [CrossRef]
43. de la Torre-González, A.; Navarro-León, E.; Albacete, A.; Blasco, B.; Ruiz, J.M. Study of Phytohormone Profile and Oxidative Metabolism as Key Process to Identification of Salinity Response in Tomato Commercial Genotypes. *J. Plant Physiol.* **2017**, *216*, 164–173. [CrossRef]
44. Boudet, A.-M. Evolution and Current Status of Research in Phenolic Compounds. *Phytochemistry* **2007**, *68*, 2722–2735. [CrossRef] [PubMed]
45. Bellaloui, N.; Brown, P.H. Cultivar Differences in Boron Uptake and Distribution in Celery (*Apium graveolens*), Tomato (*Lycopersicon esculentum*) and Wheat (*Triticum aestivum*). *Plant Soil* **1998**, *198*, 153–158. [CrossRef]
46. Malkawi, H.I.; Al-Quraan, N.A.; Owais, W.M. Acetolactate Synthase Activity and Chlorsulfuron Sensitivity of Gamma-Irradiated Lentil (*Lens culinaris* Medik.) Cultivars. *J. Agric. Sci.* **2003**, *140*, 83–91. [CrossRef]
47. Hasanuzzaman, M.; Hossain, M.A.; Fujita, M. Nitric Oxide Modulates Antioxidant Defense and the Methylglyoxal Detoxification System and Reduces Salinity-Induced Damage of Wheat Seedlings. *Plant Biotechnol. Rep.* **2011**, *5*, 353–365. [CrossRef]
48. Nahar, K.; Hasanuzzaman, M.; Alam, M.M.; Rahman, A.; Suzuki, T.; Fujita, M. Polyamine and Nitric Oxide Crosstalk: Antagonistic Effects on Cadmium Toxicity in Mung Bean Plants through Upregulating the Metal Detoxification, Antioxidant Defense and Methylglyoxal Detoxification Systems. *Ecotoxicol. Environ. Saf.* **2016**, *126*, 245–255. [CrossRef]
49. Rivero, R.M.; Ruiz, J.M.; García, P.C.; López-Lefebre, L.R.; Sánchez, E.; Romero, L. Resistance to Cold and Heat Stress: Accumulation of Phenolic Compounds in Tomato and Watermelon Plants. *Plant Sci.* **2001**, *160*, 315–321. [CrossRef]
50. Bieleski, R.L.; Turner, N.A. Separation and Estimation of Amino Acids in Crude Plant Extracts by Thin-Layer Electrophoresis and Chromatography. *Anal. Biochem.* **1966**, *17*, 278–293. [CrossRef]

51. Strasser, R.; Srivastava, A.; Tsimilli-Michael, M. The Fluorescence Transient as a Tool to Characterize and Screen Photosynthetic Samples. In *Probing Photosynthesis: Mechanism, Regulation and Adaptation*; Yunus, M., Pathre, U., Mohanty, P., Eds.; Taylor & Francis: London, UK, 2000; pp. 443–480.
52. Fu, J.; Huang, B. Involvement of Antioxidants and Lipid Peroxidation in the Adaptation of Two Cool-Season Grasses to Localized Drought Stress. *Environ. Exp. Bot.* **2001**, *45*, 105–114. [CrossRef]
53. Junglee, S.; Urban, L.; Sallanon, H.; Lopez-Lauri, F. Optimized Assay for Hydrogen Peroxide Determination in Plant Tissue Using Potassium Iodide. *Am. J. Anal. Chem.* **2014**, *05*, 730–736. [CrossRef]
54. Xiao, H.; He, W.; Fu, W.; Cao, H.; Fan, Z. A Spectrophotometer Method Testing Oxygen Radicals. *Prog. Biochem. Biophys.* **1999**, *26*, 180–182.
55. Law, M.Y.; Charles, S.A.; Halliwell, B. Glutathione and Ascorbic Acid in Spinach (*Spinacia oleracea*) Chloroplasts. The Effect of Hydrogen Peroxide and of Paraquat. *Biochem. J.* **1983**, *210*, 899–903. [CrossRef] [PubMed]
56. Moreno, D.A.; Carvajal, M.; López-Berenguer, C.; García-Viguera, C. Chemical and Biological Characterisation of Nutraceutical Compounds of Broccoli. *J. Pharm. Biomed. Anal.* **2006**, *41*, 1508–1522. [CrossRef] [PubMed]

MDPI

Article

Evaluation of the Fungicidal Effect of Some Commercial Disinfectant and Sterilizer Agents Formulated as Soluble Liquid against *Sclerotium rolfsii* Infected Tomato Plant

Rania A. A. Hussien [1,*], Mai M. A. Gnedy [2], Ali A. S. Sayed [3], Ahmed Bondok [4], Dalal Hussien M. Alkhalifah [5], Amr Elkelish [6,7,*] and Moataz M. Tawfik [8]

1 Fungicide, Bactericide and Nematicide Department, Central Agricultural Pesticides Lab (CAPL), Agriculture Research Center (ARC), Giza 11835, Egypt
2 Pesticide Formulation Research Department, Central Agricultural Pesticides Lab (CAPL), Agriculture Research Center (ARC), Giza 11835, Egypt
3 Botany Department, Faculty of Agriculture, Fayoum University, Fayoum 63514, Egypt
4 Department of Plant Pathology, Faculty of Agriculture, Ain Shams University, Cairo 11566, Egypt
5 Department of Biology, College of Science, Princess Nourah bint Abdulrahman University, Riyadh 11671, Saudi Arabia
6 Department of Biology, College of Science, Imam Mohammad Ibn Saud Islamic University (IMSIU), Riyadh 11623, Saudi Arabia
7 Botany Department, Faculty of Science, Suez Canal University, Ismailia 41522, Egypt
8 Botany Department, Faculty of Science, Port Said University, Port Said 42526, Egypt
* Correspondence: raniahussien187@arc.sci.eg (R.A.A.H.); amr.elkelish@science.suez.edu.eg (A.E.)

Citation: Hussien, R.A.A.; Gnedy, M.M.A.; Sayed, A.A.S.; Bondok, A.; Alkhalifah, D.H.M.; Elkelish, A.; Tawfik, M.M. Evaluation of the Fungicidal Effect of Some Commercial Disinfectant and Sterilizer Agents Formulated as Soluble Liquid against *Sclerotium rolfsii* Infected Tomato Plant. *Plants* **2022**, *11*, 3542. https://doi.org/10.3390/plants11243542

Academic Editors: Špela Mechora and Dragana Šunjka

Received: 2 November 2022
Accepted: 2 December 2022
Published: 15 December 2022

Abstract: Globally, root rot disease of tomato plants caused by *Sclerotium rolfsii* is a severe disease leading to the death of infected plants. The effect of some commercial antiseptics and disinfectant agents, such as chloroxylenol (10%), phenic (10%) and formulated phenol (7%) on the control of root rot pathogen and its impact on growth and chemical constituents of tomato seedlings cv. Castle Rock were investigated in vitro and in vivo. The antifungal activity was measured in vitro following the poisoned food technique at different concentrations of 1000, 2000, 3000 and 4000 μL/L. Disinfectant agents and atrio (80%) were tested in vivo by soaking 20-day-old tomato seedlings in four concentrations of 125, 250, 500 and 1000 μL/100 mL water for 5 min and thereafter planting in soil infested by *S. rolfsii*. Fresh and dry weight, shoot and root length, and chemical constituents of tomato seedlings infected by *S. rolfsii* were investigated at 35 days after planting (DAP). Experimental results indicated that chloroxylenol (10%) was the most effective on fungus in vitro, recorded an effective concentration (EC_{50} = 1347.74 μL/L) followed by phenic (10%) (EC_{50} = 1370.52 μL/L) and formulated phenol (7%) (EC_{50} = 1553.59 μL/L). In vivo, atrio (80%) and disinfectant agents at different concentrations significantly ($p \leq 0.05$) reduced disease incidence, increased shoot and root lengths and increased dry and fresh weight. Additionally, it significantly increased chlorophyll a, chlorophyll b, total carotenoids, total carbohydrates, total proteins, and total phenols. The highest reduction of root rot incidence and increase tomato growth parameters, as well as chemical compositions, were recorded on tomato seedlings treated with atrio (80%) as well as formulated phenol (7%) at different concentrations, followed by chloroxylenol (10%) at 125 and 250 μL/100 mL, whereas phenic (10%) was found to be the least effective treatment. Therefore, the application of formulated phenol (7%) could be commercially used to control tomato root rot diseases and increase the quality and quantity of tomato plants since it is promising against the pathogen, safe, and less expensive than fungicides.

Keywords: antifungal effect; chloroxylenol; phenic; phenol; *Sclerotium rolfsii*; tomato plants

1. Introduction

Tomato (*Solanum lycopersicum* L.) is one of the most important vegetable crops in the world due to increasing demand, dietary value and widespread production [1]. It is a rich source of vitamins (A and C), minerals, beta-carotene, and a high amount of water [2]. The 10 leading producers of tomato in the world are China, India, Turkey, USA, Egypt, Italy, Iran, Spain, Mexico and Brazil [3]. The world's tomato production in 2020 was 187 million tonnes, with an average yield of 37 tonnes per hectare [3].

Sclerotium rolfsii is a destructive disease for many plants, causing damping-off and root rot of nursery seedlings, wilting and blight on adult plants [4]. It is responsible for high economic losses sustained by tomato producers each year [5]. Chemical fungicides have been used for disease management, become an integral part of agriculture, and increased food production [6]. However, they are costly; their extensive use has raised concerns about residual effects and toxicity to humans, animals and the environment, and efficacy has also decreased due to the emergence of fungicide-resistant pathogens [7].

On the contrary, antiseptics and disinfectants are non-selective and anti-infective agents that can be applied topically. They are used in healthy sectors and care centers to inhibit the growth of microbes and inanimate objects [8]. Chloroxylenol or para-chloro-meta-xylenol (PCMX) is a commercially cheap liquid antiseptic with low toxicity, low metal corrosivity, an active pH of 4–9, and is yet powerful enough to use as a disinfectant due to its broad spectrum of antimicrobial effects, that is, bacteria, fungi and yeast. It can kill 98% of microbes in just 15 s [9].

Phenol (carbolic acid) is generally a protoplasmic poison and was the first antiseptic employed by Joseph Lister (1912–1927), and displays effective antimicrobial activity against a wide range of microorganisms. Recently, phenols have been widely used in the healthcare, cosmetic, food and pharmaceutical industries to prevent unwanted microbial resistance [10]. It is less potent than chloroxylenol in inhibiting microorganism activity [11].

This research aims to find a new safe characteristic of the active ingredient and formulate it in a suitable formulation type to be used as an alternative to conventional fungicides. Additionally, we determine the antifungal activity of some commercial antiseptics and disinfectant agents and recommend the best agent to control tomato root rot disease caused by *S. rolfsii*.

2. Results

2.1. Characterization of Formulation Components

The total solubility of phenol is 100%, 62.5% and 33% in acetone, xylene and water, respectively. It exhibits acidic properties as seen by its free acidity of 0.098, suggesting that it requires an acidic adjuvant for its formulation (Table 1).

Table 1. Physico-chemical properties of phenol as an active ingredient.

Solubility % (*w*/*v*)			Free Acidity as % H_2SO_4
Water	**Acetone**	**Xylene**	
33	100	62.5	0.098

Data in Table 2 show the physico-chemical properties of surfactants, that is, Sisi 6, polyethylene glycol 600 mono laurate (PEG 600 ML), and polyethylene glycol 600 dilaurate (PEG 600 DL) used for preparing phenol as soluble concentrates. According to the hydrophilic–lipophilic balance (HLB), Sisi 6 and PEG 600 ML dispersed agents since their HLB values were more than 13, whereas PEG 600 DL was ~10–12. On the other hand, these surfactants reduced the surface tension values compared to water. The critical micelle concentration (CMC) of Sisi 6 was 0.5% and possessed low surface tension of 28.5 dyne/cm, and the CMC value of PEG 600 ML was 0.3% and possessed low surface tension of 30.64 dyne/cm, and the CMC value of PEG 600 DL was 0.4% and possessed low surface tension of 30.23 dyne/cm. Further, the acidic properties were 0.245, 0.882, and 0.049 for Sisi 6, PEG 600 ML, and PEG 600 DL, respectively. These findings conclude that all surface-active agents tested were suitable for preparing phenol as soluble concentrate formulation. Since the HLB of a surfactant is related to its solubility, a surfactant having a high range (13) will tend to be water-soluble [12].

Table 2. Physicochemical characteristics of surfactants used for preparing phenol as soluble concentrates.

Surface Active Agent	Surface Tension (dyne/cm) at CMC	CMC%	HLB	Free Acidity as % H_2SO_4
Sisi 6	28.5	0.5	>13	0.245
PEG 600 ML	30.64	0.3	>13	0.882
PEG 600 DL	30.23	0.4	10–12	0.049

CMC is important in the selection of surfactants for specific applications. Generally, at a concentration greater than the CMC value, the surface tension of the solution does not decrease with a further increase in surface tension concentrations [13]. In addition, the solubility of surfactant in water is considered an approximate guide to its HLB [14]. These findings conclude that the surfactants were suitable as a spreading agent to prepare the soluble liquid formulation.

Data in Table 3 demonstrate the physicochemical properties of the commercial disinfectants before and after storage at (54 ± 3 °C) for three days. Free acidity, alkalinity, and surface tension of formulated phenol (7%) and phenic (10%) did not change. In contrast, a slight decrease in free alkalinity was shown in chloroxylenol after three days of storage. Further, these disinfectants were soluble and clear with no sedimentation in both cases, indicating the ability of the formulation to keep its properties either before or after storage conditions [13].

Table 3. Physicochemical properties of disinfectants before and after storage at (54 ± 3 °C) for three days.

Storage	Physicochemical Properties	Commercial Disinfectants		
		Phenol Formulated	Chloroxylenol	Phenic
Before storage	Surface tension (dyne/cm)	40	36.97	36.97
	Free acidity as % H_2SO_4	0.249	0	0
	Free alkalinity as % NaOH	0.0	0.72	1.84
	Solubility	Soluble	Soluble	Soluble
	Sedimentation	nil	nil	nil
After storage	Surface tension (dyne/cm)	38	36.97	36
	Free acidity as % H_2SO_4	0.249	0	0
	Free alkalinity as % NaOH	0	0.52	1.84
	Solubility	Soluble	Soluble	Soluble
	Sedimentation	nil	nil	nil

2.2. *Physicochemical Properties of Spray Solution at the Recommended Field Dilution Rate (1.5%)*

Phenol-formulated and other commercial disinfectants showed low surface tension values, high viscosity, high electrical conductivity, and a low alkaline pH value compared to water and the active ingredient (Table 4). The surface tension (dyne/cm) of the spray solution was 28, 33.44, and 34.58 recorded in formulated phenol (7%), phenic (10%), and chloroxylenol (10%), respectively. The pH of the spray solution was 7.42, 8.46, and 8.94 recorded in formulated phenol (7%), chloroxylenol (10%), and phenic (10%), respectively. The viscosity (cm/poise) of the spray solution was 1.71, 1.70, and 1.20 in formulated phenol (7%), phenic (10%), and chloroxylenol (10%), respectively. Further, the conductivity (µMHOS) of the spray solution was 585, 448, and 425 in phenic (10%), chloroxylenol (10%), and formulated phenol (7%), respectively.

Table 4. Physicochemical properties of spray solution at the recommended field dilution rate (1.5%).

Compounds	Physico-Chemical Properties			
	Surface Tension (dyne /cm)	*p*H Value	Conductivity (μMHOS)	Viscosity (cm/poise)
Water	72	9.21	350	0.89
Phenol	34.79	7.51	370	1.19
Formulated phenol (7%)	28	7.42	425	1.71
Chloroxylenol (10%)	34.58	8.46	448	1.20
Phenic (10%)	33.44	8.94	585	1.70

2.3. The Antifungal Activity of Disinfectant and Atrio (80%) on S. Rolfsii by Poisoned Food Technique In Vitro

Data in Table 5 and Figure 1 representing the application of disinfectant agents, that is, formulated phenol (7%), chloroxylenol (10%), and phenic (10%) at four concentrations of 1000, 2000, 3000, and 4000 μL/L show induced reduction on linear growth of *S. rolfsii* in vitro. No fungus growth was observed at a high concentration of all compounds. Further, the lowest EC50 value was observed by chloroxylenol (10%) (1347.74 μL/L) followed by 10% phenic (1370.52 μL/L) and 7% formulated phenol (1553.59 μL/L).

Table 5. Inhibitory effect (%) of antiseptic and disinfectant agents tested at different concentrations against *S. rolfsii* in vitro.

Compounds	Concentrations (μL/L)				EC_{50}	EC_{90}	Slope Value
	1000	2000	3000	4000			
Formulated phenol (7%)	33.33	55.55	72.22	100	3.1239 +/− 0.3266	3995.7593	1553.59
Chloroxylenol (10%)	36.66	64.44	88.88	100	3.2679 +/− 0.3469	3324.9652	1347.74
Phenic (10%)	38.88	58.88	85.55	100	3.123 +/− 0.3341	3525.7869	1370.52

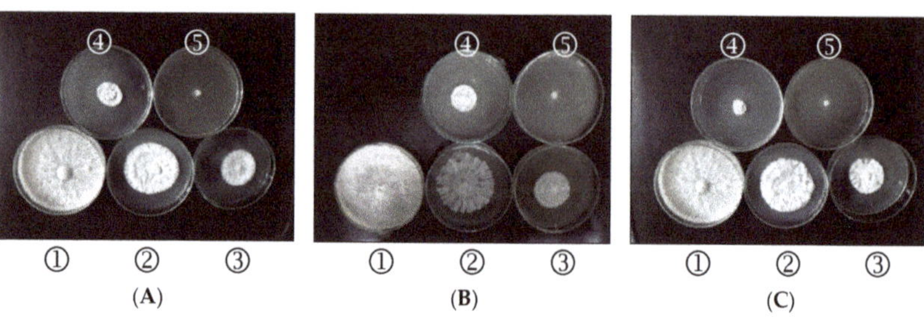

Figure 1. The antifungal activity of disinfectant agents in vitro measured by poisoned food technique. (**A**) Formulated phenol (7%); (**B**) chloroxylenol (10%); (**C**) phenic (10%). ① Control, ② 1000 μL/L, ③ 2000 μL/L, ④ 3000 μL/L, and ⑤ 4000 μL/L.

2.4. The Antifungal Activity of Disinfectant and Atrio (80%) on S. rolfsii (Greenhouse Conditions)

2.4.1. Effect of disinfectant agents at different concentrations and Atrio 80% on the incidence of tomato root rots caused by *S. rolfsii* at 35 DAP (greenhouse conditions)

Tomato seedlings planted in soil infected with *S. rolfsii* showed disease symptoms (yellowing, root rot and crown rot). The disease incidence was 93.75% at 35 DAP compared to control sterilized soil. Tomato seedlings treated with a high concentration (1%) of formulated phenol (7%) as well as atrio 80%, recorded lower disease incidence ~12.50%, whereas the concentrations 0.5% and 0.25% of formulated phenol (7%) recorded ~18.75% and 25.0%, respectively. Therefore, treating tomato seedlings with formulated phenol (7%) at different concentrations of 1, 0.5, and 0.25% caused disease reduction of ~86.7, 80.0, and 73.3%, respectively, compared to other tested agents.

The disease reduction recorded in seedlings treated with chloroxylenol (10%) at a concentration of 0.25% was 66.67%, whereas using phenic (10%) at a concentration of 0.125 reduced the disease by 53.33% (Table 6 and Figure 2). However, the high concentration (1%) of chloroxylenol and phenic antagonized the pathogen in vitro; it was toxic for the plant and killed the tomato seedlings after 5 days of treatment.

Table 6. Effect of disinfectant agents at different concentrations and atrio 80% on the incidence of tomato root rots caused by *S. rolfsii* at 35 DAP (greenhouse conditions).

Treatments	Concentrations	Disease Incidence (%)	Disease Reduction (%)
Control infected soil	N/A *	93.75	0.00
Control sterilized soil	N/A	0.00	100.00
Formulated phenol (7%)	125µL 100 mL^{-1}	43.75	53.33
	250µL 100 mL^{-1}	25.00	73.33
	500µL 100 mL^{-1}	18.75	80.00
	1000µL 100 mL^{-1}	12.50	86.67
Chloroxylenol (10%)	125µL 100 mL^{-1}	37.50	60.00
	250µL 100 mL^{-1}	31.25	66.67
	500µL 100 mL^{-1}	56.25	40.00
Phenic (10%)	125µL 100 mL^{-1}	43.75	53.33
	250µL 100 mL^{-1}	56.25	40.00
	500µL 100 mL^{-1}	66.60	28.96
Atrio (80%)	2 g L^{-1}	12.50	86.67

* N/A, not applicable.

Figure 2. Effect of disinfectant agents at different concentrations and atrio 80% on the incidence of tomato root rots caused by *S. rolfsii* at 35 DAP (greenhouse conditions). (**A**) Formulated phenol (7%); (**B**) chloroxylenol (10%); (**C**) phenic (10%); (**D**) (**I**) Atrio (80%); (**II**) control pathogen; (**III**) healthy control. ① 125 µL 100 mL^{-1}; ② 250 µL 100 mL^{-1}; ③ 500 µL 100 mL^{-1}; ④ 1000 µL 100 mL^{-1}.

2.4.2. Effect of Disinfectant Agents at Different Concentrations and Atrio 80% on Growth Parameters of Tomato Seedlings Infected by *S. rolfsii* at 35 DAP

Tomato seedlings grown in soil infected with *S. rolfsii* recorded a reduction in all growth parameters, that is, plant height, shoot and root length, and fresh and dry weight at 35 DAP compared to the control sterilized soil (Figure 3, Table S1). Compared to untreated plants, tomato seedlings treated with atrio (80%) or disinfectant agents at different concentrations significantly ($p \leq 0.05$) reduced tomato root rot incidence (Table 6) and increased all growth parameters at 35 DAP. The maximum increase in all growth parameters was recorded at a high concentration of 1% of formulated phenol (7%), atrio (80%), 0.5% of formulated phenol (7%), 0.25% of chloroxylenol (10%), and 0.125% of phenic (10%), respectively. The high concentration (1%) of chloroxylenol and phenic was toxic for the plant and killed the tomato seedlings after 5 days of treatment.

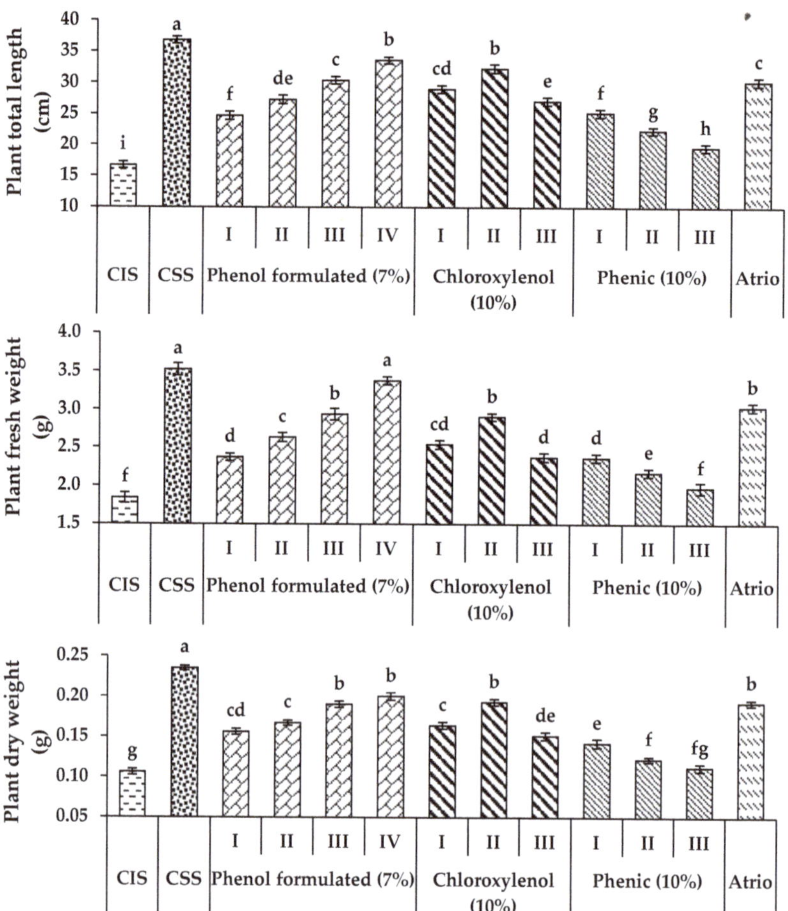

Figure 3. Effect of disinfectant agents at different concentrations and atrio 80% on growth parameters of tomato seedlings infected by *S. rolfsii* at 35 DAP. CIS; Control infected soil; CSS; control sterilized soil; (I) 125 μL 100 mL^{-1}; (II) 250 μL 100 mL^{-1}; (III) 500 μL 100 mL^{-1}; (IV) 1000 μL 100 mL^{-1}. The values shown in the figures are means ± SEM (n = 3), with different alphabetic letter/s being significantly different ($p < 0.05$) following Tukey's post hoc test.

2.4.3. Effect of Disinfectant Agents at Different Concentrations and Atrio 80% on Metabolite of Tomato Seedlings Infected by *S. rolfsii* at 35 DAP

Compared to control sterilized soil, tomato seedlings planted in soil infected with *S. rolfsii* recorded low leaf pigment concentration (Figure 4, Table S2). Tomato seedlings treated with atrio (80%) and disinfectant agents at different concentrations significantly ($p \leq 0.05$) reduced tomato root rot incidence (Table 6) and increased chlorophyll A and

B, total chlorophyll, and total carotenoids concentration in leaves of tomato seedlings at 35 DAP compared to untreated plants. The maximum increase in leaf pigments was recorded at a 1% concentration of formulated phenol (7%), atrio (80%), 0.5% of formulated phenol (7%), 0.25% of chloroxylenol (10%), and 0.125% of phenic (10%), respectively.

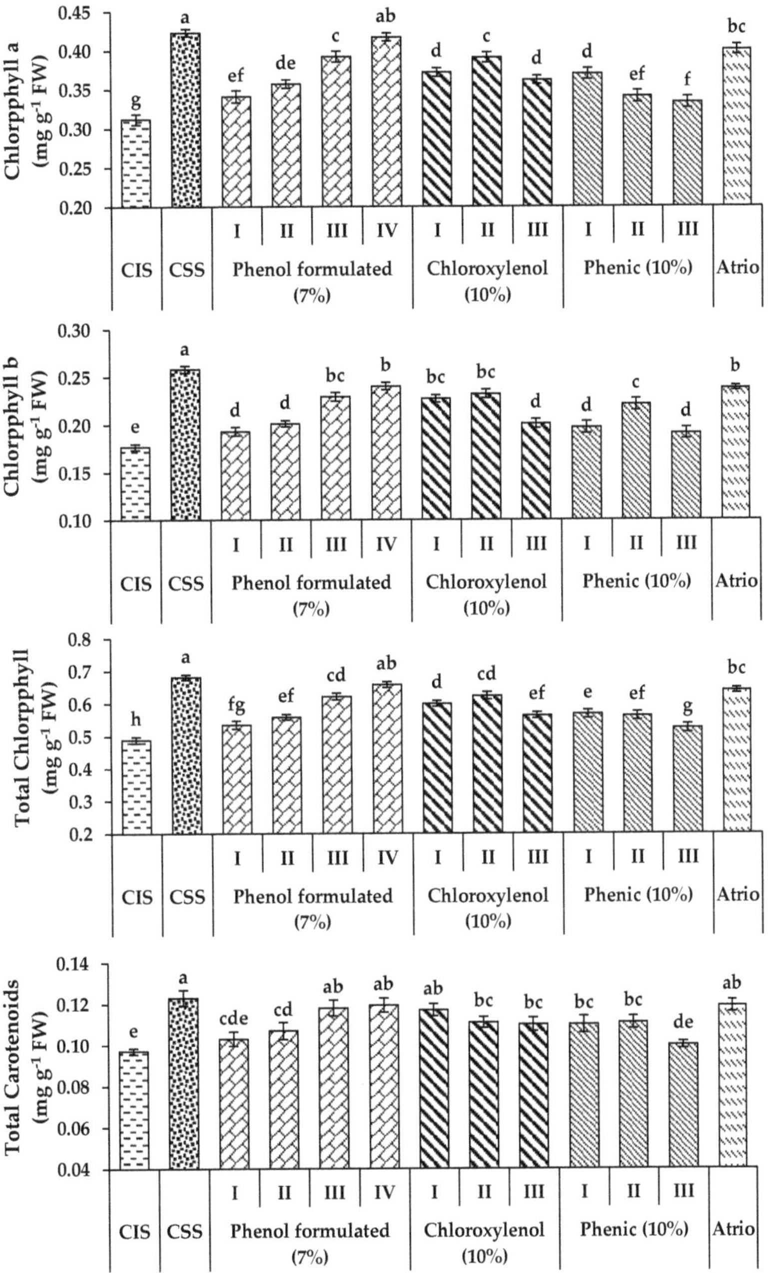

Figure 4. *Cont.*

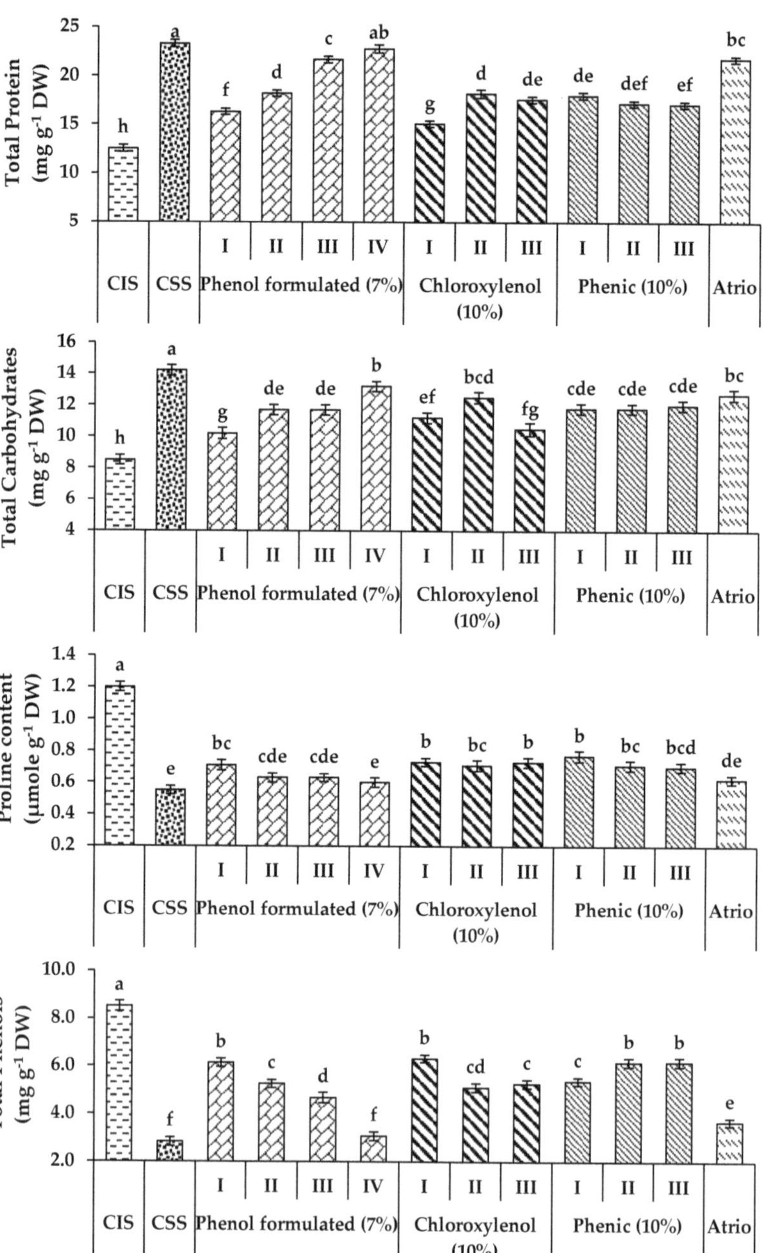

Figure 4. Effect of disinfectant agents at different concentrations and atrio 80% on chemical constituents of tomato seedlings infected by *S. rolfsii* at 35 DAP. **CIS**; Control infected soil; CSS; control sterilized soil; (I) 125 µL 100 mL^{-1}; (II) 250 µL 100 mL^{-1}; (III) 500 µL 100 mL^{-1}; (IV) 1000 µL 100 mL^{-1}. The values shown in the figures are means ± SEM (n = 3), with different alphabetic letter/s being significantly different (p < 0.05) following Tukey's post hoc test.

Further, treating with atrio (80%) and disinfectant agents at different concentrations significantly ($p \leq 0.05$) improved chemical constituents, that is, total carbohydrates, protein content, proline content and total phenols of tomato seedlings at 35 DAP compared to untreated plants. The maximum increase in chemical constituents of tomato seedlings was recorded at a 1% concentration of formulated phenol (7%), atrio (80%), 0.5% of formulated phenol (7%), 0.25% of chloroxylenol (10%), and 0.125% of phenic (10%), respectively (Figure 4, Table S3). The high concentration (1%) of chloroxylenol and phenic was toxic for the plant and killed the tomato seedlings after 5 days of treatment.

3. Discussion

The tomato is an important vegetable crop due to its economic importance and nutritional value. It is one of the most important crops in Egypt and is used for food and industrial purposes [15]. *Sclerotium rolfsii* is a necrotrophic soil-borne plant pathogen that attacks more than 500 plant species belonging to over 100 families, causing chlorosis and wilting of entire plants and finally reducing crop yield and quality [16,17]. Synthetic fungicides have been used to control plant diseases worldwide. Although synthetic fungicides are highly effective, their repeated use has led to problems, such as environmental pollution, development of resistance, and residual toxicity [18].

The antiseptics used in this study are composed of phenols and chloroxylenol compounds. Each compound can be combined or used individually to achieve an antifungal effect. Disinfectants are used to treat the surface of inanimate objects and eliminate all pathogenic microorganisms by causing the denaturation of microbial proteins or enzymes. The antimicrobial properties of the disinfectant against some pathogenic bacteria have been reported earlier [19].

The physicochemical properties of phenol were carried out to determine the appropriate formulation type (Table 1). When selecting a pesticide formulation, several factors must be addressed; that is, the feasibility of utilizing a certain formulation in a specific location to control the target pest and whether the created product will offer effective control [20]. According to the Ref. [21], the pesticide which can be formulated is limited by solubility and hydrolytics. Therefore, the soluble liquid formulation is suitable for the tested material.

Results showed that surfactants applied to soluble liquid would reduce the surface tension of spray droplets, providing more coverage for toxicants by decreasing the contact angle of the spray on a solid surface [22]. The active ingredient with various inert components, known as additives or adjuvants, enhances the active ingredient's effectiveness. Spreaders, wetting agents, stickers, foaming agents, and compatibility agents are all common additives [23]. Phenol showed free acidity, and one of the proposed surfactants (Sisi 6, PEG 600 ML, and PEG 600 DL) also showed free acidity, indicating that it might be employed in the formulation procedure of this chemical with no chemical interaction expected [23].

There were no observable changes for the soluble concentrate local formulation and commercial disinfectants before and after accelerated storage (Table 3), as it showed nearly the same values for free acidity or alkalinity, surface tension and solubility with no sedimentation in both cases, indicating the ability of the formulation to keep its properties in either normal or accelerated storage conditions with expected stability [24].

The spray solution used at a dilution rate (1.5%) had low surface tension, high viscosity, high conductivity, and low pH. Lowering the surface tension of a pesticide spray solution predicts increased wettability and spreading over the treated surface, which leads to increased pesticidal efficiency [25]. Increasing spray solution viscosity causes less drift and increases retention, sticking, and pesticidal effectiveness [26]. Increasing the electrical conductivity of the spray solution would lead to the deionization of insecticides and increase their deposit and penetration on the tested surface, resulting in an increase in insecticidal efficiency [27].

Application of selected disinfectant agents, that is, chloroxylenol (10%), phenic (10%) and formulated phenol (7%) at four concentrations 1000, 2000, 3000 and 4000 μL/L showed induced reduction in linear growth of *Sclerotium rolfsii* in vitro. Our results were in line with an earlier study [28] reporting that chloroxylenol had complete antifungal activity. Phenol and chloroxylenol cause the denaturation of proteins and inhibition of enzymes in microorganisms [29,30].

In this study, treating tomato seedlings with different disinfectant agents led to disease reduction, explaining the improvement of plant growth parameters, photosynthetic pigments and chemical constituents compared to those which were untreated [31]. However, there are no adequate studies on the effect of antiseptic and disinfectant agents against phytopathogenic fungi.

4. Materials and Methods

4.1. Tested Materials

4.1.1. Commercial Disinfectants

A. Chloroxylenol, or para-chloro-meta-xylenol (PCMX), is a mixture of 4.8% chloroxylenol + 9.9% terpineol and absolute alcohol. It was supplied by Agricultural Development Markets, Nadi El Seid St., Dokki, Giza.

B. Phenic contains more than 98% high-quality, high-impact saponified tar oils and carbonates. It has between 6.5 and 7% pure phenol, a highly effective disinfectant. It is produced by the International Company for Chemicals and Industrial Detergents (Cairo, Egypt).

4.1.2. Active Ingredient

Phenol or carbolic acid (C_6H_5OH), a white crystalline solid, was supplied by EL-Gomhoria Co., Cairo, Egypt.

4.1.3. Surface-Active Agents

A. Sisi-6, an anionic surfactant prepared by neutralizing aryl alkyl sulphonic acid with alkaline.

B. Polyethylene glycol 600 di-laurate (PEG 600 DL) (Alexandria, Egypt), a nonionic surfactant supplied by The National Company for Starch, Yeast and Detergents, Alexandria.

C. Polyethylene glycol 600 mono laurate (PEG 600 ML), a nonionic surfactant, supplied by The National Company for Starch, Yeast and Detergents, Alexandria.

4.2. Physico-Chemical Properties of Basic Formulation Constituents

4.2.1. Active Ingredient

A. Solubility is determined by measuring the volume of distilled water, acetone and xylene for complete solubility or miscibility of one gram of an active ingredient at 20°C [32]. The solubility (%) was calculated according to the following equation:

$$\text{Solubility}(\%) = \frac{W}{V} \times 100 \quad (1)$$

W = Active ingredient weight, V = Volume of solvent required for complete solubility.

B. Free acidity or alkalinity was determined according to the Refs. [33,34].

4.2.2. Surface-Active Agents

A. Surface tension was measured using a Du-Nouy tensiometer for solutions containing a 0.5% (w/v) surface-active agent following the American Society of Testing Materials [35].

B. Hydrophilic–lipophilic balance (HLB): The solubility of a surfactant in water was used to approximate its hydrophilic–lipophilic balance [14].

C. Critical micelle concentration (CMC): The concentration of the tested surfactants at which the surface tension of the solution does not decrease as the surfactant concentration increases (CMC) was determined using the technique given by the Ref. [13].

D. Free-acidity or alkalinity was determined as mentioned previously.

4.3. Preparation of Phenol as Soluble Concentrate Formulation

The formulation of phenol was carried out by combining an active ingredient and surfactant in water in three forms (7% + 5% + 88%), (7% + 7.5% + 85.5%), and (7% + 10% + 83%). These three mixes were exposed to several tests to determine the optimal composition. The surface tension of the prepared mixtures was then evaluated at a field dilution rate of (0.5%), and the combination with the lowest surface tension was regarded as successful since it demonstrated the best wetting, spreading, and pesticidal efficiency when sprayed over the treated surface.

4.4. Physicochemical Properties of Disinfectants before and after Storage

A. Surface tension was determined as mentioned before.
B. Free acidity or alkalinity was determined as mentioned previously.
C. Accelerated storage was done to check the stability of local formulations at 54 ± 30 °C for three days according to the Ref. [35].

4.5. Determination of the Physico-Chemical Properties of the Spray Solution at the Field Dilution Rate

A. Surface tension was measured using the du Nouy Tensiometer method described by the Ref. [36].
B. The pH was determined using an Adwa (AD8000) pH meter [35].
C. Viscosity was measured at room temperature with a "Brookfield DV II + PRO" digital viscometer and UL rotational adaptor (ULA) [37].
D. Electrical Conductivity was measured using Cole–Parmer pH/Conductivity following the method described by the Ref. [35].

4.6. Isolation and Identification of the Fungal Pathogen

Tomato plants showing typical root and crown rot symptoms were collected from a private field in Qaliubia Governorate, Egypt. The infected root samples were washed with tap water to remove the adhering soil particles, cut into small fragments, and surface-sterilized by dipping in 5% sodium hypochlorite solution for 5 min following the method [38]. The segments were washed several times with sterilized distilled water, dried between two folds of sterilized filter papers and transferred under aseptic conditions to sterilized Petri dishes containing potato dextrose agar medium (PDA). Thereafter, plates were incubated at 25 ± 2 °C, and developed colonies were picked up after five days, transferred onto a new PDA medium and purified using hyphal tip techniques [39]. The isolated fungus was identified microscopically according to Barnett and Hunter [40]. The identification was confirmed in the Plant Pathology Department, Faculty of Agriculture, Cairo University, Giza, Egypt. The stock culture was maintained on DPA slants and kept at 10 °C in the refrigerator for further experiments.

Antifungal Assay In Vitro

The antifungal activity of antiseptic and disinfectant agents, that is, chloroxylenol (10%), phenic (10%) and formulated phenol (7%) against *S. rolfsii* was investigated by using a food poisoning technique [41]. Disinfectant agents at concentrations of 1000, 2000, 3000, and 4000 µL/L were mixed with 50 mL of sterilized PDA medium and transferred equally into three Petri dishes. The media were allowed to solidify. Then a five-day old fungal culture disk of 9 mm diameter was taken and inoculated to the centre of the Petri dishes containing disinfectant agents. Instead of a PDA, a medium without disinfectant agents served as the control. All plates were incubated at 25 ± 2 °C, and radial growth of the colony was measured when the mycelia of control had almost filled the Petri dishes. Each test was performed in triplicate.

The fungal growth inhibition was calculated due to treatments against the control using the following formula [42]:

$$\textbf{Inhibition}(\%) = [\frac{C - T}{C} \times 100] \quad (2)$$

where C is the average of three replicates of hyphal extension (mm) of the control and T is the average of three replicates of hyphal extension (mm) of plates treated with tested material. EC_{50} and EC_{90} values were determined by the linear regression (LPD) line computer program of the tested fungus percentage inhibition probit vs. logs of the concentrations (µL/L) of the disinfectant's agents. The EC_{50} and EC_{90} values indicate the effective concentrations (µL/L) that cause 50% and 90% growth inhibition. In essence, the lower the value of EC_{50} and EC_{90}, the higher the efficacy of disinfectant agents in the test under consideration.

4.7. Greenhouse Experiment

Selected antiseptic and disinfectant agents, that is, chloroxylenol (10%), phenic (10%) and formulated phenol (7%), were evaluated for controlling root rot disease on tomato plants caused by *S. rolfsii* in the greenhouse (temperature (25 °C ± 2 °C) and humidity 65%) at the Central Agriculture Pesticide Laboratory, ARC, Giza, during the growing season 2021. Pots were filled with cornmeal sand medium at 3% (W/W), infested by *S. rolfsii* and watered regularly for 10 days before planting to ensure the distribution of inoculum. Tomato seedlings cv. Castle rock of 20 days old were soaked in the tested materials at different concentrations of 125, 250, 500, and 1000 µL/100 mL water for 5 min and fungicide (atrio 80% WP, Strchemi Industrial Chemicals, Egypt) at the recommended dose of 2 g L^{-1}, individually. Tomato seedlings were planted at the rate of four seedlings/pot and four replicated (pots) were used for each treatment. At 35 DAP, root rot disease incidence was evaluated as the number of root-rot-diseased plants relative to the number of plant seedlings in each treatment according to the Ref. [43]. Disease incidence was calculated relative to the control infected soil. Fresh and dry weight and shoot and root length, growth parameters and chemical constituents of tomato seedlings infected by *S. rolfsii* at 35 DAP were determined as follows:

1. Total carbohydrates were determined and expressed as glucose according to the Shaffer–Somogi micro-method [44].
2. Total protein content was determined indirectly using nitrogen concentration estimated by the semi-micro-Kjeldahl method, and a Kjeldahl conversion coefficient of 6.25 was used [45].
3. Total phenols were determined using the colourimetric method of Folin–Denis as described by the Ref. [46].
4. Proline content was determined according to the method described by the Ref. [47].
5. Chlorophylls (a and b) and carotenoid concentrations were determined following the Ref. [48].

4.8. Statistical Analysis

The obtained data were subject to analysis of variance (ANOVA), using Minitab Statistical Software 20 [49]. Significant differences between means were compared at $p < 0.05$ following Tukey's post hoc test.

5. Conclusions

Experimental results proved that treating tomato seedlings with high concentration (1%) of formulated phenol (7%) as well as low concentrations 0.25 and 0.5% of chloroxylenol (10%) improved growth parameters and chemical constituents of treated plants compared to those which were untreated. We also conclude that these disinfectants are more effective at controlling root rot disease; however, toxicological studies for the recommended agents are needed.

Supplementary Materials: The following supporting information can be downloaded at: https://www.mdpi.com/article/10.3390/plants11243542/s1, Table S1. Effect of antiseptic and disinfectant agents at different concentrations on growth parameters of tomato seedlings infected by *S. rolfsii* at 35 DAP: Table S2. Effect of antiseptic and disinfectant agents at different concentrations on leaf pigments of tomato seedlings infected by *S. rolfsii* at 35 DAP: Table S3. Effect of antiseptic and disinfectant agents at different concentrations on chemical constituents of tomato seedlings infected by *S. rolfsii* at 35 DAP.

Author Contributions: Conceptualization, R.A.A.H. and M.M.A.G.; methodology, R.A.A.H., M.M.A.G.; software, A.A.S.S., A.B., D.H.M.A., A.E. and M.M.T.; validation, R.A.A.H., M.M.A.G., A.A.S.S., A.B, D.H.M.A., A.E and M.M.T.; formal analysis, R.A.A.H., M.M.A.G., A.A.S.S., D.H.M.A., A.E. and M.M.T.; investigation, D.H.M.A., A.E. and M.M.T.; data curation, A.A.S.S., D.H.M.A., A.E. and M.M.T.; writing—original draft preparation R.A.A.H., M.M.A.G., A.A.S.S., D.H.M.A., A.E. and M.M.T.; writing—review and editing, R.A.A.H., M.M.A.G., A.A.S.S., A.B., D.H.M.A., A.E. and M.M.T.; visualization, A.A.S.S.; supervision, A.B., D.H.M.A., A.E. and M.M.T.; funding acquisition, A.B., D.H.M.A. and M.M.T. All authors have read and agreed to the published version of the manuscript.

Funding: This research received no external funding.

Institutional Review Board Statement: Not applicable.

Informed Consent Statement: Not applicable.

Data Availability Statement: Date is contained within the article and supplementary material.

Acknowledgments: We would like to extend our thanks to Princess Nourah bint Abdulrahman University Researchers Supporting Project number PNURSP2022R15, Princess Nourah bint Abdulrahman University, Riyadh, Saudi Arabia.

Conflicts of Interest: The authors declare no conflict of interest.

References

1. Kimura, S.; Sinha, N. Tomato (*Solanum lycopersicum*): A model fruit-bearing crop. *Cold Spring Harb. Protoc.* **2008**, *3*, pdb.emo105. [CrossRef] [PubMed]
2. Willcox, J.K.; Catignani, G.L.; Lazarus, S. Tomatoes and cardiovascular health. *Crit. Rev. Food Sci. Nutr.* **2003**, *43*, 1–18. [CrossRef]
3. FAOSTATS. 2022. Available online: https://www.fao.org/faostat/en/#data/QCL (accessed on 15 August 2022).
4. Flores-Moctezuma, H.E.; Montes-Belmont, R.; Jiménez-Pérez, A.; Nava-Juárez, R. Pathogenic diversity of *Sclerotium rolfsii* isolates from Mexico, and potential control of southern blight through solarization and organic amendments. *Crop Prot.* **2006**, *25*, 195–201. [CrossRef]
5. De Curtis, F.; Lima, G.; Vitullo, D.; De Cicco, V. Biocontrol of *Rhizoctonia solani* and *Sclerotium rolfsii* on tomato by delivering antagonistic bacteria through a drip irrigation system. *Crop Prot.* **2010**, *29*, 663–670. [CrossRef]
6. Ons, L.; Bylemans, D.; Thevissen, K.; Cammue, B.P. Combining biocontrol agents with chemical fungicides for integrated plant fungal disease control. *Microorganisms* **2020**, *8*, 1930. [CrossRef] [PubMed]
7. Fan, F.; Hamada, M.S.; Li, N.; Li, G.Q.; Luo, C.X. Multiple fungicide resistance in *Botrytis cinerea* from greenhouse strawberries in Hubei Province, China. *Plant Dis.* **2017**, *101*, 601–606. [CrossRef]
8. Mark, L.W. Overview of antiseptics and disinfectants. In *Merck Sharp and Dohme Corp*; A Subsidiary of Merck and Co. Inc.: Kenilworth, NJ, USA, 2015.
9. El Mahmood, A.M.; Doughari, J.H. Effect of Dettol® on viability of some microorganisms associated with nosocomial infections. *Afr. J. Biotechnol.* **2008**, *7*, 1554–1562.
10. Emam, M.A.; Abd EL-Mageed, A.M.; Niedbała, G.; Sabrey, S.A.; Fouad, A.S.; Kapiel, T.; Piekutowska, M.; Mahmoud, S.A. Genetic Characterization and Agronomic Evaluation of Drought Tolerance in Ten Egyptian Wheat (*Triticum Aestivum* L.) Cultivars. *Agronomy* **2022**, *12*, 1217. [CrossRef]
11. Ismail, M.A.; Amin, M.A.; Eid, A.M.; Hassan, S.E.-D.; Mahgoub, H.A.M.; Lashin, I.; Abdelwahab, A.T.; Azab, E.; Gobouri, A.A.; Elkelish, A.; et al. Comparative Study between Exogenously Applied Plant Growth Hormones versus Metabolites of Microbial Endophytes as Plant Growth-Promoting for *Phaseolus Vulgaris* L. *Cells* **2021**, *10*, 1059. [CrossRef]
12. El-Attal, Z.M.; Sawy, M.S.S.; Said, A.A.A. Laboratory studies on some Egyptian granular carrier for Linden and Endrin. *Bull. End. Sci. Egypt Econ. Ser.* **1974**, *8*, 7–11.
13. Hashim, A.M.; Alharbi, B.M.; Abdulmajeed, A.M.; Elkelish, A.; Hozzein, W.N.; Hassan, H.M. Oxidative Stress Responses of Some Endemic Plants to High Altitudes by Intensifying Antioxidants and Secondary Metabolites Content. *Plants* **2020**, *9*, 869. [CrossRef] [PubMed]

14. Lynch, M.I.; Griffin, W.C. Food Emulsions. In *Emulsion Technology*; Lissant, K.J., Ed.; Marcell Decker Inc.: New York. NY, USA, 1974; Mukerjee, P.; Mysels, K.J. *Critical Micelle Concentration of Aqueous Surfactant Systems*; National Bureau of Standards: Washington, DC, USA, 1971; pp. 1–21.
15. Giovannucci, E. Tomatoes, tomato-based products, lycopene, and cancer: Review of the epidemiologic literature. *J. Natl. Cancer Inst.* **1999**, *91*, 317–331. [CrossRef] [PubMed]
16. Billah, K.M.M.; Hossain, M.B.; Prince, M.H.; Sumon, M.D.M.P. Pathogenicity of *Sclerotium rolfsii* on different host, and its over wintering survival: A mini-review. *Int. J. Adv. Agric. Sci.* **2017**, *2*, 1–6.
17. Yaqub, F.; Shahzad, S. Pathogenicity of *Sclerotium rolfsii* on different crops and effect of inoculum density on colonization of mungbean and sunflower roots. *Pak. J. Bot.* **2005**, *37*, 175–180.
18. Erper, I.; Turkkan, M.; Karaca, G.H.; Kilic, G. Evaluation of in vitro antifungal activity of potassium bicarbonate on *Rhizoctonia solani* AG 4 HG-I, *Sclerotinia sclerotiorum* and *Trichoderma* sp. *Afr. J. Biotechnol.* **2011**, *10*, 8605–8612.
19. Mellefont, L.A.; McMeekin, T.A.; Ross, T. The effect of abrupt osmotic shifts on the lag phase duration of foodborne bacteria. *Int. J. Food Microbiol.* **2003**, *83*, 281–293. [CrossRef]
20. Fishel, F.M. *Pesticide Formulations*; EDIS: Singapore, 2010.
21. FAO/WHO. *Manual on Development and Use of FAO and WHO Specifications for Pesticides*, 1st ed.; 3rd Rev. FAO Plant Production and Protection, FAO: Rome, Italy, 2010; Volume 36, p. 3.
22. El-Sisi, A.G. Preparation of some insecticidal formulations using local constituent and testing their efficiency. Foliar pests of watermelon in Hawaii. *Tropical Manag.* **1985**, *35*, 90–96.
23. Libs, E.S.; EL Salim, R.A. Formulation of essential oil pesticides technology and their application. *Agric. Res. Tech.* **2017**, *9*, 42–60.
24. Hala, M.I.; Hanan, S.T.; Naglaa, K.Y. Effect of storage on the stability and biological effectiveness of some insecticides. *J. Biol. Chem. Environ. Sci.* **2016**, *11*, 265–282.
25. Pereira, V.J.; da Cunha, J.R.; De Morais, T.P.; Ribeiro-Oliveira, J.P.; De Morais, J.B. Physical-chemical properties of pesticides: Concepts, applications, and interactions with the environment. *J. Biosci.* **2016**, *32*, 627–641. [CrossRef]
26. Spanoghe, P.; De Schampheleire, M.; Van der Meeren, P.; Steurbaut, W. Influence of agricultural adjuvants on droplet spectra. *Pest Manag. Sci.* **2007**, *63*, 4–16. [CrossRef] [PubMed]
27. El-Sisi, A.G.; El-Mageed, A.; El-Asawi, T.F.; El-Sharkawy, R.A. Improvement the physico-chemical properties and efficiency of some insecticides formulation by using adjuvants against cotton leafworm *Spodoptera littoralis* (BOISD.). *J. Plant Prot. Pathol.* **2011**, *2*, 757–764. [CrossRef]
28. Nowrozi, H.; Kazemi, A.; MotallebiKhah, F.; Ghooshchi, F.; Sharabiani, A.K. Antifungal activity of benzalkonium chloride, Dettol, and chlorhexidine on opportunistic isolated fungi from the environment and operating rooms in private clinics of Tehran in 2011. *J. Birjand Univ. Med. Sci.* **2013**, *20*, 305–311.
29. Abed, A.R.; Hussein, I.M. In vitro study of antibacterial and antifungal activity of some common antiseptics and disinfectants agents. *Kufa J. Vet. Med. Sci.* **2016**, *7*, 148–159.
30. Ascenzi, J.M. Chloroxylenol: An old-new antimicrobial. In *Handbook of Disinfectants and Antiseptics*; M. Dekker: New York, NY, USA, 1996.
31. Hura, K.; Hura, T.; Dziurka, K.; Dziurka, M. Carbohydrate, phenolic and antioxidant level in relation to chlorophyll A content in oilseed winter rape (*Brassica napus* L.) inoculated with *Leptosphaeria maculans*. *Eur. J. Plant Pathol.* **2015**, *143*, 291–303. [CrossRef]
32. Nelson, F.C.; Fiero, G.W. Pesticide formulations, a selected aromatic fraction naturally occurring in petroleum as a pesticide solvent. *J. Agric. Food Chem.* **1954**, *2*, 735–737. [CrossRef]
33. W.H.O. *Specification of Pesticides Used in Public Health*, 5th ed.; WHO: Geneva, Switzerland, 1979.
34. Dobrat, W.; Martijn, A. *Physic-Chemical Methods for Technical and Formulated Pesticides*; Collaborative International Pesticides Analytical Council: Harpenden, UK, 1995.
35. *D133*; American Society of Testing and Materials Standard Test Method For Surface and Inter Facial Tension of Solution of Surface-Active Agents. ASTM: West Conshohocken, PA, USA, 2001.
36. *D-1331-14*; American Society of Testing and Materials Standard Test Methods for Surface and Interfacial Tension of Solution of Surface-Active Agents. ASTM: West Conshohocken, PA, USA, 2014.
37. *2196–15*; American Society of Testing and Materials Standard Test Methods for Rheological Properties of Non-Newtonian Materials by Rotational (Brookfield viscometer DV+II pro.) ASTM-D. ASTM: West Conshohocken, PA, USA, 2015.
38. Burr, T.J.; Hunter, J.E.; Ogawa, J.M.; Abawi, G.S. A root rot of apple caused by *Rhizoctonia solani* in New York nurseries. *Plant Dis. Rep.* **1978**, *62*, 476–479.
39. Dhingra, O.D.; Sinclair, J.B. *Basic Plant Pathology Method*; CRC, Boca: Raton, FL, USA, 1984.
40. Barnett, H.L.; Hunter, B.B. *Illustrated genera of imperfect fungi. Mac Millan Publishing Company*; New York and Mac Millan Publishers: London, UK, 1987; 220p.
41. Mohanty, R.C.; Ray, P.; Rath, S. In-vitro antifungal efficacy study of plant leaf extracts against three dermatophytes. *CIB Tech. J. Microbiol.* **2012**, *1*, 27–32.
42. Satya, P.R.; Manisha, S.P.; Khushboo, S. Evaluation of the antifungal activity of the potent fraction of hexane extract obtained from the bark of *Acacia nilotica*. *Int. J. Sci. Res.* **2014**, *3*, 730–738.
43. El-Mohamedy, R.S.R.; Jabnoun-Khiareddine, H.; Daami-Remadi, M. Control of root rot diseases of tomato plants caused by *Fusarium solani*, *Rhizoctonia solani* and *Sclerotium rolfsii* using different chemical resistance inducers. *Tunis. J. Plant Prot.* **2014**, *9*, 45–55.

44. Association of Official Agriculture Chemists (A.O.A.C.). *Official Methods of Analysis*, 12th ed.; A.O.A.C.: Washington, DC, USA, 1995.
45. Juliano, B.O. *Rice in Human Nutrition*; (FAO Food and Nutrition Series. No. 26); FAO: Rome, Italy, 1993; ISBN 92-5-103149-5.
46. Baba, S.A.; Malik, S.A. Determination of total phenolic and flavonoid content, antimicrobial and antioxidant activity of a root extract of *Arisaema jacquemontii* Blume. *J. Taibah Univ. Sci.* **2015**, *9*, 449–454. [CrossRef]
47. Bates, L.S.; Waldren, R.P.; Teare, I.D. Rapid determination of free proline for water-stress studies. *Plant Soil* **1973**, *39*, 205–207. [CrossRef]
48. Arnon, D.I. Copper enzymes in isolated chloroplast. Polyphenol-oxidase in *Beta Vulgaris* L. *Plant Physiol.* **1949**, *24*, 1–5. [CrossRef] [PubMed]
49. Gomez, K.A.; Gomez, A.A. *Statistical Procedures for Agriculture Research*, 2nd ed.; June Wiley & Sons. Inc.: New York, NY, USA, 1984.

MDPI
St. Alban-Anlage 66
4052 Basel
Switzerland
Tel. +41 61 683 77 34
Fax +41 61 302 89 18
www.mdpi.com

Plants Editorial Office
E-mail: plants@mdpi.com
www.mdpi.com/journal/plants

www.ingramcontent.com/pod-product-compliance
Lightning Source LLC
LaVergne TN
LVHW071559170726
843515LV00008B/2305

* 9 7 8 3 0 3 6 5 7 3 9 6 0 *